PHOTOCHEMISTRY AND PHOTOPHYSICS OF POLYMER MATERIALS

PHOTOCHEMISTRY AND PHOTOPHYSICS OF POLYMER MATERIALS

Edited by

NORMAN S. ALLEN

WILEY

A John Wiley & Sons, Inc., Publication

Published by John Wiley & Sons, Inc., Hoboken, New Jersey
Published simultaneously in Canada

Library of Congress Cataloging-in-Publication Data:

Handbook of photochemistry and photophysics of polymer materials / edited by
Norman S. Allen.
p. cm.
Includes bibliographical references and index.
ISBN 978-0-470-13796-3 (cloth)
1. Polymers–Optical properties. 2. Polymers–Effect of radiation on. 3. Photochemistry–Industrial applications. 4. Photoelectrochemistry–Industrial applications. 5. Photopolymerization. I. Allen, Norman S.
QD381.9.O66.H36 2010
620.1′920495–dc22

2009031389

10 9 8 7 6 5 4 3 2 1

CONTENTS

PREFACE

In the last four decades, the photophysics and photochemistry of polymeric materials have evolved from what were essentially esoteric, often topical, basic research specialities into what one would clearly class today as one of the most significant and important industrial fields and specializations of our modern era. In fact, our whole modern industrial world today in terms of electronic devices, computers, solar cells, printing, imaging, copying and recording systems, to name but a few, would not exist without this creative science surrounding light and its interactions with plastics. Furthermore, our modern-day applications of commercial plastics would not exist without sound knowledge and understanding of their degradation and stabilization processes.

Today, this subject embraces many fields of science and technology, all continuing to develop and change rapidly. However, we have reached a stage where now it is important to categorize and collate our knowledge achieved to date on most of the specific areas of industrial significance. Indeed, only recently, some research specializations such as OLED's and plastic solar cells have emerged into new industrial technologies that will make their mark in the forthcoming years. Our current knowledge achieved to date in most of the areas is clear and scientifically sound and this handbook hopes to lay the foundations of such an understanding for future developments. There are always controversies in all fields of research and indeed this will continue as mankind develops ideas and new technologies. The foundations laid in this book will hopefully facilitate such developments for specialists, academics, and industrialists worldwide.

Thus, the theme of this book brings together a wide range of subjects in the field of photophysics and photochemistry of polymers by underpinning the fundamental, functional, operational, developmental, and application knowledge acquired to date.

The book comprises a total of 17 chapters, each dealing with a specific topic from the basic empirical approach to specific applications in the various fields. The utilization of light energy and its effects is the main theme of the book. Indeed, each chapter interlinks well in terms of the fundamentals, mechanisms, processes, and applications. As one might expect, some of the topics in each chapter interrelate and cross boundaries as many plastics materials have numerous applications.

Chapter 1 on energy transfer and migration processes is indeed the key to the whole operational processes underpinning the interactions of light energy with polymer materials. Light absorption and its dismutation is crucial to all the technologies applicable to polymer materials in terms of their photonic applications. Thus, how the energy is utilized, degraded, and transformed plays a major role in determining the viability of a particular polymer in its application. Here the inherent structural features and morphology of the material control the nature of the photophysical processes that result in energy transformations and potential chemical reactions. All these are dealt with in subsequent chapters. Concurrent with this chapter is Chapter 2 dealing with more complex effects of polymer materials in solution, especially in polyelectrolytic media where important processes for energy transfer can be controlled and made more effective, for example, in light harvesting, enhanced photochemistry, and rheological modifications. Examination of the light emission processes is crucial to our understanding of such processes and their efficiency. In Chapter 3, an extension to light emissive processes is discussed in relation to chemiluminescence. This is light emission in a different context that arises from the inherent features of the polymer itself rather than being light induced. All polymer materials degrade in their lifetime, even during their initial manufacture, and here this technique has attracted much interest worldwide into a specialized field probing the durability of the polymer. In fact, it has become a highly sensitive method of providing rapid information on polymer life expectancies and stabilization efficiency. In the applications area, polymeric electrooptical materials incorporating nonlinear (NLO) chromophores have important commercial potential. Chapter 4 concentrates on the design features of efficient dipolar NLO chromophores via different processes for various electrooptic device applications. This includes guest–host systems, chromophore functionalization, dendritic and cross-linked materials, to name but a few. Here, state-of-the-art technologies are covered with emphasis on high NLO responses and good stability for their long-term future. As an extension to this area, we have novel exploitations in the development of smart devices based on metallodendrimers, and their structure–property relationships are described in detail in Chapter 5. Such metallated macroassemblies show enormous promise with some intriguing architectures for future development. On another front, polymers with photochromic properties have developed and become a major factor in the manufacture of optical storage media. Materials based on azobenzene functionalities are fundamental to this process and are covered in terms of the latest developments in Chapter 6. Cycloaddition reactions and the use of thyamine as new high-potential devices for high storage capacities are some of the key developments here. Metallopolymers with integrated metal complexes is a further addition to the development of molecular devices, especially with regard to enhanced applications for OLED's and metal complexation of polyconjugated materials. Their role in solar energy devices is becoming an area of significant development. Their optical and luminescence characteristics underpin their efficiency and this is covered in Chapter 7. On the application side, clearly one of the major roles of polymer materials in future optical electronics is their development in solar cell construction as one potential solution for the future energy crisis facing us all. Their development and construction and structure–efficiency relationships are described in some detail in Chapter 8. Here the cost-effective devices with high conversion efficiency that can compete with inorganic devices is the key to their success and this chapter shows the way forward. Another major area closely looked into the previous chapters is the development of OLED polymers that has grown into a major field of commercialization, especially in terms of electronic devices and displays with of course thin flat screen televisions being one of the most significant. Structure LED properties for a range of such polymers is described in depth in Chapter 9 probes the molecular device complexities with light emitting properties and how these may be maximized for effect. Moving on to some different topics, the use of light in synthesis of polymer materials is also an area of significance and has been closely allied to many commercial operations for many years. Polymerization processes being the key to the manufacture of polymeric materials involves a number of mechanisms, some requiring heat while others requiring light or radiation. The latter has some major beneficial advantages over conventional methods. These are controlled reactions, energy efficiency, polydispersity, and *in situ* production of materials, that is, in coatings applications such as electronic devices. These processes are indeed vast in terms of potential mechanisms and methods. Free radical addition polymerization through the use of active light-absorbing initiators is a major field of commercial significance and academic interest. Chapter 10 underpins all the complex mechanistic processes involved in free radical polymerization chemistry covering most of the currently available and topical initiators that have been investigated. On the other front and indeed another closely related manufacturing technology for electronic devices is the use of cationic initiators. Here there are many other advantages over that of conventional free radical initiators such as ring opening reactions, absence of oxygen inhibition, and production of what one can describe as "living polymers" for further extentions. Chapter 11 delves into some depth on the types and mechanisms of such processes in relation to new commercial developments. In the electronic world, the key to circuitry, for example, is the process and technology of imaging science through the use of light energy. Photoimaging, as it is known, is indeed a major field where various polymer materials, monomers, and prepolymers are processed and converted by light into an image. Lithographic printing, for example, is a well-established technology and continues to be developed into newer materials for processing. This takes us into the realms of nanolithography and the adaptability and use of specific lasers for nanoimaging. All these processes, new technologies, and polymer materials in photoimaging science and lithography are defined and discussed in Chapter 12. The last area in terms of the synthetic polymerizations with light is the field of photografting. Here light is utilized to graft monomers and polymers onto surfaces either to enhance the properties of materials or to develop molecular electronic devices. In this technology, free radical initiation, cationic, and cross-linking processes can all play a role depending on the end use, covered in depth in Chapter 13. In the fabrication of polymeric electrooptical devices, an another major field of significance is in the ablation of polymer materials to yield specially designed and fabricated surfaces. This area of photoablation underpins a number of important and established commercial technologies in microdevices and microfluidic channels, to name but a few. The computing and printing industries could not, for example, function well without this modern type of photoapplication. Here the mechanisms and efficacy of photoablation and its applications in the electronics field is described in detail. The key here is to produce effective clear ablative surfaces or holes without contamination, and how to design and modify photochemistry to suit is described in Chapter 14. Remaining with light and its effects on polymer materials, we come to one of the most important fields that one could argue is imperative in terms of the future use and applicability of electronic polymers and that is their stability to the actual light activation itself. Indeed, all organic polymers and for that matter inorganic types undergo some type of photochemically induced reaction in their useful lifetime that can cause a breakdown to varying degrees in their physical and mechanical properties. Such processes are complicated by the nature of the light, environmental effects, and the manufacturing and compositional nature of the polymer material itself. Such processes and their effects are clearly described in Chapter 15 for both conventional thermoplastics through to modern-day electronic and composite materials. An extension to this issue is the usefulness of light-induced degradation processes and its chemistry for the production of environmental polymers where there is widespread application in minimizing, for example, waste and pollution. There are also extensive applications in agriculture for enhanced and controlled crop growth and subsequent degradation and mulching. The area of photodegradable polymers has significant commercial development and applicability from the plastic carrier bag through to agricultural greenhouse films. These topics covering the types of polymers and modes of operation are described in some detail in Chapter 16. In the end, it is important to be able to overcome many of the detrimental processes described here in previous chapters, especially if one is to be able to secure the long-term viability of modern-day polymer materials in the environment. Thus, the role and effectiveness of stabilization technologies is crucial, and Chapter 17 thus covers the structural types and modes of operation of stabilizers and stabilization mechanisms for polymer materials. Hopefully, this will provide an opportunity for the more specialized optical electronic specialists to look into the ways and methods for extending the practical usefulness of their materials.

In conclusion, I hope this book will bring together what has been a series of major developments in the polymer world where light and its energy has been put to valuable use in enhancing the technological developments we find ourselves in today and the future. The book I am sure will provide specialists in the fields concerned with a sound understanding and basis for the future in launching new materials while at the same time having a clearer view and opportunity of related areas for problem solving.

NORMAN S. ALLEN (EMERITUS)

Dalton Research Institute Manchester
Metropolitan University
Manchester, UK

CONTRIBUTORS

Concha Abrusci, Departamento de Fotoquímica de Polímeros, Instituto de Ciencia y Tecnología de Polímeros, C.S.I.C., Madrid, Spain

Sahar Al-Malaika, Polymer Processing & Performance, School of Engineering and Applied Science, Aston University, Birmingham, UK

Norman S. Allen, Faculty of Science and Engineering, Chemistry and Materials Department, Manchester Metropolitan University, Manchester, UK

Xavier Allonas, Department of Photochemistry, UMR CNRS 7525, University of Haute Alsace, ENSCMu, Mulhouse Cedex, France

Hazel Assender, Department of Materials, University of Oxford, Oxford, UK

Aaron Barkhouse, Department of Materials, University of Oxford, Oxford, UK

Fernando Catalina, Departamento de Fotoquímica de Polímeros, Instituto de Ciencia y Tecnología de Polímeros, C.S.I.C., Madrid, Spain

Sébastien Chenais, Laboratoire de Biomatériaux et Polyméres de Spécialité, Université Paris, 13/CNRS, Villetaneuse, France

Teresa Corrales, Departamento de Fotoquímica de Polímeros, Instituto de Ciencia y Tecnología de Polímeros, C.S.I.C., Madrid, Spain

Joe A. Crayston, School of Chemistry, EaStCHEM, University of St Andrews, North Haugh, St Andrews, Fife, UK

Céline Dietlin, Department of Photochemistry, UMR CNRS 7525, University of Haute Alsace, ENSCMu, Mulhouse Cedex, France

Mirko Faccini, Laboratories of Molecular Nanofabrication, $MESA^+$ Research Institute for Nanotechnology, University of Twente, Enschede, The Netherlands; Supramolecular Chemistry and Technology, $MESA^+$ Research Institute for Nanotechnology, University of Twente, Enschede, The Netherlands

Li-Juan Fan, College of Chemistry, Chemical Engineering and Materials Science, Soochow University, Suzhou, China

Sébastien Forget, Laboratoire de Physique des Lasers, Université Paris, 13/CNRS, Villetaneuse, France

Jean P. Fouassier, Department of Photochemistry, UMR CNRS 7525, University of Haute Alsace, ENSCMu, Mulhouse Cedex, France

Grégory Franc, Department of Chemistry, McGill University, Montreal, Quebec, Canada

Jean-Luc Gardette, Laboratoire de Photochimie Moléculaire et Macromoléculaire, UMR CNRS 6505, Université Blaise Pascal (Clermont-Ferrand), Aubière Cedex, France

Pieter Gijsman, DSM Research, Geleen, The Netherlands

C. Hewitt, Polymer Processing & Performance, School of Engineering and Applied Science, Aston University, Birmingham, UK

Marius G. Ivan, Department of Chemistry, Center for Catalysis Research and Innovation, University of Ottawa, Ontario, Canada

Wayne E. Jones, Jr., Department of Chemistry, State University of New York at Binghamton, Binghamton, NY, USA

Muhammet U. Kahveci, Faculty of Science and Letters, Chemistry Department, Istanbul Technical University, Maslak, Istanbul, Turkey

Ashok K. Kakkar, Department of Chemistry, McGill University, Montreal, Quebec, Canada

Jacques Lalevée, Department of Photochemistry, UMR CNRS 7525, University of Haute Alsace, ENSCMu, Mulhouse Cedex, France

Thomas Lippert, General Energy Department, Paul Scherrer Institut, Villigen PSI, Switzerland

Avtar S. Matharu, Department of Chemistry, University of York, York, UK

Ali E. Muftuoglu, Faculty of Arts and Sciences, Department of Chemistry, Fatih University, Buyukcekmece, Istanbul, Turkey

Carmen Peinado, Departamento de Fotoquímica de Polímeros, Instituto de Ciencia y Tecnología de Polímeros, C.S.I.C., Madrid, Spain

P.S. Ramanujam, Department of Chemistry, University of York, York, UK

David N. Reinhoudt, Supramolecular Chemistry and Technology, MESA$^+$ Research Institute for Nanotechnology, University of Twente, Enschede, The Netherlands

Joanne R. Ritchie, School of Chemistry, EaStCHEM, University of St Andrews, Fife, Scotland, UK

Agnès Rivaton, Laboratoire de Photochimie Moléculaire et Macromoléculaire, UMR CNRS 6505, Université Blaise Pascal (Clermont-Ferrand), Aubière Cedex, France

Juan Cesar (Tito) Scaiano, Department of Chemistry, Center for Catalysis Research and Innovation, University of Ottawa, Ottawa, Ontario, Canada

H.H. Sheena, Polymer Processing & Performance, School of Engineering and Applied Science, Aston University, Birmingham, UK

Alain Siove, Laboratoire de Physique des Lasers, Universite Paris, 13/CNRS, Villetaneuse, France

Linda Swanson, Department of Chemistry, University of Sheffield, Sheffield, UK

Mehmet A. Tasdelen, Faculty of Science and Letters, Chemistry Department, Istanbul Technical University, Maslak, Istanbul, Turkey

Sandrine Therias, Laboratoire de Photochimie Moléculaire et Macromoléculaire, UMR CNRS 6505, Université Blaise Pascal (Clermont-Ferrand), Aubière Cedex, France

Lukas Urech, General Energy Department, Paul Scherrer Institut, Villigen PSI, Switzerland

Willem Verboom, Laboratories of Molecular Nanofabrication, MESA$^+$ Research Institute for Nanotechnology, University of Twente, Enschede, The Netherlands; Supramolecular Chemistry and Technology, MESA$^+$ Research Institute for Nanotechnology, University of Twente, Enschede, The Netherlands

Yusuf Yagci, Faculty of Science and Letters, Chemistry Department, Istanbul Technical University, Maslak, Istanbul, Turkey

Ali Gorkem Yilmaz, Faculty of Science and Letters, Chemistry Department, Istanbul Technical University, Maslak, Istanbul, Turkey

1

ENERGY TRANSFER AND ELECTRONIC ENERGY MIGRATION PROCESSES

LI-JUAN FAN AND WAYNE E. JONES JR.

1.1 INTRODUCTION

1.1.1 Foundations of Energy Transfer and Energy Migration

Photoinduced energy transfer and energy migration processes have been studied extensively in luminescent polymers for several decades and have applications in sensors, optics, and solar energy conversion [1–7]. Energy transfer refers to a photophysical process whereby the excitation energy of an excited luminophore (donor "D") moves to a chromophore (acceptor "A"). This process can occur by either

Photochemistry and Photophysics of Polymer Materials, Edited by Norman S. Allen

a nonradiative process or a radiative process and is typically thermodynamically spontaneous. Within a unimolecular system, including macromolecules and one-dimensional polymer systems, this process is called electronic energy migration. This can occur between identical molecular subunits where no thermodynamic gradient exists or between different structures resulting in an energy transfer gradient. Thus, energy migration is a specific class of energy transfer. Here, the term "energy transfer" will be used to refer to an interaction between two different species ("heterogeneous" transfer) while the term "energy migration" to refer to processes involving the same species ("homogeneous" transfer), unless otherwise noted.

Electronic excitation results in a quasiparticle, termed an "exciton" in a one-dimensional polymer system [8–11]. When an insulator or a semiconductor absorbs light, an electron from the valence band is excited into the conduction band, and the missing electron in the valence band leaves a hole behind. The exciton can be regarded as a bound state of the excited electron and hole. It is an electrically neutral, excited state of the molecule. An exciton is mobile in a homogeneous one-dimensional system resulting in "hopping" of the exciton state among identical particles with equal energy rather than localizing on one chromophore. Therefore, energy migration can also be called "exciton migration" or "exciton hopping." The energy of a migrating "exciton" is not dissipated during migration; however, the kinetic relaxation of the molecule to the ground state results in a natural lifetime unless the exciton interacts with a lower energy trap site.

There are two nonradiative energy transfer mechanisms that can be described: the Förster mechanism and the Dexter mechanism [8,10]. Both energy transfer mechanisms occur through a radiationless process. Förster energy transfer, also known as through-space or dipole–dipole energy transfer, involves the long-range coupling of the donor and acceptor dipoles. The resonance between the donor dipole moment and the acceptor dipole moment is facilitated by the presence of intervening solvent dipoles. According to Förster, this kind of energy transfer mechanism is mainly affected by three factors: (1) the spectral overlap between the absorption spectrum of the acceptor and the fluorescence spectrum of the donor; (2) the distance between donor and acceptor since both dipole–dipole interaction energy and resonance are distance dependent; and (3) the orientation of the dipoles of the donor and the acceptor molecules and the intervening medium. The Förster energy transfer is favored when donor and acceptor are rigidly held in good alignment, because resonance is maximized when the oscillating dipole of the excited donor and the transition dipole of the acceptor ground state are aligned. The energy transfer rate (k_{ET}) for the Förster mechanism is described by the following relationship:

$$k_{\mathrm{ET}}(\text{Förster}) = k\frac{\kappa^2 k_{\mathrm{D}}^0}{R_{\mathrm{DA}}^6}J \tag{1.1}$$

where k is a constant determined by experimental conditions such as solvent index of refraction and concentration, κ^2 is related to the interaction between the oscillating donor dipole and the acceptor dipole, which depends on the square of the transition dipole moments for the donor and the acceptor and the orientation of the dipoles in space, k_D^0 is the pure radiative rate of the donor, J is the spectral overlap integral, and R_{DA} is the distance between donor and acceptor.

Dexter energy transfer is known as through-bond energy transfer and takes place through a double electron exchange mechanism within the molecular orbitals of the donor and the acceptor. The electronic coupling leading to energy transfer requires significant orbital overlap. Therefore, close interaction between the excited donor and the acceptor ground state is necessary. The rate constant of Dexter energy transfer, k_{ET}(Dexter), is shown below:

$$k_{ET}(\text{Dexter}) = KJ\,e^{(-2R_{DA}/L)} \tag{1.2}$$

where R_{DA} is the distance between donor (D) and acceptor (A) relative to their van der Waals radii L, K is related to the specific orbital interactions, and J is the normalized spectral overlap integral. Thus, the Dexter transfer rate is affected by the spectral overlap and also the separation distance between D and A.

The rates of the Förster energy transfer and Dexter energy transfer depend on the separation distance between the donor and the acceptor, shown qualitatively in Fig. 1.1. The former decreases as the inverse sixth power of the distance, whereas the latter falls off exponentially as the distance increases. Therefore, the Förster energy transfer is able to occur over very large distances, while Dexter energy transfer will give much greater rates at short distances and close contacts.

There is also a radiative energy transfer mechanism termed the "trivial" mechanism of energy transfer. It is accomplished through radiative deactivation of a luminophore donor and reabsorption of the emitted photon by a chromophoric acceptor. There is no direct interaction between the excited donor and the ground

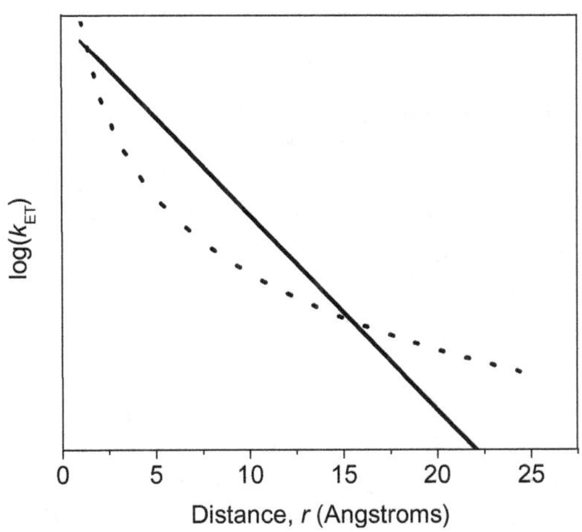

FIGURE 1.1 Plot of log(k_{ET}) versus distance r for both Dexter (solid line) and Förster (dotted line) energy transfer mechanisms, excluding any criterion other than distance. k_{ET} denotes the rate of energy transfer.

state of the acceptor. Thus, the "trivial" energy transfer is a two-step sequence as shown in Equations 1.3 and 1.4, where D^* and D denote the excited and ground states of the donor and A^* and A denote the excited and ground states of the acceptor.

$$D^* \rightarrow D + h\nu \tag{1.3}$$

$$h\nu + A \rightarrow A^* \tag{1.4}$$

The trivial mechanism is not very efficient given the statistical probability of direct emission and reabsorption. However, the efficiency of this kind of energy transfer can be maximized with high emission quantum yield of the donor, high concentration and high extinction coefficient of the acceptor, and strong overlap between the emission and absorption spectra. In most systems, the trivial process does not contribute significantly compared to the other two mechanisms.

1.1.2 Luminescent Polymer Systems

Excited-state processes in polymer materials strongly depend on their electronic and structural properties. There are several types of polymers that are capable of energy transfer and migration. A polymer with a saturated backbone and pendant chromophores such as naphthalene, anthracene, and commercial dyes can be described as type I polymer [11]. In some cases, the chromophore has also been incorporated as a component of the polymer backbone repeat unit or covalently linked to the polymer as a terminal end group. Typically, introduction of the chromophore is achieved through premodification of the monomers or postfunctionalization of the polymers.

The photophysics of type I polymers is very similar to that of the corresponding small-molecule substituent chromophores that they contain. The polymer can be described as an inert scaffold that restricts the movement and relative distance of the chromophores. As a result, the photophysics of these polymers can be strongly affected by the conformation of the polymer backbone and, to a lesser extent, by the aggregation of the polymers. The basic photophysics and energy migration processes of this type of fluorescent polymers have been studied extensively by Guillet and Webber over the past two decades and have been summarized elsewhere [1,11].

Another type of polymers involves conjugated backbones with extensive electronic coupling and has been classified as type II polymers. These so-called "molecular wire" conjugated polymers have received substantial and growing interest in recent years due to their potential wide application as light emitting diodes (LEDs), photovoltaics, and sensors [3,4,9]. Understanding energy transfer and migration in these polymer systems is critical to their success in these applications. An interesting subset of conjugated polymers is the group known as electrically conducting polymers first discovered in 1977 by MacDiarmid and Coworkers [12]. This Nobel Prize winning work has been described elsewhere and is beyond the scope of this review [13]. Here, focus will exclusively be on the fluorescent conjugated polymers that are typically semiconducting in nature.

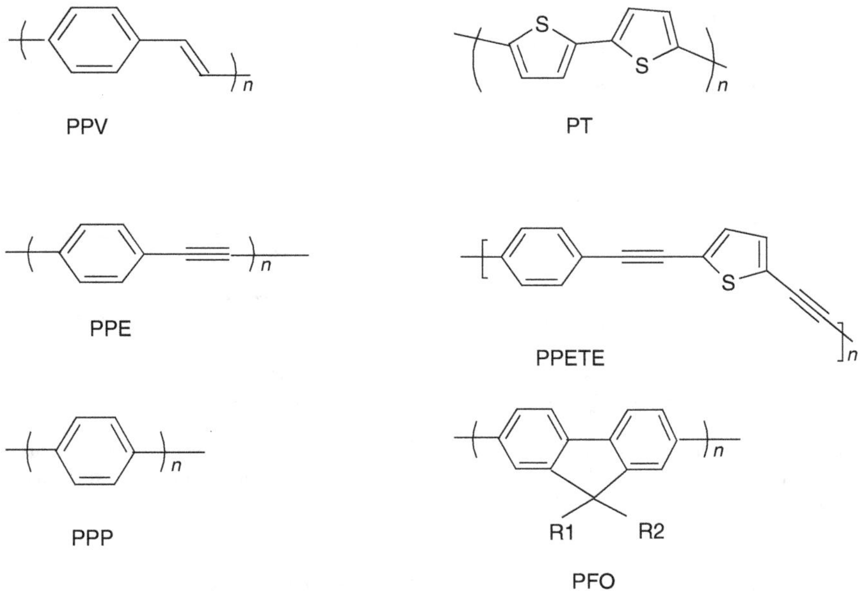

FIGURE 1.2 Basic structures of several fluorescent conjugated polymers.

There are two distinctive characteristics of conjugated polymer fluorophores relative to their small-molecule model compounds. The first is that the initial excited state formed by absorption of a photon is not a localized electron–hole pair. The excitation energy can be described as a delocalized *exciton* that spreads over several repeat units along the polymer backbone. The second characteristic is that the excitation energy can migrate along the polymer backbone. Some representative examples of conjugated polymer backbones are given in Fig. 1.2. They are poly (phenylene vinylene) (PPV), polythiophene (PT), poly[*p*-(phenylene ethynylene)] (PPE), poly[*p*-(phenyleneethylene)-*alt*-(thienyleneethynylene)] (PPETE), poly (*para*-phenylene) (PPP), and polyfluorenes (PFO).

Compared to type I fluorescent polymers, the conjugated polymer backbone is the active chromophore. The monomer units that make up the polymer might not be inherently fluorescent. The absorption and emission of photons involve electronic transitions between a ground-state singlet (S_0) and a excited-state singlet (typically S_1) as shown in Fig. 1.3. Radiative (k_r) and nonradiative (k_{nr}) transitions result in relaxation to the ground state and the observed kinetic lifetime τ of these systems is governed by the relationship

$$\tau = 1/(k_r + k_{nr}) \tag{1.5}$$

Introduction of transition metal elements might introduce significant spin–orbit coupling into the system that increases intersystem crossing, k_{isc}, and results in population of lower energy triplet states (T_1). The result is phosphorescence between the lowest energy triplet state and the ground-state singlet S_0.

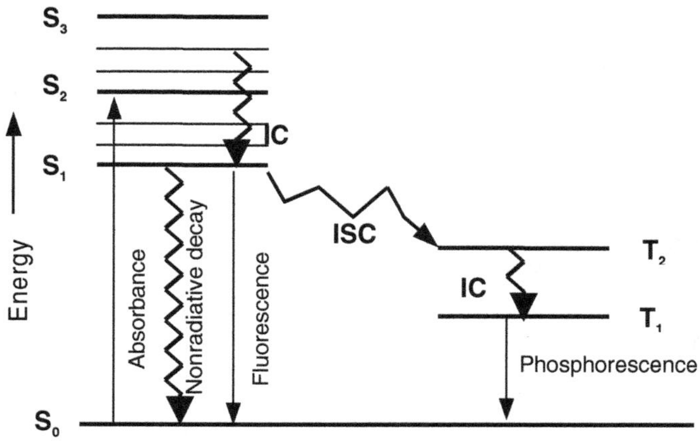

FIGURE 1.3 The Jablonski diagram. k_r, k_{nr}, and k_{isc} denote the rate constants of the deactivating processes from S_1; IC indicates the internal conversion process and ISC indicates the intersystem crossing process.

1.2 ENERGY TRANSFER AND MIGRATION PROCESSES IN CONJUGATED POLYMERS

1.2.1 Light Harvesting and Fluorescence Superquenching

Many applications of conjugated polymers take advantage of the facile energy migration that can occur through the extended π-system of the polymer backbone. The process can be represented using an electronic state diagram as shown in Figs. 1.4 and 1.5. The process of exciton migration has been described in the literature as following a random walk between energetically degenerate and equidistant chromophoric sites [9].

Here the natural lifetime of the exciton is not affected by the migration, although changes in polarization are evidence of the migration event. An exception to this

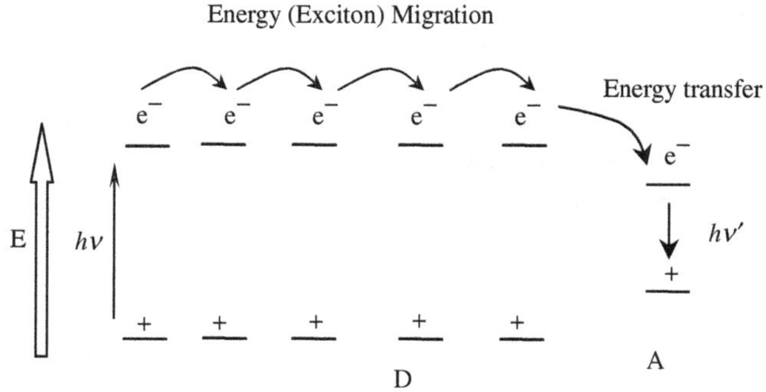

FIGURE 1.4 Illustration of the exciton (bound electron–hole pair) migration process in a conjugated polymer ending with a red-shift emission.

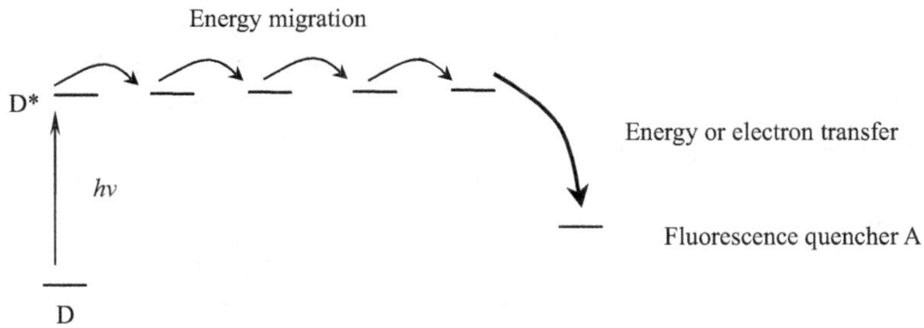

FIGURE 1.5 Illustration of energy migration from a donor state D* to a fluorescence quenching acceptor site A.

would be when the exciton comes into proximity with an energy quenching site. This site could be a narrow-bandgap chromophore with emission at a long wavelength (Fig. 1.4) or a fluorescence quencher (Fig. 1.5). Two forms of energy transfer are involved in both situations: the energy migration among the identical excitonic sites of the polymer backbone (homotransfer) followed by energy transfer or electron transfer to a quenching acceptor (heterotransfer). Energy migration could occur over very large distances along the polymer chain depending upon the lifetime of the exciton and the relative loading of the acceptor.

The energy transfer situations depicted in Figs. 1.4 and 1.5 represent different models currently discussed in the literature. In the case of Fig. 1.4, the energy migration and transfer efficiency can be characterized by the emission ratio between the polymer backbone donor and the fluorescence from the acceptor group. This phenomenon has been called "light harvesting" since the conjugated polymer captures the incident light efficiently over the entire polymer and dumps the energy out at the lower energy trap site. This is the basis for solar harvesting in photosynthetic cells [1,3,14].

Superquenching of polymer fluorescence can be described based on Fig. 1.5. In this case, energy migration in the polymer and subsequent energy transfer or electron transfer to the acceptor is very efficient. Rapid exciton mobility combined with energy transfer to a trap site can result in million-fold amplification and the "superquenching" phenomenon in sensory system [6,9,15,16]. Superquenching has also been observed with gold nanoparticles in conjugated polymer systems further extending the scope of this phenomenon [17].

1.2.2 Intramolecular Versus Intermolecular Energy Migration

In recent years, there has been growing interest in understanding the energy migration processes that occur in conjugated polymer materials [3]. Many have found that the rate and efficiency of energy migration in conjugated polymers strongly depend on the conformation, aggregation, and electronic structure properties of the polymer. This dependence is due to the fact that energy migration processes can occur both intramolecularly and intermolecularly in different assemblies of the polymer materials.

In the case of a rigid-rod polymer in dilute solution, the system is free of any specific interaction between polymer chains or chain segments. In this case, the exciton usually follows a random walk along the polymer backbone, which is called "intramolecular energy migration process" [9,16]. When the polymer is in an aggregated state, the "intermolecular energy migration process" is the more dominant mechanism, although the intramolecular energy migration could still occur at a slower rate [18,19]. The aggregation state could be in several forms, such as specific interaction induced polymer aggregation in solution, spin-coating film, LB film, and self-assembled polymers onto the surface of a sphere [14,18–23]. Intramolecular energy migration takes place in one dimension, while the intermolecular energy migration takes places in two or three dimensions. Sometimes, polymer segments can fold back to form aggregation with another segment of the same polymer chain. The energy migration between these two segments is still classified as intermolecular here since it is more close to interchain migration as discussed in greater detail below.

Most conjugated polymers form aggregates at high concentration or in the solid state when the chains are in close contact. This aggregation tends to order the polymer chains, increasing p-conjugation and resulting in a red shift in the observed absorption and emission process. In some assemblies, well-packed conjugated chains preserve electronic independence and there is no observed interaction between the polymer chromophores. In other cases, the interaction between the conjugated segments results in the formation of new excited states such as excimers, exciplexes, and charge transfer species. When two adjacent polymer chromophores have π-electron interaction in the excited state but not in the ground state, the excited state is referred to as an excimer if the segments are identical or an exciplex if they are different. These aggregates are more common in the process of energy migration in condensed polymers.

1.2.3 Probing of Energy Migration in Different Conjugated Systems

There is considerable interest in understanding the mechanism and variables related to rate and efficiency for intramolecular and intermolecular energy migration in conjugated polymer systems. Many techniques, other than the room-temperature steady-state spectroscopy, have been used, such as single-molecule spectroscopy, time-resolved fluorescence spectroscopy, and low-temperature fluorescence and emission polarization spectroscopy. In addition, some specific conjugated polymers and assemblies have been designed to facilitate the study of isolated polymers or aggregated polymers with precise conformation or controlled migration. Some theoretical modeling and simulation have also been advanced by some groups for further understanding the experimental phenomena. Poly(arylene ethynylene) (PAE), PPV, and polyphenylene are the most discussed conjugated polymer systems in the literature and will be described in detail here as representative examples.

1.2.3.1 Poly(arylene ethynylene) Systems Energy migration processes in PAE have received great interest due to their relatively rigid structure, photophysical

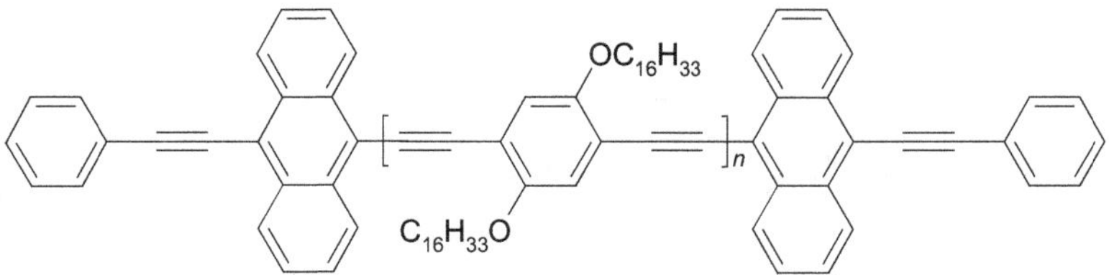

FIGURE 1.6 Structure of PPE endcapped with anthracene.

characteristics, and capability for facile long-range energy migration. Swager and coworkers have devoted considerable effort in this area, along with their outstanding research in the amplification of fluorescence-based chemosensory signal. One creative method for probing energy migration efficiency is to attach low-bandgap termini at the ends of a conjugated polymer chain. This was first demonstrated by Swager et al. in 1995 [24]. They synthesized the highly luminescent PPE polymer and incorporated anthracene into the backbone (Fig. 1.6). In this system, the polymer acted as an antenna that harvested the optical energy and transferred it to the low-energy anthracene group. The energy migration efficiency could be quantified by the emission ratio between anthracene and the polymer backbone. Figure 1.7 shows the absorption and emission spectra for this representative polymer. In this study, the absorption was a composite spectrum of the model PPE without the anthracene endcap and a model small-molecule anthracene chromophore. This suggested that the end group on the polymer produces very little perturbation in the polymer electronic structure. Interestingly, these low-energy termini were found to have a large effect on the emission behavior. Most emissions occurred in a band at 524 nm, which belonged to anthracene, while only a small amount (<5%) of emission at 478 nm was from the polymer backbone. This demonstrated that 95% of the excitation energy was

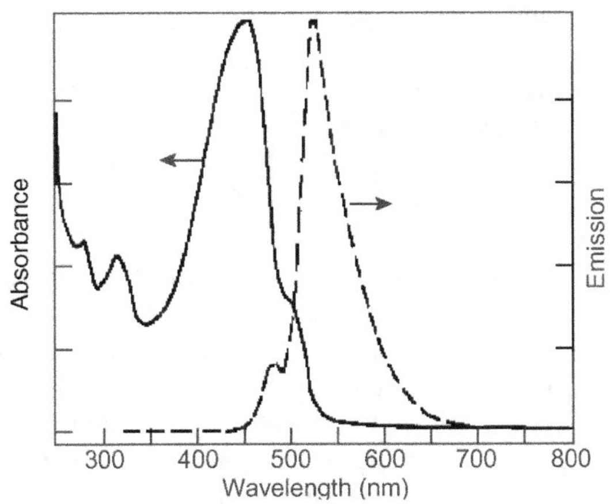

FIGURE 1.7 Absorption and emission spectra of the anthracene-endcapped PPE in dilute solution. For a detailed comparison among the spectra of the model polymer, anthracene, and this polymer, please see the original literature. (*Source*: Ref. [24].)

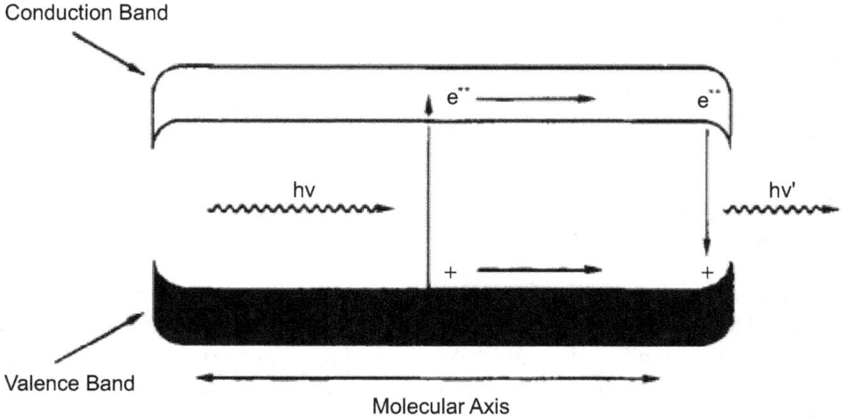

FIGURE 1.8 Schematic illustration of energy migration in a semiconductive molecular wire with a lower bandgap at the terminus. The exciton (the electron–hole pair) is initially generated by the photoexciton, migrates along the conjugated polymer backbone to the terminus, and then recombines with emitting fluorescence. (*Source*: Ref. [24].)

transferred to the end group and was attributed to facile energy migration in a semiconductive molecular wire with a decrease in the bandgap trap at the end (Fig. 1.8).

The molecular wire concept was first advanced in a chemosensory system by Swager and coworkers in 1995 [16,25]. In this case, energy migration could be classified as intramolecular, one-dimensional random walk since the photophysical study was carried out in dilute chloroform solutions. PPE is a rigid-rod polymer that suggested very little interaction between the polymer chains or segments at low concentrations. The one-dimensional energy migration along the polymer backbone can be described as involving through-bond Dexter energy transfer as a result.

One-dimensional random walks do not always provide for the most efficient energy migration. There are at least two reasons to explain the relatively sluggish one-dimensional energy migration in some systems. First, an excitation might retrace certain portions of the polymer backbone several times before quenching or relaxation. Second, the through-bond Dexter energy transfer process can be limited by orbital overlap as a result of chain defects, or conformational irregularities might exist in certain polymer backbones with more degrees of freedom.

To improve the efficiency of energy migration in chemosensory polymers, Swager et al. extended their interest in designing new conjugated polymers to assemblies capable of two- or three-dimensional random walks [14,18]. The idea was based on the consideration that increased dimensionality might decrease the probability of an excitation retracting a given segment of the polymer. In addition, multidimensional intermolecular energy transfer usually involves more facile Förster-type processes, which depend on the dipolar interaction between donor and acceptor [8,10].

The photophysics and energy transport properties of another PPE (Fig. 1.9) were investigated in highly aligned monolayer and multilayer Langmuir–Blodgett (LB) films as well as in the spin-coating film [18]. The cyclophane pendant group was

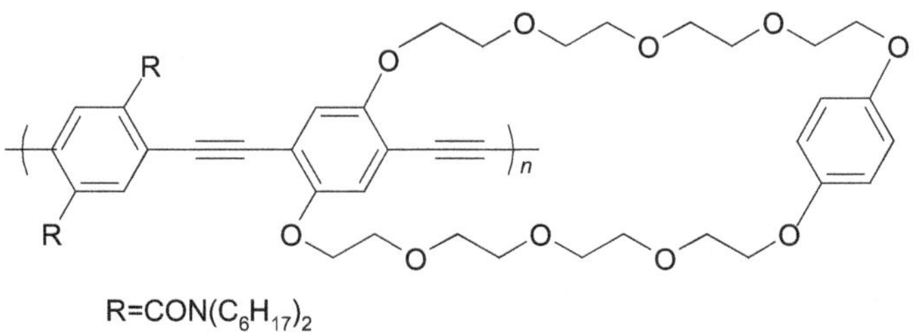

FIGURE 1.9 Structure of a PPE with cyclophane pendant group.

introduced in this case to prevent fluorescence self-quenching in the film due to intermolecular π-stacking between the fluorophores. Compared with fluid polymer solution, all the films displayed some degree of red shift. This was consistent with increased interaction (π-stacking) between the polymer backbones in the aggregated state (Fig. 1.10). The spectral profile in the case of the LB film showed a more limited change compared to the spin-cast film that was red shifted and exhibited a broadening in the spectra. A progression of red shifts in the LB film emission was observed with an increase in the number of layers, indicating more aggregation with increasing number of aligned layers in the films (Fig. 1.11).

Direct evidence of energy migration in the different films was found in polarized fluorescence emission studies. The fluorescence anisotropy indicated that intramolecular energy migration (with a Dexter hopping mechanism) existed in both the solution and spin-cast films. Further energy migration and transfer studies were

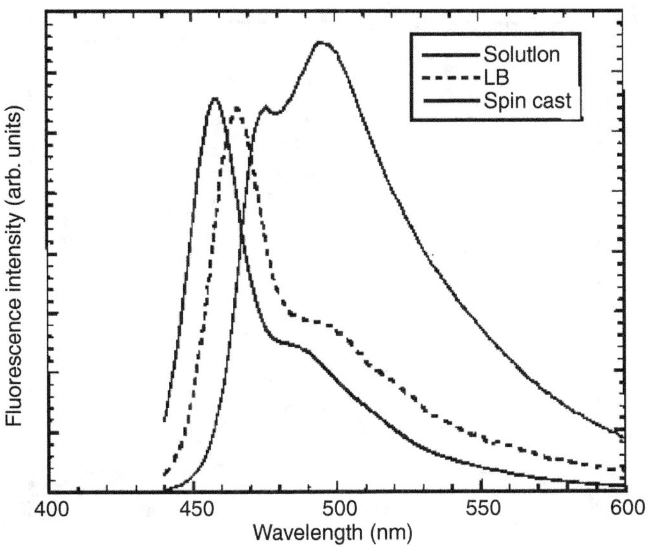

FIGURE 1.10 Comparison of the emission spectra for PPE with cyclophane pendant group (polymer in Fig. 1.9) in solution, as a highly aligned LB film (two layers) and as spin-cast from $CHCl_3$. The excitation wavelength was 420 nm. (*Source*: Ref. [18].)

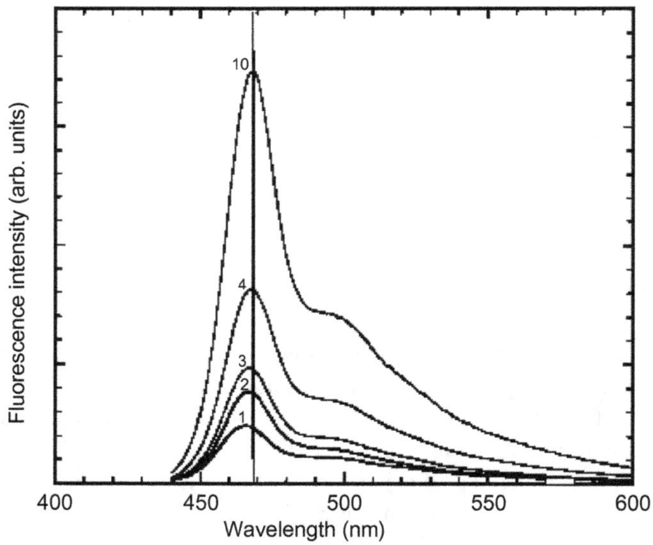

FIGURE 1.11 Emission spectra for LB films of the polymer in Fig. 1.9 containing 1, 2, 3, 5, and 10 aligned layers. The excitation wavelength was 420 nm. (*Source*: Ref. [18].)

carried out by using a luminescence trap Acridine Orange (AO) located on the polymer surface, which was prepared by dipping the LB film into the AO solution in methanol. As seen in Fig. 1.12, while selectively exciting the chromophoric polymer, the AO emission increases. This is accompanied by a decrease in the polymer

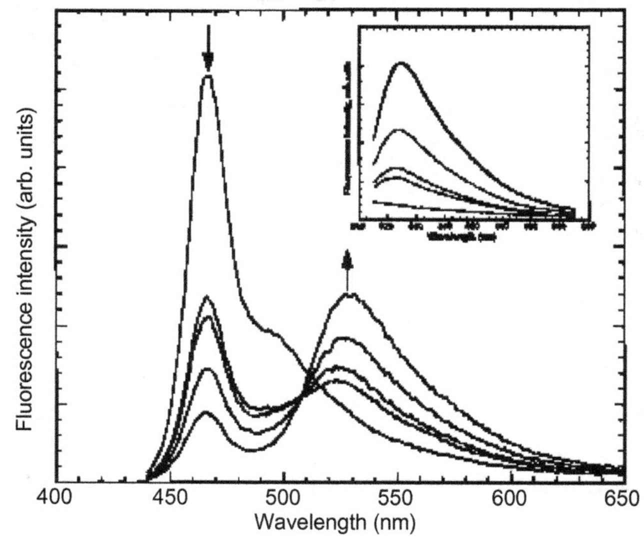

FIGURE 1.12 Emission ($\lambda_{ex} = 360$ nm) from a bilayer LB film of polymer in Fig. 1.9 treated with increasing concentrations (0, 8×10^{-8}, 1×10^{-7}, 3×10^{-7}, and 5×10^{-7} M) of AO in MeOH before dipping. The inset shows the emission spectra of the same films excited at 490 nm, where AO was selectively excited. (*Source*: Ref. [18].)

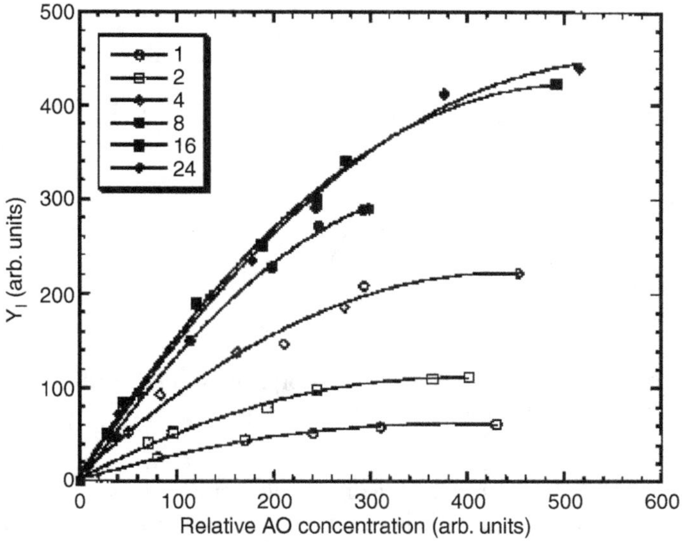

FIGURE 1.13 Summary of the AO fluorescence (excited at 360 nm, fluorescence detected at 525 nm) obtained from LB films of polymer in Fig. 1.9 containing different numbers of layers and concentrations of AO. (*Source*: Ref. [18].)

emission in the bilayer LB film. This phenomenon is consistent with that depicted in Fig. 1.4. The AO fluorescence intensity plotted against the relative trap concentration for LB films with different layers is shown in Fig. 1.13. The AO emission increases with an increase in the number of polymer layers until about 16 layers. The results indicate that energy migration takes place in a three-dimensional topology and that energy migration was very efficient. It was speculated that the energy migration saturates due to the diffusion length of the polymer exciton limiting access to all sites with higher numbers of layers.

Some theoretical work and computer modeling were also carried out in this system. The consistency between modeling and the experimental data suggested that both intralayer (in the film plane) and interlayer (normal to the LB film) energy migration have a very fast rate. The energy migration is dominated by a dipole–dipole (Förster-type) mechanism. This may be facilitated by the alignment of the rigid PPE molecules in the LB film.

Modifications to the secondary polymer structure can also be used to enhance intramolecular energy migration in dilute solution. By inducing the polymer chain into an extended conformation, there is increased π-conjugation and as a result energy migration [19,26]. For example, pentiptycene-incorporating PPE endcapped with 10-(phenylethynyl)anthracenyl has been synthesized by the Swager group (Fig. 1.14) [19]. The endcap was used as the energy trap in this case. The strategy was to introduce rigid and bulky tetra-*tert*-butylpentiptycene units in the backbone. This serves to limit the interchain interactions and at the same time makes the polymers more soluble. In addition, this bulky moiety has a substantial "internal free volume," which facilitates the polymer's alignment in nematic liquid crystalline (LC) media.

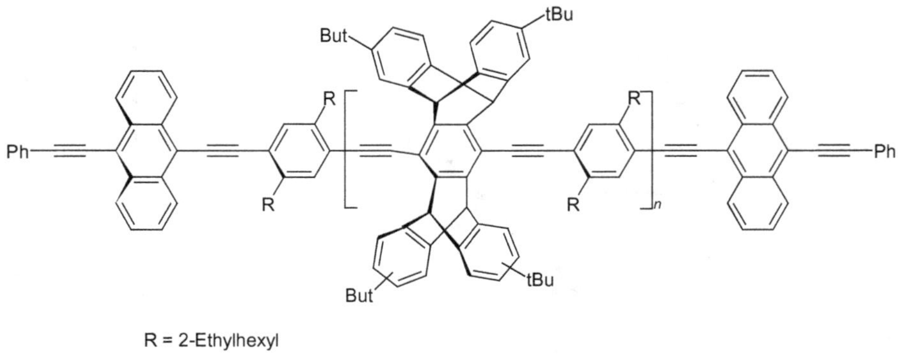

FIGURE 1.14 Structure of pentiptycene-incorporating PPE endcapped with 10-(phenylethynyl)anthracenyl groups.

The absorption and emission spectra of the polymer in CH_2Cl_2 solution, film, and LC solution are shown in Fig. 1.15. The fluorescence spectra were selectively excited into the p-conjugation of the polymer. In dilute CH_2Cl_2 solution, the polymer has an intense peak that can be attributed to the polymer backbone in both absorption (around 377 nm) and emission (around 420 nm). There is also a much weaker peak belonging to the end group at about 466 nm in the absorption and 500 nm in the emission. The red shift in absorption and the strong emission from the end group suggest facile energy migration in the polymer film. More interestingly, emission from the end group increases dramatically in LC solution. Both energy transfer efficiency (up to 36%) and quantum yield (0.90) increase significantly, compared to those in CH_2Cl_2 solution (approximately 7% and 0.60, respectively).

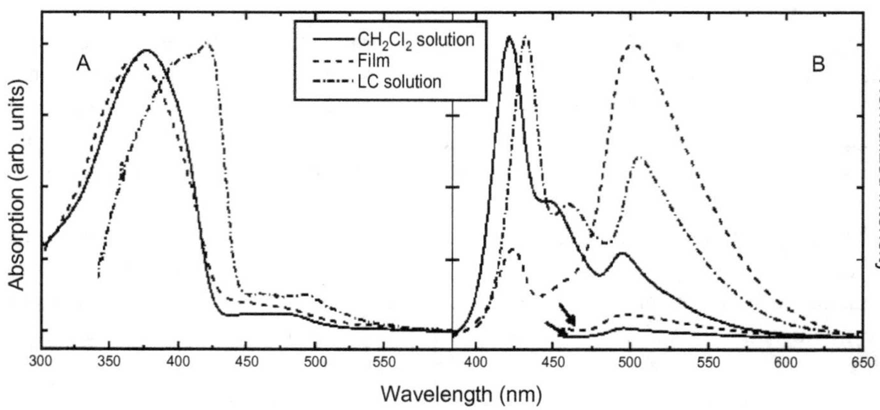

FIGURE 1.15 Normalized absorption (a) and emission (b) spectra of the polymer in Fig. 1.14. Emission spectra were acquired upon excitation at 370 nm and normalized according to the quantum yields. The lines marked with arrows correspond to the emission spectra obtained by direct excitation of the anthracenyl group at 450 nm. (*Source*: Ref. [19].)

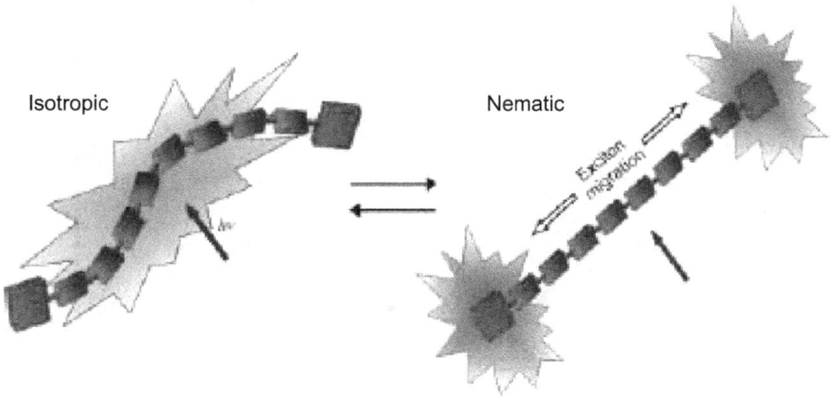

FIGURE 1.16 Simplified representation of the polymer shown in Fig. 1.14, in isotropic and LC solutions. (*Source*: Ref. [19].)

Enhancement in both energy transfer and total quantum yield was attributed to "the highly uniform alignment of the polymer chains and the increased electronic conjugation within the polymer backbone in the order LC phase" [19]. This was supported by further anisotropy experiments. The polymer exhibits much greater emission when excited parallel to the nematic LC direction than in the perpendicular direction. The rigidification and planarization of the polymer chains result in stronger electronic coupling along the polymer backbone, and thus increased conjugation (Fig. 1.16). It was concluded that intrachain Dexter energy migration was enhanced in LC solution compared with that in isotropic solution. The emission spectra in Fig. 1.17 supported the above arguments. The intramolecular Förster mechanism would not be expected to contribute to the enhanced energy migration since it increased the distance between the chromophores.

Energy migration has previously been shown to be involved in superquenching related to fluorescent conjugated polymer chemosensors. In this context, the role of different pendant groups and their structural spacing on energy migration has been studied by Jones and coworkers [15,27–30]. Energy transfer and migration in the PPETE system were studied extensively in dilute THF solution (Fig. 1.18). The study of the ttp-PPETE system involved energy migration leading to energy transfer at trap sites where transition metal cations were detected along the polymer backbone.

Stern–Volmer analysis provides a convenient method to study the binding event that leads to trapping of the energy migration. In this case, it was determined that a static quenching mechanism involving preexcitation binding of the metal cation was involved. This has also been observed in many other conjugated polymer chemosensory systems that involve energy migration [2]. Careful investigation showed that energy migration along the polymer backbone accounted for an observed upward curvature of the Stern–Volmer plot. A modified Stern–Volmer static quenching model was successfully applied to this system by incorporating an energy transfer term that takes into account different energy transfer mechanisms in the Stern–Volmer

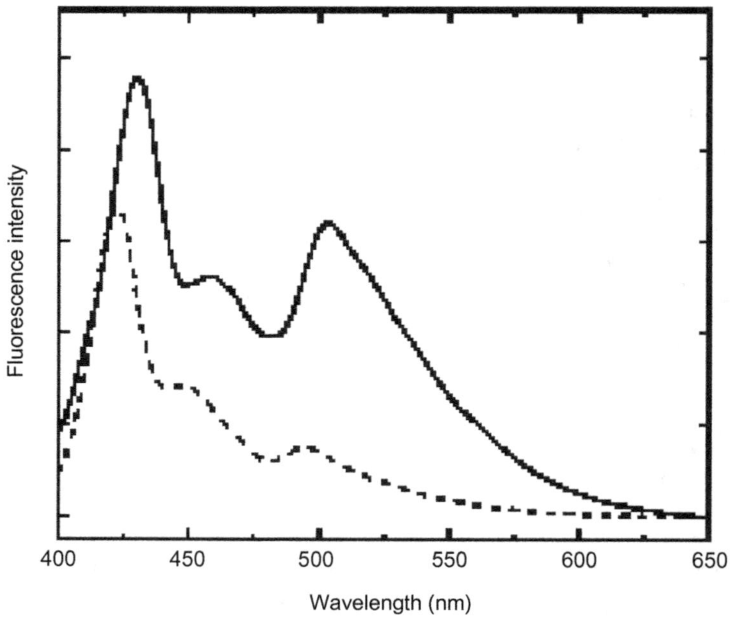

FIGURE 1.17 Fluorescence spectra of the polymer in Fig. 1.14 in the nematic LC phase at 25°C (solid line) and in the isotropic phase at 60°C (dashed line). (*Source*: Ref. [19].)

relationship. It was also demonstrated that the quenching mechanism, Förster or Dexter, was found to depend on the identity of the quencher analyte involved [27].

The (x% tmeda)-PPETE series are the polymers with PPETE as the backbone, having variable percent loadings of the *N,N,N′*-trimethylethylenediamino pendant group. The quantum yields were found to be directly related to the loading of the amino groups and can be modeled by a Stern–Volmer-type relationship. Photophysical studies related the total quenching efficiency to the inherent rate of photoinduced electron transfer from the amino group to the polymer backbone, the lifetime of the exciton, the rate of excitation energy migration along the polymer backbone, and the total loading of the receptor on the polymer. The amino group loading dependence of the fluorescence suggested that the energy migration along the polymer backbone

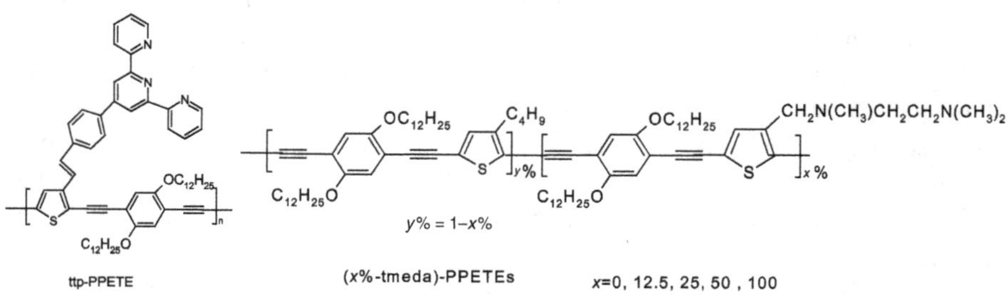

FIGURE 1.18 Structures of ttp-PPETE (left) and (x% tmeda)-PPETEs (right).

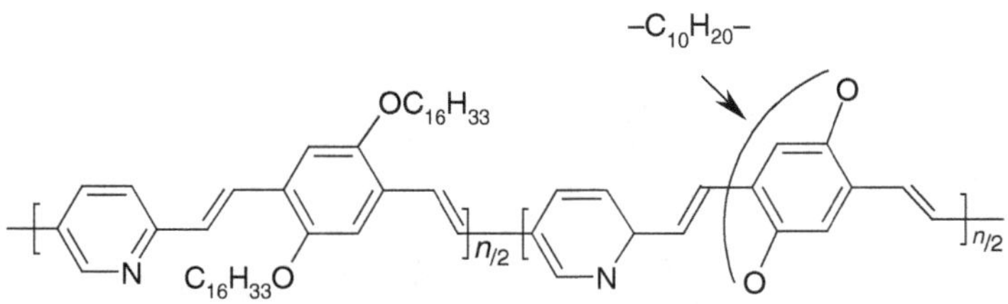

FIGURE 1.19 Structure of PPV-PPyV.

must be relatively slow. Cation titrations of this series of polymers showed no sensitivity enhancements with reduced receptor loading, which was consistent with the relatively sluggish energy migration along the polymer backbone [30].

1.2.3.2 Poly(phenylene vinylene) Systems PPV-type conjugated polymers have also received extensive interest during the past decade. The conformations of PPVs and their derivatives are more diverse than those of the PAEs since the PPVs have less rigid polymer backbone. Barbara and coworkers investigated the energy transfer processes of several PPV-conjugated polymers in a spin-cast film by single-molecule fluorescence spectroscopy (SMS) together with some other techniques [31–33]. Direct information about the individual molecules, such as energy transfer for the single polymer chain, can be obtained from SMS studies.

Barbara and coworkers reported on the copolymer poly(*p*-phenylene vinylene-*co*-*p*-pyridylene vinylene) (PPV-PPyV) (Fig. 1.19) embedded in polystyrene films [31]. The surprising single-step photobleaching kinetics and discrete jumps in single-molecule fluorescence intensity indicated that the conjugated polymer chain is a multichromophoric system. In other words, there were different absorbing and emitting species in a single conjugated polymer system. However, there was strong communication between these chromophores, which facilitate energy transfer to a localized fluorescence quenching polymer defect.

Poly[2-methoxy,5-(29-ethyl-hexyloxy)-*p*-phenylene vinylene] (MEH-PPV) (Fig. 1.20) is one of the most studied conjugated polymers in the literature. Barbara and coworkers used time-resolved single-molecule spectroscopy to investigate the energy transfer process on isolated single molecules of MEH-PPV [33]. Their experimental results suggested that the intramolecular electronic energy transfer converts the mobile excitons to trapped excitons that are located at one or more low-energy regions (energy trap) on a timescale shorter than the excited-state lifetime.

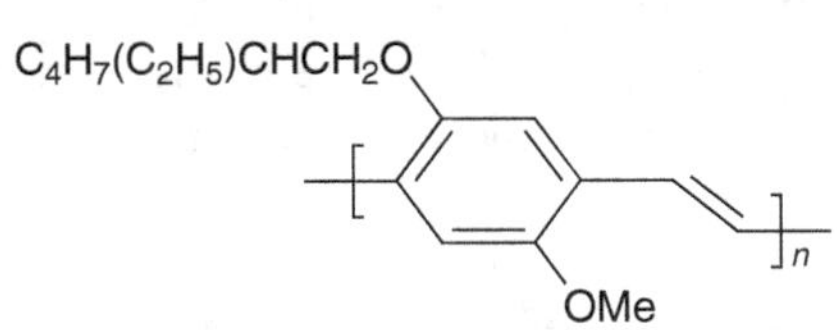

FIGURE 1.20 Structure of MEH-PPV.

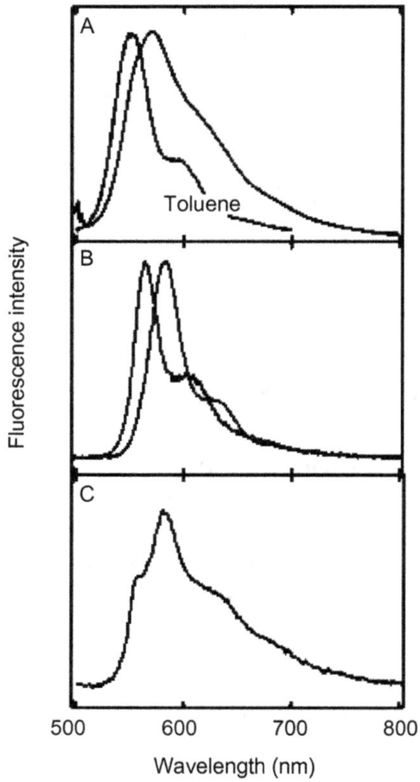

FIGURE 1.21 (a) The "ensemble" fluorescent spectrum of MEH-PPV in a pure liquid toluene solution (left peak) and in a polycarbonate matrix at ambient temperature (right peak). (b) Typical single-molecule spectra of MEH-PPV peaked at 560 and 580 nm. (c) A single molecule spectrum of MEH-PPV that has a "mixed" feature of both a 560-nm peak and a 580-nm peak. (*Source*: Ref. [33].)

Those energy traps, corresponding to the red shift peak in the SMS spectra (Fig. 1.21), come from the exciton interaction between nearby parallel oriented chromophores through the chain–chain contacts in the polymer "cylinder" conformation since different conformations exist in the polymer system (Fig. 1.22) [32]. There are some "funnels" such as the ordered-parallel-folded chain regions that favor rapid energy transfer due to the aligned transition dipole moments and short chromophore energy transfer distances. Further time-resolved emission transient indicated there are always several intermediate energy levels in the system, which also suggests that multiple energy transfer funnels exist. Thus, the whole system of the energy funnel could be envisaged as forming a "landscape for intramolecular electronic energy transfer" [33].

The picosecond time-resolved data (on a similar MEH-PPV single molecule) are also highly suggestive of rapid and efficient funneling of S_1 excitons to energy traps with oxygen quenchers [32]. The experimental result implies that a significant number

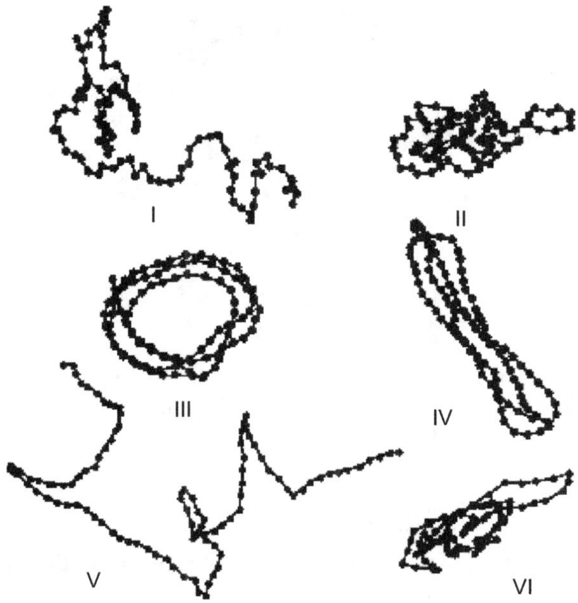

FIGURE 1.22 Typical conformations (I, random coil; II, molten globule; III, toroid; IV, rod; V, defect coil; VI, defect cylinder) of a 100-segment homopolymer generated by Monte Carlo simulations. (*Source*: Ref. [32].)

of the initially excited chromophores are associated with excitons that are transferred to the quencher in a timescale much less than 50 ps. However, measurements on a few of the single molecules reveal evidence of energy transfers (quenching) on a timescale as long as 100 ps. Therefore, slow and inefficient energy transfer still exists, perhaps in molecules with less conformational order. The intramolecular energy transfer is relatively slow. Fast energy migration in one molecule might take place among the different segments of the folded polymer chains. In addition, the nanosecond decay component associated with aggregation has not been observed in the time-resolved single-molecule fluorescence dynamics. This suggested that interchain interactions are required for MEH-PPV, as will be further confirmed by Schwartz and coworkers based on bulk spectroscopic data in the following example.

A more detailed investigation into the relative efficiencies of the interchain versus intrachain energy migration process in MEH-PPV was carried out by Schwartz and coworkers [34–37]. The ideal light harvesting or superquenching process should take place in a system where flow of energy can be controlled and directed to certain regions. To achieve the goal of controlling the energy flow, they constructed an artificial system consisting of a composite of conjugated polymer chains that have been aligned on the nanometer scale and encapsulated into the channels of a periodic mesoporous silica glass (Fig. 1.23) [34,35]. As shown in Fig. 1.23, the polymer chains within the channels are supposed to be well oriented and isolated from each other. However, the chain tails that extend out of the pores are randomly oriented and can be in contact with each other.

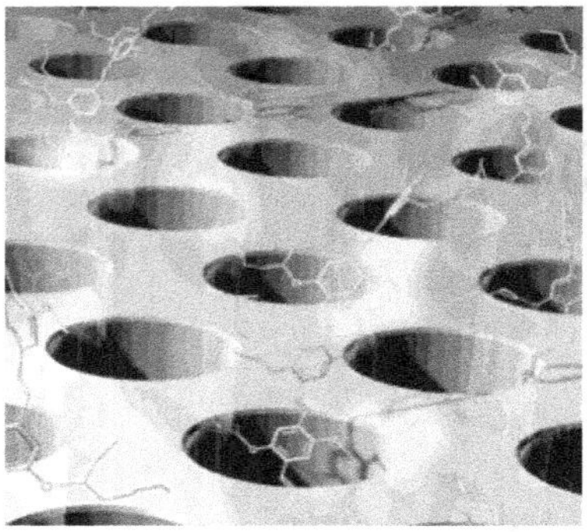

FIGURE 1.23 Schematic representation of single MEH-PPV chains embedded in the channels of an ordered mesoporous silica glass. Graphic courtesy of Daniel Schwartz, D.I.S.C. Corporation. (*Source*: Ref. [34].)

The steady-state and time-resolved luminescence were used to monitor energy transfer in the polymer/silica composites and to probe the fate of the emissive excitons created on the polymer chains after energy transfer (Figs. 1.24 and 1.25). The results showed that the aligned polymer chains inside the pores are isolated in a solution-like environment, and that the randomly oriented chains outside the pores have a degree of interchain contact similar to that in the film. In addition, the spectra indicated that excitons on conjugated polymer segments outside the pores, which are preferentially excited by the light polarized against the pore direction, migrate to lower energy segments inside the pores where they emit light that is red shifted and polarized along the pore direction.

Moreover, the role of interchain versus intrachain energy transfer in conjugated polymers was also separated in this specially designed system. The anisotropy data (Figs. 1.26 and 1.27) also suggest that although energy migration with a Förster mechanism between polymer chains (interchain energy migration) is fast, migration with the Dexter mechanism (intrachain energy migration) along a chain is unexpectedly slow. Correspondingly, the rapid anisotropy decay (Fig. 1.27) could be assigned to interchain energy transfer.

On the basis of the above experiments, the relative sequence of the interchain and intrachain migration in these composites could be described as follows. The interchain transfer takes place in a few picoseconds between the conjugated polymer segments outside the pores that are physically located within a Förster transfer radius. Once the energy has migrated sufficiently, there are no lower energy segments close enough for Förster transfer to occur, so interchain energy transfer ceases. Excitons can continue to migrate along the polymer backbone, from the coiled and nonaligned polymer segment with short conjugation length and high energy outside the pores to the straight and oriented polymer segments with longer conjugation length and lower energy, encapsulated within the channels, but as is evident from Fig. 1.26, they do so quite slowly, on a timescale of a few hundred picoseconds. Further investigation into this system led to the conclusion that the much slower energy migration along the conjugated polymer chain by roughly two orders of magnitude compared to energy

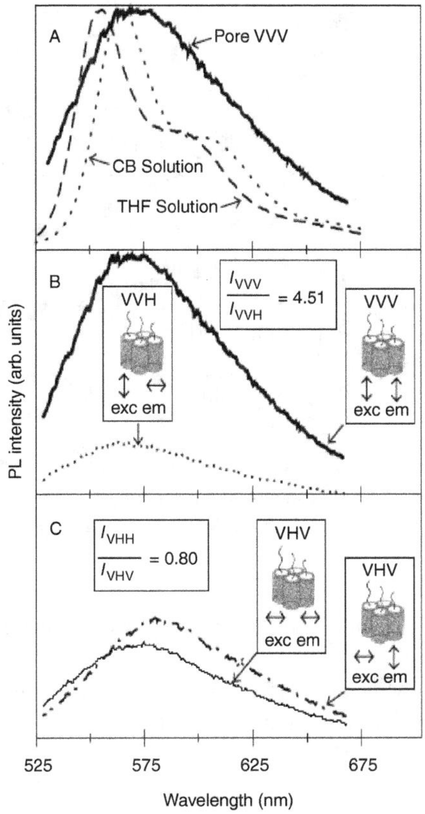

FIGURE 1.24 Emission spectra for MEH-PPV in different environments and under different excitation conditions. (a) Normalized steady-state PL spectra of MEH-PPV in chlorobenzene (CB) solution, THF solution, and the mesoporous silica composite with excitation and collection polarizations parallel to the pore direction (VVV). (b) PL of MEH-PPV in the nanostructured composite with excitation light polarized along the pore direction. The heavy solid curve [VVV, same as in (a)] shows emission collected polarized along the pore direction. The dotted curve (VVH) is for emission collected polarized perpendicular to the pore direction. (c) PL of MEH-PPV in the nanostructured composite with excitation light polarized perpendicular to the pore direction. The dot-dashed curve (VHV) shows emission collected polarized along the pore direction, and the thin solid curve is for emission collected polarized perpendicular to the pores (VHH). (*Source*: Ref. [35].)

transfer between conjugated polymer chains has important implications for devices based on these materials [35].

Structural variation through careful synthetic manipulation, such as controlling the conjugated length of the polymer chromophore or varying the chromophore distance on the polymer backbone, was an alternative way to investigate the energy transfer process. Padmanaban and Ramakrishnan prepared soluble MEH-PPV with a statistical control of the conjugation length over a very wide range by selective elimination

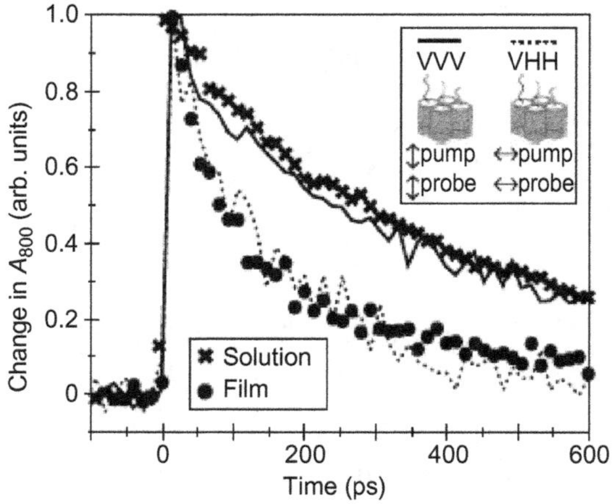

FIGURE 1.25 Ultrafast transient absorption dynamics of MEH-PPV in different environments probed at 800 nm after excitation at 490 nm, normalized to the same intensity at time 0. All four curves (solution, film, VVV, and VHH) were taken under identical excitation conditions. (*Source*: Ref. [35].)

of acetate groups in the sulfone-sulfoxide precursor while leaving the methoxy group unaffected (Fig. 1.28) [38].

The red shift in the absorption and emission with increasing statistic conjugation length was consistent with the initial expectation for the MEH-PPV-*x*s in the solution

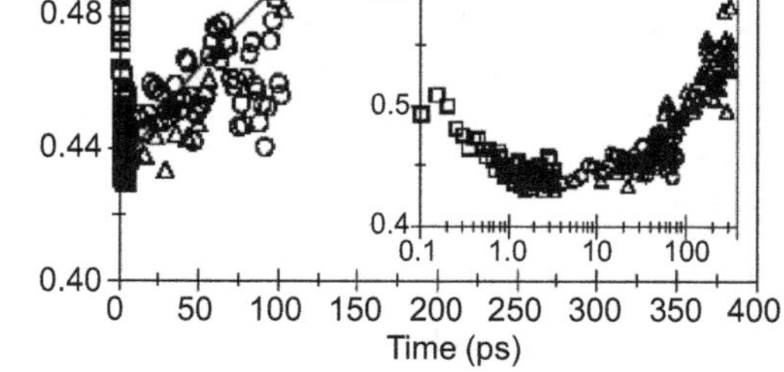

FIGURE 1.26 Ultrafast stimulated emission anisotropy of MEH-PPV in mesoporous silica glass with the excitation pulse polarization parallel to the pore direction (excited at 490 nm and probed at 590 nm). The solid curve in the figure is a 250-ps exponential rise and the different symbols represent scans taken with different amounts of time between points. □, 0.067 ps per point; ○, 1.67 ps per point; △, 6.67 ps per point. The inset shows the same data with a log scale for the time axis.

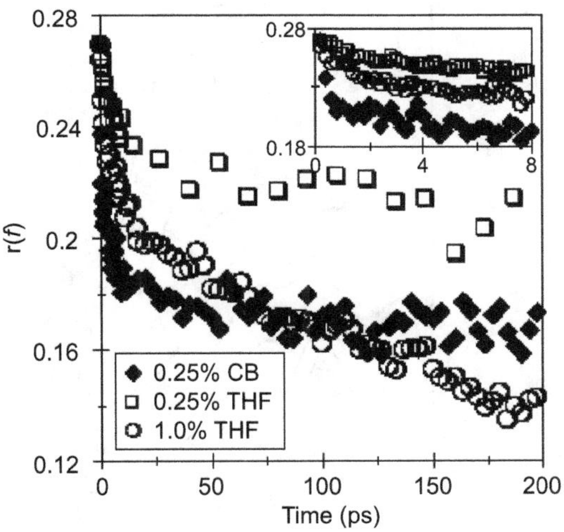

FIGURE 1.27 Ultrafast stimulated emission anisotropy of MEH-PPV in three different solution environments (excited at 490 nm and probed at 590 nm). The inset shows the early time dynamics on an expanded scale.

(Figs. 1.29 and 1.30). The fine structure of the spectra for the MEH-PPV-*x*s could be attributed to the existence of different lengths of conjugated segments. The control experiment for PPV oligomers with various conjugated lengths supported this deduction. Further comparison between the reconstructed and recorded spectra

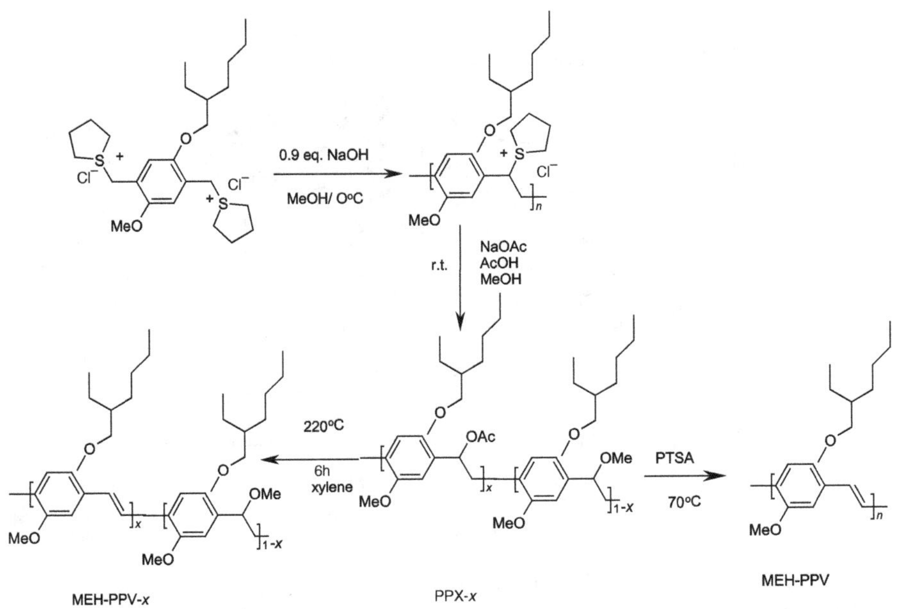

FIGURE 1.28 Synthesis of MEH-PPV-*x* and subsequent conversion to MEH-PPV-100.

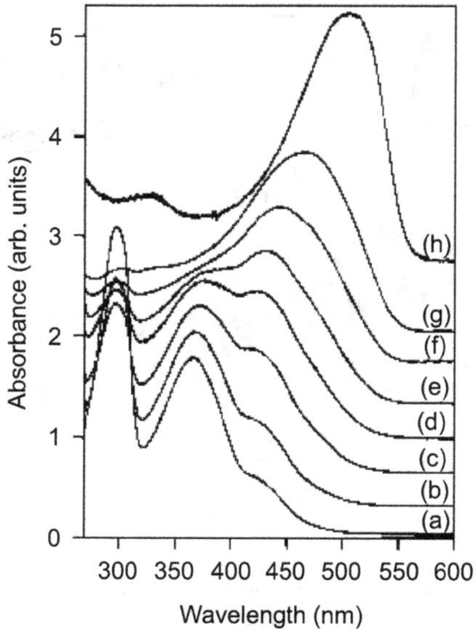

FIGURE 1.29 Normalized and Y-offset UV–visible absorption spectra of copolymers MEH-PPV-x: (a) $x = 10$; (b) $x = 20$; (c) $x = 30$; (d) $x = 45$; (e) $x = 55$; (f) $x = 69$; (g) $x = 85$; and (h) $x = 100$. (*Source*: Ref. [38].)

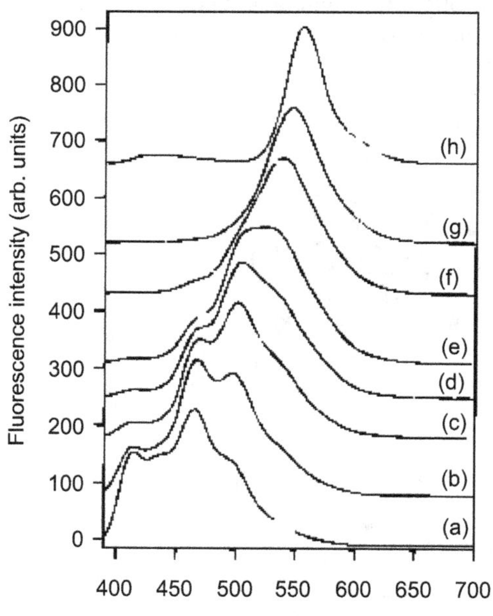

FIGURE 1.30 Normalized and Y-offset fluorescence spectra of copolymers MEH-PPV-x: (a) $x = 10$; (b) $x = 20$; (c) $x = 30$; (d) $x = 45$; (e) $x = 55$; (f) $x = 69$; (g) $x = 85$; and (h) $x = 100$. The concentration of the polymer is 10^{-6} M in CH_2Cl_2 and excited at 370 nm. (*Source*: Ref. [38].)

indicated the occurrence of intrachain energy transfer in the solution [reconstructed absorption spectra were obtained as follows: the individual probabilities of finding the various oligomer segments in each of the MEH-PPV-*x* samples were thus calculated and the absorption spectra of the OPV-*n* oligomers were factored (to reflect this expected oligomer concentrations) and summed up to generate the reconstructed absorption spectra]. In addition, in the case of the reconstructed spectra, no energy transfer is assumed to occur. The significant bathochromic shift of the spectra could be attributed to the chromophore aggregates. Devoid of the fine structure in the emission spectra of the MEH-PPV-*x* thin films indicated the presence of rapid excited energy transfer from short to longer conjugated segments and aggregates.

The extent of energy transfer was quantified by the fractional areas under the emission curves below and above 500 nm, which was calculated for both the reconstructed and the observed spectra (Fig. 1.31). The difference in the fractional areas between the observed and reconstructed spectra is a reflection of the extent of energy transfer that has occurred in the polymers. (For detailed calculation and discussion, please see the original reference.) The results show that intrachain energy migration still exists in dilute solution even when conjugated polymer chromophores were truncated to different conjugated lengths on the backbone. In addition, the extent of energy transfer increases as the average conjugation length increases since the probability of finding neighboring lower energy segments increases. However, the fluorescence quantum yields of MEH-PPV-*x* samples decrease with an increase in the extent of conjugation, which is contradictory to the model PPV oligomer. The authors attributed this phenomenon to the presence of a distribution of conjugation length within a single polymer chain that facilitates the energy transfer and lowers the emission [38].

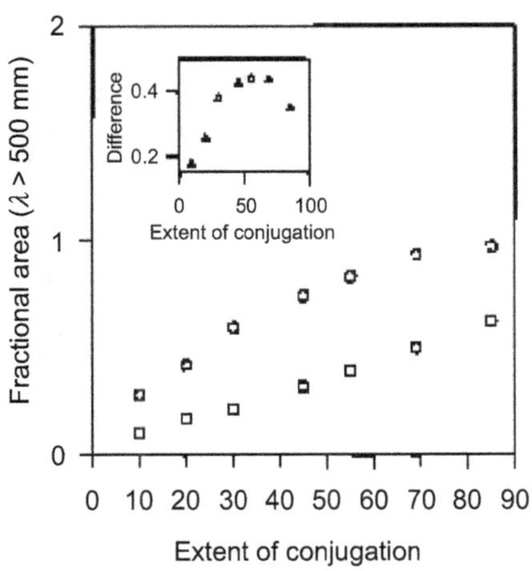

FIGURE 1.31 Plot of fractional area under the emission curve (CH_2Cl_2 solutions) for $\lambda > 500$ nm ($Area_{\lambda>500}/Area_{total}$) as a function of extent of conjugation "*x*". Open circles correspond to the recorded spectra, while open squares are for the reconstructed spectra. *Inset:* Variation of the difference of the fractional areas between the recorded and reconstructed spectra as a function of "*x*". (*Source*: Ref. [38].)

Peng et al. pointed out that energy transfer or energy migration has been extensively investigated in many artificial molecules and polymer systems [39]. However, energy transfer as a process in which excitation energy is released from one species to another has not been clearly stated in many occasions. They constructed a series of schematic diagrams of the situation in the polymer system where energy transfer might take place (Fig. 1.32) [39]. The energy transfer can be harvested and then transferred sequentially, directionally, and efficiently with multipigment array in photosynthesis (a). Energy transfer might take place between different pendent chromophores (b), or between the segments on the polymer backbone with the same chemical structures but different conjugation lengths (c). Since polymer could form aggregates with the other part of the same chain or from another chain, energy transfer could happen from the nonaggregated segments to the aggregates (d, e); Sometimes, the energy could be transferred from some pendant chromophore to the conjugated polymer backbone (f). Excited energy can also be transferred from the randomly directed conjugated polymer segments to the oriented parts (g). In some conjugated polymer systems, there exist aggregates with various extents of interactions between luminophores. Therefore, the energy could be transferred sequentially from the individual conjugated segments, through the loose aggregates, finally to the most aligned, compact aggregates (h).

Some of the energy transfer processes in Fig. 1.32 have been recognized, which were discussed in the above several paragraphs. However, the direct observation of the process itself in conjugated polymers, usually by time-resolved fluorescence spectroscopy, was rarely seen in the literature. The lack of the observation is possibly because energy migration or transfer is too rapid in fully conjugated polymers. In addition, it is

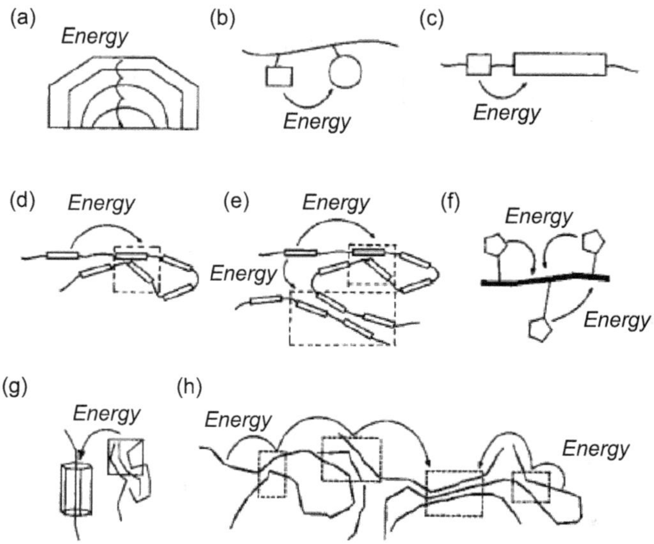

FIGURE 1.32 Schematic diagrams of the different environments for energy transfer, as discussed in the text. (*Source*: Ref. [39].)

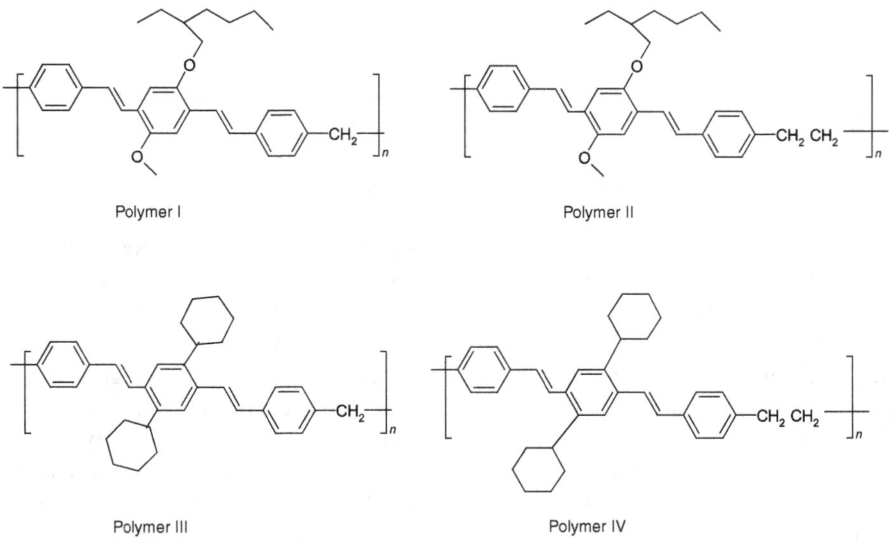

FIGURE 1.33 Structure of polydistyrylbenzene (PDSB) **I–IV**.

difficult to identify the energy transfer mechanism since it is complicated by a disordered environment. Peng et al. designed polymers with so specific structures to simplify the situation. They synthesized polymers of substituted distyrylbenzene (Fig. 1.33) with well-defined conjugation length as repeating units linked by methylene and ethylene linkages (polymers **I–IV** as shown below). The confusion in many other fully conjugated polymers, caused by the variation of conjugation lengths due to conformational distortion and tetrahedron defects, is excluded in these architectures. This series of the polymer could isolate the effects of interactions between individual luminophores (DSB units) on the photophysical properties and the effects of chemical structure on such interactions. In addition, the saturated spacer between the luminophores on the polymer backbone might slow down the energy transfer process to an extent observable by current available techniques.

They investigated the energy transfer process in the polymer chloroform solutions of various concentrations (0.01–10 mg/mL) and polymer films by UV–vis absorption and excitation, steady-state, and time-resolved photoluminescence spectroscopy. They found that the proximity between distyrylbenzene luminophores is critical to the interactions between luminophores and to the energy transfer processes. In concentrated solutions and solid films, aggregates exist besides the individual conjugated polymer chromophores. Different types of aggregates resulting from different extents of interactions, such as loose, compact, and the most aligned aggregates, between luminophores have been observed by photoluminescence.

They also directly observed the sequential energy transfer from individual DSB units to their aggregates and then from the loose aggregates to the relatively more compact aggregates by time-resolved fluorescence spectroscopy. This observation explained that the emission of the concentrated solutions and films of PDSBs **I–IV** is entirely or almost the aggregation emission in the steady-state spectroscopy.

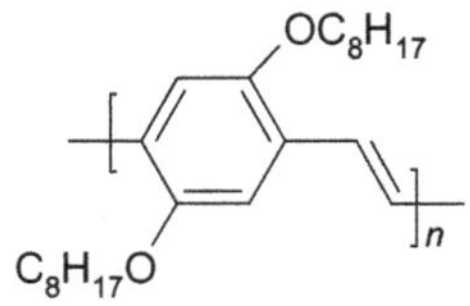

FIGURE 1.34 Structure of DOO-PPV.

However, such a sequential energy transfer process in fully conjugated polymers has never been directly observed at room temperature. The authors suggested that a similar process could take place in fully conjugated polymers with ultrafast dynamics due to the different lengths of the conjugated segments coming from the conformational distortion and tetrahedron defects, or the various extents of aggregation, although it is probably masked behind other photophysical processes [39].

The conformation of another PPV-type polymer, a long-chain 2,5-dioctyloxy-*p*-phenylene vinylene (DOO-PPV) polymer (Fig. 1.34), and photophysics and energy transfer were investigated by Chen and coworkers [40]. One important technique used by them for the study is the time-resolved fluorescence spectra, which was called fluorescence intensity–time trace in the original literature. This technique could provide valuable information on energy transfer as well as the number of emitters in the system studied. When the polymer system has only a few chromophores and luminophores, or very efficient energy transfer from many absorbing sites within the polymer to a few emitting sites, the trace would be dominated by one or two large steps. The appearance of many small steps or gradual changes in the trace would indicate that there are many insulated absorbing and emitting chromophores, lacking communication with each other.

Figure 1.35 presents a typical fluorescence time trace of an individual polymer. At short time intervals, this polymer exhibits multistep emission, while at a longer

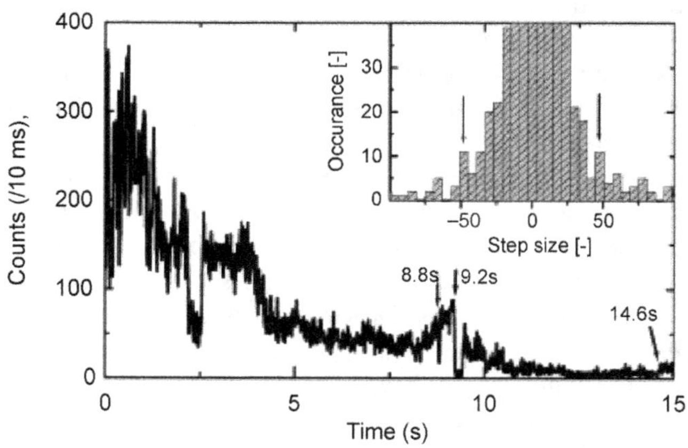

FIGURE 1.35 First 15 s of the fluorescence time trace of single long-chain DOO-PPV polymer under excitation with linearly polarized light recorded with 10 ms time resolution. The inset is a histogram of the change in intensity between each time interval. (*Source*: Ref. [40].)

timescale there is a slow smooth decay in emission intensity. There are several abrupt quantized intensity changes superimposed on small gradual changes in this trace. This phenomenon suggests that the polymer is composed of two types of regions. In the first region, which is responsible for the large jumps in intensity, three-dimensional energy transfers must be very efficient. Therefore, the excitation energy from many absorption sections was funneled to a few low-energy emission sites. The second region is responsible for the more gradual changes, indicating that many sections both absorb and emit. This could be attributed to the relatively inefficient one-dimensional energy transfer along the polymer backbone. For the polymer in Fig. 1.35, the first 3 s is characterized by both large abrupt fluctuations in emission and smaller changes in intensity on the order of the shot noise [40].

To further confirm the explanation made above for the fluorescence time trace, another new technique, the time-dependent modulation depth measurement, was performed on this polymer system. Observing the time dependence (or intensity dependence) of the modulation depth provides information on the relative orientation of the absorbing sections within the polymer. The results suggested that there is greater anisotropy in the orientation of absorption dipoles in the so-called "extended" region responsible for gradual changes, while the "core" region responsible for the abrupt changes has a slightly more isotropic arrangement of dipoles. Several conclusions can be made when combining the information from the fluorescence time traces and time-dependent modulation depth experiment. The first is that the energy transfer in the anisotropic extended region was not efficient (very likely the intrachain), leading to the existence of multiple absorbing and emitting sites. The second is the close packing in the relatively more isotropic "core" region, which allows efficient multidimensional energy migration from many absorbing to a few emitting sites. The results from the time-averaged and ensemble spectroscopic experiments on this polymer were consistent with the above conclusions [40].

Chen et al. synthesized another type of PPV, dialkoxy-substituted poly[phenylene vinylene]s (dROPPV 1/1, 0.2/1, and 0/1), consisting of two repeating units with different side chain lengths (methoxy and 3,7-dimethyloctyloxy) (see Fig. 1.36) [41].

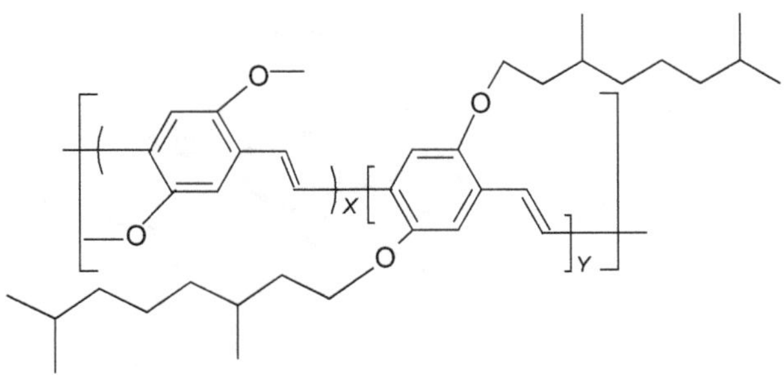

X/Y=1/1 dROPPV- 1/1

X/Y=1/5 dROPPV- 0.2/1

X/Y=0/1 dROPPV- 0/1

FIGURE 1.36 Structure of dROPPVs.

The control of the polymer conformation and aggregation state was facilitated by the fine-tuning of the copolymer composition, without altering the intrinsic properties of single polymer chains. In addition, these polymers can serve as "a model system to clarify roles of aggregates (the sites with ground-state interchain interactions) and the independent chain segments in the well-packed chains (the chain segments that are compactly packed without interaction) in the emission mechanism of conjugated polymers" [41].

The effect of the packing and aggregation of the polymer chain on the photophysics of the material was investigated by regular absorption and emission spectra, luminescence decay curve, and emission spectrum dynamics. These studies were carried out on the polymer chloroform solution as well as on the as-spin-coated films, annealed spin-coated films, as-casted films, and annealed cast films. The experimental results reveal that several kinds of polymer chain arrangement could exist in the polymer film (Fig. 1.37). One is the independent segments with random orientation in the loosely packed region. In the case of the well-packed region, sometimes the aggregates form, which are the ground-state species consisting of conjugated segments with an extension of π-system over one another. However, chains in the well-packed regions do not necessarily always form aggregates, especially for those rigid and orientated chain segments. These segments exhibit some photophysical characteristics of independent species with relatively long conjugation length. Therefore, as shown in Fig. 1.37, the energy transfer sequence could be from the independent segment in the loosely packed region to the independent segment with longer conjugated length in the well-packed chains and then to the aggregates [41].

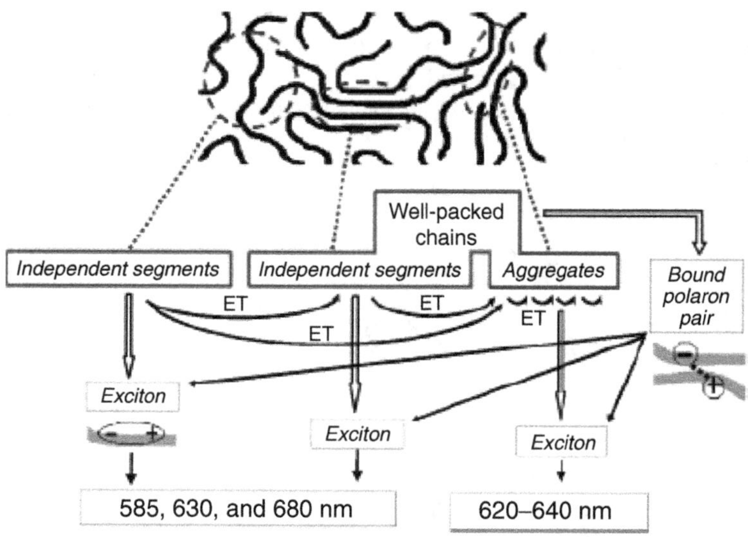

FIGURE 1.37 A proposed emission mechanism for the dROPPV polymers at the excitation of energy higher than the bandgap. (*Source*: Ref. [41].)

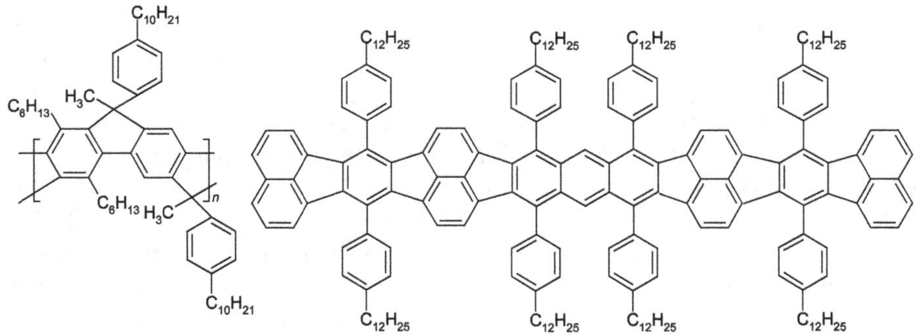

FIGURE 1.38 Chemical structure of m-LPPP (left) and RS19 (right).

1.2.3.3 ***Poly(para-phenylene)*** Another well-studied conjugated polymer is the ladder-type polymer PPP. The very extended and rigid molecular structure of this class of polymers avoids chain folding. In addition, the polymer luminophore aggregates are rarely found in these polymer materials due to their special structure. Therefore, the energy transfer process is relatively simpler than that in other conjugated polymers. Basically, it is well agreed that only intrachain energy migration along the polymer chain takes place in dilute solution, while the multidimensional interchain energy migration exists in the polymer films.

List et al. probed the energy migration EEM process in a methyl-substituted ladder-type poly(*para*-phenylene) (m-LPPP) through doping this material by small concentrations of a highly fluorescent p-conjugated macromolecule (RS19). The structures of the m-LPPP and RS19 are shown in Fig. 1.38 [42]. The experiments were carried out with various RS19 concentrations and under different temperatures. The experimental outcomes were discussed and modeled by Monte Carlo. Both the experimental and modeling results support the following picture for how the energy migration and energy transfer take place (Fig. 1.39). Energy transfer is always the sum of at least two processes in a system consisting of a polymer and a guest molecule, that is, the migration within the host, followed by a transfer to the guest.

The results from the experiment under various temperatures indicated that the first step is a temperature-dependent migration process of singlet excitons, which can be a

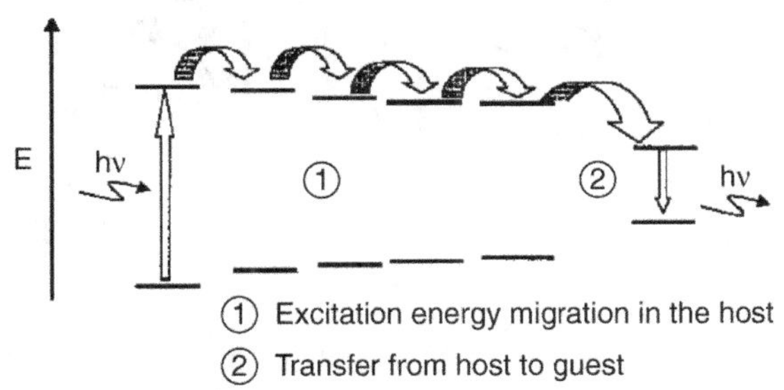

FIGURE 1.39 Schematic of the EEM process:. (*Source*: Ref. [42].)

Dexter- or a Coulomb-type energy transfer mechanism. However, the second step is a temperature-independent transfer, which could be attributed to a Förster dipole–dipole interaction type [42].

Müller et al. carried out low-temperature single-molecule spectroscopy studies on this rigid-rod ladder-type PPP in films without any other chromophore in the system [43]. The homogeneously broadened, strongly polarized emission from individual chromophore units on a single chain could be identified through this technique. Gated fluorescence spectroscopy found that the excitation energy transfer to the lowest energy chromophore is usually absent at low temperatures. The decrease in spectra overlap between the donor and acceptor units along the single rigid polymer chain accounts for the reduced energy transfer by lowering the temperature, considering the chromophoric spectral linewidth narrows with a decrease in the temperature [43].

Beljonne and coworkers explored the energy migration processes in a poly (indenofluorene) (PIF) endcapped with a perylene derivative (Fig. 1.40) by means of ultrafast spectroscopy with correlated quantum-chemical calculations [44,45]. The study of energy migration in dilute solution with a similar molecular design, a PPE-conjugated chain as the energy donor endcapped with anthracene as the acceptor, which was carried out by Swager et al., has been described in the previous paragraphs. In this system, the poly(indenofluorene) chains act as energy donors and the perylene derivatives as the acceptors. The contribution from the intrachain type or interchain type to the overall energy migration process was resolved by comparing the time-integrated luminescence with transient absorption spectra measured in solution and in films.

Basically, the slow intrachain energy migration was found to be dominating in the solution, on a timescale of 500 ps. However, the energy migration rate in a film turned out to be on a few tens of picoseconds scale, with one order of magnitude increase with respect to that in the solution. The high efficiency in films was attributed to the

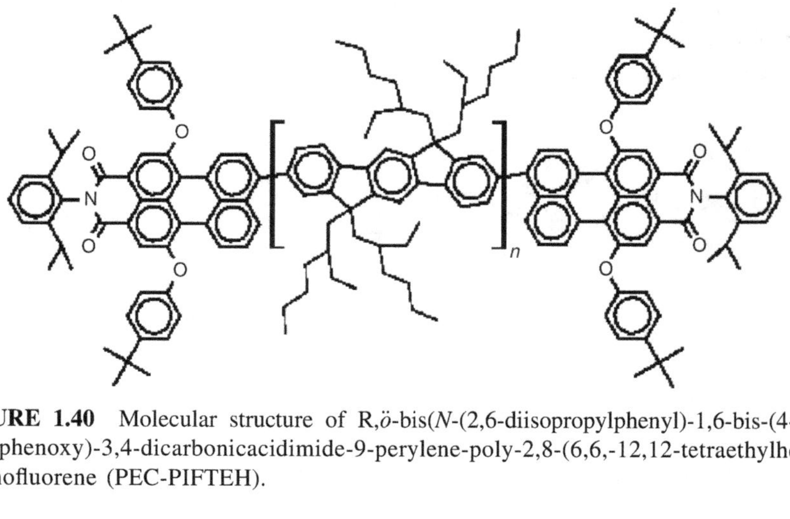

FIGURE 1.40 Molecular structure of R,ö-bis(*N*-(2,6-diisopropylphenyl)-1,6-bis-(4-*tert*-butylphenoxy)-3,4-dicarbonicacidimide-9-perylene-poly-2,8-(6,6,-12,12-tetraethylhexyl) indenofluorene (PEC-PIFTEH).

emergence of additional energy transfer channels due to the close contacts between adjacent chains. These findings were supported by computer modeling using the quantum-chemical calculations. The modeling was carried out on the basis that the whole transfer process was a two-step mechanism, which was also mentioned in the last example by List et al, that is, first a homonuclear energy migration among the conjugated polyfluorene chromophores and then a heteronuclear energy transfer to the perylene derivatives.

To rationalize the different dynamics observed in the solution and in the solid state, computer modeling was carried out step by step with gradually increasing level of sophistication [45]. The energy migration process in the solution was described as an intramolecular energy transfer process along the polymer chain, and only energy hopping to the nearest neighboring chromophores has been considered at first. Afterward, electronic couplings between the donor and acceptor chromophores, spectral overlaps between the donor emission and acceptor absorption, as well as the donor–acceptor relative orientation (which affects the electronic coupling) were taken into account. The energy transfer rates have been computed in the framework of an improved Förster model.

The homonuclear energy migration among the conjugated polyfluorene chromophores was found to be very slow due to weak electron coupling and small spectral overlaps. However, the heteromolecular energy transfer from an indenofluorene segment to an attached perylene derivative is calculated to be typically two orders of magnitude faster than that in the homonuclear process. The significant increase in the rates resulted from the higher electronic couplings and also more efficient donor–acceptor overlap in the second step of the energy transfer process as pointed out in Fig. 1.4. Thus, in situations where only intramolecular transfer takes place, the rate-determining step for energy transfer from the conjugated polymer to the end-capping perylene is the energy hopping process (Dexter energy transfer mechanism) along the conjugated main chains.

The modeling of the energy migration process in the film was a little more complicated. The intermolecular energy migration was possible since the two interacting poly(indenofluorene) chromophores can form different chains and in a cofacial arrangement. The electronic couplings and the corresponding homopolymer energy migration rates computed are always found to be larger than the corresponding intrachain values. In addition, the heteronuclear energy transfer between donor and acceptor was also found to be much faster when the perylene derivative was lying on top of the oligoindenofluorene segment compared to the situation when the donor and acceptor were lying on the same chain [45].

According to these results, it can be concluded that close contacts between molecules provide an efficient pathway for energy migration in conjugated materials as a result of increased donor–acceptor electronic couplings. That *interchain* exciton migration is faster than *intrachain* migration is consistent with the increase in transfer rate observed experimentally when going from solution to the solid state. The authors also pointed out that the rates of intrachain processes would increase significantly if the chains were perfectly conjugated and the excitations coupled coherently along the chains. Actually, the enhancement of energy migration by increasing the conjugated polymer length has been demonstrated by Swager et al. through rigidifying the PPE polymer in a liquid crystal solvent.

The understanding of the long-range intrachain energy transfer process in conjugated polymer is very important for their application as light harvesting or superquenching materials. This process was modeled on the basis of the hopping rates calculated using the improved Förster model [45]. It was found that the transfer dynamics depends on the conjugated length and the distribution. Energy transfer is faster in polymers built from the short conjugated segments since the distance-dependent, site-to-site electronic coupling is larger in this situation. The modeling also showed that energy migration could experience more steps during the exciton lifetime with an increase in the total chain length, while the number of excitations effectively reaching the perylene group decreases.

Further theoretical analysis suggested a clear picture for the energy transfer and migration processes in this system. In a single polymer chain, the first step is the fast energy transfer from the closest indenofluorene units to the covalently attached perylene. Following this is the slow energy migration along the polymer backbone, from the relatively far conjugated polymer site to the closer site, finally reaching the perylene endcap. From the above discussion, it is obvious that the rate-limiting step for the overall excitation energy transfer process is the homonuclear energy migration along the polymer backbone. The simulated transient photoluminescence spectra agree well with the experimental results, supporting the above conclusions [45].

1.2.4 Summarizing the Characteristics of Energy Transfer and Energy Migration in Conjugated Polymers

There are still more excellent examples of the energy migration and energy transfer processes in the literature. They have not been included here due to the authors' limited reading and capability and the scope of this chapter. However, from the above discussion, it can be concluded that the energy migration and transfer processes in the conjugated polymers are very complicated.

First, even in the conjugated polymer materials with the same chemical structure, there are different absorbing and emitting species that affect the ground and the excitation energy and thus the energy transfer process. These absorbing and emitting species could be *independent polymer segments* with different conjugated length or *aggregates* with different extents of aggregation. Sometimes, the polymer segments remain independent even in some well-packed regions. The aforesaid complication, as discussed in great detail throughout this section, comes from the conformation variation in different environments, different structural defects formed during the synthesis, as well as some inherent characteristics of a specific type of polymer. Generally, the energy transfer takes place from the high-energy species to the low-energy species on the premise that all conditions are the same.

Second, the energy transfer taking place between chromophores of the same chemical structure was called "homonuclear energy transfer," while the term "heteronuclear energy transfer" was used to describe the process occurring between chromophores with different chemical structures. However, "energy transfer" is the most common term sometimes used to describe all the processes mentioned above. In some cases, such as in most parts of this section, "energy migration" is used to refer to the process wherein the "homonuclear energy transfer" repeated many times, while "energy transfer" is used to refer to "heteronuclear energy transfer."

Third, the energy transfer mechanism could be the Dexter hopping mechanism, which relies more on the direct orbital overlap, and the Förster dipole–dipole energy transfer, which is more influenced by the relative orientations and the distance between the donors and acceptors. The Förster transfer is demonstrated to be much faster than Dexter transfer in the conjugated polymers.

Fourth, energy transfer or migration could take place intermolecularly or intramolecularly (between the different parts of the folded-back polymer chain classified as intermolecular transfer). It has been demonstrated experimentally and theoretically that the intermolecular energy migration is more efficient compared to the intramolecular one. Several reasons account for this difference. One is that intermolecular energy migration usually takes place via a Förster mechanism since the donor and acceptor belonging to different molecules could be in a cofacial arrangement in most cases, while the intramolecular one takes place via the Dexter mechanism since the donor and acceptor are neighboring on the same chain. Another reason is that intermolecular energy migration is always found in the close-packed region, especially in the film. Three-dimensional migration is possible and the "additional funnels" lead to more efficient migrations.

Finally, the energy migration could be "undirectional" or "directional." The unidirectional migration or "exciton random walk" usually takes place along the single polymer chain, where the conjugation length of the chromophore is considered to be identical, or different but statistically distributed. The intramolecular "random walk" leads to a slow energy migration since the exciton might revisit the sample many times. "Directional" migration takes place in specifically designed polymers constructed by chromophores with sequentially decreasing energy gap or in the polymer film or other forms where absorbing or emitting species of the same chemical structure but different energy gaps exist. The former requires more synthetic efforts, while the latter needs more control of the polymer conformation, aggregation, and packing. Obviously, the directional energy migration would be efficient and thus more useful in the application, which would be briefly discussed in the following section.

1.3 APPLICATIONS

The unique properties of the amplified fluorescent conjugated polymer due to their facile energy migration have been utilized in many applications, such as chemical sensors, biosensors, light emitting devices, and artificial light antennae. The application of conjugated polymer as sensors has been extensively reviewed by Swager et al. in 2000 and 2007 [2,5]. In their 2007 review, they have restricted their topic in sensor with fluorescence as the readout. The design of these kinds of sensors mostly took advantage of the energy migration in the polymer systems, which results in amplifying fluorescence signal output, and consequently enhanced sensitivity.

Some sensor designs make use of the superquenching phenomena (Fig. 1.5). Thus, the presence of the target analytes is indicated by the fluorescence intensity change. Other systems utilize the "light harvesting" properties (Figs. 1.5 and 1.39), and therefore, the presence of the analyte is signaled by the emission of another fluorophore.

As already detailed in Swager's reviews, some conjugated polymer sensory systems were in dilute solution where one-dimensional energy migration was predominant or in aggregation state where multidimensional energy migration was possible. The aggregation state could be achieved in concentrated solution, in films, coated on the spheres, and so on [14,18,20,21,25]. Sometimes, especially in the polyelectrolyte solution system, the analyte-induced aggregations provide additional sensitivity and selectivity [22,23,46–48]. The polyelectrolyte conjugated polymer system is especially useful in biosensors, which attracted many research efforts recently. In the case of DNA detection, the fluorescence resonance energy transfer (FRET) mechanism was used as the transduction pathway. The efficient light harvesting and emitting properties of the conjugated polymers make them excellent energy donors in these FRET-based sensing schemes. Bazan and coworkers have extensively explored these types of sensing systems [49–55]. These systems initially contain two major components: a water-soluble conjugated polymer cationic electrolyte and dye-labeled probe molecules. If the target DNA is added into the system, which has specific interaction with the probe molecules, the anionic DNA will bring the probe molecule close to the conjugated polymer, making the FRET between the polymer and the dye possible. Excitation of the polymer resulted in the emission of the dye as the signal output. Other systems were also developed with a similar mechanism.

Efficient energy migration and energy transfer were also used to tune the color emission of the fluorescence emitting devices such as polymer LEDs (PLEDs) [40,56–61]. Basically, some other emitting species such as dyes were covalently or noncovalently introduced into the conjugated polymer system. The efficient energy transfer from the polymer to the new fluorophore, or vice versa, causes the emission from the new fluorophores. Fine color tuning could be achieved by selecting different polymer and fluorophore matches, attaching the site (incorporated in the main chain, at the chain termini, or as side chain), or manipulating the morphology. In some cases, efficient white emission could be achieved through the incorporation of a low-bandgap green light emitting fluorophore and red light emitting moieties into the backbone of a blue light emitting bipolar polyfluorene copolymer. The energy transfer process, mostly the FRET process, was discussed more than migration.

Another promising application for conjugated polymer is its use in artificial antenna system. In these systems, the light harvesting was directed to an energy trap where many photoactivated processes, such as the photosynthesis, can take place. Because of their versatile physical and electronic properties as well as ease of synthetic control, porphyrins have been widely used as building blocks for such constructions. Up to now, the most studied system is dendritic or cyclic porphyrin arrays, which can direct all the absorbing lights into a core [62–68]. The goal of these studies is to mimic the antenna effect in the biological system in the nature. This is also a direction for the conjugated polymers although few reports about other polymers besides porphyrins could be found in the literature [69].

REFERENCES

1. S.E. Webber, *Chem. Rev.* **1990**, *90*, 1469.
2. S.W. Thomas, G.D. Joly, and T.M. Swager, *Chem. Rev.* **2007**, *107*, 1339.
3. J.R. Terje and A. Skotheim, *Handbook of Conducting Polymers*, 3rd edition, CRC Press, 2007.
4. T.A.E. Skoheim, R.L. Elsenbaumer, and J.R. Reynolds (Eds.), *Handbook of Conducting Polymers*, 2nd edition, Marcel Dekker, New York, 1998.
5. D.T. McQuade, A.E. Pullen, and T.M. Swager, *Chem. Rev.* **2000**, *100*, 2537.
6. L.-J. Fan, Y. Zhang, C.B. Murphy, S.E. Angell, M.F.L. Parker, B.R. Flynn, and W.E. Jones Jr., *Coord. Chem. Rev.* **2009**, *253*, 410.
7. A.P. de Silva, H.Q.N. Gunaratne, T. Gunnlaugsson, A.J.M. Huxley, C.P. McCoy, J.T. Rademacher, and T.E. Rice, *Chem. Rev.* **1997**, *97*, 1515.
8. N.J. Turro, *Modern Molecular Photochemistry*, University Science Books, Sauslito, CA, 1991.
9. T.M. Swager, *Acc. Chem. Res.* **1998**, *31*, 201.
10. J.R. Lakowicz, *Principles of Fluorescence Spectroscopy*, Kluwer Academic/Plenum Publishers, New York, 1999.
11. J. Guillet, *Polymer Photophysics and Photochemistry*, Cambridge University Press, Cambridge, UK, 1985.
12. C.K. Chiang, C.R. Fincher, Y.W. Park, A.J. Heeger, H. Shirakawa, E.J. Louis, S.C. Gau, and A.G. MacDiarmid, *Phys. Rev. Lett.* **1977**, *39*, 1098.
13. A.J. Heeger, *Angew. Chem. Int. Ed.* **2001**, *40*, 2591.
14. D.T. McQuade, A.H. Hegedus, and T.M. Swager, *J. Am. Chem. Soc.* **2000**, *122*, 12389.
15. Y. Zhang, C.B. Murphy, and W.E. Jones, *Macromolecules* **2002**, *35*, 630.
16. Q. Zhou and T.M. Swager, *J. Am. Chem. Soc.* **1995**, *117*, 12593.
17. C. Fan, S. Wang, J.W. Hong, G.C. Bazan, K.W. Plaxco, and A.J. Heeger, *Proc. Natl. Acad. Sci. USA* **2003**, *100*, 6297.
18. I.A. Levitsky, J. Kim, and T.M. Swager, *J. Am. Chem. Soc.* **1999**, *121*, 1466.
19. E.E. Nesterov, Z. Zhu, and T.M. Swager, *J. Am. Chem. Soc.* **2005**, *127*, 10083.
20. J.H. Wosnick, J.H. Liao, and T.M. Swager, *Macromolecules* **2005**, *38*, 9287.
21. R. Deans, J. Kim, M.R. Machacek, and T.M. Swager, *J. Am. Chem. Soc.* **2000**, *122*, 8565.
22. C. Tan, E. Atas, J.G. Muller, M.R. Pinto, V.D. Kleiman, and K.S. Schanze, *J. Am. Chem. Soc.* **2004**, *126*, 13685.
23. J.G. Muller, E. Atas, C. Tan, K.S. Schanze, and V.D. Kleiman, *J. Am. Chem. Soc.* **2006**, *128*, 4007.
24. T.M. Swager, C.J. Gil, and M.S. Wrighton, *J. Phys. Chem.* **1995**, *99*, 4886.
25. Q. Zhou and T.M. Swager, *J. Am. Chem. Soc.* **1995**, *117*, 7017.
26. A. Rose, C.G. Lugmair, and T.M. Swager, *J. Am. Chem. Soc.* **2001**, *123*, 11298.

27. C.B. Murphy, Y. Zhang, T. Troxler, V. Ferry, J.J. Martin, and W.E. Jones, *J. Phys. Chem. B* **2004**, *108*, 1537.
28. L.J. Fan, Y. Zhang, and W.E. Jones, *Macromolecules* **2005**, *38*, 2844.
29. L.J. Fan and W.E. Jones, *J. Am. Chem. Soc.* **2006**, *128*, 6784.
30. L.J. Fan and W.E. Jones, *J. Phys. Chem. B* **2006**, *110*, 7777.
31. D.A. Vanden Bout, W.-T. Yip, D. Hu, D.-K. Fu, T.M. Swager, and P.F. Barbara, *Science* **1997**, *277*, 1074.
32. D. Hu, J. Yu, K. Wong, B. Bagchi, P.J. Rossky, and P.F. Barbara, *Nature* **2000**, *405*, 1030.
33. J. Yu, D. Hu, and P.F. Barbara, *Science* **2000**, *289*, 1327.
34. T.-Q. Nguyen, J. Wu, S.H. Tolbert, and B.J. Schwartz, *Adv. Mater.* **2001**, *13*, 609.
35. T.-Q. Nguyen, J. Wu, V. Doan, B.J. Schwartz, and S.H. Tolbert, *Science* **2000**, *288*, 652.
36. T.-Q. Nguyen, V.S. Doan, and J. Benjamin, *J. Chem. Phys.* **1999**, *110*, 4068.
37. B.J. Schwartz, *Annu. Rev. Phys. Chem.* **2003**, *54*, 141.
38. G. Padmanaban and S. Ramakrishnan, *J. Am. Chem. Soc.* **2000**, *122*, 2244.
39. K.-Y. Peng, S.-A. Chen, and W.-S. Fann, *J. Am. Chem. Soc.* **2001**, *123*, 11388.
40. W.-Y. Sun, S.-C. Yang, J.D. White, J.-H. Hsu, K.Y. Peng, S.A. Chen, and W. Fann, *Macromolecules* **2005**, *38*, 2966.
41. K.-Y. Peng, S.-A. Chen, W.-S. Fann, S.-H. Chen, and A.-C. Su, *J. Phys. Chem. B* **2005**, *109*, 9368.
42. E.J.W. List, C. Creely, G. Leising, N. Schulte, A.D. Schlüer, U. Scherf, K. Mülen, and W. Graupner, *Chem. Phys. Lett.* **2000**, *325*, 132.
43. J.G. Müller, U. Lemmer, G. Raschke, M. Anni, U. Scherf, J.M. Lupton, and J. Feldmann, *Phys. Rev. Lett.* **2003**, *91*, 267403.
44. G.P.D. Beljonne, C. Silva, E. Hennebicq, L.M. Herz, R.H. Friend, G.D. Scholes, S. Setayesh, K. Müllen, and J.L. Bredas, *Proc. Natl. Acad. Sci. USA* **2002**, *99*, 10982.
45. E. Hennebicq, G. Pourtois, G.D. Scholes, L.M. Herz, D.M. Russell, C. Silva, S. Setayesh, A.C. Grimsdale, K. Mullen, J.L. Bredas, and D. Beljonne, *J. Am. Chem. Soc.* **2005**, *127*, 4744.
46. L. Chen, D.W. McBranch, H.-L. Wang, R. Helgeson, F. Wudl, and D.G. Whitten, **1999**, *96*, 12287.
47. S. Kumaraswamy, T. Bergstedt, X. Shi, F. Rininsland, S. Kushon, W. Xia, K. Ley, K. Achyuthan, D. McBranch, and D. Whitten, *Proc. Natl. Acad. Sci. USA* **2004**, *101*, 7511.
48. F. Rininsland, W. Xia, S. Wittenburg, X. Shi, C. Stankewicz, K. Achyuthan, D. McBranch, and D. Whitten, *Proc. Natl. Acad. Sci. USA* **2004**, *101*, 15295.
49. F. Wang and G.C. Bazan, *J. Am. Chem. Soc.* **2006**, *128*, 15786.
50. B. Liu and G.C. Bazan, *J. Am. Chem. Soc.* **2006**, *128*, 1188.
51. B. Liu, S. Wang, G.C. Bazan, and A. Mikhailovsky, *J. Am. Chem. Soc.* **2003**, *125*, 13306.
52. B. Liu, S. Baudrey, L. Jaeger, and G.C. Bazan, *J. Am. Chem. Soc.* **2004**, *126*, 4076.
53. S. Wang, B.S. Gaylord, and G.C. Bazan, *J. Am. Chem. Soc.* **2004**, *126*, 5446.
54. C. Chi, A. Mikhailovsky, and G.C. Bazan, *J. Am. Chem. Soc.* **2007**, *129*, 11134.
55. E.S. Baker, J.W. Hong, B.S. Gaylord, G.C. Bazan, and M.T. Bowers, *J. Am. Chem. Soc.* **2006**, *128*, 8484.

56. C. Ego, D. Marsitzky, S. Becker, J. Zhang, A.C. Grimsdale, K. Mullen, J.D. MacKenzie, C. Silva, and R.H. Friend, *J. Am. Chem. Soc.* **2003**, *125*, 437.

57. J. Cabanillas-Gonzalez, A.M. Fox, J. Hill, and D.D.C. Bradley, *Chem. Mater.* **2004**, *16*, 4705.

58. L. Chen, S. Xu, D. McBranch, and D. Whitten, *J. Am. Chem. Soc.* **2000**, *122*, 9302.

59. F.J.M. Hoeben, P. Jonkheijm, E.W. Meijer, and A.P.H.J. Schenning, *Chem. Rev.* **2005**, *105*, 1491.

60. J. Jacob, S. Sax, T. Piok, E.J.W. List, A.C. Grimsdale, and K. Mullen, *J. Am. Chem. Soc.* **2004**, *126*, 6987.

61. S. Zheng, J. Shi, and R. Mateu, *Chem. Mater.* **2000**, *12*, 1814.

62. I.-W. Hwang, T. Kamada, T.K. Ahn, D.M. Ko, T. Nakamura, A. Tsuda, A. Osuka, and D. Kim, *J. Am. Chem. Soc.* **2004**, *126*, 16187.

63. Y. Nakamura, I. Hwang, N. Aratani, T.K. Ahn, D.M. Ko, A. Takagi, T. Kawai, T. Matsumoto, D. Kim, and A. Osuka, *J. Am. Chem. Soc.* **2005**, *127*, 236.

64. F. Hajjaj, Z.S. Yoon, M.-C. Yoon, J. Park, A. Satake, D. Kim, and Y. Kobuke, *J. Am. Chem. Soc.* **2006**, *128*, 4612.

65. M. Park, M.-C. Yoon, Z.S. Yoon, T. Hori, X. Peng, N. Aratani, J.-I. Hotta, H. Uji-i, M. Sliwa, J. Hofkens, A. Osuka, and D. Kim, *J. Am. Chem. Soc.* **2007**, *129*, 3539.

66. I.-W. Hwang, M. Park, T.K. Ahn, Z.S. Yoon, D.M. Ko, D. Kim, F. Ito, Y. Ishibashi, S.R. Khan, Y. Nagasawa, H. Miyasaka, C. Ikeda, R. Takahashi, K. Ogawa, A. Satake, and Y. Kobuke, *Chem. Eur. J.* **2005**, *11*, 3753.

67. M.-S. Choi, T. Aida, T. Yamazaki, and I. Yamazaki, *Chem. Eur. J.* **2002**, *8*, 2667.

68. D. Holten, D.F. Bocian, and J.S. Lindsey, *Acc. Chem. Res.* **2002**, *35*, 57.

69. B. Dietzek, W. Kiefer, J. Blumhoff, L. Bottcher, S. Rau, D. Walther, U. Uhlemann, M. Schmitt, and J. Popp, *Chem. Eur. J.* **2006**, *12*, 5105.

2

OPTICAL PROPERTIES OF POLYELECTROLYTES

LINDA SWANSON

A polyelectrolyte can be defined as a macromolecule with ionizable repeat units, which can consequently display pH-dependent behavior when dissolved in aqueous solution.

As the term polyelectrolyte can also be extended to encompass natural biopolymers, which contain ionizable groups such as peptides, proteins, and DNA, it is perhaps not surprising that synthetic polyelectrolytes are often used as simple models for more complex biological systems. Indeed, many of the photophyscial techniques discussed in this chapter were developed by biologists and biophysicists interested in the conformational behavior and interactions of biopolymers [1].

Photochemistry and Photophysics of Polymer Materials, Edited by Norman S. Allen

Polyelectrolytes can be classed as either weak or strong depending on the nature of the electrolyte repeat unit. For example, strong acids containing a sulfonate group are fully ionized at all but the very extremes of pH. The ionization behavior of polyelectrolytes, which include units that display weak acid or base behavior in their small molecule analogs, on the other hand, depends markedly on the pH of the solution. Samples containing carboxylic acid repeat units fall into this latter category.

At this point, it is perhaps useful to consider how the dissociation behavior of a weak polyelectrolyte is characterized in aqueous solution.

A small molecule weak acid, for example, will dissociate in aqueous solution as follows

$$\underset{1-\alpha}{\mathrm{RCOOH}} \rightleftharpoons \underset{\alpha}{\mathrm{H}^+} + \underset{\alpha}{\mathrm{COO}^-} \tag{2.1}$$

where α is the degree of ionization (or neutralization). For example, if the pH is adjusted by the addition of NaOH, then α is defined as

$$\alpha = \frac{[\mathrm{Na}^+] + [\mathrm{H}^+] - [\mathrm{OH}^-]}{[\mathrm{RCOOH}]} \tag{2.2}$$

The titration behavior can be described by the Henderson–Hasselbalch equation

$$\mathrm{pH} = \mathrm{p}K_\mathrm{a}^\mathrm{o} + \log\{\alpha/(1-\alpha)\} \tag{2.3}$$

where $\mathrm{p}K_\mathrm{a}^\mathrm{o}$ is the dissociation constant and is the pH at which 50% ionization occurs.

In a polyacid, the presence of neighboring groups modifies the titration behavior. The first COOH group in the polymer to be ionized is similar to that of the low molar mass analog. However, as increasing numbers of units become ionized, the negative charge created serves to make the removal of a proton from any remaining carboxylic groups more difficult. Consequently, the $\mathrm{p}K_\mathrm{a}$ value decreases with increasing pH. In an attempt to describe this behavior, Katchalsky and Spitnik [2] proposed the extended Henderson–Hasselbalch equation:

$$\mathrm{pH} = \mathrm{p}K_\mathrm{a} + n\log\{\alpha/(1-\alpha)\} \tag{2.4}$$

where $\mathrm{p}K_\mathrm{a}$ and n are constants for a given titration and depend on the nature of the polyacid and the ionic strength of the solution. If a polyelectrolyte adopts the same conformation under all aqueous solution conditions then a plot of pH versus $\log\{\alpha/(1-\alpha)\}$ should produce a straight line with a slope of n and intercept of $\mathrm{p}K_\mathrm{a}$. The occurrence of nonlinear plots has been used [2–4] to confirm the existence of a conformational transition despite the fact that Equation 2.4 is only valid for a limited range of α values (around $\alpha = 0.5$).

The titration behavior of polycarboxylic acids can be described at any α value by

$$\mathrm{pH} = \mathrm{p}K_\mathrm{a}^\mathrm{o} + \log\{\alpha/(1-\alpha)\} + \Delta G/kT = \mathrm{p}K_\mathrm{app} + \log\{\alpha/(1-\alpha)\} \tag{2.5}$$

following a thermodynamic treatment of a general expression by Marcus [5]. In Equation 2.5, pK_a^o refers to the monocarboxylic analog, ΔG is the electrostatic free energy required to remove a hydrogen ion from the charged polyion, and pK_{app} is the apparent ionization constant of the polymer at a degree of ionization, α.

2.1 FLUORESCENCE OF POLYELECTROLYTES

In this chapter as far as the optical properties of polyelectrolytes are concerned, more emphasis will be placed on the fluorescence characteristics because these predominate in the literature. Phosphorescence measurements have not received as much attention in this respect although examples are available [6–15].

2.1.1 Probes and Labels

As most polyelectrolytes are nonfluorescent, two general approaches have been adopted to introduce luminescence into the sample. A luminescent species may be incorporated into the macromolecule either during polymerization (see e.g., Ref. [16]) or post-synthesis, a polyelectrolyte solution can be doped with fluorescent material [6,17–21].

To achieve the latter, spectroscopists have exploited the fact that when in a coiled conformation in aqueous solution, certain polyelectrolytes can solubilize [6,12,17–21] low molar organic molecules. On changing the pH, the chain expands releasing this material into the aqueous phase. Analysis of the resultant emission from the probe can reveal information concerning its environment and thence, the conformation of the polymer [6,12,17–21].

This strategy has proved effective in providing information pertinent to the solution behavior of polyelectrolytes at the *molecular level*. However, potential problems can arise with this approach in that the probe does not remain soluble under all solution conditions. The introduction of a covalently bound fluorophore in the form of a label [12,16,19,20,22–26] enables the study of specific regions of the polyelectrolyte over the entire pH range, removing the problem of probe insolubility. Typically, the loading level should be approximately one fluorescent monomer for [25,26] every one hundred electrolyte repeat units. These conditions are such that introduction of the fluorescent reporter does not perturb the properties of the polyelectrolyte under study.

Both of these approaches have their merits, however, use of a labeled polymer offers specific advantages over that of a solublized probe: potentially the probe can be sequestered in both the hydrophobic and the more hydrophilic regions of the polyelectrolyte, in addition to the aqueous phase (see e.g., Ref. [20]), while a label allows the tagged site to be monitored directly [12,16,20,22–26].

A third strategy has involved increasing the label content such that the luminescent species itself acts as a modifier, which subsequently creates a new class of fluorescent polyelectrolyte [20,27–42]. Copolymers of this type, containing a high degree of aromaticity have shown potential as media for efficient light harvesting (i.e. photon harvesting polymers [43]) for application in solar energy conversion and in carrying out enhanced photochemistry, [i.e. so-called photozymes (see, for example references 31, 32, 36, 38, 40, 44 and 45)]. Photon harvesting in polymers is the subject of a comprehensive review [43].

2.2 HOMOPOLYMERS

Because poly(carboxylic acid)s have received particular attention in the literature in respect of their fluorescence behavior in aqueous solution [6,12,13, 16-24,26,29-34,43-46] and have been comprehensively studied by classical methods [2-4,47-52] , this important class of polyelectrolyte will be reviewed initially before moving on to other examples.

2.2.1 Polymethacrylic Acid

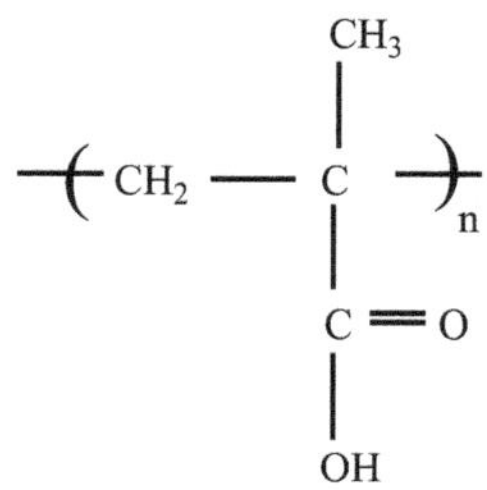

poly(methacrylic acid), PMAA

The active repeat unit in poly(methacrylic acid) (PMAA) is the carboxylic acid function, which dissociates upon increasing pH to form negatively charged carboxylate anions. Early potentiometric investigations of PMAA, following data treatment with the extended Henderson–Hasselbalch equation (see Equation 2.4), concluded that the polymer undergoes "anomalous" titration behavior as a result of a change in conformation [2–4,47–49]. PMAA can be classed as an amphiphilic polymer containing both hydrophobic (water insoluble) and hydrophilic (water soluble) units [17] dependent on the degree of ionization. The changes in the degree of hydrophobicity and hydrophilicity account for the conformational state at any given pH. The macromolecular backbone is hydrophobic at all pH values and, at higher degrees of protonation of PMAA, the carboxylic acid groups can also be considered as hydrophobic. On increasing the degree of ionization, carboxylate anions form increasing the degree of hydrophilicity of the polyelectrolyte. The amphiphilic nature of the polymer is consequently instrumental in dictating the conformational behavior of PMAA. Extensive investigations using potentiometry [2–4,47–50], viscometry [48,51,52], Raman spectroscopy [53], and scattering methods [54–59] have established that PMAA undergoes a dramatic change in conformation between pH 4 and 6. (This is equivalent to a degree of ionization, α, of 0.1 to 0.3). At pH 4, a contracted, globular structure exists to minimize unfavorable contacts between the hydrophobic units of the polyelectrolyte and the aqueous phase. The presence of the (hydrophobic) backbone methyl unit induces additional coiling, known as *hypercoiling* [48], which results in a more compact structure under these conditions [2–4,12,16–19,21,22,24,26,46–48,51–59]. As the degree of ionization of the solution is increased, carboxylate anions form resulting in increasing levels of repulsion between these charges. Between pH 5 and 6, the carboxylate anion concentration is such that repulsive interactions dominate over the hydrophobic forces to collapse the chain and a macroscopic switch occurs from a compact, globular form to an expanded water-swollen state. The fact that PMAA changes its conformation in response to an external stimulus such as pH can be viewed as a "smart response" and consequently PMAA can be classed as a "smart polymer."

As far as the fluorescence behavior of PMAA is concerned it could arguably be considered as the most extensively studied polyelectrolyte (see e.g., Refs. [6,12,16–22,24,26,46]) and consequently provides the ideal example to introduce the various fluorescence spectroscopic techniques to study polyelectrolytes in general.

2.2.1.1 Steady-State Spectroscopy Although the simplest tool available to probe the solution behavior of polyelectrolytes, data derived from steady-state spectroscopy has proven extremely informative in terms of revealing the conformational switch of PMAA in aqueous media. Steady-state measurements encompass a variety of forms of spectroscopic analysis ranging from examination of shifts in the spectral profile [6,19–22,24] to changes in the intensity of luminescence sampled at a particular wavelength [6,17,18,22–24,60–62] of a label or probe.

For example, simple fluorescence intensity measurements on dispersed hydrocarbon probes such as anthracene [17], perylene [18], pyrene [6,17,22], 9,10-dimethylanthracene (DMA) [60], and coumarin dyes [62] have confirmed that PMAA displays pH-dependent solution behavior. A marked decrease in the intensity of the probe occurs between pH 5 and 6, which coincides with the conformational transition of PMAA as determined by classical methods [2–4,47–50]. Two interrelated effects account for this behavior: the solubilizing capacity of the polymer promotes an increase in the concentration of the probe in the solution [6,17,18,60,62] and because the intensity of the fluorescence observed is proportional to the excited state population the resulting emission is enhanced. The hydrocarbons may also be considered to be preferentially solubilized within the hydrophobic domains or structures of the hypercoiled state [6,22]. This results in a degree of protection from the deactivating effects of the aqueous phase and a concomitant increase in the fluorescence observed [6,17,18,22,60,62].

A well-known characteristic of pyrene is that its fluorescence emission spectrum is sensitive to the environment in which it is dispersed [63,64]. In particular, the relative fluorescence intensities of two vibronic bands, termed 3 and 1, respectively, change markedly with solvent polarity. Estimation of the ratio of band 3 to 1 can consequently serve as a sensor of hydrophobicity: a ratio of 1.7, for example, would be expected when pyrene is dispersed in *n*-pentane [65], whereas in aqueous solution a value of 0.55 is observed [6,20]. The sensitivity of the vibrational fine structure of the fluorescence of pyrene to its environment has been exploited by spectroscopists keen to probe the polarity, of a variety of media such as micelles [66], microemulsions [67], polymer colloids [68], and microgels [69,70]. Analysis of the spectral profile of pyrene dispersed in PMAA has proven particularly informative in terms of probing the pH-triggered smart response of the polymer [6,19,21,22,71]. For example, pyrene was dispersed in PMAA and fluorescence emission spectra were sampled across the pH range. The conformational transition of the polyelectrolyte was revealed, via the resultant 3/1 intensity ratios: at low pH, a ratio of 1.1 was derived [6], which reflects an increased degree of hydrophobicity within the compact confines of the hypercoiled state. As the pH is increased, the carboxylate ion concentration also increases until a macroscopic switch in conformation occurs at pH 6 and the probe is ejected into the aqueous phase. A 3/1 ratio of 0.55 results under these conditions [6], which is similar to that observed when pyrene is dispersed in water [6,20]. A dramatic reduction in the ratio occurs between pH 5 and 6, which marks the conformational transition as determined by classical methods [2–4,47–52]. Clearly, changes in the vibrational fine structure of the fluorescence of pyrene can successfully monitor the conformational switch of PMAA.

Water-soluble probes have also been reported to undergo marked changes in their fluorescence characteristics when dispersed in polyelectrolyte solutions [18,52,61,72–78]. For example, the cationic dye, auramine O (AuO) is virtually nonfluorescent in aqueous solution but an increase in intensity is observed in the presence of PMAA at low pH as a result of enhanced binding in the hypercoiled state [72,73,75,78]. On neutralization, the fluorescence from AuO decreases as the compact structure breaks down forming the expanded state and the probe is released to the aqueous phase [52]. In a recent extension of this theme, the sensitivity of the emission spectrum of AuO to the environment in which it resides has been further exploited by covalently bonding the dye to PMAA [61]. Figure 2.1 shows the

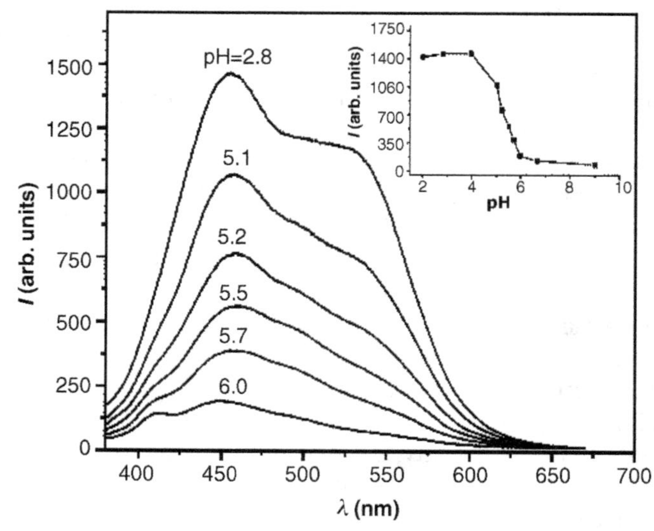

FIGURE 2.1 Fluorescence emission spectra of auramine-labeled PMAA as a function of pH. Inset: change in the maximum emission intensity with pH. [Reproduced from reference 61.]

fluorescence spectra from the PMAA bound dye as a function of pH and the corresponding variation of the maximum fluorescence intensity sampled at 450 nm (inset). Several features are apparent from consideration of the spectral data in Fig. 2.1: because the emission of the dye is sensitive to the solution viscosity [79], the high intensity of fluorescence observed at low pH reflects encapsulation of the fluorophore in a viscous microdomain and is consistent with the formation of the hypercoiled conformation of PMAA under these conditions [61]. As the pH is raised, increasing concentrations of carboxylate anions lead to increased Coulombic repulsions, which results in a gradual expansion of the chain. Between pH 4 and 6, which marks the conformational transition of PMAA from classical measurements [2–4,47–52], a dramatic reduction in the emission from the label is apparent as a consequence of two effects: expansion of the chain exposes the dye to the aqueous phase. This, coupled with a resultant decrease in the microviscosity of the environment surrounding the label serves to reduce the intensity of fluorescence observed. At pH values in excess of 6, the emission from AuO remains low and is consistent with the formation of the extended water-swollen structure under these conditions.

Morawetz and coworkers [24,80–84] have carried out an extensive series of studies of the aqueous solution behavior of polyelectrolytes using the water-soluble dye, 1-dimethylamino-5-naphthyl sulfonate (dansyl) as a covalently bound fluorescent "reporter" group. It is known [85] that the emission intensity of fluorophores of this type increases on absorption from aqueous solution into proteins. An increase in the quantum yield (accompanied by a blue shift of the emission spectrum) has been interpreted as a result of a decrease in the polarity of the medium [86]. These properties have made the dansyl label particularly attractive for probing the conformational characteristics of polymeric acids. For example, the change in the fluorescence of a dansyl label attached to PMAA [24] was used to follow the response of the polymer to a pH jump. At higher degrees of ionization, expansion of the chain leads to increased exposure of the label to the aqueous phase, which results in quenching of the emission. At low pH, on the other hand, an enhancement in fluorescence was observed because PMAA exists in a highly contracted state under these conditions, which protects the dansyl label from the aqueous phase.

2.2.1.2 Excited State Lifetime Measurements The sensitivity of the fluorescence lifetime (τ_f) of a fluorescent label or probe to its environment can often yield important information concerning the conformation of a polyelectrolyte in aqueous solution [12,19–22,26,46,60,61,87–89].

When dispersed in a homogeneous environment, the decay of fluorescence intensity [$I(t)$] from a fluorescent species might be expected to follow a simple, single-exponential model as described by Equation 2.6

$$I(t) = I_0 \exp(-t/\tau_f) \tag{2.6}$$

where I_0 is the initial intensity and τ_f, the excited state lifetime.

For example, Fig. 2.2 shows the transient fluorescence from pyrene (10^{-6} *M*) dispersed in PMAA at pH 5.7 [90] following excitation from a hydrogen discharge

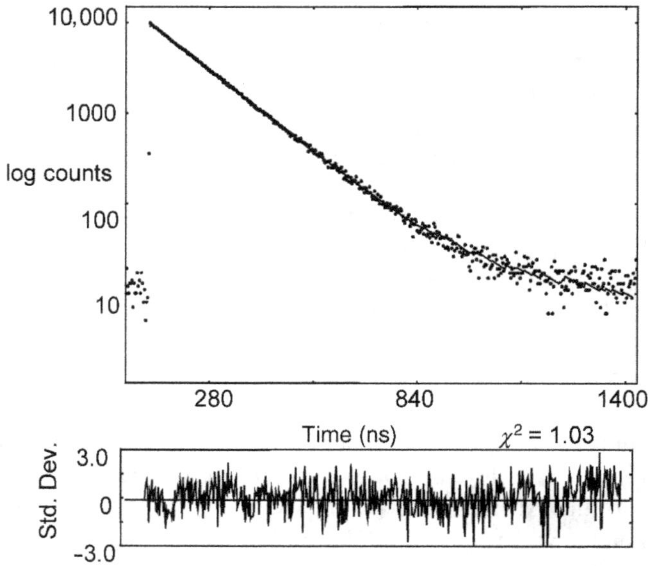

FIGURE 2.2 Fluorescence decay from pyrene (10^{-6} *M*) dispersed in PMAA at pH 5.7 modeled by a single-exponential function ($\lambda_{ex} = 340$ nm; $\lambda_{em} = 380$ nm; air saturated).

lamp and acquisition via the single photon counting method [91]. Classical measurements indicate [2–4,47–52] that the polyelectrolyte adopts an extended water-swollen conformation and is incapable of solubilizing low molar mass organic material [6,17,18,22,60] under these conditions. The probe consequently resides in a largely aqueous environment [6] as reflected by the transient fluorescence in Fig. 2.2 [90]. The decay is well-described by a monoexponential function by consideration of the fitting statistics used to judge the quality of the model: the reduced χ^2 is 1.03 (χ^2 should be close to unity for a good fit) and the residuals are randomly distributed around zero. (The residuals represent the deviation of the fitted function to each datum point.) These factors taken together are indicative of a statistically adequate fit. The lifetime derived from this form of analysis is 132 ns, which implies that a single fluorescent excited state exists in a homogeneous aqueous environment.

At pH 3.2, on the other hand, the decay becomes markedly nonexponential (see Fig. 2.3a). Modeling by a function of the form of Equation 2.6 was not statistically adequate [90]: the χ^2 value is 5.75 and a marked deviation in the residuals is apparent. This infers that additional fluorescent excited states are present under these conditions. Analysis of the data in terms of a double-exponential function as described by Equation 2.7

$$I(t) = I_1 \exp(-t/\tau_{f1}) + I_2 \exp(-t/\tau_{f2}) \tag{2.7}$$

did not prove adequate on statistical grounds: the χ^2 value was 2.1 and a nonrandom distribution of residuals was apparent [90].

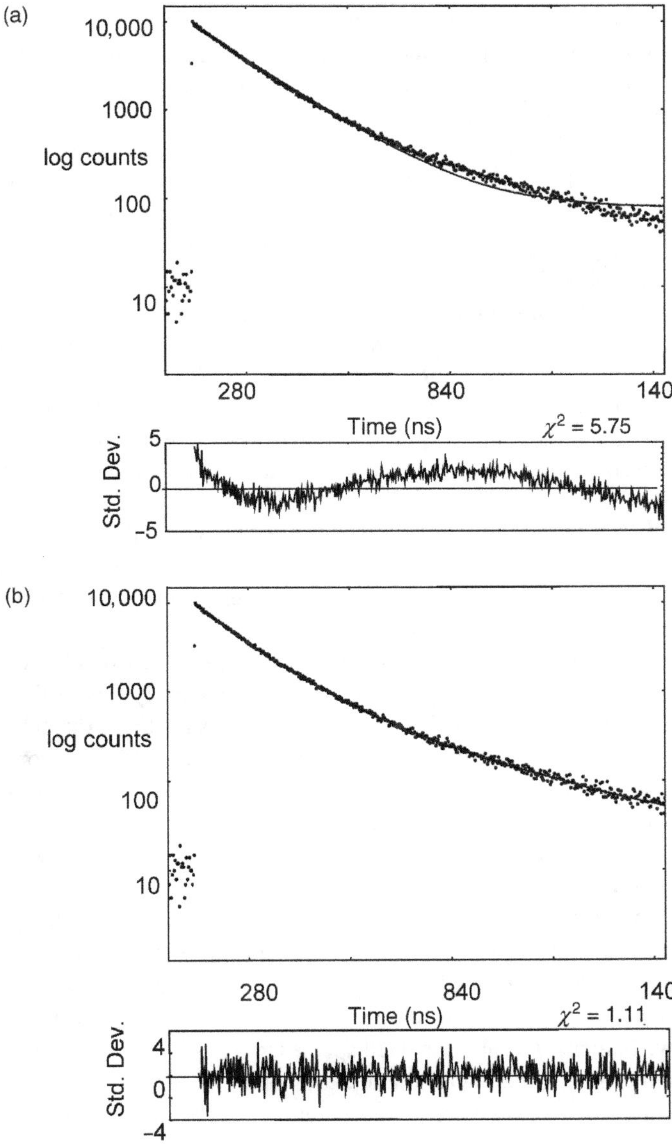

FIGURE 2.3 Fluorescence decay from pyrene ($10^{-6}\,M$) dispersed in PMAA at pH 3.2 modeled by (a) a single-exponential function ($\lambda_{ex} = 340$ nm; $\lambda_{em} = 380$ nm; air saturated) and (b) a triple-exponential function ($\lambda_{ex} = 340$ nm; $\lambda_{em} = 380$ nm; air saturated).

The decay of the pyrene/PMAA dispersion at pH 3.2 could only find adequate statistical description [90] through use of a triple-exponential function of the following form

$$I(t) = I_1 \exp(-t/\tau_{f1}) + I_2 \exp(-t/\tau_{f2}) + I_3 \exp(-t/\tau_{f3}) \tag{2.8}$$

where I_i represents the amount of each component (τ_{fi}) present.

Figure 2.3b shows the improvement in fitting statistics as follows: χ^2 is close to unity and the residuals are randomly distributed around zero. Three lifetimes are derived as a consequence of this form of analysis: 109, 174, and 303 ns, respectively. The shorter-lived component is similar to that derived at pH 5.7 in the expanded conformation and reflects probes which reside in the aqueous phase. On the other hand, τ_2 is presumed [90] to reflect fluorophores, which populate the more polar regions of the PMAA coil. The longest-lived component, τ_3, is associated with a distribution of probes, which reside in the protective domains of the hypercoiled conformation. The triple-exponential nature of the decay is indicative of a heterogeneity of environments in the hypercoiled conformation at pH 3.2 which is a consequence of varying degrees of water penetration [26].

Comparison of the transient fluorescences at pH 5.7 (Fig. 2.2) and pH 3.2 (Fig. 2.3) reveals that the duration of the decay has increased at low degrees of ionization. This is consistent with solubilization of pyrene in the protective domains of the hypercoiled conformation.

Labeled PMAA samples also display complex decay kinetics in aqueous media (see e.g., Refs. [12,19–22,26,46,60–62,87–89]). Selected data from various labels, namely 1-vinyl naphthalene (1VN) [12], acenaphthylene (ACE) [12], dansyl [88], and vinyl pyrene (VPy) [87], at different pH conditions are listed in Table 2.1.

The data listed in Table 2.1 follow the same general trends irrespective of the label. The lifetime is shorter at a higher pH when the polyelectrolyte is in the expanded water-swollen conformation, which exposes the excited state to the quenching effects of the aqueous phase. In addition, the labels reside in a largely homogeneous environment, which results in fluorescence, which can be successfully described by a monoexponential function of the form of Equation 2.6.

More complex decay kinetics are generally encountered at low pH in the hypercoiled conformation requiring double-exponential analyses (see Equation 2.7). This reflects the fact that the fluor exists in a heterogeneous medium [26] in the compact state due to differing degrees of water penetration and quenching. In the more hydrophobic regions of the hypercoil, a longer-lived component (τ_2) is derived from decay analyses, which can be associated with species, which reside in a highly protected environment. The shorter-lived species (τ_1) has its origins in fluorophores, which are in intimate contact with the aqueous phase.

Calculation of an average fluorescence lifetime, via Equation 2.9

$$\langle\tau\rangle = \frac{\sum I_i\tau_i^2}{\sum I_i\tau_i} \tag{2.9}$$

has proven informative [26,61,89] in investigation of the aqueous solution behavior of PMAA and allows simplification of the time-resolved fluorescence to reveal information pertinent to the conformational switch. For example, analysis of the transient emission from ACE–PMAA in terms of a double-exponential function of the form of Equation 2.7 and treatment of the resultant decay parameters via Equation 2.9,

TABLE 2.1 Fluorescence Lifetime Data for Various Labeled PMAA Samples in Aqueous Media

Label	τ_1 (ns)	τ_2 (ns)
1VN (pH 2.0) [12]	7.4	44.4
1VN (pH 12.0) [12]	24.0	—
ACE (pH 2.0) [12]	12.4	33.4
ACE (pH 12.0) [12]	17.9	—
Dansyl (pH 3.9) [88]	7.9	21.9
Dansyl (pH 9.0) [88]	3.7	—
VPy (pH 4.0) [87]	15.5	177
VPy (pH 11.0) [87]	100	

allowed derivation [26] of $\langle\tau\rangle$ as a function of pH. A plot of the data is shown in Fig. 2.4. At low degrees of ionization, the ACE label is preferentially located within the protective confines of the hypercoiled conformation. The resulting $\langle\tau\rangle$ value is ca. 30 ns which indicates that a majority of the fluors are protected from the deactivating effects of the aqueous phase. At high pH, $\langle\tau\rangle$ decreases to ca. 17 ns as the expanded water-swollen structure forms allowing increased ACE–water contacts. Further examination of Fig. 2.4 reveals that between pH 5 and 6, a dramatic reduction in $\langle\tau\rangle$ is apparent, which is similar to the transition region estimated by classical techniques [2–4,47–52]. Clearly these data can be considered as the spectroscopic equivalent of a titration curve allowing the conformational behavior of the polyelectrolyte to be monitored in aqueous solution.

2.2.1.3 Fluorescence Quenching Measurements Addition of an external species, known as a quencher (Q), which is capable of deactivating the excited state through collision can provide information concerning the extent/degree of exposure of a fluorescent species. The process is outlined below

$$F^* + Q \rightarrow F + Q^* \tag{2.10}$$

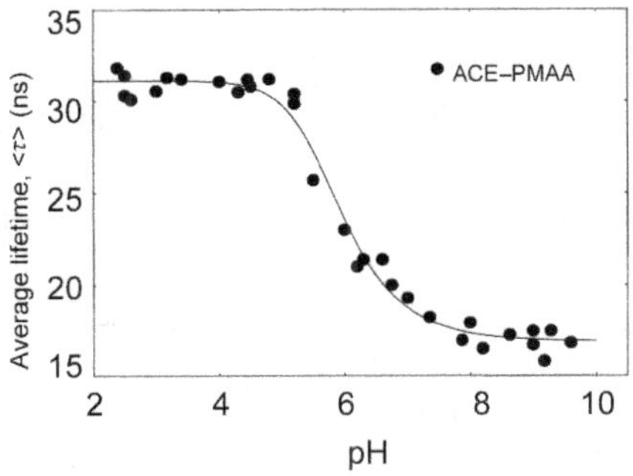

FIGURE 2.4 Average lifetime $\langle\tau\rangle$ for ACE–PMAA as a function of pH ($\lambda_{ex} = 290$ nm; $\lambda_{em} = 340$ nm; air saturated).

The fluorescent excited state (F^*) can transfer its energy to Q (which is usually nonfluorescent) over very short (i.e., contact) distances and is reflected by a decrease in the fluorescence properties such as the quantum yield (Φ), intensity of fluorescence (I), and lifetime (τ).

If the process is dynamic then it can be described by the Stern–Volmer equation

$$\Phi^o/\Phi = I^o/I = \tau^o/\tau = 1 + k_q\tau^o[Q] \tag{2.11}$$

where Φ^o and Φ are the quantum yields in the absence and presence of some concentration of Q, whereas I^o and I and τ^o and τ are the corresponding intensity and lifetime data, respectively. The bimolecular quenching constant (k_q) reflects the efficiency of the process and, in effect, the degree of exposure of the excited state to Q. k_q for a diffusion-controlled reaction of unit efficiency involving spherical molecules may be estimated [92] from

$$k_q = \frac{8RT}{300\eta} \tag{2.12}$$

for a solution of viscosity, η, at a temperature, T, with R as the gas constant.

Under such ideal conditions, k_q would be of the order of $10^{10}\,\text{mol}^{-1}\,\text{dm}^3\,\text{s}^{-1}$ with water as the solvent at 298 K. Comparison of the bimolecular quenching constant for a fluorescent PMAA sample in solution with this ideal diffusion-controlled value forms the basis of interpretation of quenching data [6,12,19,22,46,93–96]. Consequently, Stern–Volmer experiments have featured prominently in the investigation of the aqueous solution behavior of PMAA because information regarding the degree of "openness" or "compactness" of the polymer coil can be derived. Researchers have utilized both probes dispersed in, and labels covalently attached to PMAA to monitor the pH-dependent solution behavior [6,12,19,22,46,93–96]. A variety of quenchers have been used in this context including oxygen [6,19,22], copper(II) [22,94], methylviologen [94], thallium(I) [6,12,19,22,95,96], nitromethane [6,12,19,22,93,94], and iodide [6,12,94].

One of the limitations of quenching experiments on probe/PMAA dispersions is that at high pH, the low molar mass organic material is ejected into the aqueous phase. Under these conditions Stern–Volmer experiments can only report *indirectly* on the polyelectrolyte conformation from examination of the quenching behavior of the probe. The use of labeled polyelectrolytes [6,12,19,22,46,93–96] allows the tagged site to be monitored *directly* across the pH range, including the expanded form, via quenching experiments. Selected data from a variety of label/quencher combinations are listed in Table 2.2.

Examination of the k_q values listed in Table 2.2 clearly reveals the conformational transition of the polyelectrolyte: for example, PMAA has been labeled with 1-pyrene acrylic acid (PyAA) [22], ACE [12], 1VN [12], and vinyl diphenylanthracene (VyDPA) [94] and quenched with nitromethane in aqueous media. The k_q values derived from lifetime data for each of these labels are of the order of 0.5×10^9 $\text{mol}^{-1}\,\text{dm}^3\,\text{s}^{-1}$ at low degrees of ionization. The formation of the hypercoiled conformation provides a degree of protection for the excited state from the quencher, and a low value of k_q results.

TABLE 2.2 Biomolecular Quenching Constants Derived from Various Labeled PMAA Samples

Quencher	Label	k_q (acidic) × 10^{-9} mol^{-1} dm^3 s^{-1}	k_q (basic) × 10^{-9} mol^{-1} dm^3 s^{-1}
CH_3NO_2	1VN [12]	0.40 (pH 3.6)	
CH_3NO_2	ACE [12]	0.60 (pH 4.6)	
CH_3NO_2	VyDPA [94]	0.48 (pH 3.4)	3.70 (pH 6.4)
CH_3NO_2	PyAA [22]	0.54 (pH 4.0)	7.00 (pH 7.0)
I^-	1VN [12]	—	0.24 (pH 6.5)
I^-	ACE [12]	—	0.05 (pH 6.5)
I^-	VyDPA [94]	0.03 (pH 3.45)	0.02 (pH 7.4)
I^-	PyAA [22]	0.007 (pH 2.5)	0.05 (pH 11.8)
Tl^+	1VN [12]	0.30 (pH 3.1)	700 (pH 11.4)
Tl^+	PyAA [22]	0.043 (pH 2.5)	24 (pH 11.8)
Cu^{2+}	VyDPA [94]	0.09 (pH 2.3)	58 (pH 9.0)
Cu^{2+}	PyAA [22]	0.11 (pH 2.2)	12 (pH 7.2)

Further examination of Table 2.2 reveals that extremely efficient quenching occurs by nitromethane at pH values in excess of 6. This behavior is consistent with the formation of the expanded water-swollen form: under these conditions, quenching is diffusion-controlled and is dictated by the viscosity of the solvent (water).

A dramatic reduction in the quenching efficiency is apparent from the labeled PMAA samples (see Table 2.2) on use of I^- at a low pH. This reflects the fact that the anionic quencher is inhibited from accessing the hydrophobic hypercoil. At pH values in excess of 6, although PMAA is in an extended conformation, repulsion occurs between the I^- and the carboxylate ions which results in reduced k_q values (ca. 0.2–0.05 × 10^9 mol^{-1} dm^3 s^{-1}) compared to that observed while using CH_3NO_2 under similar pH conditions (see Table 2.2).

When a positively charged species such as Tl^+ [12,22,94] or Cu^{2+} [22,94] is used to quench the fluorescence from the labeled PMAA samples, on the other hand, the resultant k_q values at high pH greatly exceed that of a diffusion-controlled reaction (see Table 2.2). Such behavior is considered [12,22,94,95] to reflect the fact that the cation clusters around the polysalt form, which creates a high local concentration of quencher in proximity to the label (much higher than the nominal bulk quencher concentration in solution). This results in highly efficient quenching by Tl^+ and Cu^{2+} as reflected by the large k_q values listed in Table 2.2. At low pH, on the other hand, the quencher suffers repulsion due to the hydrophobic nature of the hypercoil, which results in a very low bimolecular quenching constant ($k_q \sim 0.04$–0.1 mol^{-1} dm^3 s^{-1}).

More sophisticated quenching models have been proposed [95,96] in an attempt to describe the complex quenching behavior observed when a cation such as thallium(I) interacts with fluorescently labeled PMAA samples as outlined below.

The Hindered Access Model Webber and coworkers have invoked [95,96] a hindered access model [1] to account for the complex quenching behavior of

PMAA and Tl^+ from samples labeled at various cites within the macromolecule. They reasoned that as a population of fluorophores is isolated from the quencher ions, this requires use of a two-state model in which the probes are divided into an accessible fraction, f_a, and an inaccessible fraction, f_b, ($f_b = 1 - f_a$). If it is assumed that following addition of the quencher, the fluorescence intensity of the accessible species decreases according to the Stern–Volmer relation (Eq. 2.11), where $K_{SV} = k_q\tau^o$, while the intensity of the f_b probes remains unaffected, then the following expression holds

$$\frac{I_o}{(I_o - I)} = 1/f_a + 1/(K_{SV}^a f_a[Q]) \tag{2.13}$$

where $I_o = I_{oa} + I_{ob}$. Consequently, a plot of $I_o/(I_o - I)$ versus 1/[Q] should be linear with an intercept $= f_a$. K_{SV}^a can subsequently be derived from the ratio of the intercept and the slope. As far as the PMAA/Tl^+ system is concerned, the model was only deemed appropriate [95,96] at high ionic strength conditions.

The Combined Stern–Volmer and Perrin Model A model has been proposed by Morishima et al. [97] which takes account of Manning's theory [98] of polyelectrolytes and introduces a modification into the Stern–Volmer equation to describe sphere-of-action (Perrin) quenching: this has been termed combined Stern–Volmer and Perrin Analysis and has been adopted [95,96] in an effort to describe quenching of fluorescence from labeled PMAA by Tl^+ ions, for example.

Manning [98] developed a theory that characterizes polyelectrolytes in aqueous solution in terms of the dimensionless charge density, ξ

$$\xi = \frac{e^2}{\in kTb} \tag{2.14}$$

where b is the average spacing between the charged groups on the polyelectrolyte, e is the proton charge, $\in$ is the solvent dielectric constant, k is the Boltzmann constant, and T is the temperature. The critical charge density is defined as $\xi_o = N^{-1}$ where N is the charge of the counterion. When $\xi > \xi_o$, the counterions are condensed on to the polyelectrolyte until the effective charge density equals ξ_o. At the other extreme, when $\xi < \xi_o$, no condensation occurs and the counterions are completely dissociated. ξ can consequently be considered as a measure of the counterions that condense to the polyelectrolyte and those that remain free in the solution. (Some authors refer to these latter species as atmospheric ions [95–97].)

As far as deactivation by Tl^+ is concerned, a fluorescent label attached to PMAA is considered [95,96] to undergo a mixture of static and dynamic quenching. (Static quenching [1] can be defined as a process that occurs too fast to resolve within the timescale of the experiment. In other words, a ground-state interaction or complex forms between the quencher and the fluorophore before excitation. Such a situation would perhaps be not unexpected when counterions condense in high concentrations to a polyelectrolyte backbone in close proximity to a fluorescent label.)

The Perrin Model describes an "active sphere" as the volume around a fluorophore such that a quencher present within this volume will deactivate fluorescence with unit efficiency. Quenchers outside of this zone do not influence the excited state.

If the exchange of the condensed and atmospheric Tl^+ ions is slow compared to the lifetime of the excited state, a mixture of dynamic and static quenching can be described by a function combining both the Stern–Volmer and Perrin terms as in

$$\frac{I_o}{I} = (1 + K_{SV}[Tl^+]_A)\exp(n) \tag{2.15}$$

where K_{SV} is the dynamic quenching constant ($K_{SV} = k_q\tau^o$), $[Tl^+]_A$ is the concentration of atmospheric Tl^+ ions, and n is the average number of condensed Tl^+ ions in the active sphere.

A degree of success has been achieved through the use of this model to describe, for example, the quenching by Tl^+ ions of PMAA samples labeled at various sites [95,96].

2.2.1.4 Nonradiative Energy Transfer Fluorescence nonradiative energy transfer (NRET) measurements require the presence of two distinct fluorescent species in the system and have the potential to yield detailed information regarding both the magnitude and the mechanism of the conformational change in polyelectrolytes. The process can be distinguished from Stern–Volmer kinetics in that NRET can occur over much larger distances (typically up to 10 nm).

The presence of a fluorescent donor species, D, which is capable of transferring energy to an acceptor, A, forms the basis of NRET measurements as outlined below:

$$D^* + A \rightarrow D + A^* \tag{2.16}$$

The spectroscopic conditions which lead to an enhancement of this from of energy transfer include the following:

(1) Good spectral overlap of the fluorescence of D^* and the absorbance of A.
(2) A large Φ for D^* (in the absence of A).
(3) A large extinction coefficient, ε_A, for A.

Condition (1) can be quantified in terms of the spectral overlap integral, J. This is the integrated overlap of the experimental absorption and emission bands and is defined as

$$J \equiv \int_0^\infty I_D \varepsilon_A \, d\bar{\nu} \tag{2.17}$$

where I_D is the fluorescence spectrum of D and ε_A the absorption spectrum of A. Both the emission and absorbance spectra are usually plotted in wavenumbers and normalized so that complete overlap corresponds to a J value of unity.

As the degree of energy transfer is also dependent on the separation distance between D and A, Förster [99] developed the following expression for the rate constant for energy transfer (k_{ET})

$$k_{ET} = \frac{k\kappa^2 k_D^o J(\varepsilon_A)}{r^6} \tag{2.18}$$

where r is the separation distance between D and A. k is a term that accounts for experimental conditions such as the concentration and the refractive index of the solvent. The orientation of the transition dipoles of D and A is described by the quantity, κ, and is usually assigned a value of 2/3 for randomly oriented groups [1]. k_D^o is the rate constant for the donor in the absence of acceptor.

In practice, as k_{ET} is inversely proportional to r^6, it is often more convenient to measure the efficiency of energy transfer (ET) rather than k_{ET}. Determination of the critical separation distance, R_o, provides a means of quantifying, experimentally, the ET efficiency.

R_o is defined [99] as

$$R_o = \left[\frac{900 \ln 10\kappa^2 \phi_{DA}}{125\pi^5 n^4 N_o}\right]^{1/6} \tag{2.19}$$

where ϕ_{DA} is the donor fluorescence quantum yield in the absence of energy transfer, n is the refractive index of the solvent at the wavelength of excitation, and N_o is Avogadros constant.

R_o is termed [99] as the distance at which the concentration of acceptor is such that the rate of energy transfer equals the rate of deactivation of D. Under these conditions, the efficiency of ET is 50%, allowing modification of Equation 2.18 to

$$k_{ET} = \frac{k_D^o R_o^6}{r^6} \tag{2.20}$$

As the efficiency of ET is defined as

$$ET = \frac{k_{ET}}{k_D^o + k_{ET}} \tag{2.21}$$

then combination of Equations 2.20 and 2.21 allows the efficiency to be expressed in terms of R_o and r

$$ET = \frac{R_o^6}{R_o^6 + r^6} \tag{2.22}$$

Once R_o has been determined for a particular D and A pair, it is possible to estimate, experimentally, the actual separation distance (r) in a particular system, via Equation 2.22, prompting the use of the term Spectroscopic Ruler Technique to describe such measurements.

In an extension of the premise of the Spectroscopic Ruler Technique, if the donor and acceptor species are covalently bound to a macromolecule, information concerning distances within polymer chains [26,100–104] and polyelectrolytes, in particular [26,100,104], can be determined. For example, Guillet and coworkers [104] investigated a PMAA sample labeled internally with a naphthalene donor and an anthryl acceptor, which was sited at the chain terminus. Examination of the steady-state spectra [104] provided an indication of the proximity of D to A and hence offered a simple yet effective means of monitoring the degree of expansion or collapse of the polyelectrolyte under various pH conditions. Figure 2.5 shows the fluorescence emission resultant on exciting the naphthyl donor at 290 nm at pH 2.0 and 10.54, respectively, for the double-labeled PMAA sample. The enhancement in intensity observed from the anthryl acceptor between 400 and 450 nm provides clear evidence of efficient energy transfer in the compact, hypercoiled conformation. Further examination of Fig. 2.5 reveals that less emission from A is apparent on expansion of the labeled PMAA at higher degrees of ionization. The authors [104] concluded from this investigation that energy transfer was enhanced due to the hypercoiled conformation and also as a result of the increased mobility of the chain ends. (Both conditions bring D and A to within the critical R_o distance.)

In a time-resolved experiment, the occurrence of NRET can be confirmed by examination of the transient fluorescence of D: the donor fluorescence lifetime will be quenched (i.e., the donor emission will decay faster) as its energy is transferred to A [26].

An alternative approach for estimation of the efficiency of ET to that from Equation 2.22 can be achieved by consideration of the excited state lifetime data as described by Equation 2.23

$$\mathrm{ET} = 1 - \frac{\langle \tau_{\mathrm{ET}} \rangle}{\langle \tau_{\mathrm{D}} \rangle} \tag{2.23}$$

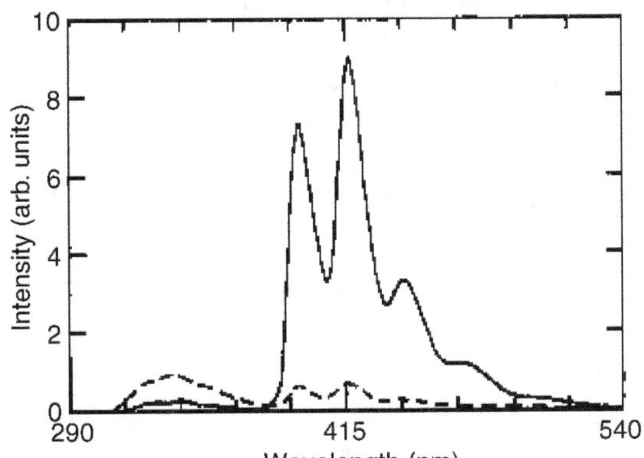

FIGURE 2.5 Steady-state fluorescence emission spectra of a PMAA sample labeled with naphthalene and anthracene with $\lambda_{ex} = 290$ nm at pH 10.54 (dashed line) and pH 2.0 (continuous line). [Reproduced from reference 104.]

where $\langle \tau_{ET} \rangle$ and $\langle \tau_D \rangle$ are the average fluorescence lifetime of the donor in the presence and absence of energy transfer, respectively. Estimation of $\langle \tau \rangle$ involves analysis of the donor decay curve in terms of multiexponential functions (e.g., Eq. 2.7) and subsequent treatment of the fitting parameters by Equation 2.9.

A comprehensive study of the conformational behavior of PMAA in aqueous solution via both fluorescence spectroscopy and dynamic light scattering (DLS) has revealed [26] subtle information concerning the mechanism of the pH response of the polyelectrolyte. Investigation of a double-labeled sample containing ACE as the donor and anthryl methyl methacrylate (AMMA) as the acceptor via NRET proved particularly informative in this respect. Figure 2.6 shows the average donor lifetime $\langle \tau \rangle$ for ACE–AMMA–PMAA as a function of pH. Also shown for comparison is the corresponding single-labeled sample, ACE–PMAA. Energy transfer clearly occurs at pH values below 6 in the hypercoiled conformation: $\langle \tau \rangle$ for ACE–PMAA is always superior to that of ACE–AMMA–PMAA as would be expected because emission from the donor is quenched in the presence of NRET. At high degrees of ionization (greater than pH 6), the two plots are superimposable indicating that expansion of the chain occurs and D and A are out with the critical transfer distance. Use of Equation 2.23 allows the energy-transfer efficiency to be estimated: the nonradiative transfer efficiency remains constant (ET $\approx$ 0.14) between pH 2 and 5 but decreases to zero at pH $>$ 5 on expansion of the coil. As the R_o value for this donor–acceptor pair has been determined as 2.3 nm [105] and the coil dimensions derived [26] from DLS are of the order of 15 nm at pH 6, the degree of energy transfer would be expected to be minimal under these conditions which is in agreement with the experimental observations [26]. The authors [26] concluded that by pH 5.6 for PMAA, the macroscopic switch in conformation is complete. In addition, the energy-transfer study [26] resolved to some extent the controversy regarding the exact nature of the mechanism of the smart response in PMAA: potentiometric experiments have been performed [47], which suggest that the polyelectrolyte has a broad and ill-defined neutralization region which has been attributed [4] to a gradual breakdown in short-range interactions between methyl units and carboxylic acid groups. Experimental evidence suggests that the change is cooperative and occurs in one step, analogous to

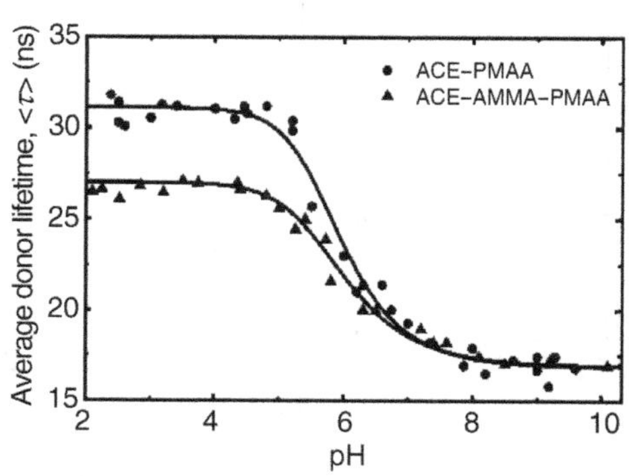

FIGURE 2.6 Average donor lifetime $\langle \tau \rangle$ for ACE–AMMA–PMAA and ACE–PMAA as a function of pH ($\lambda_{ex} = 290$ nm; $\lambda_{em} = 340$ nm; air saturated). [Reproduced from reference 26.]

the helix-coil transition of polypeptides [106]. Whereas data from Raman spectroscopy indicate the presence of a multiplicity of structures, suggesting that the transition occurs as a progressive rather than a cooperative change [53]. Energy-transfer measurements may be considered ideal for interrogating such short-range phenomena because the technique is responsive to interactions, which occur on the nm-length scale. As a consequence of NRET experiments [26], the breakdown in short-range interactions between methyl units and carboxylic acid and carboxylate anions as the pH increases toward neutralization was monitored. Steady-state and time-resolved measurements [26] revealed that a gradual rearrangement of the coil occurs over several pH units as mutual repulsion between increasing concentrations of carboxylate anion begins to dominate. At a critical pH (confirmed by DLS measurements [26]), repulsive interactions involving anions serve to expand the coil into a water-swollen structure breaking down any remaining hydrophobic forces of attraction in the process. The authors [26] concluded as a consequence of information from both NRET and DLS measurements that small-scale rearrangements in structure occur between pH 2 and 5, rather than a large-scale expansion. This is subsequently followed by a macroscopic change in dimension of PMAA at a pH of ~5.

2.2.1.5 Excimer Formation Interaction of a ground-state species (M) with an excited state (M^*) results in the formation of an excited state dimer, known as an excimer. The process is outlined below

$$M^* + M \rightleftarrows D^* \tag{2.24}$$

Excimer formation requires a sandwich-type overlap between M and M^* and can occur either intramolecularly or intermolecularly. The short-range nature of the phenomenon (<2 nm) lends the technique to probing the solution behavior of polyelectrolytes. Pyrene is often cited as the classical excimer-forming species and so provides the ideal example of this spectroscopic method: the experimental approaches adopted have included the use of both probes [6,71] and labels [87] to reveal the smart response of PMAA.

Section 2.2.1.1 highlighted how the vibrational fine structure of pyrene, when dispersed under relatively dilute conditions, in the absence of excimer formation, can be used to reveal the pH-dependent response of PMAA. An alternative approach has used the occurrence of intermolecular excimer formation between dispersed pyrenyl probes under relatively concentrated conditions to monitor the conformational switch of the polymer [6,71]. When dispersed in PMAA, at concentrations higher than its solubility limit in water, partitioning of the probe between the hydrophobic polymer coil and the aqueous phase results [6,71]. Dependent on the pH of the solution and consequently the conformational state of the polyelectrolyte, the balance between these two distinct populations can be affected. For example, Fig. 2.7 shows the fluorescence emission spectra for the probe 1,3-bis(1-pyrenyl)propane (1,3PP) when dispersed in aqueous solutions of PMAA [71]. At low pH, an emission spectrum consistent with that of monomeric unassociated pyrene is observed centered around

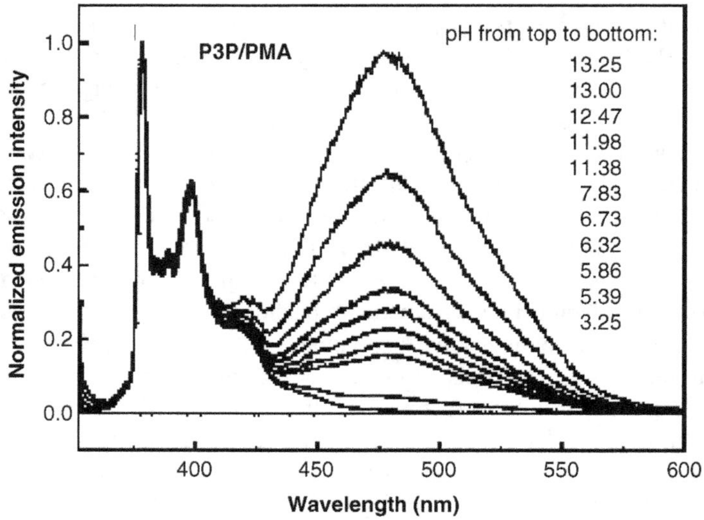

FIGURE 2.7 Normalized fluorescence emission spectra of 1,3PP dispersed in PMAA as a function of pH ($\lambda_{ex} = 348$ nm; air saturated). [Reproduced from reference 71.]

400 nm. Under these conditions, 1,3PP is sequestered into the hydrophobic domains of the compact conformation, which depletes the population in the aqueous phase. Furthermore, each pyrene moiety is solubilized within its own distinct hydrophobic cavity in isolation from other probes and thus, preventing excimer formation within the hypercoil. At pH 5.86, a broad structureless emission, centered at ca. 480 nm, is apparent which can be assigned to that of the excimer. Fluorescence from this species increases at the expense of monomer emission (at ca. 400 nm) as the pH is raised further to highly basic conditions. This reflects increased intermolecular interactions between probes following ejection into the aqueous phase: from Equation 2.24, excimer formation results in quenching of the excited state monomer, M^* population.

Plotting the ratio of the emission intensity of the excimer sampled at 480 nm (I_E) [71] against that analyzed in the monomer region (I_M) at ca. 400 nm provided another means by which the conformational switch of PMAA could be monitored. Between pH 3 and 5, I_E/I_M remained low and constant which implies that very little excimer formation occurs in the system. This reflects on the fact that the majority of the probes are sequestered in the hypercoil leaving a population approaching the solubility limit in water which remains in the aqueous phase. As the 1,3PP probes occupy their own individual cavities in the hypercoil, intermolecular excimer formation is minimal. As the conformational switch occurs between pH 5 and 6, a fraction of the probes is released into the aqueous phase exceeding the solubility limit and consequently increasing intermolecular interactions between the 1,3PP moieties to minimize contact with the aqueous phase. This is reflected by a sharp increase and then a plateau in the I_E/I_M plot at pH values beyond 6. In addition to the intermolecular interactions discussed above, if 1,3PP is dispersed in a fluid medium, intramolecular excimer formation is also possible, because rotation of the pyrenyl units about the propyl spacer can occur in a sandwich-type configuration. As a consequence, the 1,3PP probe has been used as a microviscosity sensor [107,108]. It was, therefore, acknowledged [71] that the 1,3PP probe also reports on the microviscosity of the PMAA hypercoil via the extent of excimer emission observed. At a low pH, the microviscosity is such that rotation of the pyrenyl units into an excimer-forming conformation is prevented and 1,3PP is fully extended. At a high pH, on the other hand, ejection of 1,3PP into the lower-viscosity aqueous phase allows facile rotation of the aromatic units resulting in a high degree of excimer formation.

2.2.1.6 Steady-State Anisotropy Following continuous excitation with vertically polarized light, a distribution of fluorophores whose transition vectors for the absorption process are vertically aligned will be photoselected, creating an excited state population, which possesses a degree of anisotropy (r) or optical order, in an otherwise isotropic distribution of fluorophores. Measurement of the intensity of fluorescence, via an emission polarizer in planes parallel ($i_{\parallel}$) and perpendicular ($i_{\perp}$) to the vertical plane allows estimation of r from

$$r = \frac{i_{\parallel} - iG_{\perp}}{i_{\parallel} + 2Gi_{\perp}} \tag{2.25}$$

where G is the factor that accounts [109] for the wavelength dependence of the transmission of radiation by the monochromator. G can be estimated, for example, by comparison of the emission intensities from the sample under study analyzed in planes parallel and perpendicular, respectively, to the plane of *horizontally* polarized excitation. Experimentally, the optical order within the excited state population may also be expressed as a degree of polarization, p, of the sample

$$p = \frac{i_{\parallel} - iG_{\perp}}{i_{\parallel} + Gi_{\perp}} \tag{2.26}$$

This alternative form of expression has been favored by some researchers in the literature [16,18,46]. The quantities p and r can be simply interchanged via the following relationship

$$r = 2p/(3-p) \tag{2.27}$$

If no significant rotation of the fluorophores occurs within the excited state fluorescence lifetime, then a high degree of anisotropy will be retained within the sample reflected by a value of r close to r_{o} (r_{o} is a spectroscopic parameter and is the value of r the instant the excited state population is created). If significant rotation of the photoselected excited states occurs, on the other hand, the estimate of r from Equation 2.25 is consequently small (<0.01) because the optical order created at the instant of excitation is rapidly lost.

Use of the Perrin equation

$$r^{-1} = r_{\mathrm{o}}^{-1}(1 + \tau_{\mathrm{f}}/\tau_{\mathrm{c}}) \tag{2.28}$$

allows the quantification of the speed of rotation of the photoselected population within the excited state lifetime, τ_f, through the correlation time, τ_c. τ_f can be measured independently by the single photon counting method [91], for example. The problem in the steady-state experiment lies in the estimation of r_o. Two approaches have been adopted in the literature. r_o may be estimated as the value of τ_c in the limit as τ_c tends to infinity at a constant τ_f. As τ_c is related to the molar volume, V, of the fluor and viscosity, η, of the medium, via the Stokes–Einstein equation

$$\tau_{\mathrm{c}} = \frac{\eta V}{RT} \tag{2.29}$$

then r_o (p_o) may be estimated [16,46] as the intercept of a plot of r^{-1} (p^{-1}) against T/η. (This extrapolation procedure assumes proportionality between τ_c and η/T.)

As an alternative approach, r_o (p_o) may be derived by extrapolation of r^{-1} (p^{-1}) to zero τ_f while τ_c is held constant. This can be achieved by the addition of relatively small concentrations of a quencher capable of dynamically deactivating the excited state [110]. r^{-1} is subsequently plotted against I/I^o or τ/τ^o to allow derivation of r_o^{-1} from the intercept. For example, this form of extrapolation for ACE-labeled PMAA at pH 12.7 using iodide as a quencher results in an r_o^{-1} value of 8.7 [111].

In the context of probing the conformational behavior of PMAA and the use of steady-state polarization measurements in particular, we should turn our attention at this point to the pioneering work of Anufrieva and Gotlib [16]. In a comprehensive investigation of PMAA labeled with anthryl derivatives and 1-anilino-8-naphthalene sulfonate (ANS), the conformational transition of the polyelectrolyte was monitored [16] in aqueous solution via the relaxation time (τ_w). (τ_w is an alternative means of expressing the macromolecular motion and is simply the equivalent of $3\tau_c$.) A dramatic increase in the mobility of the labeled polyelectrolyte was observed as the degree of ionization was increased. Within the hypercoiled conformation, hydrophobic interactions between methyl groups create what the authors describe as an *internal structure*. This structure is subsequently broken down as charge repulsions from increasing concentrations of carboxylate anions induce expansion into a flexible fast-moving chain.

The polarization of a DMA-labeled PMAA sample was monitored [18] as a function of pH, and τ_c was later derived [46] at various degrees of ionization, via Equation 2.28. τ_c varies from ca. 32 ns at $\alpha = 0$ to ca. 6 ns at $\alpha = 0.8$. Not surprisingly, the authors [46] offered a similar explanation for the pH dependence of τ_c to that of Anufrieva and Gotlib [16]: essentially, a breakdown in the hypercoil structure occurs as α increases and the polyelectrolyte expands allowing increased mobility of the chain segments.

2.2.1.7 Time-Resolved Anisotropy Measurements Time-resolved anisotropy measurements (TRAMs) offer a distinct advantage over their steady-state

$$r^{-1} = r_o^{-1}(1 + \tau_f/\tau_c) \tag{2.28}$$

counterparts in that they allow probing of more complex motions: ultimately, two or more correlation times can be derived from these data. The premise of the Perrin relationship (Eq. 2.28) is that a single motion occurs within the system under study. Consequently, if a more complex relaxation behavior is operative then only an average anisotropy can be monitored from the various motions via steady-state measurements.

In a time-resolved experiment, the time-dependent intensities, $i_{\|}(t)$ and $i_{\perp}(t)$, are measured in planes parallel and perpendicular, respectively, to the plane of polarization of the incident radiation. These parameters are subsequently combined, via Equation 2.30, to generate the decay of anisotropy, $r(t)$

$$r(t) = \frac{i_{\|}(t) - iG_{\perp}(t)}{i_{\|}(t) + 2Gi_{\perp}(t)} = \frac{d(t)}{s(t)} \tag{2.30}$$

where $s(t)$ and $d(t)$ are the sum and difference functions, respectively. As in the steady-state experiment, G corrects [109] for the instrumental differences in the transmission and detection efficiencies in the determination of $i_{\|}(t)$ and $i_{\perp}(t)$.

The loss of r will result through molecular motion within the excited state fluorescence lifetime (τ_f) until the photoselected population achieves an isotropic orientation. If the anisotropy decays following a simple, single relaxation mechanism, it will be described by Equation 2.31

$$r(t) = r_o \exp(-t/\tau_c) \tag{2.31}$$

where τ_c characterizes the rate of motion under study.

When TRAMs are made on dispersed probes, τ_c will reflect the speed of rotation of the fluorophore [20,46,60,76], which can be related to the microviscosity of the medium. In the context of probing the structure of PMAA, the fluorescent dyes have been occluded in the hypercoiled conformation [46,60,76] allowing an estimate of the size of the rotating, solubilizing cluster to be derived from the resultant τ_c. Alternatively, if the fluorophore is covalently attached to the polyelectrolyte in the form of a label then, depending on its mode of attachment, information concerning motion of the chain ends [46,60,76], the backbone [26,88,112,113], and chain substituents [26,88] can be derived from τ_c.

There are several analytical procedures available for derivation of relaxation information from time-resolved anisotropy experiments, the merits of which have been discussed at length elsewhere [25,112,114]. The salient points are covered here: direct analysis of $r(t)$ using a function such as Equation 2.31 is the most straightforward method but can become particularly problematic if the motion under study is comparable to the width of the excitation pulse [25,112,114]. Furthermore, as $r(t)$ can suffer contamination from the polarizing effects of stray excitation from the source, particularly in weakly fluorescent samples, other methods are required to overcome such artifacts. Impulse reconvolution [115] allows mathematical "removal" of the instrumental pulse from the experimental data and involves an analysis of $s(t)$ by a statistically adequate model function (e.g., Eq. 2.8). The "best fit" to $s(t)$ is

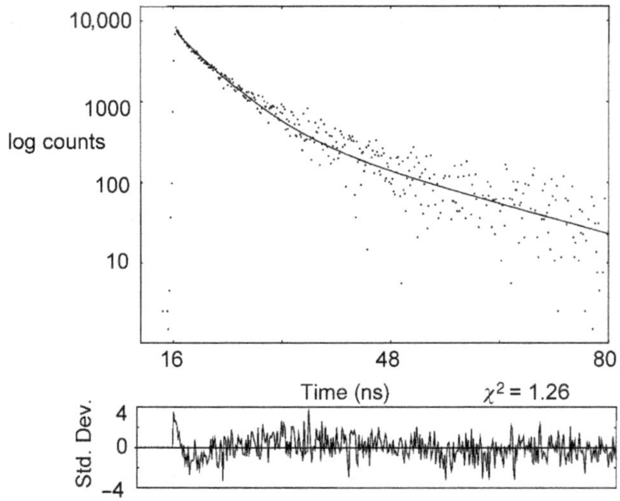

FIGURE 2.8 Difference function, $d(t)$, and impulse reconvolution fit for ACE–PMAA (10^{-3} wt%) at pH 11 ($\lambda_{ex} = 290$ nm).

represented by an impulse response function which, in combination with an assumed model (e.g., Eq. 2.31) for the anisotropy decay, may be used in an iterative least squares reconvolution procedure to fit $d(t)$. This fit is optimized by varying the fitting parameters, which allows derivation of relaxation data for the process under investigation.

Figure 2.8 shows an example of a difference function, $d(t)$, from an ACE–PMAA sample at pH 11 and associated fit to a single-exponential model of the form of Equation 2.31 following impulse reconvolution [115] analysis. The statistical criteria used to judge the quality of fit to the data are similar to those described in Section 2.2.1.3: χ^2 should be close to unity for a good fit and the residuals (which estimates the deviation of each datum point to the model) should be randomly distributed around zero. Clearly, from consideration of the χ^2 value and distribution of the residuals, a single-exponential model is appropriate for ACE–PMAA under these conditions (see Fig. 2.8). The τ_c derived as a consequence of this form of analysis is 7.9 ns [112]. This can be contrasted with the value (ca. 1.3 ns) observed for a similarly labeled poly(methyl methacrylate) sample in dichloromethane [116]. Given the slight difference in viscosity of the two solvents (and accounting for this), the τ_c observed for ACE–PMAA is greater than that of the parent ester. These observations imply that the ACE label experiences hindered mobility even in the expanded water-swollen state of PMAA at high degrees of ionization.

Early time-resolved anisotropy experiments on PMAA used a combination of anthryl-based labels and probes [18,46,60,76] in an effort to fully characterize the conformational switch of the polyelectrolyte in aqueous solution. In their study of probes dispersed in and labels incorporated into PMAA, Treloar and coworkers [60,76] derived information from anisotropy experiments pertinent not only to the cluster size of the rotating units, but also to the structure of the hypercoil itself.

For the probes 9-methyl anthracene (MA) and DMA solubilized in the hypercoiled conformation of PMAA, the anisotropy decays could best be described by invoking double-exponential analysis of the form described by Equation 2.32

$$r(t) = r_1 \exp(-t/\tau_{c1}) + r_2 \exp(-t/\tau_{c2}) \tag{2.32}$$

Such complexity in anisotropy decay kinetics was considered to be consistent with the existence of two rotating units. As a consequence of these measurements, a model was proposed [60,76] for the structure of PMAA at low degrees of ionization, later termed [46] the *String of Pearls Model.* This model suggests that the hypercoiled conformation does not exit as a single homogeneous entity but, instead, forms a series of globules linked by faster-moving segments, which are exposed to the aqueous phase.

Further TRAMs on PMAA labeled with dansyl [88], 1VN [112], ACE [112] and more recently AMMA [26] and aminoacridine derivatives [113] have also revealed the pH-dependent conformational behavior. For example, Fig. 2.9 shows a plot of τ_c versus pH for various aminoacridine derivatives (see Scheme 2.1 for the structures) covalently bound to the polyelectrolyte.

The relaxation data follow the same general trend irrespective of the label [113]: at a low pH, τ_c is long, which is indicative of restricted motion in the hypercoil. At pH 4, τ_c maximizes [113], which is presumed to reflect a tightening of the hypercoil [24] and a resultant increase in microviscosity under these conditions. This behavior has been attributed to the presence of carboxylate and carboxylic acid groups, which form strong hydrogen bonding interactions [24] and have been observed [26,88,112] in other labeled PMAA systems following anisotropy measurements. Further examination of Fig. 2.9 reveals that τ_c decreases between pH 4 and 6 as expansion of the coil

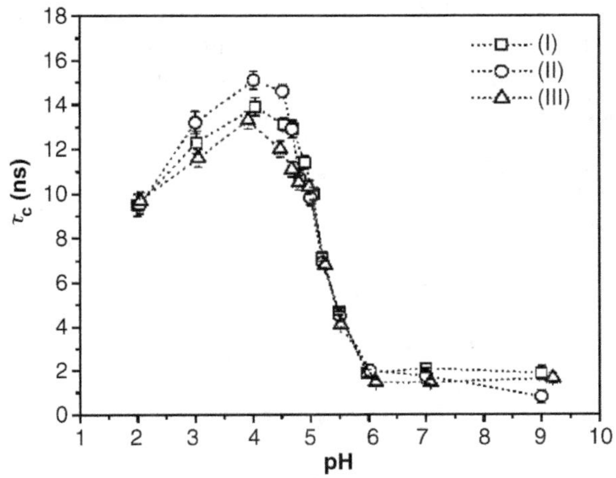

FIGURE 2.9 τ_c for the 9-aminoacridine derivatives bound to PMA (see Scheme 2.1) as a function of the pH of the solution at 298 K ($\lambda_{ex} = 400$ nm; $\lambda_{em} = 460$ nm). [Reproduced from reference 113.]

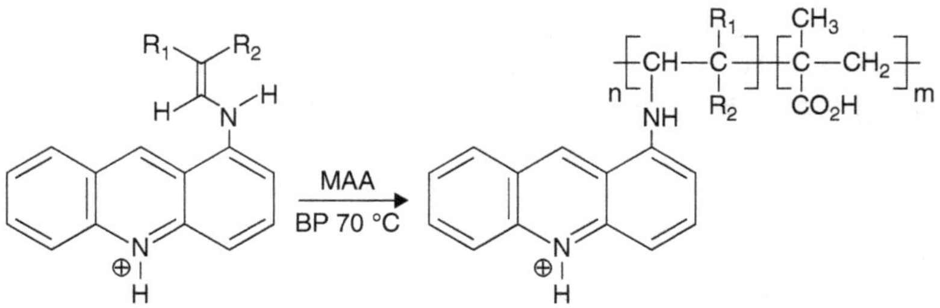

SCHEME 2.1 Copolymerization of the 9-aminoacridine derivatives with MAA using benzyl peroxide (BP). Compound **I**, R1 = R2 = COOEt; compound **II**, R1 = CN, R2 = COOEt; compound **III**, R1 = R2 = CN.

occurs due to ionic charge repulsions. Beyond pH 6, τ_c plateaus at ca. 2 ns, which is indicative of fast label relaxation in the expanded polysalt form.

2.2.1.8 ***Summary*** Fluorescence techniques have confirmed that PMAA undergoes a change in conformation between pH 4 and 6. At a low pH, the polyelectrolyte forms a compact structure (or *hypercoil* [48]) due to the presence of the backbone methyl substituent, which induces additional coiling to minimize contact with the aqueous phase. Anisotropy data indicate [46,60,76] that the hypercoil is not a single homogeneous entity but, instead, is a series of globules linked by faster-moving segments, which are exposed to the aqueous phase. At pH 4, a "tightening" of the collapsed state occurs [26,88,112,113] as a consequence of hydrogen bonding interactions between carboxylic acid units and carboxylate anions [24]. NRET measurements [26] have revealed that the conformational transition occurs over a broad pH range and involves subtle rearrangements of the coil as a consequence of the breakdown in hydrophobic interactions between methyl substituents and carboxylic units in response to increasing concentrations of carboxylate anions. At a critical pH, the concentration of anions is such that a macroscopic switch in conformation occurs between pH 4 and 6 to the expanded water-swollen state.

2.2.2 Poly(acrylic acid)

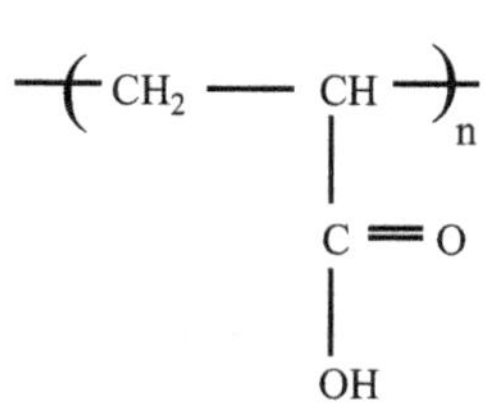

poly(acrylic acid), PAA

The lack of a backbone methyl group (which is present in PMAA) has a profound effect on the aqueous solution behavior of PAA: potentiometric [3,4,47,49] and viscometric measurements [3,117] indicate that PAA undergoes a relatively smooth change from a statistical coil at low pH to an extended state at high degrees of ionization as repulsive interactions between carboxylate anions dominate. Consequently, PAA has been regarded [118] as the "ideal" synthetic polyelectrolyte due to the absence of a conformational change.

2.2.2.1 Steady-State Spectroscopy The fact that PAA adopts a random conformation at low pH [3,4,47,49,117] results in the inability of the polyelectrolyte to solubilize low molar mass organic material [17]. This is in stark contrast to the behavior observed for PMAA [6,17,18,60] as highlighted in Section 2.2.1.1: PMAA is known [6,17,18,60] to solubilize low-molecular weight hydrocarbons in its compact form in aqueous media. Further evidence for the absence of any significant conformational change has been revealed by examination of the vibrational fine structure of the emission spectrum of pyrene when dispersed in PAA: the relative intensity of bands 3 and 1 of the probe change in different solvents and can consequently serve as a polarity indicator [63,64] for a variety of systems [66–70]. For example, pyrene (10^{-6} *M*) was dispersed in a dilute aqueous solution of PAA and the fluorescence spectra obtained as a function of pH [20]. An I_3/I_1 ratio of ca. 0.55 was calculated which was invariant with the degree of ionization. The fact that this value is the same as that determined for pyrene (10^{-6} *M*) dispersed in water [6,20] was considered [20] to reflect the fact that if the pyrene is solubilized at all by the PAA, it experiences a microenvironment very similar to that of the aqueous phase. This suggests that PAA adopts an expanded water-swollen conformation across the entire pH range.

Changes in the fluorescence spectral characteristics of water-soluble probes have also been used to provide information concerning the aqueous solution behavior of PAA [72–75,77]. For example, the fluorescence intensity of AuO is low in water but is enhanced in the presence of PMAA, particularly at low degrees of ionization [52,73]. This enhancement was attributed to a higher local viscosity at the molecular level created by the polyelectrolyte chains. The emission from AuO dispersed in PAA, on the other hand, was low across the complete pH range [73] and implies that the probe resides in a water-rich phase, which is consistent with an expanded polymer conformation.

The fluorescence of a dansyl-labeled PAA has been used to examine the response of the polyelectrolyte to a pH jump [24]. At higher degrees of ionization, expansion of the PAA chain leads to the exposure of the label to the aqueous phase resulting in fluorescence quenching. However, this effect was observed to be greater for a similarly labeled PMAA sample, which exists in a highly contracted state at low pH. These investigations provide further evidence for not only the presence of the hypercoiled PMAA structure, but also the absence of a significant conformational change in PAA.

2.2.2.2 Excited State Lifetime Measurements Corroborative evidence for an expanded water-swollen PAA structure across the entire pH range has been derived from excited state lifetime measurements [20,65,87,119]. For example, pyrene 10^{-5} *M* was dispersed as a probe in a dilute PAA aqueous solution (10^{-2} wt% w/v) to monitor the intramolecular interactions of the polyelectrolyte [20]. The fluorescence decay from the pyrene/PAA dispersion was well described by a monoexponential function of the form of Equation 2.6 resulting in a lifetime of ca. 132 ns, which was invariant with pH [20]. The fact that the lifetime is comparable to that of pyrene in water [20] was considered strong evidence that the probe resides in a microenvironment analogous to that of the aqueous phase. Similar trends in fluorescence lifetime data were reported [20] for the probe MA when dispersed in PAA: at pH 3.4, τ was estimated as 10.9 ns, which is identical to that of the probe in water. These observations are consistent with the existence of an expanded water-swollen PAA conformation across the entire pH range in agreement with steady-state spectral measurements [20,24,72–75,77].

2.2.2.3 Fluorescence Quenching Measurements Data as a consequence of spectral and lifetime measurements [20,24,72–75,77] discussed in the preceding sections have inferred that PAA forms a water-swollen and expanded conformation across the entire pH range. Fluorescence quenching experiments offer the opportunity of probing the "degree of openness" of a polyelectrolyte coil in aqueous solution and, as such, can provide information concerning the magnitude of any conformational change [6,12,19,22,46,80,93–96]. Researchers have recognized this fact as far as PAA is concerned [22,65,93,119] and Table 2.3 lists the bimolecular quenching constants as a result of Stern–Volmer experiments on naphthyl-labeled [65,93] and pyrenyl-labeled [22] samples at various pH conditions.

Examination of the quenching constants listed in Table 2.3 clearly reveals that, irrespective of the label adopted, or the pH investigated, when nitromethane is used as a quencher, the resultant k_q value lies between 4 and $5 \times 10^9\ \mathrm{mol^{-1}\,dm^3\,s^{-1}}$. This is similar to that for a diffusion-controlled process and is consistent with PAA adopting an expanded state under these conditions: the quencher is unhindered by the water-swollen polymer coil as it diffuses toward the excited state. The fact that the k_q value observed is constant with pH implies that either PAA is unresponsive to pH in terms of its conformational behavior or the Stern–Volmer experiment is insensitive to any change in conformation that may occur. Further consideration of the data listed in Table 2.3 demonstrates that the k_q values at low pH are 10 times greater than that of PMAA using similar labels and solution conditions (see Table 2.2). This provides

TABLE 2.3 Biomolecular Quenching Constants Derived from Various Labeled PAA Samples

Quencher	Label	k_q (acidic) $\times$ $10^{-9}\ \mathrm{mol^{-1}\,dm^3\,s^{-1}}$	k_q (basic) $\times$ $10^{-9}\ \mathrm{mol^{-1}\,dm^3\,s^{-1}}$
CH_3NO_2	ACE [93]	4.9 (pH 3.7)	—
CH_3NO_2	Pyrene [22]	3.5 (pH 2.2)	3.9 (pH 7.2)
Tl^+	Pyrene [22]	20 (pH 2.0)	100 (pH 7.2)
Tl^+	ACE [65]	3.3 (pH 2.3)	760 (pH 11)
Cu^{2+}	Pyrene [22]	2.1 (pH 2.1)	45 (pH 7.6)

evidence for the existence of the compact conformation of PMAA and further illustrates the importance of the backbone methyl group in inducing the hypercoiled state.

Further examination of the data listed in Table 2.3 reveals that the Tl^+ [22,65] and Cu^{2+} [22] ions quench with a similar efficiency to that of CH_3NO_2 at low pH. This reflects the fact that the cations experience a largely aqueous environment as they diffuse toward the excited state and would be consistent with an expanded polyelectrolyte coil under these conditions. In contrast, at higher pH values (see Table 2.3), more efficient quenching occurs than that of a diffusion-controlled process. Similar behavior has been observed [12,22,94,95] for PMAA (see Table 2.2) and signals the existence of carboxylate anions under these conditions: the cation clusters around the polysalt form, which creates a high local concentration of quencher around the label (higher than the nominal bulk quencher concentration in solution). Extremely efficient quenching results under these conditions as reflected by the large k_q values for PAA (see Table 2.3).

In summary, Stern–Volmer experiments infer that PAA adopts an expanded water-swollen conformation across the pH range.

2.2.2.4 Nonradiative Energy Transfer Unique information concerning the conformational behavior of aqueous solutions of PAA bearing naphthyl (donor) and pyrenyl (acceptor) species has been derived from NRET measurements [120]. A plot of the ratio of pyrene to naphthalene emission intensity (I_{PY}/I_{NP}) upon excitation of the naphthalene fluorophores at 290 nm as a function of pH is shown in Fig. 2.10. The

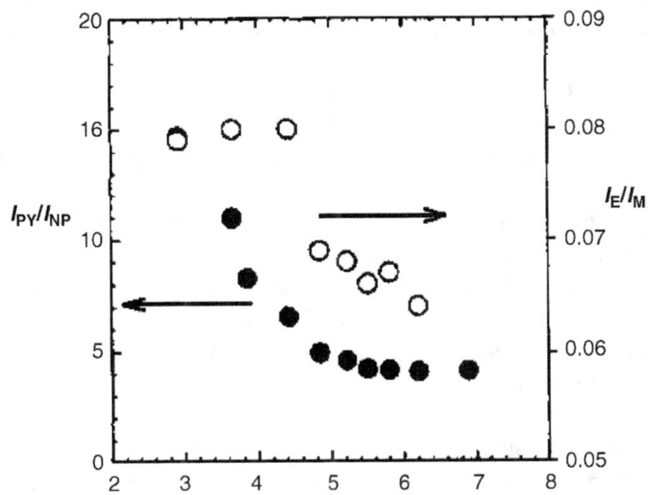

FIGURE 2.10 Plots of the ratio of pyrene to naphthalene emission intensities (I_{PY}/I_{NP}) and the ratio of pyrene excimer to monomer emission intensities (I_E/I_M) for 0.1 *M* NaCl aqueous solutions of PAA-Py-N as a function of solution pH. Polymer concentration 0.1 g l^{-1}; λ_{ex} 290 nm (for I_{PY}/I_{NP}) and λ_{ex} 343 nm (for I_E/I_M). [Reproduced from reference 120.]

large value derived for I_{PY}/I_{NP} at pH 3.8 (ca. 11) is indicative of a high degree of energy transfer and is consistent with the donor and acceptor being in close proximity to one another. This would be the case in a collapsed (or partially collapsed) polyelectrolyte coil, for example. The ratio drops to ca. 4 at pH values in excess of 4.7 which implies that D and A are out with the critical transfer distance and is consistent with the formation of the expanded polysalt form. Clearly, the change in the degree of energy transfer as a function of pH as monitored by the I_{PY}/I_{NP} ratio reveals [120] that some form of conformational change occurs in the labeled PAA sample. This is in contrast to that observed from steady-state spectral measurements [20,24,72–75,77] and Stern–Volmer quenching experiments [22,65,93].

The pK_a of the double-labeled sample was estimated [120] as 4.5 via potentiometric titration. Interestingly, it was reported [120] that the change in energy-transfer efficiency (as revealed from the I_{PY}/I_{NP} ratios in Fig. 2.10) does not coincide with the pK_a of the polymer but occurs at much lower pH (3.8 from the midpoint of the graph in Fig. 2.10) and over a much broader range. However, further reference to Fig. 2.10 reveals that the I_{PY}/I_{NP} plot reaches a plateau at pH 4.7, which is identical to the pK_a. The authors [120] presumed that the NRET interactions occur predominately between naphthalene and isolated pyrene groups, which are themselves hydrophobic. At low pH, attractive interactions keep a small number of hydrophobic groups in close proximity (including the donors and acceptors), which results in a high degree of energy transfer. When the solution reaches the critical pH, the electrostatic repulsive interactions between the carboxylate units overcome these forces of attraction and the degree of energy transfer decreases and plateaus at a pH corresponding to the pK_a of the labeled sample. The authors [120] concluded from these measurements that the overall conformational reorganization of PAA takes place over a wide pH range. A similar observation has been made for PMAA [26] following NRET experiments.

2.2.2.5 Excimer Formation The short-range nature (ca. 2 nm) of the interaction between a ground-state chromophore and an excited state species during excimer formation has been exploited by researchers to monitor the aqueous solution behavior of PAA [23,87,120–122].

For example, the photophysical behavior of a double-labeled PAA sample containing pyrene and naphthalene has been investigated in aqueous solution via both NRET and excimer formation [120]. To examine the excimer, selective excitation of the pyrene was carried out at a wavelength of 343 nm and the emission intensity of the monomer and excimer was sampled. A plot [120] of the ratio of the intensity of the pyrene excimer to monomer (I_E/I_M) as a function of pH is shown in Fig. 2.10. (Also shown in Fig. 2.10 is a measure of the degree of energy transfer via a plot of the ratio of pyrene to naphthalene emission intensity [I_{PY}/I_{NP}] on excitation of the naphthalene fluorophores at 290 nm, across the same pH range.) Clearly, on examination of Fig. 2.10, the large I_E/I_M ratio below pH 4 (ca. 15) is indicative of a high level of excimer formation: because short-range (<2 nm) interactions are required to from an excimer, this behavior can be associated with a collapsed state. At pH values above 5, on expansion of the chain, the ratio decreases because the chromophores cannot form an excimer within the excited state lifetime. The transition in the curve representing the variation in I_E/I_M is sharp and its midpoint (pH 4.7) coincides with the pK_a of the polyelectrolyte [120]. The estimation of the degree of excimer formation of a fluorescently labeled PAA sample as a function of pH in conjunction with NRET measurements [120] has revealed that the two spectroscopic techniques report on different fluorophore populations: the changes in energy-transfer efficiency via the I_{PY}/I_{NP} intensity ratio occur over a wider pH range resulting in a more diffuse transition (see Fig. 2.10). To account for these observations, the authors [120] proposed that only a small number of excimer-forming sites exist in the double-labeled polymer, which are stabilized by hydrophobic interactions. These excimers do not break down until the polyelectrolyte chain expands to a certain size and is then followed by dissociation in a cooperative manner. The NRET and the excimer experiments indicate, however, that PAA undergoes a conformational change at the pK_a of the polymer which is in contrast to the observations made from both steady-state [20,24,72–75,77] and Stern–Volmer quenching experiments [22,65,93].

An analysis of the time-resolved fluorescence in excimer-forming polymer systems can also prove informative in elucidating the conformations of macromolecular chains in solution: these data are particularly relevant in terms of the structures adopted by polyelectrolytes in aqueous media. In this latter context, it is useful at this stage to briefly overview the development of the kinetic schemes to describe excited state interactions in fluorescent polymer solutions.

The inability of the classical Birks scheme [123] to describe excimer formation in polymer systems in organic media (see Scheme 2.2) has long been recognized.

At the heart of this mechanism, mathematical functions describe the time-dependent decay of the monomer fluorescence [$I_M(t)$] in terms of the sum of two exponentials

$$I_M(t) = I_1 \exp(-t/\tau_1) + I_2 \exp(-t/\tau_2) \tag{2.33}$$

which reflects monomer emission and that of quenched monomer following dissociation of the excimer.

Emission of the excimer [$I_D(t)$], on the other hand, is modeled in terms of a growth and decay term as in

$$I_D(t) = I_3 \exp(-t/\tau_3) - I_4 \exp(-t/\tau_4) \tag{2.34}$$

The growth term represents an association of M and M* to form D*, whereas the excimer emission itself is described by the decay term.

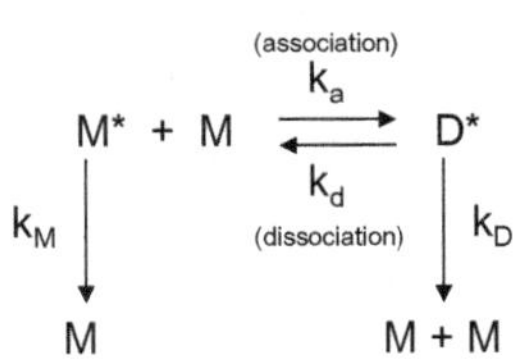

SCHEME 2.2

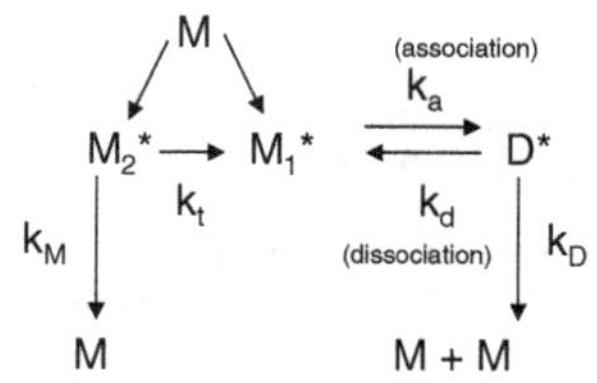

SCHEME 2.3

Satisfactory statistical analyses of emission resultant from copolymers containing the luminophore of interest and a spectroscopically inactive comonomer could only be described [124–126] if three exponentials are invoked as in Equation 2.8.

As a consequence of such investigations, it has been proposed [124–126] that two kinetically distinct monomer species exist in addition to excimer in aromatic polymers. Two independent kinetic schemes have been proposed [124–127], which are descriptive of the heterogeneity of monomer sites in such polymers (see Scheme 2.3).

One monomer (M_2) is kinetically isolated and is unable to form excimers within its excited state lifetime, whereas the second monomer (M_1) can readily take part in excimer formation.

The complexity observed in polymer fluorescence decay kinetics is further exacerbated when fluorescent polyelectrolytes are dissolved in aqueous media [29,30,33,35,37,43,120,122,128–132]: segregation of the macromolecular structure into hydrophobic and hydrophilic-rich domains results in differing degrees of water penetration which further complicates the time-resolved fluorescence [26]. Within this context, more recent attempts to describe time-resolved polymer photophysical data include use of the "blob model" [133,134], which accounts for the range of environments encountered in heterogeneous systems by invoking a distribution of rate constants for excimer formation.

As far as the excimer decay kinetics of PAA in aqueous media is concerned, de Melo and coworkers [122,130,131] have investigated the time-resolved fluorescence from a series of samples modified with various amounts of pyrene and naphthalene, respectively. Even when the aromatic content was as low as 2 mol%, excimer formation was evident in the steady-state spectra. The fluorescence decays were complex irrespective of the label and were best modeled by a triple-exponential function (as in Eq. 2.8) both when emission was sampled in the monomer and excimer regions. In contrast to the distribution of rate constants in the "blob model" [133,134], the authors favored a scheme that describes the decay kinetics in terms of discrete rate constants. The data were also consistent with previous schemes [124–127] that account for the presence of two distinct types of monomer in addition to that of excimer in macromolecular systems: one monomer enjoys kinetic isolation and is unable to form excimers, whereas the second is able to participate in excimer formation within its fluorescence lifetime. The authors [130] concluded from both steady-state and time-resolved data that PAA undergoes a conformational change from a compact form in acidic solution to an open coil at high pH. Furthermore, as the number of isolated monomers was observed to increase [130] with increasing degrees of ionization in the PAA sample, this is also consistent with a switch to an extended chain.

2.2.2.6 Steady-State Anisotropy In contrast to the excimer experiments discussed in the previous section, steady-state polarization [16,46] and anisotropy measurements [20] on various labeled PAA samples as a function of pH indicate the absence of a significant conformational transition. For example, for PAA labeled with DMA [46], the polarization varies from ca. 0.11 at low degrees of neutralization to ca. 0.085 at $\alpha = 0.4$. These data suggest that PAA exists as a flexible expanded coil, which undergoes rapid segmental motion across the pH range. In contrast, the polarization of a similarly labeled PMAA sample at $\alpha = 0$ is ca. 0.26 which infers that the label mobility is restricted under these conditions. This is to be expected due to the presence of a backbone methyl substituent that induces *hypercoiling* [48] in PMAA at low pH.

Steady-state anisotropy measurements on pyrene dispersed in PAA reveal [20] that the emission from the probe is totally depolarized at pH 3 (i.e., the anisotropy created on photoselection is lost instantaneously). This behavior indicates that pyrene undergoes rapid rotation in a medium with a viscosity similar to that of water and is consistent with observations as a consequence of spectral and lifetime experiments [20,24,72–75,77]: the probe is solubilized to a minimal extent in PAA under these conditions.

2.2.2.7 Time-Resolved Anisotropy Measurements Time-resolved anisotropy experiments on PAA [20,46,76,88] have played a pivotal role in providing information, at the molecular level, concerning the aqueous solution behavior of PAA. For example, TRAMs on PAA samples labeled with ACE and 1VN, respectively, have confirmed [20] that the polyelectrolyte undergoes a slight change in conformation with pH. The anisotropy decay was analyzed in terms of a single-exponential function of the form of Equation 2.31. The resultant τ_c decreased from ca. 6 ns in acidic solution to ca. 2 ns at pH values in excess of 6 for both the ACE- and 1VN-labeled samples, respectively. These data imply that PAA exists in a relatively expanded state even at low degrees of ionization. The pH of the midpoint of the conformational transition as a consequence of TRAMs for PAA was estimated as ca. 4 irrespective of the label. This value coincides with the pK_a of PAA derived by independent measurement [23,49,120] and infers, not surprisingly, that the conformation of the backbone is instrumental in affecting the solution behavior of the polyelectrolyte. By way of contrast, the pH-induced response of PAA is not nearly as dramatic as that of PMAA. For example, time-resolved anisotropy experiments on an ACE-labeled PMAA sample reveal [112] that τ_c decreases from ca. 50 ns at a low pH, which is consistent with the motion of a large slow-moving globule, to ca. 8 ns in the polysalt form. The latter value is indicative of a fast-moving water-swollen structure. These data further highlight the effect of the backbone methyl group in inducing *hypercoiling* [48] in PMAA.

2.2.2.8 ***Summary*** Amassing and summarizing the evidence from the photophysical experiments described in this section, it is clear that PAA undergoes a change in conformation [20,88,120,130] close to the pK_a of the polymer. However, this change is nowhere near as dramatic as that in PMAA and is a consequence of the absence of a backbone methyl group. (The hydrophobicity of the methyl substituent induces additional coiling or *hypercoiling* [48] in PMAA.) PAA forms a fast-moving flexible coil [20,88] at high degrees of neutralization, which is water-swollen and expanded [20,22,24,65,72–75,77,93,119]. On protonation, the carboxylic acid units become hydrophobic and to reduce unfavorable contacts with the aqueous phase, the polyelectrolyte coil partially contracts [20,23,87,88,120–122] decreasing the segmental motion of the backbone [20]. However, the degree of water penetration and expansion in this conformation is such that PAA is unable to solubilize probes [17] or certainly presents the low molar mass material with a microenvironment very similar to that of the aqueous phase [20].

2.2.3 Polystyrene Sulfonic Acid

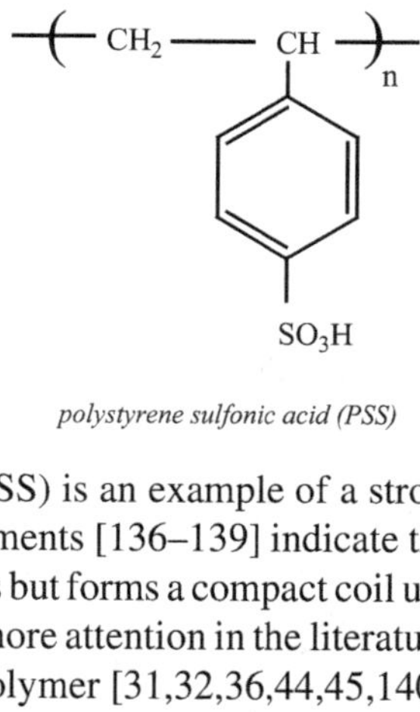

polystyrene sulfonic acid (PSS)

Polystyrene sulfonic acid (PSS) is an example of a strong polyacid [135]. Viscosity and light scattering measurements [136–139] indicate that PSS exists in an extended conformation at all pH values but forms a compact coil under high salt concentrations. Although PSS has received more attention in the literature in terms of its fluorescence behavior when part of a copolymer [31,32,36,44,45,140–142], nevertheless, interest has been shown [87,143–150] in the homopolymer due, in part, to the fact that it is inherently fluorescent as a consequence of the phenyl repeat unit. The high degree of aromaticity within PSS means that it readily forms excimers in aqueous solution via a sandwich-type overlap between a ground state and an electronically excited phenyl unit: the resulting monomer fluorescence is centered at ca. 287 nm [150], whereas that of excimer occurs at ca. 325 nm [150]. Researchers have taken advantage of this fact and used the intensity of monomer fluorescence (I_M) to excimer (I_E) to probe the aqueous solution behavior of PSS and in the presence of various salts [143,145,150].

For example, Major and Torkelson [150] examined the impact of coil size and conformation on the fluorescence properties of PSS through the addition of low-molecular weight electrolyte. A dramatic increase in excimer formation was observed for solutions of high ionic strength as a consequence of chain contraction. In contrast, little effect on the excimer population was reported on addition of low salt concentrations. These observations were consistent with viscosity experiments [139] and imply that as the salt concentration is increased, continued reduction in coil size occurs until the θ dimensions are reached.

Alternative approaches to monitoring the solution behavior of PSS through the use of water-soluble probes [147–149] and fluorescently labeled samples [87,144,146] have also appeared in the literature. For example, the fluorescence decay of a VPy-labeled PSS sample [87] was well described by a single-exponential function as in Equation 2.6. This contrasts to the behavior observed for a similarly labeled PMAA [87], which could only be modeled through use of a double-exponential function of the form of Equation 2.7. The authors concluded that a distribution of label environments exists in PMAA, which results in complex decay kinetics. The fact that only one excited state species is present in the PSS sample is indicative that the coil is more homogeneous in nature which is consistent with an extended conformation [136–139] under salt-free conditions.

When the fluorescence from a DPA-labeled PSS sample was quenched with Tl^+, deviation from Stern–Volmer behavior was observed [146]. A measure of success was achieved in description of the data, however, through use of the Hindered Access Model [1] (see Equation 2.13). This complex behavior was considered to reflect the fact that quenching can occur via Tl^+ ions which are condensed onto the polyelectrolyte in addition to deactivation by those ions which diffuse through solution. The data were also consistent with an extended PSS conformation in the presence of low concentrations of salt, which is in agreement with other fluorescence data [87,145,150] and light scattering and viscosity measurements [136–139].

2.2.4 Poly[2-(dimethylamino)ethyl methacrylate]

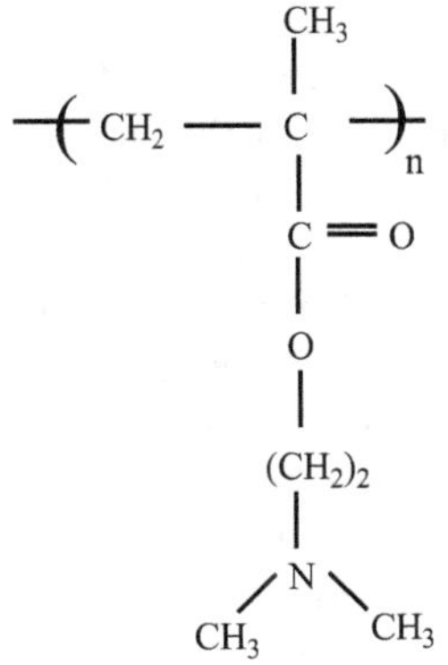

poly (2-dimethylamino ethyl methacrylate)

Potentiometric [50] and viscometric [151] measurements on poly[2-(dimethylamino) ethyl methacrylate] (PDMAEMA) indicate that the polyelectrolyte undergoes a conformational change from an expanded structure at low pH when the amine unit is fully protonated, to a collapsed structure at high pH. Potentiometric [50] data reveal a sharp discontinuity at an α value of ca. 0.15, whereas viscometric measurements [151] infer that a gradual expansion of the polymer coil occurs with increasing degrees of ionization. Following independent investigations of the titration behavior of PDMAEMA, there is a general consensus [152–154] that intramolecular association occurs between the amine unit and the carbonyl group, which effectively lowers the pK_a of the polyelectrolyte compared to that in small molecule analogs.

Corroborative evidence that PDMAEMA undergoes a change in conformation has resulted from fluorescence spectroscopic measurements using several water-soluble probes dispersed in aqueous solutions of the polyelectrolyte [151,155,156]. At low pH [50,151], mutual repulsion between the positive charges in the protonated form induces expansion of the coil. At pH values in excess of 8, on the other hand, deprotonation of the amine units occur allowing hydrophobic forces of attraction to dominate which serve to collapse the chain. Binding experiments using methyl orange (MO) have indicated [151] that PDMAEMA undergoes a transition between pH 7 and 9 from an extended structure to a compact coil which coincides with the pK_a of the polymer from independent measurement [152]. The fact that the binding strength of the dye was observed to maximize at pH 8 was considered to reflect two separate effects: (1) at pH 7, the polyelectrolyte is protonated which results in strong binding to the negatively charged MO and (2) by pH 8.5, the degree of protonation reduces significantly, whereas the hydrophobicity increases and the PDMAEMA undergoes a coil collapse. Under these conditions, hydrophobic domains are created which promote attraction (or solubilization) between the MO and the polyelectrolyte. (The fact that dyes can be solubilized by the compact conformation of polyelectrolytes has been recognized in other systems in the literature, most notably PMAA [72,73,75,78].) The reduction in electrostatic attraction and the increase in the contribution from hydrophobic association between the MO and PDMAEMA as the pH is increased toward the pK_a results in maximum binding at pH 8 [151]. A similar behavior in fluorescence characteristics of the negatively charged probe, TNS, was observed when dispersed in PDMAEMA [151]: the emission intensity from TNS maximized at pH 8.5. As an enhancement in the fluorescence of TNS occurs when the solvent polarity in which it is dissolved decreases [157], the maximum observed for the TNS/PDMAEMA dispersion was also considered to reflect enhanced binding in the protonated form below pH 8 and solubilization in the compact state above pH 8.5 [151]. Further evidence for the existence of a conformational transition in PDMAEMA has been derived from fluorescence measurements from a dansyl-labeled sample [152]. Dansyl derivatives are known to be polarity-sensitive [86] showing a shift in maximum intensity to shorter wavelength with decreasing polarity. A wavelength shift from ca. 520 nm at low pH to ca. 500 nm at pH values in excess of 8 was observed. When the wavelength maximum was plotted as a function of pH, a sharp transition was evident between 6 and 8 [152]. Analysis of this spectroscopic equivalent of a titration curve by a modification to the Henderson–Hasselbalch equation (Eq. 2.4) resulted in a pK_a of 7.8, which was in agreement with that derived from potentiometry [152].

2.3 COPOLYMERS

In the proceeding sections, emphasis has been placed on pH-responsive homopolymers whose conformational behavior is dictated by the hydrophobic to hydrophilic balance within the system. The switch in conformation is marked by a change in macroscopic properties, such as solution viscosity, over a narrow and characteristic pH range specific to that polymer which is important from a technological viewpoint. Manipulation of the switch to any desired pH has attracted much interest in the literature because this could lead to many more potential applications for the modified polyelectrolyte. Synthetic strategies adopted to achieve this end have included simple copolymerization of an ionizable repeat unit with a hydrophobic monomer [20,27–43,129] to form statistical copolymers and formation of block copolymers [71,158–180]. (A block copolymer consists of sequences of monomers A and B such as $-(A)_n-(B)_m$ or $-(A)_n-(B)_m-(A)_n$, where n and m represent the number of repeat units.)

Chemical modification by simple copolymerization, for example, creates a new class of polyelectrolyte [20,27–43,129] allowing exploitation of their amphiphilic nature: the ability of such polymers to form macromolecular aggregates with a micellar-type structure has been recognized [28,33,128] and their capacity to solubilize organic species in aqueous media [31,32,38,40,44] is of importance in consideration of potential applications which range from controlled release of materials [165,181–183] to photochemical conversion and storage [31,32,36,38,39,44,45].

When the hydrophobic modifier is an aromatic species, this creates amphiphilic polyelectrolytes which display complex fluorescence decay behavior [29,30,33,35,37,43,120,122,128,129,131,184,185] on two counts: preferential self-assembly of the hydrophobic units into domains segregated from the more hydrophilic regions of the macromolecule and the aqueous phase creates a microheterogeneous environment [28,33,128]. This in itself will result in complex photophysics due to varying degrees of water penetration [26] if the excited state is sensitive to the environment in which it resides. Secondly, high local concentrations of aromatic species have the potential to lead to excited state interactions such as energy migration as outlined below

$$D^* + D \rightarrow D + D^* \tag{2.35}$$

which is essentially an energy-transfer process between similar chromophores. The energy continues to migrate until it encounters a trap such as an acceptor species. Excimer formation, which also acts as a trap for the migrating energy, can further complicate the fluorescence decay behavior. Consequently, the excited state decay kinetics in amphiphilic polyelectrolytes containing an appreciable aromatic content, for example, require a minimum of three exponential terms as in Equation 2.8 for a statistically adequate fit [29,30,33,35,37,43,122,128–131,185].

In spite of the inherent complexity, the photophysical behavior of these polyelectrolytes has attracted significant interest in the literature over a number of years (see e.g., Refs. [20,27–43,122,128–131,184,185]). Much effort has been directed to elucidate the mechanisms of energy transfer [35,37,40,41,43] and trapping [32,35,38,41,43,129] yielding not only information directly relevant to considerations of solar energy conversion, but also information that is pertinent to the use of polymeric surfactants in solubilization phenomena. As the literature is quite extensive in this area, the copolymers will be grouped and discussed according to their architecture in the subsequent sections.

2.3.1 Random and Alternating Copolymers

In a seminal series of publications, Guillet and coworkers [27,31,32,36,38,40,44,45,128,140–142,186–191] have utilized hydrophobically modified polyelectrolytes based on acrylic acid (AA) [27,38] and sulfonic acid [31,32,36,44,128,142,188,189] in the field of energy harvesting and enhanced photochemistry. The polymer is presumed to form a "pseudo micellar"-type structure with the hydrophobic light-absorbing groups forming hypercoiled domains in the interior surrounded by ionic hydrophilic units on the exterior of the "micelle" thereby promoting the *antenna effect* [27]. The *antenna effect*, shown schematically in Fig. 2.11, is the basis for the complex photophysics observed in these systems [20,27–43,128,129] but is also the rationale behind potential applications [31,32,36,38,40,44,45]. In simple terms, the phenomenon results from the capture of a photon by one donor species (D) bound to the polymer. If D is present in high concentrations, extensive energy migration occurs until the energy is trapped by an acceptor (A) at another site within the macromolecule. Energy transfer then occurs to the trap. In the hypercoiled conformation, this is an extremely efficient process [128]. However, as the light-absorbing units are concentrated into the hydrophobic domains in intimate contact with one another, this often results in extensive excimer formation, which competes [30,31,38,39,41] with the accepter species for the migrating energy. The fact that "antenna" polyelectrolytes can solubilize low molar mass material in

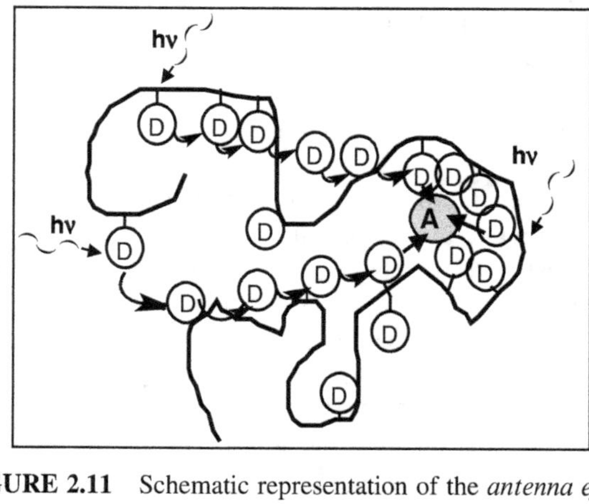

FIGURE 2.11 Schematic representation of the *antenna effect*.

aqueous media has been exploited by Guillet and coworkers [31,32,38,40,44] allowing development of so-called photozymes [31,32,36,38,40,44,45]. Photozymes are amphiphilic polyelectrolytes that absorb UV radiation and efficiently channel the harvested light into solubilized molecules, which can subsequently undergo photochemical conversions [32,36,44]. Examples of the photochemical reactions successfully performed in photozymes include the photooxidation of polynuclear aromatics [32], the photodechlorination of polychlorinated biphenyls [44], and the photosynthesis of previtamin D_3 [36].

Over a number of years, Webber and coworkers have been interested in developing "photon-harvesting"[43] polymers based on carboxylate derivatives [29,37,41,129] modified with aromatic light-absorbing species. (Photon-harvesting polymers operate on a similar premise to the antenna effect: extensive energy migration occurs between donor chromophores, which "harvests" and transfers the energy to an acceptor.) Early investigations were concerned with the photophysical behavior of a series of 2-vinyl naphthalene-based methacrylic acid (MAA) copolymers containing covalently bound anthracene groups in aqueous media [129]. The effect of the anthryl traps on the naphthyl fluorescence was studied in an attempt to elucidate the mechanism and to assess the efficiency of anthracene sensitization or "photon harvesting" in both alternating and random copolymers. It was found that anthracene sensitization for the alternating sample was more efficient than for the random copolymer for a comparable anthryl loading. One contributing factor to this enhancement in aqueous solution is that excimer formation is essentially absent in the alternating copolymer such that one photophysical pathway that competes with photon harvesting is eliminated. Later work [41] was concerned with investigating the role of electronic energy transport and polymer microstructure in a series of statistical and alternating copolymers of vinyl phenanthrene and MAA labeled with anthryl traps. Evidence for very efficient energy transfer in acidic solution was derived from both steady-state and time-resolved data. In the collapsed conformation at low pH, energy migration and transfer to the anthryl label was more efficient than at high degrees of ionization. The authors [41] concluded that there was little difference in the efficiency of trapping between the statistical and alternating sample. In terms of design of efficient photon-harvesting polymers, it was suggested that the most important component was the chromophore: the Förster R_o value [99] should be as large as possible to facilitate energy transfer and self-quenching or excimer formation should be repressed by sterically modifying the chromophore, for example.

Modification of poly(2-vinylpyridine) with a long-chain *n*-dodecyl bromide was the first published example of a "polysoap" in the literature [192]. (The term polysoap arises from the fact that the resulting modified polyelectrolyte displays similar properties to that of an ordinary soap [192,193].) Dependent on the aqueous solution conditions and the nature of the hydrophobe, modified polyelectrolytes can associate either intramolecularly or intermolecularly [194], and this has created interest from a technological perspective: potential applications include use as associative thickeners and rheology modifiers [195].

If association occurs intramolecularly between hydrophobic substituents on the same chain, then a unimolecular micelle or "unimer" is formed. Hydrophobic interactions between several polymer chains, on the other hand, can result in the formation of large aggregates, which can phase separate. An element of control can be effected over these two types of interactions by varying the concentration of polyelectrolyte [33,184,196–199]: for example, intramolecular association predominates in dilute solution, whereas intermolecular association dominates at higher polymer concentrations. Varying the salt content [184,196,198–200], addition of surfactants [121,201–204] to the solution and polymer architecture [197] can play an influential role in dictating the nature of the hydrophobic association. In the latter context, block copolymers, for example, tend to from intermolecular aggregates, whereas random copolymers associate largely intramolecularly.

The sensitivity of luminescence spectroscopy allows the examination of a wide concentration range from the ultra-dilute regime through to the bulk phase. Fluorescence spectroscopic techniques would appear ideally suited to probing the nature of hydrophobic association of polysoaps in aqueous media because intramolecular interactions can be examined in isolation to intermolecular effects. Consequently, luminescence methodologies have enjoyed increasing exposure to such amphiphilic systems over the last two decades [20,33–35,38,93,120,121,130–132,173,184,185,194,196–206]. For example, the ability of hydrophobically modified polyelectrolytes to form micelles through macromolecular association in aqueous solution has allowed probing of these structures by examination of changes in the vibrational fine structure of solubilized pyrene [20,93,173,199,201]. The conformation of the macromolecular chain has been determined under a variety of conditions via the extent of excimer formation [33,184,197,198] and nonradiative energy transfer [35,120,196,197,200,206] while TRAMs [20,205] on solubilized probes has provided information concerning the microviscosity of the "pseudo micellar" core.

In a comprehensive investigation to assess the effect of molecular architecture on the solution properties of hydrophobically modified polyelectrolytes, fluorescently labeled copolymers based on 2-(acrylamido)-2-methylpropane sulfonate (AMPS) and *N*-octadecyl have been synthesized [196]. The fluorophore (either a naphthyl or a pyrenyl derivative) was an integral part of the *N*-octadecyl modifier and was introduced through copolymerization with AMPS. Both single- and double-labeled samples were prepared allowing use of a variety of fluorescence techniques to probe the aqueous solution properties in the presence and absence of NaCl. Excimer formation and energy-transfer experiments, in particular, provided valuable information concerning the nature of macromolecular association in aqueous solution in this system. These measurements revealed that intramolecular interactions between the hydrophobic modifiers dominate promoting formation of unimolecular micelles in aqueous solution, which adopts a "flowerlike structure" irrespective of the salt concentration. Stern–Volmer experiments using nitromethane as a quencher provided evidence for contraction of the chain under high NaCl concentrations. Whereas quenching with Tl^+ in the *N*-octadecyl-modified systems was best described by the hindered access model [1] as in Equation 2.13. The fact that 94% of the labels are accessible to Tl^+, from this model, was considered to reflect a high degree of hydration in the micellar core [196].

Swanson and coworkers [20,93,205] have been concerned with hydrophobically modifying polyacids in order to tune the conformational transition to higher pH values. This has been achieved by copolymerization of acrylic acid and methacrylic acid with styrene and various *n*-alkyl methacrylates. ACE was introduced as a label in trace amounts to fully characterize the solution behavior of these amphiphilic polyelectrolytes by various spectroscopic techniques such as fluorescence quenching [93], excited state lifetime measurements [20], and TRAMs [20,93,205]. TRAMs have revealed [20,93,205] that the conformational transition of the modified acids is shifted to higher pH and is a direct consequence of the manipulation of the hydrophobic to hydrophilic balance within the macromolecule. In unmodified polyacids, there is a competition between a hydrophobic desire to collapse the chain to minimize contact with the aqueous phase and a drive to expand through electrostatic charge repulsions. It is this balance, which determines the pH at which the change in conformation occurs. Modification increases the propensity for hydrophobic interactions and shifts the balance. Consequently, increased concentrations of carboxylate anion are required to overcome these attractive forces and expand the polyelectrolyte, which can only be achieved at higher pH. Evidence for intramolecular association in the modified polyelectrolytes was confirmed through solubilization of probes such as pyrene [20,93]. Solubility of the probe was retained to pH values well above that of the conformational change of the unmodified polyacids from examination of the I_3 to I_1 ratio [20,93], excited state lifetime [20], and the rotational correlation time [20].

2.3.2 Block Copolymers

Synthesis of block copolymers containing hydrophobic or water-insoluble units with hydrophilic, solvated blocks allows exploitation of the amphiphilic nature of these systems. In many cases, spherical aggregates are formed from self-assembly of the insoluble blocks into micelle-type structures consisting of a hydrophobic core surrounded by a shell or corona of solvated blocks. Dependent on the constituent components within the block, stimuli responsive micelles can be formed which are critically dependent on pH and salt content. The use of luminescence spectroscopy has featured prominently in the study of block copolymer micelles: for a representative sample of this extensive literature see, for example, references [71,158–180] and references therein. Valuable information concerning the nature and stability of block copolymer micelles has been derived from fluorescence spectroscopic techniques such as excimer formation [162,163], energy transfer [164,168,169,172,175], and TRAMs [161,165,180]. The sensitivity of the fluorescence spectrum of pyrene to its environment [63,64] has been exploited to determine [158,159,171,174,176,178,179] the critical micelle concentration (CMC) following aggregation of the block copolymer chains: at low concentrations, only molecularly dissolved chains (unimers) are present which are unable to solubilize pyrene. This results in a high I_1/I_3 ratio. Above the CMC, self-assembly of the copolymers occurs, the probe migrates into the hydrophobic core (or the core–corona interface) and the I_1/I_3 ratio decreases. The vibrational fine structure of the emission spectrum of pyrene has also been used to

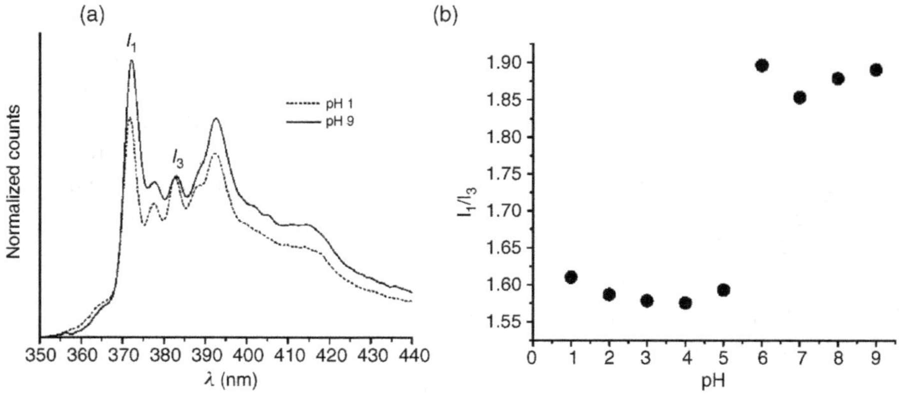

FIGURE 2.12 (a) Fluorescence spectra of pyrene ($9.00 \times 10^{-6}\,M$) in aqueous solutions of sodium p(2-(acrylamido)-2-methylpropane sulfonate)$_{70}$-*block*-(3-(acrylamido)-3-methylbutanoate)$_{62}$) (1% w/v) at pH 1 and 9. The spectra are baseline-adjusted and normalized to give equal peak heights at I_3. (b) I_1/I_3 as a function of pH for the sample in (a). [Reproduced from reference 173.]

probe the behavior of micelles formed in a variety of diblock copolymer systems in response to changes in the pH [166,173,174] and salt content [158,174].

For example, McCormick and coworkers [173] have dispersed pyrene in block copolymers of AMPS and 3-(acrylamido)-3-methylbutanoate (AMBA). The pH response of the system was investigated by using the sensitivity of the fluorescence spectrum of the probe to the polarity of the medium. Figure 2.12a shows an example of the spectra at two pH extremes: the decrease in the intensity of band 1 from pH 9.0 to 1.0 is indicative of pyrene, which resides in a more hydrophobic environment and is consistent with the formation of micelles under acidic conditions. Figure 2.12b shows a plot of the resultant I_1 to I_3 ratio across the pH range. A dramatic decrease in the ratio is observed between pH 5.0 and 6.0. This supports NMR data, which indicates that dehydration of the PAMBA block occurs at pH 5.5, which serves to create near monodisperse micelles, which can solubilize low molar mass material.

Researchers have also used pyrene dispersed as a probe to provide information regarding the nature of the micellar core formed from block copolymers of 2-(dimethylamino)ethyl methacrylate (DMAEMA) and methacrylic acid [167] and DMAEMA with 2-(diethylamino)ethyl methacrylate (DEAEMA) [166]. A similar degree of hydrophobicity in the micellar cores, as sensed by the probe, was determined in both types of block copolymers.

In low-molecular weight surfactants, above the CMC, a constant exchange occurs between aggregated material and that which is "free" (i.e., unimers) in the aqueous phase. Similarly, with block copolymers, a dynamic equilibrium exists between micelles and nonaggregated chains. The average residence time of a single block copolymer chain within a micelle is of importance because this will dictate the stability of the system. Researchers have used fluorescence techniques to probe such characteristics as the exchange rate between unimers and polymeric micelles [168] based on, for example, block copolymers of styrene and MAA and *tert*-butyl styrene and MAA: 2-vinyl naphthalene was covalently attached to the styryl block and acted as the energy donor, whereas the acceptor, pyrene, was dispersed in a unlabeled block copolymer sample. On mixing, pyrene exchange between nonlabeled and labeled block copolymer micelles was monitored via the extent of energy transfer: increased naphthyl quenching occurs as the pyrene acceptor enters the labeled core. The conclusion from this study was that the probe exchanges extremely slowly between micelles.

In a similar manner, the exchange kinetics have also been investigated for various block copolymers based on dimethylaminoakyl methacrylate and MAA via energy transfer from a donor-labeled core to a dispersed acceptor probe [160]. It was concluded from these measurements that there is a direct link between the hydrophobicity of the core and the exchange rate constant: increasing the hydrophobicity of the copolymer slows down the exchange rate due to less favorable interactions with the aqueous phase.

Energy-transfer measurements have also provided an insight into the structure of amphiphilic block copolymer micelles with hydrophobically modified polyelectrolyte shells [172].

The pH-induced micellization of a DMAEMA-*b*-DEAEMA diblock copolymer has been studied in detail using dynamic light scattering, small-angle neutron scattering, and fluorescence spectroscopy [166]. The DMAEMA constitutes the corona of the micelle, whereas the DEAEMA forms the core. Pyrene was used as a probe to determine the nature of the DEAEMA blocks. It was shown that the hydrophobicity of the micellar cores increased progressively as the solution pH was adjusted from pH 7 to 9. In the presence of an electrolyte, it was possible to observe both individual chains (unimers) and micelles under certain conditions. The critical micellization pH depended on both the copolymer concentration and also the background electrolyte concentration.

In an extension of the work using a pyrene probe [166], TRAMs have been performed [180] on a fluorescently labeled DMAEMA-*b*-DEAEMA block copolymer to detect the onset of micellization. A pyrenyl derivative, located at the DEAEMA block end, allowed motion of this site to be monitored. A significant reduction in the mobility of the label (as reflected by τ_c) was apparent at concentrations in excess of the CMC following single-exponential analyses, of the form of Equation 2.31, of the anisotropy decay. This was considered to reflect the inclusion of the labeled DEAEMA block into the core of the micelle. Assuming that a spherical geometry is adopted above the CMC, it was possible to estimate the micellar diameter from time-resolved anisotropy experiments. The resultant τ_c, when treated by Equation 2.29, provided an estimate of the micelle diameter as ca. 12 nm, which is approximately half that derived from dynamic light scattering measurements. The discrepancy between the two values is consistent with a model, which suggests that the chain ends are not "frozen" into position but that limited motion occurs due to fluidity within the micellar core.

TRAMs [161] have also been performed on styrene MAA block copolymers tagged either at the end (in the styrene block) or at the block boundary and have revealed that the styryl core is very compact in aqueous media. The corona/core interface, on the other hand, was considered to be not a distinct boundary but rather more heterogeneous in nature.

2.4 CONCLUDING REMARKS

Fluorescence methodologies have proven invaluable in studying the aqueous solution properties of polyelectrolytes.

Judicious choice of label allows specific sites within a macromolecule to be monitored, whereas dispersion of probes can reveal information concerning the microviscosity and hydrophobicity of the coil. Hydrophobic modification creates amphiphilic polyelectrolytes, which can self-assemble to from media of importance as rheology modifiers and in solubilization phenomena concerned with light harvesting and enhanced photochemistry. The sensitivity of fluorescence spectroscopy allows probing of self-assembly and aggregation across a wide concentration range allowing isolation of purely intramolecular effects from intermolecular association. This is of importance in effecting control over the nature of the interaction for subsequent applications.

REFERENCES

1. J.R. Lakowicz, *Principles of Fluorescence Spectroscopy*, Kluwer Academic, New York, 1999.
2. A. Katchalsky and P. Spitnik, *J. Polym. Sci.* **1947**, *2*(4), 432–446.
3. V. Crescenzi, *Adv. Polym. Sci.* **1968**, *5*, 358.
4. J.C. Leyte and M. Mandel, *J. Polym. Sci. Part A: Gen. Papers* **1964**, *2*(4PA), 1879–1891.
5. R.A. Marcus, *J. Chem. Phys.* **1955**, *23*(6), 1057–1068.
6. T.S. Chen and J.K. Thomas, *J. Polym. Sci. Part A: Polym. Chem.* **1979**, *17*(4), 1103–1116.
7. T. Okubo and N.J. Turro, *J. Phys. Chem.* **1981**, *85*(26), 4034–4038.
8. N.J. Turro and T. Okubo, *J. Phys. Chem.* **1982**, *86*(9), 1535–1539.
9. N.J. Turro, G. Caminati, and J. Kim, *Macromolecules* **1991**, *24*(14), 4054–4060.
10. N.J. Turro, J. Kim, and G. Caminati, *Macromolecules* **1993**, *26*(8), 1930–1935.
11. J.R. Ebdon, D.M. Lucas, I. Soutar, and L. Swanson, *Anal. Proc.* **1993**, *30*, 431–433.
12. I. Soutar and L. Swanson, *Eur. Polym. J.* **1993**, *29*(2–3), 371–378.
13. I. Soutar, L. Swanson, N.J. Flint, and R. Haywood, *J. Fluoresc.* **1998**, *8*, 327–334.
14. M. Flores, R. Rodriguez, and R. Arroyo, *Mater. Lett.* **1999**, *39*(6), 329–334.
15. K. Haskins-Glusac, M.R. Pinto, C. Tan, and K.S. Schanze, *J. Am. Chem. Soc.* **2004**, *126*, 14964–14971.
16. E.V. Anufrieva, and Y.Y. Gotlib, *Adv. Polym. Sci.* **1981**, *40*, 1.
17. G. Barone, V. Crescenz, A.M. Liquori, and F. Quadrifo, *J. Phys. Chem.* **1967**, *71*(7), 2341–2345.

18. F.E. Treloar, *Chem. Scripta* **1976**, *10*(5), 219–224.
19. A.F. Olea, and J.K. Thomas, *Macromolecules* **1989**, *22*(3), 1165–1169.
20. J.R. Ebdon, B.J. Hunt, D.M. Lucas, I. Soutar, L. Swanson, and A.R. Lane, *Can. J. Chem.: Rev. Can. Chim.* **1995**, *73*(11), 1982–1994.
21. A.F. Olea, H. Rosenbluth, and J.K. Thomas, *Macromolecules* **1999**, *32*(24), 8077–8083.
22. D.Y. Chu and J.K. Thomas, *Macromolecules* **1984**, *17*(10), 2142–2147.
23. N.J. Turro, and K.S. Arora, *Polymer* **1986**, *27*(5), 783–796.
24. B. Bednar, H. Morawetz, and J.A. Shafer, *Macromolecules* **1985**, *18*(10), 1940–1944.
25. I. Soutar, L. Swanson, R.E. Imhof, and G. Rumbles, *Macromolecules* **1992**, *25*(17), 4399–4405.
26. L. Ruiz-Pérez, A. Pryke, M. Sommer, G. Battaglia, I. Soutar, L. Swanson, and M. Geoghegan, *Macromolecules* **2008**, *41*, 2203–2211.
27. D.A. Holden and J.E. Guillet, *Ann. N.Y. Acad. Sci.* **1981**, *366*, 11–23.
28. Y. Morishima, Y. Itoh, and S. Nozakura, *Makromol. Chem.: Macromol. Chem. Phys.* **1981**, *182*(11), 3135–3147.
29. Y. Morishima, T. Kobayashi, S.I. Nozakura, and S.E. Webber, *Macromolecules* **1987**, *20*(4),807–813.
30. F. Bai and S.E. Webber, *Macromolecules* **1988**, *21*(3), 628–633.
31. M. Nowakowska, B. White, and J.E. Guillet, *Macromolecules* **1988**, *21*(12), 3430–3437.
32. M. Nowakowska, B. White, and J.E. Guillet, *Macromolecules* **1989**, *22*(5), 2317–2324.
33. C.L. McCormick, C.E. Hoyle, and M.D. Clark, *Macromolecules* **1990**, *23*(12), 3124–3129.
34. C.L. McCormick, C.E. Hoyle, and M.D. Clark, *Macromolecules* **1991**, *24*(9), 2397–2403.
35. Y. Morishima, Y. Tominaga, S. Nomura, M. Kamachi, and T. Okada, *J. Phys. Chem.* **1992**, *96*(4), 1990–1994.
36. M. Nowakowska, V.P. Foyle, and J.E. Guillet, *J. Am. Chem. Soc.* **1993**, *115*(14), 5975–5981.
37. Y. Itoh, H. Satoh, T. Yasue, A. Hachimori, H. Satozono, S. Suzuki, and S.E. Webber, *Macromolecules* **1994**, *27*(6), 1434–1439.
38. D.M. Gravett and J.E. Guillet, *Macromolecules* **1995**, *28*(1), 274–280.
39. Y. Itoh, O. Yamashita, A. Hachimori, M. Kojima, and S. Suzuki, *J. Polym. Sci. Part A: Polym. Chem.* **1995**, *33*(1), 137–142.
40. D.M. Gravett and J.E. Guillet, *Macromolecules* **1996**, *29*(2), 617–624.
41. D.J. Kiserow, Y. Itoh, and S.E. Webber, *Macromolecules* **1997**, *30*(10), 2934–2940.
42. R. Rutkaite, G. Buika, and J.V. Grazulevicius, *J. Photochem. Photobiol. A: Chem.* **2001**, *138*(3), 245–251.
43. S.E. Webber, *Chem. Rev.* **1990**, *90*(8), 1469–1482.
44. M. Nowakowska, E. Sustar, and J.E. Guillet, *J. Am. Chem. Soc.* **1991**, *113*(1), 253–258.
45. M. Nowakowska, A. Karewicz, N. Loukine, and J.E. Guillet, *Polymer* **2002**, *43*(7), 2003–2009.
46. K.P. Ghiggino and K.L. Tan, in: D. Phillips (Ed.), *Polymer Photophysics: Luminescence, Energy Migration and Molecular Motion in Synthetic Polymers*, Chapman and Hall, London, 1985.

47. R. Arnold, *J. Colloid Sci.* **1957**, *12*, 549–556.
48. A. Katchalsky, *J. Polym. Sci.* **1951**, *7*(4), 393–412.
49. M. Mandel, *Eur. Polym. J.* **1970**, *6*(6), 807–822.
50. M. Pradny, J. Holata, and S. Sevcik, *Makromol. Chem.: Macromol. Chem. Phys.* **1989**, *190*(5), 1079–1088.
51. H. Eisenberg, *J. Polym. Sci.* **1958**, *30*(121), 47–66.
52. E.V. Anufrieva, T.M. Birshtei, T.N. Nekrasov, O.B. Ptitsyn, and T.V. Shevelev, *J. Polym. Sci. Part C: Polym. Symp.* **1968** (16PC), 3519–3531.
53. J.L. Koenig, A.C. Angood, J. Semen, and J.B. Lando, *J. Am. Chem. Soc.* **1969**, *91*(26), 7250.
54. C. Heitz, M. Rawiso, and J. Francois, *Polymer* **1999**, *40*(7), 1637–1650.
55. M. Moan and C. Wolff, *Polymer* **1975**, *16*(11), 776–780.
56. M. Moan, C. Wolff, J.P. Cotton, and R. Ober, *J. Polym. Sci. Part C: Polym. Symp.* **1977** (61), 1–8.
57. M. Moan, *J. Appl. Crystallogr.* **1978**, *11*(Oct), 519–523.
58. J. Plestil, D. Hlavata, J. Labsky, Y.M. Ostanevich, and V.Y. Bezzabotnov, *Polymer* **1987**, *28*(2), 213–216.
59. J. Plestil, Y.M. Ostanevich, V.Y. Bezzabotonov, D. Hlavata, and J. Labsky, *Polymer* **1986**, *27*(6), 839–842.
60. K.L. Tan and F.E. Treloar, *Chem. Phys. Lett.* **1980**, *73*(2), 234–239.
61. R.V. Pereira and M.H. Gehlen, *J. Phys. Chem. B* **2006**, *110*(13), 6537–6542.
62. G. Jones and M.A. Rahman, *J. Phys. Chem.* **1994**, *98*(49), 13028–13037.
63. A. Nakajima and H. Baba, *Bull. Chem. Soc. Jpn.* **1970**, *43*(3), 967.
64. K. Kalyanasundaram and J.K. Thomas, *J. Am. Chem. Soc.* **1977**, *99*(7), 2039–2044.
65. D.M. Lucas, PhD dissertation, Lancaster University, **1993**.
66. J. Jay, L.J. Johnston, and J.C. Scaiano, *Chem. Phys. Lett.* **1988**, *148*(6), 517–522.
67. P. Lianos, J. Lang, and R. Zana, *J. Phys. Chem.* **1982**, *86*(24), 4809–4814.
68. I. Soutar and L. Swanson, *Langmuir* **2006**, *22*(13), 5904–5910.
69. S. Pankasem, J.K. Thomas, M.J. Snowden, and B. Vincent, *Langmuir* **1994**, *10*(9), 3023–3026.
70. N.J. Flint, S. Gardebrecht, and L. Swanson, *J. Fluoresc.* **1998**, *8*, 343–353.
71. E. Szajdzinska-Pietek, M. Pinteala, and S. Schlick, *Polymer* **2004**, *45*, 4113–4120.
72. M. Mandel, and W.H.J. Stork, *Biophys. Chem.* **1974**, *2*(2), 137–143.
73. C. Braud, *Eur. Polym. J.* **1977**, *13*(11), 897–901.
74. G. Muller and J.C. Fenyo, *J. Polym. Sci. Part A: Polym. Chem.* **1978**, *16*(1), 77–87.
75. B. Erny and G. Muller, *J. Polym. Sci. Part A: Polym. Chem.* **1979**, *17*(12), 4011–4020.
76. M.J. Snare, K.L. Tan, and F.E. Treloar, *J. Macromol. Sci.: Chem.* **1982**, *A17*(2), 189–201.
77. J. Jager and J.B.F.N. Engberts, *Eur. Polym. J.* **1987**, *23*, 295–299.
78. M.G. Krakovyak, E.V. Anufrieva, E.A. Sycheva, and T.V. Sheveleva, *Macromolecules* **1993**, *26*(26), 7375–7378.
79. G. Oster and Y. Nishijima, *J. Am. Chem. Soc.* **1956**, *78*, 1581–1584.
80. H.L. Chen and H. Morawetz, *Macromolecules* **1982**, *15*(5), 1445–1447.

81. H.L. Chen and H. Morawetz, *Eur. Polym. J.* **1983**, *19*(10–11), 923–928.
82. B. Bednar, H. Morawetz, and J.A. Shafer, *Macromolecules* **1984**, *17*(8), 1634–1636.
83. B. Bednar, Z.M. Li, Y.H. Huang, L.C.P. Chang, and H. Morawetz, *Macromolecules* **1985**, *18*(10), 1829–1833.
84. H. Morawetz, Photophysical and Photochemical Tools in Polymer Science, NATO ASI Ser. 1986, CI82, p. 85–95.
85. G. Weber and D.J.R. Laurence, *Biochem. J.* **1954**, *56*(4), R31–R31.
86. L. Stryer, *J. Mol. Biol.* **1965**, *13*, 482–495.
87. R.D. Stramel, C. Nguyen, S.E. Webber, and M.A.J. Rodgers, *J. Phys. Chem.* **1988**, *92*(10), 2934–2938.
88. B. Bednar, J. Trnena, P. Svoboda, S. Vajda, V. Fidler, and K. Prochazka, *Macromolecules* **1991**, *24*(8), 2054–2059.
89. I. Soutar, L. Swanson, T. Annable, J.C. Padget, and R. Satgurunathan, *J. Colloid Interf. Sci.* **2006**, *303*(1), 205–213.
90. Y. Fang, PhD dissertation, Lancaster University, **1998**.
91. D.V. O'Connor and D. Phillips, *Time-correlated Single Photon Counting*, Academic Press, New York, 1984.
92. E.J. Bowen, *Trans. Faraday Soc.* **1954**, *50*, 97–102.
93. J.R. Ebdon, D.M. Lucas, I. Soutar, and L. Swanson, *Macromol. Symp.* **1994**, *79*, 167–177.
94. J.A. Delaire, M.A.J. Rodgers, and S.E. Webber, *J. Phys. Chem.* **1984**, *88*(25), 6219–6227.
95. J.H. Clements and S.E. Webber, *J. Phys. Chem. A* **1999**, *103*, 2513.
96. J.H. Clements and S.E. Webber, *J. Phys. Chem. B* **1999**, *103*(43), 9366–9377.
97. Y. Morishima, H. Ohgi, and M. Kamachi, *Macromolecules* **1993**, *26*(16), 4293–4297.
98. G. Manning, *J. Chem. Phys.* **1969**, *51*, 924.
99. T. Förster, *Naturwissenschaften* **1946**, *6*, 166.
100. G. Liu, J.E. Guillet, M. Vlegels, and E.J. Goethals, *Macromolecules* **1991**, *24*(14), 4094–4100.
101. G.J. Liu and J.E. Guillet, *Macromolecules* **1990**, *23*(5), 1388–1392.
102. G.J. Liu, J.E. Guillet, E.T.B. Altakrity, A.D. Jenkins, and D.R.M. Walton, *Macromolecules* **1990**, *23*(5), 1393–1401.
103. G.J. Liu and J.E. Guillet, *Macromolecules* **1990**, *23*(11), 2973–2977.
104. G.J. Liu, J.E. Guillet, E.T.B. Altakrity, A.D. Jenkins, and D.R.M. Walton, *Macromolecules* **1991**, *24*(1), 68–74.
105. I.B. Berlman, *Energy Transfer Parameters of Aromatic Compounds*, Academic Press, New York, 1973.
106. G. Schwarz, *J. Mol. Biol.* **1965**, *11*, 64–77.
107. K.A. Zachariasse, W. Kuhnle, U. Leinhos, P. Reynders, and G. Striker, *J. Phys. Chem.* **1991**, *95*(14), 5476–5488.
108. K.A. Zachariasse, G. Duveneck, and R. Busse, *J. Am. Chem. Soc.* **1984**, *106*(4), 1045–1051.
109. T. Azumi and S.P. McGlyn, *J. Phys. Chem.* **1962**, *37*, 2413.
110. G. Weill, *C. R. Acad Sci. Paris B* **1971**, *272*, 116.
111. I. Soutar and L. Swanson, *Polymer* **1994**, *35*, 1942.

112. I. Soutar and L. Swanson, *Macromolecules* **1994**, *27*, 4304–4311.
113. R.V. Pereira and M.H. Gehlen, *Macromolecules* **2007**, *40*, 2219–2223.
114. A.J. Marsh, G. Rumbles, I. Soutar, and L. Swanson, *Chem. Phys. Lett.* **1992**, *195*(1), 31–36.
115. M.D. Barkley, A.A. Kowalczyk, and L. Brand, *J. Chem. Phys.* **1981**, *75*(7), 3581–3593.
116. I. Soutar, L. Swanson, R.L. Christensen, R.C. Drake, and D. Phillips, *Macromolecules* **1996**, *29*(14), 4931–4936.
117. G. Barone, V. Crescenzi, F. Quadrifoglio, and V. Vitagliano, *Ric. Sci.* **1966**, *36*, 447.
118. H. Morawetz, *Macromolecules in Solution*, 2nd edition, Wiley, New York, Chapter 7, 1975.
119. K.S. Arora and N.J. Turro, *J. Polym. Sci. Part B: Polym. Phys.* **1987**, *25*(2), 243–262.
120. D.F. Anghel, V. Alderson, F.M. Winnik, M. Mizusaki, and Y. Morishima, *Polymer* **1998**, *39*(14), 3035–3044.
121. K. Schillen, D.F. Anghel, M.D. Miguel, and B. Lindman, *Langmuir* **2000**, *16*(26), 10528–10539.
122. T. Costa, M.D. Miguel, B. Lindman, K. Schillen, and J.S.S. de Melo, *J. Phys. Chem. B* **2005**, *109*(23), 11478–11492.
123. J.B. Birks, *Photophysics of Aromatic Molecules*, Wiley-Interscience, London, 1970.
124. D. Phillips, A.J. Roberts, and I. Soutar, *Polymer* **1981**, *22*(3), 293–298.
125. D. Phillips, A.J. Roberts, and I. Soutar, *Polymer* **1981**, *22*(4), 427–429.
126. I. Soutar, D. Phillips, A.J. Roberts, and G. Rumbles, *Sci. J. Polym Part B: Polym Phys.* **1982**, *20*(10), 1759–1770.
127. D.A. Holden, P.Y.K. Wang, and J.E. Guillet, *Macromolecules* **1980**, *13*(2), 295–298.
128. J.E. Guillet, J.Y. Wang, and L.Y. Gu, *Macromolecules* **1986**, *19*(11), 2793–2798.
129. F. Bai, C.H. Chang, and S.E. Webber, *Macromolecules* **1986**, *19*(10), 2484–2494.
130. J.S. de Melo, T. Costa, M.D. Miguel, B. Lindman, and K. Schillen, *J. Phys. Chem. B* **2003**, *107*, 12605–12621.
131. J.S. de Melo, T. Costa, A. Francisco, A.L. Macanita, S. Gago, and I.S. Goncalves, *Phys. Chem. Chem. Phys.* **2007**, *9*, 1370–1385.
132. G.L. Smith and C.L. McCormick, *Macromolecules* **2001**, *34*(16), 5579–5586.
133. A.K. Mathew, H. Siu, and J. Duhamel, *Macromolecules* **1999**, *32*(21), 7100–7108.
134. H. Siu and J. Duhamel, *J. Phys. Chem. B* **2005**, *109*(5), 1770–1780.
135. G.E. Boyd and K. Bunzl, *J. Am. Chem. Soc.* **1974**, *96*, 2054.
136. R.A. Mock and C.A. Marshall, *J. Polym. Sci.* **1954**, *13*(69), 263–277.
137. R.A. Mock, C.A. Marshall, and T.E. Slykhouse, *J. Phys. Chem.* **1954**, *58*(6), 498–503.
138. U.P. Strauss and Y.P. Leung, *J. Am. Chem. Soc.* **1965**, *87*(7), 1476–1480.
139. A. Takahash, T. Kato, and M. Nagasawa, *J. Phys. Chem.* **1967**, *71*(7), 2001.
140. J.E. Guillet, N.A.D. Burke, M. Nowakowska, H. Reese, and D.M. Gravett, *Macromol. Symp.* **1995**, *98*, 53–72.
141. M. Nowakowska, E. Sustar, and J.E. Guillet, *J. Photochem. Photobiol. A: Chem.* **1994**, *80*(1–3),369–376.
142. E. Sustar, M. Nowakowska, and J.E. Guillet, *J. Photochem. Photobiol. A: Chem.* **1992**, *63*(3),357–365.

143. P. Ander and M.K. Mahmoudhagh, *Macromolecules* **1982**, *15*(1), 213–214.
144. R.D. Stramel, S.E. Webber, and M.A.J. Rodgers, *J. Phys. Chem.* **1989**, *93*(5), 1928–1934.
145. N.J. Turro and T. Okubo, *J. Phys. Chem.* **1982**, *86*(9), 1485–1487.
146. M.E. Morrison, R.C. Dorfman, W.D. Clendening, D.J. Kiserow, P.J. Rossky, and S.E. Webber, *J. Phys. Chem.* **1994**, *98*(21), 5534–5540.
147. C. Peyratout, E. Donath, and L. Daehne, *J. Photochem. Photobiol. A: Chem.* **2001**, *142* (1), 51–57.
148. A.K. Chibisov and H. Gorner, *Chem. Phys. Lett.* **2002**, *357*(5–6), 434–439.
149. I. Moreno-Villoslada, M. Jofre, V. Miranda, R. Gonzalez, T. Sotelo, S. Hess, and B.L. Rivas, *J. Phys. Chem. B* **2006**, *110*, 11809–11812.
150. M.D. Major and J.M. Torkelson, *Macromolecules* **1986**, *19*(11), 2801–2806.
151. T. Takagishi, T. Hosokawa, and Y. Hatanaka, *J. Polym. Sci. Part A: Polym. Chem.* **1989**, *27*(1), 1–13.
152. P. van de Wetering, N.J. Zuidam, M.J. van Steenbergen, O. van der Houwen, W.J.M. Underberg, and W.E. Hennink, *Macromolecules* **1998**, *31*(23), 8063–8068.
153. M. Pradny and S. Sevcik, *Makromol. Chem.: Macromol. Chem. Phys.* **1985**, *186*(1), 111–121.
154. M. Pradny, J. Lokaj, M. Novotna, and S. Sevcik, *Makromol. Chem.: Macromol. Chem. Phys.* **1989**, *190*(9), 2229–2234.
155. H. Kozuka, T. Hosokawa, and T. Takagishi, *J. Polym. Sci. Part A: Polym. Chem.* **1989**, *27*(2),555–563.
156. T. Takagishi and T. Hosokawa, *J. Polym. Sci. Part A: Polym. Chem.* **1989**, *27*(6), 1925–1933.
157. W.O. McClure and G.M. Edelman, *Biochemistry* **1966**, *5*(6), 1908–1919.
158. D.V. Pergushov, E.V. Remizova, M. Gradzielski, P. Lindner, J. Feldthusen, A.B. Zezin, A.H.E. Muller, and V.A. Kabanov, *Polymer* **2004**, *45*(2), 367–378.
159. L.E. Bromberg and D.P. Barr, *Macromolecules* **1999**, *32*(11), 3649–3657.
160. S. Creutz, J. van Stam, F.C. De Schryver, and R. Jerome, *Macromolecules* **1998**, *31*(3), 681–689.
161. J. Chan, S. Fox, D. Kiserow, C. Ramireddy, P. Munk, and S.E. Webber, *Macromolecules* **1993**, *26*(25), 7016–7023.
162. T. Cao, P. Munk, C. Ramireddy, Z. Tuzar, and S.E. Webber, *Macromolecules* **1991**, *24* (23), 6300–6305.
163. K. Prochazka, D. Kiserow, C. Ramireddy, Z. Tuzar, P. Munk, and S.E. Webber, *Macromolecules* **1992**, *25*(1), 454–460.
164. T. Rager, W.H. Meyer, G. Wegner, and M.A. Winnik, *Macromolecules* **1997**, *30*(17), 4911–4919.
165. S.E. Webber, *J. Phys. Chem. B* **1998**, *102*(15), 2618–2626.
166. A.S. Lee, A.P. Gast, V. Butun, and S.P. Armes, *Macromolecules* **1999**, *32*(13), 4302–4310.
167. J.F. Gohy, S. Creutz, M. Garcia, B. Mahltig, M. Stamm, and R. Jerome, *Macromolecules* **2000**, *33*(17), 6378–6387.
168. J. van Stam, S. Creutz, F.C. De Schryver, and R. Jerome, *Macromolecules* **2000**, *33*(17), 6388–6395.

169. P. Matejicek, F. Uhlik, Z. Limpouchova, K. Prochazka, Z. Tuzar, and S.E. Webber, *Macromolecules* **2002**, *35*(25), 9487–9496.

170. S. Yusa, Y. Shimada, Y. Mitsukami, T. Yamamoto, and Y. Morishima, *Macromolecules* **2003**, *36*(11), 4208–4215.

171. Y. Li and K. Nakashima, *Langmuir* **2003**, *19*(3), 548–553.

172. P. Matejicek, J. Humpolickova, K. Prochazka, Z. Tuzar, M. Spirkova, M. Hof, and S.E. Webber, *J. Phys. Chem. B* **2003**, *107*(32), 8232–8240.

173. B.S. Sumerlin, A.B. Lowe, D.B. Thomas, and C.L. McCormick, *Macromolecules* **2003**, *36*(16), 5982–5987.

174. M. Pinteala, V. Epure, V. Harabagiu, B.C. Simionescu, and S. Schlick, *Macromolecules* **2004**, *37*(12), 4623–4634.

175. S. Holappa, L. Kantonen, F.M. Winnik, and H. Tenhu, *Macromolecules* **2004**, *37*(18), 7008–7018.

176. C. Giacomelli, L. Le Men, R. Borsali, J. Lai-Kee-Him, A. Brisson, S.P. Armes, and A.L. Lewis, *Biomacromolecules* **2006**, *7*(3), 817–828.

177. J.Y. Zhang, Y.T. Li, S.P. Armes, and S.Y. Liu, *J. Phys. Chem. B* **2007**, *111*, 12111–12118.

178. O. Colombani, M. Ruppel, F. Schubert, H. Zettl, D.V. Pergushov, and A.H.E. Muller, *Macromolecules* **2007**, *40*(12), 4338–4350.

179. M. Burkhardt, N. Martinez-Castro, S. Tea, M. Drechsler, I. Babin, I. Grishagin, R. Schweins, D.V. Pergushov, M. Gradzielski, A.B. Zezin, and A.H.E. Muller, *Langmuir* **2007**, *23*, 12864–12874.

180. R. Rutkaite, L. Swanson, Y. Li, and S.P. Armes, *Polymer* **2008**, *49*, 1800–1811.

181. M. Stepanek, K. Krijtova, K. Prochazka, Y. Teng, S.E. Webber, and P. Munk, *Acta Polym.* **1998**, *49*(2–3), 96–102.

182. M. Stepanek, K. Krijtova, Z. Limpouchova, K. Prochazka, Y. Teng, P. Munk, and S.E. Webber, *Acta Polym.* **1998**, *49*(2–3), 103–107.

183. Y. Teng, M.E. Morrison, P. Munk, S.E. Webber, and K. Prochazka, *Macromolecules* **1998**, *31*(11), 3578–3587.

184. K.D. Branham, G.S. Shafer, C.E. Hoyle, and C.L. McCormick, *Macromolecules* **1995**, *28*(18),6175–6182.

185. T. Costa, M.G. Miguel, B. Lindman, K. Schillen, J.C. Lima, and J.S. de Melo, *J. Phys. Chem. B* **2005**, *109*(8), 3243–3251.

186. M. Nowakowska, H. Bakhtiyari, E. Osselton, M.R. Steele, and J.E. Guillet, *J. Photochem. Photobiol. A: Chem.* **1992**, *64*(3), 329–342.

187. J.E. Guillet and W.A. Rendall, *Macromolecules* **1986**, *19*(1), 224–230.

188. M. Nowakowska, B. White, and J.E. Guillet, *Macromolecules* **1989**, *22*(10), 3903–3908.

189. E. Sustar, M. Nowakowska, and J.E. Guillet, *J. Photochem. Photobiol. A: Chem.* **1990**, *53*(2),233–250.

190. N.A.D. Burke, M. Templin, and J.E. Guillet, *J. Photochem. Photobiol. A: Chem.* **1996**, *100*(1–3), 93–100.

191. M. Nowakowska, J. Storsberg, S. Zapotoczny, and J.E. Guillet, *New J. Chem.* **1999**, *23*(6), 617–623.

192. U.P. Strauss and E.G. Jackson, *J. Polym. Sci.* **1951**, *6*(5), 649–659.

193. U.P. Strauss, in: E.D. Goddard and K.P. Ananthapadmanabhan (Eds.), *Interactions of Surfactants with Polymers and Proteins*, CRC Press, Boca Raton, FL, 1993.

194. H. Yamamoto, M. Mizusaki, K. Yoda, and Y. Morishima, *Macromolecules* **1998**, *31*(11), 3588–3594.

195. D.N. Schultz and J.E. Glass (Eds.), *Polymers as Rheology Modifiers*, ACS SymSer. 462, ACS, Washington, DC, 1991.

196. M. Mizusaki, Y. Morishima, and F.M. Winnik, *Macromolecules* **1999**, *32*, 4317–4326.

197. Y. Morishima, in: M.W. Urban, and T. Provder (Eds.), *Multidimensional Spectroscopy of Polymers*, ACS. SymSer. 598, Chapter 29, 1995.

198. M.C. Kramer, J.R. Steger, C.L. McCormick, in: M.W. Urban and T. Provder (Eds.), *Multidimensional Spectroscopy of Polymers*, ACS. SymSer. 598, Chapter 22, **1995**.

199. S. Yusa, A. Sakakibara, T. Yamamoto, and Y. Morishima, *Macromolecules* **2002**, *35*, 5243–5249.

200. Y.X. Hu, M.C. Kramer, C.J. Boudreaux, and C.L. McCormick, *Macromolecules* **1995**, *28*, 7100–7106.

201. M. Seki, Y. Morishima, and M. Kamachi, *Macromolecules* **1992**, *25*(24), 6540–6546.

202. D.F. Anghel, J.L. Toca-Herrera, F.M. Winnik, W. Rettig, and R. von Klitzing, *Langmuir* **2002**, *18*(14), 5600–5606.

203. D.F. Anghel, S. Saito, and A. Baran, *Langmuir* **1998**, *14*(19), 5342–5346.

204. M. Vasilescu, D.F. Anghel, M. Almgren, P. Hansson, and S. Saito, *Langmuir* **1997**, *13*(26), 6951–6955.

205. I. Soutar, C. Jones, D.M. Lucas, and L. Swanson, *J. Photochem. Photobiol. A: Chem.* **1996**, *102*(1), 87–92.

206. M. Mizusaki, N. Kopek, Y. Morishima, and F.M. Winnik, *Langmuir* **1999**, *15*, 8090–8099.

3

CHEMILUMINESCENCE PROCESSES IN POLYMERIC MATERIALS

Teresa Corrales, Carmen Peinado, Concha Abrusci, Norman S. Allen, and Fernando Catalina

3.1 INTRODUCTION

In the last few decades, many investigations have been undertaken to elucidate the degradation mechanism of polymers, which can be induced by exposure to various factors such as heat, oxygen, UV light, humidity, and mechanical stress, and results in flaws such as brittleness, cracking, and fading. Many authors have carried out studies to follow the oxidative aging of polymeric materials by means of several techniques. These studies need much time and are not sensitive enough in certain applications.

Photochemistry and Photophysics of Polymer Materials, Edited by Norman S. Allen

Chemiluminescence (CL), due to its low quantum yield of emission [1] ($\phi_{CL} = 10^{-9}$) and the special equipment required for its determination, has not been recognized as an important industrial analytical method until recently [2,3]. New improved equipments have been employed for chemiluminescence detection, including multicell imaging chemiluminescence instruments, which make it possible to determine the CL of up to 23 samples at the same time, effective gas flow distribution, and a constant temperature in the cells of the sample chamber [4]. In this regard, very sensitive charged coupling device (CCD) cameras have been used [5,6] and coupled with microchannel plate (MCP) photon amplifier to have image-intensified devices [7]. The most common is a conventional photomultiplier tube (PMT) used in photon-counting mode, and commercial equipments have been developed in the past few years. These improved devices have contributed to the acceptance of chemiluminescence as a broad technique for research and a quality control tool for the polymer industry, because CL is sensitive to detect the very first stage of oxidation of polymers, and does not require long periods of time to obtain information.

This technique of chemiluminescence has become a useful tool for the study of polymer degradation [8], oxidation mechanisms [9], and kinetics [10,11], and stabilizer efficiency [12,13]. This fact is due to its advantages with respect to other well-established techniques [14]. It is well known [15,16] that the chemiluminescence in polymers is due to the light emission that accompanies the thermal decomposition of the thermooxidative degradation products (hydroperoxides), which are formed during processing or in-service-life of the material under ambient conditions [6,17]. This bimolecular reaction promotes ketone products to its lowest triplet state and the radiative deactivation gives chemiluminescence emission in the visible region.

The chemiluminescence emission can be related to the hydroperoxide (POOH) content of a polymer [18]. The relationship of several factors such as temperature, molar mass [19,20], oxygen concentration [21], and antioxidants [22,23] to changes of the CL intensity in time or with the temperature has been established.

The first step of the hydrocarbon polymer degradation processes is the formation of polymeric alkyl radicals ($P^{\bullet}$) and its reaction with oxygen to give peroxy radicals

$$\mathrm{PH} \xrightarrow{\Delta} \mathrm{P\cdot} \qquad \text{(a)}$$

$$\left.\begin{array}{ll} \mathrm{P\cdot} + \mathrm{O_2} \longrightarrow \mathrm{POO\cdot} & \text{(b)} \\ \mathrm{POO\cdot} + \mathrm{PH} \longrightarrow \mathrm{POOH} + \mathrm{P\cdot} & \text{(c)} \\ \mathrm{POOH} \longrightarrow \mathrm{PO\cdot} + \mathrm{\cdot OH} & \text{(d)} \end{array}\right\} \text{Autooxidation cyclic process}$$

$$\mathrm{PO\cdot} + \mathrm{PH} \longrightarrow \mathrm{POH} + \mathrm{P\cdot} \qquad \text{(e)}$$

$$\mathrm{P\cdot} + \mathrm{P\cdot} \longrightarrow \mathrm{P\text{-}P} \qquad \text{(f)}$$

$$\mathrm{PO\cdot} + \mathrm{PO\cdot} \longrightarrow \mathrm{POOP} \qquad \text{(g)}$$

$$\mathrm{POO\cdot} + \mathrm{POO\cdot} \longrightarrow \mathrm{POH} + {}^1\mathrm{O_2} + \mathrm{P}\,{}^3(\mathrm{C{=}O})^* \dashrightarrow \mathrm{PCO} + \mathbf{h\nu} \qquad \text{(h)}$$

$$\mathrm{POO\cdot} + \mathrm{POO\cdot} \longrightarrow 2\,\mathrm{PO\cdot} + \mathrm{O_2} \qquad \text{(i)}$$

SCHEME 3.1 Simplified "autooxidation mechanism" of polyolefins considering the chemiluminescence pathway of reaction.

($POO^{\bullet}$) ((a) and (b) in Scheme 3.1). These radicals centered in oxygen can react in different ways giving a variety of product species in an autooxidation catalytic process ((c)–(e) in Scheme 3.1). Among the various reactions that can take place, such as hydrogen atom abstraction of the peroxy radicals from another polymer macromolecule (c), or homolytic decomposition of hydroperoxides (d), or radical recombination (f and g), the bimolecular termination reaction of two peroxy radicals (h) has a smaller activation energy [24] ($\Delta E = 12\,\mathrm{kJ\,mol^{-1}}$) being strongly exothermic compared to the other reactions. The involved energy gain in this reaction is $462\,\mathrm{kJ\,mol^{-1}}$. Hence, this bimolecular reaction (h) is the only one that can promote ketone products to its lowest triplet state and the further radiative deactivation by chemiluminescence emission in the visible region [25].

Otherwise, the heterochain polymers that contain heteroatom as part of the chemical constitution of the polymer along the main chain, besides degradation described for homopolymers, and also other mechanisms may be considered in their degradation mode, that is, depolymerization or elimination reactions.

The intensity of chemiluminescence, I_{CL}, as it is stated in Equation (h) of Scheme 3.1, is proportional to the rate of the bimolecular reaction of ($POO^{\bullet}$) species, which has a rate constant, k_b, and considering an efficiency factor, f. This factor of proportionality depends on different factors such as the yield of formation and emission of the excited state responsible for the chemiluminescence and the inherent apparatus constant.

$$I_{CL} = fk_b[POO^{\bullet}]^2 \tag{3.1}$$

The intensity of chemiluminescence can be determined as the peak-top intensity $I_{CL\text{-}max}$ or the area of the peak, $A_{CL\text{-}max}$. In the absence of antioxidants, the rate formation of peroxy radicals, $d[POO^{\bullet}]/dt$, can be expressed in terms of the initiation rate of oxidation, R_i.

$$d[POO^{\bullet}]/dt = R_i - k_b[POO^{\bullet}]^2 \tag{3.2}$$

The term R_i includes the rate of generation by heat of $P^{\bullet}$ (Eqs. (a) and 3.5) and the possible ways of disappearance of $P^{\bullet}$, for example, hydrogen atom abstraction and recombination.

Under nitrogen, the emission of chemiluminescence from polymer samples corresponds to the initial concentration of peroxy radicals ($POO^{\bullet}$) and it is proportional to the hydroperoxide content formed during the processing of the material, this being related to its thermal oxidation history. Under these conditions, $R_i = 0$. Hence, Equation 3.2 can be simplified as follows [26]:

$$d[POO^{\bullet}]/dt = -k_b[POO^{\bullet}]^2 \tag{3.3}$$

If $[POO^{\bullet}]_0$ is the initial concentration of peroxide radicals, the differential of Equation 3.3 gives the following equation:

$$1/[POO^{\bullet}] = k_b t + 1/[POO^{\bullet}]_0 \tag{3.4}$$

Substituting Equation 3.4 into the expression of the intensity of chemiluminescence, Equation 3.1, gives Equation 3.5:

$$\frac{1}{\sqrt{I_{CL}}} = t\sqrt{\frac{k_b}{f}} + \frac{1}{([POO^\bullet]_0\sqrt{fk_b})} \tag{3.5}$$

The reciprocal of the square root of the chemiluminescence intensity versus time gives a linear relationship as published by Kihara and Hosoda [26]. In the region where the plot gives a straight line, the slope is the value of $\sqrt{k_b/f}$. This parameter is named CL-decay rate and may be used to evaluate the thermal history of different polymers.

In the presence of oxygen, the chemiluminescence intensity (I_{CL}) is significantly enhanced with respect to the emission produced under nitrogen. As the samples are highly oxidized in a diffusion-controlled reaction simultaneous to the emission, reaction (b) in Scheme 3.1 is very fast and the relative concentration of [POO$^\bullet$] will be larger in proportion to that of [P$^\bullet$]. The rate of oxidation (R_i) in Equation 3.2 increases under these conditions, the bimolecular termination of peroxy radical, reactions (f) and (g) in Scheme 3.1, is, therefore, predominant. All these parameters can be used to evaluate the degradation in different materials and the effectiveness of antioxidants in the polymer stability.

In this chapter, some of the recent advances on the application of chemiluminescence to the study of several types of polymers performed by the authors are reviewed, together with the contributions of other authors in the field.

3.2 CHEMILUMINESCENCE OF POLYOLEFINS

3.2.1 Pristine Polyolefins

In the past 50 years, many authors have studied the degradation mechanism of polyolefins, because aging of polymers constitutes the main reason of detrimental changes in their properties. A large number of studies have been dedicated to describe the chemiluminescence from oxidation of polypropylene [27,28]. Under nitrogen, CL has been found to be proportional to peroxide content accumulated during accelerated aging of PP for all levels of oxidation. It was concluded that CL measured is a function of the geometric collection efficiency of the apparatus and the sample form, and also the temperature of oxidation. The effect of exposing peroxidized PP to dimethylsulfide (DMS) was reported by Zahradnickova et al. [29]. The destruction of peroxides when oxidized PP was exposed to DMS was shown to consist of two stages, an initial high reaction and a subsequent low reaction rate. The nature of the fast decomposing fraction of titratable peroxides is still under discussion, although CL measurements in oxygen clearly showed that exposure to DMS had a great stabilizing effect.

The maximum chemiluminescence intensity has been seen to be proportional to the oxygen pressure, which supported the hypothesis that light emission comes from

POOH decomposition than from bimolecular disproportionation of peroxyl radicals [30]. The results indicated that the concentration of oxygen in the polymer was not sufficient to scavenge all the polymer radicals, so that the termination reactions of alkyl radicals cannot be neglected. And even when working with pure oxygen at atmospheric pressure, the conditions are far from the critical concentration of oxygen in the polymer at which the oxidation rate becomes independent of oxygen. The work undertaken by Matisova-Rychla and Rychly [31] demonstrated that the chemiluminescence from the thermal oxidation of PP was linked to the decomposition of associated hydroperoxides, and that the kinetics of isothermal CL intensity versus time were affected by the kinetics of the appearance of potential light emitters having changing quality from single to α,β-unsaturated carbonyl groups.

Several models postulate that oxidation of polymers is an inhomogeneous process, where weak sites oxidize and initiate failure of the bulk. However, factors such as the effect of catalyst residues and the presence of crystalline and amorphous regions in polymers cannot be disregarded as other causes of inhomogeneous degradation. The imaging chemiluminescence has been revealed as a very useful technique to demonstrate the role of physical spreading in the oxidation of stabilized PP, and also it has been employed to study the effects of artificial recycling of PP waste films [32].

PP exhibits higher oxidation susceptibility compared with polyethylene. This well-known behavior is ascribed to tertiary hydrogen atoms in the structure of PP. Recently, the thermal and photooxidative behavior of polyethylenes with different manufacturing histories (linear low-density polyethylene [LLDPE], metallocene polyethylene [mPE], and high-density polyethylene [HDPE]) has been investigated by means of CL technique, which suggested a relationship between CL and some structural characteristics of the polymers. The results were compared to those obtained by thermogravimetric analysis (TGA), in terms of determination of the frequency factor and activation energy for the overall photooxidation process, FTIR, and derivative UV spectroscopy [33]. It was found that the metallocene polymer exhibited the lowest CL intensity that corresponded to the lowest concentration of oxidized products determined by FTIR. For LLDPE and mPE, the intensity of CL showed a gradual increase with aging time, whereas for HDPE, an autocatalytic oxidation process was observed, and the emission was detected at a lower temperature than for LLDPE and mPE, which showed an inhibition period (Fig. 3.1).

These results indicated that HDPE was more susceptible to oxidation, followed by LLDPE and mPE. A similar order for the stability was found through the measurement of hydroperoxide concentration and carbonyl index growth. The mPE and LLDPE samples showed an initial autoretarding effect, whereas HDPE film exhibited a shorter induction period and the highest oxidized species levels after heating time. The results correlated with those obtained by TGA; activation energies for the thermal degradation of polyethylenes were determined at different oven aging times. HDPE gave rise to a significant decrease in the activation energies with aging, and showed the highest $\Delta E_a/\Delta t$ values. This study showed that CL provides useful data for the characterization of the stability of several polyethylenes having

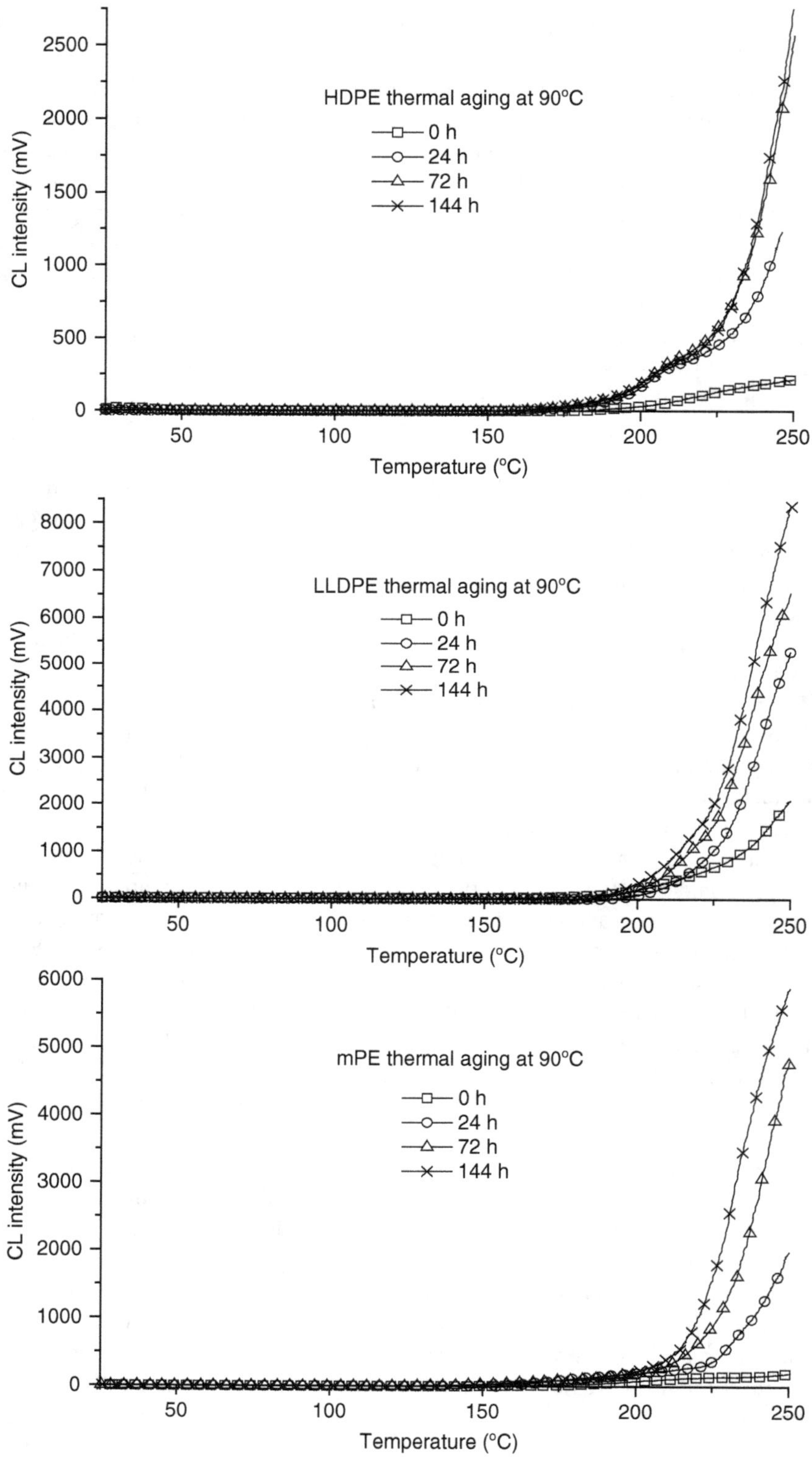

FIGURE 3.1 Chemiluminescence spectra under nitrogen of HDPE, LLDPE, and mPE polymers at 170°C.

various structures and to establish a certain correlation between structure and properties.

Generally, polyethylene exhibits a double sigmoidal shape on the isothermal chemiluminescence curve. Setnescu et al. interpreted the behavior as oxidation of very active sites (double bonds, branches, etc.) of PE and oxidative crystalline-phase conversion [34]. Zlatkevich suggested that overlaid sigmoids reflect two oxidation processes related to two kinds of peroxides with different stability assumed to exist in the oxidized polyolefins [35].

The study carried out by Broska and Rychly [36] revealed that the first stage is related to oxidation started in the vicinity of vinyl bonds, representing the vulnerable site of the polymer chain due to easily abstracted allylic hydrogens, and their concentration decreased at the beginning of the thermooxidative degradation. Otherwise, the second peak was ascribed to CL of decomposition of peroxides accumulated in the polymer during oxidation. T_{max} of the low temperature peak decreases on oxidation, because the oxidative groups formed, such as carbonyls, hydroxyls, and unsaturation, are introduced to the polymer chain as defects, resulting in a lower melting temperature. T_{max} of the second peak was seen to decrease as well, associated with decreasing stability of peroxides.

Also, the dependence of polyethylene oxidation on oxygen concentration and cross-linking has been evaluated. In oxygen-rich conditions, the structure of the polymer is the main factor affecting the efficiency of hydroperoxide formation. However, in the case of poor oxygen atmosphere, the oxidation is governed by the oxygen diffusion. When the polymer oxidation is undertaken in a relatively high concentration of oxygen and at high temperature, the system can reach the diffusion-controlled mode of oxidation. In such conditions, the rate of autooxidation increases to the level where the diffusion and solubility of oxygen are limited. It resulted in a reduction of the oxidation reaction, and it is reflected as a break in the isothermal CL curve of oxidation as double-stage oxidation, oxygen independent and oxygen diffusion-controlled reactions. On the other hand, the oxidation is affected by cross-linking, by the increase of the rate of oxidation due to chain branching, and by changing the rate of diffusion of oxygen to the system. The correlation of integrated CL under dynamic analysis and insoluble fraction of PE clearly showed that cross-linking reaction on the oxidized polymer is important at the stage of oxygen diffusion-controlled oxidation, where the termination starts to involve the reaction of alkyl radicals leading to cross-linking.

Thermal oxidation of ultrahigh molecular weight polyethylene (UHMWPE) has been studied by chemiluminescence analysis, to obtain a better understanding of the mechanism responsible for the emission during degradation [37]. UHMWPE exhibited dual sigmoidal CL curves typical of PE. The first chemiluminescence peak coincided with an initial build-up of hydroperoxides determined by FTIR, and the second CL peak followed the build-up of carbonyls. These results indicated that during the first stage of the oxidation, when the concentration of hydroperoxide is high and the carbonyl content is low, the chemiluminescence emission is due to decomposition of hydroperoxide. At longer aging times, the CL becomes more associated with the accumulation of carbonyls.

3.2.2 Stabilized Polymer Systems

Polyolefin stability is an important field of research with enormous industrial interest, with particular emphasis on the study of antioxidant efficiency. The chemiluminescence emission of LDPE films including some commercial antioxidants based on hindered phenols, varying in structure (Scheme 3.2) has been studied under different conditions, to evaluate their stabilization activity in these films and related their effects to hydroperoxide levels [22]. Hindered phenols, in an oxygen-deficient environment (e.g., in melt processing), can act as chain-breaking donors (CB-D) and acceptors (CB-A) simultaneously in a catalytic way [38,39]. Both mechanisms for these antioxidant additives will give rise to stable products contributing to a depletion in the concentration of $P^{\bullet}$ and $POO^{\bullet}$ species and subsequently the value of I_{CL}. In the presence of this hindered phenol antioxidant, peroxy radicals are trapped by the antioxidant during the induction period of oxidation (OIT_{CL}), the rate of formation of peroxy radicals is nearly similar to their rate of termination by recombination, and the CL signal is low. When the antioxidant is consumed, the radical concentration increases due to autoacceleration of the oxidation, and it is reflected by an increase in the chemiluminescence intensity.

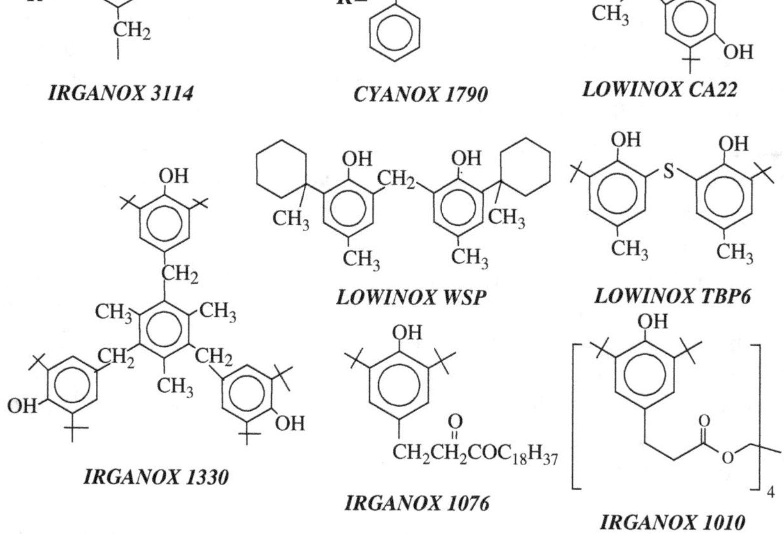

SCHEME 3.2 Structures of hindered phenolic antioxidants.

TABLE 3.1 Chemiluminescence Under Nitrogen at 170°C of LDPE films (120 μm) That Contain 0.1% w/w of Different Hindered Phenols and Their Initial Hydroperoxide Content

Stabilizer (0.1% w/w)	$I_{CL\text{-}max}$ (mV)	$A_{CL\text{-}peak} \times 10^{-3}$ (mV)	CL-decay rate $\times 10^4$ ($mV^{-1/2}\,s^{-1}$)	$[POOH]_0 \times 10^6$ ($mol\,g^{-1}$)[a]
Free additive	196/303	5.0	0.18	3.98
Cyanox 1790	40	0.6	1.42	2.40
Irganox 3114	111	1.9	0.66	2.98
Irganox 1330	90	1.8	0.62	2.82
Lowinox CA22	115	1.5	0.70	2.70
Lowinox TBP6	79	1.0	0.73	2.86
Lowinox WSP	58	0.7	1.57	2.46
Irganox 1010	76	1.5	0.43	2.94
Irganox 1076	47	0.5	2.47	2.26

[a]Determined by iodometric titration.

The measurement of the chemiluminescence intensity allows to obtain information about the rate of degradation and the efficiency of the additives because the presence of antioxidant stabilizers reduces the rate formation of excited ketones responsible for the emission of light. Under nitrogen, the oxidation is inhibited and the emission is a measure of the peroxides present in the material formed during its processing at high temperatures. As expected, the intensity of chemiluminescence and the area of the emission curve decrease for the films containing hindered phenols compared with free additive polyethylene thus demonstrating their protective effect (Table 3.1).

The presence of antioxidants clearly increased the CL-decay rate (Fig. 3.2) together with the intensity of chemiluminescence in the peak-top and integrated area of CL-signal. A good correlation is seen between the chemiluminescence parameters, CL-decay rate, $I_{CL\text{-}max}$, and $A_{CL\text{-}peak}$ and the initial hydroperoxide contents of the films. Irganox 1076 exhibits the best stabilization effect as expected due to its well-known good performance under processing conditions. This is associated with its low melting point and good compatibility due to the long alkyl chain present in its structure. In contrast, Irganox 1010, which is claimed to extend the long-term stability, has the poorest antioxidant efficiency under processing conditions. The order of antioxidant efficiency of the hindered phenols measured by chemiluminescence under nitrogen, which corresponds with the stabilization during processing, are compared as follows: Irganox 1076 > Lowinox WSP ≈ Cyanox 1790 > Lowinox TBP6 > Lowinox CA22 > Irganox 3114 ≈ Irganox 1330 > Irganox 1010.

Cyanox 1790 and Lowinox WSP exhibited the next best stabilization efficiencies (of Irganox 1076). This can be related with the combination of substitution in both *ortho* positions of a methyl-phenyl group and a *tert*-butyl group, which increases the radical trapping rate constant [40]. These two products are similar from the point of view of their substitution in the *para* position of the phenolic nucleus [41] and they will give the same type of stable quinone methide. Lowinox TBP6 exhibited a lower

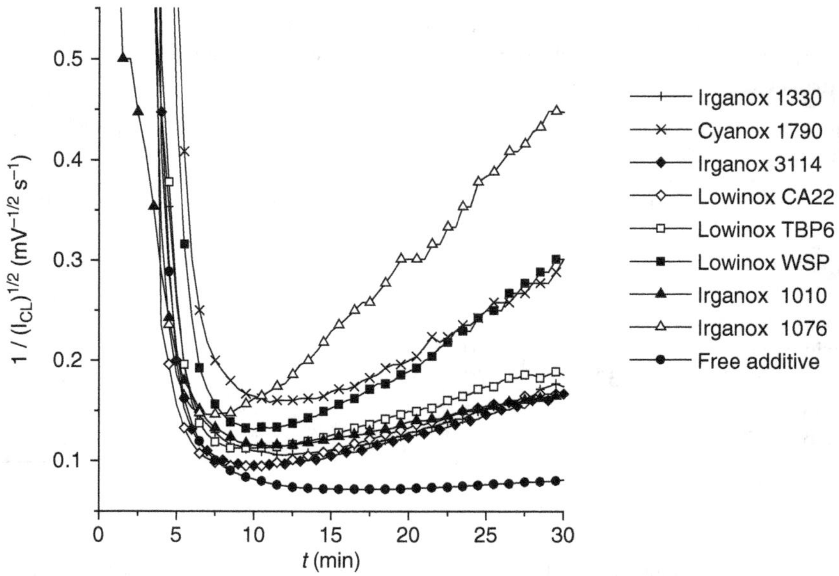

FIGURE 3.2 Plot of the reciprocal of the square root of chemiluminescence intensity versus time obtained at 170°C under nitrogen for free and stabilized low-density polyethylene films with the different phenolic antioxidants (0.1% w/w).

stabilization efficiency from the chemiluminescence data under nitrogen and this can be related with the incorporation of sulfur-containing functional groups, which enhances its radical trapping effectiveness. The initial hydroperoxide content of the film possessing this additive does not appear to be sensitive to the commented behavior and the value is in the same order as the remainder of the antioxidants. Also, a small difference, not detected in the hydroperoxide content, has been found in the chemiluminescence parameters between the phenol Lowinox CA22 and the rest of the phenols Irganox 3114, Irganox 1330, and Irganox 1010, the latter showing the lowest antioxidant efficiency, as commented on before. Lowinox CA22 possesses an *ortho* mono-*tert*-butyl phenol and its corresponding phenoxyl radical has a lower lifetime and higher radical reactivity compared with the other structures. The lower stability of the *ortho* di-*tert*-butylic phenols can be explained, as well as, Irganox 1010, in terms of their lower compatibility in the polyethylene melt. These two products have the highest melting point, above 200°C and the processing of the films was carried out at 190°C.

Phenolic antioxidants are well known for being melt processing stabilizers as well as long-term thermal stabilizers. In the chemiluminescence measurements on the polyethylene films under oxygen (Fig. 3.3), the antioxidant effect of the phenols is clear when these induction times are compared with those of the free additive polyethylene film (0.73 h) (Table 3.2). The results showed that the structure of the phenolic moiety will be a crucial factor influencing the stabilization performance.

The rate of oxidations, ν_{ox}, has a similar value for all the polymer films. This is consistent with the inherent susceptibility to oxidation of the material [42] and the

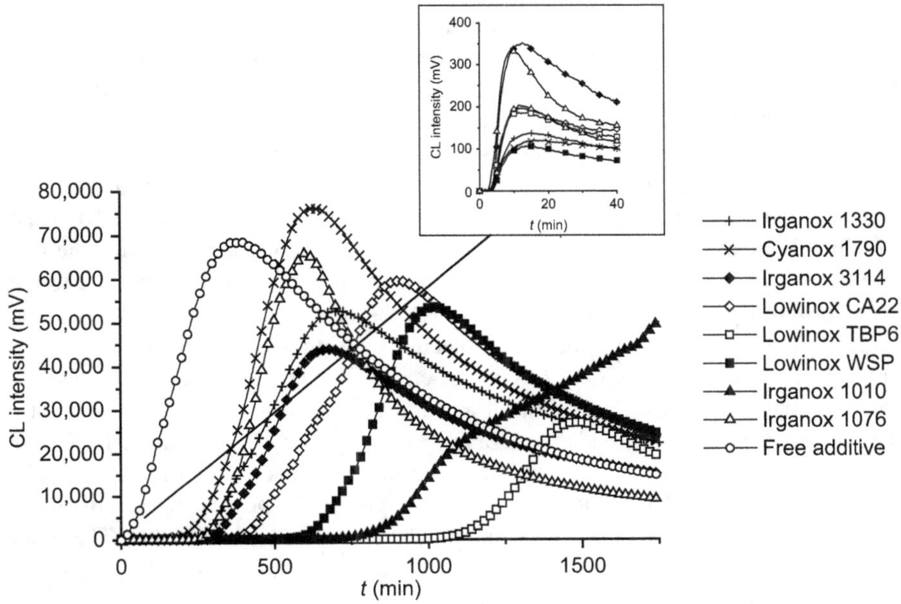

FIGURE 3.3 Chemiluminescence emission profiles versus time of low-density polyethylene films free and stabilized with the different phenolic antioxidants (0.1% w/w), determined at 170°C under oxygen.

low-density polyethylene must be oxidized at a similar rate when the antioxidants are consumed. In general, a good correlation between the other chemiluminescence parameters $I_{CL\text{-}max2}$, t_{max2}, $t_{½}$, and OIT_{CL170} were found. The oxidation induction time gives the following order in the antioxidant efficiency: Lowinox TBP6 > Irganox 1010 > Lowinox WSP > Lowinox CA22 > Irganox 3114 ≈ Irganox 1076 > Cyanox 1790 ≈ Irganox 1330.

TABLE 3.2 Chemiluminescence Under Oxygen at 170 and 200°C of LDPE Films

Stabilizer	$I_{CL\text{-}max1}$ (mV)	I_s (mV)	t_{max2} (h)	$I_{CL\text{-}max2} \times 10^{-4}$ (mV)	$t_{1/2}$ (h)	v_{ox} (h^{-1})	OIT_{CL170} (h)	OIT_{CL200} (h)	OIT_{DSC} (min)
Free additive	—	—	6.6	6.86	2.6	1.6	0.73	0.17	0
Cyanox 1790	120	115	10.5	7.63	6.8	1.3	5.1	0.69	10.3
Irganox 3114	352	148	11.2	4.40	7.7	1.4	6.1	0.73	9.6
Irganox 1330	137	117	11.8	5.28	7.6	1.2	5.1	0.64	16.4
Lowinox CA22	196	144	15.2	5.97	10.5	1.1	8.4	0.97	11.3
Lowinox TBP6	186	40	24.6	2.72	21.2	1.3	19.8	3.9	65.9
Lowinox WSP	106	58	16.9	5.37	13.3	1.3	11.7	1.02	49.8
Irganox 1010	203	59	31.9	5.34	22.5	1.3	14.1	1.27	16.7
Irganox 1076	333	116	10.1	6.59	7.2	1.6	5.7	0.53	3.0

This order in efficiency as determined by chemiluminescence is in agreement with the order in the oxidation induction time measured by calorimetry, $OIT_{DSC.}$, (Table 3.2). Lowinox TBP6 exhibits the best behavior in a melt medium attained in oxygen, with results that can be correlated with the stabilization during the product lifetime. This molecular structure contains a thiophenol group, which acts as a strong $POO^{\bullet}$ scavenger and hydroperoxide decomposer due to the chemistry of both the phenolic and activated sulphidic antioxidants being involved [43]. Irganox 1010 and Lowinox WSP are the next additives that follow in the stabilization activity, the first due to its high number of phenolic-OH groups per mole and the second due to the close proximity of the hydroxyl groups. The latter will cause internal hydrogen bonding with little steric strain on the molecule, causing the molecule to be a strong antioxidant [44]. Lowinox CA22 does not equate to the performance of Lowinox WSP, probably due to the substitution at the *ortho* positions, as explained before. The remainder of the antioxidants, Irganox 3114, Irganox 1076, Cyanox 1790, and Irganox 1330, exhibits low efficiency of chemiluminescence signal inhibition. The reference, Irganox 1076, which was very efficient in the processing stabilization, shows a low performance under oxygen, maybe due to its high volatility. The latter phenomenon is important in considering performance as measured by chemiluminescence activity in the sample chamber.

It has been observed that the annealing of stabilized polypropylene prolonged the inhibition periods of oxidation. This effect has been associated to better homogenization of the antioxidant at a molecular level in the polymer, and so the stabilizer would be more ready to react with polymer peroxy radicals and to suppress the formation of associated hydroperoxides [45]. Also, it has been related to the existence of respective domains with predominating low and high molar fractions in the amorphous part of the polymer, which is the most vulnerable part with respect to oxidation. The stabilizer tends to dissolve better in the lower molar mass regions, then, the domains of higher molar mass should be less protected. Annealing tended to distribute the molecules of the stabilizer into higher molar mass domains, and the induction period of oxidation increased. Other authors suggested that during annealing in an inert atmosphere, other reactive sites such as hydroperoxides and terminal unsaturations were removed, which allowed polymer of a better quality to be obtained [46].

Since the discovery of [C_{60}]fullerene, a new allotropic state of carbon, several applications including thermal stabilization of polymers have been described [47]. Chemiluminescence measurements on high-density polyethylene stabilized with fullerene revealed the very low efficiency of that additive in the oxidative protection of polyethylene in comparison with the methyl[C_{60}]fullerepiramate, which behaved as a moderately good antioxidant. The oxidation induction time obtained in the presence of the adduct with levopimaric acid methyl ester was longer than the induction period of polyethylene stabilized with fullerene. However, the maximum oxidation rate and the maximum chemiluminescence intensity were similar. These results indicated that methyl[C_{60}]fullerepiramate was a good radical scavenger in the induction period, and exhibited a poor antioxidant efficiency over the propagation stage of thermal aging. Other adducts similar to fullerene did not exhibit antioxidant activity, and the stabilization efficiency was dependent on the nature of functional groups of substituted derivatives.

3.2.3 Effect of Micron and Nanoparticle Titanium Dioxide

Nowadays, titanium dioxide is the most important white pigment used in the plastics industry. The main criteria for selecting a pigment are opacity and tinting strength. It has been observed that pigment particle size and surface area play an important role in dispersion. If the number of aggregates and agglomerates is few, good dispersion will be obtained and high opacity and tinting strength will develop [48]. Consequently, particle size selection has become one of the principal developments in pigment technology in recent years. It has been shown that nanoparticles of TiO_2 have a significant effect on the properties of coating materials [49]. The small size is responsible for the different properties, making these materials attractive for many applications [50–56].

In many cases, pigments can influence the thermal and photochemical stability of a polymer material. Titanium dioxide absorbs much of the ultraviolet radiation, protecting the polymers against UV light. However, the UV energy may be converted to heat, and creates radical sites on the surface of the pigment particle, which accelerates the breakdown of the polymer. Generally, the pigment is surface-coated with alumina, silica, or siloxanes to minimize photochemical activity at the polymer/pigment interface, and to improve the dispersibility of the pigment. However, high levels of surface treatment reduce the proportion of the titanium dioxide in the pigment, reducing opacity and tinting strength.

The thermal activity of a series of nano- and micron-particle grade anatase and rutile titanium dioxide pigments, with various densities of surface treatments, particle size, and surface area, have been determined by chemiluminescence in monomodal metallocene polyethylene [57].

The chemiluminescence analysis under nitrogen was undertaken for metallocene polyethylene films including 0.5% pigment, and compared to unpigmented polyethylene, stabilized with Irganox 1010 and not stabilized (PE Control) (Table 3.3). Similar values of $I_{CL\text{-}max}$ for metallocene polyethylene with and without antioxidant were obtained, which indicated the low antioxidant efficiency under processing of Irganox 1010. However, marked suppression of chemiluminescence emission was observed for the anatase-pigmented polymer, which correlated well with the lower initial hydroperoxide content determined in the material. In general, this effect was enhanced for coated grades, where the S.D. nano-anatase grade exhibited the lowest value of hydroperoxide indicating the protective influence of the hydroxyapatite coating. For uncoated pigments, the chemiluminescence varied with particle size, PC50 < PC500 < PC105. Apparently, the greater surface particle pigment allows the higher contact with the polymer, and induces more oxidation during processing as measured by the hydroperoxide analysis. A similar trend was observed for the rutile grade. Untreated nano-rutile showed the highest emission, and the thermal sensitization decreased with the alumina content, RCL 696 < Chlo.

TABLE 3.3 Chemiluminescence Under Nitrogen at 170°C of Metallocene Polyethylene Films Containing the Nano and Micron Grade Titania Pigments at 0.5% w/w Concentrations with 0.05% w/w Irganox 1010 Antioxidant Together with Control Unpigmented Samples

Pigment (0.5% w/w)	Pigment Type	$I_{CL\text{-}max}$ (mV)	$A_{CL\text{-}peak} \times 10^{-2}$ (mV)	$[ROOH]_0$ (mg/g)
None (PE control—unstabilized)		200	46.8	172
None (stabilized Irganox 1010)		154	41.3	61
PC50	Nano-anatase (20 nm)	54	12.1	35
PC105	Nano-anatase (15 nm)	135	38.4	67
PC500	Nano-anatase (5 nm)	105	24.4	125
S.D.	Nano-anatase coated (70 nm)	47	10.3	19
AT-1	Micron-anatase (0.24 μm)	65	21.1	26
Nano-rutile	Nano-rutile (25 nm)	407	154.1	112
Chlo	Micron-rutile 1% alum. (0.28 μm)	115	37.8	43
RCL 696	Micron-rutile 3.4% alum. (0.28 μm)	66	16.3	32

In general, the stabilization effect observed in the presence of titania was enhanced with the content of pigment, and lower values of $I_{CL\text{-}max}$ and $A_{CL\text{-}peak}$ were obtained for metallocene polyethylene with 2% w/w pigment in comparison with 0.5% content.

Under oxygen conditions, two emission peaks were detected for stabilized polymer films, a weak emission due to the thermal history of the film and a second intense peak observed when the oxidation induction period is attained (Fig. 3.4).

The high efficiency of Irganox 1010, as a long-term thermal stabilizer in polyethylene film, was evidenced by the increased oxidation induction time (OIT) (Table 3.4).

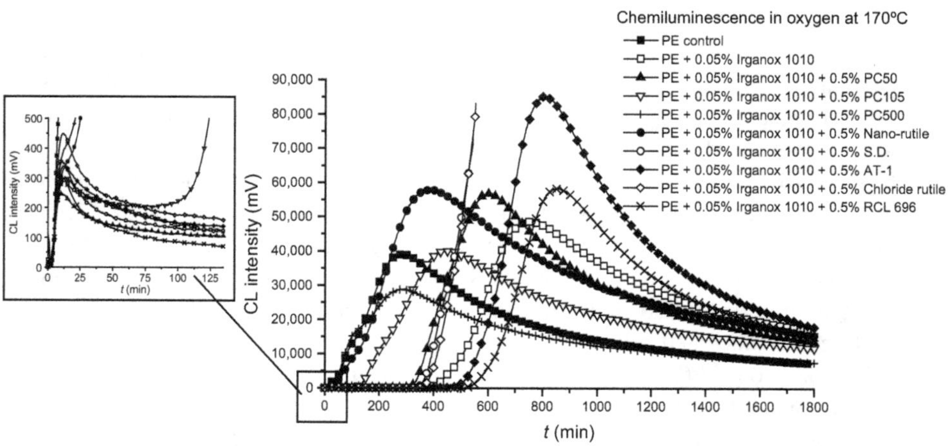

FIGURE 3.4 First and second chemiluminescence peak curves versus time of metallocene polyethylene films free, stabilized with Irganox 1010 (0.05% w/w), and pigmented with micron- and nanoparticle titanium dioxide (0.5%), determined at 170°C under oxygen.

TABLE 3.4 Chemiluminescence under Oxygen in Melt Condition at 170°C of Metallocene Polyethylene Films Containing the Nano and Micron Grade Titania Pigments at 0.5% w/w Concentrations with 0.05% w/w Irganox 1010 Antioxidant Together with Control Unpigmented Samples

Pigment (0.5% w/w)	Pigment Type	$I_{CL\text{-}max1}$ (mV)	I_s (mV)	t_{max2} (h)	$I_{CL\text{-}max2} \times 10^{-4}$ (mV)	OIT_{CL} (h)	OIT_{DSC} (min)
None (PE control—unstabilized)		—	—	4.7	3.9	0.7	0
None (Irganox 1010)		289	118	12.5	4.9	8.1	7.2
PC50	Nano-anatase (20 nm)	246	99	10	5.7	5.6	11.3
PC105	Nano-anatase (15 nm)	302	205	7.5	4.0	2.3	0
PC500	Nano-anatase (5 nm)	—	—	4.8	2.9	0.6	0
S.D.	Nano-anatase coated (70 nm)	354	113	—	>9	6.8	2.2
AT-1	Micron-anatase (0.24 μm)	341	131	13.5	8.5	9.7	9.1
Nano-rutile	Nano-rutile (25 nm)	—	—	6.4	5.8	1.8	0
Chlo	Micron-rutile (0.28 μm) 1% alum.	449	150	—	>9	8.2	3.6
RCL 696	Micron-rutile (0.29 μm) 3.4% alum.	299	60	14.3	5.8	10.4	6.2

In the presence of pigments, the activity of the antioxidant was modified. Nanoparticles of the rutile pigment exhibited a thermal sensitizing effect and lower t_{max2}, $I_{CL\text{-}max2}$, and OIT were observed. This effect was suppressed for micron-pigmented samples. Chlo and RCL 696 exhibited longer induction periods than for the unpigmented polymer, and as it was observed under nitrogen, the protective effect was enhanced with coating. A similar trend was observed for anatase-grade titania. The OIT was reduced in the presence of nanoparticles with respect to the micron particles. For the untreated nanoparticles, the OIT decreased in the order PC50 > PC105 > PC500, which followed the decrease in particle size. The greater surface area of particle would allow greater contact with the polymer and induce more oxidation. In general, a similar order in the oxidation of polymer films was found with all the chemiluminescence parameters. And a good correlation was found between the thermal activity of pigments determined by means of chemiluminescence OIT_{CL} and the OIT_{DSC} obtained by calorimetry.

For all polyethylene samples containing 2% pigment, higher oxidation was observed. OITs were greatly reduced, and for nanoparticles the first emission peak was not well defined appearing as a shoulder on the intense second peak. Due to the higher oxidation rates of the pigmented polyethylene samples, OITs were not detected by DSC. In contrast, CL provides useful data for the characterization of the stability of the polyethylene [7]. CL exhibits higher sensitivity and it allows one to make measurements at lower temperature, closer to the real degradation conditions in the solid state. Whereas DSC is insensitive to detect the OIT below the melting point, and the curve obtained is the sum of many different exothermic oxidation reactions, which makes difficult the determination of the OIT.

The work undertaken by Fearon et al. compared DSC and CL techniques for studying polymer oxidation [58]. It was evidenced the higher sensitivity of CL to detect traces of peroxides, under conditions where DSC was insensitive, and the advantages of CL to study semicrystalline polymers where overlapping peaks from peroxide decomposition and melting complicated the study by DSC.

3.3 CHEMILUMINESCENCE OF OTHER POLYMERS

3.3.1 Natural Polymers

Gelatines are a class of proteinaceous materials that are derived from the parent protein structure of collagen [59,60], by a procedure that involves the destruction of the collagen secondary structure and, in some aspects of the primary and tertiary structures. The gelatine manufacture is carried out by acid pretreatment (type A gelatine) or alkali pretreatment (type B gelatine) from bone and skin collagen fibrils in a reproducible way, but always, gelatine is a degraded protein. Gelatine from successive extractions has different physical and chemical properties due to the hydrolytic degradation that takes place during the manufacturing process. Its variety of properties have focused much scientific interest due to the large number of commercial applications such as edible, pharmaceutical, and photographic uses.

After the pretreatment, the first extract (with hot water) is obtained at the lowest temperature and always exhibits a higher bloom value or gel strength, which is the common parameter to measure gelatine quality [61].

Gelatine is composed of long chains of amino acids joined through a peptide linkage. The structure of the commercial type B gelatines is heterogeneous with respect to the compositional amino acids, with a higher content of glycine (33 mol%), proline (10 mol%), and hydroxyproline (5 mol%). Proline and its hydroxy derivatives are nitrogen heterocyclic organic compounds that can suffer, as well as glycine, an oxidation reaction in the α-carbon to the nitrogen of the N-alkyl amide of their structure in the gelatine. Such reactions have been observed in the autooxidation of N-alkyl amide [62–64] model compounds at temperatures lower than 100°C and also photochemically at short (254 nm) and long ($\lambda > 300$ nm) wavelengths. This oxidation reaction has also been observed in polymers such as nylon and polyamides [65,66].

Recently, the chemiluminescence analysis has been undertaken for two type B gelatines, which are the most employed photographic gelatine as components in silver halide photographic films. The two gelatine raw materials selected differed in molecular weight (bloom value) to study the effect on their thermal stability [67]. The non-isothermal measurements of chemiluminescence under nitrogen of the two bloom value gelatine grades were undertaken (Fig. 3.5). The samples exhibited weak intensity emission, confirming the thermal stability of dry gelatines. The emission of chemiluminescence started in the region of their glass transition temperatures, due to loss of rigidity, which favors the emission reaction. The gelatine of bloom

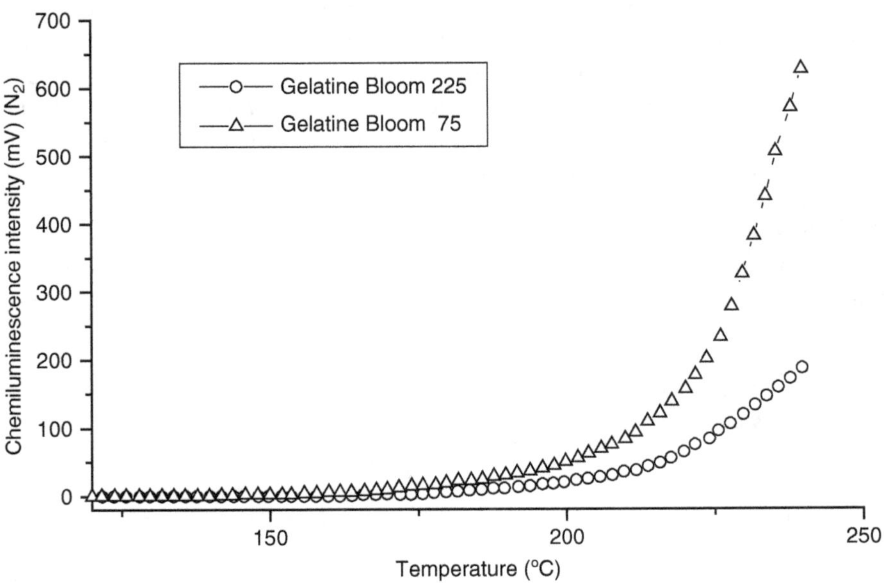

FIGURE 3.5 CL temperature ramping curves obtained for gelatine films under nitrogen.

TABLE 3.5 Chemiluminescence Data of Gelatines Type B Obtained Under Nitrogen at 225°C

Sample	T (°C)	Test Atm.	$I_{CL\text{-}max}$ (mV)[a]	$A_{CL\text{-}peak} \times 10^3$ (mV)[b]	I_s (mV)[c]	$[POOH]_0$ (mol g^{-1})[c]
G-75	160	N_2	58	20.5	—	3.50×10^{-6}
G-225	160	N_2	20	6.4	—	2.75×10^{-6}
G-75	225	N_2	660	—	640	
G-225	225	N_2	170	—	200	

[a]Intensity of the chemiluminescence maximum.
[b]Area of the chemiluminescence peak.
[c]Determined by iodometric titration.

75 exhibited an early and more intense emission than the gelatine of bloom 225. This result was in accordance with the more degraded structure of gelatine bloom 75 because of its extraction at higher temperatures on the manufacturing process. The higher quality of the bloom 225 gelatine is evidenced by the behavior of the films in isothermal chemiluminescence experiments under nitrogen (Table 3.5).

In the case of gelatine bloom 75, the emission started during the pretests ($T < 225$° C) in contrast with the bloom 225 gelatine. This behavior confirmed its lower rigidity and the higher oxygen mobility in the material. The CL-emission at 225°C, temperature above the glass transition, was enhanced by the loss of rigidity of the film and showed the same behavior, that is, higher emission of the lower bloom gelatine. The initial hydroperoxide contents determined for the gelatine films were low, but were in agreement with the order of stability.

Under oxygen, both materials behave with the same order of stability commented before, again high bloom gelatine exhibits the lower emission. The graphs of the ramping temperature CL-emission in Arrhenius form showed two different temperature intervals. The process of low activation energy predominated at low temperatures, and that of high activation energy predominated at higher temperatures in the glass transition region (Table 3.6). This high activation energy value was an indication that bond cleavage and the subsequent formation and reaction of peroxyl radicals shown before can contribute to the emission. For this mechanism, high temperatures (T_g region) and mobility in the material are required. The higher bloom gelatine exhibited higher activation energy of the CL-emission in the T_g region confirming

TABLE 3.6 Activation Energies[a] of the CL-Emission for Commercial Gelatines

Material	T_g Region $175 < T < 225$°C	Low T Region $T < 160$°C
Gelatine bloom 75	89.8	42.3
Gelatine bloom 225	105.6	43.2

[a]E_a in kJ mol^{-1}.

TABLE 3.7 Chemiluminescence Data of Gelatines Type B Obtained Under Oxygen

Sample	T (°C)	Test Atm.	$I_{CL\text{-}max}$ (mV)[a]	$A_{CL\text{-}peak} \times 10^3$ (mV)[b]	I_s (mV)[c]	$[POOH]_0$ $(mol\ g^{-1})$[c]
G-75	160	O_2	860	—	900	3.50×10^{-6}
G-225	160	O_2	550	—	500	2.75×10^{-6}
G-75	225	O_2	17,000	219.9	—	
G-225	225	O_2	13,000	207.1	—	

[a]Intensity of the chemiluminescence maximum.
[b]Area of the chemiluminescence peak.
[c]Determined by iodometric titration.

again its higher rigidity. The obtained values of the temperature coefficients from Arrhenius plot were in the range of those obtained for other polymers [68]. For example, with cellulose in sheet form, Matisova-Rychla and Rychly [45] founded 48 and 72 kJ mol^{-1} for the intervals below and above the T_g of the cellulose, which, as in gelatine, strongly depends on the content of moisture.

In the presence of oxygen, at higher temperatures than T_g, the emission was clearly enhanced due to the higher oxygen mobility in the materials (Table 3.7). The emission profiles at 225 and 160°C are shown in Fig. 3.6, a stationary emission (I_s) was

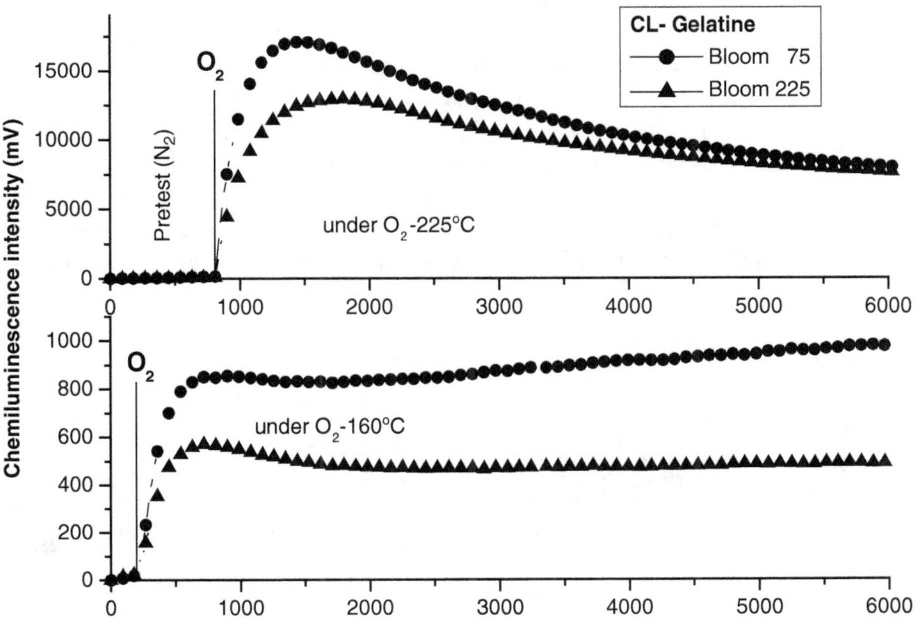

FIGURE 3.6 CL emission profiles for gelatine films obtained at 160 and 225°C under oxygen. The pretest (heating rate 10°C/min) data under nitrogen are also included.

observed at 160°C indicating the lower oxidation rate, in contrast with that at 225°C, where the emission gave a peak with much higher intensity. The gelatine grade of lower bloom showed the higher degradation rate in agreement with its lower quality.

Photographic gelatine was exposed to bacterial and fungal degradations, in water and under controlled conditions, and the biodegradation extent was studied by chemiluminescence and viscosimetry [69]. Bacteria and fungi were isolated and identified from black and white cinematographic films in a previous work [70]. It was seen that autoclave sterilization of gelatine solution drastically reduces the viscosity of the gelatine. This fact confirms that hydrolytic degradation mechanism is through a cleavage of the peptide bond of the protein, as no significant oxidation of the gelatine was detected by chemiluminescence. In contrast, biodegradation by bacteria, at low temperatures, decreases the viscosity of the gelatine via enzymatic degradation, and also the metabolic activity of microorganism generates reactive oxygen species (ROS), as peroxide radicals or hydroperoxides, hydrogen superoxide, and radical anion superoxide, which induces an important oxidation in the gelatine structure. These species are intermediates in the microbial aerobic metabolism [71] and are frequent in physiological processes [72], and some microorganisms can produce a high concentration of ROS, for example, in the case of lichens [73] and some bacteria [74]. The oxidation of gelatine was detected by the drastic increase in the chemiluminescence emission of the materials. It is important to remark that the intensity of CL did not correlate quantitatively with the viscosity decay indicating that hydroperoxide formation depends on the metabolism of each bacterium. In general, much higher CL emission intensities were observed for samples biodegraded by fungi with respect to those obtained for gelatine biodegraded by bacteria. The obtained CL-profiles versus time for gelatines biodegraded by *Penicillium chrysogenum* and *Alternaria alternata*, as illustrations, are plotted in Fig. 3.7.

The obtained results on chemiluminescence emission from gelatine materials confirmed that this technique can be a useful tool to characterize the solid state of gelatines. It can be applied to different commercial grades and applications for oxidation control during the manufacture and use of gelatines, and is able to quantify the oxidation that is produced by microorganism in the structure of gelatine due to

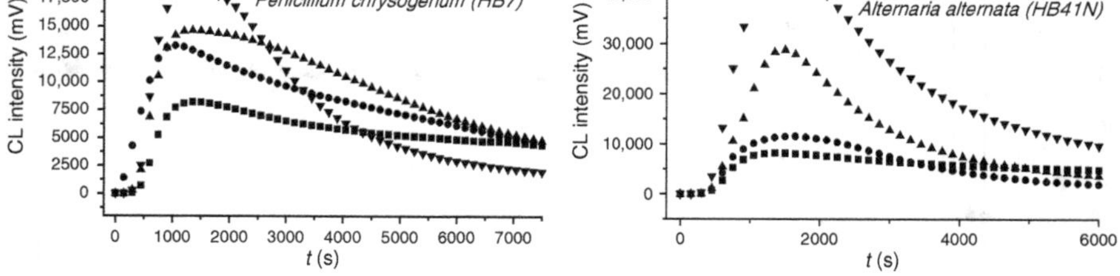

FIGURE 3.7 Chemiluminescence of gelatine films under oxygen at 225°C. Biodegradation assay at 25°C in aqueous solution with *Penicillium chrysogenum* (*HB7*) and *Alternaria alternata* (*HB41N*) during (■) 0, (●) 24, (▲) 48 and (▼) 72 h.

their metabolism. The differences observed in the chemiluminescence under different conditions are sensitive to the molecular weight and gel resistance value.

Cellulose is the most important polymer found in nature. Cellulose is the structural component of the primary cell wall of green plants, many forms of algae, and the oomycetes, about 33% of all plant matter is cellulose (the cellulose content of cotton is 90% and that of wood is 50%). Cellulose is a polysaccharide consisting of a linear chain of several hundred to over ten thousand 1,4-β linked D-glucose units. This polymer has many applications as emulsifier, stabilizer, dispersing agent, thickener, and gelling agent, but these are generally subsidiary to its most important use of holding on to water, as water cannot penetrate crystalline cellulose, but dry amorphous cellulose absorbs water becoming soft and flexible.

The mechanism of degradation of cellulose is complex. It was proposed [75] that the carbon atoms in positions 1 and 4 of glucopyranosyl units are preferential sites of the primary oxygen attack, but other positions such as carbon atoms 2 and 3 linked to alcoholic groups and carbon atom 5 linked to methylhydroxyl group should also be considered. Several works have been devoted to investigate the thermal degradation of cellulose by chemiluminescence [76–78]. Three initiating processes have been postulated as responsible for emission (1) thermal decomposition of charge-transfer complexes formed between cellulose and oxygen, (2) thermal decomposition of peroxides, and (3) chain scission process at elevated temperatures.

The studies showed that cellulose degradation was influenced by absorbed oxygen, chemisorbed water, and metal ion impurities. In acid media, cellulose degrades by an ionic mechanism, and free radical degradation takes place in an alkaline environment. Under oxygen, both mechanisms may be involved in initiation degradation in dependence on conditions of media, although free radical process was predominant. From kinetics analysis of non-isothermal runs, faster and slower processes could be distinguished. This is based on the slope change observed at 170°C for CL intensity versus temperature plot, which is in agreement with the abrupt start of the mass loss determined by thermogravimetric analysis during oxidation of cellulose. It has been attributed to the σ-transition determined by measurements of temperature dependence of dielectric properties, and related to migration of protons in cellulose [79]. Polymerization degree was determined, and plots DP_0/DP versus time also showed the two regions of decomposition. A correlation between non-isothermal CL and decrease of polymerization degree kinetics was found, indicating that chemiluminescence is related to main chain scissions.

The oxidative degradation of cellulose has been compared with that of other polysaccharide polymers, such as dextran and pullulan [80]. Pullulan is a polysaccharide polymer consisting of maltotriose units. Three glucose units in maltotriose are connected by an 1,4-α-glycosidic bond, whereas consecutive maltotriose units are connected to each other by an 1,6-α-glycosidic bond. Pullulan is produced from starch by the fungus *Aureobasidium pullulans*. Dextran is a complex, branched polysaccharide made of many glucose molecules joined into chains of varying lengths, used as an antithrombotic (antiplatelet), and to reduce blood viscosity. The straight chain consists of 1,6-α-glycosidic linkages between glucose molecules, whereas branches begin from 1,4-α linkages (and in some cases, 1,2-α and 1,3-α linkages as well).

Dextran is synthesized from sucrose by certain lactic acid bacteria, the best-known being *Leuconostoc mesenteroides* and *Streptococcus mutans*. Dextran is also formed by the probiotic *Lactobacillus brevis* to create the crystals of tibicos or water kefir-fermented beverage with reported health benefits.

Under oxygen, the rate constants from non-isothermal chemiluminescence analysis and from kinetics of the decay of polymerization degree were determined, and a discrepancy was found for pullulan compared to cellulose and dextran. It was attributed to the different mechanisms of degradation due to the different structures of polysaccharides. The 1,4-α-glycosidic bond in pullulan is more sensitive to split thermally in comparison to 1,4-β linkages in cellulose and 1,6-α linkages in dextran, where first, the free radical oxidation of carbon 6 of glucopyranose unit takes place and leads to the formation of carboxyl groups and water contributing to the subsequent heterolytic cleavage of C–O–C bonds linking glucopyranosyl units.

In the presence of methoxy magnesium methyl carbonate, the chemiluminescence emission in pullulan and cellulose increased significantly. It may be attributed to the decomposition of dioxetanes, with higher quantum yield of emission, which were formed from peroxidation of terminal semiacetal groups in the presence of MgO.

The use of chemiluminescence technique has expanded to others natural polymers. Recently, Millington et al. reported studies on polymers such as the fibrous proteins wool and feather keratin, silk fibroin, and reconstituted collagen from bovine skin, which revealed new information about their degradation [81]. The potential of CL to analyze the effectiveness of treatments to protect the polymer during processing could contribute to diminish the loss of properties of the types of polymers widely used in textile industry.

3.3.2 Acrylic Polymers

Chemiluminescence has been applied to monitoring the network structure formation under UV irradiation of epoxy acrylate resins [82]. These types of polymers are used as protective coatings, and a series of requirements (temperature resistance, good mechanical properties, etc.) for these materials must be satisfied. The degree of cross-linking plays an important role on their properties, and has been studied by several techniques. The CL emission was measured after exposure to the light of epoxy acrylate films, and activation energies calculated from Arrhenius plots. E_a is related to the degree of formation of polymer network. The activation energy increased with irradiation dose, which was correlated to the enhancement of the rigidity of the system. CL ramping test for the films exhibited a maximum in the temperature range of glass transition, and the maxima shifted to higher temperature with increasing irradiation dose. The behavior observed for T_g, similar to E_a, described the network formation of epoxy acrylate resin during UV irradiation.

With regard to adhesive technology, the requirement for shortening the curing times has brought a new approach to use of formulations that will polymerize under UV or visible radiation. Photochemical polymerization has allowed the development of one-component adhesives that cure in a few minutes (or seconds) at room temperature. Recently, society has become aware of the necessity to reduce or eliminate volatile organic component (VOC) emissions from coating applications. These reasons together with the growing demand for adhesive use in technologies and specialty applications have lead chemists to look for alternatives to the traditional solvent-based adhesives. Radiation-cured adhesives, by means of electron beam (EB) or UV light, show the potential to replace conventional solvent-based systems that are ozone-depleting.

The prediction of the useful life of this type of materials is a major challenge for industry. Recently, the photooxidation of an UV-cured polyurethane–acrylate adhesive (Loctite L350) has been monitored through the use of chemiluminescence and fluorescence techniques [83]. Homogeneous mixtures of the adhesive formulation and two fluorescent probes were prepared, and irradiated with polychromatic light until limiting conversion was reached. Then, the cured adhesives were photoaged and the fluorescence and chemiluminescence measured at different irradiation times. For chemiluminescence studies, samples were prepared without fluorescent probes. CL data as a function of UV irradiation time are summarized in Table 3.8.

Under nitrogen, the unexposed sample exhibited a weak CL at 100°C, indicating a low concentration of hydroperoxides ($t_{irr} = 0$). After UV irradiation, the CL-integrated intensity and maximum intensity increased with exposure time. From I_{max} data, it was seen that the photooxidation of the L350 adhesive films started only after a few hours of UV irradiation, and then, remained constant for at least the first 70 h. This kind of study may be considered to provide a reliable evaluation of the early stages of photooxidation.

The non-isothermal analysis was undertaken for nonirradiated sample of the polyurethane–acrylate-based adhesive. And from the Arrhenius plot of the CL intensity over the temperature range 60–150°C, the activation energy for CL emission was determined to be 85 kJ mol^{-1}.

Two commercially available pH-sensitive fluorescent probes were used to monitor the photooxidation of the UV-cured polyurethane–acrylate adhesive L350: Oregon Green (OG) and *p*-dimethylaminosalicylic acid (*p*-DASA). During the period of UV-exposure, the photostability of these fluorescent compounds was assessed by UV-absorption monitoring and the fluorescence emission was measured at several

TABLE 3.8 Kinetic Data of Chemiluminescence from the UV-Cured Aliphatic Polyurethane–Acrylate-Based Adhesive L350, at 100°C Under Nitrogen, After UV Irradiation

t_{irr} (h)	I_{max} (mV)	t_{max} (s)	$I_{residual}$ (mV)	Area (V s)	$v_{CL} \times 10^5$ (mV$^{-1/2}$ s^{-1})
0	36	405	0	298	12.3
1	226	400	—	355	3.0
3	417	474	222	644	1.6
6	504	522	277	816	1.2
12	510	565	340	900	0.8
24	728	670	367	1289	0.6
48	779	680	—	1013	0.6
63	791	749	—	1444	0.4

irradiation times. During photooxidation, a rapid decrease of the emission was observed. Different arguments might be put forward for the observed fluorescence intensity decrease, such as the formation of oxidation products (peroxide, etc.) generated during photochemical degradation of polyurethanes [84,85], that are able to quench the excited states of emitting fluorophores.

To analyze the oxidation level of the adhesive by fluorescence, the fluorescence of the two probes as a function of the CL-parameters, maximum intensity, and decay rate, for the polyurethane–acrylate-based adhesive, were plotted (Fig. 3.8). Quenching of fluorescence followed the same kinetics as the chemiluminescence increase during UV-exposure. The two fluorescent probes used were sensitive to the formation of oxidation products. Therefore, in this work, good correlations between fluorescence and chemiluminescence were established, which allow monitoring photodegradation by using extrinsic emission fluorescent probes and/or CL emission from bulk material.

3.3.3 Poly(styrene-*b*-ethylene-*co*-butylene-*b*-styrene) (SEBS) Block Copolymers

Styrenic block copolymers, SEBS, are triblock copolymers where polystyrene is the end segment and the elastomer segment is based on poly(ethylene-butylene). These materials are members of the family of thermoplastic elastomers, exhibit many of the physical properties of rubber, improved mechanical properties due to the presence of the hard styrenic segments, and in comparison to SBS, enhanced resistance to oxygen and ozone attack as a result of hydrogenation of the olefinic double bonds. The influence of two commercial antioxidants, Irganox 1330 and Irgafos 168, on their thermal stability has been evaluated on the basis of chemiluminescence analysis [86]. The efficiency of the stabilizer is a useful source of information for assessing the consumption of stabilizers by different processes during SEBS production, storage, weathering exposure, and exposure under application conditions.

A kinetic study has been carried out to determine the order of the reaction and to establish a correlation between the structure and composition of the materials with their stability properties. Integration methods, lifetimes, and initial rate analysis were used to determine the order of the reaction [87]. A plot of $\ln(I/I_0)$ versus time is lineal, confirming first-order kinetics for oxidation of SEBS. A pseudo-first order of the reaction is also confirmed, because half lifetime remains constant for different I_0, and the double logarithmic plot of initial rates versus intensity maxima (which are proportional to the initial peroxides concentration) gave a straight line.

Chemiluminescence analysis was recorded under nitrogen, and all the samples exhibited a weak CL indicating a low concentration of hydroperoxides [87]. A rapid decrease of the CL-decay rate was observed as Irgafos 168 concentration increased (Fig. 3.9) and the chemiluminescence emission area increased. This feature has been related to the mechanism of stabilization of the Irgafos 168, which mainly acts as a secondary antioxidant peroxide decomposer, giving rise to stable products and preventing further degradation of the SEBS samples. In this work, it was first reported for both primary and secondary stabilizing efficiency of a hindered aryl phosphate in

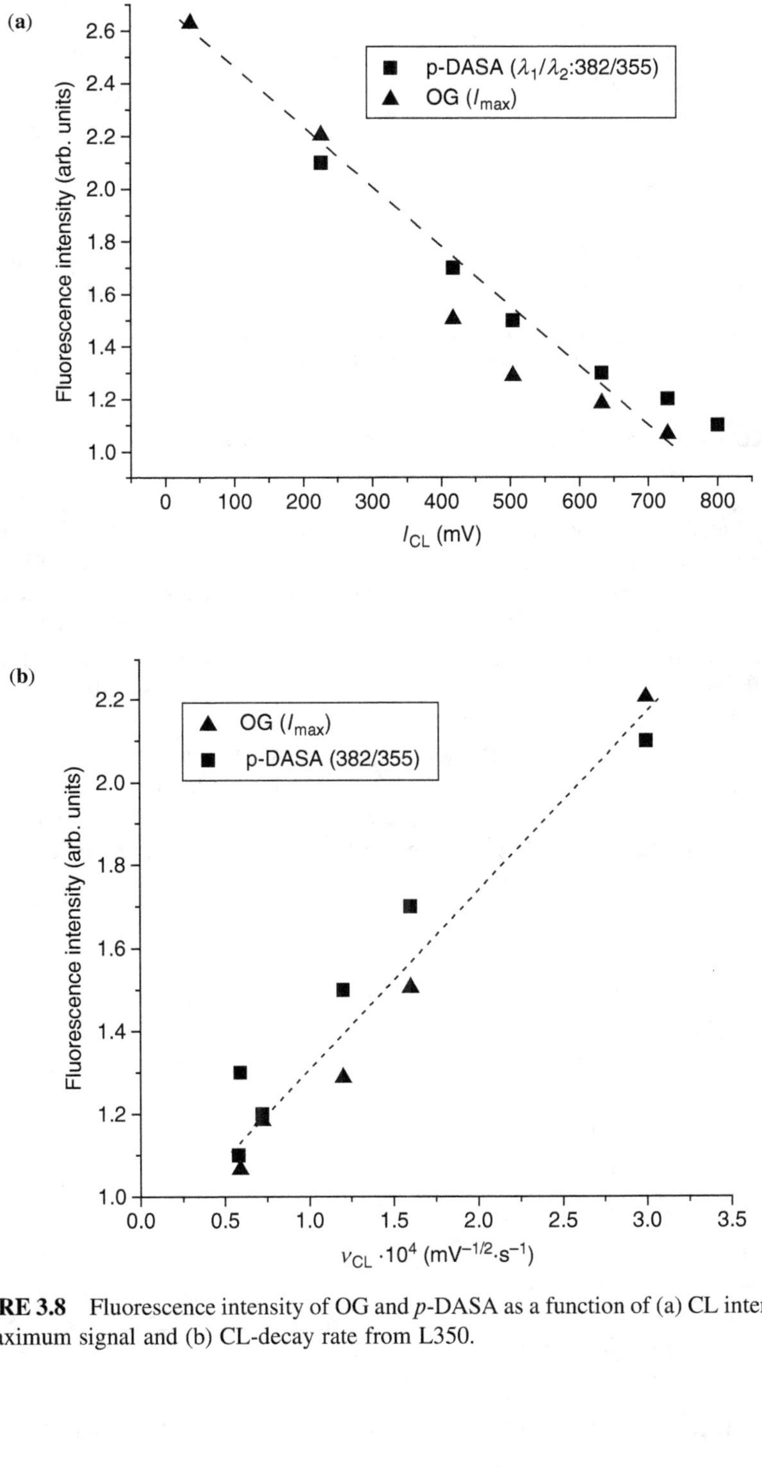

FIGURE 3.8 Fluorescence intensity of OG and *p*-DASA as a function of (a) CL intensity at the maximum signal and (b) CL-decay rate from L350.

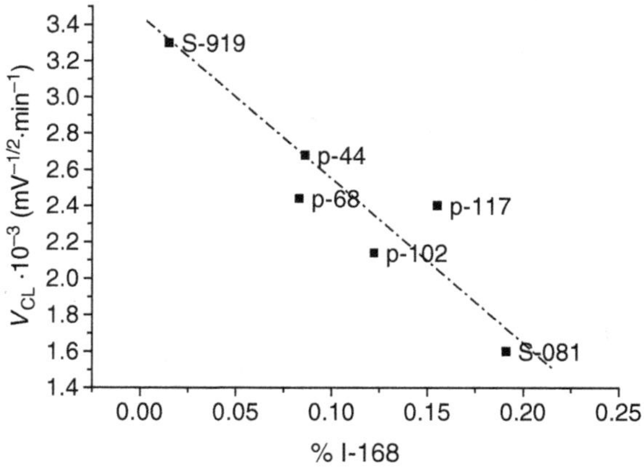

FIGURE 3.9 CL-decay rate, under nitrogen at 185°C, versus concentration of Irgafos 168.

hydrogenated block copolymers SEBS. In the binary mixture of stabilizers, the hindered phosphite has a synergistic effect, protecting the phenol Irganox 1330 from excessive consumption during processing of the polymer sheets. Whereas phosphites alone provide no long-term stabilization at lower temperatures during service-life, their combination with sterically hindered phenols also contributes to the improvement of long-term behavior preserving the concentration of the hindered phenol Irganox 1330 during processing at elevated temperatures.

The isothermal measurement under oxygen showed good correlations between chemiluminescence parameters and the concentration of stabilizers. The oxidation induction times have been seen to increase with the ratio of Irgafos 168/Irganox 1330 (Fig. 3.10). In general, the OITs increased with the transition temperature determined

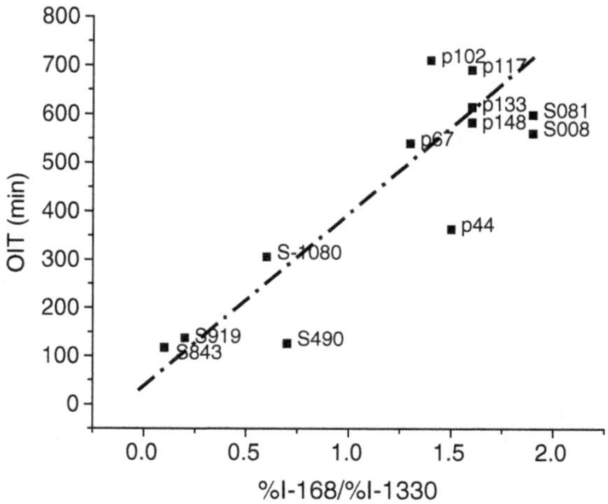

FIGURE 3.10 Plot of OIT versus the amount ratio between I-168 and I-1330.

in the nitrogen chemiluminescence analysis, associated to the glass temperature of polystyrene domains. This behavior has been attributed to changes in bulk morphology of the SEBS, which lead to a modification of elastomer–polystyrene interphase. These results indicated that for these copolymers the oxidation started in the interfacial region.

The relation between thermal stability of SEBS and their structure has also been investigated by differential scanning calorimetry. An order–disorder transition was observed at temperatures around 40–45°C, depending on the thermal history of the sample. Similar transitions were attributed to the existence of aggregates that behaves as paracrystalline regions in ethylene–propylene elastomers, responsible for the nonsimple viscoelastic behavior of the samples involving morphological differences between samples. The OITs decreased as the area of this transition peak increased, which revealed that thermooxidation stability of SEBS was influenced by the presence of paracrystalline aggregates. In this work, a new potential of the chemiluminescence technique was revealed because it opened the possibility to investigate the mechanism of oxidation apart from kinetics.

3.3.4 Condensation Polymers

3.3.4.1 Poly(ethylene-co-1,4-cyclohexane dimethylene terephthalate) The thermal and photochemical oxidation processes of a polyester based on poly(ethylene-*co*-1,4-cyclohexanedimethylene terephthalate) (PECT) with approximately 30% of 1,4-cyclohexanedimethanol have been studied by means of chemiluminescence [88]. Also, the stabilization activity of some commercial antioxidants has been evaluated, and it has been related to hydroperoxide levels in the polymer.

The chemiluminescence analysis of UV-degraded PECT samples (free additive), under nitrogen, was undertaken as a function of UV aging time (Table 3.9). The initial sample exhibited a weak chemiluminescence emission, which was in accordance with the low concentration of oxidized products (hydroperoxides) determined in the sample by UV absorbance at 420 nm. The emission intensity was seen to increase during the degradation process. This was due to the higher oxidation levels reached during UV degradation and its further decomposition.

TABLE 3.9 Chemiluminescence of PECT Free of Additives as a Function of UV Aging Time, Under Nitrogen at 150°C

Sample	Aging time (weeks)	I_{max1} (mV)	Integrated CL[a] (mV)	[POOH] (μg/g)
PECT	0	25	846	35
	13	78	2358	120
	43	143	3191	135
	56	194	4649	900
	76	343	4784	2500
	85	489	6004	3100

[a]Integration time 1 h.

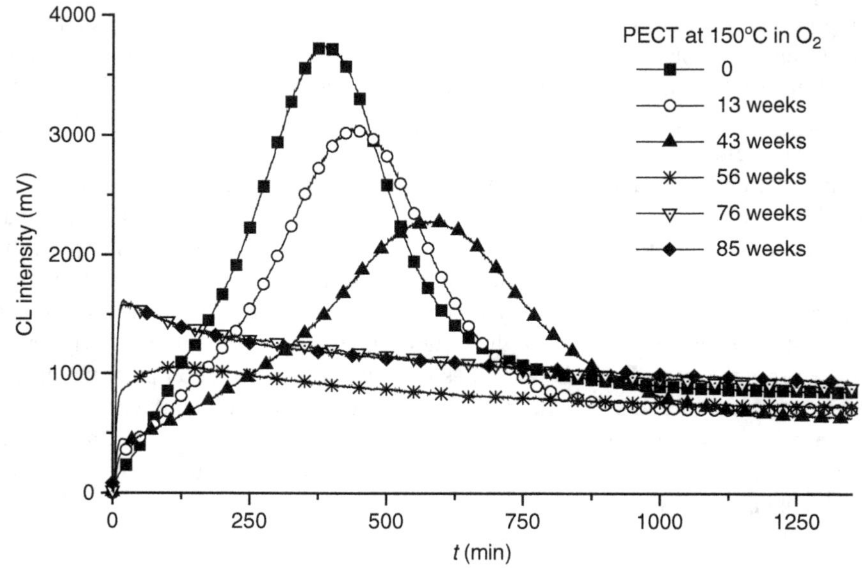

FIGURE 3.11 Chemiluminescence spectra under oxygen of UV-degraded PECT samples (free additive) at different aging times.

Under oxygen, a much higher intensity of chemiluminescence was observed compared to that obtained under nitrogen atmosphere (Fig. 3.11). Two emissions (I_{max1} and I_{max2}) were detected with different evolutions of their intensities depending on the UV degradation of the samples (Table 3.10).

Initial PECT sample showed the highest intensity of chemiluminescence. As the aging time is increases, the peak-top time is longer and a decrease on the intensity of CL is observed. It has been associated with the different oxygen diffusions in the initial and degraded samples of the material, because during the UV aging, cross-linked polymer has been detected. This cross-linking was seen to increase with time, making difficult the diffusion of oxygen into the polymer and the formation of new

TABLE 3.10 Chemiluminescence of PECT Free of Additives as a Function of UV Aging Time, Under Oxygen at 150°C

Sample	Aging time (weeks)	t_1 (min)	I_{max1} (mV)	t_2 (min)	I_{max2} (mV)
PECT	0	383	3743	—	—
	13	444	3057	25	360
	43	589	2287	25	444
	56	—	—	25	1056
	76	—	—	24	1581
	85	—	—	23	1607

peroxides was restricted. The physical spreading of the oxidation from an initial highly reactive center to other particles would be more hindered [6,89]. After 13 h of aging time, a new increasing peak appears at a shorter time (25 min of heat treatment in the CL analyzer). This new early emission has been related to the formation of peroxides of different stabilities, as already described by Gijsman et al. [90]. The chemiluminescence was similar to that observed for poly(ethylene terephthalate) film under nitrogen and oxygen. PET showed a peak-top time similar to the observed maximum for long-range aged PECT films, and the CL emission increased significantly with UV degradation.

These facts indicated that the ethyleneglycol segments should probably be the oxidation sites. The initial high CL emission, which decreases with UV degradation time, must be due to the cyclohexanedimethylene units. Hydroperoxides formed in these aliphatic segments are responsible for this difference with the homopolymer PET. Two possibilities of different peroxy radical formation were postulated, in the ethylene units, as it was observed for PET, and in the cyclohexanedimethylene units.

The thermally degraded PECT samples (free additive and with stabilizers Irganox 3114 and Alkanox 28) were analyzed by chemiluminescence under nitrogen atmosphere, at 150°C. Under this condition, only one emission was observed as it was described for UV degraded samples. In general, a correlation can be established between the chemiluminescence intensity and the hydroperoxide content determined by the iodometric method at different aging times, in agreement with the results obtained by other authors [36].

The CL of thermally degraded PECT samples was undertaken under oxygen at 150°C, to evaluate the effectiveness of antioxidants incorporated into the copolymer (Fig. 3.12). The maximum of the initial intensity of CL emission (I_{max1}) and the oxidation induction time (OIT) were determined (Table 3.11).

The presence of antioxidant decreases the CL emission drastically compared with the free additive sample, confirming the general good efficiency of the employed antioxidants, which work as radical scavengers and peroxide decomposers decreasing the oxidation during film processing, and hence, the CL emission. The free additive sample showed the lowest OIT value compared with the stabilized samples. The stability of the copolymer was highly increased when Alkanox 28 was added in combination with Cyanox 1790 or Irganox 3114. This fact indicates a synergistic effect of the antioxidants and longer oxidation induction time was observed for these samples.

3.3.4.2 Poly(ethylene terephthalate) Poly(ethylene terephthalate) is a widely used semicrystalline polymer. The macroscopic properties of PET such as thermal, mechanical, optical, and permeation properties depend on its specific internal morphologies and microstructure arrangement. It can be quenched into the completely amorphous state, whereas thermal and thermomechanical treatments lead to partially crystallized samples with easily controlled degrees of crystallinity. The crystallization behavior of thermoplastic polymers is strongly affected by processing conditions [91–93].

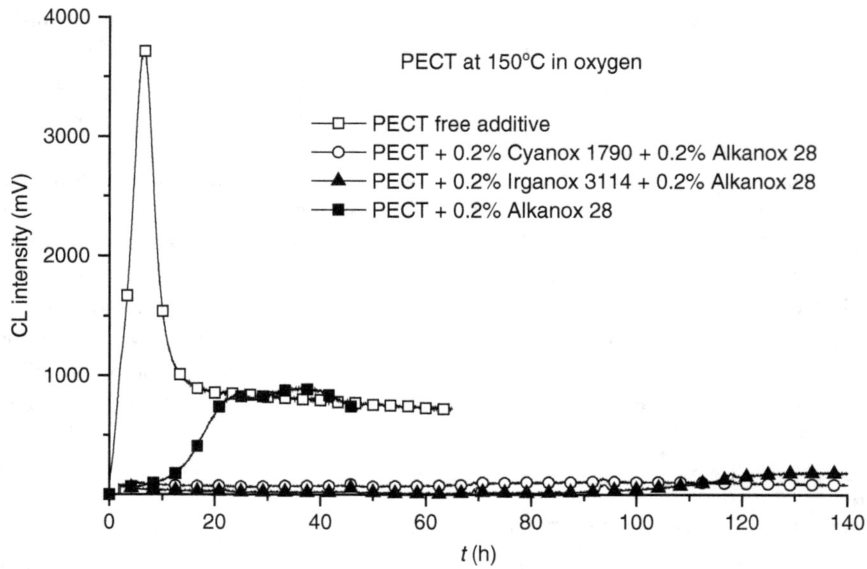

FIGURE 3.12 Chemiluminescence evolution under oxygen of PECT samples, free additive, and stabilized.

The chemiluminescence emission of several poly(ethylene terephthalate) samples (initial and annealed at different temperatures) has been studied [94]. The Coumarin probes (C152, C153, and C337) have been used as physical enhancer of chemiluminescence in PET, as this polymer exhibited low emission intensity in the whole range of temperatures [95–97]. These sensitizers offer nonintrusive methods, amplifying the CL signal without altering the reaction kinetics. Amplified chemiluminescence emission intensity was found for poly(ethylene terephthalate) doped with Coumarin C337, PET_{C337} compared to those obtained for the other samples (Fig. 3.13).

It is proposed that in the presence of fluorescent probes, the energy transfer from excited carbonyl (donor molecule) to an originally unexcited acceptor molecule can take place. Because the energy of excitation of the acceptor comes from the excited donor, the donor is radiationlessly deactivated to its ground electronic state. The acceptor molecule, which has become excited at the expense of the donor, may return

TABLE 3.11 Chemiluminescence of Initial PECT Samples Under Oxygen at 150°C

Simple	t_{max} (h)	I_{max} (mV)	OIT (h)
PECT free additive	6.4	3743	1.5
PECT + 0.2% Cyanox 1790 + 0.2% Alkanox 28	85.0	107	59.0
PECT + 0.2% Irganox 3114 + 0.2% Alkanox 28	136.0	187	100.0
PECT + 0.2% Alkanox 28	37.9	883	13.0

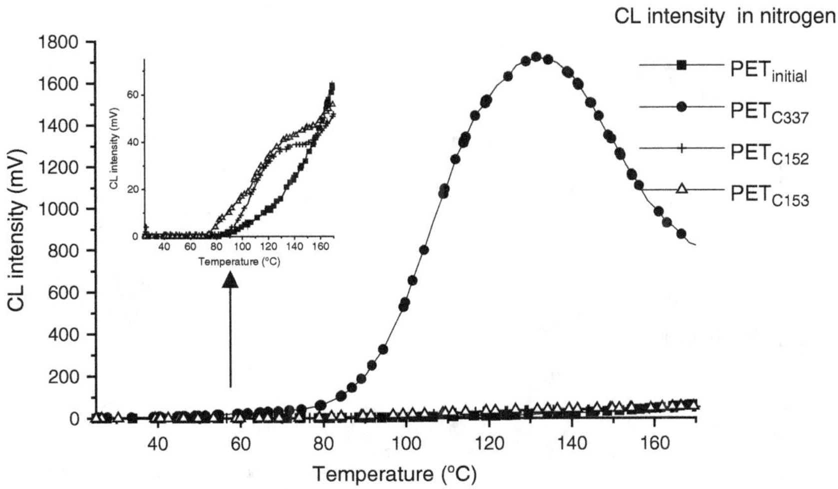

FIGURE 3.13 Chemiluminescence temperature ramping curves obtained for $PET_{initial}$, PET_{C337}, PET_{C152} and PET_{C153}, under nitrogen.

to its ground state and fluorescence emission from the acceptor is detected (Scheme 3.3). A requirement for energy transfer is the overlap of the emission spectrum of the donor and the absorption spectrum of the acceptor. The energy transfer from excited triplet state of carbonyl to the singlet state may be more favorable for C337, which exhibited the lowest energy of singlet excited state and would act as the most effective physical enhancer of the chemiluminescence emission.

The Arrhenius plots exhibited pronounced slope changes in the chemiluminescence emission with temperature, and different zones are distinguished (Fig. 3.14). A slope change was observed around 75°C assigned to the α-transition of poly(ethylene terephthalate). This transition involves motion of long segments, which favors the mobility of hydroperoxides and their bimolecular termination reaction. Another slope change is observed around 113°C, attributed to the exothermic recrystallization peak during heating (T_c) and the corresponding decrease of the free volume fraction, which would restrain the peroxyl recombination, which originates excited carbonyl moieties and the energy transfer to the fluorescent probe.

The sensitization of fluorescence of an acceptor molecule is considered a concentration-dependent process. The enhancement of chemiluminescence with the Coumarin C337 content was observed in the range of concentration studied. This result would indicate that energy transfer is a diffusion-controlled process. The requirement for donor and acceptor to diffuse to near or actual physical contact with

$$\mathrm{POO\cdot} + \mathrm{POO\cdot} \longrightarrow \mathrm{POH} + {}^{1}\mathrm{O}_2 + \mathrm{P}\,{}^{3}(\mathrm{C{=}O})^{*} \dashrightarrow \mathrm{PCO} + h\nu_{PET} \qquad (1)$$

$$\mathrm{P}\,{}^{3}(\mathrm{C{=}O})^{*} + \mathrm{Coumarin} \longrightarrow \mathrm{P(C{=}O)} + \mathrm{Coumarin}^{*} \dashrightarrow \mathrm{Coumarin} + h\nu_{Coumarin} \qquad (2)$$

SCHEME 3.3 Chemiluminescence emission in polymers doped with Coumarins as enhancer.

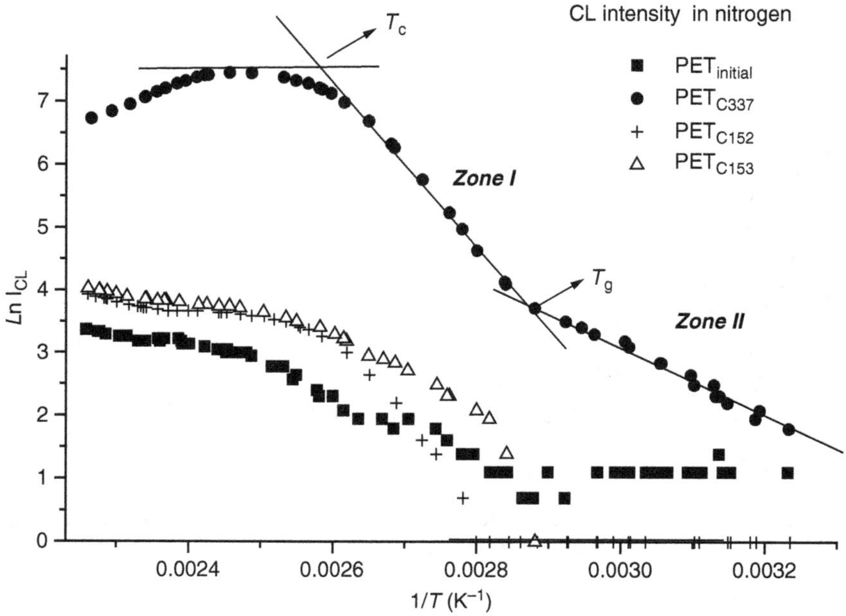

FIGURE 3.14 Arrhenius plots of chemiluminescence intensity versus 1/T for $PET_{initial}$, PET_{C337}, PET_{C152}, and PET_{C153}, under nitrogen.

each other suggests that rigidity of the medium would greatly influence the energy transfer process from excited carbonyl moieties to Coumarin fluorescent probes. Then, the chemiluminescence emission decrease at temperatures above 130°C would be related to the decrease of energy transfer process during heating. As it is a diffusion-controlled process, this fact may be explained in terms of the more restricted mobility of hydroperoxides and Coumarin to diffuse near each other in the polymer after the crystallization process during heating.

The influence of the degree of crystallinity in the bimolecular decomposition of hydroperoxides was studied by non-isothermal chemiluminescence tests under nitrogen of PET_{C337}, PET_{C152}, and PET_{C153} after annealing at various temperatures (Fig. 3.15). The annealed PET samples followed a similar trend to their corresponding not-annealed PET samples. In general, a lower chemiluminescence emission for annealed samples with respect to the initial sample was observed. In this case, several factors may affect the chemiluminescence emission. The increasing crystallinity of annealed poly(ethylene terephthalate) samples may be considered, which would restrict the mobility of hydroperoxides to react due to the higher rigidity of the medium. Furthermore, as the energy transfer from the triplet excited state of carbonyl to the singlet state of Coumarin is considered a diffusion-controlled process, it would be highly affected by the degree of order in the medium. The translocation of probe may also influence the probability of the energy transfer from the donor to the fluorescent probe, which would be less favored because the ability of the Coumarin

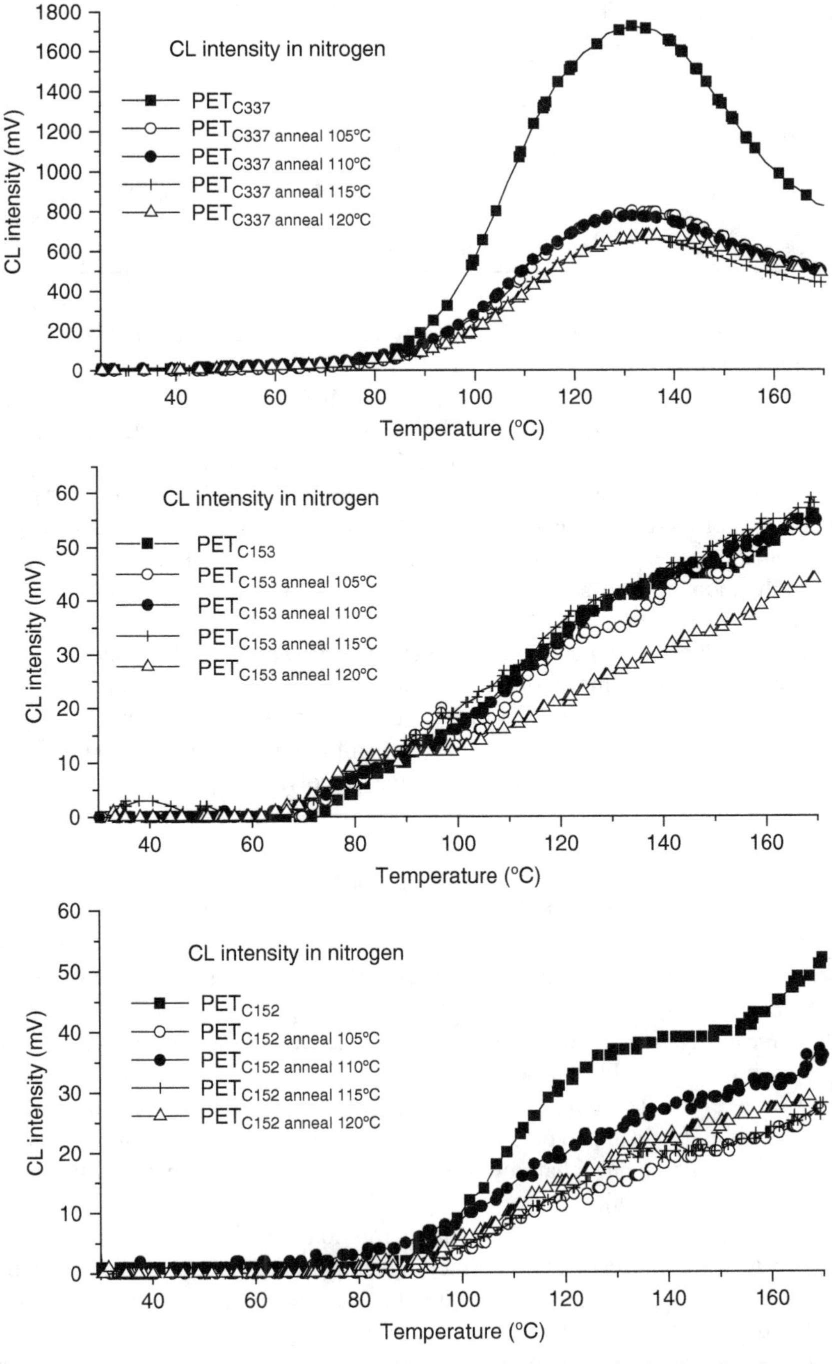

FIGURE 3.15 Chemiluminescence temperature ramping curves obtained under nitrogen, for PET$_{C337}$, PET$_{C152}$, and PET$_{C153}$, and after annealing at different temperatures.

TABLE 3.12 Thermal Transition and Activation Energies Determined by Chemiluminescence Under Nitrogen

Sample	E_{actI} (kJ mol^{-1})	E_{actII} (kJ mol^{-1})	T_g (°C)	T_c (°C)
PET_{C337}	116	42	75	113
$PET_{C337\ anneal\ 105°C}$	90	43	75	113
$PET_{C337\ anneal\ 110°C}$	92	55	76	113
$PET_{C337\ anneal\ 115°C}$	81	65	78	112
$PET_{C337\ anneal\ 120°C}$	78	73	—	112

C337 to undergo diffusional motion in poly(ethylene terephthalate) matrix may be more restricted.

In general, the glass transition temperature increases slightly with annealing temperature, as well as the E_{actII} associated with the bimolecular decomposition of hydroperoxide below T_g increases (Table 3.12). This phenomenon may be associated with the higher crystallinity after annealing process, as it was observed by differential scanning calorimetry. However, lower values of activation energies above glass transition (E_{actI}) were found for annealed samples than for the initial sample, as the recrystallization of amorphous regions during heating is reduced with annealing temperature increase.

This study showed that chemiluminescence method is a useful technique to sense temperature-dependent morphological changes, that is, annealing processes, and to determine relaxation temperatures and exothermic recrystallization peaks. The use of a fluorescent molecule to amplify the chemiluminescence emission allows to application of the innovative method to all types of polymers including those with low intensity emission.

3.3.4.3 Poly(ε-caprolactone) Poly(ε-caprolactone) (PCL) is a biodegradable, biocompatible, and semicrystalline polymer, which has expanded greatly its field of applications, most of them related with tissue engineering, due to its good biocompatibility, ability to form compatible blends and copolymers with a wide range of other polymers, improving its mechanical properties, processability, and high permeability. The stability of biopolymers used as substrate for tissue engineering is an important field of research, with particular emphasis on their durability during service-life. The long-term degradation behavior of PCL films obtained by two different methods in two biologically related media, DMEM, employed in many cell cultures, and a phosphate-buffered solution (PBS), has been evaluated by chemiluminescence [98].

Under oxygen, no chemiluminescence emission was detected at temperatures below 65°C, which corresponds to the melting temperature of the polymer determined by DSC. At temperatures above T_m, the mobility of the hydroperoxides is favored and their bimolecular termination reaction takes place. Replotting the ramping temperature CL-emission profile under oxygen in Arrhenius form yields two different temperature intervals: region I and region II. It would indicate that poly(ε-caprolactone) undergoes thermal degradation by a two-step mechanism. These results were in good agreement with those of other authors [99]. It has been reported that thermal degradation of PCL showed a two-step mechanism. The first stage was attributed to the random chain scission and the second step corresponded to specific chain end scission.

The isothermal chemiluminescence analysis under oxygen at 120°C revealed a weak emission due to the thermal history of the film at the onset of heating, and a second peak when the oxidation induction period was attained. The initial sample showed the highest intensity of CL second peak. For samples degraded in PBS, as the aging time is increases, the peak-top time ($t_{max2)}$ is longer and a decrease in the intensity of CL is observed (Fig. 3.16). It was attributed to the different oxygen diffusion in the initial and degraded samples of the material, that is, during the aging, degradation of amorphous phase takes place reducing the diffusion of oxygen in the polymer and the formation of new peroxides would be restricted. The physical spreading of the oxidation from an initial highly reactive center to other particles would be more hindered [100,101]. As the degradation process developed, a new increasing peak appears at shorter times (t_{max1}), which could be due to the formation of new peroxides in different active sites of poly(ε-caprolactone). The different stability of the species formed will be related with their mobility, which depends on the microenvironment and the molecular weight of the polymer segment.

A different evolution of the CL emission peaks was observed depending on the media and time of degradation (Table 3.13). The chemiluminescence results were in

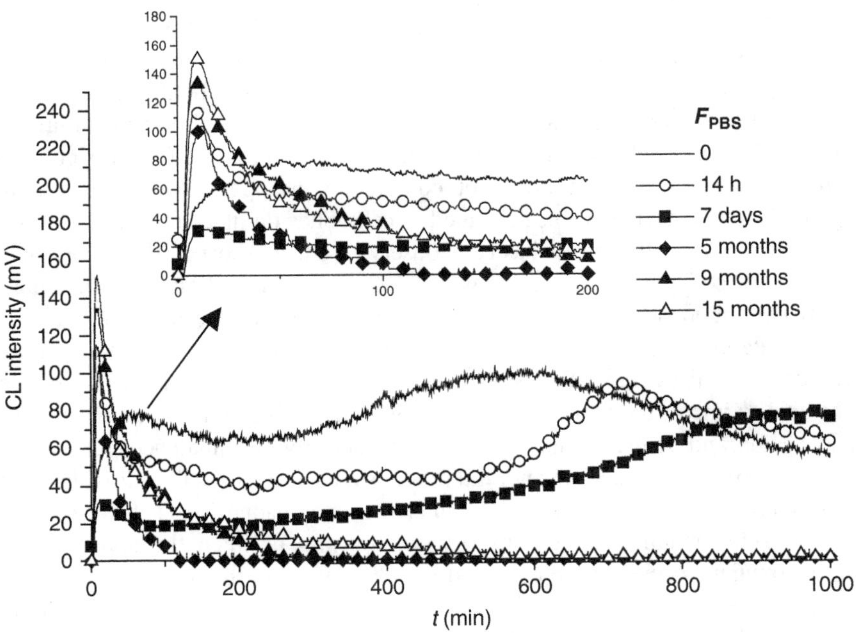

FIGURE 3.16 Chemiluminescence intensity versus time at 120°C under oxygen of PCL in PBS.

TABLE 3.13 CL Parameter of Poly(ε-caprolactone) Films at 120°C Under O_2 Atmosphere

	PBS media				DMEM media			
Degradation time	t_{max1} (min)	$I_{CL\text{-}max1}$ (mV)	t_{max2} (min)	$I_{CL\text{-}max2}$ (mV)	t_{max1} (min)	$I_{CL\text{-}max1}$ (mV)	t_{max2} (min)	$I_{CL\text{-}max2}$ (mV)
0	66	80	582	100	66	80	582	100
14 hours	9	114	727	92	10	32	—	—
7 days	11	32	966	77	8.5	57	—	—
1 month	9	119	924	53	8.5	81	—	—
5 months	12	103	—	—	7.5	117	—	—
6 months	9.5	123	—	—	11	126	—	—
9 months	10	133	—	—	8	167	—	—
15 months	9.5	152	—	—	9.5	241	—	—
18 months	7.5	183	—	—	9.5	539	—	—

good agreement with those obtained by DSC and GPC, where a higher degradation as a consequence of the ester groups hydrolysis, for DMEM, was observed with respect to PBS media.

PCL films were alkaline-treated to enhance the surface hydrophilicity and cell adhesion, and the degradation behavior of untreated (PCL_{cells}) and alkaline-treated ($APCL_{cells}$) samples in the presence of fibroblasts was studied by chemiluminescence [102]. Considering that CL detects specifically the surface emission, this technique results are of great interest to analyze the surface phenomena that conducts to a change in the hydrophilic degree and, consequently, to a better cell adhesion.

For alkaline-treated membranes, high-intensity signals appear in samples immersed for short times (up to 4 days). In the case of nontreated membranes, this type of emission prologues longer time, and does not completely disappear even after 1 month (Fig. 3.17). These facts have been related to cell adhesion mechanism: a transitory, but significant, oxidation stress has been observed, during short culture times, in the same type of cells employed in this work [103]. The stimulation of mitochondrial activity can induce the synthesis of ROS, which are formed in mitochondria when energy metabolism is compromised. ROS includes singlet oxygen and hydrogen peroxide, as well as free radicals such as superoxide anion and hydroxyl radical, being able to impair the function of several cellular components including proteins, nucleic acids, and lipids. Oxidative stress is produced by the imbalance created when exposure to oxidant changes the normal red-ox status of major cells antioxidants and has been related with apoptosis [104], but also with cell adhesion [105,106].

The results obtained showed a clear relationship between the appearance of reactive oxygen species in the cells and the CL signals in the longer time regions, that is, these species contribute to the peroxide generation. In the case of alkaline-treated membranes, $APCL_{cells}$, which in turn show a higher degree of adhesion and proliferation of different cell lines, these emissions disappear after 4 days, whereas for nontreated membranes they remain for a more prolonged period of time.

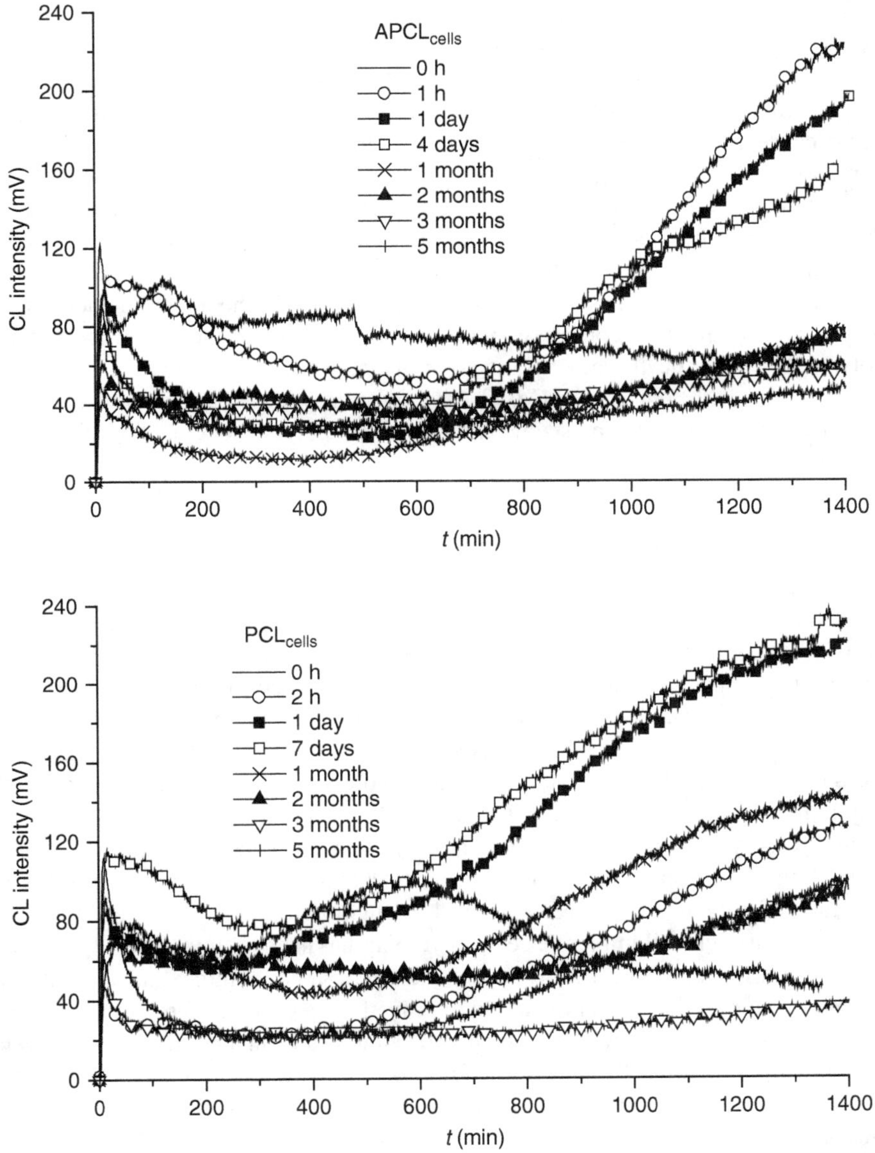

FIGURE 3.17 Chemiluminescence intensity versus time at 120°C under oxygen of APCL$_{cells}$ and PCL$_{cells}$ degraded at different times.

Other condensation polymers as polyamides should also be mentioned. For polyamides prepared by hydrolytic polymerization, the rate of thermooxidation is highly influenced by the content and ratio of carboxylic and amino end groups. In the case of polyamides with only carboxylic end groups, loss of molecular weight is observed from the beginning of degradation [107,108]. These polymers have been extensively studied, and the oxidation behavior of stressed polyamides, stabilized polyamides, and the influence of reactive end groups have been analyzed by chemiluminescence technique [109–112]. For polyamide 6,6, a bimolecular hydroperoxide decomposition model describes the chemiluminescence runs, which are composed of three interdependent stages related to (1) termination of peroxyl radicals generated in processing and storage, (2) depletion of terminal amino groups due to their direct oxidation or to the condensation reactions with oxidation products, and (3) oxidation of CH_2 groups close to an amido group. The effect of metal salts on oxidation in polyamides has been recently analyzed [113]. The transition metal salts may form coordination complexes with the amido groups, and the properties of polyamides, including thermooxidative stability, may be modified. In fact, it has been described that in the presence of Co, Ni, Cu, and Zn chlorides, alone or combined with KI, the copper salts are the most efficient stabilizing species, reducing the emission of the third stage.

3.4 CONCLUDING REMARKS

Chemiluminescence emission technique is very useful for analyzing the degradation in polymers, mainly due to its high sensitivity for detecting the early stages of the process in comparison with other methods. In the past decade, the use of chemiluminescence technique has extensively grown, and the oxidation behavior of a great variety of polymers has been investigated. The results from the chemiluminescence measurements can be correlated with those obtained from other conventional techniques, and new applications of CL are developing to obtain valuable information related to polymer structure, interactions, morphology, dynamic changes, rigidity, crystallinity, and orientation. Some of the recent advances in the application of CL to the study of polymers have been reviewed through representative examples.

The origin of chemiluminescence in polyolefins has been profoundly analyzed and it has contributed to the better understanding of their complex mechanism of thermooxidation The thermal oxidation of polyethylenes with different manufacturing histories has been compared, which allowed to establish a relationship between CL and some structural characteristics of the polymers. Modification of their stability in the presence of antioxidants, or other additives such as the activity of nano- and micron particles of pigments has been evaluated.

Section 3.3 included the work dedicated to a great variety of materials with different nature, and reflects the importance of chemiluminescence to study the degradation processes in several areas. The stability of natural polymers, such as type B gelatines, which are the most photographic employed material, has been analyzed by means of chemiluminescence, and the influence of the molecular weight determined. Also, gelatines have been exposed to bacterial and fungal degradations, to quantify by CL the deterioration produced by microorganism. The oxidative degradations of polysaccharide polymers, such as cellulose, dextran, and pullulan have been compared. A discrepancy in their behavior was found, attributed to the different mechanisms of degradation due to the glycosidic bonds connecting the glucose units in the structure of polysaccharides. The use of CL technique has been expanded to study the degradation of other natural polymers used in the textile industry, such as fibrous proteins wool and feather keratin, silk fibroin, which could contribute to the development of protective treatments to diminish the loss of properties during processing.

CL has been employed to study the mechanism of degradation in polymers such as UV-cured aliphatic polyurethane–acrylate-based adhesives. In styrenic block copolymers, SEBS, the order of the reaction was determined, and the efficiency of antioxidants was evaluated. Also, the relationship between thermooxidation of SEBS and morphology was established, indicating that for these copolymers the oxidation started in the interfacial region.

The comparison of the evolution of chemiluminescence emission in thermally and photochemically aged PECT and PET indicated the formation of hydroperoxides with different stabilities in the structure of PECT, in the ethylene units as it was observed in PET, and in the cyclohexanedimethylene units.

As some polymers exhibit low emission intensity, a nonintrusive method based on sensitizers amplifying the CL signal has been proposed. The method was employed to sense temperature-dependent morphological changes, and to determine relaxation temperatures and exothermic recrystallization peaks in PET.

The long-term degradation of PCL has been evaluated by means of chemiluminescence. PCL is a biopolymer used as substrate for tissue engineering. It was clear that degradation depended on the media and time of degradation. The degradation of PCL in the presence of fibroblasts showed a clear relationship between the appearance of ROS in the cells and the CL signals in the longer time regions.

The prediction of the useful lifetime of polymer materials is a major challenge for the industry. New research efforts have been focused to understand the complex mechanism of polymer degradation by analyzing the chemiluminescence emission, due to its capabilities as well as commercially developed instrumentation. In the next few years, it would be expected that chemiluminescence will earn a place as a well-established research technique in many laboratories, and will contribute to the better understanding of the molecular structural level and the relationship with the macroscopic properties and the behavior in the use of the polymeric materials.

ACKNOWLEDGMENT

The authors thank the Spanish Ministerio de Educación y Ciencia for financial support (MAT2006-05979).

REFERENCES

1. R.E. Kellogg, *J. Am. Chem. Soc.* **1969**, *91*, 5433.
2. D.R. Kohler and C. Krohnke, *Polym. Degrad. Stab.* **1998**, *62*, 385.

3. D.R. Kohler and C. Krohnke, *Polym. Degrad. Stab.* **1999**, *63*, 173.
4. M. Hamskog, G. Ahlblad, G. Färnet, P. Gijsman, and B. Terselius, *Polym. Test.* **2003**, *22*, 363.
5. M. Celina, G.A. George, D.J. Lacey, and N.C. Billingham, *Polym. Degrad. Stab.* **1995**, *47*, 311.
6. D.J. Lacey and V. Dudler, *Polym. Degrad. Stab.* **1996**, *51*, 101.
7. S. Hosoda, Y. Seki, and H. Kihara, *Polymer* **1993**, *34*, 4602.
8. J.M. Gromek and M.R. Derrik, *Polym. Degrad. Stab.* **1993**, *39*, 261.
9. M. Celina and G.A. George, *Polym. Degrad. Stab.* **1993**, *40*, 323.
10. G.A. George, in: *Developments in Polymer Degradation*, Vol. 3, N. Grassie, Ed. Applied Science Publishers, 1981, p. 173.
11. M. Celina, G.A. George, and N.C. Billingham, in: R.L. Clough, N.C. Billingham, and K.T. Gillen (Eds.), *Polymer Durability: Degradation, Stabilization and Lifetime Prediction, Advances in Chemistry Series 249*, American Chemical Society, Washington, DC, 1996, Chapter 11, p. 159.
12. L. Matisova-Rychla and J. Rychly, in: R.L. Clough, N.C. Billingham and K.T. Gillen (Eds.), *Polymer Durability: Degradation, Stabilization and Lifetime Prediction, Advances in Chemistry Series 249*, American Chemical Society, Washington, DC, 1996, Chapter 12, p. 185.
13. V. Dudler, D.J. Lacey, and Ch. Kröhnke, *Polym. Degrad. Stab.* **1996**, *51*, 115.
14. D.R. Kohler and C. Krohnke, *Polym. Degrad. Stab.* **1999**, *63*, 165.
15. G.E. Ashby, *J. Polym. Sci.* **1961**, *50*, 99.
16. M.P. Schard and C.A. Russel, *J. Appl. Polym. Sci.* **1964**, *8*, 995.
17. A. Kron, B. Stenberg, T. Reitberg, and N.C. Billingham, *Polym. Degrad. Stab.* **1996**, *53*, 119.
18. N.C. Billingham, E.T.H. Then, and P.H. Gijman, *Polym. Degrad. Stab.* **1991**, *42*, 263.
19. H. Mori, T. Hatanaka, and M. Terano, *Macromol. Rapid Commun.* **1997**, *18*, 157.
20. M. Kato and Z. Osawa, *Polym. Degrad. Stab.* **1999**, *65*, 457.
21. M. Strlic, J. Kolar, B. Pihlar, L. Matisova-Rychla, and J. Rychly, *Eur. Polym. J.* **2000**, *16*, 2351.
22. F. Catalina, C. Peinado, N.S. Allen, and T. Corrales, *J. Polym. Sci. Part A: Polym. Chem.* **2002**, *40*, 3312.
23. F. Gugumus, *Polym. Deg. Stab.* **1999**, *63*, 41.
24. L. Reich and S.S. Stivala, *Autooxidation of Hydrocarbons and Polyolefins*, Dekker, New York, **1969**.
25. S.K. Brauman and J.G. Pronkp, *J. Polym. Sci., Part B: Polym. Phys.* **1988**, *26*, 1205.
26. H. Kihara and S. Hosoda, *Polym. J.* **1990**, *22*, 763.
27. A. Kron, B. Stenberg, T. Reitberger, and N.C. Billingham, *Polym. Degrad. Stab.* **1996**, *53*, 119.
28. D.J. Lacey and V. Dudler, *Polym. Degrad. Stab.* **1996**, *51*, 101.
29. A. Zahradnickova, J. Sedlar, and D. Dastych, *Polym. Degrad. Stab.* **1991**, *32*, 155.
30. L. Achimsky, L. Audouin, J. Verdu, L. Rychla, and J. Rychly, *Eur. Polym. J.* **1999**, *35*, 557.
31. L. Matisova-Rychla and J. Rychly, *Polym. Degrad. Stab.* **2000**, *67*, 515.

32. G. Ahlblad, P. Gijsman, B. Terselius, A. Jansson, and K. Moller, *Polym. Degrad. Stab.* **2001**, *73*, 15.
33. T. Corrales, C. Peinado, N.S. Allen, M. Edge, G. Sandoval, and F. Catalina, *J. Photochem. Photobiol. Part A: Chem.* **2002**, *147*, 213.
34. R. Setnescu, S. Jipa, and Z. Osawa, *Polym. Degrad. Stab.* **1998**, *60*, 377.
35. L. Zlatkevich, *Polym. Degrad. Stab.* **1983**, *50*, 83.
36. R. Broska and J. Rychly, *Polym. Degrad. Stab.* **2001**, *72*, 271.
37. K. Jacobson, *Polym. Degrad. Stab.* **2006**, *91*, 2126.
38. J. Pospísil, G. Scott (Ed.), in: *Developments in Polymer Stabilisation—1*, Applied Science Publishers Ltd, London, 1979.
39. J. Pospisil, *Adv. Polym. Sci.* **1989**, *36*, 69.
40. J.A. Howard and K.U. Ingold, *Can. J. Chem.* **1963**, *41*, 2800.
41. J. Pospisil, in: G. Pritchard (Ed.), *Plastics Additives*, Chapman and Hall, London, 1998.
42. R. Setnescu, S. Jipa, and Z. Osawa, *Polym. Degrad. Stab.* **1998**, *60*, 377.
43. J. Lucki, *Polym. Degrad. Stab.* **1985**, *11*, 75.
44. Yu.A. Shyaprikov, S.G. Kiryushkin, and A.P. Marin, Antioxidative stabilisation of polymers, Taylor and Frances Ed, London, 1996.
45. L. Matisova-Rychla and J. Rychly, *J. Polym Sci., Part A: Polym. Chem.* **2004**, *42*, 648.
46. L. Zlatkevich, *J. Polym. Sci., Polym. Phys. Ed.* **1985**, *23*, 2633.
47. S. Jipa, T. Zaharescu, B. Gigante, C. Santos, R. Setnescu, T. Setnescu, M. Dumitri, L.M. Girghiu, W. Kappel, and I. Mihalcea, *Polym. Degrad. Stab.* **2003**, *80*, 209.
48. J. Murphy, *Additives for Plastics Handbook*, Elsevier Advanced Technology, Oxford, 1996, Chapter 7.
49. C.E. Vignolo, *Eur. Coatings J.* **1995**, *Part 5*, 359.
50. N.S. Sariciftci, D. Braun, C. Zhang, V.I. Srdanov, A.J. Heeger, G. Stucky, and F. Wudl, *Appl. Phys. Lett.* **1993**, *62*, 585.
51. V.L. Colvin, M.C. Schlamp, and A.P. Alivisatos, *Nature* **1994**, *370*, 354.
52. L. Merhari, K.E. Gonsalves, Y. Hu, W. He, W.S. Huang, M. Angelopoulos, W.H. Bruenger, C. Dzionk, and M. Torkler, *Microelectron. Eng.* **2002**, *63*, 391–403.
53. D. Godovsky, *Adv. Pol. Sci.* **2000**, *153*, 163.
54. Y.N. Konan, R. Gurny, and E. Allemann, *Int. J. Pharm.* **2002**, *233*, 239.
55. T. Yonezawa, N. Toshima, and H.S. Nalwa (Eds.), *Advanced Functional Molecules and Polymers*, Vol. 2, Overseas Publishers Association, 2001.
56. N.S. Allen, M. Edge, G. Sandoval, A. Ortega, C.M. Liauw, J. Stratton, and R.B. McIntyre, *Polym. Degrad. Stab.* **2002**, *76*, 305.
57. T. Corrales, C. Peinado, N.S. Allen, M. Edge, G. Sandoval, and F. Catalina, *J. Photochem. Photobiol. Part A: Chem.* **2003**, *156*, 151.
58. P.K. Fearon, S.W. Bigger, and N.C. Billingham, *J. Therm. Anal. Calorim.* **2004**, *76*, 75.
59. K.H. Gustavson, *The Chemistry and Reactivity of Collagen*, Academic Press, New York, 1956.
60. G.N. Ramachandram, (Ed.), *Treatise on Collagen*, Academic Press, Inc., New York, 1967.
61. A.G. Ward and A. Court (Eds.), *The Science and Technology of Gelatin*, Academic Press, Inc., New York, 1977.

62. M.V. Lock, and B.F. Sagar, *J. Am. Chem. Soc., B* **1966**, 690.
63. B.F. Sagar, *J. Chem. Soc.* **1967**, 1047.
64. W.H. Sharley and W.E. Mochel, *J. Am. Chem. Soc.* **1959**, *81*, 3000.
65. L. Tang, D. Sallet, and J. Lemaire, *Macromolecules* **1982**, *15*, 1432.
66. T. Karstens and V. Rossbach, *Makromol. Chem.* **1990**, *191*, 757.
67. C. Abrusci, A. Martin-Gonzalez, A. Del Amo, F. Catalina, P. Bosch, and T. Corrales, *J. Photochem. Photobiol. Part A: Chem.* **2004**, *163*, 537.
68. L. Matisová-Rychlá, J. Rychlý, and K. Slovák, *Polym. Degrad. Stab.* **2003**, *82*, 173.
69. C. Abrusci, D. Marquina, A. Santos, A. Del Amo, T. Corrales, and F. Catalina, *J. Photochem. Photobiol. Part A: Chem.* **2007**, *185*, 188.
70. C. Abrusci, A. Martin-Gonzalez, A. Del Amo, F. Catalina, J. Collado, and G. Platas, *Int. Biodeterior. Biodegrad.* **2005**, *56*, 58.
71. I. Fridovich, *Ann. N.Y. Acad. Sci.* **1999**, *873*, 13.
72. S.-X. Chen and P. Schopfer, *Eur. J. Biochem.* **1999**, *260*, 726.
73. R.P. Beckett, F.V. Minibayena, N.N. Vylegzhanina, and T. Tolpysheva, *Plant Cell Environ.* **2003**, *26*, 1827.
74. M.M. Huycke, D.M. Moore, W. Joyce, L. Shepard, Y. Kotache, and M. Gilmore, *Mol. Microbiol.* **2001**, *42*, 729.
75. F. Shafizadeh and A.G.W. Bradbury, *J. Appl. Polym. Sci.* **1979**, *23*, 1431.
76. M. Strlic, J. Kolar, B. Pihlar, J. Rychly, and L. Matisova-Rychla, *Polym. Degrad. Stab.* **2001**, *72*, 157.
77. J. Rychly, M. Strlic, L. Matisova-Rychla, and J. Kolar, *Polym. Degrad. Stab.* **2002**, *78*, 357.
78. J. Rychly, L. Matisova-Rychla, M. Lazar, K. Slovak, M. Strlic, D. Kocar, and J. Kolar, *Carbohydr. Polym.* **2004**, *58*, 301.
79. A.M. Emsley, A.M. Herman, and H. Heywood, *Polymer* **2001**, *42*, 2893.
80. M. Pletenikova, L. Matisova-Rychlá, J. Rychly, and I. Lacik, *Carbohydr. Polym.* **2007**, *69*, 50.
81. K.R. Millington, G. Maurdev, and M.J. Jones, *Polym. Degrad. Stab.* **2007**, *92*, 1504.
82. O.L. Kramer, E.M. Battalov, A.I. Voloshin, and Y.A. Prochukhan, *Russ. J. Appl. Chem.* **2001**, *74*, 1190.
83. C. Peinado, N.S. Allen, E.F. Salvador, T. Corrales, and F. Catalina, *Polym. Degrad. Stab.* **2002**, *77*, 523.
84. C.S. Schollenberg and F.D. Stewart, in: K.C. Frisch, S.L. Reegen (Eds.), *Advances in Urethane Science and Technology*, Vol. 4, Technomic Publishing, Westport, CT, 1976, p. 68.
85. J.L. Gardette, and J. Lemaire, *Makromol Chem.* **1982**, *183*, 2415–2425.
86. C. Peinado, T. Corrales, M.J. García-Casas, F. Catalina, V. Ruiz, and M.D. Parellada, *Polym. Degrad. Stab.* **2006**, *91*, 862.
87. N.C. Billingham and M.N. Grigg, *Polym. Degrad. Stab.* **2004**, *83*, 441.
88. N.S. Allen, G. Rivalle, M. Edge, T. Corrales, and F. Catalina, *Polym. Degrad. Stab.* **2002**, *75*, 237.
89. M. Celina, G.A. George, D.J. Lacey, and N.C. Billingham, *Polym. Degrad. Stab.* **1995**, *47*, 311.

90. P. Gijsman, J. Hennekens, and J. Vincent, *Polym. Degrad. Stab.* **1993**, *42*, 95.
91. M. Avrami, *J. Chem. Phys.* **1941**, *9*, 177.
92. D. Turnbull and J.C. Fisher, *J. Chem. Phys.* **1949**, *17*, 71.
93. J.D. Hoffman, J.J. Weeks, and W.M. Murphey, *J. Res. Natl. Bureau Stand.*, **1959**, *63A*, 67.
94. T. Corrales, C. Abrusci, C. Peinado, and F. Catalina, *Macromolecules* **2004**, *37*, 6596.
95. L. Matisová-Rychlá, J. Rychlý, and K. Slovák, *Polym. Degrad. Stab.* **2003**, *82*, 173.
96. Y.A. Vladimirov, A. Arroyo, J. Michael Taylor, Y. Y. Tyurina, T. Matsura, V. A. Tyurin, V.E. Kagan, *Arch. Biochem. Biophys.* **2000**, *376*, 154.
97. D. Slawinska and J. Slawinski, *J. Biolumin. Chemilumin.* **1989**, *4*, 226.
98. J. Peña, T. Corrales, I. Izquierdo-Barba, A.L. Doadrio, and M. Vallet-Regí, *Polym. Degrad. Stab.* **2006**, *91*, 1424.
99. G. Sivalingam, R. Karthik, and G. Madras, *J. Anal. Appl. Pyrolysis* **2003**, *70*, 631.
100. V. Dudler, Th. Bolle, and G. Rytz, *Polym. Degrad. Stab.* **1998**, *60*, 351.
101. M. Celina, G.A. George, and N.C. Billingham, *Polym. Degrad. Stab.* **1993**, *42*, 335.
102. J. Peña, T. Corrales, I. Izquierdo-Barba, M.C. Serrano, T. Portolés, R. Paggani, and M. Vallet-Regí, *J. Biomed. Mater. Res.* **2006**, *76A*, 788.
103. A.G.A. Coombes, S.C. Rizzi, M. Williamson, J.E. Barralet, S. Downes, and W.A. Wallace, *Biomaterials* **2002**, *23*, 315.
104. N. Takeyama, S. Miki, A. Hirakawa, and T. Tanaka, *Exp. Cell Res.* **2002**, *274*, 16.
105. M. Aikawa, S. Sugiyama, C.C. Hill, S.J. Voglic, E. Rabkin, Y. Fukumoto, F.J. Schoen, J.L. Witztum, and T. Libby, *Circulation* **2002**, *106*, 1390.
106. A.R. Parrish, J.M. Catania, J. Orozco, and A.J. Gandolfi, *Toxicol. Sci.* **1999**, *51*, 80.
107. J. Sebenda and B. Lanska, *J. Macromol. Sci.* **1993**, *A30*, 669.
108. B. Lanska, *Eur. Polym. J.* **1994**, *30*, 197.
109. G. Ahlblad, D. Forsstrom, B. Stenberg, B. Terselius, T. Reitberger, and L.-G. Svensson, *Polym. Degard. Stab.* **1997**, *55*, 287.
110. J. Jacobson, B. Stenberg, B. Terselius, and T. Reitberger, *Polym. Degrad. Stab.* **1999**, *64*, 17.
111. B. Lanská, L. Matisova-Rychlá, and J. Rychlý, *Polym. Degrad. Stab.* **2001**, *72*, 249.
112. P. Cerruti, C. Carfagna, J. Rychlý, and L. Matisova-Rychlá, *Polym. Degrad. Stab.* **2003**, *82*, 477.
113. P. Cerruti, J. Rychlý, L. Matisova-Rychlá, and C. Carfagna, *Polym. Degrad. Stab.* **2004**, *84*, 199.

4

NONLINEAR OPTICAL POLYMERIC MATERIALS

Mirko Faccini, David N. Reinhoudt, and Willem Verboom

Photochemistry and Photophysics of Polymer Materials, Edited by Norman S. Allen

4.1 INTRODUCTION

Polymeric electrooptic (EO) materials incorporating nonlinear optical (NLO) chromophores have shown commercial potential as active media in high-speed broadband waveguides for optical switches, optical sensors, and information processors. Meanwhile, several reviews appeared describing the different aspects of this topic. Some of them focus more on the different classes of chromophores [1] (e.g., charge-transfer molecules, octopolar compounds, ionic materials, multichromophore systems, and organometallics), on their design and synthesis, and on the optimization of their intermolecular interactions to obtain large macroscopic EO activities [2,3]. Others describe chromophore orientation techniques [4] (e.g., static field poling, photoassisted poling, all optical poling, and contact and corona poling), or material characterization and the steps and techniques required for EO device fabrication and operation [5,6].

This chapter concentrates on the design of efficient dipolar NLO chromophores and the different approaches for their incorporation in non-centrosymmetric materials, including guest–host polymer systems, chromophore-functionalized polymers (side-chain and main-chain), cross-linked chromophore–macromolecule matrices, dendrimers, and intrinsically acentric self-assembled chromophoric superlattices. The different architectures will be compared together with the requirements (e.g., large EO coefficient, low optical absorption, high stability, and processability) for their incorporation into practical EO devices. First, a brief introduction to nonlinear optics is presented.

4.2 GENERAL BACKGROUND

Nonlinear optics originates from the ability of matter to respond in a nonlinear way to the interaction with external electromagnetic fields (e.g., that associated with light) [7,8]. When an external forcing field is applied to a material, it causes a displacement of the charges in molecules and atoms. In case of a low-intensity field, this induced polarization (P) is linearly proportional to the field strength (E). However, under sufficiently intense fields (e.g., laser light), the relationship is no longer linear, and the polarization can be expressed as a power series expansion. Thus, the molecular polarization p can be written as (neglecting quadrupolar terms):

$$p = \alpha E + \beta EE + \gamma EEE + \cdots \tag{4.1}$$

where α is the molecular polarizability, whereas β, γ, and so forth are the molecular hyperpolarizabilities of the first order, second order, and so forth, corresponding to second-, third-, and higher-order nonlinearities, respectively. On a macroscopic scale, the polarizability can be expressed as:

$$P = \chi^{(1)} E + \chi^{(2)} EE + \chi^{(3)} EEE + \cdots \tag{4.2}$$

where χs are the macroscopic susceptibilities.

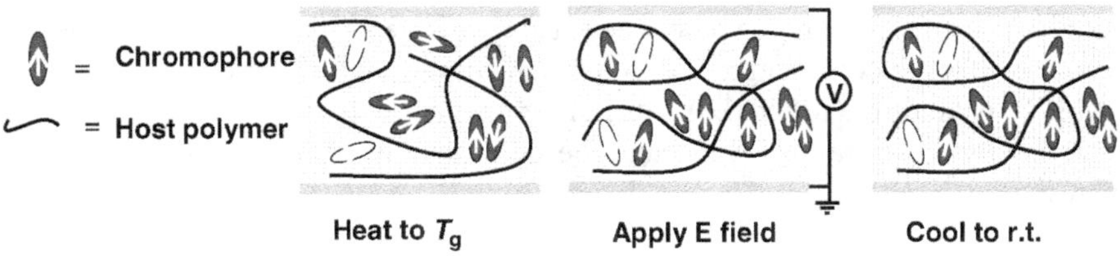

FIGURE 4.1 Schematic representation of the electric field poling process.

Unlike the first- and third-order terms of the above equations, a requirement to induce second-order nonlinear optical activity is the non-centrosymmetry or acentric symmetry, both at the molecular and macroscopic levels [9]. At the molecular level, highly polarizable materials with the absence of symmetry can be easily obtained by connecting strong electron donors and acceptors by a conjugated π bridge, giving strong charge-transfer molecules. The application of an electric field will effect the mixing of a neutral ground state and a charge-separated excited state, thus changing the molecular polarization of the dipole.

To achieve electrooptic activity at the macroscopic level, chromophores must display non-centrosymmetry. To achieve this condition, dipolar molecules should be oriented to yield acentric chromophore lattices. The most commonly used method is electric field poling (corona poling or contact poling) of NLO polymers [9] (Fig. 4.1). The polymer containing NLO chromophores must first be converted into thin films by spin-coating on conductive substrates. Then, by heating the film close to its glass transition temperature (T_g), the material becomes softer, allowing dipoles to increase their mobility. Subsequently, an electric field is applied and the chromophores in the matrix can reorient toward the electric field. After some time, the temperature can be decreased until below the T_g of the material, while keeping the orienting field applied, eventually resulting in an acentric ordered lattice. Other orientation methods used include non-centrosymmetric crystallization [10], Langmuir–Blodgett (LB) film formation [11], layer-by-layer (LbL) growth on solid supports [12–14] and liquid crystals [15], and the incorporation of chromophores into inclusion compounds (supramolecular alignment) [16].

Electrooptic activity arises from the ability of a material to change its refraction index on application of an external electric field. As the refractive index relates with the speed of the light transiting a material, electrooptic activity can be defined as a voltage-controlled phase shift of light. Therefore, the application of an electric field will cause a change in the charge distribution of the material (change of mixing of ground and excited forms) and thus, alter the speed of light propagating through the material. In the case of Boltzmann-distributed chromophores, the macroscopic electrooptic activity (r_{33}) of the material is given by

$$r_{33} = \frac{2N\beta f(\omega)\langle\cos^3\theta\rangle}{n^4} \tag{4.3}$$

where N is the chromophore number density, $f(\omega)$ is the local optical field correction factor from the dielectric nature of the environment surrounding the chromophore, n is the refractive index, and $\langle\cos^3\theta\rangle$ is the order parameter. If electrostatic interactions between chromophores are neglected, the order parameter in Equation 4.3 can be expressed as:

$$\langle\cos^3\theta\rangle = \frac{\mu F}{5kT} \tag{4.4}$$

where k is the Boltzmann's constant, T is the poling temperature, μ is the dipole moment of the chromophore, and F $(= f_0 E_p)$ is the electric poling field felt by the chromophore. By combining Equations 4.3 and 4.4, the EO coefficient becomes

$$r_{33} = \frac{2N\mu\beta f(\omega)f(o)E_p}{5kTn^4} \tag{4.5}$$

A variety of techniques are being employed to measure the EO activity, r_{33}, including ellipsometry [17,18], attenuated total reflection (ATR) [19], and two-slit interference modulation [20].

From Equation 4.5, it is clear that for a given $\mu\beta$ chromophore, the EO activity can be maximized either by increasing the poling field strength E_p or by increasing the chromophore loading (concentration) in the polymer matrix. However, this linear relationship is only valid at low chromophore concentration, where $\langle\cos^3\theta\rangle$ and N are independent. Differently, at higher concentrations, intermolecular electrostatic interactions among dipoles cannot be neglected, and the relationship between $\langle\cos^3\theta\rangle$, N, and E_p becomes more complex. Therefore, to achieve the maximum EO activity, the product $\langle\cos^3\theta\rangle N$ has to be optimized.

4.3 APPLICATIONS OF EO MATERIALS

During the past two decades, great efforts from scientists have led to a good fundamental knowledge of the relationship between chemical structure and NLO properties, and thus, to the production of materials with ever-increasing performances. Based on these advances, several devices incorporating polymeric electrooptic materials have been fabricated for different types of applications, including electrical to optical signal transduction, optical switching in local area networks (LANs), millimeter wave signal generation, optical beam steering (for addressing phosphors in flat panel displays), backplane interconnections for high-speed personal computers, or wavelength division multiplexing (WDM) [3,5]. Among all these devices, electrooptic modulators play a fundamental role in many growing areas of broadband telecommunication and to increase the diffusion of multimedia services such as high-quality cable television, telephone, real-time videoconferencing, telemedicine, distance learning, video-on-demand, and ultra-fast Internet.

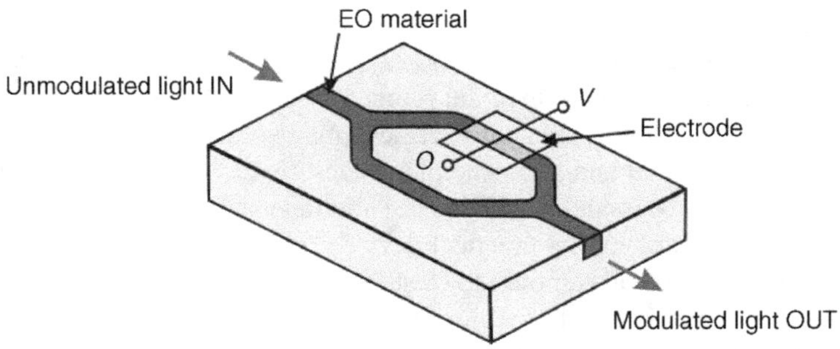

FIGURE 4.2 Schematic representation of a MZ modulator.

A simple device configuration for light modulation based on a second-order NLO phenomenon is the Mach–Zehnder (MZ) interferometer, which acts as an electrical to optical switch (Fig. 4.2).

If no electric field is applied, the input light is split into two beams propagating in separate arms of the MZ modulator to then recombine at the end of the Y-junction ("on"-position). By applying an electric field of a proper intensity to one arm, the refractive index of this channel will change, resulting in a phase retardation of π relative to the signal traversing the other arm and thus to destructive interference ("off"-position). Using this principle, the MZ modulator can transduce the applied electrical signal onto the optical beam as an amplitude modulation.

The relative phase shift $\Delta\varphi$ of light passing through an electrooptic material when an electric field is applied is given by Equation 4.6, where n is the material refractive index, r is its EO coefficient, L is the propagation length (waveguide length), E is the strength of the applied electric field, and λ is the operation wavelength.

$$\Delta\varphi = \frac{\pi n^3 rLE}{\lambda} \tag{4.6}$$

The minimum voltage required for a π-phase shift, called half-wave voltage V_π, can be expressed as:

$$V_\pi = \frac{\lambda h}{n^3 rL\Gamma} \tag{4.7}$$

where h is the electrode spacing and Γ is the overlap integral of the electrical and optical wave (close to 1).

4.4 MATERIAL REQUIREMENTS

In 2000, an encouragingly low half-wave voltage V_π of 0.8 V was achieved in a polymer waveguide modulator using highly nonlinear organic chromophores [21].

This breakthrough, together with a more recent demonstration of a device with exceptional bandwidths (up to 200 GHz) [22], has provided a solid foundation for applying polymeric EO materials in the future generations of telecommunications.

However, apart from large nonlinearities, many other essential parameters, including a good thermal and photochemical stability, low optical loss (high transparency), and good processability, need to be simultaneously optimized for the active material to be incorporated in a practical device.

For the nonlinear optical response to be stable during processing and operation of chromophore/polymer materials, the chromophores need to be chemically stable at all temperatures that the system encounters in electric field poling, and should withstand the fabrication steps needed for device fabrication. Usually, during device processing, the temperature can rise up to 250°C, whereas during operation, the material is subject to temperatures around 100°C for long periods of time.

Moreover, after removal of the poling field, the electrooptical response of the material should be stable over time. For this reason, the glass transition temperature (T_g) of the polymer should be high enough for the chromophore acentric order to be kept frozen over device operation.

EO devices are expected to be operational for several years. Therefore, chromophores should possess a sufficient photostability to withstand a high-intensity illumination for long periods of time, without structural degradation, resulting in a loss of nonlinearity.

Optical loss of chromophore-containing materials is a key performance parameter in EO devices. In general, EO polymers must possess a good optical transparency at datacom wavelengths (840 nm) and telecom wavelengths (1310 and 1550 nm). Optical loss can be due either to vibrational absorption or to electronic absorption. Vibrational absorption is mainly due to the C–H overtones coming from the polymer backbone (especially at low chromophore loadings), whereas electronic absorption is mainly caused by the charge-transfer (HOMO–LUMO) band of the chromophores. High β red-shifted chromophores could cause some long-wavelength tailing at operative wavelengths [23]. In general, the optical loss due to chromophore absorption should remain below 1 dB/cm [24]. Moreover, both the chromophores and the host polymer must exhibit a good solubility in spin-casting solvents to be converted into optical quality thin films.

4.5 CHROMOPHORE DESIGN

In general, second-order NLO materials can be considered as dipolar chromophores non-centrosymmetrically aligned in poled polymers. Therefore, the most intuitive way of achieving a bulky EO response is by optimizing the molecular first hyperpolarizability β of the active component. For organic molecules, β is generally determined in solution, using methods such as electric field-induced second harmonic (EFISH) generation [25,26] or hyper-Rayleigh scattering (HRS) [27,28]. Commonly used figures of merit for comparing chromophores are $\mu\beta$ or $\mu\beta/M_w$ (M_w is the chromophore molecular weight).

Oudar and Chemla proposed a simple two-state quantum mechanical model as a powerful tool to predict the molecular first hyperpolarizability β in the design of the second-order NLO chromophores [29]:

$$\beta = (\mu_{ee} - \mu_{gg})\,(\mu_{ge})^2/(\Delta E_{ge})^2 \tag{4.8}$$

where $(\mu_{ee} - \mu_{gg})$ is the difference between the dipole moments of the exited and ground states, μ_{ge} is the transition dipole moment (transition matrix element between ground and exited states), and ΔE_{ge} is the HOMO–LUMO energy gap. In the early 1990s, based on this model, Marder and coworkers developed a structure/function relationship that illustrates how β, for a given conjugation bridge, can be maximized through an optimal combination of donor and acceptor strengths, which can be viewed as tuning the degree of mixing between the neutral and the charge separated form [30,31]. They have shown that by structural modifications of donor–acceptor polyenes, β increases in a sinusoidal manner with molecular parameters such as the bond length alternation (BLA), which is defined as the average of the difference between carbon single and double bond lengths in the conjugated chromophore core. For a given conjugation bridge, there is an optimal combination of donor and acceptor strengths to maximize the molecular first hyperpolarizability, and a further increment of the strength will only reduce β.

Quantum mechanical analysis based on these theories has resulted in an incredibly useful tool for designing chromophores with a large molecular hyperpolarizability. Some representative examples of chromophores with an ever-improved molecular optical nonlinearity developed over the past decade are reported in Table 4.1.

The search for new chromophores has focused on the development of various types of acceptor moieties and conjugated bridges. Mainly, all chromophores developed for EO applications contain invariably amine-based donors [32]. Structural modifications have been explored to improve the solubility, and the thermal or photostability of the chromophores. It has been noted that the use of a 4-(diarylamino)phenyl electron donor results in a significant improvement in thermal stability compared with the 4-(dialkylamino)phenyl-substituted derivatives [33–37]. A theoretical comparison between aminophenyl and aminothienyl donor systems reported recently provides useful guidelines for the design of improved NLO chromophores [32].

Although very large hyperpolarizabilities have been obtained with long and unprotected polyene bridges, these materials are chemically and photochemically unstable or it is impossible to prepare them in adequate yield and purity, and therefore, they are not suitable for practical applications. On the other hand, chromophores based on fused ring systems, such as naphthalene benzimidazoles [38] or phthalocyanines [39,40] possess good nonlinearity and thermal stabilities over 350°C, but they are poorly soluble in spin-casting solvents and they cannot be effectively poled.

Improvements in thermal and photochemical stability have been achieved by using conformationally locked polyenes [41,42] by incorporating the polyene chain into ring systems such as isophorone [43] or by introducing heterocyclic conjugating units such as thiazole [44], furan, and thiophene [45]. Many studies have investigated the

TABLE 4.1 ***μβ* Values for Representative NLO Chromophores[a] and r_{33} Values for Their Guest-Host Polymers.[b]**

NLO Chromophore	λ_{max} (nm)	$\mu\beta$ (10^{-48} esu)	NLO Chromophore	λ_{max} (nm)	$\mu\beta$ (10^{-48} esu)
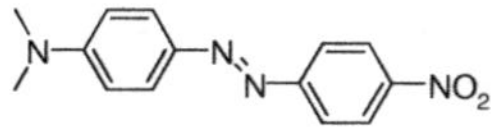	475	580	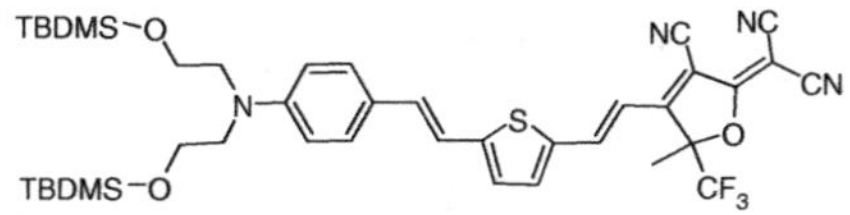	711	N/A
DR	$r_{33} = 13$ pm/V (30% in PMMA)		**AJL-8**	$r_{33} = 42$ pm/V (25% in APC)	
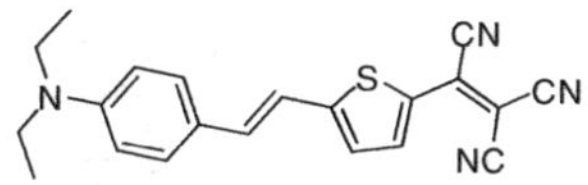	640	6200	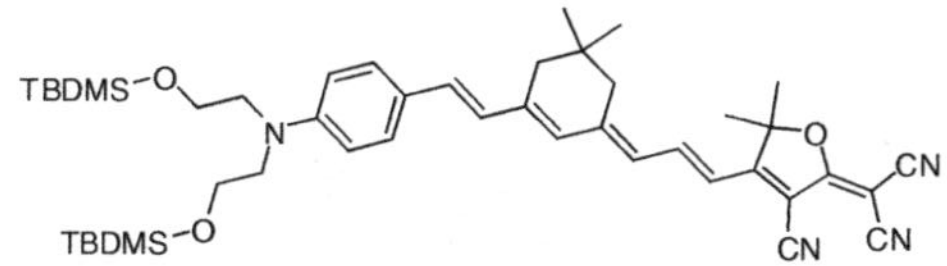	695	35000
	$r_{33} = 30$ pm/V (25% in PI)		**CLD**	$r_{33} = 92$ pm/V (25% in APC at 1.06 μm)	
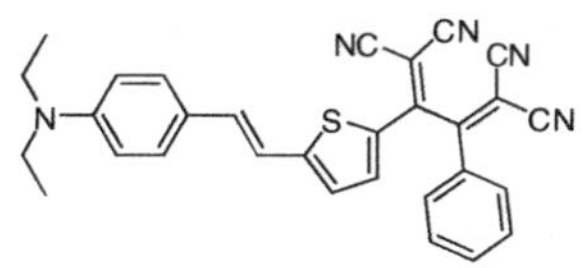	604	N/A	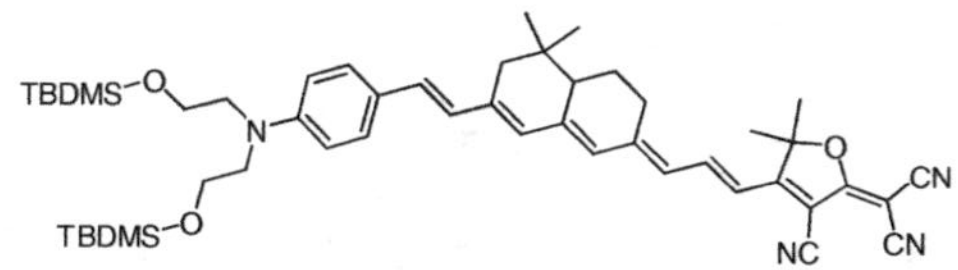	N/A	N/A
Ph-TCBD	$r_{33} = 36$ pm/V (25% in PQ100)		**GLD**	$r_{33} = 105$ pm/V (17.5% in PMMA)	

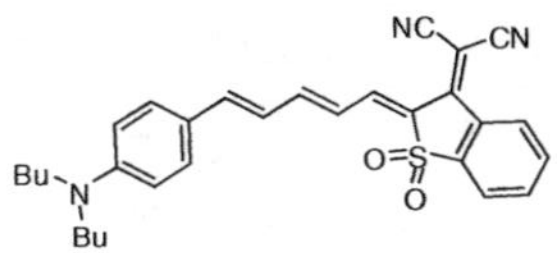	770	13500	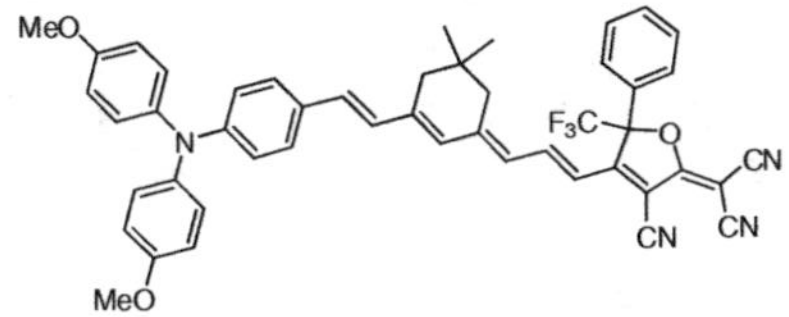	770	N/A
	$r_{33} = 55$ pm/V (20% in PC)			$r_{33} = 106$ pm/V (25% in APC)	
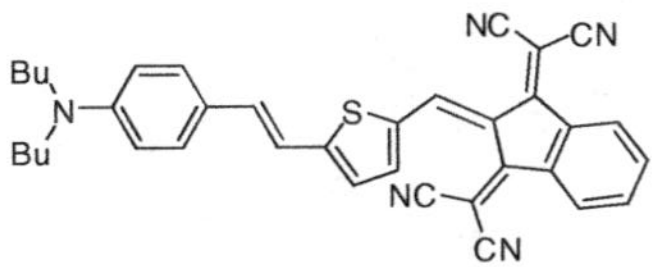 **TCI**	770	6100	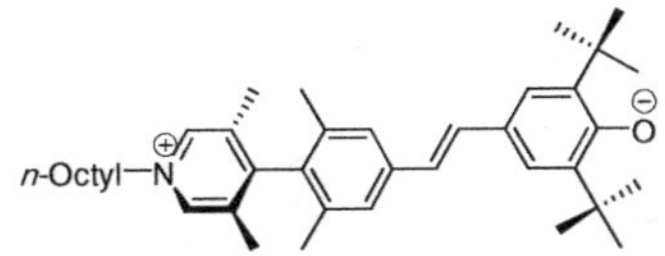 **TM-2**	462	−315000
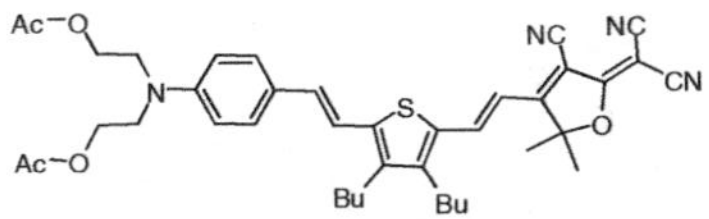 **FTC**	650	18000	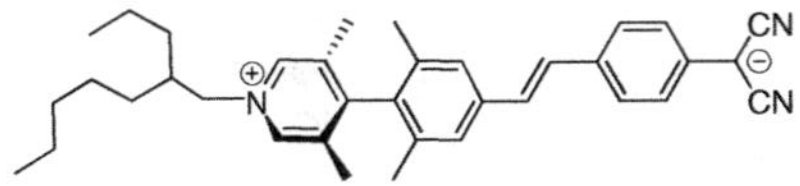 **TMC-3**	540	−488000
	$r_{33} = 57$ pm/V (20% in PMMA at 1.06 μm)			$r_{33} = 330$ pm/V (10% in PVP)	

[a] Measured at 1907 nm;
[b] Measured at 1.3 μm, otherwise indicated

role of heteroaromatics and clarified the effect of the nature and location of the heterocyclic ring on the charge-transfer (CT) transitional energies of the chromophores, leading to the development of the "auxiliary donor and acceptor" model [46–48]. This theory correlates the molecular hyperpolarizability β with the electron density of the π-conjugation, arguing that electron-excessive/deficient heterocyclic bridges act as auxiliary donors/acceptors giving larger β values.

Many different acceptor groups have been investigated. Very large nonlinearities have been achieved using, for instance, nitro, cyanovinyl, thiobarbituric acid moieties [49], or strong multicyano-containing heterocyclic electron acceptors [35,50–52].

Highly hyperpolarizable chromophores have been reported using the tricyanovinyl (TCV) moiety as electron acceptor and various conjugating moieties ($\mu\beta$, as high as 9800×10^{-48} esu) [53]. However, the TCV group is very sensitive to chemical attack. To overcome this, Jen and coworkers [54] prepared a series of thermally and chemically stable chromophores containing a 2-phenyl-tetracyanobutadienyl acceptor (**Ph-TCBD**), which have been demonstrated to be much more stable toward amine nucleophiles than their tricyanovinyl counterparts. An enormous progress has been achieved since the introduction of the tricyano-derivatized furan (TCF) acceptor in the late 1990s [55–57]. Its incorporation in chromophores such as **FTC**, **CLD**, and **GLD** (Table 4.1) resulted in chromophore figures-of-merit, $\mu\beta$, as high as $35{,}000 \times 10^{-48}$ esu [21]. Moreover, these chromophores display a good solubility in spin-casting solvents, a good processability, and a thermal stability of approximately 300°C.

More recently, Marks and coworkers developed a totally different approach for obtaining very high β values. Their strategy does not focus on extensive planar π-conjugation, which is prone to chemical, thermal, and photochemical instabilities [58], but on twisted π-electron system chromophores. These unconventional twisted π zwitterionic structures (TM and TMC) exhibit unprecedented hyperpolarizabilities as large as 15 times greater than those reported previously ($\mu\beta$ as high as $-488{,}000 \times 10^{-48}$ esu) [59,60].

4.6 GUEST–HOST SYSTEMS

Guest–host systems have been the first NLO polymer systems investigated, because they can be easily obtained by simply dissolving an EO chromophore in a compatible amorphous polymeric matrix, to form a solid solution. The selection of polymer host should be based on a good optical transparency, high thermal stability, and good solubility in spin-casting solvents. To obtain a high EO response, the chromophore should be able to dissolve in the polymer matrix at high loadings, without phase separation to occur. A high T_g polymer is desirable to maintain the electrically induced non-centrosymmetric order stable over time at device operating temperatures. Although the T_g of commonly used host polymers is 150–250°C, the incorporation of a chromophore will induce plasticization, considerably lowering the T_g of the composite material, and therefore, reducing the temporal stability of the EO response.

A number of different chromophores have been investigated in various low-T_g polymer lattices, such as polycarbonate (PC) or poly(methyl methacrylate) (PMMA), obtaining very large EO coefficients. An r_{33} value of 21 pm/V at 1.06 μm was achieved by Sun et al. in PMMA thin films doped at 30–40 wt% with a strong heteroaromatic electron-acceptor chromophore [51]. A polycarbonate film, containing 20 wt% of a highly NLO-active compound yielded a very large r_{33} value of 55 pm/V at 1.06 μm [50]. However, the limited thermal stability of these materials, suggested the use of high-T_g polymers such as polyimides [61,62] and polyquinolines [63,64]. On the other hand, as the chromophore is not covalently connected to the polymer backbone, it can sublimate out of the blend when high poling and processing temperatures are required. Much improved long-term stabilities at elevated temperatures have been obtained using highly thermally stable chromophores such as **Ph-TCBD**, incorporated as a guest (20 wt%) in the rigid-rod polyquinoline PQ-100 ($T_g = 265°C$). An r_{33} value of 36 pm/V was recorded at 1.3 μm after poling, which remained at ~80% of its original value at 85°C for over 1000 h [54].

Recently, a very large EO coefficient ($r_{33} = 169$ pm/V at 1.3 μm) and an excellent long-term alignment stability at 85°C under vacuum for more than 500 h have been demonstrated by incorporating a large $\mu\beta$ chromophore bearing a 2-dicyanomethylidene-3-cyano-4,5-dimethyl-5-trifluoromethyl-2,5-dihydrofuran (CF_3-TCF) electron acceptor into PQ-100 as a guest (25 wt%) [64].

Together with the development of highly NLO-active molecules such as **CLD** and **FTC**, amorphous polycarbonate (APC) has been extensively investigated as a host polymer, because of its low crystallization tendency, good solubility in halogenated solvents and high glass transition temperature ($T_g = 205°C$). Its good compatibility with large $\mu\beta$ chromophores and its high dielectric constant allow EO coefficients as large as 92 pm/V (25% **CLD-1**/APC composite, at 1.06 μm) to be routinely obtained. MZ modulators fabricated from **CLD-1**/APC showed a good thermal stability at 50°C, an optical loss of 1.7 dB/cm, and a low modulation voltage (V_π) of 3.7 V [65,66]. EO modulators have also been fabricated from 30% **CLD-1**/PMMA material, demonstrating a V_π value of 0.8 V [21]. However, due to the low T_g of this composite, the dynamic thermal stability of poling-induced alignment was only 75°C, that is, 40°C lower than for the corresponding APC material. More recently, really large r_{33} values (as high as 169 pm/V at 1.31 μm) have been recorded for a series of thermally stable chromophores with a 4-(diarylamino)phenyl donor and a strong CF_3-TCF electron acceptor incorporated at 25 wt% in APC [67].

Furthermore, Garner et al. [68] demonstrated the practicality of using polysulfone as a novel host material. A chromophore–polysulfone system showed a similarly high EO performance and a lower V_π value, compared with the analogous polycarbonate system. Moreover, this material combines a reasonably high T_g of 190°C and a high refractive index of 1.63 and is a better host than polycarbonate with respect to the photostability [69].

Very recently, it has been reported that a guest–host system, consisting of twisted π-zwitterionic chromophores **TM** and **TMC** in polyvinylphenol (PVP) provided very large EO responses (r_{33} as high as 330 pm/V at 1.31 μm; 10 wt%), 3–5 times greater than that ever-reported [59].

4.7 SIDE-CHAIN SYSTEMS

Different from that for guest–host systems, in the side-chain polymers, the NLO chromophores are covalently attached to the polymer backbone, rather than being simply dissolved into it. These systems have the advantage that high chromophore loadings (and therefore, high NLO responses) can be obtained, without phase separation, crystallization, or chromophore sublimation. In general, the glass transition temperatures of side-chain polymers are considerably higher than of a guest–host system with comparable chromophore loading (no plasticization effect occurs) [70]. Therefore, an improved thermal and temporal stability of the poled order is observed, because the chromophore rotational freedom is restricted by the chemical connection to the polymer.

Many different NLO chromophore-functionalized polymers have been investigated, including polymethacrylates, polystyrenes, poly(acrylamides), polyurethanes (PU), polyquinolines, polyesters, polyethers, and polyamides [4,70,71]. In the next sections, more attention will be paid to high-T_g polymers such as polyimides and polycarbonates.

4.7.1 Polyimides

NLO chromophore-functionalized polyimides have attracted a lot of interest thanks to their high T_g and excellent temporal stability. Verbiest et al. [72,73] and Miller et al. [74] reported several highly stable aromatic polyimides (Fig. 4.3), including the synthesis of the **PI-1** polymer, which is the first reported example of a processable "donor-embedded" side-chain polyimide having a very high T_g of 350°C, and a chemical stability at temperatures as high as 350°C. Poled samples of polymer **PI-1** have an EO coefficient of 4–7 pm/V (at 1.3 μm) and a much higher orientational

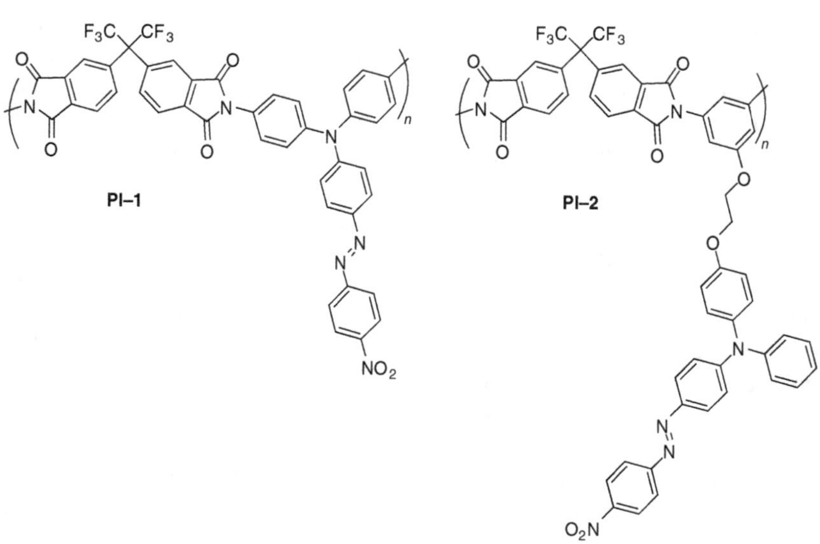

FIGURE 4.3 Processable polyimides **PI-1** and **PI-2**.

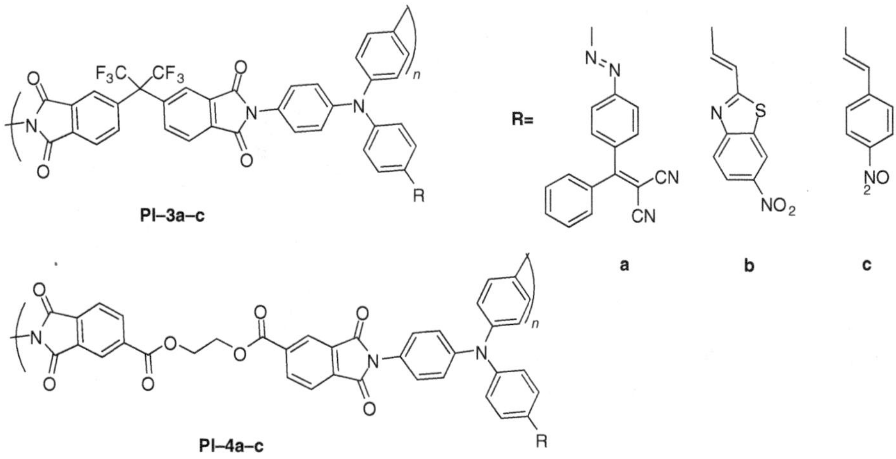

FIGURE 4.4 Polyimides **PI-3a–c** and **PI-4a–c.**

stability (up to 225°C) compared with a true side-chain polyimide (**PI-2**) in which a similar chromophore is covalently linked to the polymer backbone by a flexible spacer.

Davey et al. reported a general, convergent approach for the synthesis of protected diamine NLO chromophores, which allows both acid- and base-sensitive chromophores to be incorporated, using either alkaline or acidic deprotection, into polyimide backbones [75]. The obtained chromophores have been condensed into two types of high-T_g/high thermal stability polyimide chain structures (Fig. 4.4). The polymers based on the more rigid 6F-subunit (**PI-3a–c**) exhibit T_g values in excess of 300°C, however, due to the lower inherent mobility, they do not show high nonlinearities ($\chi^{(2)} = 16$–48 pm/V). In contrast, polyimides based on the more flexible TMEG-DA subunit (**PI-4a–c**), containing the same chromophore, have lower T_gs (225–265°C), but an increased NLO response ($\chi^{(2)} = 44$–82 pm/V), due to the greater structural mobility and poling efficiency.

However, the synthetic methods for aromatic side-chain polyimides include tedious procedures and the fact that not all chromophores can survive the harsh imidization conditions limits their application. To alleviate this problem, Chen et al. developed a two-step, generally applicable synthetic approach for the synthesis of NLO side-chain aromatic polyimides, which consists of a one-pot preparation of a preimidized, hydroxyl-containing polyimide, followed by covalent attachment of a chromophore onto the backbone of the polyimide via a post-Mitsunobu reaction [76]. Using this facile methodology, NLO side-chain polyimides with a wide variety of pendant NLO chromophores have been synthesized (**PI-5a–c**, Fig. 4.5) with fine control of the chromophore loading. The resulting NLO polyimides possess high T_gs (>200°C), large EO coefficients (up to 34 pm/V at 0.63 μm and 11 pm/V at 0.83 μm), and a long-term stability of the dipole alignment (>500 h at 100°C).

A new synthetic method for the effective attachment of a wide variety of chromophores, even highly active ones such as **FTC** and **CLD**, to polyimide

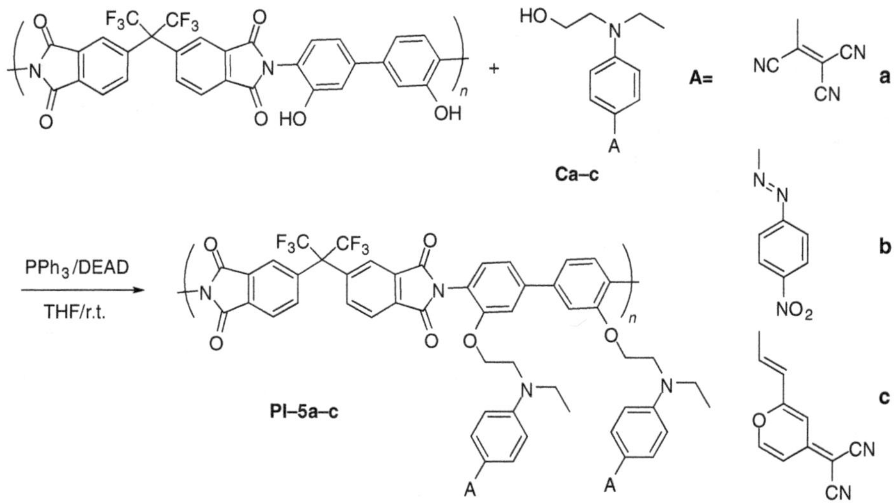

FIGURE 4.5 Synthesis and structures of polyimides **PI-5a–c.**

backbones has recently been developed by Wright et al. [77] and Guenthner et al. [78]. They have shown that the benzyl alcohol pendant group in the polymers can be chemically modified with a bis-functional linking agent, such as bis(isocyanates) or bis(acid chlorides), to afford reactive side-chain polyimides. The reactive isocyanate or acid chloride pendant group can then be easily linked to an alcohol-containing dye (Fig. 4.6). The resulting materials possess r_{33} values as high as those achieved in guest–host systems (e.g., 60 pm/V at 1300 nm, for an **FTC**-functionalized polyimide), whereas having substantially higher T_gs (>170°C), and an enhanced stability of the poling-induced order. Mach–Zehnder optical interferometers have been fabricated with polymers that contained CLD- and FTC-type chromophores. Their long-term aging performance (for months at four temperatures ranging from ambient to 110°C) has been determined from the increase of the V_π value of the modulator [79]. Multiyear high-temperature stability was predicted by fitting the data to a newly developed aging model.

4.7.2 Polycarbonates

Although polycarbonates have been widely employed as a host material for NLO chromophores (APC in particular), thanks to good optical properties, high T_g, and good processability, there are only few reports about the incorporation of chromophores as a side-chain in a polycarbonate backbone.

To cope with the growing demand for highly stable materials, we synthesized and investigated the EO properties of a series of side-chain polycarbonates, in which the highly photostable **TCVDPA** chromophore is either incorporated as part of the polymer backbone or connected through a flexible linker (Fig. 4.7) [80]. The materials were synthesized by postreaction of a polymer containing the triphenylamine donor. All obtained materials possess good solubility, thermal stability as high as 370°C,

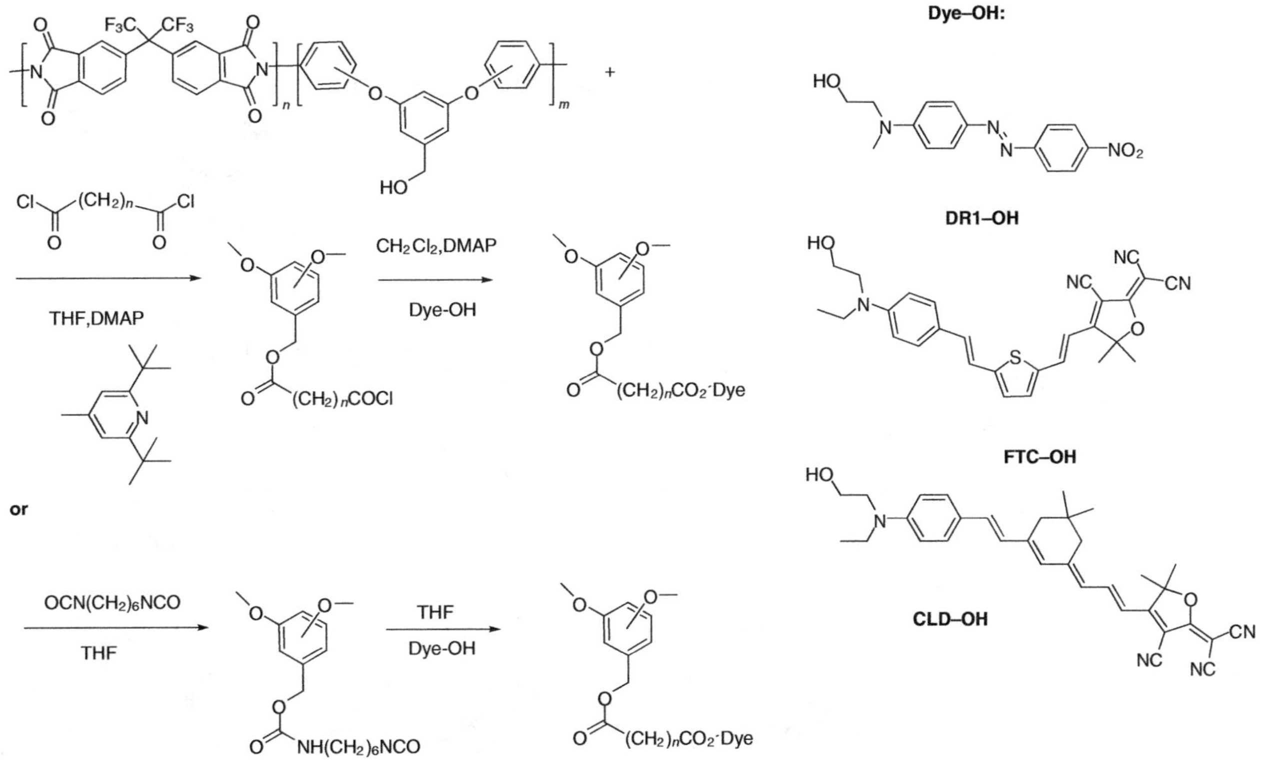

FIGURE 4.6 Synthesis of high-$\mu\beta$ chromophore-functionalized polyimides.

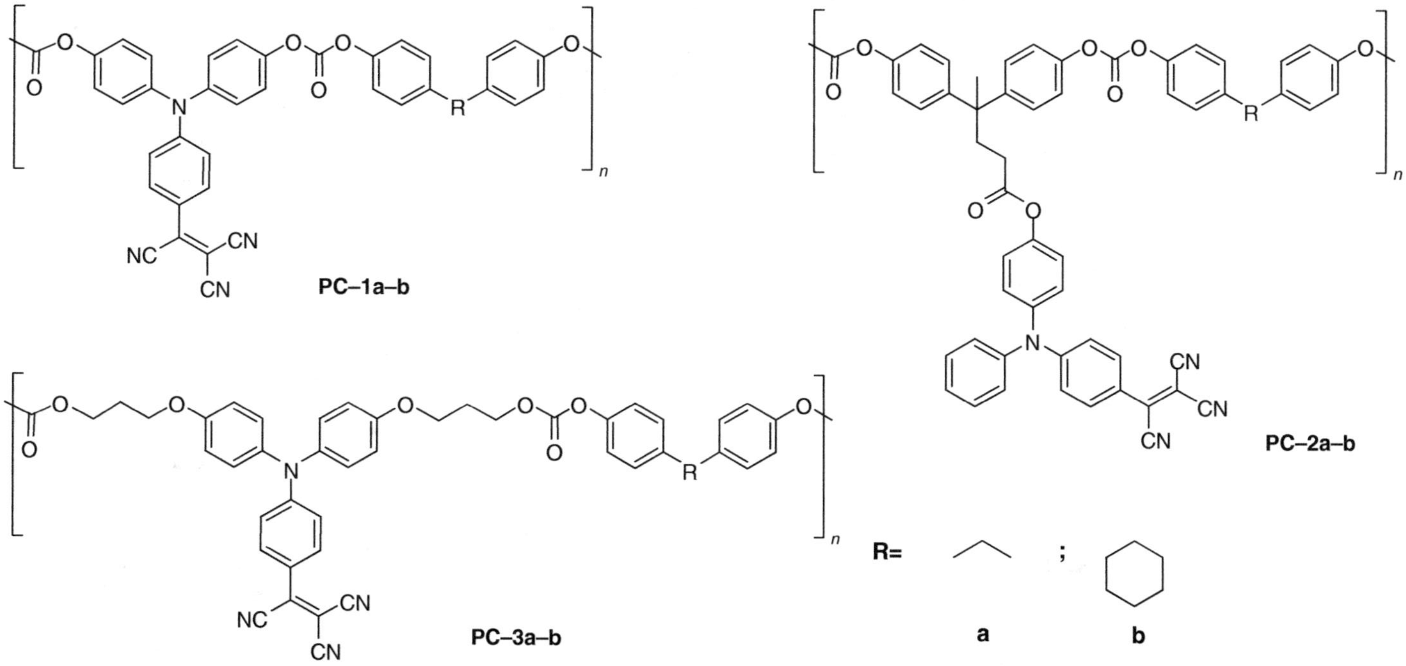

FIGURE 4.7 **TCVDPA**-functionalized polycarbonates.

significantly higher T_g, and stability of the EO activity, compared with guest–host systems containing the same chromophore. The polymer **PC-1b** in which the electron-donor part of the chromophore is donor-embedded showed the highest T_g (215°C), but its EO activity was only 15 pm/V due to the inherent structural rigidity that limits the rotational freedom of the chromophores. However, the alkoxy spacers connecting the chromophore to the PC structure in polymer **PC-2a** provide the flexibility required for efficient poling, leading to an r_{33} value of 33 pm/V, in spite of the lower T_g (170°C).

A more versatile approach for the synthesis of NLO side-chain polycarbonates involves the preparation of a perfluorinated PC polymer [81]. It contains free pendant carboxylic acid groups, to which any hydroxyl-functionalized molecule can be easily and efficiently incorporated by coupling (Fig. 4.8). In addition, keeping the chromophore out of the system until the last reaction step allows the greatest flexibility, both in forming the polymer backbone and in controlling the amount of chromophore incorporated. The T_g of these polymers varied from 180 to 220°C, depending on the structure and the degree of functionalization. Good thermal stabilities and EO activities as high as 40 pm/V were also obtained.

4.8 MAIN-CHAIN SYSTEMS

Another approach to attenuate the poled-order relaxation comprises the use of main-chain polymer systems, in which the chromophores are chemically incorporated in the polymer backbone itself, rather than being attached as pendant groups. The main difference between the main-chain and the side-chain approach is that large segmental motion of the polymer backbone is needed for poling and relaxation [82]. Main-chain NLO polymers can be divided into three categories (Fig. 4.9): (1) head-to-tail [83]; (2) random [84], where the chromophore dipole moments are pointing along the polymer backbone; and (3) accordion polymers [85], where the dipole moments are nearly perpendicular to the main chain.

With the purpose of improving processability, thermal stability, and alignment stability, a wide variety of main-chain chromophoric polymers have been investigated, including polyurethanes [86], polyimides [87], polyamides [88], carbazoles [89], and polyesters [90]. However, to date, most main-chain polymers show relatively poor processabilities (including solubility and poling efficiency) and/or low NLO responses. In addition, the choice of the chromophores suitable for main-chain incorporation is limited and high loadings are difficult to achieve. For these reasons, the current researches in the NLO polymer field are mainly focused on the side-chain and cross-linked types. Therefore, main-chain systems will not be discussed in detail in this chapter.

4.9 CROSS-LINKED SYSTEMS

Covalently attaching chromophores to the polymer backbone or incorporating them into the backbone, as described in Sections 4.7 and 4.8, can effectively increase the

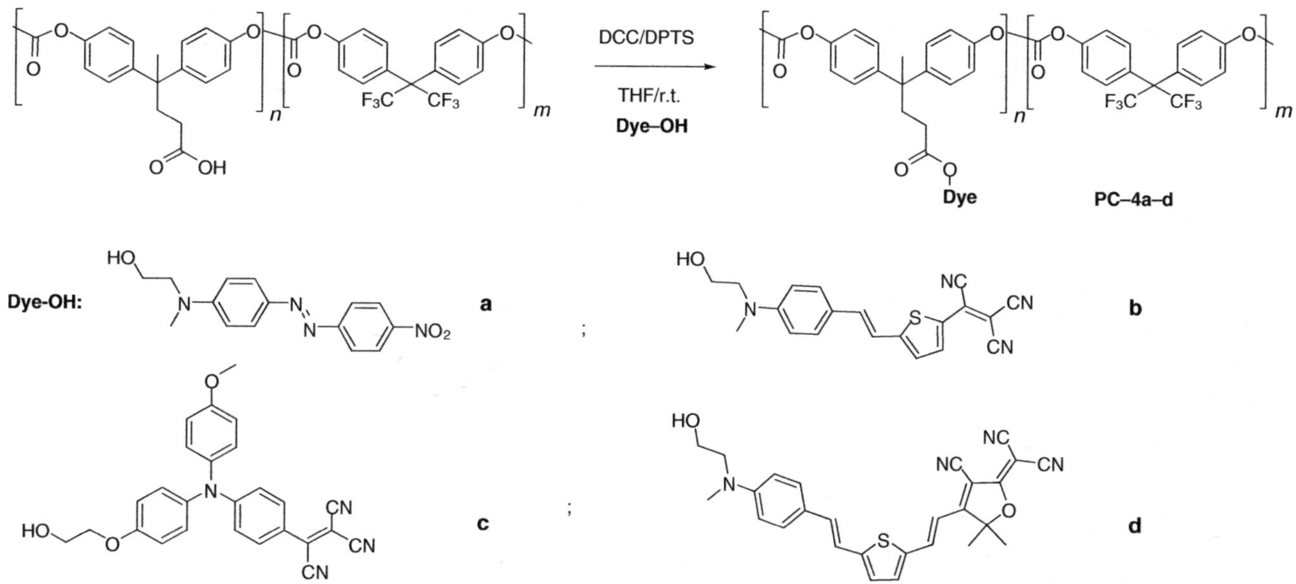

FIGURE 4.8 Synthesis of chromophore-functionalized polycarbonates.

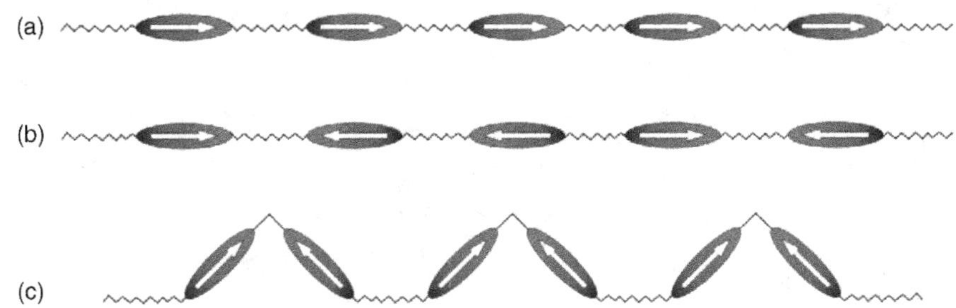

FIGURE 4.9 Different types of main-chain NLO polymers: (a) head-to-tail, (b) random, and (c) accordion.

chromophore loading, and prevent phase separation, and thermal relaxation of the chromophore dipole moment. However, such high-T_g materials require high temperature poling, where chromophore decomposition may occur. Moreover, this process lacks flexibility, because the screening of the potential host polymers and varying the polymer/chromophore compositions require tedious batch-to-batch production, which makes it difficult to control the composition and the properties of the final material precisely.

The use of low-T_g cross-linkable materials offers the advantage that the poling process can be conducted at relatively lower temperatures during the hardening process, ending up with a high-T_g non-centrosymmetric material. In general, the cross-linking process must not degrade the optical quality of the films, such as causing defects and poor uniformity, which could increase the optical loss. The general approaches to lattice hardening are photo-induced cross-linking and thermally induced cross-linking [91]. Photo-induced cross-linking has the advantage that the lattice hardening process can be completely separated from the temperature-dependent poling process. However, the UV or visible light applied to activate the photoinitiator could be preferably absorbed by the NLO chromophore, making the exposure ineffective and promoting chromophore degradation. This drawback has so far limited this development, hence thermally induced cross-linking has been applied more widely [9].

As both lattice hardening and poling are temperature-dependent processes, fine-tuning of the conditions should be done to simultaneously achieve a high poling efficiency and a high T_g. Increasing the temperature permits a higher chromophore mobility to reorient along the poling field. It also antagonistically drives the hardening process, hence, reducing the mobility. On the other hand, the application of a too high electric field to a soft film can result in material breakdown. This often leads to a trade-off between poling efficiency and material stability. Thus, optimum conditions can be achieved using stepped poling protocols (where temperature and electric field are increased in a series of steps) [92].

Different thermally induced cross-linking approaches have been investigated, such as the formation of sol–gel networks [93,94] or reactions resulting in polyimides [95] and maleimides [96]. Thermoset PU have been widely studied for EO applications [92,97]. Zhang et al. synthesized a high $\mu\beta$-derived isophorone-derived

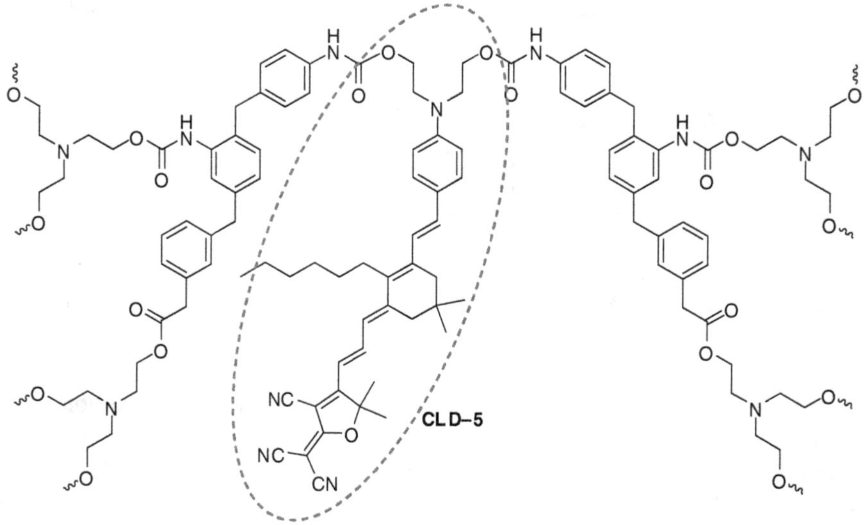

FIGURE 4.10 Example of a **CLD-5** containing polyurethane.

phenyltetraene chromophore (**CLD-5**) modified with a hexyl group at the middle of the π-conjugate bridge to improve the solubility and two hydroxyl terminal groups for covalent incorporation into cross-linked PU polymer systems [98] (Fig. 4.10). The chromophore was incorporated into a PU matrix based on triethanolamine (TEA) and toluene-2,4-diisocyanate (TDI) and, after curing and poling, an EO coefficient of 57.6 pm/V at 1.06 μm and a dynamic stability till about 80°C was obtained for the resulting material. By using more rigid monomeric cross-linkers, the thermal stability could be enhanced up to 133°C. This gain, however, does not come without a sacrifice in the EO activity. In addition, from a study on the cross-linking density, it is clear that excessive cross-linking is harmful to electrical poling of a polyurethane material and that cross-linking by itself is not enough to provide a very high thermal stability of electrical field-induced chromophore alignment.

Sol–gel and PU oligomerization reactions require a strict control of the reaction conditions, because atmospheric moisture can negatively influence the reaction, causing phase separation and optical loss. To tackle this problem, a new cross-linking unit, trifluorovinyl ether (TFVE), has been introduced. TFVE-containing monomeric units can be converted into perfluorocyclobutane (PFCB) containing polymers by a radical-mediated thermal cyclopolymerization reaction. These polymers have excellent properties such as a low dielectric constant, good thermal stability, and optical transparency [99].

A new synthetic strategy for incorporating a wide variety of NLO chromophores into PFCB polymers has been developed by Ma et al. [100]. The chromophore loading can be tuned by varying the ratio of chromophore-substituted di-TFVE monomer and the tri-TFVE inert monomer. The obtained mixture is then prepolymerized at 150°C, spin-coated to obtain high-quality films, and efficiently cross-linked at 180–250°C. All resulting NLO PFCB (Fig. 4.11) thermosets possess excellent solvent resistance,

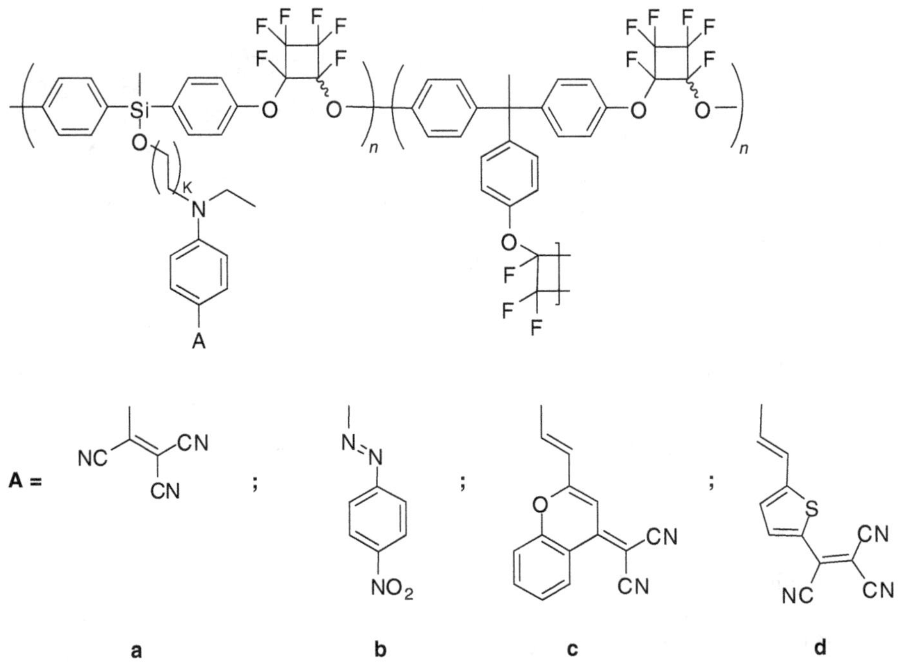

FIGURE 4.11 Structures of the PFCB polymers.

large r_{33} values, and can retain ~80% of their original values at 85°C for more than 1000 h.

PFCB chemistry, thanks to its versatility, has been applied to different kinds of material architectures, including side-chain polymers, NLO dendrimers, and dendronized polymers, as will be described in the next section.

More recently, another lattice-hardening approach has been demonstrated using the thermally reversible Diels–Alder [4 + 2] cycloaddition reaction (Fig. 4.12) to provide significant advantages over the conventional NLO thermosets, such as high poling efficiency and fine-tuning of the processing temperatures [101,102]. This procedure has mainly been applied to obtain NLO dendrimers and dendritic polymers, which are described in the next sections.

4.10 DENDRITIC SYSTEMS

In the past decade, much effort has been made to develop polymeric materials possessing simultaneously large EO coefficients, high thermal and photostabilities, and low optical losses, which could be suitable for incorporation into practical EO devices. One major obstacle that limits the development in this area is to efficiently translate high nonlinearities into large macroscopic EO activities. In fact, although the $\mu\beta$ values of chromophores have been improved more than 250-fold, only several

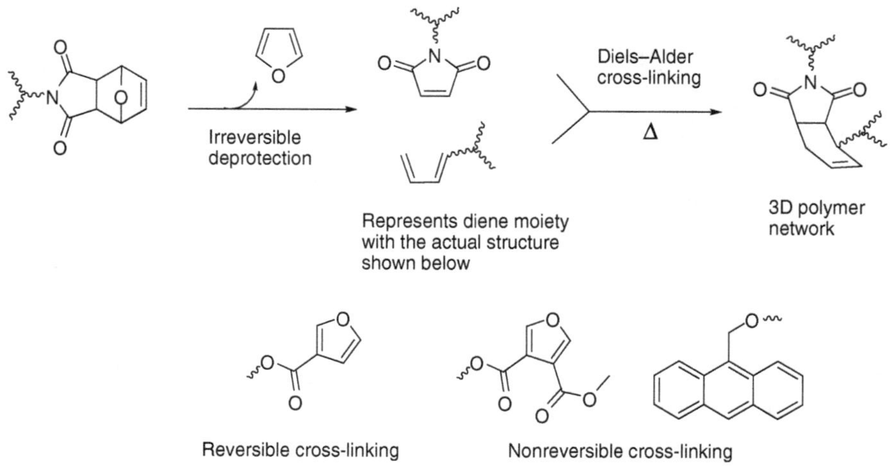

FIGURE 4.12 Representation of the Diels–Alder cross-linking reactions.

times of enhancement of the r_{33} value could be achieved. From the ideal-gas model, the EO coefficient should increase linearly with the chromophore number density (loading) in the polymer matrix, making an r_{33} of several hundreds of pm/V theoretically obtainable [103]. However, molecules with large dipole moments cannot be treated as noninteracting, and strong intermolecular dipole–dipole interactions, especially at high chromophore loading levels, become competitive with the poling induced non-centrosymmetric alignment [104–106]. Recently, theoretical and experimental results by Robinson and coworkers demonstrated that a logical approach to improve the maximum achievable EO activity is to modify the shape of the chromophores by introducing bulky substituents [107]. Derivatization of chromophores with these inert groups makes them more spherical-shaped, limiting intermolecular electrostatic interactions, and hence, antiparallel clustering, therefore, enabling higher poling efficiency. Disappointingly, only a slight increase of the r_{33} could be achieved using this method.

Chromophore-containing dendritic structures have emerged as an alternative solution to achieve spherical shape modification of chromophores [108]. In spite of any conventional EO polymer, the void-containing structure of dendrimers provides the site isolation needed for chromophores to independently reorient under the external poling field [109]. Moreover, these dendritic materials possess a monodisperse and well-defined globular geometry. Their structure is synthetically controllable in size and shape, allowing wide control over solubility, processability, viscosity, and stability.

4.10.1 3D-Shaped Dendritic NLO Chromophores

One of the very first examples of the spontaneous, non-centrosymmetric organization of NLO chromophores in dendritic structures has been reported by Yokoyama et al. [110]. They studied the conformational and NLO properties of a series of azobenzene-containing dendrons, synthesized by introducing 1–15 azobenzene branching units and by placing aliphatic functionalities at the end of the dendritic chains (Fig. 4.13). HRS measurements showed that the synthesized dendrons had a cone-shaped conformation, with each chromophore contributing coherently to the macroscopic EO activity with no need of application of an external field. In fact, the β measured for the azobenzene dendron **D1** with 15 chromophoric units was 3010×10^{-30} esu, a value which is more than 20 times larger than that of the individual azobenzene monomer (150×10^{-30} esu).

To explore the dendritic effect on the poling efficiency of dipolar NLO chromophores, Ma and coworkers [111] modified the highly stable **Ph-TCBD** chromophore with three highly fluorinated aromatic dendrons (Fig. 4.14). In comparison with the pristine analog, the resulting dendritic chromophore **D2** exhibits a 20°C higher decomposition temperature and a large blue shift (29–42 nm) of the charge-transfer absorption maximum (λ_{max}), indicating the influence of the fluoro-rich dendrons on the microenvironment of the core chromophore in solid films. When **D2** and **Ph-TCBD** were incorporated with the same amount of active component into APC (12 wt %), the poled films of **D2** showed a three times larger EO coefficient (30 pm/V at 1.3 μm) than the pristine chromophore, providing a clear evidence of the improved poling efficiency due to the dendrons.

More recently, Dalton and coworkers investigated the EO properties of trifunctional dendritic chromophores and the effectiveness of the theoretical analysis as a guide in the bottom-up design of these molecular architectures [112]. The analyzed structures consisted of three thiophene-containing **FTC**-type chromophores, connected to the tribranched inert core through flexible spacers. Four triarm EO dendrimers were prepared and evaluated to explore the effects on the EO behavior of end-on relative to the side-on chromophore attachment geometry as well as varying chromophore-to-dendrimer core tether groups (Fig. 4.15).

These dendritic materials were dispersed into APC and tested as thin film composites. The side-on geometry provided a more stable EO signal, requiring larger activation energies to induce dipole randomization than the end-on type. The differences in average r_{33} between side-on and end-on geometries were small, but consistent. The EO behavior depended heavily on the length and rigidity of the moieties used to covalently anchor the chromophore to the inert host or core. A nearly threefold enhancement in EO coefficient was noted when the short diester tether was replaced by a longer, more aliphatic system, however, a faster thermal decay of r_{33} was observed.

These experimental findings, together with quantum mechanical modeling, were then employed to guide the design and synthesis of triarm dendritic structures with extended outer peripheral functionalities [113] (Fig. 4.16). The resulting materials were used to fabricate stand-alone thin films, without the addition of an inert polymer host. These all-dendrimer films exhibit a high poling efficiency (r_{33}/E_p) and a stunning linear relationship between r_{33} and N. The linear dependence holds even at very high chromophore concentrations ($N = 6.45 \times 10^{20}$ chromophores/cm^3), yielding a maximum EO coefficient of 140 pm/V (at 1.31 μm).

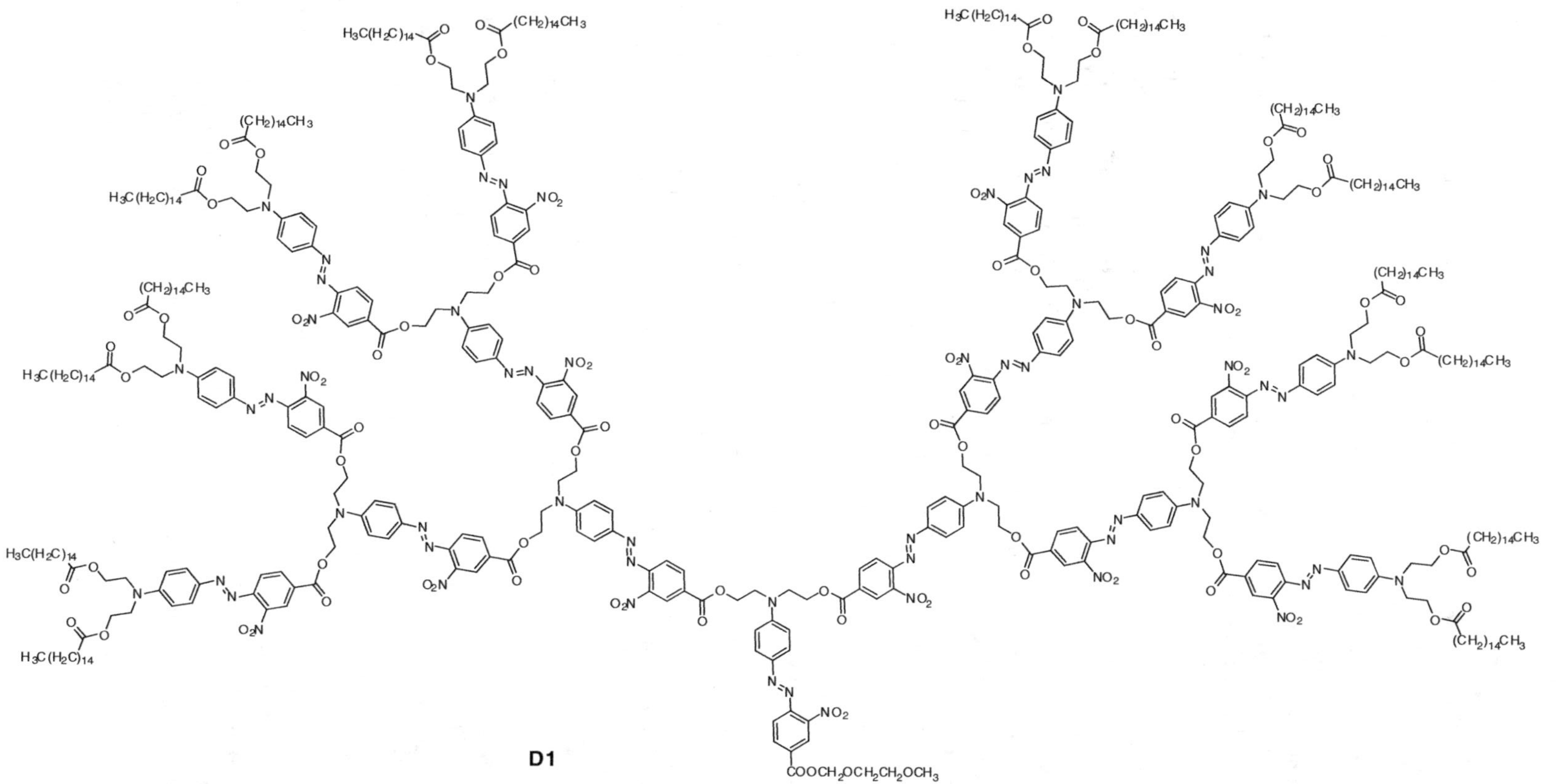

D1

FIGURE 4.13 Structure of the azobenzene dendron **D1.**

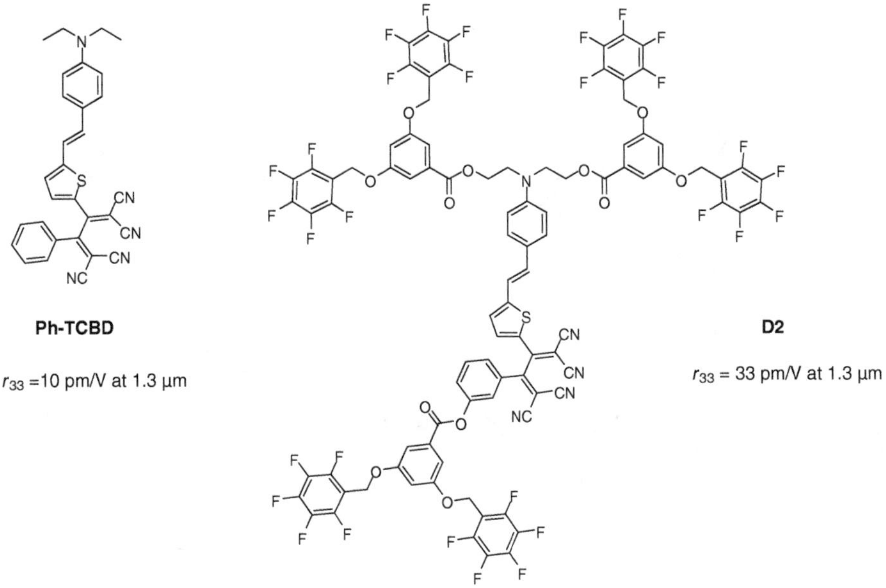

FIGURE 4.14 Comparison between dendritic chromophore **D2** and its pristine NLO chromophore **Ph-TCBD.**

4.10.2 Cross-Linkable NLO Dendrimers

Low-molecular weight multichromophore containing dendrimers (described in the previous section) have very large r_{33} values, which is very promising for the next generation EO materials. However, the materials obtained with this approach have intrinsically low T_gs, which translates into a poor thermal stability of the poling order, and a low solvent resistance (high solubility in spin-casting solvents), which limits their incorporation into multilayer polymer optical devices.

To alleviate this problem, Ma et al. developed an NLO dendrimer, having the center core connected to a **Ph-TCBD**-containing chromophore and thermally cross-linkable TFVE-containing dendrons at the periphery (Fig. 4.17) [114]. The resulting dendrimer **D5**, thanks to its relatively high molecular weight (4664 Da), can be directly spin-coated, with no need of a prepolymerization process. Moreover, this approach offers the advantage that the sequential hardening/cross-linking process can be efficiently conducted during electric field poling, making it possible that large EO coefficients ($r_{33} = 60$ pm/V at 1.55 μm) and long-term alignment stability (over 90% at 85°C for more than 1000 h) can be simultaneously obtained.

Recently, Sullivan et al. introduced a novel, thermally curable, tricomponent dendrimer glass, which takes advantage of the Diels–Alder cycloaddition reaction to achieve efficient cross-linking during poling [115]. This tricomponent material consists of a multichromophore dendrimer functionalized with a diene-containing outer periphery (**D6**), a guest chromophore that is bis-functionalized with a

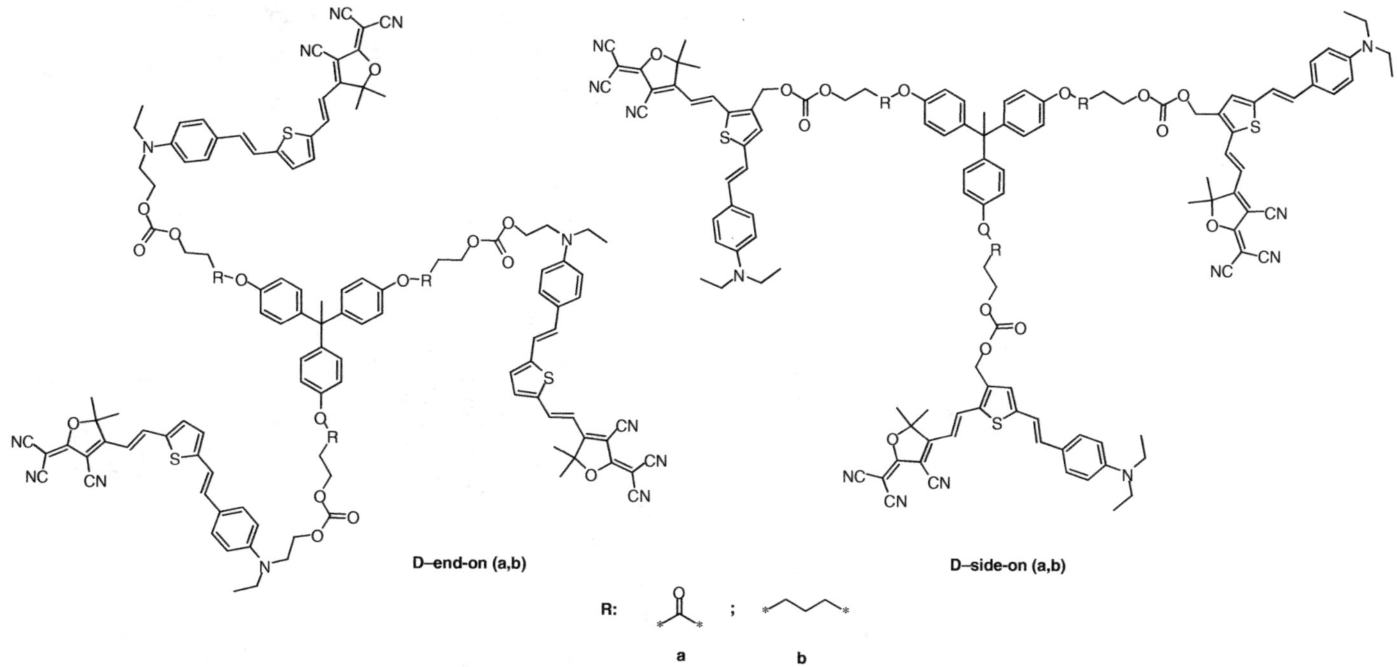

FIGURE 4.15 Structures of the FTC-containing dendrimers.

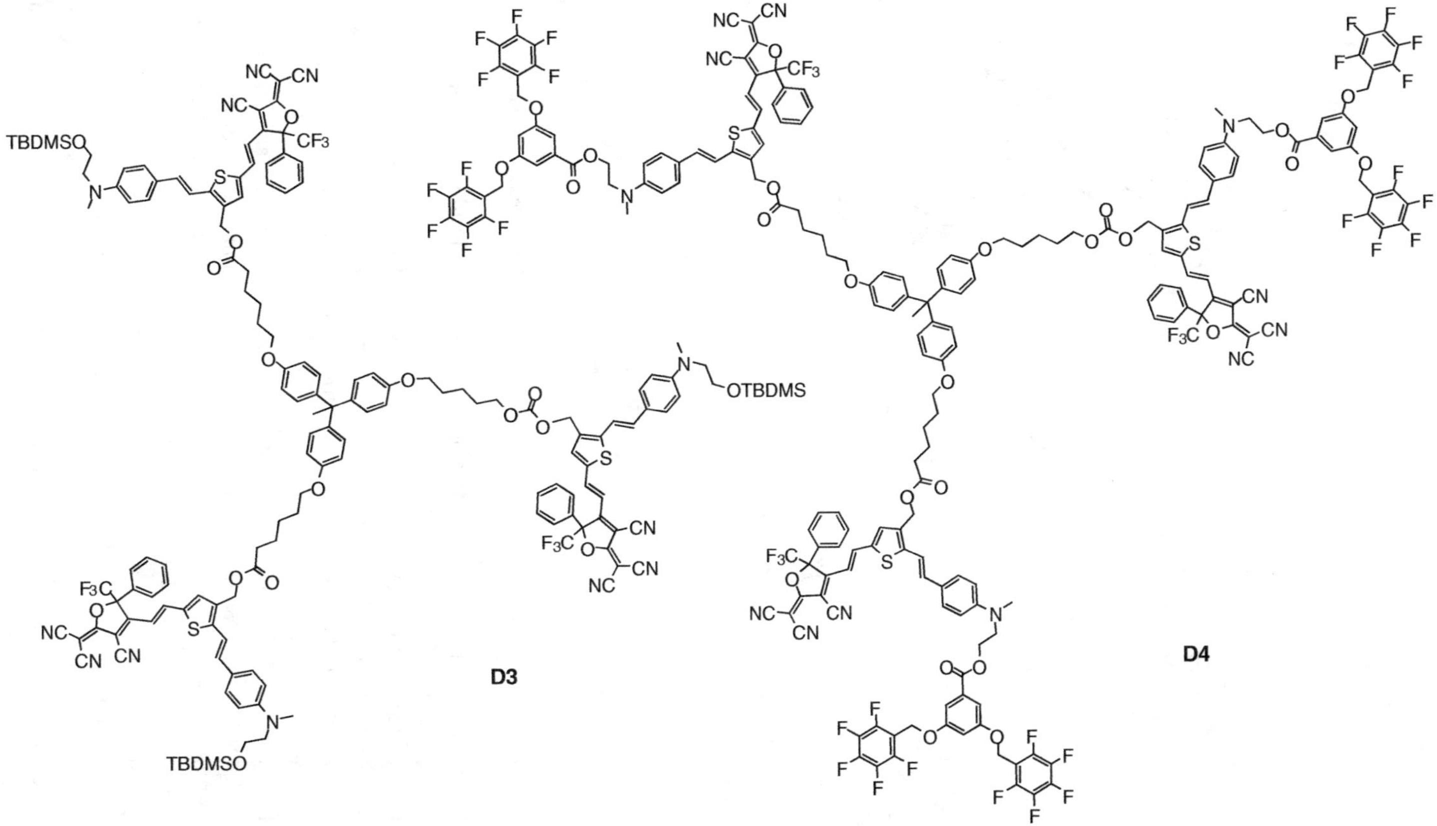

D3

D4

FIGURE 4.16 Structures of triarm EO dendrimers **D3** and **D4.**

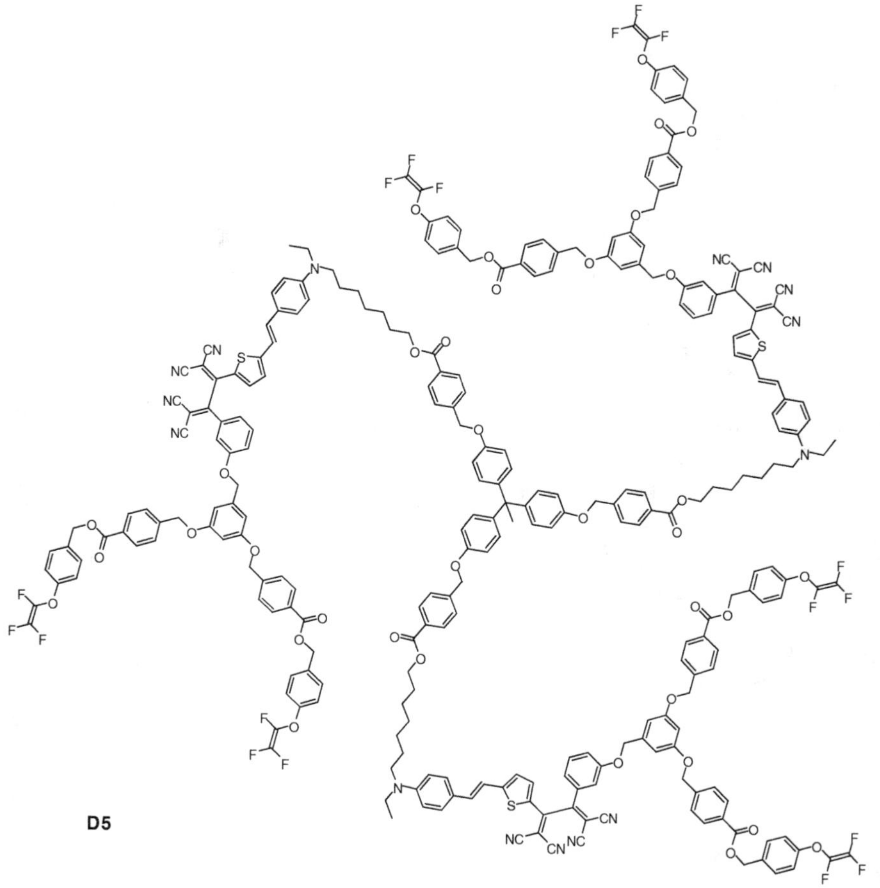

D5

FIGURE 4.17 Structure of the triarm cross-linkable NLO dendrimer **D5.**

furan-protected dienophile (**GC**), and an optically inert maleimide-based dienophile cross-linking agent (**MD**) (Fig. 4.18).

The optimized tricomponent mixture has an r_{33} of 150 pm/V (at 1.31 μm) and a thermal stability of up to 130°C (a 48°C improvement over similar uncross-linked materials). The materials were insoluble in acetone and retained 90% of their original r_{33} value after 15 months at room temperature.

Using a similar Diels–Alder cycloaddition-based approach, Jen and coworkers developed a series of cross-linkable dendrimers by functionalization of the **AJL8**-type chromophore with diene-containing dendrons, which could be cross-linked at a later stage by using a trismaleimide dienophile (Fig. 4.19) [116].

The high EO activity and solvent resistance allowed these materials to be processed through multiple lithographic and etching steps to fabricate a race-track-shaped microring resonator. By coupling this resonator with a side-polished optical fiber, a broadband electric-field sensor with a high sensitivity of 100 mV/m at 55 MHz has been demonstrated [116].

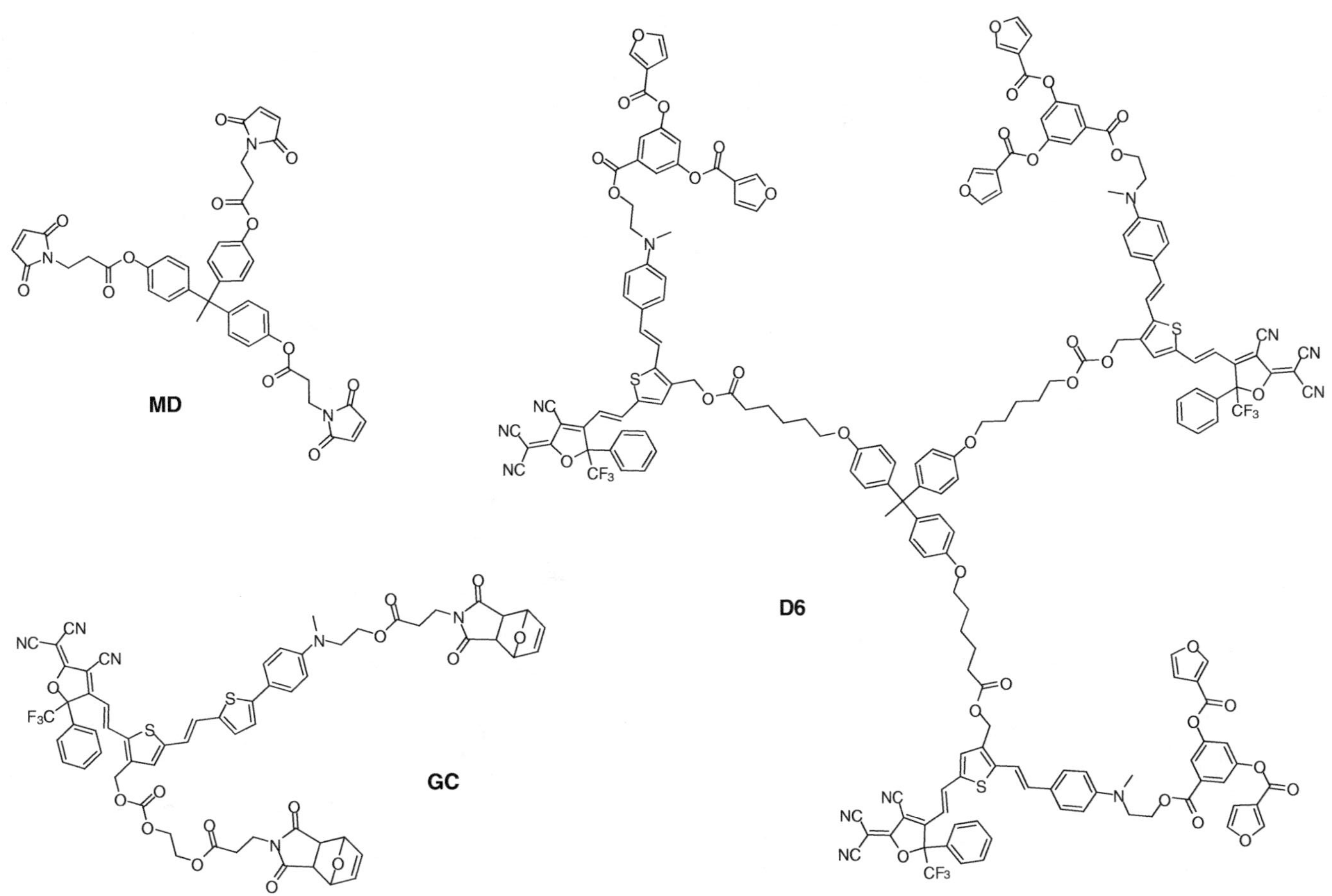

FIGURE 4.18 Structures of the cross-linkable dendrimer **D6**, the chromophore **GC**, and optically inert precross-linker **MD**.

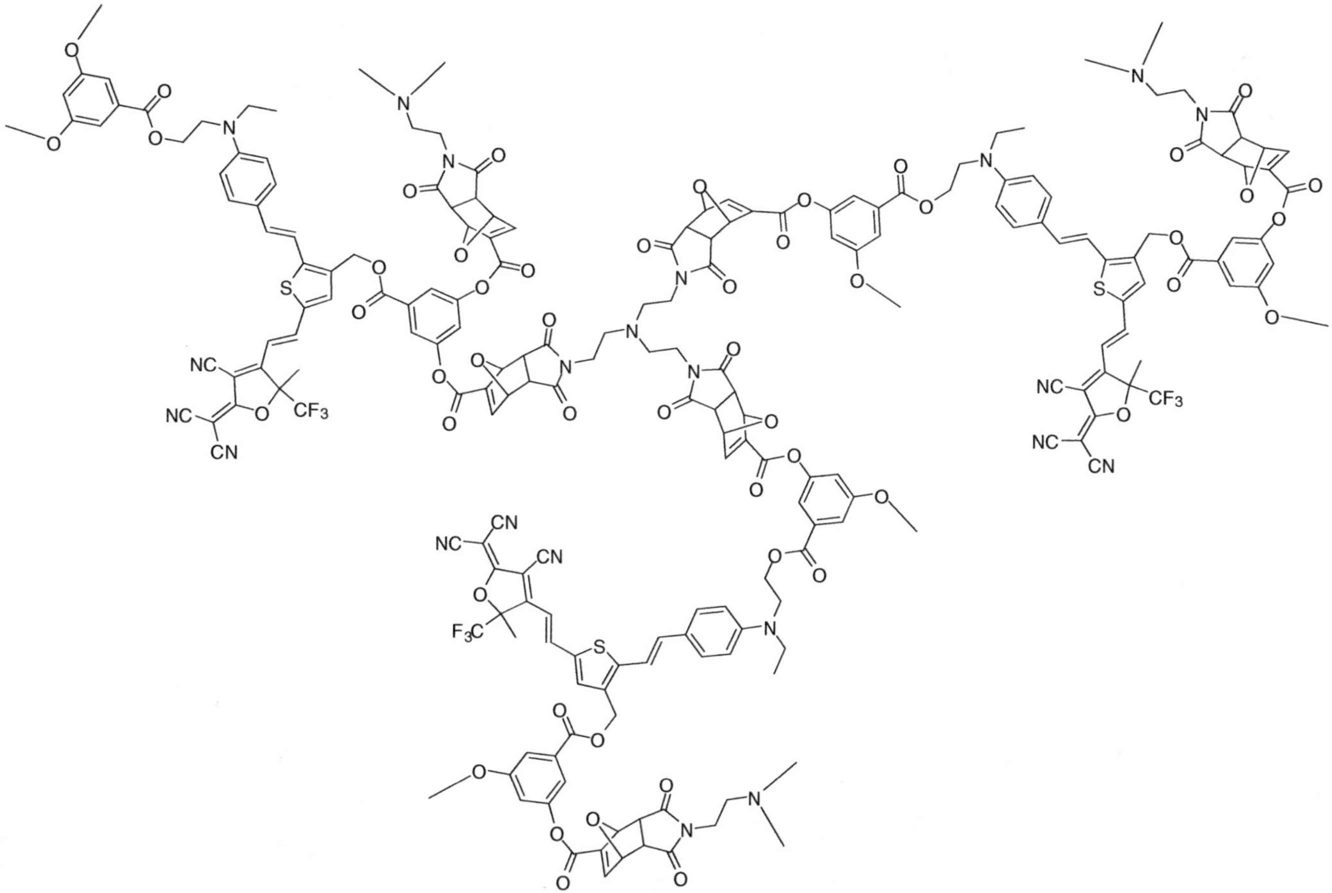

FIGURE 4.19 Example of an **AJL-8**-containing cross-linked EO dendrimer.

4.10.3 Side-Chain Dendronized NLO Polymers

Although stand-alone EO dendritic glasses with high, ever-improving poling efficiencies (EO activities) can be obtained, a drawback is the long and tedious synthesis needed to produce sufficiently high molecular weight dendrimers with good film forming properties.

Another approach is to produce dendron-substituted polymers (or dendronized polymers) by combining the site-isolation effect of dendrimers with the good processability of linear polymers. This strategy provides a greater flexibility in designing suitable molecular structures for realizing high-performance EO materials.

Making use of this principle, Jen et al. synthesized the first side-chain dendronized polymer by attaching TFVE-containing dendritic moieties to a chromophore-functionalized polystyrene-based backbone **DP-1a** (Fig. 4.20) [117]. The poled films of this polymer showed an r_{33} of 81 pm/V (at 1.31 μm), a value that is about 2.5 times larger than that of the corresponding pristine side-chain NLO polymer. A similar modification has also been applied to high $\mu\beta$ **CLD** type of chromophores, obtaining dendronized polymers **DP-1b** and **DP-2** with r_{33} values of 97 and 111 pm/V (at 1.31 μm), respectively [118]. Polymer **DP-2** assembles in a pseudo-cylindrical rigid-rod conformation, which may explain the high poling efficiency. In fact, in this rigid polymer, the chain entanglement may be reduced, allowing the chromophores a higher freedom to reorient in the channels of the cylindrical structure.

To improve the thermal stability, a high-T_g cardo-type polyimide with a dendronized CLD-type chromophore has been developed [119]. A high poling efficiency was achieved to afford a very large EO coefficient (71 pm/V at 1.3 μm); more than 90% of this value can be retained at 85°C for more than 650 h.

The synthetic/processing scheme based on the Diels–Alder cycloaddition reaction described in the previous section has also been employed to generate chromophore-functionalized side-chain EO polymers *in situ* during the poling process [101] (Fig. 4.21). The advantages of using this approach are the very mild heating conditions, the specificity, and the absence of ionic species and catalysts. A series of highly efficient EO materials has been produced by covalent attachment of maleimide-containing NLO chromophores onto PMMA-based polymers substituted with pendant anthracene diene groups (PMMA-AMA) [120]. Different macromolecular architectures were created by changing the attaching mode of the chromophore onto the polymer and onto the fluorinated dendritic units. The obtained polymers showed a good optical quality and processability, fairly high T_gs (~150°C), and large r_{33} values (as high as 60 pm/V at 1.31 μm). Addition of a bismaleimide (BMI) cross-linker to the polymer blend resulted in an increase of the temporal stability and the solvent resistance, without a significant decrease of the poling efficiency.

4.11 SELF-ASSEMBLED SYSTEMS

The main requirement for molecule-based second-order NLO materials is a non-centrosymmetric organization of the constituent active species. Furthermore, organic

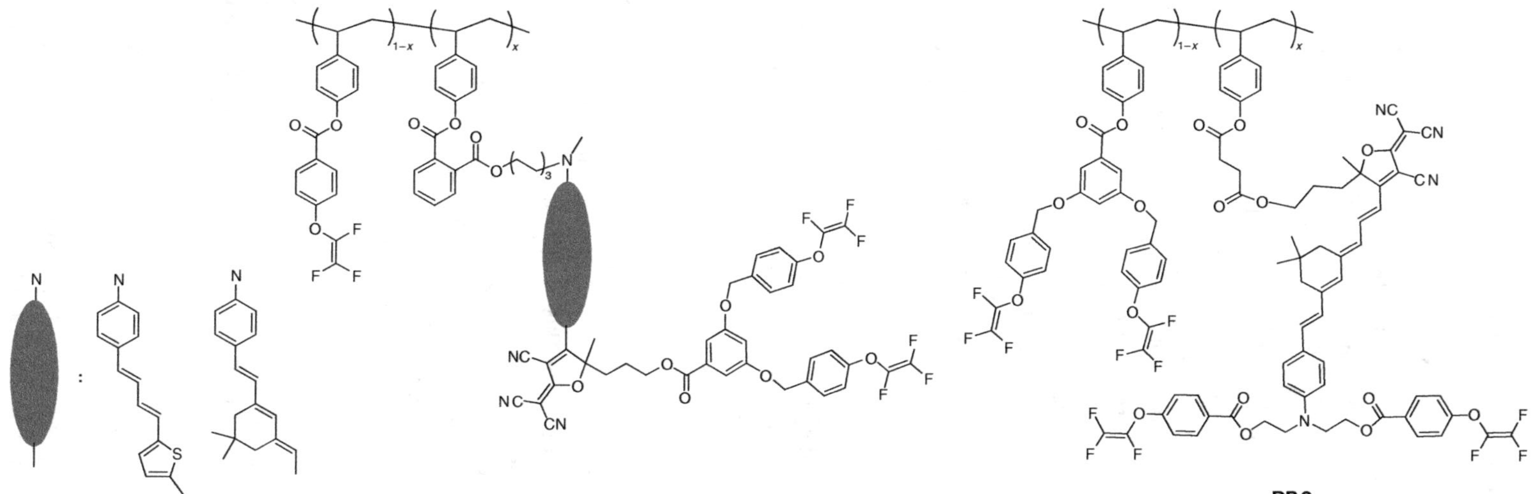

a b

DP 1-a r_{33} = 81 pm/V at 1.3 μm

DP 1-b r_{33} = 97 pm/V at 1.3 μm

DP 2

r_{33} = 111 pm/V at 1.3 μm

Cylindrically shaped polymer

FIGURE 4.20 Side-chain dendronized polymers with thermoset TFVE groups.

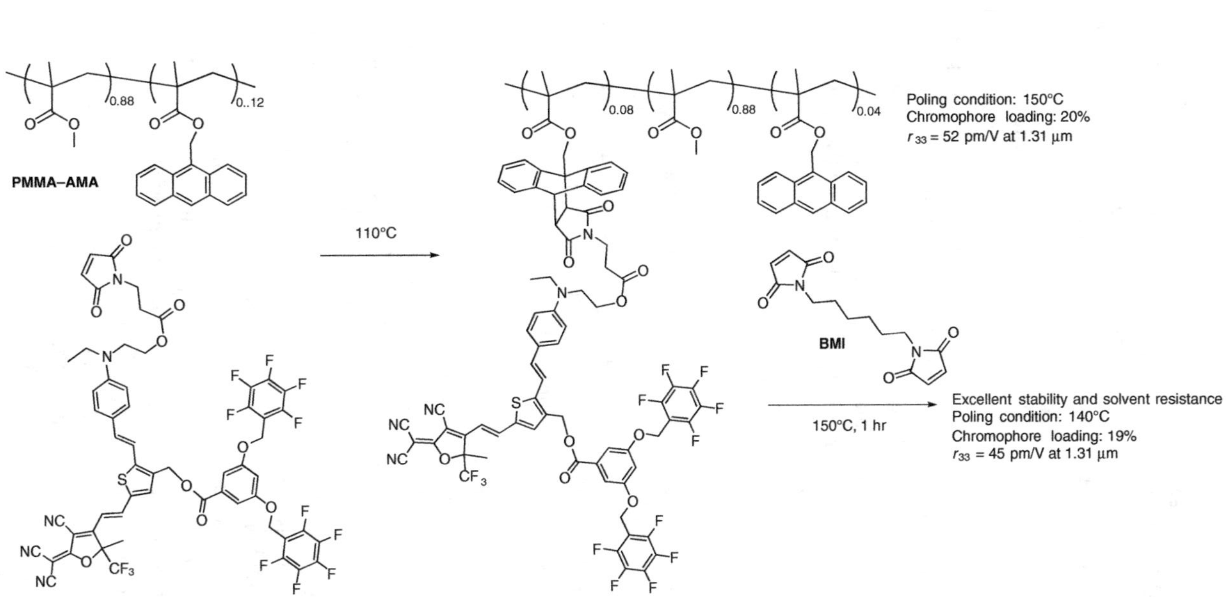

FIGURE 4.21 Example of the synthesis of a side-chain polymer using Diels–Alder postfunctionalization.

materials should be effectively fabricated in device-applicable (waveguiding) films of micrometer-scale thickness, good optical quality, and homogeneity over areas as large as centimeter squares.

Electric field poling of polymers, as described in the previous sections, is the most commonly used methodology to achieve polar order of chromophores within the selected matrixes. Although this method makes use of strong external forces, it does not take full advantage of the large nonlinear optical properties (β) of organic chromophores. Therefore, the preparation of bulk materials in which the dipoles are well aligned is still a hard challenge.

A totally different strategy to achieve acentric film architectures is by self-assembly. Self-assembled chromophore multilayer structures tend to have an intrinsic molecular dipolar alignment. It does not require electric field poling to achieve a highly acentric film, thus eliminating a possible cause of surface damage and defects. Based on this principle many design strategies have been explored. The most prominent examples are liquid crystals [15], chirality [121,122], Langmuir–Blodgett film growth, head-to-tail hydrogen bonding, covalent layer-by-layer chemisorption from solution, and vapor deposition.

4.11.1 Langmuir–Blodgett Films

The LB technique takes advantage of amphiphilic molecules, having a hydrophilic head and a hydrophobic tail, to achieve alignment at the air/water interface. This technique allows films to be deposited with great control at the molecular level, obtaining well-ordered structures with high chromophore number densities and homogeneous thicknesses [11]. As during deposition, the majority of amphiphilic molecules tend to adopt a head-to-head or a tail-to-tail arrangement (Y-type), which is centrosymmetric, it is necessary to manipulate the molecular structure to impose non-centrosymmetric arrangements (X-type or Z-type) (Fig. 4.22). This issue has been overcome in several ways: alternating optically active layers with inactive spacers, using complementary dyes with chromophores hydrophobically substituted at opposite ends, the tail attached to the donor (C_nH_{2n+1}–D–π-A) and acceptor (D–π-A–C_nH_{2n+1}) in adjacent layers, or chromophores with two hydrophobic end groups [123]. Some molecules show a particular non-centrosymmetric Y-type structure called herringbone arrangement, in which the dipoles are arranged in a plane parallel to the substrate and the alignment is retained in all subsequent layers [124]. The second harmonic intensity increases quadratically with the number of bilayers. Despite the success of the LB technique in providing waveguiding NLO-active thin films consisting of more than 100 monolayers, several important drawbacks are related to the fragility of the films, such as the low temporal stability of the dipole order (even at room temperature), the scattering from microdomains, and the tedious thick films growth. Recently, the use of alternative approaches for the reduction of molecular mobility, such as polyelectrolyte complexation [125,126], hydrogen-bonding [127], and photopolymerization [128] has led to LB films with improved thermal stabilities.

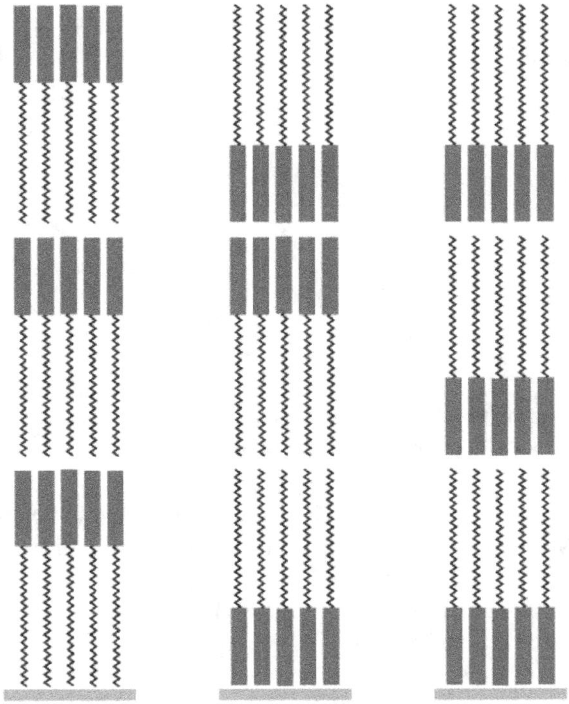

FIGURE 4.22 Schematic representation of centrosymmetric (Y-type) and non-centrosymmetric (X-type and Z-type) LB structures.

4.11.2 Covalent Layer-by-Layer Assemblies

Another approach to make thin-film NLO materials is based on the sequential construction of covalently self-assembled chromophore-containing multilayer structures as first introduced by Marks and coworkers [129,130]. The general strategy for superlattice construction employs three steps (Fig. 4.23a). Chemisorption of alkyl or benzyl halide-containing trichlorosilane coupling agents (step i) onto flat, hydroxy-terminated surfaces (e.g., glass, silicon, and organic polymers) provides functionalized surfaces for the polar anchoring of a bifunctional chromophore precursor. The quaternization/anchoring process (step ii) converts the NLO-inactive precursor into the NLO-active chromophore. It also creates surface hydroxyalkyl functionalities that are subsequently used to "lock in" the polar structure with a capping agent via the three-dimensional siloxane network formation. This "capping" reaction (step iii) planarizes the structure, exposes silanol functionalities mimicking the original SiO_2 interface, and thus, allows superlattice construction by iteration of steps i–iii.

This layer-by-layer chemisorptive siloxane-based self-assembly approach is particularly attractive because it offers a greater net chromophore alignment and

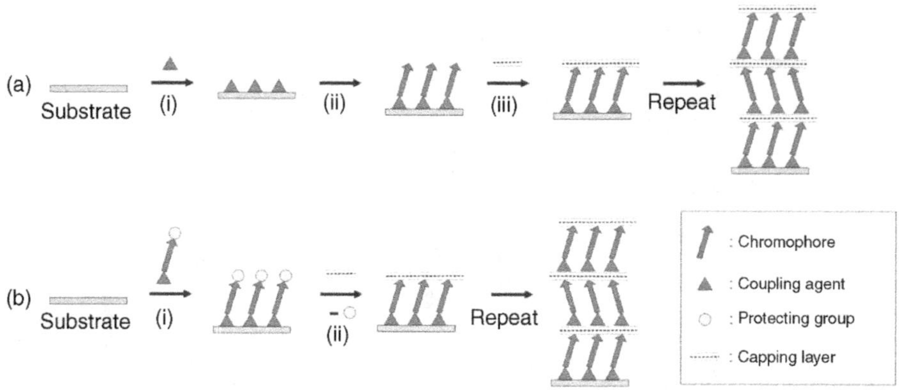

FIGURE 4.23 Schematic representation of (a) the three-step and (b) the two-step LbL assembly processes for chromophoric superlattices.

number densities than poled films and a much better structural control and stability than LB films. Molecular orientation is intrinsically acentric. In fact, chromophores are covalently linked to the substrates, and further locked into place with strong covalent cross-links. Hence, the microstructural orientation is very stable and the films are closely packed and robust.

This three-step assembly method is suitable for a wide range of donor–acceptor chromophore precursors, such as stilbazole, acetylenic, azobenzene [131,132], or pyrrole [133] resulting in films with very large NLO/EO responses ($\chi^{(2)}$ of 150–220 pm/V and r_{33} values as high as ~80 pm/V).

Recently, a novel type of X-shaped 2D chromophore with extended conjugation has been used. The resulting self-assembled chromophoric films exhibit a dramatically blue-shifted optical maximum (325 nm), while maintaining a large EO response ($\chi^{(2)}$ ~232 pm/V at 1064 nm; r_{33} ~ 45 pm/V at 1310 nm) [14,134] (Fig. 4.24).

However, this three-step process is really time-consuming and tedious, especially due to the quaternization reaction that involves inefficient spin-coating followed by vacuum treatment in an oven at 110°C, to obtain full chromophore coverage. To enhance the process efficiency and speed, a greatly improved all-"wet-chemical"

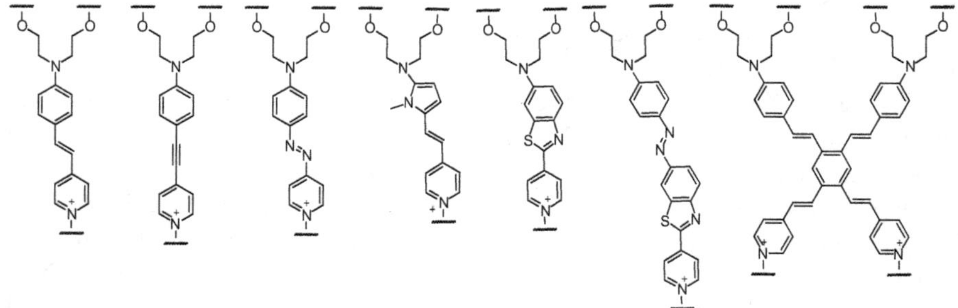

FIGURE 4.24 High-β chromophoric building blocks used in self-assembled superlattices.

two-step approach has been introduced [135]. Specifically, the deposition technique (Fig. 4.23b) employs the iterative combination of only two steps: (1) self-limiting polar chemisorption of a protected high-β chromophore monolayer ($t \sim 15$ min) and (2) *in situ* selective trialkylsilyl protecting group removal, plus capping of each "deprotected" chromophore layer using octachlorotrisiloxane ($t \sim 25$ min).

This all-"wet-chemical" two-step process can be efficiently implemented in a vertical dipping procedure to yield polar films consisting of more than 80 alternating chromophore and capping layers. Each nanoscale bilayer (chromophore + polysiloxane layer are ~3.26 nm thick) can be grown at least one order of magnitude more rapidly than the previously used siloxane-based solution deposition methodologies [136]. The thermally and photochemically robust superlattices exhibit very large EO responses (up to $\chi^{(2)}$ 370 pm/V and $r_{33} \sim 120$ pm/V determined by second harmonic generation (SHG) measurements at $\lambda = 1064$ nm) [137] and high chromophore surface densities and have been integrated in frequency doubling devices [138], ultrafast optical switches [139], and EO phase modulators [140,141].

Several alternative self-assembly approaches for producing thermally stable, acentric chromophoric multilayers have been reported [142–144]. The most prominent example is that developed by Katz et al. [145,146], which takes advantage of the zirconium phosphonate/phosphate coordinative bonding to fix layers of a dye to one another producing films with a good structural regularity and stability to orientational randomization of up to 150°C. Another example utilizes the electric field-induced LbL assembly technique of ionic species, followed by UV irradiation to convert the ionic bonds between layers into covalent bonds [147].

4.11.3 Hydrogen-Bonded and Supramolecular Assemblies

An attractive way of self-assembling highly EO-active organic materials is by using supramolecular interactions, such as intermolecular hydrogen bonding, to achieve highly ordered, noncovalently bound, acentric structures under mild conditions [148–150].

Saadeh et al. successfully applied this approach to generate a series of polymer-like materials exhibiting large optical nonlinearities [151]. These assemblies were built through the spontaneous formation of three parallel hydrogen bonds between a NLO chromophore bearing two 2,6-diacetamido-4-pyridone moieties (**D1–3**) and a monomer with a diimide group or two uracil groups (**A1–3**) (Fig. 4.25). These NLO supramolecular self-assemblies can form amorphous films with a good optical quality, r_{33} as high as 70 pm/V, and a long-term stability (4000 h) of the SHG signal at room temperature.

Generally, self-assembled organic thin films are formed in solution. However, the solution techniques have the disadvantage of being time-consuming and limited by the diffusion and aggregation of molecules that interferes with the deposition process. To tackle this problem, Gunter and coworkers have developed a novel vapor phase deposition technique, to produce supramolecular assemblies with well-defined polar ordering [152]. This technique, the so-called organic molecular beam deposition (OMBD), consists of evaporation of chromophores onto a substrate under high or

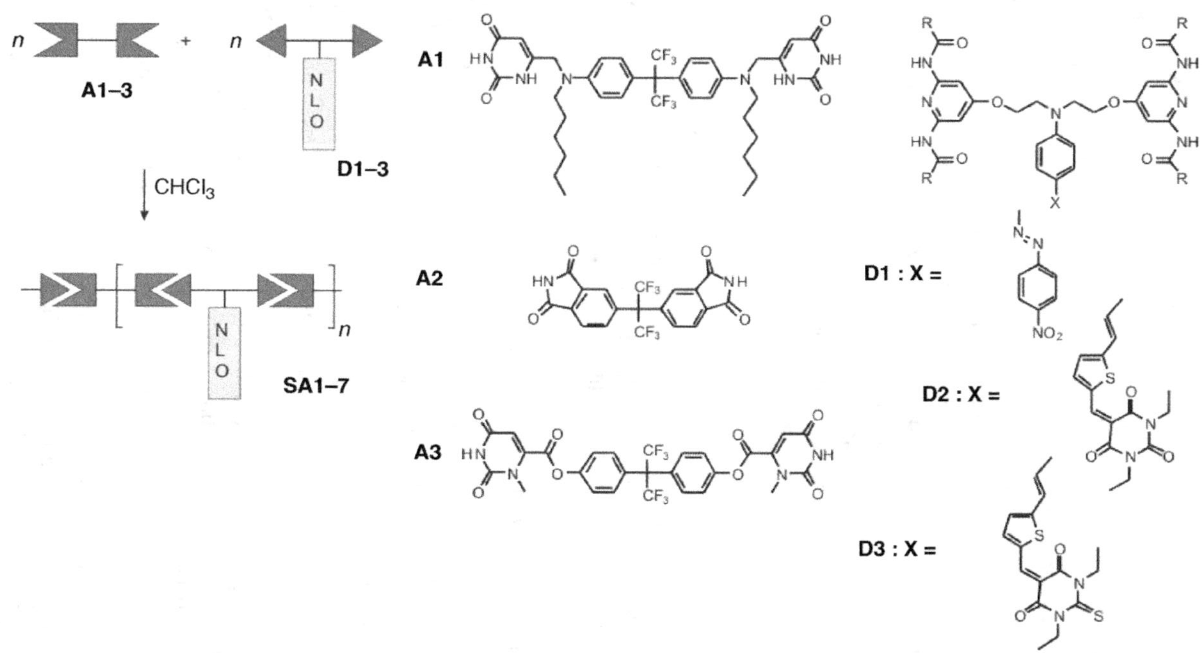

FIGURE 4.25 Structures of monomers **A1–3** and **D1–3** and synthesis of their supramolecular self-assemblies **SA1–7**.

ultrahigh vacuum (UHV) conditions. Suitable molecules must be strongly hydrogen-bonded to one another in a head-to-tail fashion to form linear supramolecular assemblies in the solid state. The vapor phase approach offers many advantages over the solution phase one, such as high chemical purity, faster growing rates, and superior control over growth parameters. Films can be grown homogeneously over large areas and with a high degree of molecular ordering and tight molecular packing over thicknesses of about 1 μm. However, only low nonlinearities ($d_{11} = 5$ pm/V) could be obtained so far [153].

Significantly larger EO responses have been achieved by Facchetti et al. applying a similar vapor deposition process to heteroaromatic-based chromophores [154]. These high-response molecules contain one pyridine ring as hydrogen-bond acceptor at one end and at the opposite end either a hydrazobenzoic acid or hydrazophenyl group as hydrogen-bond donor, making them capable of self-organizing in a head-to-tail fashion from the vapor phase (Fig. 4.26). High-quality films, as thick as 700 nm, could be grown in a few hours either on siloxane-modified substrates or on hydroxyl functionalized substrates. The EO responses at 1.30 and 1.55 μm ($r_{33} \sim$ 14–26 pm/V; $\chi^{(2)}$ 136–320 pm/V) are exceeding by about two orders of magnitude to those of previously reported H-bonded EO films [155].

Recently, Jen and coworkers developed a new class of molecular glasses based on the reversible self-assembly of aromatic/perfluoroaromatic (Ar–Ar^F) dendron-substituted nonlinear high-$\mu\beta$ chromophores [156]. In these molecules (Fig. 4.27), both phenyl and pentafluorophenyl rings are incorporated as peripheral dendrons on the π-bridge and the donor-end of the chromophores **C1**, **C2**, and **C3**.

Despite being nondirectional, and therefore, needing an external electric field to achieve acentric order, complementary Ar–Ar^F interactions have been demonstrated to improve the poling efficiency of the self-organized molecular glasses. In fact, chromophore **C3** gave the highest r_{33} value (108 pm/V at 1310 nm) among all monolithic molecular glasses, due to the favorable Ar–Ar^F interactions. This value is more than two times larger than those obtained for chromophores **C1** or **C2** that do not have such interactions. Moreover, the binary 1 : 1 composite of **C1** and **C2** showed an r_{33} value of 130 pm/V. These poled thin films could retain over 90% of their original r_{33} values at room temperature for more than 2 years. On the contrary, the temporal

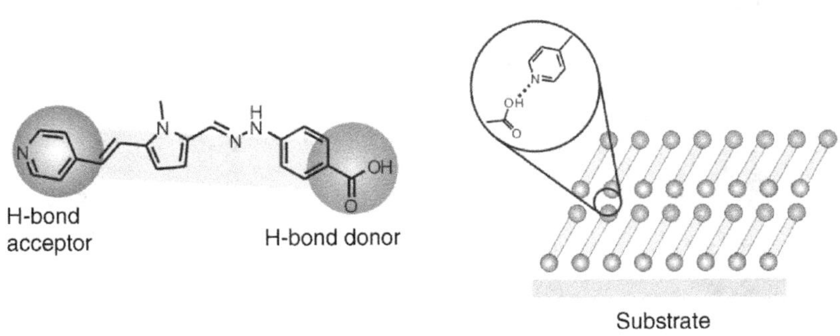

FIGURE 4.26 Graphical illustration of the head-to-tail self-assembly of chromophores by vapor deposition.

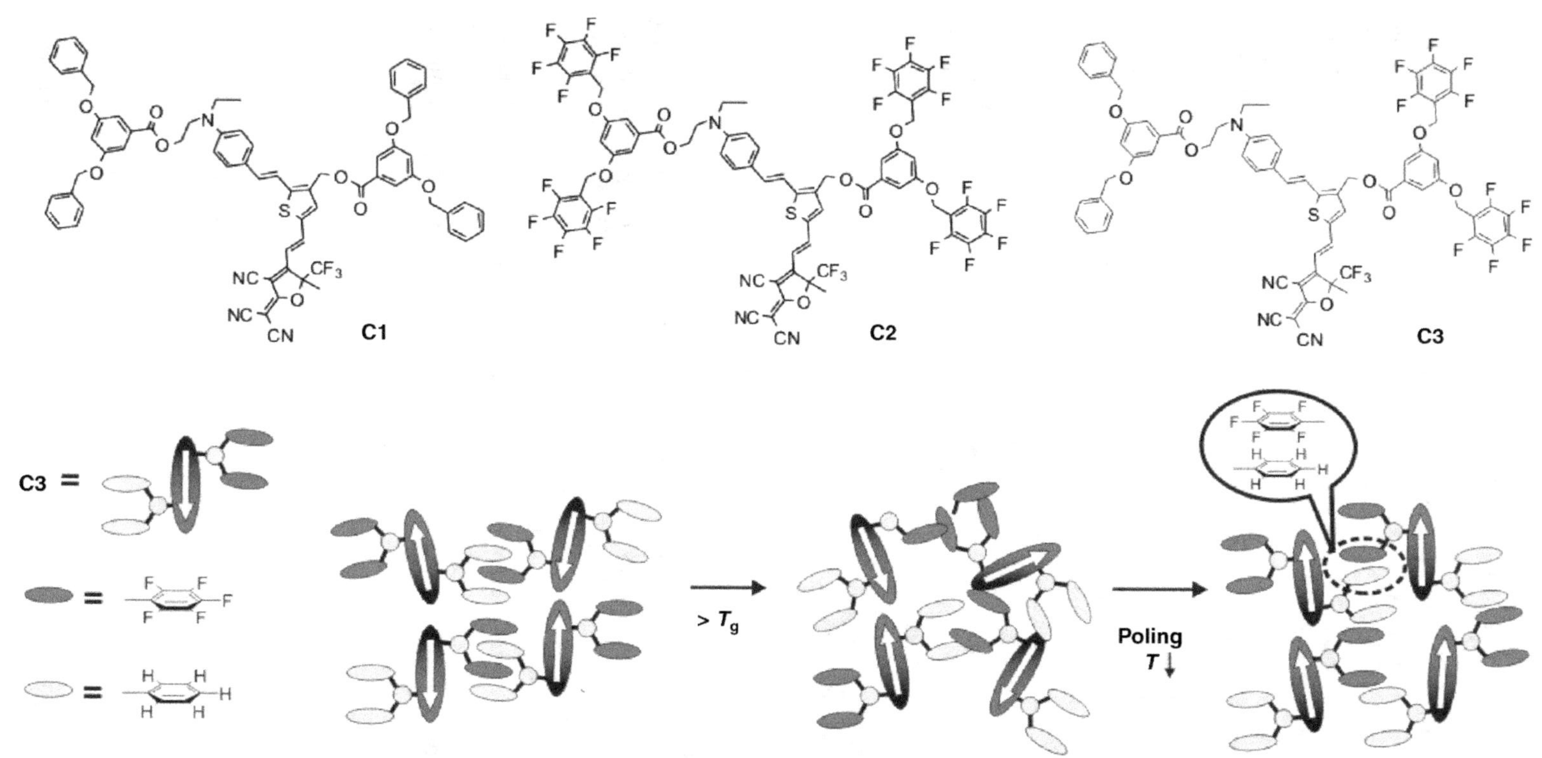

FIGURE 4.27 Structures of the glass-forming chromophores and a graphical illustration of the alignment of self-assembled chromophore **C3** by Ar–Ar^F interactions.

stability of the glasses without the Ar–Ar^F interactions deteriorated dramatically within 1 month. Although the chromophore number density in these molecular glasses are already very high (~2 times that of the typical guest–host polymers), it is still possible to further enhance the r_{33} values of these materials by doping a second chromophore with higher β in these glasses, obtaining stunning r_{33} values (up to 327 pm/V).

4.12 CONCLUSIONS AND OUTLOOK

In this chapter, an overview of the state-of-the-art polymeric NLO materials is given. The great effort made by scientists during the past 20 years has led to a better knowledge of the influence of chromophore design on material structure–function relationships. In particular, the theories developed by Dalton and coworkers have provided a better fundamental understanding of the role of chromophore shape and structure in defining the maximum achievable EO activity [107]. This has led to the development of a new generation of chromophores, with improved acceptors and conjugated bridges giving better stabilities, reduced dipole–dipole interactions, and large nonlinearities. Recently, an alternative approach for obtaining high β values, based on twisted π-zwitterionic structures, instead of on extensive planar conjugation, has been reported by Marks and coworkers [60]. This new molecular design resulted in unprecedented molecular hyperpolarizabilities, and thus, provides new paradigms for future molecule-based EO materials [59].

Several diverse synthetic approaches to obtain EO materials with acentric symmetry have been pursued. Among them, poling of guest–host polymers and chromophore-functionalized polymers have been investigated most intensely. Nevertheless, innovative designs similar to that involving the development of multichromophore-containing dendritic materials investigated by Dalton and coworkers have been introduced [113]. These dendritic materials are of particular interest because of the opportunities for molecular-scale architectural control [109]. Moreover, a polymer host is not required and stand-alone thin films can be fabricated with extremely high loading densities and larger EO activities than those that have been found in conventional EO materials. These properties, in combination with a good processability, high stability, and solvent resistance, makes NLO dendrimers a good candidate for next generation high-performance EO devices [116].

Self-assembly methods proceeding from substrates, such as layer-by-layer chemisorptive siloxane-based self-assembly [14] and vapor deposition techniques [154], have experienced a substantial growth in the past 5 years. Thanks to intrinsic molecular dipolar alignment, thermally robust and highly EO-active films can be achieved without the use of external forces. Although prototype devices integrating self-assembly superlattices have been fabricated, some crucial aspects, such as the ability of growing optical quality films in the order of micrometers thickness, still needs to be optimized.

Although the state-of-the-art organic electrooptics has advanced considerably during the past 5 years in terms of getting materials with large NLO responses and induced temporal stabilities of the polar order, the problem of chemical (aging) and photochemical stability (light-induced chromophore degradation) requires more systematic study and better understanding to be solved. The solution to this issue is critical for organic EO modulators and other devices to become a reality. Therefore, more intensive research efforts will be required to realize the full potential of organic EO materials.

REFERENCES

1. T. Verbiest, S. Houbrechts, M. Kauranen, K. Clays, and A. Persoons, *J. Mater. Chem.* **1997**, *7*, 2175–2189.
2. L.R. Dalton, *J. Phys.: Condens. Matter* **2003**, *15*, R897–R934.
3. L.R. Dalton, W.H. Steier, B.H. Robinson, C. Zhang, A. Ren, S. Garner, A. Chen, T. Londergan, L. Irwin, B. Carlson, L. Fifield, G. Phelan, C. Kincaid, J. Amend, and A. Jen, *J. Mater. Chem.* **1999**, *9*, 1905–1920.
4. F. Kajzar, K.S. Lee, and A.K.Y. Jen, *Adv. Polym. Sci.* **2003**, *161*, 1–85.
5. L.R. Dalton, *Adv. Polym. Sci.* **2002**, *158*, 1–86.
6. L. Dalton, A. Harper, A. Ren, F. Wang, G. Todorova, J. Chen, C. Zhang, and M. Lee, *Ind. Eng. Chem. Res.* **1999**, *38*, 8–33.
7. P.N. Prasad and D.J. Williams, *Introduction to Nonlinear Optical Effects in Molecules and Polymers*, John Wiley and Sons, New York, **1991**.
8. R.W. Boyd, *Nonlinear Optics*, Academic Press, San Diego, CA, **1992**.
9. D.M. Burland, R.D. Miller, and C.A. Walsh, *Chem. Rev.* **1994**, *94*, 31–75.
10. O.R. Evans and W. Lin, *Acc. Chem. Res.* **2002**, *35*, 511–522.
11. G.J. Ashwell, *J. Mater. Chem.* **1999**, *9*, 1991–2003.
12. J.R. Heflin, M.T. Guzy, P.J. Neyman, K.J. Gaskins, C. Brands, Z. Wang, H.W. Gibson, R.M. Davis, and K.E. VanCott, *Langmuir* **2006**, *22*, 5723–5727.
13. E.H. Kang, T. Bu, P. Jin, J. Sun, Y. Yang, and J. Shen, *Langmuir* **2007**, *23*, 7594–7601.
14. H. Kang, G. Evmenenko, P. Dutta, K. Clays, K. Song, and T.J. Marks, *J. Am. Chem. Soc.* **2006**, *128*, 6194–6205.
15. G. Hennrich, A. Omenat, I. Asselberghs, S. Foerier, K. Clays, T. Verbiest, and J. L. Serrano, *Angew. Chem., Int. Ed.* **2006**, *45*, 4203–4206.
16. L. Wang, J. Yoshida, N. Ogata, S. Sasaki, and T. Kajiyama, *Chem. Mater.* **2001**, *13*, 1273–1281.
17. C.C. Teng and H.T. Man, *Appl. Phys. Lett.* **1990**, *56*, 1734–1736.
18. F. Michelotti, G. Nicolao, F. Tesi, and M. Bertolotti, *Chem. Phys.* **1999**, *245*, 311–326.
19. V. Dentan, Y. Levy, M. Dumont, P. Robin, and E. Chastaing, *Opt. Commun.* **1989**, *69*, 379–383.
20. S. Kalluri, S. Garner, M. Ziari, W.H. Steier, Y. Shi, and L.R. Dalton, *Appl. Phys. Lett.* **1996**, *69*, 275–277.
21. Y.Q. Shi, C. Zhang, H. Zhang, J.H. Bechtel, L.R. Dalton, B.H. Robinson, and W.H. Steier, *Science* **2000**, *288*, 119–122.

22. M. Lee, H.E. Katz, C. Erben, D.M. Gill, P. Gopalan, J.D. Heber, and D.J. McGee, *Science* **2002**, *298*, 1401–1403.
23. R.R. Barto, C.W. Frank, P.V. Bedworth, S. Ermer, and R.E. Taylor, *J. Phys. Chem. B* **2004**, *108*, 8702–8715.
24. H. Ma, A.K.Y. Jen, and L.R. Dalton, *Adv. Mater.* **2002**, *14*, 1339–1365.
25. B.F. Levine and C.G. Bethea, *J. Chem. Phys.* **1975**, *63*, 2666–2682.
26. A. Willetts, J.E. Rice, D.M. Burland, and D.P. Shelton, *J. Chem. Phys.* **1992**, *97*, 7590–7599.
27. K. Clays and A. Persoons, *Phys. Rev. Lett.* **1991**, *66*, 2980–2983.
28. K. Clays and A. Persoons, *Rev. Sci. Instrum.* **1992**, *63*, 3285–3289.
29. J.L. Oudar and D.S. Chemla, *J. Chem. Phys.* **1977**, *66*, 2664–2668.
30. S.R. Marder, D.N. Beratan, and L.T. Cheng, *Science* **1991**, *252*, 103–106.
31. C.B. Gorman and S.R. Marder, *Proc. Natl. Acad. Sci. U.S.A.* **1993**, *90*, 11297–11301.
32. O. Kwon, S. Barlow, S.A. Odom, L. Beverina, N.J. Thompson, E. Zojer, J.L. Bredas, and S.R. Marder, *J. Phys. Chem. A* **2005**, *109*, 9346–9352.
33. C.R. Moylan, R.J. Twieg, V.Y. Lee, S.A. Swanson, K.M. Betterton, and R.D. Miller, *J. Am. Chem. Soc.* **1993**, *115*, 12599–12600.
34. P.V. Bedworth, Y. Cai, A. Jen, and S.R. Marder, *J. Org. Chem.* **1996**, *61*, 2242–2246.
35. C.R. Moylan, S. Ermer, S.M. Lovejoy, I.H. McComb, D.S. Leung, R. Wortmann, P. Krdmer, and R.J. Twieg, *J. Am. Chem. Soc.* **1996**, *118*, 12950–12955.
36. S. Thayumanavan, J. Mendez, and S.R. Marder, *J. Org. Chem.* **1999**, *64*, 4289–4297.
37. G. Koeckelberghs, T. Verbiest, M. Vangheluwe, L. DeGroof, I. Asselberghs, I. Picard, K. Clays, A. Persoons, and C. Samyn, *Chem. Mater.* **2005**, *17*, 118–121.
38. R.F. Shi, M.H. Wu, S. Yamada, Y.M. Cai, and A.F. Garito, *Appl. Phys. Lett.* **1993**, *63*, 1173–1175.
39. G. Rojo, G. De La Torre, J. Garcia-Ruiz, I. Ledoux, T. Torres, J. Zyss, and F. Agullo-Lopez, *Chem. Phys.* **1999**, *245*, 27–34.
40. M.O. Senge, M. Fazekas, E.G.A. Notaras, W.J. Blau, M. Zawadzka, O.B. Locos, and E.M. N. Mhuircheartaigh, *Adv. Mater.* **2007**, *19*, 2737–2774.
41. B. Kippelen, S.R. Marder, E. Hendrickx, J.L. Maldonado, G. Guillemet, B.L. Volodin, D.D. Steele, Y. Enami, Sandalphon, Y.J. Yao, J.F. Wang, H. Rockel, L. Erskine, and N. Peyghambarian, *Science* **1998**, *279*, 54–57.
42. Y.C. Shu, Z.H. Gong, C.F. Shu, E.M. Breitung, R.J. McMahon, G.H. Lee, and A.K.Y. Jen, *Chem. Mater.* **1999**, *11*, 1628–1632.
43. C. Zhang, A.S. Ren, F. Wang, J. Zhu, L.R. Dalton, J.N. Woodford, and C.H. Wang, *Chem. Mater.* **1999**, *11*, 1966–1968.
44. C.W. Dirk, H.E. Katz, M.L. Schilling, and L.A. King, *Chem. Mater.* **1990**, *2*, 700–705.
45. S. Gilmour, S.R. Marder, J.W. Perry, and L.T. Cheng, *Adv. Mater.* **1994**, *6*, 494–496.
46. P.R. Varanasi, A.K.Y. Jen, J. Chandrasekhar, I.N.N. Namboothiri, and A. Rathna, *J. Am. Chem. Soc.* **1996**, *118*, 12443–12448.
47. I.D.L. Albert, T.J. Marks, and M.A. Ratner, *J. Am. Chem. Soc.* **1997**, *119*, 6575–6582.
48. C.F. Shu and Y.K. Wang, *J. Mater. Chem.* **1998**, *8*, 833–835.
49. S. Gilmour, R.A. Montgomery, S.R. Marder, L.T. Cheng, A.K.Y. Jen, C. Yongming, J. W. Perry, and L.R. Dalton, *Chem. Mater.* **1994**, *6*, 1603–1604.

50. M. Ahlheim, M. Barzoukas, P.V. Bedworth, M. Blanchard-Desce, A. Fort, Z.Y. Hu, S. R. Marder, J.W. Perry, C. Runser, M. Staehelin, and B. Zysset, *Science* **1996**, *271*, 335–337.
51. S.S. Sun, C. Zhang, L.R. Dalton, S.M. Garner, A. Chen, and W.H. Steier, *Chem. Mater.* **1996**, *8*, 2539–2541.
52. Z.Y. Hu, A. Fort, M. Barzoukas, A.K.Y. Jen, S. Barlow, and S.R. Marder, *J. Phys. Chem. B* **2004**, *108*, 8626–8630.
53. A.K.Y. Jen, Y. Cai, P.V. Bedworth, and S.R. Marder, *Adv. Mater.* **1997**, *9*, 132–135.
54. X. Wu, J. Wu, Y. Liu, and A.K.Y. Jen, *J. Am. Chem. Soc.* **1999**, *121*, 472–473.
55. F. Wang, A.S. Ren, M. He, A.W. Harper, L.R. Dalton, S.M. Garner, H. Zhang, A. Chen, and W.H. Steier, *Polym. Mater. Sci. Eng.* **1998**, *78*, 42.
56. M. He, T.M. Leslie, and J.A. Sinicropi, *Chem. Mater.* **2002**, *14*, 2393–2400.
57. M. He, T.M. Leslie, J.A. Sinicropi, S.M. Garner, and L.D. Reed, *Chem. Mater.* **2002**, *14*, 4669–4675.
58. H. Kang, A. Facchetti, H. Jiang, E. Cariati, S. Righetto, R. Ugo, C. Zuccaccia, A. Macchioni, C.L. Stern, Z.F. Liu, S.T. Ho, E.C. Brown, M.A. Ratner, and T. J. Marks, *J. Am. Chem. Soc.* **2007**, *129*, 3267–3286.
59. A. Galvan-Gonzalez, K.D. Belfield, G.I. Stegeman, M. Canva, S.R. Marder, K. Staub, G. Levina, and R.J. Twieg, *J. Appl. Phys.* **2003**, *94*, 756–763.
60. H. Kang, A. Facchetti, P.W. Zhu, H. Jiang, Y. Yang, E. Cariati, S. Righetto, R. Ugo, C. Zuccaccia, A. Macchioni, C.L. Stern, Z.F. Liu, S.T. Ho, and T.J. Marks, *Angew. Chem., Int. Ed.* **2005**, *44*, 7922–7925.
61. J.W. Wu, J.F. Valley, S. Ermer, E.S. Binkley, J.T. Kenney, G.F. Lipscomb, and R. Lytel, *Appl. Phys. Lett.* **1991**, *58*, 225.
62. K.Y. Wong and A.K.Y. Jen, *J. Appl. Phys.* **1994**, *75*, 3308–3310.
63. Y.M. Cai and A.K.Y. Jen, *Appl. Phys. Lett.* **1995**, *117*, 7295.
64. S. Liu, M.A. Haller, H. Ma, L.R. Dalton, S.H. Jang, and A.K.Y. Jen, *Adv. Mater.* **2003**, *15*, 603–607.
65. C. Zhang, L.R. Dalton, M.C. Oh, H. Zhang, and W.H. Steier, *Chem. Mater.* **2001**, *13*, 3043–3050.
66. M.C. Oh, H. Zhang, C. Zhang, H. Erlig, Y. Chang, B. Tsap, D. Chang, A. Szep, W. H. Steier, H.R. Fetterman, and L.R. Dalton, *IEEE J. Sel. Top. Quantum Electron.* **2001**, *7*, 826–835.
67. Y.J. Cheng, J. Luo, S. Hau, D.H. Bale, T.D. Kim, Z. Shi, D.B. Lao, N.M. Tucker, Y. Tian, L.R. Dalton, P.J. Reid, and A.K.Y. Jen, *Chem. Mater.* **2007**, *19*, 1154–1163.
68. S.M. Garner, J.S. Cites, M. He, and J. Wang, *Appl. Phys. Lett.* **2004**, *84*, 1049–1051.
69. M. He, T. Leslie, S. Garner, M. DeRosa, and J. Cites, *J. Phys. Chem. B* **2004**, *108*, 8731–8736.
70. C. Samyn, T. Verbiest, and A. Persoons, *Macromol. Rapid Commun.* **2000**, *21*, 1–15.
71. C.C. Chang, C.P. Chen, C.C. Chou, W.J. Kuo, and R.J. Jeng, *J. Macromol. Sci., Polym. Rev.* **2005**, *45*, 125–170.
72. T. Verbiest, D.M. Burland, M.C. Jurich, V.Y. Lee, R.D. Miller, and W. Volksen, *Science* **1995**, *268*, 1604–1606.
73. T. Verbiest, D.M. Burland, M.C. Jurich, V.Y. Lee, R.D. Miller, and W. Volksen, *Macromolecules* **1995**, *28*, 3005–3007.

74. R.D. Miller, D.M. Burland, M. Jurich, V.Y. Lee, C.R. Moylan, J.I. Thackara, R.J. Twieg, T. Verbiest, and W. Volksen, *Macromolecules* **1995**, *28*, 4970–4974.
75. M.H. Davey, V.Y. Lee, L.M. Wu, C.R. Moylan, W. Volksen, A. Knoesen, R.D. Miller, and T.J. Marks, *Chem. Mater.* **2000**, *12*, 1679–1693.
76. T.A. Chen, A.K.Y. Jen, and Y. Cai, *Macromolecules* **1996**, *29*, 535–539.
77. M.E. Wright, S. Fallis, A.J. Guenthner, and L.C. Baldwin, *Macromolecules* **2005**, *38*, 10014–10021.
78. A.J. Guenthner, M.E. Wright, S. Fallis, G.A. Lindsay, B.J. Petteys, G.R. Yandek, D. Y. Zang, M. Sanghadasa, and P.R. Ashley, *Proc. SPIE-Int. Soc. Opt. Eng.* **2006**, *6331*.
79. G.A. Lindsay, A.J. Guenthner, M.E. Wright, M. Sanghadasa, and P.R. Ashley, *Polymer* **2007**, *48*, 6605–6616.
80. M. Faccini, M. Balakrishnan, M.B.J. Diemeer, R. Torosantucci, A. Driessen, D.N. Reinhoudt, and W. Verboom, *J. Mater. Chem.* **2008**, *18*, 5293–5300.
81. M. Faccini, M. Balakrishnan, R. Torosantucci, A. Driessen, D.N. Reinhoudt, and W. Verboom, *Macromolecules* **2008**, *41*, 8320–8323.
82. M.E. Wright, S. Mullick, H.S. Lackritz, and L.Y. Liu, *Macromolecules* **1994**, *27*, 3009–3015.
83. F. Fuso, A.B. Padias, and H.K. HallJr., *Macromolecules* **1991**, *24*, 1710–1713.
84. C. Xu, B. Wu, L.R. Dalton, P.M. Ranon, Y. Shi, and W.H. Steier, *Macromolecules* **1992**, *25*, 6716–6718.
85. G.A. Lindsay, J.D. Stenger-Smith, R.A. Henry, J.M. Hoover, R.A. Nissan, and K. J. Wynne, *Macromolecules* **1992**, *25*, 6075–6077.
86. Z. Li, S. Dong, G. Yu, Z. Li, Y. Liu, C. Ye, and J. Qin, *Polymer* **2007**, *48*, 5520–5529.
87. N. Tsutsumi, M. Morishima, and W. Sakai, *Macromolecules* **1998**, *31*, 7764–7769.
88. M. Dobler, C. Weder, O. Ahumada, P. Neuenschwander, U.W. Suter, S. Follonier, C. Bosshard, and P. Gunter, *Macromolecules* **1998**, *31*, 7676–7681.
89. Y.D. Zhang, L.M. Wang, T. Wada, and H. Sasabe, *Macromolecules* **1996**, *29*, 1569–1573.
90. G. Sheeren, A. Persoons, H. Bolink, M. Heylen, M. Vanbeylen, and C. Samyn, *Eur. Polym. J.* **1993**, *29*, 981–986.
91. L.R. Dalton, A.W. Harper, R. Ghosn, W.H. Steier, M. Ziari, H. Fetterman, Y. Shi, R. V. Mustacich, A.K.Y. Jen, and K.J. Shea, *Chem. Mater.* **1995**, *7*, 1060–1081.
92. S.S.H. Mao, Y. Ra, L. Guo, C. Zhang, L.R. Dalton, A. Chen, S. Garner, and W.H. Steier, *Chem. Mater.* **1998**, *10*, 146–155.
93. L. Chen, G. Qian, X. Jin, Y. Cui, J. Gao, Z. Wang, and M. Wang, *J. Phys. Chem. B* **2007**, *111*, 3115–3121.
94. M. Guglielmi, G. Brusatin, and G. Della Giustina, *J. Non-Cryst. Solids* **2007**, *353*, 1681–1687.
95. Y. Bai, N. Song, J.P. Gao, X. Sun, X. Wang, G. Yu, and Z.Y. Wang, *J. Am. Chem. Soc.* **2005**, *127*, 2060–2061.
96. R.-J. Jeng, C.-C. Chang, C.-P. Chen, C.-T. Chen, and W.-C. Su, *Polymer* **2003**, *44*, 143–155.
97. C. Zhang, C. Wang, L.R. Dalton, H. Zhang, and W.H. Steier, *Macromolecules* **2001**, *34*, 253–261.

98. C. Zhang, C. Wang, J. Yang, L.R. Dalton, G. Sun, H. Zhang, and W.H. Steier, *Macromolecules* **2001**, *34*, 235–243.
99. D.W. Smith and D.A. Babb, *Macromolecules* **1996**, *29*, 852–860.
100. H. Ma, J. Wu, P. Herguth, B. Chen, and A.K.Y. Jen, *Chem. Mater.* **2000**, *12*, 1187–1189.
101. M. Haller, J. Luo, H. Li, T.D. Kim, Y. Liao, B.H. Robinson, L.R. Dalton, and A.K.Y. Jen, *Macromolecules* **2004**, *37*, 688–690.
102. J. Luo, M. Haller, H. Li, T.D. Kim, and A.K.Y. Jen, *Adv. Mater.* **2003**, *15*, 1635–1638.
103. R.D. Nielsen, H.L. Rommel, and B.H. Robinson, *J. Phys. Chem. B* **2004**, *108*, 8659–8667.
104. S.D. Bella, M.A. Ratner, and T.J. Marks, *J. Am. Chem. Soc.* **1992**, *114*, 5842–5849.
105. L.R. Dalton, A.W. Harper, and B.H. Robinson, *Proc. Natl. Acad. Sci. U.S.A.* **1997**, *94*, 4842–4847.
106. I. Liakatas, C. Cai, M. Bösch, M. Jäger, C. Bosshard, P. Günter, C. Zhang, and L.R. Dalton, *Appl. Phys. Lett.* **2000**, *76*, 1368–1370.
107. A.W. Harper, S. Sun, L.R. Dalton, S.M. Garner, A. Chen, S. Kalluri, W.H. Steier, and B. H. Robinson, *J. Opt. Soc. Am. B* **1998**, *15*, 329–337.
108. H. Ma, S. Liu, J.D. Luo, S. Suresh, L. Liu, S.H. Kang, M. Haller, T. Sassa, L.R. Dalton, and A.K.Y. Jen, *Adv. Funct. Mater.* **2002**, *12*, 565–574.
109. Y.V. Pereverzev, O.V. Prezhdo, and L.R. Dalton, *Chem. Phys. Lett.* **2003**, *373*, 207–212.
110. S. Yokoyama, T. Nakahama, A. Otomo, and S. Mashiko, *J. Am. Chem. Soc.* **2000**, *122*, 3174–3181.
111. J.D. Luo, H. Ma, M. Haller, A.K.Y. Jen, and R.R. Barto, *Chem. Commun.* **2002**, 888–889.
112. P.A. Sullivan, A.J.P. Akelaitis, S.K. Lee, G. McGrew, S.K. Lee, D.H. Choi, and L. R. Dalton, *Chem. Mater.* **2006**, *18*, 344–351.
113. P.A. Sullivan, H. Rommel, Y. Liao, B.C. Olbricht, A.J.P. Akelaitis, K.A. Firestone, J. W. Kang, J. Luo, J.A. Davies, D.H. Choi, B.E. Eichinger, P.J. Reid, A. Chen, A.K.Y. Jen, B.H. Robinson, and L.R. Dalton, *J. Am. Chem. Soc.* **2007**, *129*, 7523–7530.
114. H. Ma, B. Chen, T. Sassa, L.R. Dalton, and A.K.Y. Jen, *J. Am. Chem. Soc.* **2001**, *123*, 986–987.
115. P.A. Sullivan, B.C. Olbricht, A.J.P. Akelaitis, A.A. Mistry, Y. Liao, and L.R. Dalton, *J. Mater. Chem.* **2007**, *17*, 2899–2903.
116. Z. Shi, S. Hau, J. Luo, T.D. Kim, N.M. Tucker, J.W. Ka, H. Sun, A. Pyajt, L. Dalton, A. Chen, and A.K.Y. Jen, *Adv. Funct. Mater.* **2007**, *17*, 2557–2563.
117. A.K.Y. Jen, J. Luo, S. Liu, M. Haller, L. Liu, and H. Ma, *Adv. Mater.* **2002**, *14*, 1763–1768.
118. J. Luo, M. Haller, H. Ma, S. Liu, T.D. Kim, Y. Tian, B. Chen, S.H. Jang, L.R. Dalton, and A.K.Y. Jen, *J. Phys. Chem. B* **2004**, *108*, 8523–8530.
119. J. Luo, M. Haller, H. Li, H.Z. Tang, A.K.Y. Jen, K. Jakka, C.H. Chou, and C.F. Shu, *Macromolecules* **2004**, *37*, 248–250.
120. T.D. Kim, J. Luo, Y. Tian, J.W. Ka, N.M. Tucker, M. Haller, J.W. Kang, and A.K.Y. Jen, *Macromolecules* **2006**, *39*, 1676–1680.
121. S. Sioncke, T. Verbiest, and A. Persoons, *Mater. Sci. Eng., R.* **2003**, *42*, 115–155.
122. T. Verbiest, S. Van Elshocht, M. Kauranen, L. Hellemans, J. Snauwaert, C. Nuckolls, T. J. Katz, and A. Persoons, *Science* **1998**, *282*, 913–915.
123. G.J. Ashwell, P.D. Jackson, and W.A. Crossland, *Nature* **1994**, *368*, 438–440.

124. T.L. Penner, H.R. Motschmann, N.J. Armstrong, M.C. Ezenyilimba, and D.J. Williams, *Nature* **1994**, *367*, 49–50.

125. K. Rajesh, M.S. Chandra, S. Hirakawa, J. Kawamata, and T.P. Radhakrishnan, *Langmuir* **2007**, *23*, 8560–8568.

126. G. Panambur, Y. Zhang, A. Yesayan, T. Galstian, C.G. Bazuin, and A.M. Ritcey, *Langmuir* **2004**, *20*, 3606–3615.

127. Q. Huo, S. Russev, T. Hasegawa, J. Nishijo, J. Umemura, G. Puccetti, K.C. Russell, and R. M. Leblanc, *J. Am. Chem. Soc.* **2000**, *122*, 7890–7897.

128. Y. Wang, C. Wang, X. Wang, Y. Guo, B. Xie, Z. Cui, L. Liu, L. Xu, D. Zhang, and B. Yang, *Chem. Mater.* **2005**, *17*, 1265–1268.

129. D. Li, M.A. Ratner, T.J. Marks, C. Zhang, J. Yang, and G.K. Wong, *J. Am. Chem. Soc.* **1990**, *112*, 7389–7390.

130. A.K. Kakkar, S. Yitzchaik, S.B. Roscoe, F. Kubota, D.S. Allan, T.J. Marks, W. Lin, and G. K. Wong, *Langmuir* **1993**, *9*, 388–390.

131. T.J. Marks, *Angew. Chem., Int. Ed.* **1995**, *34*, 155–173.

132. S. Yitzchaik and T.J. Marks, *Acc. Chem. Res.* **1996**, *29*, 197–202.

133. A. Facchetti, A. Abbotto, L. Beverina, M.E. van der Boom, P. Dutta, G. Evmenenko, G. A. Pagani, and T.J. Marks, *Chem. Mater.* **2003**, *15*, 1064–1072.

134. H. Kang, P. Zhu, Y. Yang, A. Facchetti, and T.J. Marks, *J. Am. Chem. Soc.* **2004**, *126*, 15974–15975.

135. M.E. Van der Boom, A.G. Richter, J.E. Malinsky, P.A. Lee, N.R. Armstrong, P. Dutta, and T.J. Marks, *Chem. Mater.* **2001**, *13*, 15–17.

136. P. Zhu, M.E. van der Boom, H. Kang, G. Evmenenko, P. Dutta, and T.J. Marks, *Chem. Mater.* **2002**, *14*, 4982–4989.

137. M.E. van der Boom, G. Evmenenko, C. Yu, P. Dutta, and T.J. Marks, *Langmuir* **2003**, *19*, 10531–10537.

138. P.M. Lundquist, W. Lin, H. Zhou, D.N. Hahn, S. Yitzchaik, T.J. Marks, and G.K. Wong, *Appl. Phys. Lett.* **1997**, *70*, 1941–1943.

139. G. Wang, P. Zhu, T.J. Marks, and J.B. Ketterson, *Appl. Phys. Lett.* **2002**, *81*, 2169.

140. Y.G. Zhao, S. Chang, A. Wu, H.L. Lu, S.T. Ho, M.E. Van der Boom, and T.J. Marks, *Opt. Eng.* **2003**, *42*, 298–299.

141. Y.G. Zhao, A. Wu, H.L. Lu, S. Chang, W.K. Lu, S.T. Ho, M.E. Van Der Boom, and T. J. Marks, *Appl. Phys. Lett.* **2001**, *79*, 587–589.

142. J.R. Heflin, M.T. Guzy, P.J. Neyman, K.J. Gaskins, C. Brands, Z. Wang, H.W. Gibson, R. M. Davis, and K.E. Van Cott, *Langmuir* **2006**, *22*, 5723–5727.

143. W.C. Flory, S.M. Mehrens, and G.J. Blanchard, *J. Am. Chem. Soc.* **2000**, *122*, 7976–7985.

144. G.A. Neff, M.R. Helfrich, M.C. Clifton, and C.J. Page, *Chem. Mater.* **2000**, *12*, 2363–2371.

145. H.E. Katz, G. Scheller, T.M. Putvinski, M.L. Schilling, W.L. Wilson, and C.E.D. Chidsey, *Science* **1991**, *254*, 1485–1487.

146. H.E. Katz, W.L. Wilson, and G. Scheller, *J. Am. Chem. Soc.* **1994**, *116*, 6636–6640.

147. Y. Wang, X. Wang, Y. Guo, Z. Cui, Q. Lin, W. Yu, L. Liu, L. Xu, D. Zhang, and B. Yang, *Langmuir* **2004**, *20*, 8952–8954.

148. F. Wuerthner, J. Schmidt, M. Stolte, and R. Wortmann, *Angew. Chem., Int. Ed.* **2006**, *45*, 3842–3846.

149. M.S. Johal, Y.W. Cao, X.D. Chai, L.B. Smilowitz, J.M. Robinson, T.J. Li, D. McBranch, and D.Q. Li, *Chem. Mater.* **1999**, *11*, 1962–1965.

150. Y.W. Cao, X.D. Chai, W.S. Yang, R. Lu, Y.S. Jiang, T.J. Li, M. Blanchard-Desce, and J. M. Lehn, *Thin Solid Films 284-285* **1996**, 859–862.

151. H. Saadeh, L. Wang, and L. Yu, *J. Am. Chem. Soc.* **2000**, *122*, 546–547.

152. C. Cai, M.M. Bosch, Y. Tao, B. Muller, Z. Gan, A. Kundig, C. Bosshard, I. Liakatas, M. Jager, and P. Gunter, *J. Am. Chem. Soc.* **1998**, *120*, 8563–8564.

153. A.N. Rashid, C. Erny, and P. Gunter, *Adv. Mater.* **2003**, *15*, 2024–2027.

154. A. Facchetti, E. Annoni, L. Beverina, M. Morone, P. Zhu, T.J. Marks, and G.A. Pagani, *Nat. Mater.* **2004**, *3*, 910–917.

155. P. Zhu, H. Kang, A. Facchetti, G. Evmenenko, P. Dutta, and T.J. Marks, *J. Am. Chem. Soc.* **2003**, *125*, 11496–11497.

156. T.D. Kim, J.W. Kang, J. Luo, S.H. Jang, J.W. Ka, N. Tucker, J.B. Benedict, L.R. Dalton, T. Gray, R.M. Overney, D.H. Park, W.N. Herman, and A.K.Y. Jen, *J. Am. Chem. Soc.* **2007**, *129*, 488–489.

5

METALLODENDRIMERS: PHOTOPHYSICAL PROPERTIES AND RELATED APPLICATIONS

GRÉGORY FRANC AND ASHOK K. KAKKAR

Photochemistry and Photophysics of Polymer Materials, Edited by Norman S. Allen

5.1 INTRODUCTION

Hyperbranched and monodispersed tree-like macromolecules that are commonly referred to as dendrimers [1–6] have been attracting tremendous attention due to their significant potential in designing novel materials for applications in material science, medicine, and catalysis. Dendrimers (Fig. 5.1) can be synthesized using two different iterative step-by-step synthetic methodologies: (1) a divergent approach that starts from a plurifunctional core and extends outward in a layer-by-layer fashion, as initially described by Vögtle and coworkers [7] and subsequently extended by Tomalia et al. [8] or (2) alternatively, using a less common convergent strategy that involves grafting dendrons onto the core [9–11].

The flurry of activity that followed these initial reports has generated a wealth of these macromolecules [12], and much emphasis has subsequently shifted to understanding their structure–property relationships, and then, applying these principles to design goal-oriented applications. In addition, during the past few years, intriguing fluorescence properties offered by dendrimers have been studied in detail, and have been extended to a wide range of applications such as light harvesting [13–15], two-photon absorption [16,17], labeling agents for biology [18–21], and in the design of efficient and competitive OLEDs [22–25]. Furthermore, investigations regarding dendrimers and their interactions with metals have also expanded at a fast pace. Metallodendrimers [4,26–28], as they are generally called, are now an integral part of the main-stream dendritic macromolecules. These well-defined architectures include metal complexes of varied forms as illustrated in Fig. 5.2. Metallodendrimers can be synthesized using a stepwise build-up approach in which coordination ligands bind to metals of a wide variety that can be placed at the core, the periphery, and even as divergent points or a combination

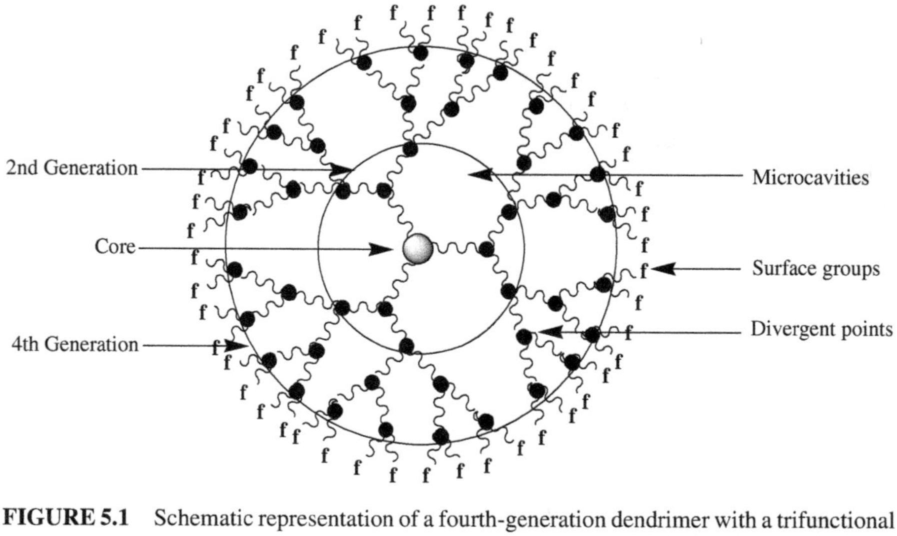

FIGURE 5.1 Schematic representation of a fourth-generation dendrimer with a trifunctional core.

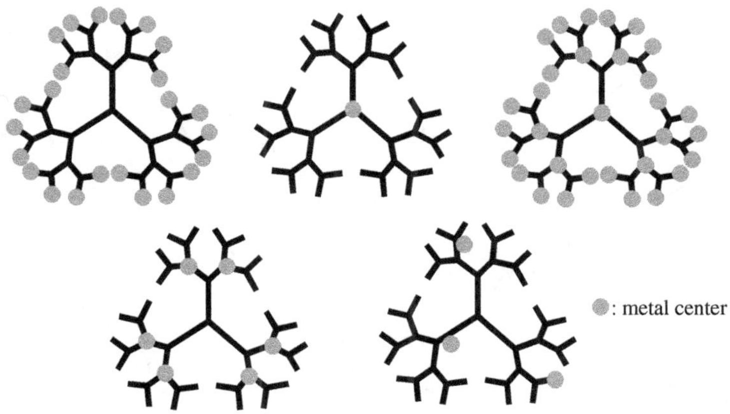

FIGURE 5.2 Different possible locations of metal centers in the dendritic architecture.

thereof. Metal centers can also be integrated "randomly" without any particular preferred site, which is ideal for sensing applications (Fig. 5.2).

Metal-functionalized dendritic assemblies have been demonstrated to be a fertile area with competitive potential for purely organic structures for their photonic properties [29]. This chapter aims at providing on overview of the different outstanding possibilities that dendrimers advantageously offer, first from a fundamental perspective by describing the main routes to build these nanometer-sized species, and then, by describing their photophysical properties, and eventually to the applications developed for these materials in various areas.

5.2 METALLODENDRIMERS WITH A CHELATING SITE AT THE CORE

The inspiration to use metals to form complex architectures has led to the captivating and attractive design of nanodevices including a photochemically driven molecular machine [30]. Coordination chemistry [31] has also been extensively used for the construction of dendrimers and has allowed the incorporation of metals including Zn and lanthanides in these unimolecular macrostructures to form emissive compounds. A global review of these architectures is provided below.

5.2.1 An Efficient Protective Dendritic Shell to Enhance Photophysical Properties

Metal centers can typically be included in the core or as the core of a dendrimer via metal–ligand coordination bonds. Complexes ($M(L)_n$) of metals (M) containing ligands (L) that can readily be replaced are generally employed for synthesizing these metallodendrimers. Due to its remarkable ability to form stable coordination compounds, Ru(II) metal ion is an excellent candidate for such a build-up process. One of the first examples of Ru(II)-based metallodendrimers was prepared by reacting

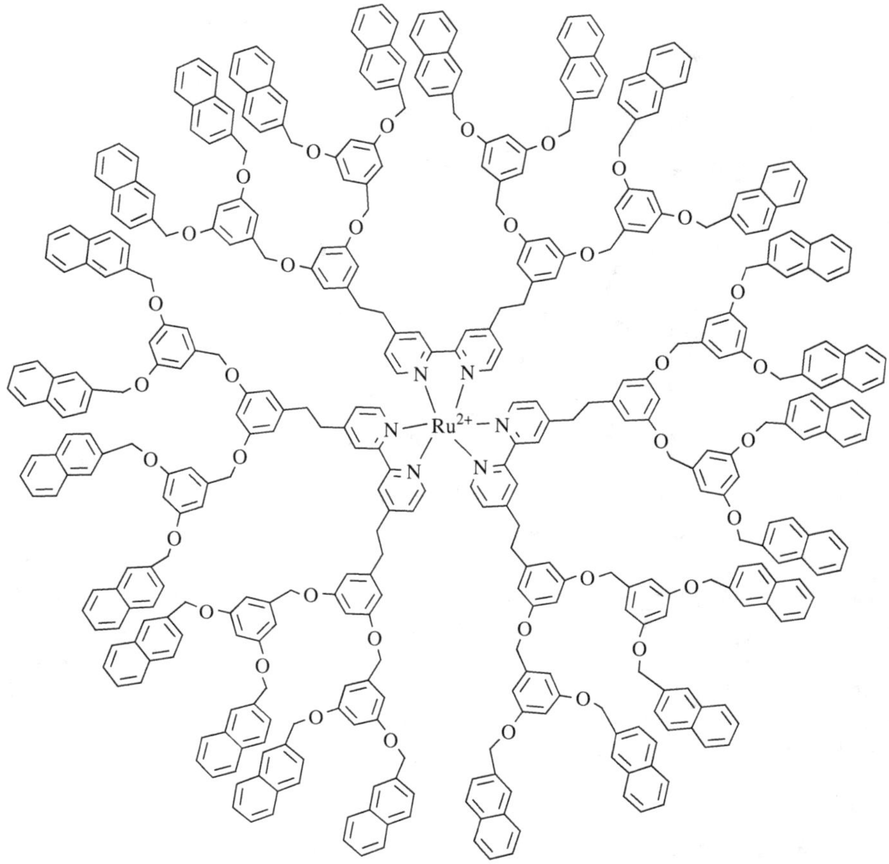

FIGURE 5.3 Ru(II)-cored dendrimer with excellent energy-transfer properties based on an antenna system from the periphery to the metal center.

$RuCl_3$ and $[Ru(bpy)_2Cl_2]$ with bipyridine (bpy) ligands containing dendritic wedges in the 4,4′-positions with up to 48 benzyl moieties (Fig. 5.3) [32]. Photophysical and luminescent properties of these metallodendrimers were found to be similar to the bipyridine mononuclear complexes, but the dendritic scaffold offered a strong barrier to quenching by dioxygen (O_2). Other salient features included a decrease of the quenching rate constant through a dynamic mechanism toward three different electron acceptors (methyl viologen dication, tetrathiafulvalene, and anthraquinone-2,6-disulfonate anion) with increasing number and generation of dendritic wedges. In another study, the influence of the peripheral groups on the possible energy transfer (ET) process in these dendrimers was reported. For this purpose, naphthalene was placed at the extremities of the dendritic wedges [33], and the photophysical properties were found to be very different leading to a light-harvesting antenna system. On irradiation of the peripheral naphthyl groups in acetonitrile, phosphorescence emission corresponding to the $[Ru(bpy)_3]^{2+}$ focal moiety ($\lambda = 610$ nm) with twice the intensity was observed. This efficient process was attributed to the high absorbance of the aromatic peripheral groups that can transfer their energy transiently to the luminescent metallated core, which can then exhibit a strong visible emission. During this process, fast fluorescence emission from the peripheral groups is transformed into long emitting phosphorescent emission of the core.

For phosphorescence, lanthanide metal ions can be used in a similar manner. Lanthanide ions have very interesting photophysical properties, but often exhibit weak absorption bands, and aggregate to form clusters, which limit their applications. Thus, a dendrimer that can provide a protective shell to isolate a cation and at the same time enhance the emission by transfer from the periphery to the lanthanide ion at the core could be of great interest. Self-assembled lanthanide-cored dendrimers have been prepared to prove such an assumption: synthesis was carried out by mixing three equivalents of polyaryl ether dendrons bearing carboxylic acid entity at the focal point with Ln(III) cations [Er(III), Tb(III), and Eu(III)] (Fig. 5.4) [34]. The authors demonstrated that the enhancement of the lanthanide cation emission associated with the dendritic core shell was observed, and an antenna effect from the periphery to the core was shown to promote this process.

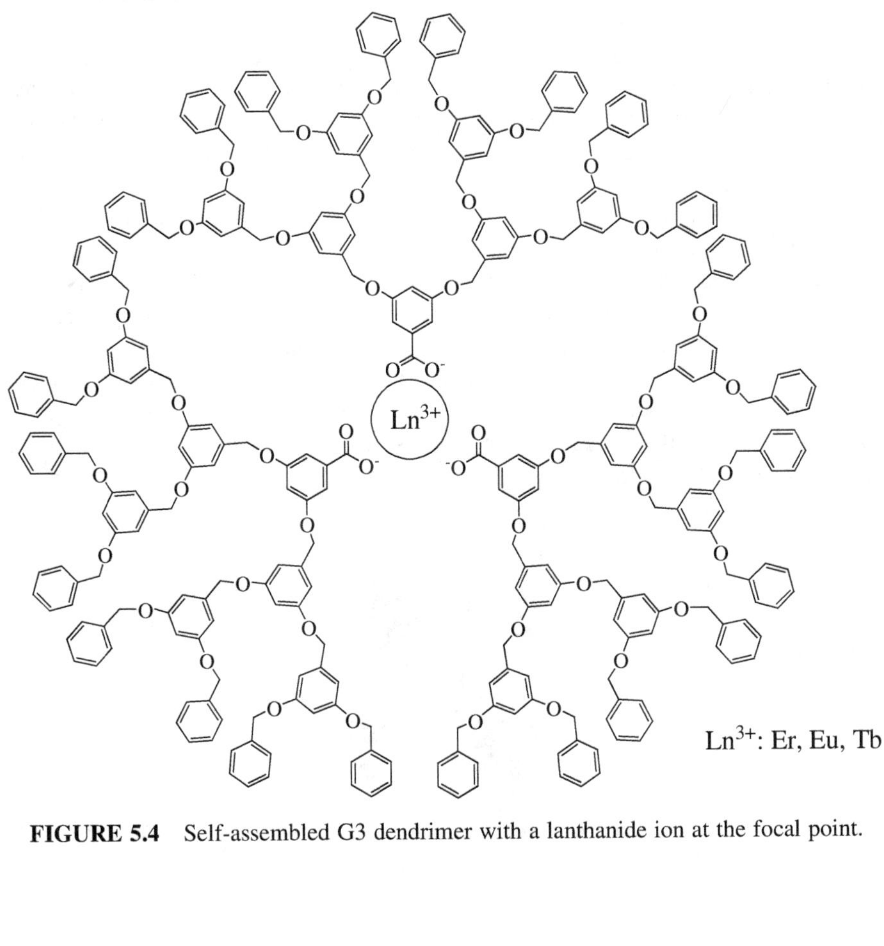

FIGURE 5.4 Self-assembled G3 dendrimer with a lanthanide ion at the focal point.

The modification of the dendritic shell with perfluorinated aromatic end groups along with an aliphatic bis-MPA scaffold was then described to make this phenomenon possible in water for Eu(III) and Tb(III) [35]. In addition, a Förster mechanism was shown to occur between the perfluorinated donor groups and lanthanide cations as acceptors. A study at the air–water surface interface was also reported with these dendrimers [36].

5.2.2 Porphyrin and Phthalocyanine-Based Dendrimers

Another approach using metallated porphyrin or phthalocyanine as a core was first reported in 1996 with a porphyrin-cored dendrimer [37]. Since then, such fluorescent macrocycles have been used as cores for a large number of macroassemblies [38,39]. A noteworthy and outstanding feature of having a dendritic shell is the prevention of aggregation of these macromolecules and allowing them to emit efficiently, a concept that was coined "site isolation" [40]. Complexation of different metals such as Zn, Co, or Pd with porphyrin afforded materials with very appealing properties. An important feature of porphyrin includes efficient photon-induced electron transfer (PET) between Zn-metallated porphyrin to free-base porphyrin. Based on the latter principle, and by mixing in a 1 : 4 ratio, free-base porphyrin containing dendritic cationic end groups with negatively charged Zn-metallated porphyrin counterparts, electron transfer leading to the single emission of the free porphyrin (Fig. 5.5) was observed [41]. This demonstrates that the dendrimers offer an intriguing architecture that hampers aggregation of the macrocycles and in which a balance between electrostatic repulsive interactions and the distance between different donors and acceptors can be achieved, to finally afford total disappearance of the signal coming from the metallated centers.

The first example of phthalocyanine-cored dendrimers with bulky dendritic substituents was reported in 1997 [42]. Subsequently, Majoral and coworkers examined phthalocyanine-cored dendrimers containing eight dendritic arms that were prepared using their classical divergent methodology [43]. Dendrimers of up to the eighth generation complexed with Co(II) were prepared to study the progressive influence of the branches on the core (Fig. 5.6), and it was demonstrated that it promotes a bathochromic shift of the Q-band and an increase in its intensity up to G5. The addition of subsequent generations did not show any further change, indicating that the influence of the branches on the core is limited up to this specific generation probably due to steric hindrance [44].

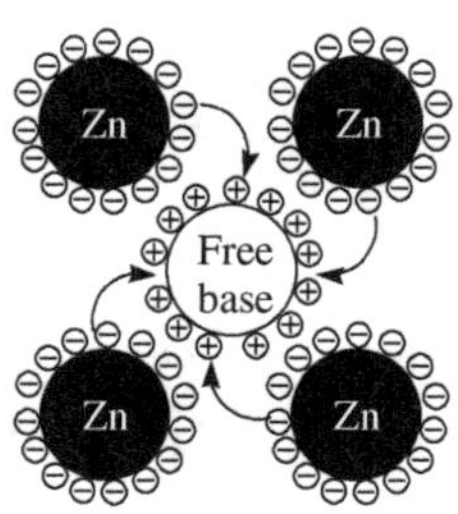

FIGURE 5.5 Enhanced electron transfer between porphyrins with protective dendritic shells.

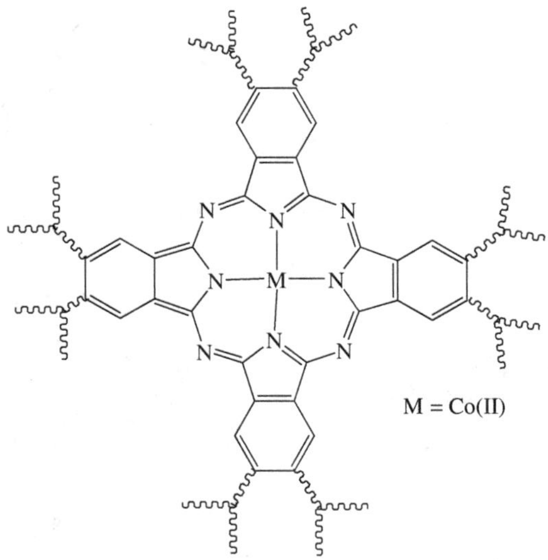

FIGURE 5.6 Metallated phthalocyanine-cored dendrimers with eight arms as described by Majoral and coworkers.

Axial dendritic substituents are an alternative way to functionalize these hyperbranched macromolecules and to avoid stacking. McKeown's group synthesized dendritic silicon phthalocyanines based on Fréchet's poly(aryl ether) dendrons, and used them to create emissive thin films by spin-coating. The disappearance of stacking rearrangement was clearly seen in the UV–vis spectra [45].

5.2.3 Metallodendrimers for Light-Emitting Diodes

Dendrimer light-emitting diodes (DLEDs) have tremendous potential due to the ability of the chemist to fine-tune dendrimers with different end groups and metals. In this regard, iridium and europium metal-based dendrimers have offered a new perspective toward phosphorescent materials (Fig. 5.7). Phosphorescent iridium complexes have shown great promise as highly efficient LEDs for emission in green, red, and blue [46–48]. Samuel's group elaborated iridium(III)-based dendrimers for LED applications by using *fac*-tris(2-phenylpyridine)iridium(III)-cored dendrimers to obtain solution-processed uniform and homogenous light-emitting layers [49]. The spin-coating process to obtain good quality thin films was subsequently optimized offering a new perspective compared to vapor deposition. These thin films exhibited interesting phosphorescent emission, with second-generation dendrimers depicting much better results [50]. The carbazole dendrons that could act as excellent hole-transporters were also included in this study, and the second-generation dendrimer gave a quantum yield of 45% in the thin film. Incorporation of metal-based dendrons seemed to confer salient improvement to the formation of LED, and was followed recently by an example with red light-emitting DLED. Using a convergent methodology, a series of Ir(III)-cored dendrimers with two carbazole dendrons were

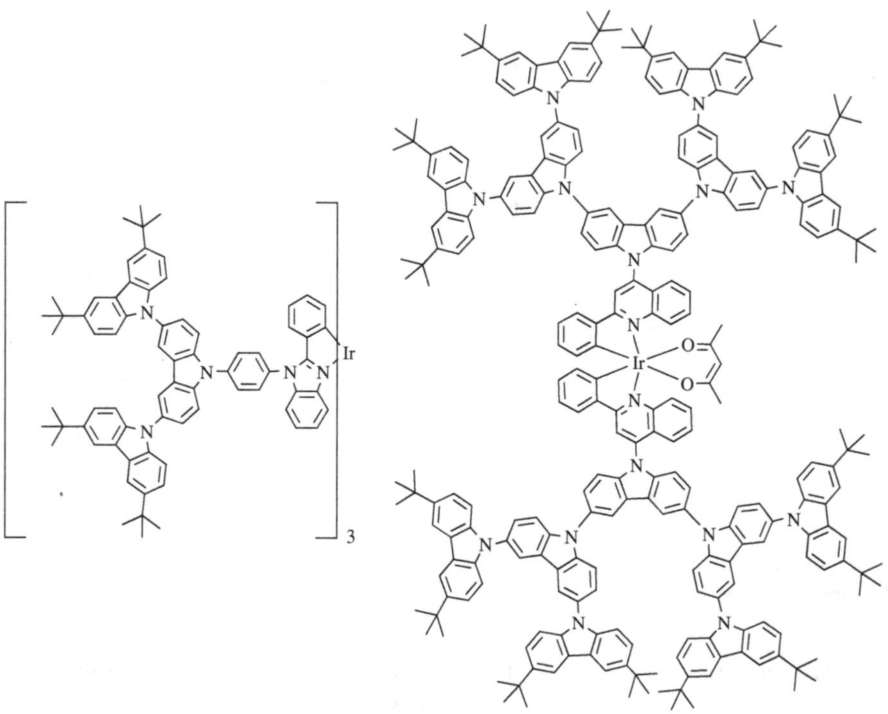

FIGURE 5.7 Second-generation irido-dendrimers with carbazole dendrons used for the formation of highly efficient green light- (*left*) and red light (*right*)-emitting diodes.

obtained. The excellent photoluminescent behavior after an easy preparation made these suitable candidates for solution-processible macromolecules in low-cost technology [51].

Eu-cored dendrimers with carbazole dendrons appended at the periphery have also been reported that exhibited very interesting properties (Fig. 5.8). A characteristic red-emission upon excitation of peripheral carbazole groups was observed after efficient electron transfer. Site-isolation and light-harvesting properties of this system afforded an approximately eightfold increase of the emission for one of their dendritic wedges, and upon adequate doping white-light emission could be observed. This enhanced the potential of these systems for future development [52,53].

5.2.4 Self-Assembled Structures with Cyclam-Cored Dendrimers

Cyclam as a chelating agent offers a new and intriguing system to complex metals compared to previous examples. Coordination ligands such as 1,4,8,11-tetraazacyclotetradecane lead to stable 1 : 1 stoichiometric complexes with divalent ions such as Co(II), Ni(II), Zn(II), or Cu(II). Vögtle and coworkers had focused their work on this well-defined core with appended 16 naphthyl groups at the periphery for complexation with Zn(II). The presence of zinc ions prevented quenching of the naphthyl units

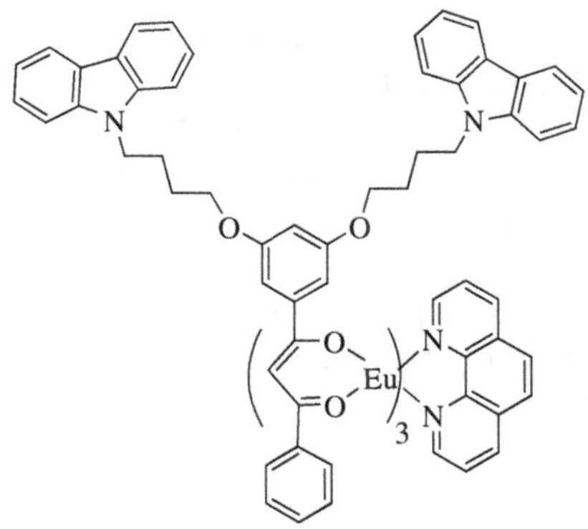

FIGURE 5.8 Eu-based dendritic complex containing carbazole units at the periphery for efficient photoluminescent applications.

by exciplex formation as reported earlier [54]. In addition, in a DCM/acetonitrile 1 : 1 v/v solution, they observed a strong emission for the aromatic end groups after titration with only 0.5 eq metal ion compared to their model compound. This was attributed to an unexpected formation of a self-assembled 1 : 2 complex with the cyclam core pointing "inward" and the dendritic wedges extending "outward" (Fig. 5.9) [55]. Another example by Vögtle and coworkers described how a [Ru(bpy)(CN)$_4$]$^{2-}$ complex can interact with cyclam-cored dendrimers and provoke major changes in the luminescence properties. Electron transfer in these metallodendrimers occurred between the naphthyl units and the metal complex that could easily be tailored by the addition of DABCO or triflic acid in the solution [56].

Lanthanide ions are also excellent candidates to include in cyclam-cored dendrimers. These systems were investigated using ions such as Nd(III), Eu(III), Gd(III), Tb(III), or Dy(III) (Fig. 5.9) [57]. ^{1}H NMR titration confirmed the presence of [M(dendrimers)$_2$]$^{2+}$ self-assembled macrostructures, and was accompanied, as

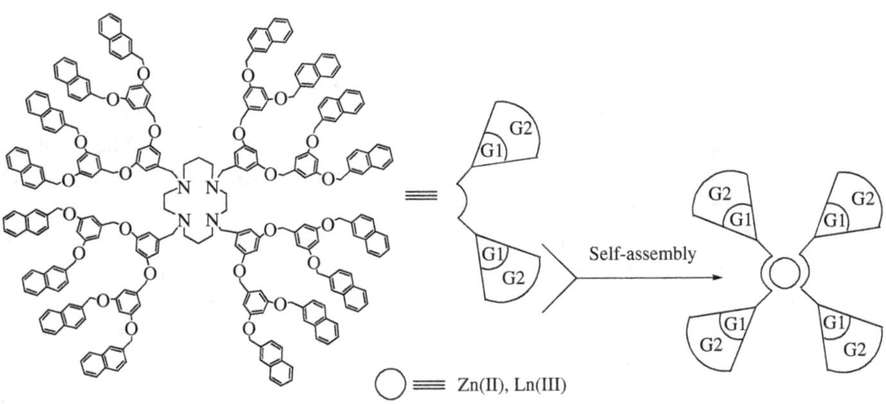

FIGURE 5.9 Self-assembly of cyclam-cored dendrimers with Zn(II) and Ln(III) ions.

expected, by the reappearance of the naphthyl-based fluorescence. Furthermore, no phosphorescence emission of the lanthanides was observed. The nearest coordination sphere seemed to play a key role in the electron transfer efficiency compared to Kawa and Fréchet's dendrimers [34]. These systems have attracted great attention as one could envisage to self-assemble different dendritic wedges around a metal to confer multifunctionalization.

The ease of synthesis and the remarkable properties that core-metallated dendrimers exhibit make them competitive candidates for a large variety of applications. In addition to the examples described above, other strategies that incorporate metals at different locations such as the periphery have been demonstrated and are summarized below.

5.3 DENDRITIC MACROMOLECULES WITH METAL CENTERS AT THE PERIPHERY

For the formation of metallodendrimers of precise nature, a second favorable position in the overall structure for complexation can obviously be at the periphery. Excellent examples of such systems have been reported that include a silicon dendrimer decorated with 243 ferrocenyl units at the periphery with stable redox activity [58]. Catalytic activity of dendrimers with metals located at the periphery has also proven to be of great interest as it has been recently reviewed by several authors [59,60]. Placing photoactive centers at this specific location can nonetheless be more intricate in this case, as demonstrated by the limited number of reported examples.

5.3.1 Ruthenium Complexes

Dendrimers have displayed great potential to convert solar energy into electricity [60,61]. In addition, a Ru(II) dye molecule can typically be used as a source of electron upon light excitation, as initially demonstrated for applications in solar cells with an overall conversion efficiency of $\eta = 7.12\%$ [62]. This study was a great source of motivation, and since then, dendrimers with Ru(II) polypyridyl moieties at the periphery (e.g., the one depicted in Fig. 5.10) have revealed great potential when investigated for charge transfer photosensitizers in nanocrystalline TiO_2 surfaces [63]. More interestingly, in this study, the G1 dendrimer displayed best photovoltaic performance ($\eta = 1.80\%$) compared to G2 ($\eta = 0.95\%$). This is probably due to an uneven absorption of the metal centers in the thin film for the larger macromolecule.

Reports of the studies with small PAMAM dendrimers functionalized with polypyridyl complexes, synthesized by starting from dendritic poly-bipyridyl ligands and using well-known coordination chemistry of Ru(II) complexes as before, have also appeared [64]. Electrochemiluminescent properties of different adducts were found to be dependent on the length of the spacer, and an increase in the length provoked an increase in the luminescence, and the structure in Fig. 5.11 was shown to give the best results.

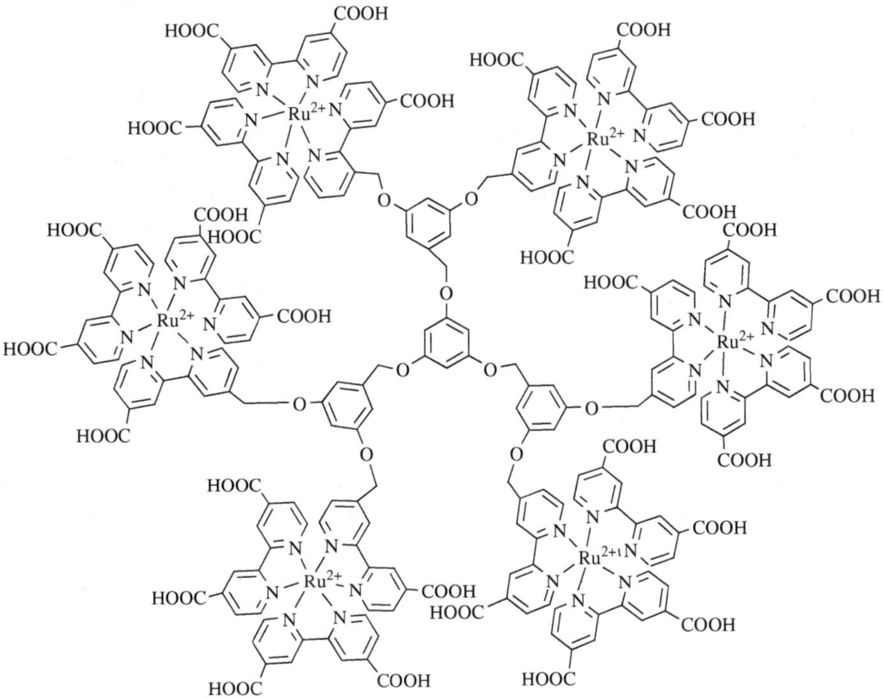

FIGURE 5.10 Second-generation Ru(II)-based dendrimer for solar cell purposes.

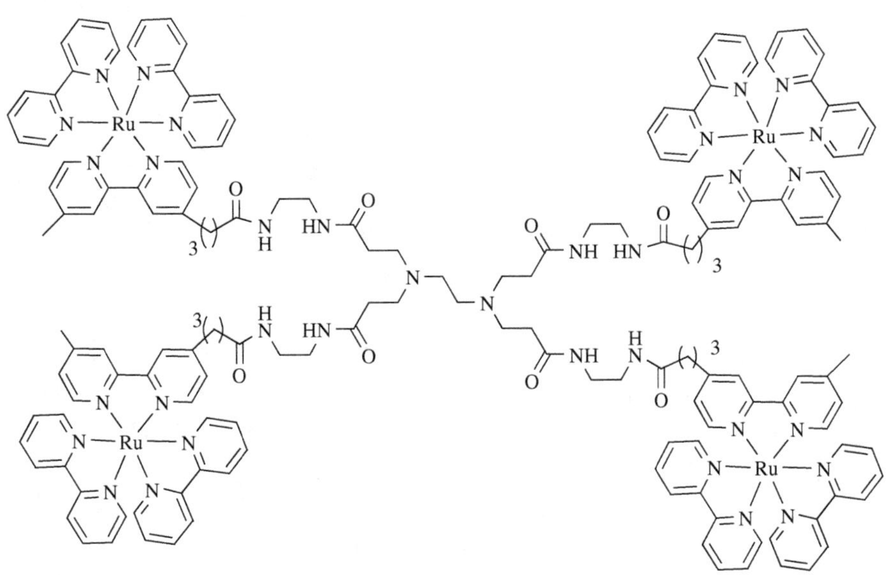

FIGURE 5.11 Polynuclear Ru(II) complex with brilliant luminescent properties.

5.3.2 Multiporphyrin Systems

As core of dendrimers, porphyrins have demonstrated valuable properties including their use for artificial light-harvesting based on an antenna effect. Porphyrins as end groups of regularly branched nanosized macromolecules posed significant challenges due to tremendous difficulties in their syntheses with poor yields. In the years including and following 2000, reports by Aida and coworkers gave some evidence of the potential of dendrimers appended with porphyrins for applications in a diverse range of fields. For example, construction of supramolecular assemblies with Zn-metallated multiporphyrin flexible dendrimers and bipyridyl guest molecules with various chain length spacer was reported [65]. They observed an association pattern using multiple and various spectroscopic studies, indicating that the selective intercalation of the guests into the dendrimers by self-assembly between the zinc atoms and the pyridine moieties led to an increase in the quantum yields. More complexed self-assembled structures were obtained with the construction of segregated arrays formed by the same dendrimers and intercalated bipyridine guests with multiple C_{60}-fullerene units [66]. Photoexcitation of the zinc porphyrin donors resulted in an electron transfer (PET) to the fullerene acceptors to generate charge separation, which was dependent on the generation and the number of C_{60}-units, with the recombination process remaining similar for all. Self-assembled systems, including a maximum of porphyrin and fullerene units (see Fig. 5.12) provided best results for charge separation, in this case, for a remarkable dendritic effect.

Other groups have also studied such macrocycles. For example, extending the molecular chain with a single carbon that connects the dendrimers with the porphyrins on the outer shell (Fig. 5.13) resulted in a significant improvement in the electron transfer process [67]. The authors reported that only ~10% of the excitation intensity was needed for 64 Zn(II) porphyrin-based structure with a longer connecting chain to

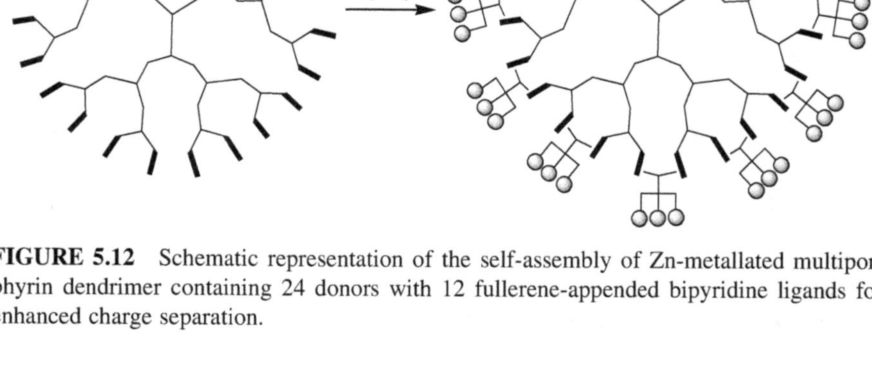

FIGURE 5.12 Schematic representation of the self-assembly of Zn-metallated multiporphyrin dendrimer containing 24 donors with 12 fullerene-appended bipyridine ligands for enhanced charge separation.

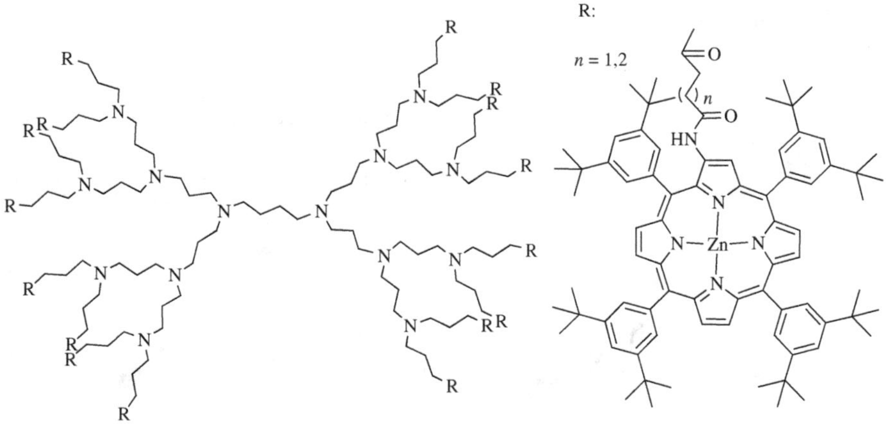

FIGURE 5.13 Poly(propylene imine) dendrimers with peripheral Zn-metallated porphyrins.

observe the same extent of annihilation throughout a complex two-step delayed process.

5.3.3 Gadolinium at the Periphery as a Contrast Agent

Magnetic resonance imaging (MRI) is a noninvasive diagnostic way to visualize tumor, and this technique has been developed at a fast pace in the recent past. Gadolinium has become one of the most studied candidates for contrast agents because it can be easily detected at concentrations ~100-fold compared to iodine. For efficiency, a well-defined dendrimer conjugated with a target moiety could be of interest for monitoring tumor efficiently. To demonstrate the potential of such a system, nanoprobes based on bifunctional dual-labeled dendrimers with covalently attached Gd(III)-DTPA chelates and near-infrared (NIR) fluorescent dye units (Cy 5.5 succinimidyl amide) have been developed for tumor mapping using fluorescence imaging [68].

Gd is also known to generate Auger and internal conversion electrons efficiently by irradiation with a neutron beam. Based on this, an increase of the local concentration of Gd should enhance the properties for tumor cell detection, and Gd neutron capture therapy could be more efficient. For this purpose, a novel water-soluble PAMAM-based dendrimers terminated with DPTA extremities covered with 254 Gd atoms were prepared with excellent results [69].

5.4 METAL CENTERS AT VARIED LOCATIONS

5.4.1 Homonuclear and Heteronuclear Dendritic Complexes

Coordination chemistry also offers an excellent way to create divergent points, and some groups have taken advantage of this specificity to synthesize large metallated

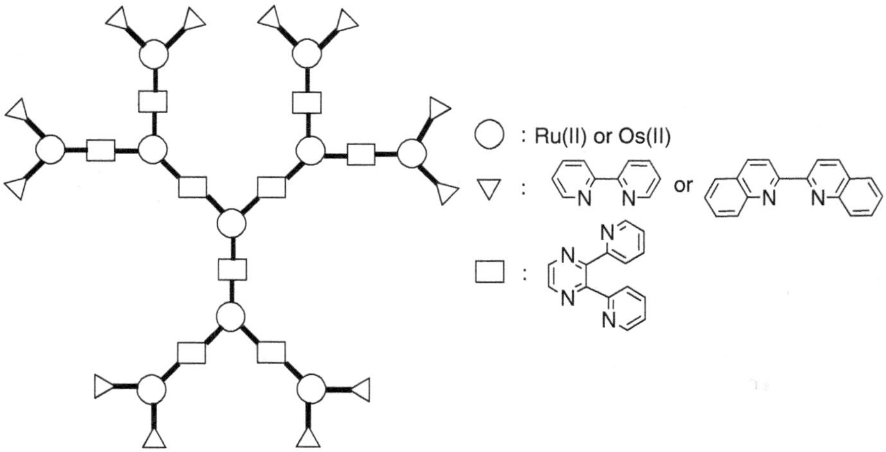

FIGURE 5.14 Decanuclear homo-polypyridine and hetero-polypyridine complexes.

macroassemblies. For example, the construction of a planar and hexagonal fractal-based nanoassembly with a size of 12.5 nm, including 36 Ru(II) and 6 Fe(II) atoms stabilized with polypyridine moieties has been reported [70]. AFM study of the latter gave a very clear clue on how self-assembly occurred during the synthesis leading to hexagonal structures. Other similar examples were reported [71] in the literature including the one with Ru(II) and/or Os(II) connectivity to form patterns encompassing 12 metal atoms at predetermined positions (Fig. 5.14). These polynuclear oligomers with transition metals can lead to unique luminescent properties with energy transfer driven by light. A suitable choice of the metal location can indeed channel the photons into a specific direction (from the center to the periphery or vice versa) resulting in potential photochemical molecular devices.

5.4.2 Phenylazomethine Dendrimers

A study of the photophysical properties of phenylazomethine dendrimers (PAD) [72] demonstrated an excellent behavior toward tailor-made metal complexation [73]. During the titration of a G4 dendrimer with $SnCl_2$, it reported that four different isosbestic points could be distinguished: after addition of 2, 4, 8, and finally 16 equivalents (Fig. 5.15). The latter corresponds to a stepwise radial complexation of each layer of the dendrimer starting from the innermost ones to the periphery. These controlled and specific complexation occurs due to a gradient of basicity of the imines of each generation accompanied with a strengthening of the visible absorption around $\lambda = 420$ nm. Modification of the core with the electron-withdrawing fluorine imines confirmed this assumption by changing the order of the assembly, the core being complexed at the end by the addition of two final equivalents of Sn^{2+} as monitored by UV–vis [74]. Experiments were also conducted with a different Lewis acid ($FeCl_3$) and proved to be as successful as for tin [75].

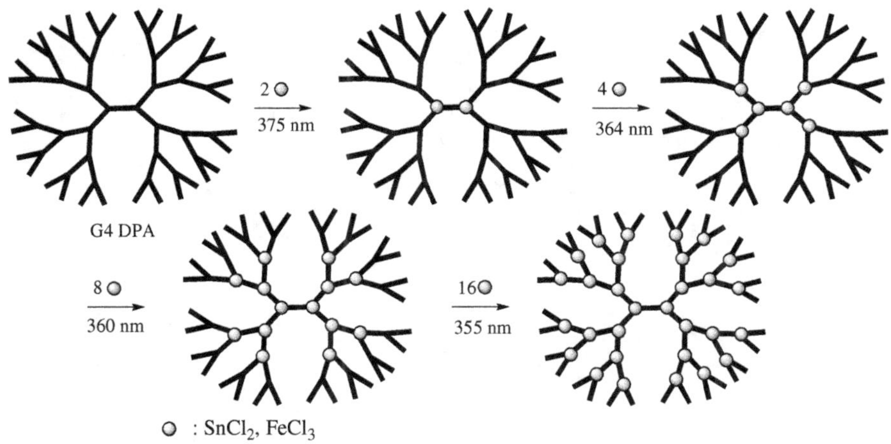

FIGURE 5.15 Schematic representation of the radial step-by-step complexation of a G4 PAD dendrimer with the wavelength of the isosbestic point for each layer on addition of tin chloride.

The above-mentioned macrostructures demonstrated really good potential as DLEDs with an excellent performance. It was demonstrated that the presence of tin chloride layer in a two-layer device improved the light-emission efficiency more than 20 times and reduced the threshold voltage by almost 5 eV [76]. In addition, solar cells were also developed with a TiO_2-based device. The Ru sensitized dye was coated with G1–G5 dendrimers and a generational increase was noted in the electron transfer, resulting in an increase of the current density, especially when the dendrimers were complexed with tin chloride. Through a simple manipulation, such as coating, metallated dendrimers afforded a 30% enhancement in the energy conversion efficiency by providing a shielding effect for the dye [77].

5.4.3 Giant Multiporphyrin Arrays

Although examples of phthalocyanine assemblies have not been reported, giant multiporphyrin arrays represent an equally attractive field of research [78]. These dendritic arrays can be prepared in a convergent or divergent manner. An example of shape-persistent convergently synthesized macroassembly is depicted in Fig. 5.16, and it was noted that yields become moderate with an increase in the number of porphyrin units [79]. The design consists of a porphyrin linked to Zn-metallated multiporphyrin branches. This provides an efficient platform for energy migration among the metallated donor centers before transfer to the acceptor with a slightly decreasing efficiency with an increasing number of metallated centers (yield is 92% in toluene for the example described in Fig. 5.16 with 20 donor centers). The efficiency of the energy transfer was proven to be dramatically shape-dependent. A star-shaped series showed a decrease in electron transfer in THF from 87% to 71% for G1–G3, respectively, whereas for a triangle-shaped series, the decrease was much more significant with 86% for G1, but only 7% for G3. As a result, electron transfer seems to

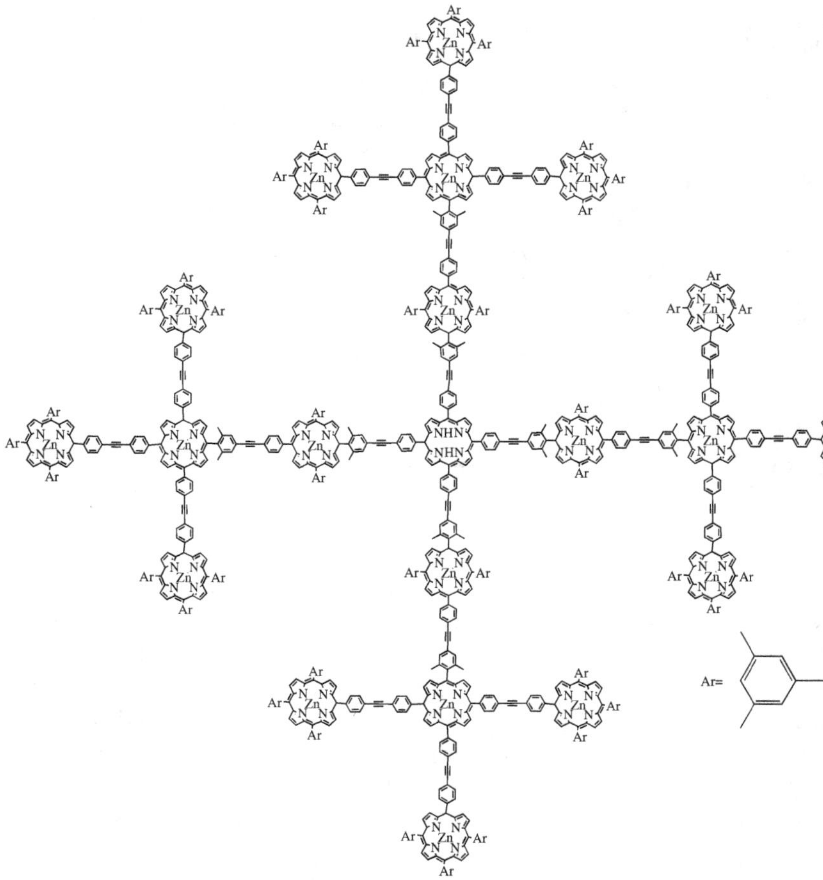

FIGURE 5.16 Multiporphyrin array with suitable geometry for high energy transfer between the metallated donors and the free-base acceptors.

be more efficient with a suitable geometry for the donors around the acceptor. The central moiety needs to be able to trap the energy and a highly congested architecture proved to be more useful for this process [80,81].

These examples demonstrate how flexible the dendritic architecture is in terms of elaborating extremely complex but still well-defined metallated macroassemblies. Assembly components can easily be tailored to perform specific tasks such as mimicking light-harvesting systems present in nature.

5.5 DENDRIMERS FOR METAL SENSING APPLICATIONS

Supramolecular chemistry that invokes weak and noncovalent interactions between a dendrimer and molecules of interest is of great significance, and has been investigated using a diverse range of methodologies. For example, dendrimers can be designed to

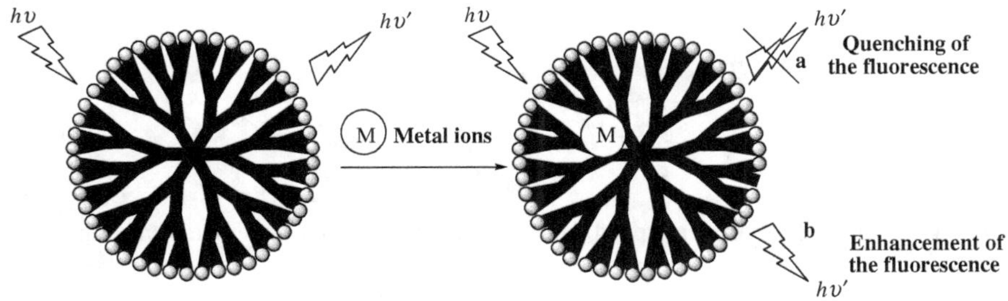

FIGURE 5.17 Fluorescent dendrimers for metal sensing by (a) quenching or (b) enhancement of the emission.

allow interactions similar to host–guest type interaction [82] with desired species, to form micelle-type assemblies under specific conditions. In addition, suitable complexation moieties can be incorporated into the scaffold as the dendrimers provide a dense platform of interaction with external agents. Dendrimers containing organic photoactive groups at the periphery are excellent candidates for metal sensor applications via signal amplification (Fig. 5.17). Many reports have been described recently that offer a broad range of possibilities for fast and efficient detection of metals [83]. However, could dendrimers enlarge this library by bringing higher selectivity or sensitivity? The latter remains to be demonstrated.

Balzani et al. carried out detailed studies related to pH sensing of various macromolecules [84] including metal ions. After reporting the modification of PPI dendrimers with dansyl groups at the periphery, Balzani's group in collaboration with Vögtle demonstrated that for a G4 PPI dendrimer bearing 32 dansyl chromophores, a 1 : 1 ratio with Co^{2+} was sufficient at low concentration to completely quench the emission (disappearance of the dansyl emission band at $\lambda = 514$ nm) [85,86]. They suggested that the quenching phenomenon upon excitation of the dansyl units was due to complexation of the cations with one of the 30 aliphatic amines of the backbone preventing any emission to take place.

This pioneering example for sensor applications was followed by many other systems in which dendritic fluorescent chemosensors were studied [87]. PAMAM dendrimers containing 1,8-naphthalimide chromophores with linked allyamine or diamine pincers were synthesized. The allyamine pincers stitched to a G2 dendrimer were found to be excellent candidates for Zn(II) detection by providing almost complete disappearance of the fluorescence emission (7% remaining after reaching $[Zn(II)] = 10^{-4}\,M$ in the solution) [88]. The case of the diamine pincers (Fig. 5.18) was even more interesting in which the presence of cations such as Zn(II), Fe(III), or Ni(II) enhanced the emission (with a 95% increase in the case of Ni(II)) [89,90].

More recently, studies using small PAMAM dendrimers functionalized with naphthyl units at the periphery have demonstrated their ability toward detection of cationic species such as Cd(II) [91] or Al(III), Zn(II), and Cu(II) [92]. Analytical studies demonstrated a decrease in the exciplex emission during the titration process.

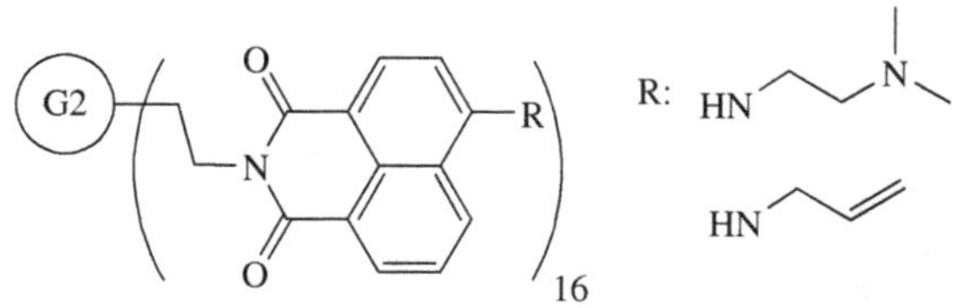

FIGURE 5.18 G2 dendrimer capped with 1,8-naphthalimide chromophores encompassing allyamine or diamine pincers for metal sensing.

Sensing has also been extended to luminescent metal ions including rare-earth Nd(III). The latter ion emits in the NIR region, and became of interest for applications dealing with long-range optical data transport, compared to commonly studied Eu(III) or Tb(III) ions that exhibit properties in the visible spectral region. For this purpose, dansyl-based dendrimers with 18 amide groups were synthesized and examined by titration experiments. A 1:1 ratio was found to be enough to observe complete disappearance of the dansyl emission with the Nd(III) being coordinated with the amide. However, sensitization of the Nd(III) NIR emission was also observed at $\lambda = 1064$ nm. The light-harvesting antenna system was found to occur in which the dendritic backbone prevented quenching by dioxygen and allowed luminescence to occur through a Förster-type mechanism. More recently, a three-component self-assembled device comprising a cyclam-cored dendrimer appended with 16 naphthyl units, a $[Ru(bpy)_2(CN)_2]$ complex, and a Nd(III) luminescent (Fig. 5.19) ion was

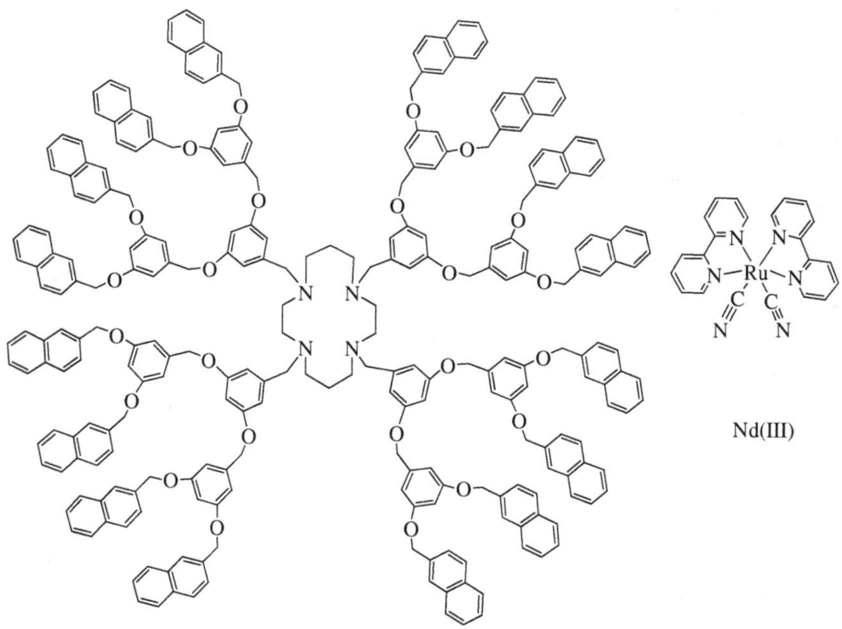

FIGURE 5.19 Three-component self-assembled device for sensitization of Nd(III) with emission in the NIR region.

investigated [93] for its light-harvesting and light-conversion properties, and an efficient electron transfer was achieved under specific conditions. The authors described that upon excitation of the naphthyl units in the UV region and the Ru(II) complex in the UV–vis region, sensitization of the Nd(III) ion was observed in the NIR region. It was noted that the Ru(II) complex serves as a mediator in the energy transfer. Studies were also carried out using Gd(III) for comparison, but selectivity of the system toward lanthanide ions having low-lying excited states hampered such a phenomenon from taking place in this case.

A more common strategy with PAMAM dendrimers appended with peripheral 1,8-naphthalimide was subsequently reported in which it was demonstrated that six different rare-earth cations can enhance the fluorescence by PET without modification of the wavelength of emission at $\lambda \cong 384$ nm and $\lambda \cong 478$ nm [94]. In addition, during their investigation, sensitivity was achieved with this system for Yb(III) being the best candidate (~3.1-fold increase), followed by Er(III) and Gd(III), Nd(III). On treatment with trifluoroacetic acid, and a detailed investigation of the photophysical properties [95], it was concluded that emission can be tailored by combining signals from an acid and lanthanide metal ion through interactions with the amide moieties of the backbone to create a rearrangement of the orbitals creating a ~8–10-fold increase in the emission for the six cations, even in the case of Nd(III) this time.

5.6 CONCLUSIONS AND PERSPECTIVES

To summarize, interactions between metals and dendrimers represent clearly a fertile, productive, and expandable domain of research. As this chapter has attempted to demonstrate, dendrimers offered an excellent and useful medium to build numerous and diversified nanodevices for a variety of applications in a diverse range of fields due to their versatility and adaptability. These two properties of dendrimers are the significant key elements to remember. In fact, the dendritic architecture is extremely "malleable" and their exploitation by chemistry is the only limiting point to the endless possibilities these macromolecules offer in the development of novel photochemically driven nanodevices. Metallodendrimers as such offer a rich area in terms of their photophysical properties. Their potential is continuously being exploited and constitutes a high topical area of research. It is hoped that in the near future, fluorescent dendrimers could probably be tailored on demand for specific interaction. This will allow further developments in the area of metal detection with a very wide range of probes, providing excellent selectivity and/or sensitivity toward a specific metal. However, fundamental research still needs to be done to obtain a better understanding of the structure–property relationships in metallated macroassemblies. It is expected that we shall continue to see this area flourish, and these novel and intriguing architectures will find ever increasing synergy with numerous fields including material science and photonics.

REFERENCES

1. D.A. Tomalia and J.M.J. Fréchet, *J. Polym. Sci., Part A: Polym. Chem.* **2002**, *40*, 2719–2728.
2. G.R. Newkome, C.N. Moorefield, and F. Vögtle, *Dendrimers and Dendrons: Concepts, Syntheses, Applications*, Wiley-VCH, Weinheim, **2001**.
3. J.-P. Majoral and A.-M. Caminade, *Chem. Rev.* **1999**, *99*, 845–880.
4. G.R. Newkome, E. He, and C.N. Moorefield, *Chem. Rev.* **1999**, *99*, 1689–1746.
5. A.W. Bosman, H.M. Janssen, and E.W. Meijer, *Chem. Rev.* **1999**, *99*, 1665–1688.
6. D.A. Tomalia, A. Naylor, and W.A.I. Goddard, *Angew. Chem., Int. Ed. Engl.* **1990**, *29*, 138–175.
7. E. Buhleier, W. Wehner, and F. Vögtle, *Synthesis* **1978**, 155–158.
8. D.A. Tomalia, H. Baker, J. Dewald, M. Hall, C. Kallos, S. Martin, J. Roeck, J. Ryder, and P. Smith, *Polym. J.* **1985**, *17*, 117–132.
9. S.M. Grayson and J.M.J. Fréchet, *Chem. Rev.* **2001**, *101*, 3819–3868.
10. K.L. Wooley, C.J. Hawker, and J.M.J. Fréchet, *J. Am. Chem. Soc.* **1991**, *113*, 4252–4261.
11. C.J. Hawker and J.M.J. Fréchet, *J. Am. Chem. Soc.* **1990**, *112*, 7638–7647.
12. G. Franc and A. Kakkar, *Chem. Commun.* **2008**, 5267–5276.
13. T. Weil, E. Reuther, C. Beer, and K. Müllen, *Chem. Eur. J.* **2004**, *10*, 1398–1414.
14. T. Weil, E. Reuther, and K. Müllen, *Angew. Chem. Int. Ed.* **2002**, *41*, 1900–1904.
15. A. Adronov, S.L. Gilat, J.M.J. Fréchet, K. Ohta, F.V.R. Neuwahl, and G.R. Fleming, *J. Am. Chem. Soc.* **2000**, *122*, 1175–1185.
16. W.R. Dichtel, J.M. Serin, C. Edder, J.M.J. Fréchet, M. Matuszewski, L.-S. Tan, T.Y. Ohulchansskyy, and P.N. Prasad, *J. Am. Chem. Soc.* **2004**, *126*, 5380–5381.
17. D.W. Brousmiche, J.M. Serin, J.M.J. Fréchet, G.S. He, T.-C. Lin, S.J. Chung, and P.N. Prasad, *J. Am. Chem. Soc.* **2003**, *125*, 1448–1449.
18. A. Myc, I.J. Majoros, T. Thomas, and J.R. Baker, *Biomacromolecules* **2007**, *8*, 13–18.
19. K.M. Kitchens, R.B. Kolhatkar, P.W. Swaan, N.D. Eddington, and H. Ghandehari, *Pharm. Res.* **2006**, *23*, 2818–2826.
20. M. Poupot, L. Griffe, P. Marchand, A. Maraval, O. Rolland, L. Martinet, F.E. L'Faqihi-Olive, C.-O. Turrin, A.-M. Caminade, J.-J. Fournie, J.-P. Majoral, and R. Poupot, *FASEB J.* **2006**, *20*, 2339–2351.
21. I.J. Majoros, A. Myc, T. Thomas, C.B. Mehta, and J.R. Baker, *Biomacromolecules* **2006**, *7*, 572–579.
22. S.-H. Hwang, C.N. Moorefield, and G.R. Newkome, *Chem. Soc. Rev.* **2008**, *37*, 2543–2557.
23. P.L. Burn, S.-C. Lo, and I.D.W. Samuel, *Adv. Mater.* **2007**, *19*, 1675–1688.
24. L. Brauge, G. Vériot, G. Franc, R. Deloncle, A.-M. Caminade, and J.-P. Majoral, *Tetrahedron* **2006**, *62*, 11891–11899.
25. P. Furuta, J. Brooks, M.E. Thompson, and J.M.J. Fréchet, *J. Am. Chem. Soc.* **2003**, *125*, 13165–13172.
26. D. Astruc, C. Ornelas, and J. Ruiz, *Acc. Chem. Res.* **2008**, *41*, 841–856.
27. D. Astruc and F. Chardac, *Chem. Rev.* **2001**, *101*, 2991–3024.

28. M. Petrucci-Samija, V. Guillemette, M. Dasgupta, and A.K. Kakkar, *J. Am. Chem. Soc.* **1999**, *121*, 1968–1969.
29. R. Hourani, M.A. Whitehead, and A. Kakkar, *Macromolecules* **2008**, *41*, 508–510.
30. T. Muraoka, K. Kinbara, and T. Aida, *Nature* **2006**, *440*, 512–515.
31. B.J. Holliday and C.A. Mirkin, *Angew. Chem. Int. Ed.* **2001**, *40*, 2022–2043.
32. F. Vögtle, M. Plevoets, M. Nieger, G.C. Azzellini, A. Credi, L. de Scola, V. de Marchis, M. Venturi, and V. Balzani, *J. Am. Chem. Soc.* **1999**, *121*, 6290–6298.
33. M. Plevoets, F. Vögtle, L. de Cola, and V. Balzani, *New. J. Chem.* **1999**, *23*, 63–69.
34. M. Kawa and J.M.J. Fréchet, *Chem. Mater.* **1998**, *10*, 286–296.
35. C. Pitois, A. Hult, and M.J. Lindgren, *J. Lumin.* **2005**, *111*, 265–283.
36. W.R. Glomm, M.-H. Glomm Ese, S. Volden, C. Pitois, A. Hult, and J. Sjöblom, *Physicochem. Eng. Aspects* **2007**, *299*, 186–197.
37. R. Sadamoto, N. Tomioka, and T. Aida, *J. Am. Chem. Soc.* **1996**, *118*, 3978–3979.
38. O. Finikova, A. Galkis, V. Rozkov, H. Cordero, C. Hagerhall, and S. Vinogradov, *J. Am. Chem. Soc.* **2003**, *125*, 4882–4893.
39. V. Rozhkov, D. Wilson, and S. Vinogradov, *Macromolecules* **2002**, *35*, 1991–1993.
40. S. Hecht and J.M.J. Fréchet, *Angew. Chem., Int. Ed.* **2001**, *40*, 74–91.
41. N. Tomioka, D. Takasu, T. Takahashi, and T. Aida, *Angew. Chem., Int. Ed.* **1998**, *37*, 1531–1534.
42. M. Kimura, K. Nakada, Y. Yamaguchi, K. Hanabusa, H. Shirai, and N. Kobayashi, *Chem. Comm.* **1997**, 1215–1216.
43. J. Leclaire, R. Dagiral, S. Fery-Forgues, Y. Coppel, B. Donnadieu, A.-M. Caminade, and J.-P. Majoral, *J. Am. Chem. Soc.* **2005**, *127*, 15762–15770.
44. J. Leclaire, R. Dagiral, A. Pla-Quintana, A.-M. Caminade, and J.-P. Majoral, *Eur. J. Inorg. Chem.* **2007**, *18*, 2890–2896.
45. M. Brewis, G.J. Clarkson, V. Goddard, M. Helliwell, A.M. Holder, and N.B. McKeown, *Angew. Chem. Int. Ed.* **1998**, *37*, 1092–1094.
46. M.A. Baldo, S. Lamansky, P.E. Burrows, S.R. Forrest, and M.E. Thompson, *Appl. Phys. Lett.* **1999**, *75*, 4–6.
47. C. Adachi, M.A. Baldo, S.R. Forrest, S. Lamansky, M.E. Thompson, and R.C. Kong, *Appl. Phys. Lett.* **2001**, *78*, 1622–1624.
48. C. Adachi, R.C. Kong, P. Djurovich, V. Adamovich, M.A. Baldo, M.E. Thompson, and S. R. Forrest, *Appl. Phys. Lett.* **2001**, *79*, 2082–2084.
49. S.-C. Lo, N.A.H. Male, J.P.J. Markham, S.W. Magennis, P.L. Burn, O.V. Salata, and I.D.W. Samuel, *Adv. Mater.* **2002**, *14*, 975–979.
50. J.P.J. Markham, S.-C. Lo, S.W. Magennis, P.L. Burn, and I.D.W. Samuel, *Appl. Phys. Lett.* **2002**, *80*, 2645–2647.
51. J. Ding, J. Lü, Y. Cheng, Z. Xie, L. Wang, X. Jing, and F. Wang, *Adv. Funct. Mater.* **2008**, *18*, 2754–2762.
52. S. Li, G. Zhong, W. Zhu, F. Li, J. Pan, W. Huang, and H. Tian, *J. Mater. Chem.* **2005**, *15*, 3221–3228.
53. S. Li, G. Zhong, W. Zhu, F. Li, J. Pan, W. Huang, and H. Tian, *Chem. Lett.* **2005**, *34*, 688–689.

54. C. Saudan, V. Balzani, P. Ceroni, M. Gorka, M. Maestri, V. Vicinelli, and F. Vögtle, *Tetrahedron* **2003**, *59*, 3845–3852.
55. C. Saudan, V. Balzani, M. Gorka, S.-K. Lee, M. Maestri, V. Vicinelli, and F. Vögtle, *J. Am. Chem. Soc.* **2003**, *125*, 4424–4425.
56. G. Bergamini, C. Saudan, P. Ceroni, M. Maestri, V. Balzani, M. Gorka, S.-K. Lee, J. van Heyst, and F. Vögtle, *J. Am. Chem. Soc.* **2004**, *126*, 16466–16471.
57. C. Saudan, P. Ceroni, V. Vicinelli, M. Maestri, V. Balzani, M. Gorka, S.-K. Lee, J. van Heyst, and F. Vögtle, *Dalton Trans.* **2004**, 1597–1600.
58. S. Nlate, J. Ruiz, V. Sartor, R. Navarro, J.-C. Blais, and D. Astruc, *Chem. Eur. J.* **2000**, *6*, 2544–2553.
59. A.-M. Caminade, P. Servin, R. Laurent, and J.-P. Majoral, *Chem. Soc. Rev.* **2008**, *37*, 56–67.
60. S.-H. Hwang, C.D. Shreiner, C.N. Moorefield, and G.R. Newkome, *New J. Chem.* **2007**, *31*, 1192–1217.
61. L. Shih-Chun and P.L. Burn, *Chem. Rev.* **2007**, *107*, 1097–1116.
62. B. O'Regan and M. Grätzel, *Nature* **1991**, *353*, 737–740.
63. A.H. Younes and T.H. Ghaddar, *Inorg. Chem.* **2008**, *47*, 3408–3414.
64. D.N. Lee, J.K. Kim, H.S. Park, Y.M. Jun, R.Y. Hwang, W.-Y. Lee, and B.H. Kim, *Synth. Met.* **2005**, *150*, 93–100.
65. J. Yang, S. Cho, H. Yoo, J. Park, W.-S. Li, T. Aida, and D. Kim, *J. Phys. Chem. A* **2008**, *112*, 6869–6876.
66. W.-S. Li, K.S. Kim, D.-L. Jiang, H. Tanaka, T. Kawai, J.H. Kwon, D. Kim, and T. Aida, *J. Am. Chem. Soc.* **2006**, *128*, 10527–10532.
67. J. Larsen, B. Brüggeman, T. Khoury, J. Sly, M.J. Crossley, V. Sundström, and E. Åkesson, *J. Phys. Chem. A* **2007**, *111*, 10589–10597.
68. V.S. Talanov, C.A.S. Regino, H. Kobayashi, M. Bernardo, P.L. Choyke, and M.W. Brechbiel, *Nano Lett.* **2006**, *6*, 1459–1463.
69. H. Kobayashi, S. Kawamoto, T. Saga, N. Sato, T. Ishimori, J. Konishi, K. Ono, K. Togashi, and M.W. Brechbiel, *Bioconjug. Chem.* **2001**, *12*, 587–593.
70. G.R. Newkome, P. Wang, C.N. Moorefield, T.J. Cho, P.P. Mohapatra, S. Li, S.-H. Hwang, O. Lukoyanova, L. Echegoyen, J.A. Palagallo, V. Iancu, and S.-W. Hla, *Science* **2006**, *312*, 1782–1785.
71. G. Denti, S. Campagna, S. Serroni, M. Ciano, and V. Balzani, *J. Am. Chem. Soc.* **1992**, *114*, 2944–2950.
72. M. Higuchi, S. Shiki, and K. Yamamoto, *Org. Lett.* **2000**, *2*, 3079–3082.
73. K. Yamamoto, M. Higuchi, S. Shiki, M. Tsuruta, and H. Chiba, *Nature* **2002**, *415*, 509–511.
74. M. Higuchi, M. Tsuruta, H. Chiba, S. Shiki, and K. Yamamoto, *J. Am. Chem. Soc.* **2003**, *125*, 9988–9997.
75. T. Imaoka, R. Tanaka, S. Arimoto, M. Sakai, M. Fujii, and K. Yamamoto, *J. Am. Chem. Soc.* **2005**, *127*, 13896–13905.
76. T. Imaoka, H. Horiguchi, and K. Yamamoto, *J. Am. Chem. Soc.* **2003**, *125*, 8104–8105.
77. N. Satoh, T. Nakashima, and K. Yamamoto, *J. Am. Chem. Soc.* **2005**, *127*, 13030–13038.
78. H. Imahori, *J. Phys. Chem. B* **2004**, *108*, 6130–6143.

79. M.R. Benites, T.E. Johnson, S. Weghorn, L. Yu, P.D. Rao, J.R. Diers, S.I. Yang, C. Kirmaier, D.F. Bocian, D. Holten, and J.S. Lindsey, *J. Mater. Chem.* **2002**, *12*, 65–80.
80. M.-S. Choi, T. Aida, T. Yamazaki, and I. Yamazaki, *Angew. Chem. Int. Ed.* **2001**, *40*, 3194–3198.
81. M.-S. Choi, T. Yamazaki, I. Yamazaki, and T. Aida, *Angew. Chem. Int. Ed.* **2004**, *43*, 150–158.
82. F.G.A. Jansen, E.M.M. de Brabander-van den Beng, and E.W. Meijer, *Science* **1994**, *226*, 1226–1229.
83. C. Kaewtong, G. Jiang, Y. Park, T. Fulghum, A. Baba, B. Pulpoka, and R. Advincula, *Chem. Mater.* **2008**, *20*, 4915–4924.
84. V. Balzani, P. Ceroni, S. Gestermann, M. Gorka, C. Kauffmann, M. Maestri, and F. Vögtle, *Chem. Phys. Chem.* **2000**, *4*, 224–227.
85. F. Vögtle, S. Gestermann, C. Kauffmann, P. Ceroni, V. Vicinelli, and V. Balzani, *J. Am. Chem. Soc.* **2000**, *122*, 10398–10404.
86. V. Balzani, P. Ceroni, S. Gestermann, C. Kauffmann, M. Gorka, and F. Vögtle, *Chem. Comm.* **2000**, 853–854.
87. U. Hahn, M. Gorka, F. Vögtle, V. Vicinelli, P. Ceroni, M. Maestri, and V. Balzani, *Angew. Chem. Int. Ed.* **2002**, *41*, 3595–3598.
88. S. Sali, I. Grabchev, J.-M. Chovelon, and G. Ivanova, *Spectrochim. Acta Part A* **2006**, *65*, 591–597.
89. I. Grabchev, D. Staneva, and R. Betcheva, *Polym. Degrad. Stab.* **2006**, *91*, 2257–2264.
90. I. Grabchev, J.-M. Chovelon, and A. Nedelcheva, *J. Photochem. Photobiol. A: Chem.* **2006**, *183*, 9–14.
91. J. Kawakami, T. Isobe, Y. Sasaki, S. Ito, M. Nagaki, and H. Kitaharan, *Anal. Sci.* **2005**, *21*, 729–730.
92. J. Kawakami, T. Mizuguchi, and S. Ito, *Anal. Sci.* **2006**, *22*, 1383–1384.
93. C. Giansante, P. Ceroni, V. Balzani, and F. Vögtle, *Angew. Chem. Int. Ed.* **2008**, *47*, 5422–5425.
94. Q.-Q. Chen, L. Lin, H.-M. Chen, S.-P. Yang, L.-Z. Yang, and X.-B. Yu, *J. PhotoChem. Photobiol. A* **2006**, *180*, 69–74.
95. S.-P. Yang, L. Lin, L.-Z. Yang, J.-M. Chen, Q.-Q. Chen, D. Cao, and X.-B. Yu, *J. Lumin.* **2007**, *126*, 515–530.

6

PHOTOCHROMIC POLYMERS FOR OPTICAL DATA STORAGE: AZOBENZENES AND PHOTODIMERS

AVTAR S. MATHARU AND P.S. RAMANUJAM

6.1 INTRODUCTION

This chapter discusses photochromic azobenzene polymers and photochromic molecules capable of undergoing photodimerization as materials for potential use in optical data storage applications. A brief introduction to photochromism, including recent examples of some photochromic systems, other than azobenzenes, based on variety of applications is presented at the outset. Thereafter, we shall focus on azobenzene photoaddressable polymers (PAPs) and [4 + 4]-photocycloaddition as materials for ultrahigh capacity optical data storage.

Photochromism is defined as a light-induced reversible change of color, and photochromic compounds are chemical species having photochromic properties [1].

Photochemistry and Photophysics of Polymer Materials, Edited by Norman S. Allen

$$\text{Isomer A} \underset{h\nu_1 \text{ or } \Delta}{\overset{h\nu_1}{\rightleftharpoons}} \text{Isomer B}$$

FIGURE 6.1 Photochromic principle.

Perhaps the first observed report of *photochromism* was by Fritzsche who noted a color change in a solution of tetracene and a reversal of the process in darkness [2]. However, it was almost one hundred years later that Hirshberg suggested the name *photochromism* to describe the phenomenon [3].

Photochromic reactions are usually based on unimolecular processes such as intramolecular ring opening/closing and *cis-trans*-photoisomerization. In photochromic experiments, a thermodynamically stable form of a species **A**, on irradiation with light is transformed to corresponding species **B** (see Fig. 6.1). The species **B** can return to the original state, species **A**, through thermal or photochemical processes. Usually, the stable species **A** absorb short wavelength light and is converted to species **B**, which absorb light at longer wavelengths. This effect is called *positive photochromism*; in some cases, the opposite occurs in which case, the effect is termed *negative photochromism*. If the photogenerated isomers are unstable and revert thermally to their initial state in the dark, then they are termed *T-type* (thermally reversible type, e.g., spiropyrans and azobenzenes). Photogenerated isomers that are thermally irreversible, but photochemically reversible are termed *P-type* (e.g., furylfulgides and diarylethenes) and applicable to optical memory, photooptical switching, and displays [4].

Thus, photochromic materials are of significant scientific, technological, and commercial interest because photoconversion and photoreversion modulates a multitude of physical properties not just color: geometrical shape, dipole moment, refractive index, birefringence, conductivity, magnetism, hydrophilicity, hydrophobicity, adhesion, and so forth [5–8]. The most obvious and widespread application of photochromic materials is in the use of sunglasses that darken on exposure to bright sunlight, and regain their transparency in darkness or low light intensity. New interesting applications include converting light into mechanical energy allowing for the possibility of bending, flexing, swelling, contraction, and motion [9–14]. Interesting polymeric, glass, and gel-like color display applications are possible through the combination of liquid crystals and photochromic materials affording a range of colors [15–18].

Photochromic materials exhibit their properties either as single crystal or in a variety of media: solution phase (liquid); liquid crystalline; sol–gel, amorphous glass, and polymeric. Single crystals afford higher quantum efficiency, but are difficult to engineer. Photochromic polymers have come to the fore because of their potential to be mass-produced on a roll-to-roll basis or can be spin-coated making them commercially attractive. The properties of any photochromic polymer are dependent upon the nature of the photochromic molecule and the nature of the matrix. The photochromic molecule may reside as a *guest* via noncovalent interactions within a *host* polymer matrix or may be covalently bound to the polymer matrix forming an inherently *homogenous* system. The polymer matrix may either increase or decrease reversibility and this information is important as it dictates the application. It is not our intention to critically appraise polymer matrix effects as they have been discussed in several key texts [6,19].

6.1.1 Examples of Photochromic Systems

Perhaps the largest class of organic photochromic molecules are those based on symmetry-allowed electrocyclic ring closure and opening. A variety of examples exist: spiropyrans, naphthopyrans, fulgides, and diarylethenes.

6.1.1.1 Reactions Involving Carbon–Oxygen Bond Cleavage: Spiropyrans, and Naphthopyrans (Chromenes) Hirshberg first reported photochromism in spiropyrans in 1950s ([1], references cited within). The spiropyran-merocyanine photochromic switch relies on UV-induced (approximately 360 nm) photolysis of the spiro-carbon–oxygen (C_{spiro}–O) bond in spiropyran (colorless closed form) to generate the intensely colored open form, merocyanine. A noticeable bathochromic shift from the UV-region to the visible region occurs as the open form of merocyanine favors improved planarity, orbital overlap, and electronic conjugation. Additionally, a large dimensional change occurs because an approximately 90° rotation of one half of the molecule occurs during conversion from ring-open (merocyanine) to ring-closed (spiropyran) forms. The reverse process may be mediated either thermally or optically by irradiation with visible light, but may be slowed by careful selection of substituents on the spiropyran. For example, 6-NO_2-substituted spiropyrans upon irradiation give a *pseudo*-stable merocyanine form (see Fig. 6.2). The presence of a strong electron-withdrawing effect of the nitro-substituent increases polarizability and dipole moment of the open form inducing aggregate formation.

Formation of aggregates may be evidenced by X-ray and electron diffraction patterns as aggregation induces formation of crystalline domains. Solvent-induced polymer swelling or plastification enhances aggregate formation even thermally without irradiation. Crystallization proceeds via a *cooperative effect*: photochromic groups aggregate and promote thermal spiropyran-merocyanine conversion sustaining crystallization. Krongauz termed this process *zipper crystallization*. However, certain vinyl polymers restrict aggregate formation because of the rigidity of the polymer backbone [20].

Polymer types and polymer matrix effects will obviously affect stability of both states. Recently, spiropyran has been attached to polystyrene Wang resin, which serves as a solid support, to minimize thermal back reaction and improve bistability (see Fig. 6.3). The resultant photochromic microbeads suspended in toluene greatly

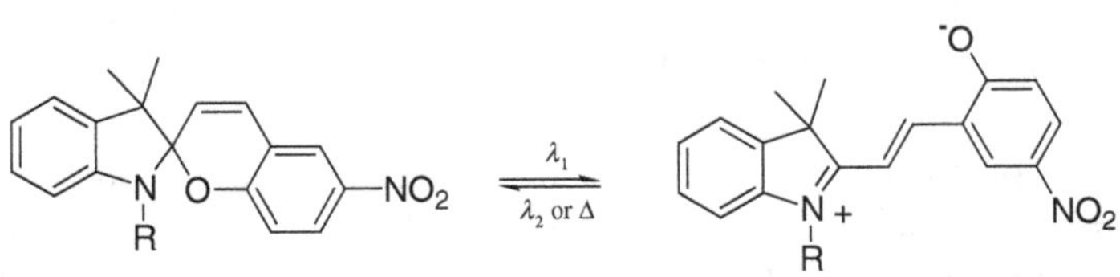

FIGURE 6.2 Photochromic switching in 6-NO_2-spiropyrans.

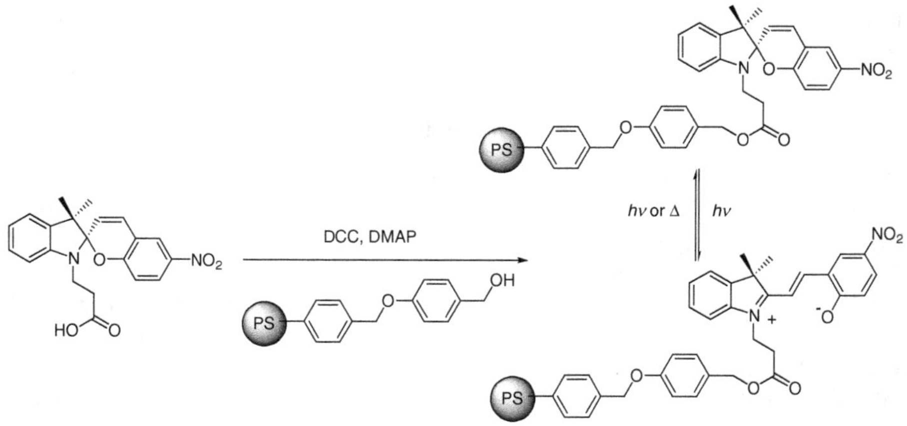

FIGURE 6.3 Photochromic switching of spiropyran attached to a Wang resin.

enhanced thermal bistability and photostability compared with unbound photochromic molecules dissolved in solution. The increased stability is attributed to the steric constraints imposed by the polystyrene beads, which prevent electrocyclic ring closure [21].

Spiropyran molecules have also been utilized as reverse wettable surfaces, whereby, a hydrophilic surface may be photomodulated to a hydrophobic surface and vice versa. Lygeraki et al. utilized the polar (mercocyanine, zwitterionic open-form) and nonpolar (spiropyran closed-form) properties of the spiropyran-merocyanine photochromic transformation to photomodulate surface wettability. 6-NO_2 BIPS (10 wt %) was added to a poly(methyl methacrylate) host matrix and patterned to produce reverse wettable surfaces. However, reversibility is limited to 6–8 irradiation cycles because of degradative photooxidation [22]. Reverse wettable surfaces are commercially important as they may enhance drug delivery or function as a self-cleaning surface.

Spiropyran ring opening and closure is also associated with reversible volume or dimensional change. Surface relief gratings can be formed for potential optical signaling and storage purposes. Ubukata et al. utilized changes in dimensionality with spiropyrans doped in a PMMA matrix to produce deep surface relief gratings upon irradiation at 366 nm, which coincide with the back and forth photoisomerization between the open and closed forms [23]. Similarly, Fragouli et al. have exploited reversible changes in dimensionality to produce optically switchable gratings for use in all-optic signal processing systems. Spiropyran was dissolved (10 wt %; best concentration for homogeneity) in a polyethylmethacrylate-*co*-methacrylate (PEM-MA) polymer matrix to produce thin polymeric gratings that change their diffraction efficiency upon laser irradiation due to changes in dimensionality rather than refractive index [24].

The coloration efficiency of photochromic spirocompounds may be enhanced by careful selection of additional chromophoric substituents. Tomasulo et al. reported amplification of coloration efficiency of photochromic oxazines using a bichromophoric approach [25].

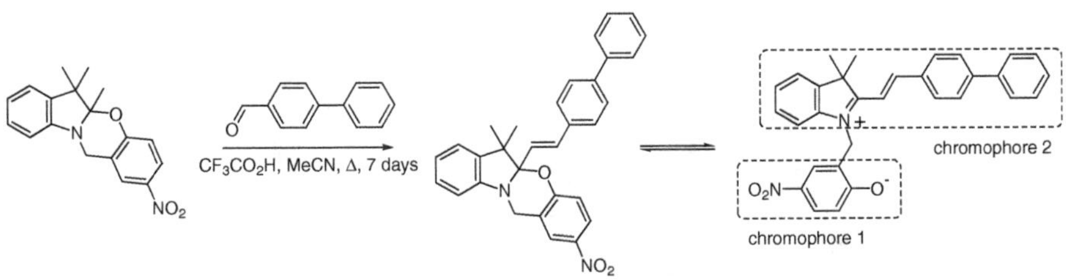

FIGURE 6.4 Bichromophoric spiropyran photochromic system.

The UV-mediated photolysis of the C–O bond produced simultaneously two chromophores (see Fig. 6.4): phenolate anion (chromophore 1) and indolium cation (chromophore 2), which absorb in the same region of the visible spectrum resulting in an amplification (doubling) of the coloration efficiency.

The photochromic behavior of naphthopyrans (chromene), such as spiroxazine, is based on one half of the molecule undergoing a 90° internal rotation following cleavage of the spiro-carbon–oxygen. Naphthopyrans are known to possess good fatigue resistance and represent one of the best sources of yellow and red colors. Evans' Group have exploited the use of photochromic naphthopyran-polymer conjugates to control the switching performance of the photochromic moiety through its positioning within the polymer (see Fig. 6.5). The covalent attachment of a polymer directly to a dye allowed control of the immediate local environment surrounding each individual dye molecule without compromising the properties of the bulk polymeric matrix in which it is contained.

Photochromic performance was enhanced when the naphthopyran moiety was located in the *middle* rather than at the terminus. A *middle* disposition of the naphthopyran provides better encapsulation by the polymeric arms giving improved coloration and decoloration rates and higher optical density in the colored open-state [26].

6.1.1.2 Reactions Involving Carbon–Carbon Bond Cleavage: Diarylethenes (DTEs), Fulgides, and Fulgimides The photogenerated isomers of diarylethenes (DTEs) and fulgides are thermally irreversible (P-type) and lend themselves to being very useful photochromic switches because they exhibit bistability and high quantum efficiencies for forward and reverse reactions [7,27,28]. DTEs exhibit better fatigue resistance and have higher durability than fulgides.

Diarylethenes (DTEs) The photochromic behavior of DTEs works on the principle of electrocyclic ring closure and ring opening via carbon–carbon bond formation and cleavage, respectively. The colorless ring-open form absorbs in the UV-region while ring-closure affords the colored form. Two main conformers of the ring-open form of DTE can be identified (see Fig. 6.6), parallel (p) and antiparallel (a-p), only the latter is photoreactive [27].

The ability of the closed form of DTE to fluoresce has led to significant interest in the development of dual-action or multiresponsive fluorescent molecular sensors for

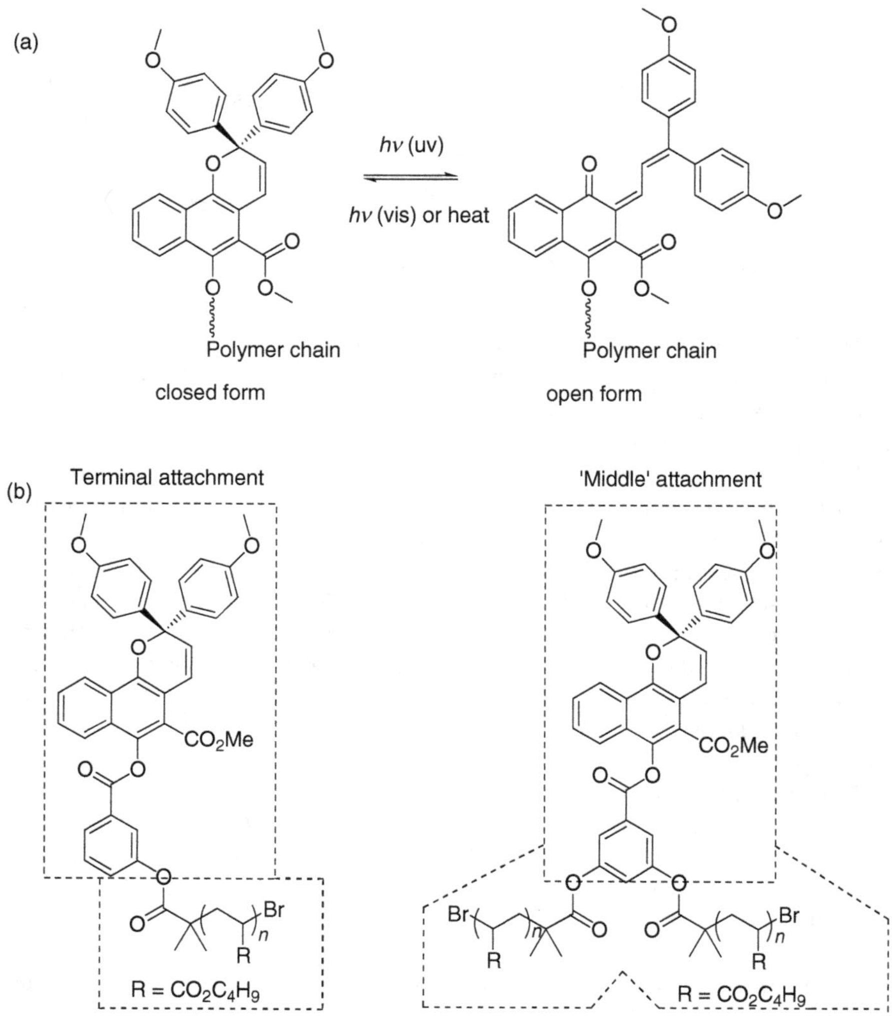

FIGURE 6.5 Evans' naphthopyran polymer conjugate systems.

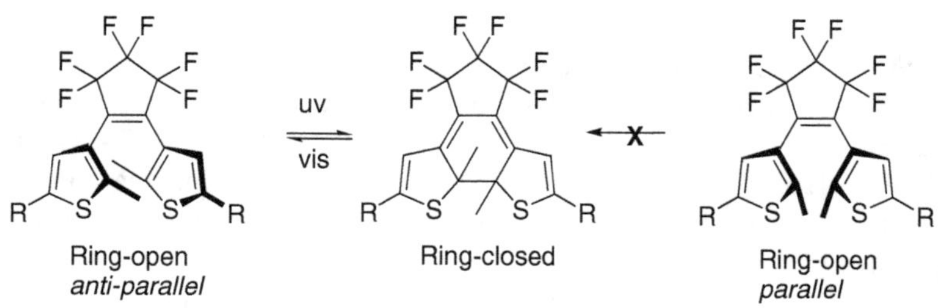

FIGURE 6.6 Dithienylethenes: photochromism and conformers.

secure data storage applications. Jiang et al. report a photo- and proton-dual responsive fluorescence switch based on DTE-bridged naphthalimide dimer impregnated in a PMMA film and its application in security data storage [29]. Suitably substituted 1,8-naphthalimide derivatives show good light stability and high fluorescent quantum yield. The latter may be fine-tuned depending on the nature and disposition of additional substituents on the 1,8-naphthalimide chromophore. Complementary electron donor substituents at the 4-position promoted fluorescence emission due to enhanced conjugation of the resultant *push–pull* system. Jiang et al. showed that the push–pull electronic properties, and thus, fluorescence emission, could also be tuned by a proton source providing a suitable substituent (e.g., morpholino) was located at the 4-position (see Fig. 6.7). Protonation of the morpholino group prevented conjugative effects with the naphthalimide ring, as the lone pair of electrons on the morpholino-nitrogen atom was no longer available, and fluorescence was not observed. Alternatively, fluorescence was also quenched by irradiation at 365 nm, converting the open form of DTE (fluorescent) to the closed form (nonfluorescent) [29].

Similarly, Kawai et al. [30] use the concept of changes in electrical conductivity associated with a photochromic response. This polymer (see Fig. 6.8) was found to be soluble in common organic solvents. UV irradiation of a thin film of the polymer turned the film blue. For the measurement of photoinduced conductivity changes, an approximately 100 nm thick polymer film was deposited on an indium tin oxide electrode. A gold electrode about 20 nm thick was then evaporated on top of the polymer film. Before the gold deposition, the polymer film was irradiated with UV

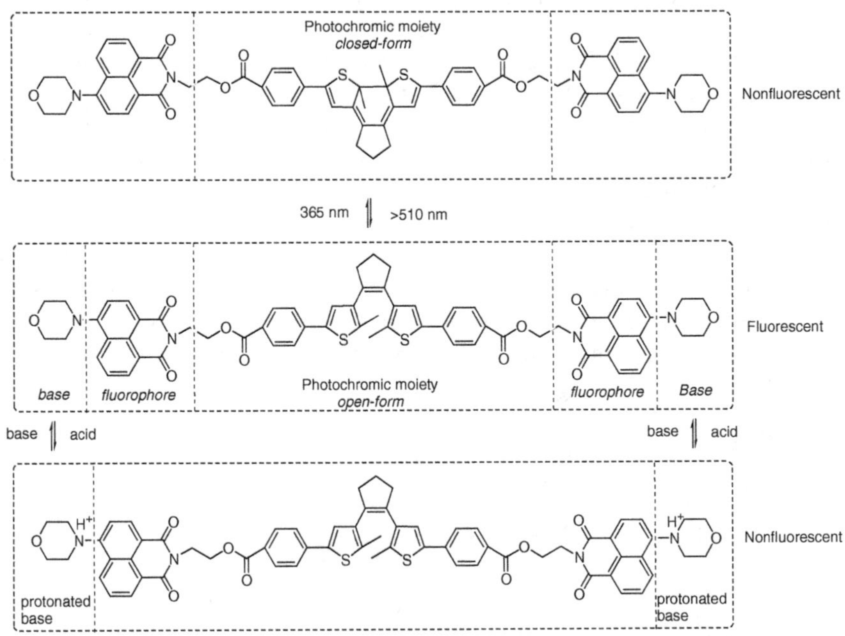

FIGURE 6.7 Photo- and proton-dual responsive DTE.

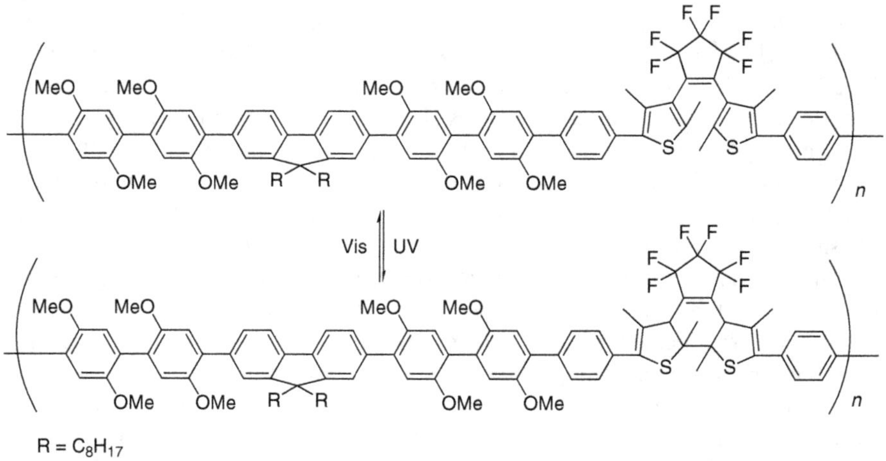

FIGURE 6.8 Photochromic and electrically conducting DTE.

light till a photostationary state was obtained. The electrical conductivity at room temperature varied between 1×10^{-15} and 2×10^{-15} S cm^{-1} between the colored and photobleached states. The latter was obtained through irradiation with visible light. As the conductivity was stable under nonirradiated conditions, it was concluded that the changes in the conductivity was the result of irradiation and not due to thermal processes. The change in the conductivity may be due to the photoinduced modulation of the π-conjugated connections in the diarylethene sites.

Wigglesworth and Branda have designed a series of copolymers (see Fig. 6.9), via ring-opening metathesis (ROMP) chemistry, whose colors were determined through appropriate blending of three diarylethene derivatives [31], which have potential application in the area of multibit optical storage.

The optical and electronic properties of the polymers can be fine tuned through the choice of appropriate substituents at the R position. Varying the length of the linearly π-conjugated chain extending along the molecular backbone can modify the color of the ring-closed form. In the ring-open form, the copolymers are colorless; on irradiation in the UV at 313 nm, they isomerize to the ring-closed form displaying blue, red, or yellow colors. Because of the unique absorption characteristics in both the ring-open and ring-closed states, the blend can display eight possible colored states in a single copolymer containing three different diarylethenes.

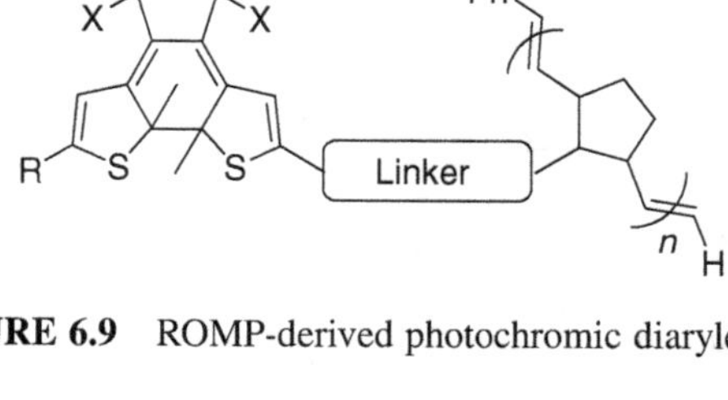

FIGURE 6.9 ROMP-derived photochromic diarylethene.

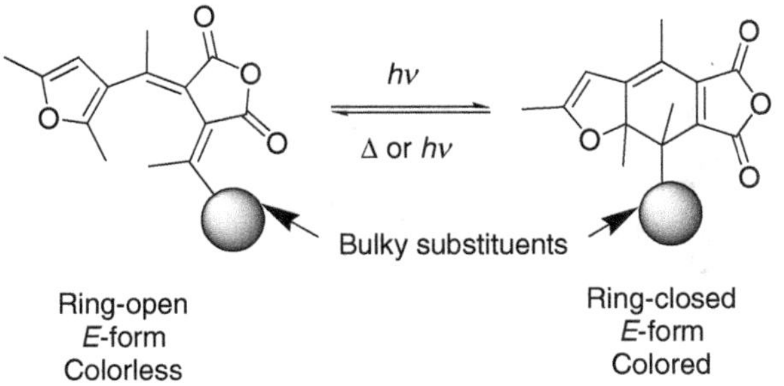

FIGURE 6.10 Photochromic fulgides with pendant bulky substituents.

Fulgides and Fulgimides Fulgide is the generic name given to derivatives of 1,3-butadiene-2,3-dicarboxylic acid (fulgenic acid) and its acid anhydride (fulgide) [32]. In fulgides, for example, furylfulgide, the stable *E*-form upon UV-irradiation undergoes conrotatory electrocyclic ring-closure to give the highly colored, ring-closed *C*-form or dihydrobenzofuran derivative, which absorbs light in the visible region. The *E*-form refers to the isomer in which one of the carbonyl groups of the acid anhydride moiety and the furan ring are in the *E* configuration (e.g., see Fig. 6.10). The observed bathochromic shift is due to enhanced mesomeric relay of the lone pair of electrons on the furan oxygen atom to the carbonyl oxygen atom inferring a rigid, planar structure of the closed *C*-form. The latter reverts to the original *E*-form upon irradiation with visible light. Heller and coworkers [33,34] reported the first thermally stable fatigue resistant near-colorless photochromic fulgides (ring-open), which underwent near quantitative conversion into their colored forms (ring-closed) on exposure to UV light. They showed it was possible to record 1000 lines mm^{-1} at 16 mW cm^{-2} in a sample of the ring-closed form dissolved in poly(methyl methacrylate) (50% w/v). However, the recording rate was slow due to the poor quantum efficiency for bleaching as rate of recording is directly proportional to rate of bleaching, that is, transformation from colored to colorless (original) state [34].

The bleaching rate can be increased by careful choice of bulky, inflexible substituents, such as adamantylidene or dicyclopropylidene [35], which increase steric strain of the closed *C*-form weakening the newly formed carbon–carbon σ-bond making it easier to ring-open (see Fig. 6.10).

Yu et al. [36] have reported applications of pyrryl fulgides with respect to optical data storage. More than 500 UV-irradiation and photobleaching (He–Ne laser, 632.8 nm) cycles of an optical disk coated with PMMA thin film doped with pyrryl fulgide without fatigue resistance of formation photodecomposed species were demonstrated.

As with all photochromic polymers, the polymer matrix effect is very important in fulgides and fulgimides as the polymer free volume affects rate of photocoloration and photobleaching. For example, Bahajaj and Asiri recently reported that the rate of photochemical reactions of fulgide doped in a variety of polymer matrices decreased in the order: polystyrene > poly(methyl methacrylate) > epoxy resin [37]. Polystyrene possesses a larger free volume than poly(methyl methacrylate) and epoxy resin

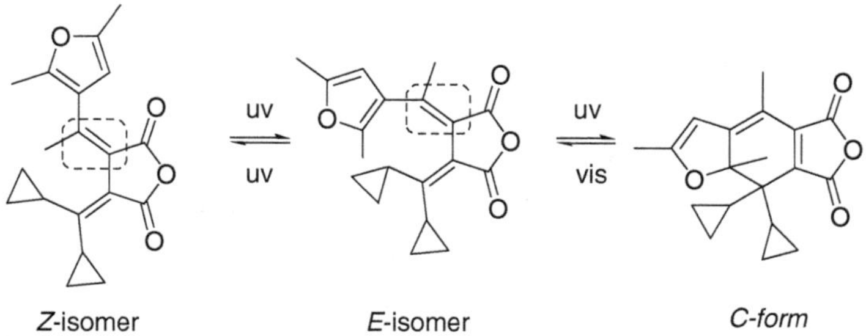

FIGURE 6.11 Bahajaj and Asiri's photochromic fulgides.

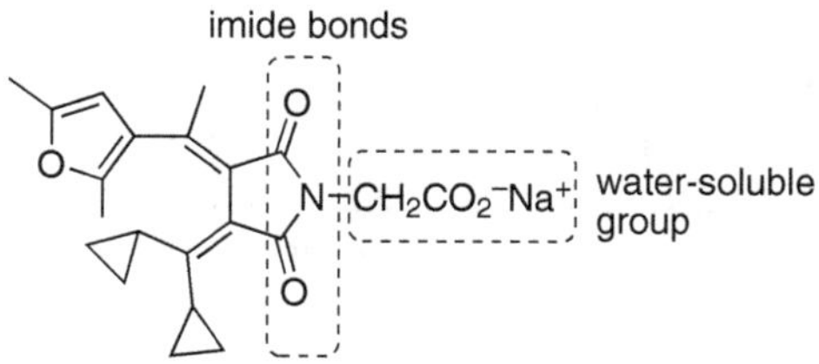

FIGURE 6.12 Water-soluble fulgimide.

allowing fast forward and backward reactions, so much so, that fulgide doped in polystyrene showed poor resistance to photochemical fatigue. Two first-order rate constants for photocoloration were determined: one at initial reaction time and the second at late reaction times, due to trapping of *E*- and *Z*-conformers of the fulgide in the polymer matrix. The initial first-order rate constant reflected ring closure of the *E*-isomer. The latter rate constant corresponded to initial conversion of *Z*-conformer to *E*-conformer, which subsequently undergoes ring closure (see Fig. 6.11).

Fulgimides are photochromic imides based on fulgides. Synthesis of fulgimides is low yielding and their reduced fatigue resistance compared with fulgides makes them commercially unattractive. However, Heller et al. have reported the synthesis of water-soluble fulgimides (see Fig. 6.12), which are more amenable to large-scale manufacture and environment friendly [38].

We shall now embark on the main focus of this chapter: photochromic azobenzene polymers and photochromic materials capable of undergoing photodimerization for potential use in optical data storage applications. Herein, the photochromic polymers described are those in which the photochromic molecule is covalently linked, either as part of the main chain or as a side-chain, to the polymeric system.

6.2 HOLOGRAPHIC DATA STORAGE USING *TRANS-CIS*-ISOMERIZATION: AZOBENZENES

The volume of information produced and stored annually increases exponentially. In 2002, Lyman and Varian calculated that print, film, magnetic, and optical storage media produced approximately five exabytes (5×10^{18} bytes) of information globally [39]. Magnetic disk is still the best medium for storage of large amounts of information, but now has severe limitations because of the *superparamagnetic effect*, that is, bits cannot be compressed further without destroying data integrity. If, the size of a magnetic domain falls below 10 nm, then thermal self-erasure occurs because the magnetized bit flips randomly finding it difficult to attain a stable state [40].

Holography [41–44] derived from the Greek words *holos* meaning *whole* and *graphein* meaning *to write*, may provide a near-term solution to the information needs of modern day society. Holography offers an alternative 3D approach that utilizes the whole volume of the recording medium rather than just the surface. Holographic storage provides the potential of storing in excess of a terabyte of information with transfer rates exceeding 1 GB/s and data access time of less than 100 μs. Gabor, in 1948, pioneered the concept of holography, but at the time complementary optics, imaging, and suitable materials were not available [41]. Commercial holographic storage media are now available as exemplified by InPhase Technologies' Tapestry™ media (see http://www.inphase-technologies.com), capable of storing and retrieving 200 GB on a standard 120 mm CD format at high speed, equivalent to a near 10-fold improvement on current consumer state-of-the-art BluRay DVD technology.

Many different materials for holography have been investigated, ranging from inorganic crystals to glasses to polymers. Any suitable material for holographic data storage applications must possess the following characteristics:

(1) Fast optical switching between the two states
(2) High thermal stability over a wide temperature range
(3) Nondestructive read-out

6.2.1 Photoaddressable Polymers Based on Azobenzenes

Photochromic behavior of azobenzene is well known and its incorporation or conversion into polymeric systems has been extensively studied and highlighted [44]. Azobenzenes undergo reversible *cis-trans*-photoisomerization (see Fig. 6.13). *trans*-Azobenzene has a strong π–π^* transition peaking at approximately 365 nm and a weak n–π^* band between 400 and 550 nm. *cis*-Azobenzene also possesses an n–π^* transition in the same range. Thus, azobenzene can be optically pumped between *trans*- and *cis*-states with radiation around 450 nm. Appropriate irradiation of the *trans*-state affords the less stable *cis*-state, which reverts back to the *trans*-state either thermally or photochemically. The resultant interaction effects change in geometry, dipole moment, and importantly, for optical storage induces dichroism and birefringence.

The first demonstration of azobenzene polymers for holographic storage was made by Wendorff and coworkers who utilized side-chain liquid crystalline polyacrylates (see Fig. 6.14) comprising terminal cyano-substituted azobenzene connected via a hexyloxy tether to an acrylate backbone [45,46].

The liquid crystalline polymer (see Fig. 6.14) was filled and aligned in a cell consisting of rubbed polyimide surfaces. Making use of conventional intensity holography, they showed that rather high diffraction efficiency (~50%) could be achieved with very low laser intensities (1 mW cm^{-2}).

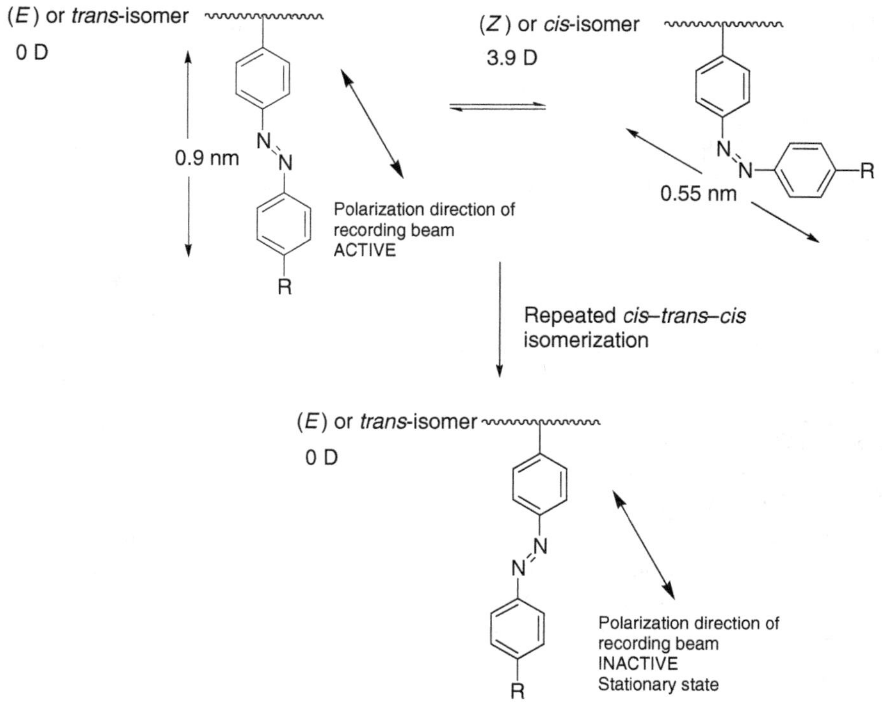

FIGURE 6.13 Azobenzene photoisomerization and interaction with polarized light.

In the case of azobenzene containing materials, it is preferable to use another type of holography called polarization holography [47]. The difference between intensity and polarization holography is illustrated in Fig. 6.15. In Fig. 6.15a, the two beams that are incident on the film (reference and object beams) have the same polarization

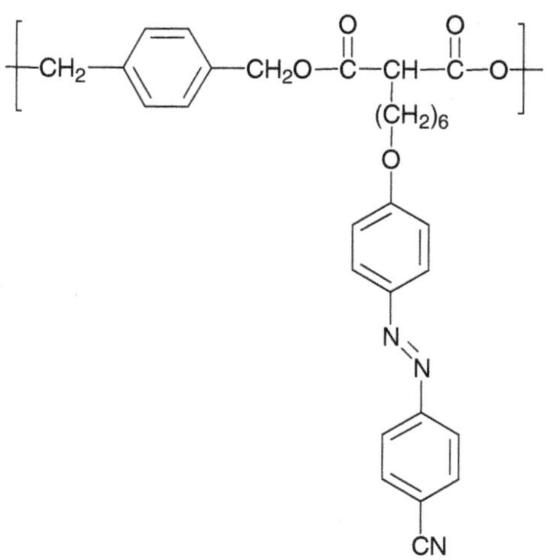

FIGURE 6.14 Eich and Wendorff's polymer.

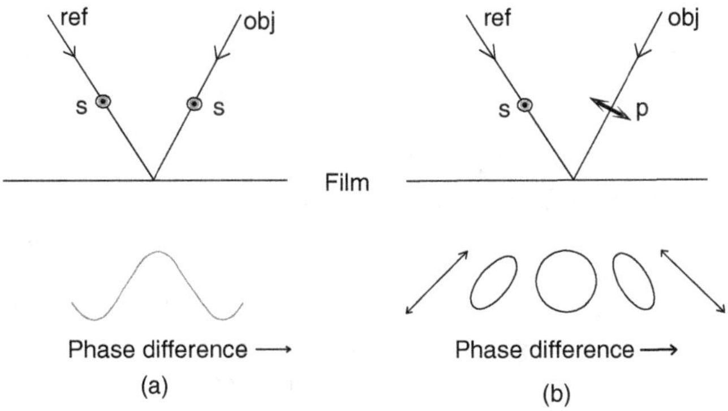

FIGURE 6.15 Principle of (a) intensity holography and (b) polarization holography.

(s-polarization shown). As the polarizations are the same and the beams are coherent, they can interfere to produce dark and bright fringes on the film. The lower half of Fig. 6.15a, shows the resultant intensity as a function of the phase difference between the two beams. This type of holography is not very efficient, resulting in only approximately 6% diffraction efficiency [48]. If, instead, phase holograms can be made (which do not absorb light), a diffraction efficiency of approximately 33% can be achieved for holograms recorded in thin films, and 100% for thick volume holograms. In Fig. 6.15b, the two beams that overlap have orthogonal polarizations. In this case, the beams do not interfere. The resulting polarization pattern is shown in the lower half of Fig. 6.15b. If the material is sensitive to polarization, then this pattern can be recorded as a hologram. If the two overlapping beams have orthogonal polarizations, then a diffraction efficiency of 100% can be achieved for even thin films. Azobenzene is remarkable in the sense that it orients perpendicular to the polarization of incident light, causing a birefringence. It is worth mentioning that azobenzene can also orient parallel to the polarization of incident light as demonstrated by Kempe et al. in an azobenzene side-chain polymer, which had been pretreated with nonpolarized UV light followed by exposure with polarized red light (633 nm). However, prolonged exposure reorients the molecules perpendicular with respect to the polarization plane, thus, implying a fast initial orientation (parallel) followed by a slow, more conventional perpendicular reorientation [49].

The advantage of a polymer with attached azobenzenes over that of *guest–host* systems is that below the glass-transition temperature, the azobenzenes are locked into the oriented positions. A birefringence of 0.35 has been obtained for azobenzene polymers, and is thermally and temporally stable. Thus, azobenzene polymers provide a powerful medium to record polarization holograms. For example, in actual experiments, an argon ion laser 488 and 515 nm or a frequency doubled YAG laser at 532 nm are used. A twisted nematic liquid crystal spatial light modulator with an image of a bitmap spatially modulates the signal beam. This beam then overlaps with a beam of constant intensity, and is of opposite circular polarization (reference beam). The hologram is recorded in an amorphous azobenzene polyester [50], which has a glass transition temperature exceeding 100°C. Investigation on the induced birefringence in such films shows that the information is stable up to a temperature of 160°C. The hologram is read with a small red-diode laser that does not alter the states of azobenzene, and which is also circularly polarized. Holograms with a size of 0.3 mm and with a data density of about 1 bit μm^{-2} have been made. Even higher data densities could be achieved, if recording lenses with high numerical aperture and shorter wavelengths could be used. Another means of increasing the data capacity is to use multiplexing techniques either using thicker films or confocal filtering techniques.

Following the seminal work of Eich and Wendorff [45], PAPs based on azobenzene have been instrumental in the development of holography [44,51]. Hvilsted and Ramanujam have extensively studied azobenzene polymers comprising a variety of scaffolds [52]. They concluded that liquid crystalline polyesters responded rapidly to laser light. A single nanosecond pulse was able to induce high permanent birefringence, stable for more than 15 years with a storage capacity of 5000 lines/mm, and showed erasability.

A PAP may be either amorphous or liquid crystalline and usually comprises suitably functionalized photochromic azobenzene and mesogenic moieties as side groups. The latter play an important cooperative or synergistic role with the photochromic group enhancing the overall birefringence (photoinduced anisotropy).

The effect of a combination of photochromic azobenzene with mesogenic units within the same polymer on cooperative motion and anisotropic enhancement is exemplified by the CP1 and K1 polymer systems of Zilker et al. [53] shown in Fig. 6.16. In the CP1-50 polymer, a clear distinction between chromophore and cooperative mesogen (CP1-50 polymer) existed. There was no apparent distinction between chromophore and mesogen in the K1-40 polymer because mesogen also contained a photoactive azo linkage. The boundary between fast-chromophore and slow-mesogen reorientation is blurred and an effective enhancement of reorientation occurs.

Molecular anisotropy of polarizability and electronic contributions may be enhanced through incorporation of terminal electron-withdrawing substituents (e.g., cyano-) to promote higher photo-induced anisotropy. Polymer K1-40 has a larger dipole moment and higher birefringence than CP1-50 polymer. In the absence of or reduction in dipolar contributions (CP1-50 polymer), the driving force for reorientation of the mesogenic units is due to steric considerations. Unlike the K1-50 polymer, in which both chromophore and mesogen are photoresponsive, the reorientation in CP1-50 polymer is slower and sequentially driven, that is, the chromophore reorients first followed by the mesogen. The need for chromophore and mesogen for cooperative motion and enhancement is important. Polymer K1-0 (not shown) comprising of 100% mesogenic groups exhibited poor birefringence (Δn 0.10). Similarly, low birefringence ($\Delta n = 0.11$) was shown by a homopolymer comprising 100% photochromic units (not shown nominally and known as K1-100). However, a combination of mesogen and photochrome, as in KI-40 polymer, gives a higher birefringence confirming the need for both moieties working in a synergistic or cooperative manner (K1-40, $\Delta n = 0.23$). Small angle X-ray scattering reveals the formation of small domains consisting of supermolecular aggregates with a high degree of orientational order. As the illumination time is increased, the degree

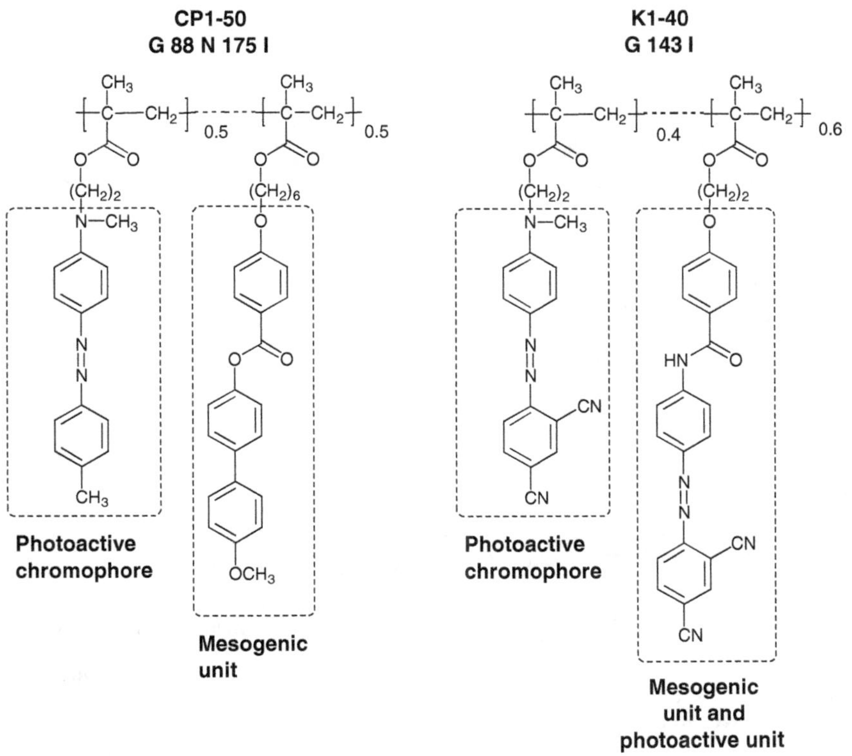

FIGURE 6.16 CP1-50 and K1-40 polymeric systems.

of ordering increases, while the domain size remains constant. Thus, scattering is due to reordering and not density fluctuations as in guest–host systems where diffusion processes are dominant.

Many different azobenzene–mesogen combinations have been investigated. Ikeda and coworkers incorporated high birefringence tolane-based mesogenic side-chains (diphenylacetylenes, see Fig. 6.17a and b) to enhance diffraction efficiency (η) and modulation of the refractive index ($\Delta\eta'$) [54]. The liquid crystalline polymer was heated 60°C into the nematic phase and gratings were formed (intensity holography). Compared with a cyanobiphenyl mesogenic side-chain, which took one hour to attain a maximum diffraction efficiency of 4%, its tolane-based counterpart reached a maximum value of 20% after only 4 minutes irradiation. Combining azobenzene photochrome with the tolane-based mesogen (see Fig. 6.17b) into a single entity further improved photosensitivity [55]. Side-chain spacer effects were evident. Short alkyl tethers afford low packing density, greater flexibility, and are more prone to surface relief grating formation. Long alkyl tethers decouple photochromic units from the main polymer backbone allowing highly ordered, tightly packed regions less prone to movement and surface grating formation.

Another polymer material (see Fig. 6.18) that has been employed for the 3D storage is based on a urethane–urea copolymer with azobenzene side-chains [56].

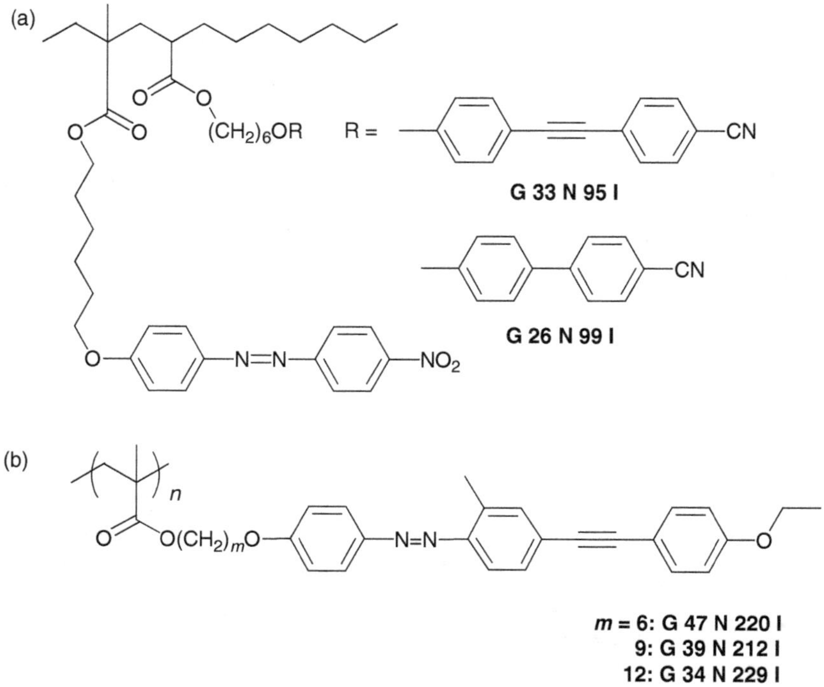

FIGURE 6.17 Azobenzene and tolane-based high birefringence polymers.

A refractive index change on the order of 10^{-2} could be induced in the polymer on irradiation with blue light. This technology can also be utilized in the fabrication of multilayer or thick films. By alternately coating urethane–urea copolymer with azobenzene and nonphotosensitive polyvinylalcohol multilayer films could be obtained. In this case, the bitmaps are recorded as images on individual layers.

Unfortunately, PAPs have not been commercially successful for high-capacity volumetric storage because of their inability to form thick films without the aid of diluents. The maximum thickness of a homopolymer film should be approximately

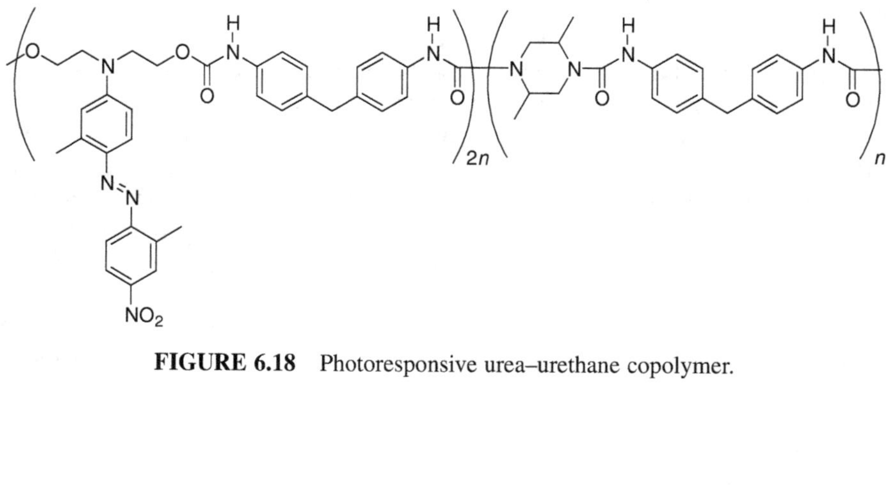

FIGURE 6.18 Photoresponsive urea–urethane copolymer.

4/absorptivity, or about 320 nm, otherwise thicker films will completely absorb the incoming light beam. It is not our intention to discuss thick film formation, but there are some interesting concepts being developed. For example, thick films with multiplexing capability, read-write longevity, and acceptable levels of optical density have been prepared using block copolymers [57]. Azobenzene photochromic side-chains comprise one block (minor) are combined with a block of mesogen material (major). Spherical microphases (10–20 nm diameter) are formed of azobenzene block concentrations of 15% or less. The chromophores become confined in a narrow environment aiding stability of the inscribed grating and minimizing surface relief grating formation.

6.2.2 Photoaddressable Cholesteric Photochromic Polymers

Photoaddressable cholesteric polymers combine the photochromic properties of azobenzene with the supermolecular order of cholesteric liquid crystals [58]. Dual responsive systems are possible (see Fig. 6.19), which absorb differing parts of the electromagnetic spectrum. For example, as shown in Fig. 6.18, the mesogenic methoxyphenylbenzoate groups (**a**) are expected to induce liquid crystallinity and are present with photoresponsive azobenzene moieties (**b**). Additionally, a third chiral moiety is included, benzylidene-*p*-menthanone fragment (**c**), capable of exhibiting a cholesteric phase and hence, a helical structure. The absorption band of moiety (**c**) occurs in the UV region, widely different from that for moieties (**a**) and (**b**). Thus, for example, UV irradiation (313 or 366 nm) induces *E*-*Z*-photoisomerization of the menthanone groups causing the linear *E*-isomer to change shape to the nonlinear structure of the *Z*-isomer, destroys the cholesteric phase (see Fig. 6.19). A significant change in the helical twisting power, β, is observed as a result of helix unwinding with the pitch tending toward longer wavelengths. Unfortunately, this process is irreversible and the image cannot be rewritten.

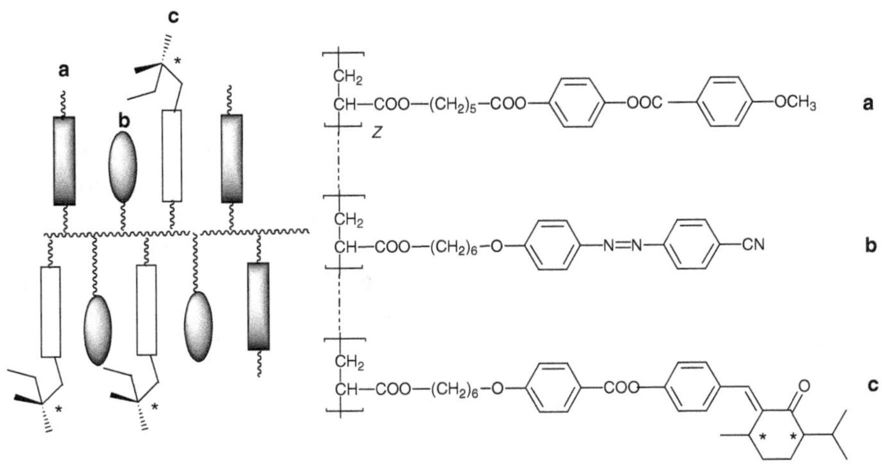

FIGURE 6.19 Shibaev's ternary photochromic cholesteric polymer.

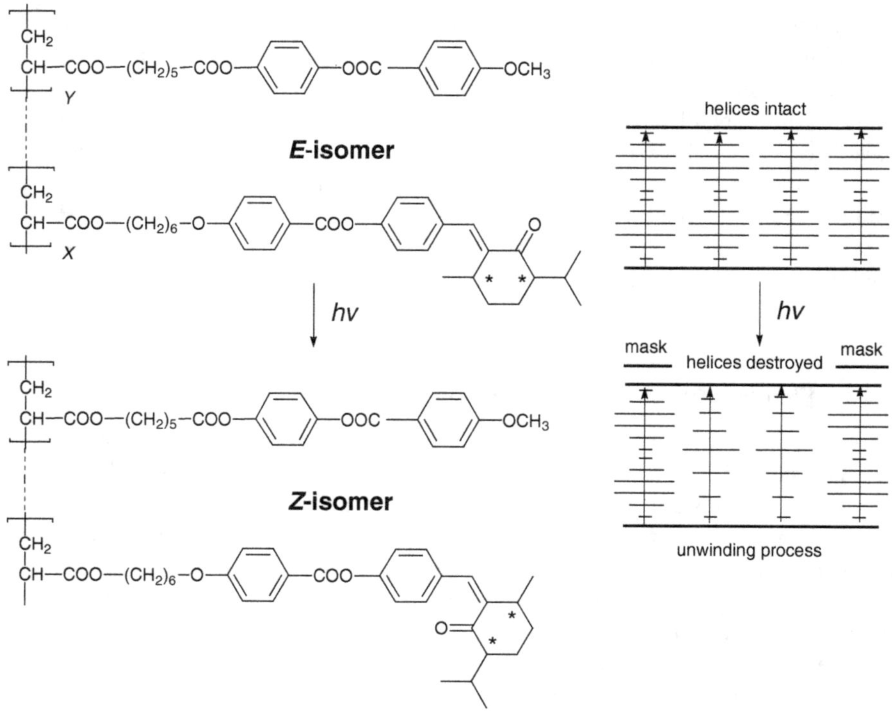

FIGURE 6.20 Selective photoinduced helix unwinding in a photochromic cholesteric polyester.

However, irradiation at 514 nm induces *cis-trans-cis*-photoisomerization within the azobenzene moiety, without affecting the menthanone group (see Fig. 6.20). Helix unwinding occurs and a bathochromic shift is detected. However, this process is reversible and heating the sample causes twisting of the helix. Many record-erase cycles can be performed with minimal fatigue or bleaching.

6.2.3 Cycloaddition Reactions for Optical Storage

Tomlinson et al. [59,60] investigated cycloaddition reactions in organic materials for holographic storage of information. A cycloaddition reaction is a ring-forming addition of m atoms of one group to n atoms of another group. A cycloadduct is the sum of the individual components undergoing cycloaddition. In this case, no groups are eliminated in the ring-forming step. Typically, new σ bonds are formed in a cycloaddition reaction. Usually, reactions involve [2 + 2], [4 + 2], or [4 + 4] cycloaddition, involving exciplex or excimer formation.

$$^{1}\mathrm{A} + \mathrm{A} \Leftrightarrow {}^{1}(\mathrm{AA})^{*} \rightarrow \mathrm{A}_2$$

However, direct evidence for excimer formation is scanty, as fast radiationless transitions tend to compete with excimer emission. For effective cycloaddition to

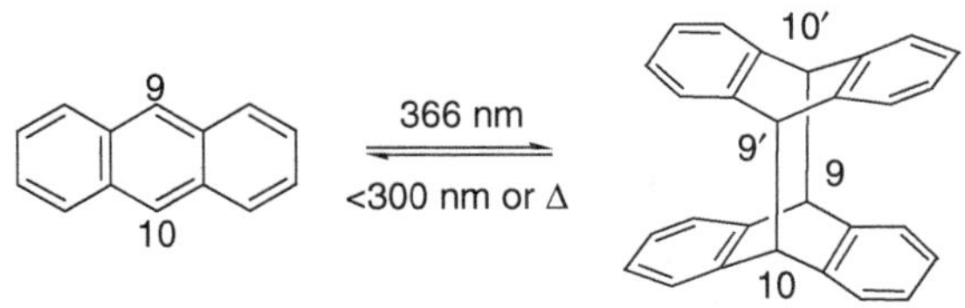

FIGURE 6.21 Anthracene monomer to dimer conversion.

occur, the reactive partners should be in close proximity and orientation relative to each other. The importance of cycloaddition reactions for optical storage applications lies in the fact that the photogenerated products are kinetically stable. The activation/deactivation energy between monomers and dimers can be on the order of several electron-volts. Competing degrading processes include photooxidation, as well as reconversion due to thermal processes.

Anthracene photodimerization (see Fig. 6.21) is well known to occur in a variety of solvents and in the solid state [61]. The photodimers are stable at room temperature and can revert to their monomeric state by photoirradiation ($\lambda < 300$ nm).

Unfortunately, anthracene photodimers are generally insoluble in common organic solvents making them unattractive for commercial applications. The dimerized structure consists of four benzene rings attached together by single carbon–carbon bonds. When the dimer is cleaved through irradiation at the appropriate wavelength, the bonds between 9,9′ and 10,10′ carbon atoms are broken, leading to a re-establishment of the aromatic structure of the middle ring in the two monomers. This increases the conjugation length, and hence, the wavelength of absorption by the monomers. When the monomers are excited at their appropriate wavelength of absorption, one of the monomers is excited forming an excimer, and thence, in association with another monomer dimerize. In solution, dimer formation occurs via diffusion of the excited monomers and collisions with monomers in the ground state, and, therefore, the quantum efficiency for the process is low. However, in crystals, it is expected that because the monomers are held rigid, the quantum efficiency can be high.

To ensure that the monomers are in proximity and oriented properly, Tomlinson et al. first dimerized the chromophores in solution, which because of their low solubility could be filtered out and the resultant precipitate was dissolved in a compatible polymer matrix such as PMMA. On excitation with a cleaving light, the molecules broke into monomers retaining their position and orientation. Thus, they were ideally situated for further experiments in optical storage. In some cases, single crystals of concerned salts were utilized. Tomlinson et al. examined several chromophores: thymine, anthracene, 2-aminopyridinium, acridizinium, and benzacridizinium. Conventional intensity holographic gratings were fabricated and the diffraction efficiencies were measured. Acridizinium compounds were found to be the best, with diffraction efficiencies between 0.01% and 5%, and could be reversibly used several times, without fatigue.

Similarly, Desvergne et al. exploited the importance of close proximity of anthracene chromophores by linking the two anthracene chromophores with an oxymethyleneoxy (e.g., A–X–A, see Fig. 6.22) tether [61]. The dynamics and

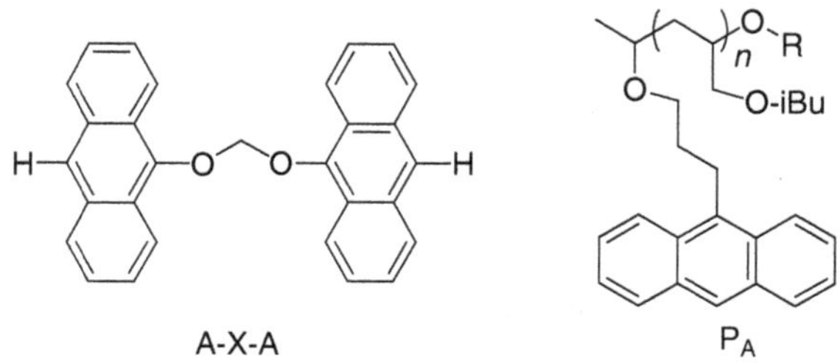

FIGURE 6.22 Desvergne's bisanthracenes.

effectively the length of the interchain tether in bisanthracenes are important as it can increase the efficiency of photodimerization. In addition, suitable substituents at the 10 and 10′-positions increased solubility of the photodimers, but reduced thermal stability. However, the unsubstituted photodimer was thermally stable and several cycles (closure at 366 nm, thermal opening) could be performed without apparent degradation. Photodimerization also proceeded efficiently in polymer matrices and was dependent on the type of matrix used. High molecular weight, high T_g (>100°C) matrices gave lower quantum yield than matrices with low molecular weight and low T_g (<80°C) suggesting a high degree of segmental or chain motion in the latter. Desvergne et al. extended their work to isobutylvinylether polymers with pendant anthracene moieties (P_A) showing that dimer-formation and dimer-breakage was possible at 366 and 280 nm, respectively. However, photochromic processes were not as efficient compared with either free anthracene monomer dissolved in a certain solvent or diluted within a polyisobutylvinyl ether matrix. Once again, the polymer matrix has a significant influence on photochromic processes, and in this case, the low mobility of the polymer chain may have caused poor efficiency [61].

Heller and Ottaway combined the photochromic properties of fulgides to promote photodimerization between pendant anthryl moieties (see Fig. 6.23) in toluene solution [62]. *E,E*-Fulgide is severely overcrowded as the two anthryl moieties are in close proximity such that on exposure to white light facile [4 + 4] cycloaddition occurs. The reverse process was induced by UV-irradiation or heat and giving the

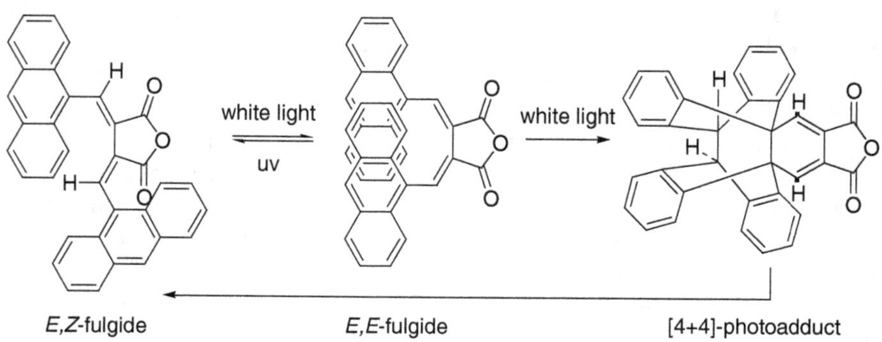

FIGURE 6.23 Anthracenyl fulgides.

E,Z-fulgide. Irradiation of the latter with white light reformed the photoadduct presumably via the *E,E*-isomer.

Ramanujam and Berg chose peptidic linkers between anthracenyl moieties to ensure that the chromophores are always close together. The attachment to a peptide also improved the solubility, and hence, made it possible to make good quality thin films. One could in this case, make use of the enormous development in the area of digital optical storage. While a conventional CD has a capacity of approximately 700 MB, a DVD has a storage capacity of 4.2 GB. This increase is possible due to application of shorter wavelength laser and optics with higher numerical aperture. In the case of digital optical storage, the spot-size (bit-size) depends on the wavelength of the laser and the optics (s approximately equal to λ/NA, where λ is the wavelength of the writing source, and NA is the numerical aperture of the optics). Thus, the shorter the source wavelength, and the larger the numerical aperture, then the larger the number of bits that can be stored on a given surface. The best commercial medium is based on Blu-Ray, which employs a blue laser at 405 nm with a numerical aperture of 0.85 to provide a storage capacity on the order of 25 GB. However, as new compact lasers are under continuous development, the possibility of a laser operating at approximately 250 nm is not far away. Hence, the development of materials and the process of storage of information in UV wavelengths are imperative. Thymine or uracil that dimerizes on irradiation at 250 nm is a well-known example. The photodimerization of thymine and uracil on exposure to strong sunlight, leading to a misreading of the DNA code in human beings, is a known cause for skin-cancer. However, the process can also provide useful technology. The experiments described below utilize a 257-nm frequency-doubled argon ion laser and quartz optics for measurements. Optical storage of information in this case would involve the measurement of the change in absorption, as the accompanying refractive index change is too small to accommodate a multibit storage process.

The structure of a thymine dipeptide (Fig. 6.24) and its absorption spectrum before and after irradiation at 257 nm is shown below (Fig. 6.25).

Initially, it was possible to achieve a transmission of 45% after 300 s of irradiation at 257 nm with a laser power of 300 mW cm^{-2}. Efforts were then undertaken to optimize the chromophores and the backbones for efficiency and fast response. Changing the chromophore from thymine to uracil, the transmission increased to 65% at an intensity of 30 mW cm^{-2} [63]. At this juncture, an optimization of the scaffold

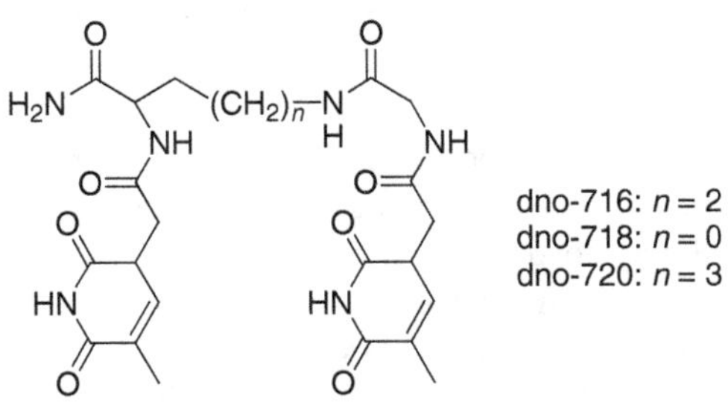

FIGURE 6.24 Chemical structure of a thymine dipeptide.

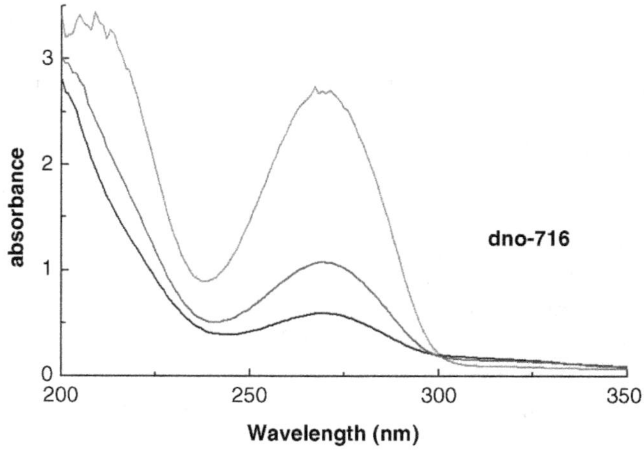

FIGURE 6.25 Absorption spectrum of thymine-peptide dimer before and after irradiation.

was undertaken and several backbones were tried: peptide scaffolds, alkyl chains, and two dendrimers based on 3-hydroxy-2-hydroxymethyl-2-methylpropionic acid (bis-MPA) and polyamidoamine (PAMAM). The dependence of the photodimerization efficiency on the number of amino acid residues, number of methylene units, and generations was investigated. In the case of dipeptides, the backbone based on ornithine-glycine was found to be optimal [64]. In the case of oligopeptides, an ornithine-glycine hexamer gave a maximum transmission of 68% in 1500 s for an intensity of 30 mW cm^{-2}. An estimation of the absorption cross-section for photons based on the theory developed by Tomlinson et al. [59] gave a value of 9×10^{-20} cm^2 far below the theoretical maximum of 1.2×10^{-18} cm^2 [65]. The reason being that a peptide backbone was too rigid to permit a molecular adjustment.

To simplify the synthesis and to provide a certain amount of flexibility of the backbone, a short alkyl tether was chosen as the backbone [66]. In this case, a maximum transmission of 65% was achieved within 30 s with an intensity of 130 mW cm^{-2}, for a film of 1-(6-bromohexyl)uracil, giving an absorption cross-section of 1×10^{-19} cm^2. Best results were obtained for the case of 1,1′(1,8-octanediyl)-bis[uracil] in which a maximum transmission of 71% was obtained in less than 20 s, for an intensity of 20 mW cm^{-2}. The absorption cross-section obtained with theoretical simulation was 0.85×10^{-18} cm^2, close to the maximum theoretical value [67].

The question of whether it was possible to increase the efficiency even further was investigated next by having as many chromophores as possible close to each other, to promote intramolecular cycloaddition. In this case, it should not be necessary to find another monomer lying around per chance for dimerization. The obvious choice of backbone in this case was a dendrimer. Two series of uracil-functionalized dendrimers based on bis-MPA and PAMAM were looked into. Dendrimers up to fourth generation were synthesized and their dimerization efficiency compared with both peptide and alkyl backbones. Dendrimers have an added advantage in the increased solubility, and hence, better quality of films. The best photodimerization efficiency was found for a first-generation uracil-PAMAM dendrimer with a maximum transmission of 73% in 100 s, at a laser intensity of $30\,\mathrm{mW\,cm^{-2}}$, giving an absorption cross-section of $1.7 \times 10^{-18}\,\mathrm{cm^2}$. For the case of bis-MPA, the best result was a transmission of 73% in 300 s with the same intensity; however, the film quality was considerably poorer [68].

This very large change in the transmission of thin films (0% to 70%) can, in principle, be exploited to store multivalued bits. Conventionally, bit storage involves the binary numbers "0" or "1." However, if eight bits can be stored and read-out from the same area, the storage capacity could reach 200 GB on a single disc. By changing exposure time at a given point, different transmission values can be achieved. This kind of a gray scale recording was demonstrated in a film of 1,1′-(1,8-octanediyl)bis [uracil] [67]. After recording the grayscale, the intensity of the laser was reduced, and the transmission of the film could be read-out constantly and accurately.

In a further development, materials that undergo efficient reversible cycloaddition in the blue for use with Blu-Ray have been investigated. It was found by Tomlinson et al. that acridizinium undergoes dimerization efficiently in the blue. Therefore, we have synthesized [69] a second-generation acridizinium-bis-MPA dendrimer, which undergoes $[4\pi + 4\pi]$ cycloaddition reaction on exposure to blue-violet light at 405 nm. The compound was tested as a solution in ethanol, showing good dimerization efficiency. A film of the dendrimer in a polyvinyl-pyrrolidone matrix (10% v/v) was tested for reversible cycloaddition reaction. Dimerization was achieved by irradiation at 405 nm and cleavage back to monomers, at 257 nm. It has been possible to obtain satisfactory results for optical storage.

6.3 CONCLUDING REMARKS

Photochromic polymers have played a pivotal role in the development of novel materials for optical data storage. Advances in holography have benefited from early research into PAPs. Although not based on PAPs, the first commercial holographic data storage devices are now available.

In the near future, UV-lasers may become economically viable such that new photochromic materials will be needed that respond to UV irradiation. Photodimerizing nucleotides, such as thymine tethered via a short peptide chain or in the form of a dendrimer, look like promising candidates. The future will certainly depend on the past, as many materials investigated previously for purely academic interest may return to the fore.

REFERENCES

1. H. Bouas-Laurent and H. Dürr, Organic Photochromism (IUPAC Technical Report). *Pure Appl. Chem.* **2001**, *73*, 639–665.
2. J. Fritzsche, *C. R. Acad. Sci. Paris* **1867**, *69*, 1035.

3. Y. Hirshberg, *C. R. Acad. Sci. Paris* **1950**, *231*, 903.
4. M. Irie, Photoswitchable molecular systems based on diarylethenes, in: B.L. Feringa (Ed.), *Molecular Switches*, Wiley-VCH GmbH, Weinheim, Germany, 2007.
5. J.C. Crano and R.J. Guglielmetti, *Organic Photochromic and Thermochromic Compounds. Volume 2: Physicochemical Studies, Biological Applications, and Thermochromism*, Kluwer Academic/Plenum Publishers, New York, 1999.
6. W. Schnabel, *Polymers and Light: Fundamental Technical Applications*, Wiley-VCH Verlag GmbH, Weinheim, Germany, 2007.
7. B.L. Feringa (Ed.), *Molecular Switches*, Wiley-VCH GmbH, Weinheim, Germany, 2007.
8. M. Irie, *Chem. Rev.* **2000**, *100*, 1683–1890.
9. M. Yamada, M. Kondo, J.-H. Mamiya, Y. Yu, M. Kinoshita, C.J. Barrett, and T. Ikeda, *Angew. Chem. Int. Ed.* **2008**, *47*, 4986–4988.
10. T.K. Mudiyanselage and D.C. Neckers, *Soft Matter* **2008**, *4*, 768–774.
11. N. Katsonis, M. Lubomska, M.M. Pollard, B.L. Feringa, and P. Rudolf, *Prog. Surf. Sci.* **2007**, *82*, 407–434.
12. Y.L. Yu and T. Ikeda, *Angew. Chem. Int. Ed.* **2006**, *45*, 5416–5418.
13. Y. Li, Y. He, X. Tong, and X. Wang, *J. Am. Chem. Soc.* **2005**, *127*, 2402–2403.
14. Y.L. Yu, M. Nakano, and T. Ikeda, *Nature* **2003**, *425*, 145.
15. R.A. van Delden, M.B. van Gelder, N.P.M. Huck, and B.L. Feringa, *Adv. Funct. Mater.* **2003**, *13*, 319–324.
16. T. Yoshioka, T. Ogata, T. Nonaka, M. Morisugu, S.-N. Kim, and S. Kurihara, *Adv. Mater.* **2005**, *17*, 1226–1229.
17. A.V. Mallia and N. Tamaoki, *Chem. Soc. Rev.* **2004**, *33*, 76–84.
18. K. Matsubara, M. Watanabe, and T. Takeoka, *Angew. Chem. Int. Ed.* **2007**, *46*, 1688–1692.
19. K. Ichimura, Photochromic polymers, in: J.C. Crano and R.J. Guglielmetti (Eds.), *Organic Photochromic and Thermochromic Compounds. Volume 2: Physicochemical Studies, Biological Applications, and Thermochromism*, Kluwer Academic/Plenum, New York, 1999, pp. 9–63.
20. V. Krongauz, *Mol. Cryst. Liq. Cryst.* **1994**, *246*, 339–346.
21. J. Whelan, J.T.C. Wojtyk, E. Buncel, *Chem. Mater.* **2008**, *20*, 3797–3799.
22. M.I. Lygeraki, E. Tsiranidou, S.H. Anastasiadis, C. Fotakis, D. Pisignano, C. Cingolani, A. Athanassiou, *Appl. Phys. A* **2008**, *91*, 397–401.
23. T. Ubukata, K. Takahashi, Y. Yokoyama, *J. Phys. Org. Chem.* **2007**, *20*, 981–984.
24. D. Fragouli, L. Persano, G. Paladini, D. Pisignano, F. Carzino, F. Pignatelli, R. Cingolani, and A. Athanassiou, *Adv. Funct. Mater.* **2008**, *18*, 1617–1623.
25. M. Tomasulo, S. Sortino, and F.M. Raymo, *Adv. Mater.* **2008**, *20*, 832–835.
26. N. Malic, J.A. Campbell, and R.A. Evans, *Macromolecules* **2008**, *41*, 1206–1214.
27. T.J. Wigglesworth, A.J. Myles, and N.R. Branda, *Eur J Org Chem.* **2005**, 1233–1238.
28. H. Tian, and Y. Feng, *J. Mater. Chem.* **2008**, *18*, 1617–1622.
29. G. Jiang, S. Wang, W. Yuan, Z. Zhao, A. Duan, C. Xu, L. Jiang, Y. Song, and D. Zhu D, *Eur. J. Org. Chem.* **2007**, *2007*, 2064–2067.
30. T. Kawai, Y. Nakashima, and M. Irie, *Adv. Mater.* **2005**, *17*, 309–314.
31. T.J. Wigglesworth and N.R. Branda, *Chem. Mater.* **2005**, *17*, 5473–5480.
32. Y. Yokoyama, *Chem. Rev.* **2000**, *100*, 1717–1739.

33. P.J. Darcy, H.G. Heller, P.J. Strydom, and J. Whittall, *J. Chem. Soc. Perkin. Trans.* **1981**, *1*, 202–205.
34. A.P. Glaze, H.G. Heller, and J. Whittall, *J. Chem. Soc. Perkin. Trans.* **1992**, 591–594.
35. Z. Sun, R.S. Hosmane, and M. Tadros, *Tetrahedron Lett.* **1995**, *36*, 3453–3456.
36. L. Yu, Y. Ming, W. Zhao, and M. Fan, *Mol. Cryst. Liq. Cryst.* **1994**, *246*, 49–58.
37. A.A. Bahajaj, and A.M. Asiri, *Chin. J. Polym. Sci.* **2008**, *26*, 1–9.
38. H.G. Heller, K. Koh, C. Elliot, and J. Whitall, *Mol. Cryst. Liq. Cryst.* **1994**, *246*, 79–86.
39. H. Lyman and P. Varian, How much information 2003?. Retrieved from http://www.sims.berkeley.edu/how-much-info-2003 on 14/08/08.
40. D.A. Thompson and J.S. Best, *IBM J. Res. Dev.* **2000**, *44*, 311–322.
41. D. Gabor, *Science* **1972**, *177*, 299–313.
42. M. Haw, *Nature* **2003**, *422*, 556–558.
43. J. Ashley, M.P. Bernal, G.W. Burr, H. Coufal, H. Guenther, J.A. Hoffnagle, C.M. Jefferson, B. Marcus, R.M. Macfarlane, R.M. Shelby, and G.T. Sincerbox, *IBM J. Res. Dev.* **2000**, *44*, 341–368.
44. A.S. Matharu, S. Jeeva, and P.S. Ramanujam, *Chem. Soc. Rev.* **2007**, *36*, 1868–1880.
45. M. Eich and J.H. Wendorff, *Makromol. Chem. Rapid Commun.* **1987**, *8*, 467–471.
46. M. Eich, J.H. Wendorff, B. Reck, and H. Ringsdorf, *Makromol. Chem. Rapid Commun.* **1987**, *8*, 59.
47. Sh.D. Kakichashvili, *Opt. Spectrosk.* **1972**, *33*, 324.
48. R.J. Collier, C.B. Burckhardt, and L.H. Lin, *Optical Holography*, 1st edition, Academic Press, London, **1971**.
49. C. Kempe, M. Rutloh, and J. Stumpe, *J. Phys. Condens. Matter* **2003**, *15*, S813–S823.
50. E. Lörincz, G. Szarvas, P. Koppa, G. Erdei, A. Süto, F. Ujhelyi, and P.S. Ramanujam, in: H.S. Nalwa (Ed.), *Handbook of Organic Electronics and Photonics, Part II*, American Scientific Publishers, USA, 2007.
51. A. Natansohn and P. Rochon P, *Chem. Rev.* **2002**, *102*, 4139–4176.
52. S. Hvilsted and P.S. Ramanujam, *Monatsh. Chem.* **2001**, *132*, 43–51.
53. S.J. Zilker, T. Bieringer, D. Haarer, R.S. Stein, J.W. van Egmond, and S.G. Kostromine, *Adv. Mater.* **1998**, *10*, 855–859.
54. S. Yoneyama, T. Yamamoto, O. Tsutsumi, A. Kanazawa, T. Shiono, and T. Ikeda, *Macromolecules* **2002**, *35*, 8751–8758.
55. K. Okano, A. Shishido, O. Tsutsumi, T. Shiono, and T. Ikeda, *J. Mater. Chem.* **2005**, *15*, 3395–3401.
56. M. Ishikawa, Y. Kawata, C. Egami, O. Sugihara, N. Okamoto, M. Tsuchimori, and O. Watanabe, *Opt. Lett.* **1998**, *23*, 1781–1783.
57. M. Haeckel, L. Kador, D. Kropp, and H.-W. Schmidt, *Adv. Mater.* **2007**, *19*, 227–231.
58. V.P. Shibaev, A. Yu Bobrovsky, and N.I. Boiko, Macromol. Symp., **2001**, *174*, 319–332.
59. W.J. Tomlinson, E.A. Chandross, R.L. Fork, C.A. Pryde, and A.A. Lamola, *Appl. Opt.* **1972**, *11*, 533–548.
60. W.J. Tomlinson, *Appl. Opt.* **1972**, *11*, 823–831.
61. J.-P. Desvergne, H. Bouas-Laurent, and A. Deffieux, *Mol. Cryst. Liq. Cryst.* **1994**, *246*, 111–118.
62. H.G. Heller and M.J. Ottaway, *J. Chem. Soc. Chem. Commun.* **1995**, 479–480.

63. P.S. Ramanujam, B. Lohse, and R.H. Berg, *Proc. SPIE* **2004**, *5351*, 144–149.
64. B. Lohse, P.S. Ramanujam, S. Hvilsted, and R.H. Berg, *J. Pept. Sci.* **2005**, *11*, 499–505.
65. B. Lohse, R.H. Berg, S. Hvilsted, and P.S. Ramanujam, *Jpn. J. Appl. Phys.* **2006**, *45*(1B), 488–492.
66. P.S. Ramanujam, B. Lohse, and R.H. Berg, *Proc. SPIE* **2006**, *6252*, 1.
67. B. Lohse, S. Hvilsted, R.H. Berg, and P.S. Ramanujam, *Chem. Mater.* **2006**, *18*, 4808–4816.
68. B. Lohse, R. Vestberg, M.T. Ivanov, S. Hvilsted, R.H. Berg, P.S. Ramanujam, and C.J. Hawker, *J. Polym. Sci. Part A* **2007**, *45*, 4401–4412.
69. B. Lohse, R. Vestberg, M.T. Ivanov, S. Hvilsted, R.H. Berg, C.J. Hawker, and P.S. Ramanujam, *Chem. Mater.*, **2008**, *20*, 6715–6720.

7

OPTICAL AND LUMINESCENCE PROPERTIES AND APPLICATIONS OF METAL COMPLEX-BASED POLYMERS

JOE A. CRAYSTON AND JOANNE R. RITCHIE

7.1 INTRODUCTION

In this chapter, the fascinating spectroscopic properties of metallopolymers containing metal complexes are considered. In some cases, a new chromophore is created when the metal and the polymer are brought together, whereas in others it is the

Photochemistry and Photophysics of Polymer Materials, Edited by Norman S. Allen

interplay between the metal and polymer chromophores that is the focus of the investigation. There are many possible applications of these materials, beginning with one of the oldest and most successful processes: photography. Indeed, the interaction of light with both metals and polymers could be used as one definition of the photographic process. Certainly, since the discovery of the light-sensitive silver halide reaction in photography, the interface between silver and the polymeric medium (e.g., gelatin) is an important consideration. In parallel with the development of the dominant silver halide process, work progressed on a variety of alternative inorganic and organic photographic processes, including gold, platinum, and "blue" (e.g. Prussian Blue) processes, some of which involve the interaction with a polymeric environment, such as the cellulose contained in paper [1,2]. Unfortunately, the wet photographic industry is now being displaced by digital methods, but similar processes are still vital in the printing industry, and over the years the interaction of light with polymers has re-emerged in completely new processes such as photolithography for the electronics industry. There is, of course, great interest in applications exploiting the photoconductivity, photochromism, photothermal, nonlinear optical, and photoluminescence properties of metallopolymers. Some of these applications arise from the conversion of a proven non-polymeric system to a polymer-based system is attractive to improve stability, to prevent leaching of the metal complex, to improve mechanical robustness, and to facilitate rapid and reproducible manufacturing. Finally, another very large area, sadly not covered in this review, is the use of light to initiate polymerization or, on a preformed polymer, to initiate cross-linking/curing or other modifications, such as laser ablation.

Perhaps one of the more enticing and inspiring ambitions of those researching in this area of metallopolymer photochemistry is the use of such polymers [3] in artificial photosynthesis for solar energy conversion [4]. Solar energy research mushroomed in the 1970s when, following the oil crisis, governments looked to alternative energy sources, and is today once again the focus of intense research. The discovery of metal complex excited state deactivation by *electron* transfer in addition to energy transfer at about this time was a spur to the study of metal complex photochemistry. Many proposed solar energy systems used $[Ru(bpy)_3]^{2+}$ as photosensitizer. However, it was quickly realized that to prevent the energy-wasting electron transfer between the oxidized ground-state species and the acceptor (the so-called back-reaction), it would be necessary to control the direction of electron transfer more carefully than was possible between simple molecules in solution. A variety of structured and heterogeneous systems was explored and some of these included polymeric systems, such as micellar solutions and a modified polystyrene backbone as support for the sensitizer. The study of these systems has revealed much about metallopolymer photochemistry. In recent years, advances in the synthesis of assemblies of metal complexes (*supramolecular* assemblies) have led to the study of the photochemistry of such systems in solar energy conversion.

More recently, a significant improvement in efficiency in solar energy systems was realized using nanocrystalline, high surface area TiO_2 as the electron acceptor and a special ruthenium sensitizer **1**, which contained carboxylated bipyridine groups that could bind directly to the surface of the oxide. The first example was the Gratzel cell

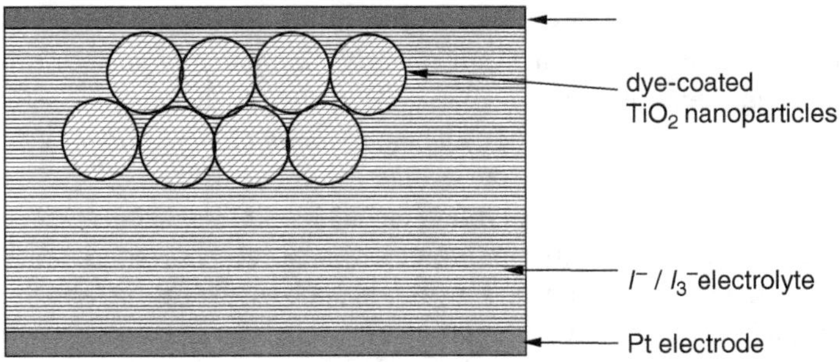

FIGURE 7.1 Dye-sensitized TiO_2 nanoparticle solar cell.

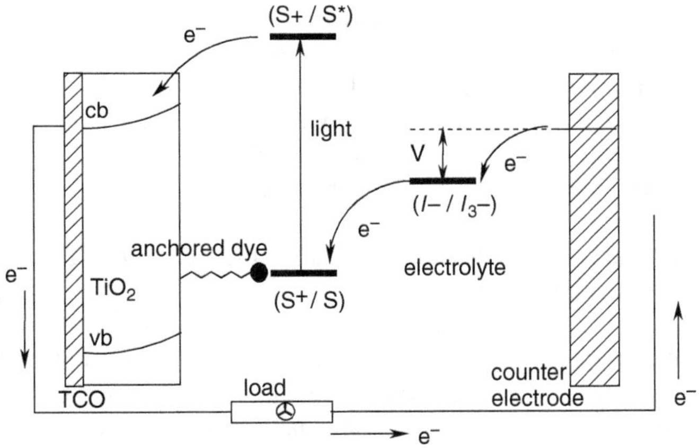

SCHEME 7.1 Current generation in the dye-sensitized TiO_2 solar cell.

(Fig. 7.1), which uses a ruthenium complex as a photosensitizer to inject electrons into a nanocrystalline TiO_2 support. Scheme 7.1 shows how current is generated in such cells. Light absorbed by the sensitizer dye S excites an electron into the LUMO in the singlet state. This electron is injected into the TiO_2 nanoparticle and emerges after tunneling through the nanoparticles to reach the transparent conducting electrode. The oxidized ground state dye remains, which then accepts an electron from the I^- in solution. The iodide is in turn oxidized to I_3^- and reduced back to I^- at the Pt electrode.

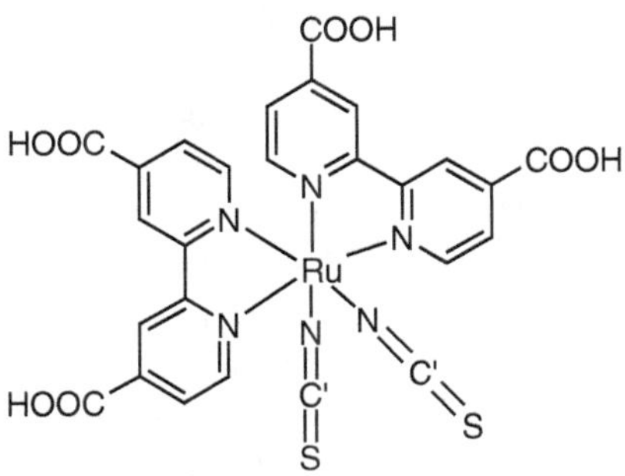

1

The disadvantage of this design is that it uses a liquid electrolyte, for example, acetonitrile, which means that the cell must be hermetically sealed. More recent Gratzel cell designs have used a nonvolatile electrolyte such as a polymer electrolyte or ionic liquid [5]. The possibility of using a polymeric photosensitizer has also been explored, and examples will be discussed later in this chapter.

The interaction of light with polymers has its own unique characteristics. Most obviously, the constraints of the polymer chain force chromophores into close proximity, leading to interactions in the ground state (aggregation) and between the excited state and a ground-state molecule to form an excimer. Excimers often emit at lower energy and without vibrational structure, so their emission is quite easily distinguished from the monomer emission. The study of excimer emission decay provides valuable information about polymer dynamics and energy migration, which involves energy transfer through resonance (Förster) or exchange (Dexter) energy transfer between adjacent chromophores on the polymer chain, or through quenching reactions with quencher molecules on the chain. Resonance energy transfer is favored over long distances due to the r^{-6} dependence on the rate (compared to exponential for exchange energy transfer). Information on polymer motion can also be obtained from exciting a chromophore with polarized light, and then, monitoring the monomer or excimer or probe molecule fluorescence decay in directions both parallel to and perpendicular to the excitation. This technique is called fluorescence depolarization.

Most of the 20th century work in polymer photochemistry concerned work on *nonconjugated* polymers, largely as an adjunct to the vast amount of work being carried out on purely organic polymers, in turn driven partly by the needs of the polymer industry to improve the photostability of polymers (or, conversely, to enhance the photodegradation of polymers for waste disposal or recycling purposes). Of course, many commercial, nonconjugated polymers only exhibit σ–σ^* absorptions in the high-energy UV region (<250 nm), leading to bond-breaking. Also, relatively weak metal–ligand bonds are clearly vulnerable to photodissociation. For applications involving light, the interest is in the absorption and luminescence phenomena at considerably lower energy. This requires chromophores absorbing light at lower energy; and such chromophores may be provided by the polymer backbone (e.g., polystyrene, poly(methyl methacrylate)) or the metal complex itself. A further distinction in these systems is whether the chromophore is part of the repeat unit of the polymer or a minor component (dopant), which is present adventitiously (e.g., from a metal complex free radical initiator) or as an end group.

The serious study of *conjugated* polymers began relatively recently, and the discovery of electric field-induced light emission from conjugated polymers was made less than 20 years ago. The discovery in 1977 that polyacetylene, when doped, showed conductivities of the same magnitude as some metals [6] sparked an interest into conjugated polymers that resulted in an enormous array of uses and applications for organic semiconductors, including artificial muscles [7], radiation-absorbing coatings [8], sensors and biosensors [9–11], and electrochromic (smart) windows. But it was not until 1990 that conjugated polymers were shown to emit light [12].

Since then the area of light-emitting polymers has become one of the fastest growing domains in terms of the number of scientific papers published in recent times. This is due to the potential applications that these materials have for flat-panel displays such as laptops and mobile phone displays. As devices can be synthesized from polymers, cheap solution-based technology such as screen printing or inkjet printing can be used to fabricate devices instead of vacuum deposition required in the traditional inorganic light-emitting diode (LED) fabrication. The mechanical properties of polymers also allow for the design of flexible displays, electronic newspapers, and roll up road signs have been envisaged. Additionally, light-emitting polymer displays have low power consumption and greater viewing angles than rival liquid crystal displays (LCDs).

The potential of the polymer light-emitting diodes (PLEDs) was enormous, not only could flexible displays be made, but also the PLEDs would combine light emission with the advantageous mechanical properties of the polymers. The previously studied small molecule organic light-emitting diodes (OLEDs) suffered from the thin organic films recrystallizing during their use, whereas polymers are much more capable of withstanding the harsh operating conditions found in an LED device. The need for expensive microfabrication equipment is a drawback in both the traditional inorganic LEDs and the small molecule organic LEDs, as it is both costly and time-consuming. Polymer LEDs can be fabricated using the relatively simple technique of spin-coating, and in the future, the use of high vacuums in processing may be eradicated completely. It is also possible to produce large area displays with polymer devices as they can be easily fabricated using solution processable techniques such as spin-coating, screen printing, or even through conventional inkjet printing techniques. Since the initial discovery in 1990, a large number of families of light-emitting polymer have been reported [13].

Light is emitted from conjugated polymers following excitation of electrons from the highest occupied molecular orbital (HOMO) to the lowest unoccupied molecular orbital (LUMO) by ultraviolet light or an applied voltage. The excited polymer then returns to the ground state by either a nonradiative process or the emission of light. Emission occurs from the lowest energy bandgap and therefore, energy is transferred from high energy states to lower energy states. This means that for a device containing two different polymers, emission will generally be observed from the long-wavelength emitter. Similarly for a polymer with a range of conjugation lengths (i.e., segments of coplanar rings), emission will occur from the longest conjugated segment. This gives a larger Stokes shift to the emission and prevents reabsorption of the emitted light by the polymer. Also, the emission often displays vibrational structure characteristic of a well-defined single emitting segment. Two kinds of energy transfers are known to exist for conjugated polymers. Firstly, energy can be transferred along a polymer chain over relatively short distances, referred to as Dexter transfer. The second type of energy transfer is that between different chains (interchain energy transfer) or different chain segments. This can occur over larger distances than for Dexter transfer and is referred to as Förster transfer [14]. Both of these processes have different selection rules.

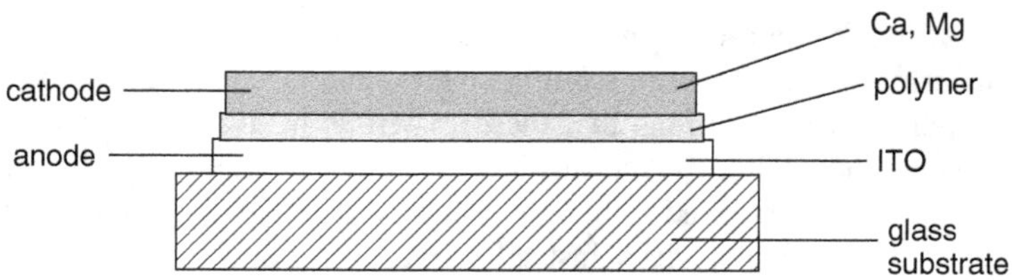

FIGURE 7.2 Structure of an OLED.

To produce electroluminescence, the organic light-emitting material is sandwiched between two electrodes, and a voltage is then applied. This forms the basis of a single-layer device, the structure of which is shown in Fig. 7.2. To allow for the emission of light, one of the electrodes must be transparent as well as conducting. Indium-tin oxide (ITO) is commonly used as it is transparent and has a relatively high work function making it a suitable anode.

A number of different metals have been used successfully as the cathode in LEDs. Calcium and magnesium are frequently used due to their low work function, which results in a lower energy barrier to electron injection than for other metals. Unfortunately, calcium metal is extremely air sensitive and it is necessary to apply another conducting layer to protect the calcium from the atmosphere. Typically, aluminum is used for this purpose due to its inertness in air and suitable work function. The cathode injects electrons into the LUMO while the anode creates holes in the HOMO creating a radical anion and radical cation, respectively (Fig. 7.3). The applied electric field then causes these charges to migrate through the polymer until two opposite charges meet. When this occurs the radicals combine to form either a singlet or a triplet excited state.

High efficiencies for conversion of current to light output can be achieved by forming bilayer devices with electron-transporting/hole-blocking layers or through

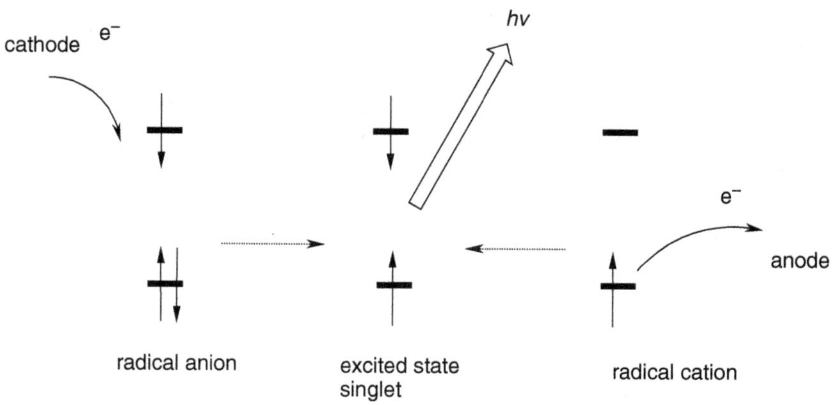

FIGURE 7.3 Generation of radical anions and cations at the electrodes of an OLED. The ions migrate to the center of the material where they meet on the same chain segment to form emissive singlet (shown here) or triplet excited states.

blending of the polymer layer with a host such as 4,4,′-bis(*n*-carbazolyl)biphenyl (CBP) and an electron transporter such as 1,3,5-tris(2-*N*-phenylbenzimidazolyl) benzene (TBPI). The technology has rapidly improved to the extent that several products with light-emitting polymer displays have entered the market. These include electric razors, car stereos with polymer backlights, and now consumer monitors and displays, including a Kodak digital camera with full color OLED display. For those interested in the applications of metallopolymers as electroluminescent materials, it is to be hoped that the potential advantages of metal complex emission such as control of wavelength and narrowness of emission lines and greater efficiencies, can be exploited fully.

This chapter will focus on the discussion of polymer photochemistry on a rather broad interpretation of metallopolymers, further subdivided into conjugated and nonconjugated types. However, developments in dendrimer photochemistry will not be covered: readers are encouraged to consult the recent review of metallodendrimers by Hwang and Newkome [15], and the dendrimers section of the excellent review on the photochemical conversion of solar energy by Balzani et al. [4]. This chapter will also not cover the photochemistry of metal–organic framework (MOF) materials or the polymers linked by hydrogen bonding.

Early studies of metallopolymers were frequently hampered by poor solubility and incomplete characterization. Although great strides have been made in the last 20 years on the efficient synthesis of soluble metallopolymers, as witnessed by the two excellent collections of expert reviews [16,17], the applications of such polymers as photoactive materials is only just beginning.

7.2 COMPLEXED POLYLIGANDS WITH NONCONJUGATED BACKBONES

As mentioned in Section 7.1, nonconjugated polymers may have the chromophore as part of the repeat unit of the polymer or a minor component (dopant) that is present adventitiously or as an end group. This chapter will be concerned with the former type. Nonconjugated polymers such as polystyrene, based on a vinyl monomer, may have different *tacticities* depending on the way the pendant groups are configured with respect to the backbone. This not only has implications for the physical properties of the polymer, such as crystallinity and glass transition temperature, but also affects the rates of energy migration within the polymer after photoexcitation. Another factor to consider is that the steric bulk (and possible charge) of the pendant metal complexes will tend to promote extended rather than coiled conformations of these polymers. This section will consider only covalently bound metal complexes, although the reader should be aware that a substantial literature exists on electrostatically immobilized metal complexes in ion-exchange polymers such as Nafion. Many of the fundamental issues concerning energy transfer and the fate of excited states in these polymers will not be covered here; instead, the reader is referred to the excellent review by Meyer and coworkers [3].

7.2.1 Polystyrene Systems

The attraction of polystyrene is that it can be easily derivatized to introduce ligand or metal complex groups. For example, following polymerization of chloromethylstyrene, ruthenium complexes can be attached to give **2** by the displacement of the chloride with the deprotonated carboxylate group of $[Ru(bpy)_2(4,4'-(CO_2H)_2bpy)]^{2+}$. Meyer and coworkers [18] used this material as a sensitizer, with the electron acceptor (e.g., methyl viologen) on a separate polystyrene chain. Due to the much-reduced diffusional mobility of the polymers, the rate of back-reaction was reduced by a factor of 500 compared to the solution value. Subsequent work by Meyer's group has favored the use of *ether* links between the bipyridyl groups and the polychloromethylstyrene [19]. Fully loaded Ru polymers show intensity-dependent, nonexponential decay due to multiphoton polarization effects, which alter the local environment around each excited chromophore. Partially loaded polymers could be prepared by using the metal complex as the limiting reagent, permitting the unreacted groups to be derivatized with a sensitizer or a different metal complex. Thus, time-resolved emission studies on a mixed Ru–Os chain shows evidence for energy transfer from Ru^* to Os^*, but unless the Ru and Os complexes are adjacent to each other, the rate of energy transfer is limited by the slow Ru^*–Ru energy transfer. More rapid energy transfer was achieved using anthracene groups to mediate energy transfer between the Ru and Os metal centers (Scheme 7.1).

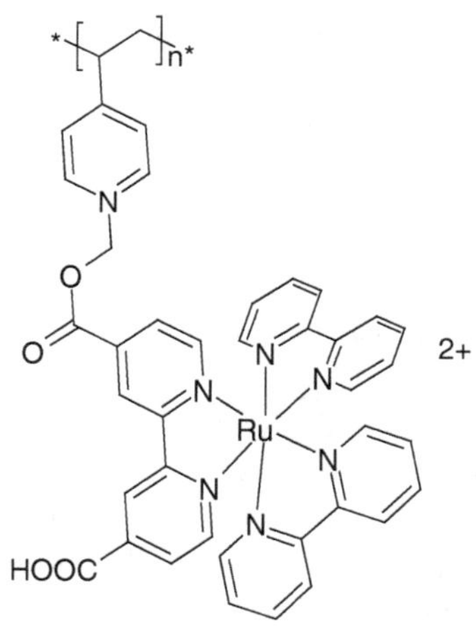

2

Much faster intrastrand energy transfer rates were observed in the homopolymer prepared by *amide* linking to the polystyrene chain. Based on results from the time-resolved resonance Raman spectroscopy, the higher rate was attributed to the localization of the excited electron on the bpy linked to the polystyrene backbone [20]. In contrast, the ether-linked polymer, due to directing effects, has the electron localized on the bpy pointing away from the polymer backbone, and thus, further away from adjacent Ru centers. Energy transfer is thought to occur through resonance energy transfer due to the large distances involved. The spin conservation condition is met even though the ground states are singlets and the excited states are triplets because the lowest energy MLCT state actually contains up to 30% singlet character. Based on these results, an antenna-sensitization-electron transfer mimic of photosynthesis was constructed within the polystyrene polymer **3**. The redox-separated state following excitation, PTZ^{+}—Ru–MV^{+}, decayed relatively slowly ($k = 15\ s^{-1}$), suggesting that it is feasible for long-lived redox equivalents to be stored on polymer chains.

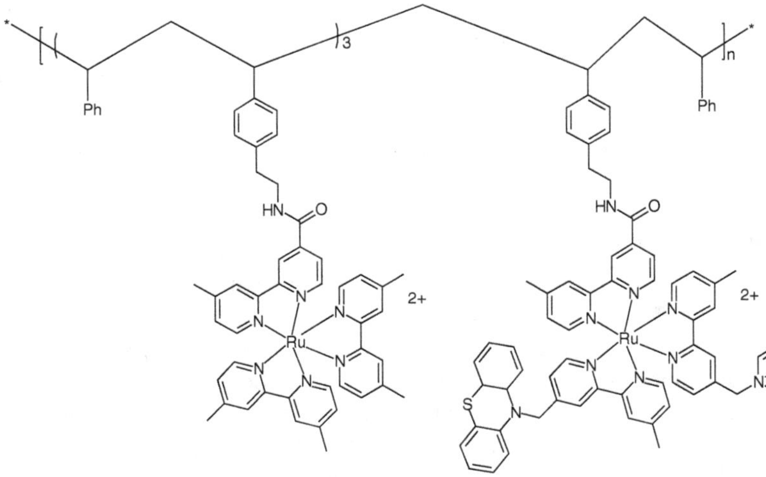

3

Polystyrenes have also been used to support chromophores useful in organic light-emitting diodes (OLEDs). Weck and coworkers have attached tris(2-phenylpyridine) iridium complexes to aminomethylated polystyrene using a Schiff base reaction, **4** [21]. There was no major diminution of the desirable luminescence properties of the iridium complexes (high emission quantum yields of 0.23 and lifetimes of about a microsecond). Similar results have been reported for aluminum and boron 8-hydroxyquinoline complexes tethered to polystyrene using Schiff base condensation [22].

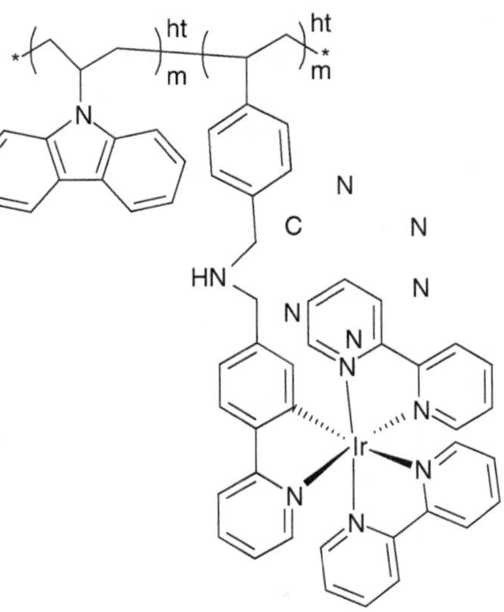

4

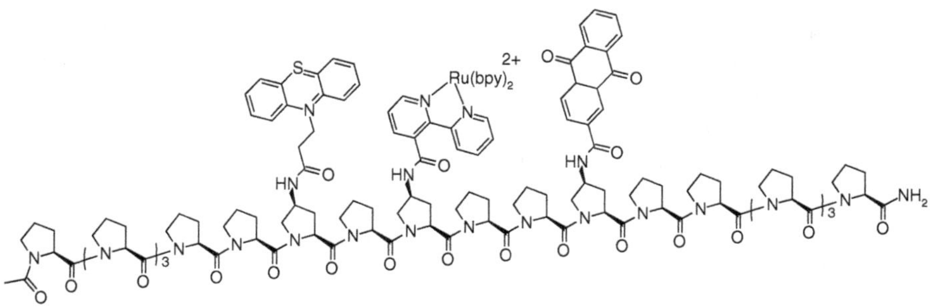

FIGURE 7.4 Oligoproline donor (PTZ)–Ru(bpy)$_3{}^{2+}$–acceptor (anthraquinone) assembly.

7.2.2 Polyprolines

In an effort to exert more control over the spatial arrangement of the various sensitizer and acceptor groups, Meyer's group have turned to solid-state peptide (Merrifield) synthesis. The amino acid selected was proline, which has a secondary amine and a cyclic structure, forcing polyprolines to adopt a rigid, helical conformation. This means that the distance between substituents on the chain is much more well defined than in polystyrene. As described for the polystyrene-based systems, it is possible to set up a Donor–Ru–Acceptor assembly on a polyproline backbone that is capable of storing redox equivalents (Fig. 7.4). The distance dependence of back electron transfer between Ru–bpy* and the oxidized donor (PTZ$^+$) was measured and shown to occur via through-space interactions [23].

7.2.3 Polyvinylpyridines

Much of the original work on polyvinylpyridines was carried out by Meyer and coworkers. They reported that in mixed RuOs polymers there was evidence for quenching of the Ru emission by the Os centers. However, recent work by Dennany et al. has found that although dynamic quenching by soluble $[Os(bpy)_3]^{2+}$ of polymer-bound Ru centers is very efficient, suggesting that one Os complex is capable of quenching several bound Ru* complexes, there is much less efficient quenching by Os complexes that are bound to the same polymer as the Ru in **5**, $[Ru(bpy)_2(PVP)_{10}Os(bpy)_2]_n{}^{4n+}$. They also found that when the polymer was immobilized on a surface the quenching by solution species occurred at a much lower rate [24].

Forster and coworkers have also prepared polyvinylpyridine complexed to a mixture of $[Ru(dpp)_2Cl]^+$ and (minority) $[Ru(dpp)_2]^{2+}$ groups, where dpp = 4,7-diphenyl-1,10-phenanthroline. In solution, dual emission from both chromophores is observed, whereas in the solid state, energy transfer to the lowest energy chromophore occurs [25]. The $Re(CO)_3Cl$(diimine) group can also be attached to polyvinylpyridine. The polymer can serve as a template for growing CdS nanoparticles [26].

The Dublin group has also reported a three-color electrochromic metallopolymer **6** based on a ruthenium phenolate complex bound to poly(4-vinylpyridine). The reversible color changes (wine red to green) are associated with the Ru(II) oxidation, whereas the mixed redox state produces a red-orange. The charge transport parameters of the polymer were thoroughly characterized [27,28].

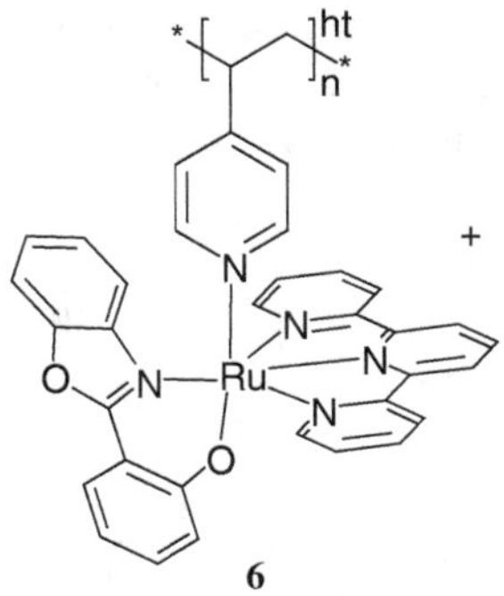

6

7.2.4 Electropolymerized Thin Films

Electropolymerization represents a useful method to control precisely the deposition of thin films of conducting polymers. These thin films have a wide variety of applications in optical devices [29]. Walder and coworkers showed that photocurrents could be generated using a trilayer thin film. The three different electropolymerized polymers were cleverly prepared from a common precursor, 4-methyl-4′-(2-pyrrolyl-1-ethyl)-2,2′-bipyridine **7**: two were alkylated, diquat derivatives and one was the Ru metal complex of this ligand, which acted as a sensitizer. The appropriate ordering of the triad in terms of their redox potentials led to the maximizing of the photocurrent (Scheme 7.2) [30].

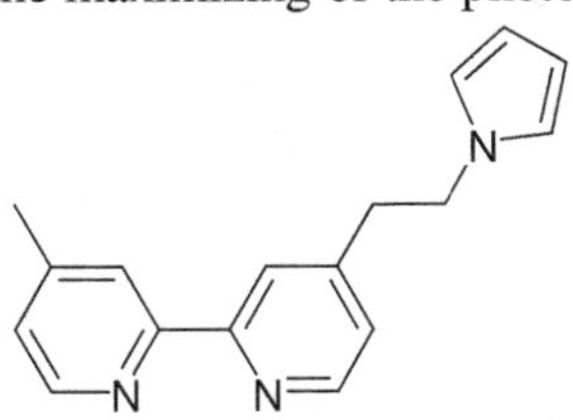

7

Hagfeldt and coworkers reported the oxidative electropolymerization of styryl-substituted Ru bipyridyl complexes, including **8**, designed to act as new sensitizer dyes in Gratzel-type cells. In contrast to other polymers, the λ_{max} of the polymer actually blue-

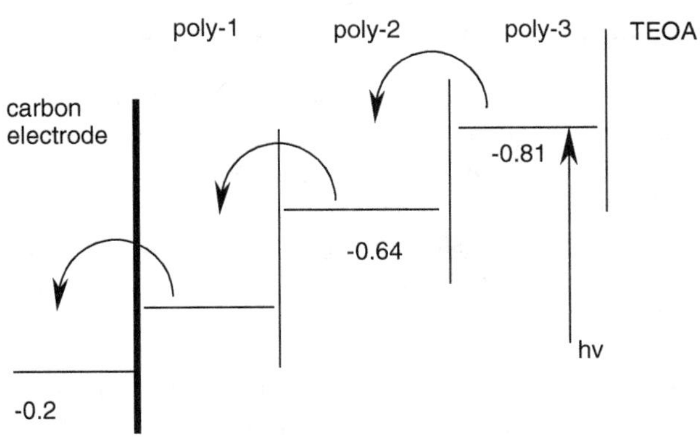

SCHEME 7.2

shifted compared to the monomer because the conjugation was interrupted by the reaction of the double bond [31]. In a later paper, they reported a vinyl thiophene-substituted complex **9**, which, surprisingly, oxidatively electropolymerized through the vinyl groups rather than the thiophene groups [32].

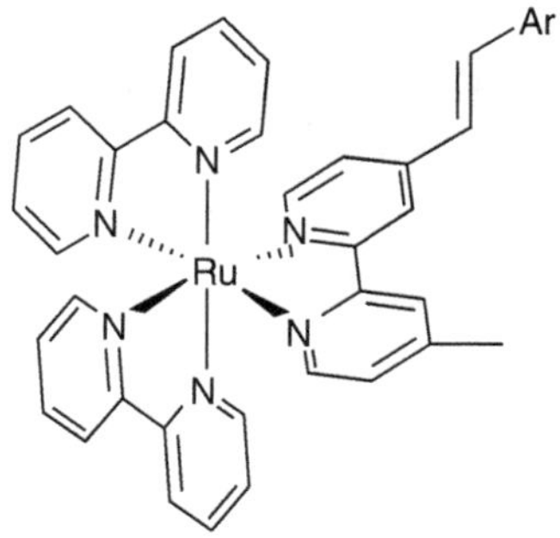

8 (Ar= -methoxyanisole) 9 (Ar=2-thienyl)

Crayston and coworkers have shown that vinylpyridine complex **10** may be electropolymerized efficiently in DMSO to yield an orange, luminescent film [33,34]. The method is much more controllable than that used in the traditional electropolymerization method for vinylbipyridine [35] and vinylterpyridine complexes [36].

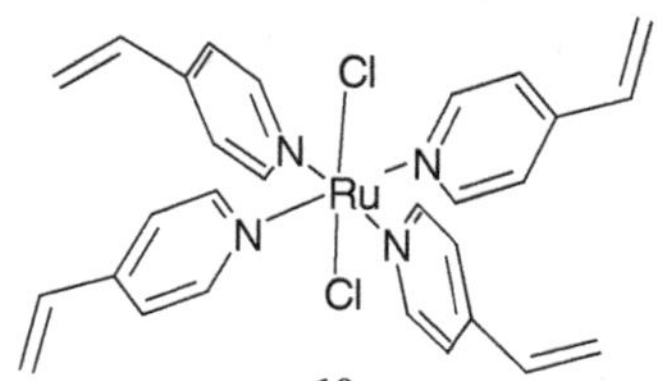

10

Improved electropolymerization was also found by Forster and coworkers using ionic liquids as the electrolyte. Electropolymerization of $Ru(aphen)_3^{2+}$ (aphen = 5-amino-1,10-phenanthroline) **11** in this medium gave thicker films with improved conductivity. The films showed electrochemiluminescence (ECL) in the presence of tripropylamine [37].

7.2.5 Metallodendrimers in Phosphorescent Light-Emitting Diodes

Although significant advances have been made in both OLEDs and PLEDs, there is an intrinsic limit of 25% internal efficiency for OLEDs due to these materials being purely fluorescent (light is emitted from the singlet excited state). For polymers, any triplet states formed will reduce the quantum efficiency as these states undergo nonradiatve processes to return to the ground state. Using one of the many phosphorescent metal complexes that have radiative triplet excited states would make internal efficiencies of 100% a realistic target. Several different phosphorescent metal complexes exist, for example, those of platinum and iridium. For optimum device efficiency, a metal with a short triplet lifetime is desired so that the excited state does not become quenched, which in turn would reduce the quantum efficiency of the device. The triplet lifetimes of several iridium complexes are suitable and, therefore, iridium complexes are the preferred choice for use in phosphorescent devices [38]. One of the most frequently used iridium complexes is $Ir(ppy)_3$**12**, which emits green light.

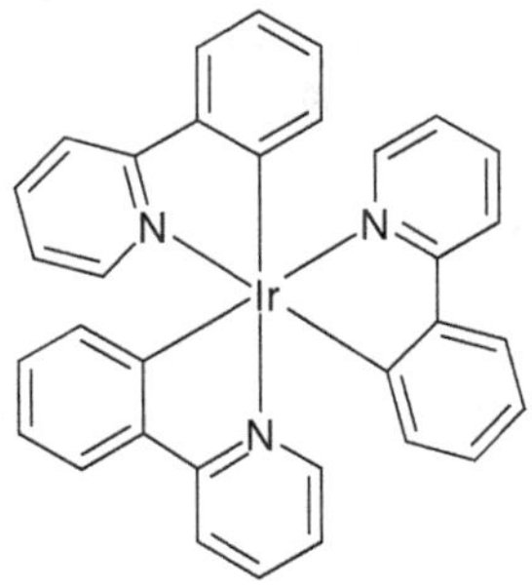

12

By altering the structure of the ligand attached to the iridium, the color of the emission can be tuned, and red, green, and blue phosphorescent devices have been prepared [39].

Triplet–triplet annihilation is prevented by doping the iridium complexes into polymers such as polyvinylcarbazole (PVK). PVK is a conducting polymer and acts as a spacer to separate the iridium chromophores. This also has the dual advantage of allowing the iridium complexes to be solution processable [40]. Light emission occurs by an exothermic or Förster transfer of energy from the host (often a polymer) to the dopant (the phosphorescent material). The relative energies of the two materials ensure that this energy transfer is not reversible and, therefore, a large external quantum efficiency can be obtained for a relatively small amount of iridium complex [41]. However, devices may suffer from phase separation of the individual components, which is detrimental to the device performance.

Another very effective method of preventing triplet–triplet annihilation was reported by Burn and coworkers using dendrimer ligands to surround and protect the iridium core. The color of the emission can be tuned by altering the core of the molecule and the properties of the material controlled by variation of the dendritic structure [42]. For example, the iridium dendrimer **13** has an external quantum efficiency of 10.4% in a single layer device. These phosphorescent dendrimers also have the additional advantage that they are sufficiently soluble and massive enough to be spin-coated from solution.

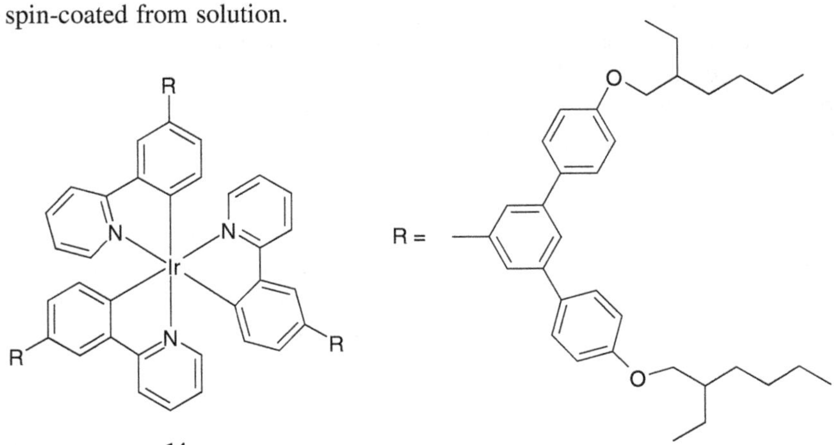

14

7.3 COMPLEXES OF POLYLIGANDS WITH CONJUGATED BACKBONES

The simplest way to turn these molecular complexes into polymers is to prepare the polymer backbones containing pure ligand, such as polybipyridine homopolymers. These tend to be rather intractable, hence, the emphasis in the succeeding sections on mixed polymer chains. Nevertheless, interesting examples of this approach include polybenzimidazoles **14**, which can coordinate to Ru centers [43]. Despite complications in the emission spectra from the formation of aggregates, LEDs prepared from this material exhibited 0.1% external quantum efficiency.

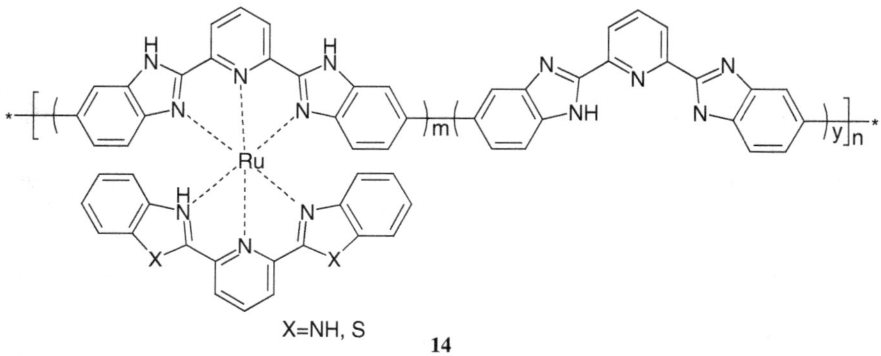

14

Cameron et al. [44] have investigated in some detail the charge transport properties of metal-coordinated polybenzimidazoles, such as **15**.

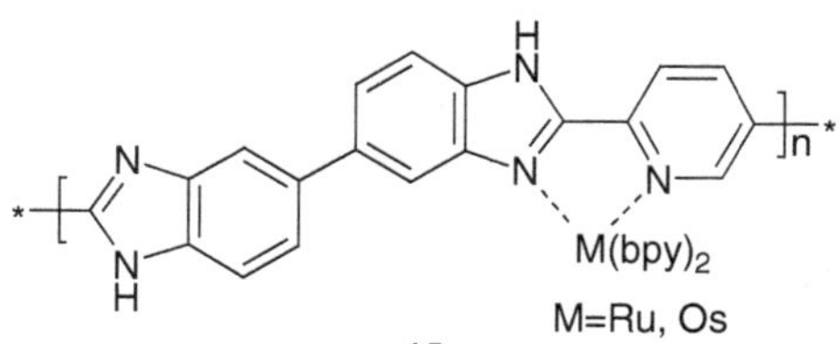

15

Strong electronic communication between the metal centers is indicated by the presence of IVCT bands and the $M(II)_2/M(II)M(III)$ and $M(II)M(III)/M(III)_2$ redox waves with large separations [44–47]. Note that a recent review of conjugated metallopolymers is available [48] and there is also an earlier review by Pickup [49].

7.3.1 Polyfluorene-Based Polymers

Polyfluorenes are sometimes referred to as stepladder poly(*para*-phenylene) (PPP) polymers as they have a backbone with a planar, ladder-like structure. The first polyfluorene LED was made by Yoshino and coworkers using poly(9,9-dihexylfluorene) **16** [50].

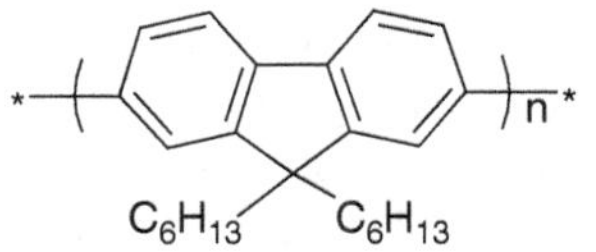

16

Polyfluorenes represent a promising family of LEPs for blue emission for several reasons. First, they display good thermal and oxidative stability with good quantum conversion. These polymers are a type of rigid-rod polymer suffering from poor solubility. In the case of polyfluorenes, however, solubility can be readily improved by the introduction of alkyl side chains at the C9 position. As the aromatic rings are held at the C9 position, the phenylene–phenylene remains virtually planar. The alkyl side chains lie perpendicular to the aromatic rings and, therefore, do not cause any steric hindrance, preserving the conjugation of the backbone [51].

The Yamamoto method can be used to synthesize fluorene homopolymers (Scheme 7.3A) as well as random alternating copolymers [52]. Alternating copolymers can be synthesized through Suzuki coupling (Scheme 7.3B), which tolerates a wide variety of different functional groups [53].

The Suzuki coupling is particularly powerful, since the polymer properties can be easily tuned by varying the structure of the monomers, and indeed a large number of different polyfluorene copolymers have been made in this way [54].

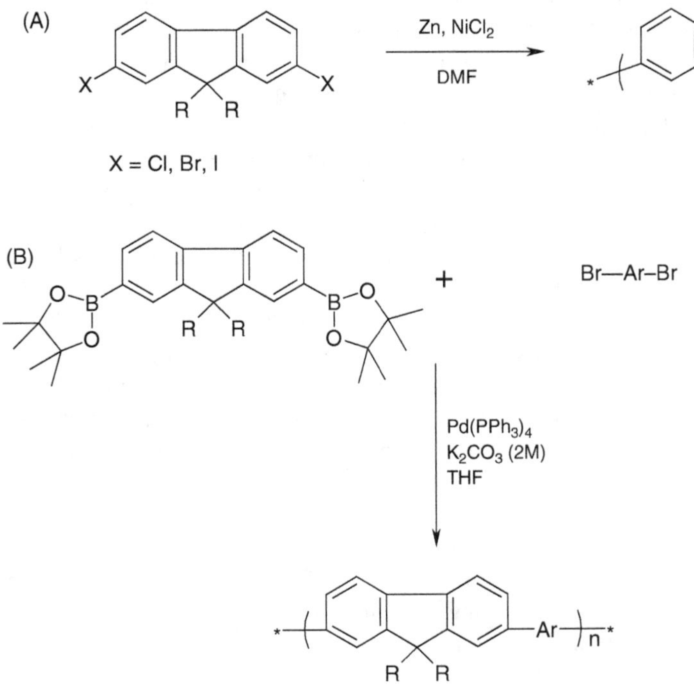

SCHEME 7.3 Synthesis of polyfluorenes.

Clearly, if metal-binding sites are required, one strategy is to prepare alternating copolymers where one monomer is based on a well-known ligand (e.g., bipyridine, phenanthroline, and acetylacetonate), which is capable of coordinating to a metal ion. One such example was a study carried out by Liu et al. In this case, three similar conjugated polymers **17** (**R** = Oct), **18**, and **19** based on fluorene and bipyridine were prepared and tested for their metal ion sensing capabilities [55].

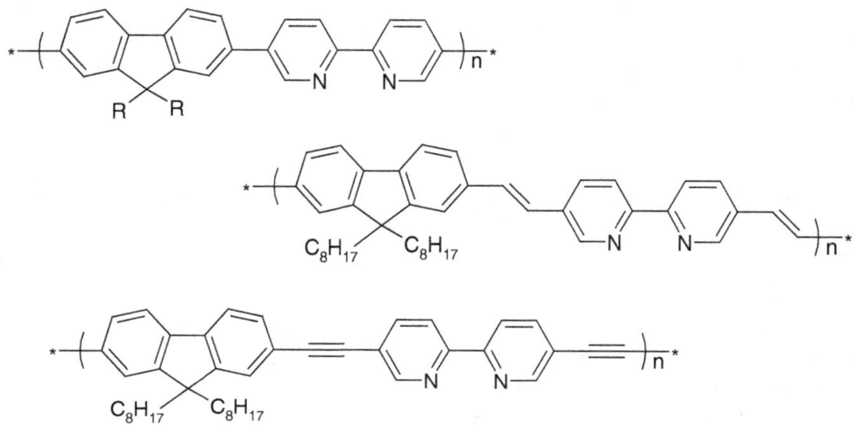

17, 18, 19

None of the three polymers showed any ion response to the alkali metal ions or the alkaline earth metals except for magnesium. Transition metals, however, produce a very sensitive ionochromic effect in all three polymers. The metal ions could be divided into three distinct groups: metals that completely quenched the luminescence (e.g., nickel), those that partially quenched the luminescence without altering the emission maxima (e.g., aluminum), and those that partially quenched the luminescence and also displayed a red shift in the emission spectra (e.g., manganese). The authors also report differing sensitivities to the metal ions between the polymers, which is explained by differences in the stiffness in the backbone of the polymer. Polymer **17** containing a single C–C bond has the most flexible backbone and also displays the most sensitive ionochromic effect. Polymer **19** with the ethyne link and, consequently, the least flexible backbone displays the weakest ionochromic effect. The authors have explained this phenomenon in terms of the flexibility of the backbone permitting the bipyridine units to become planar. The two aromatic rings of the bipyridine unit lie at an angle to each other and as a result the bipyridine is not planar and acts as a conjugation break. However, on coordination of a metal ion such as palladium, nickel, or zinc, the bipyridine unit is forced into a planar conformation and the conjugation length of the polymer is increased resulting in a bathochromic shift in the absorption and photoluminescence spectra. The ease of forming the planar bipyridine is related to the ease of binding metal ions and, therefore, accounts for the differing sensitivities of the three polymers studied.

Zhang et al. describe the properties of two alternating copolymers based on fluorene and a bipyridyl unit **17** (R = Hexyl) or a phenanthroline unit **20** to coordinate to metal ions [56]. Both of these polymers display bathochromic shifts in the absorption and luminescence spectra on coordination to some metal ions. This throws into doubt the validity of the argument given above that such red shifts are accounted for by the induced planarity of the bipyridyls increasing the conjugation length. Zhang et al. propose, instead, that each metal ion and its polymer complex have a distinct electronic structure and energy levels.

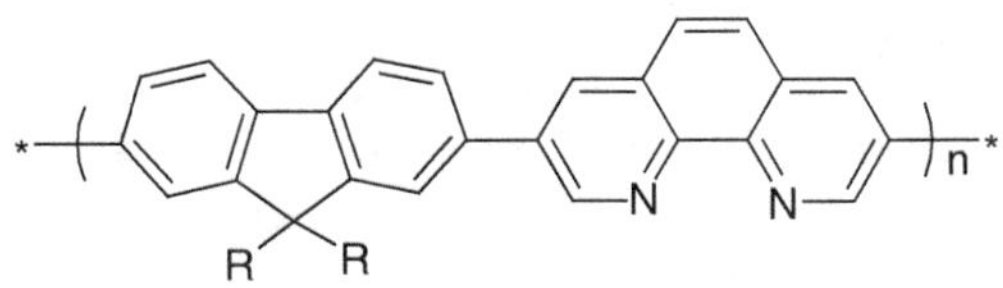

20 (R=Hexyl)

Later, a Ru complex **21** was prepared [57], and the effect of Zn, Eu, and Ir coordination on the polymer was investigated. But, in these cases, the complexes were not isolated [58]. Crayston and coworkers prepared Eu (15% doped) and Tb complexes (100% doped) of **20** and related copolymers **22** [59]. Whereas the Eu complex displayed f–f emission in solution at high concentrations and in thin films via energy transfer, the Tb complex showed only a broad green emission. Time-resolved photoluminescence experiments revealed that the Tb emission did arise from a new, red-shifted singlet state. The red shift of the singlet state energy helped to explain why the energy of the corresponding triplet was not high enough to transfer energy to the Tb f states.

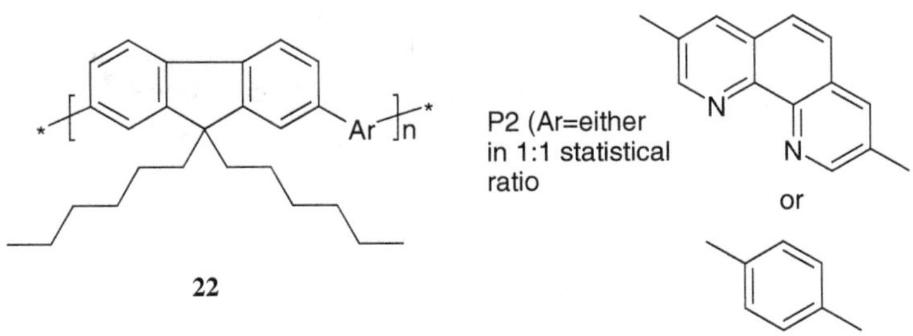

22

A benzoate–fluorene copolymer was bound to a europium complex to yield **23** and showed the characteristic red emission of this metal, demonstrating efficient energy transfer from the polymer chain to the metal complex [60].

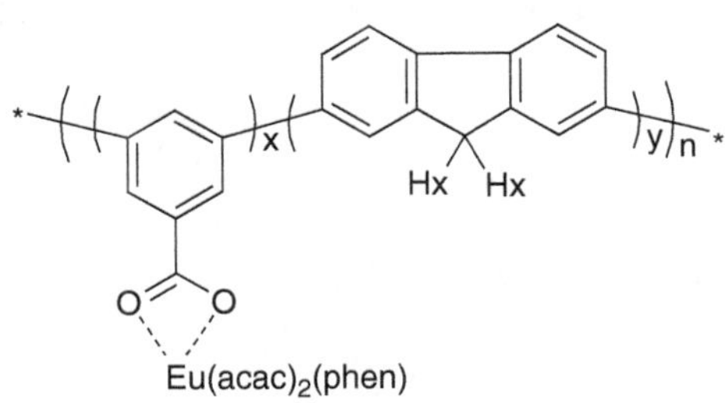

23

7.3.2 Poly(phenylene)-Based Polymers

The phenanthroline moiety is a widely used chelating ligand capable of binding to a large number of metal ions. However, difficulties have been encountered in the synthesis of the high-molecular weight polymers because of the insolubility of the phenanthroline moiety. Yasuda et al. [61] demonstrate that it is possible to synthesize a relatively high-molecular weight polymer based on phenanthroline if the second monomer has long alkyl groups attached to enhance the solubility. The polymer **24** was synthesized by Suzuki coupling with a molecular weight of 7000. In contrast to polymers **16–18**, polymer **24** displayed good sensitivity to alkaline and alkali earth metals as well as the transition metals. This can be attributed to the high coordinating ability of the phenanthroline relative to bipyridyl.

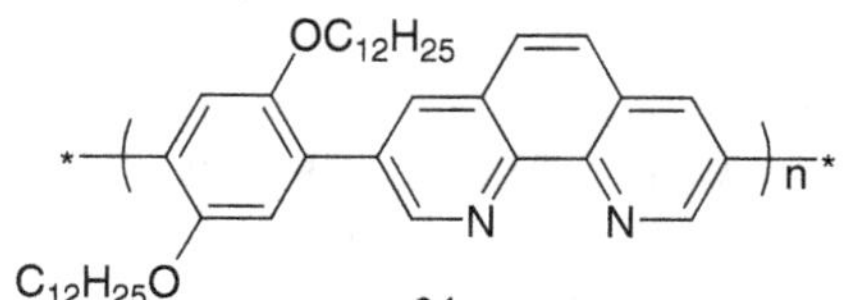

24

Bipyridine may be incorporated into a poly(*para*-phenylene vinylene) chain by Wittig reactions, yielding **25**. Solubilizing groups are necessary to characterize the polymer, whose absorption spectrum responds to a wide variety of metal ions in solution (ionochromism) [62]. The emission was also affected by different metal ions in different ways. It was proposed that metal ions that bind to both pyridine groups lead to a planarization of the chain and a quenching of the emission, whereas those that bind to only one pyridine group merely red shifted the emission.

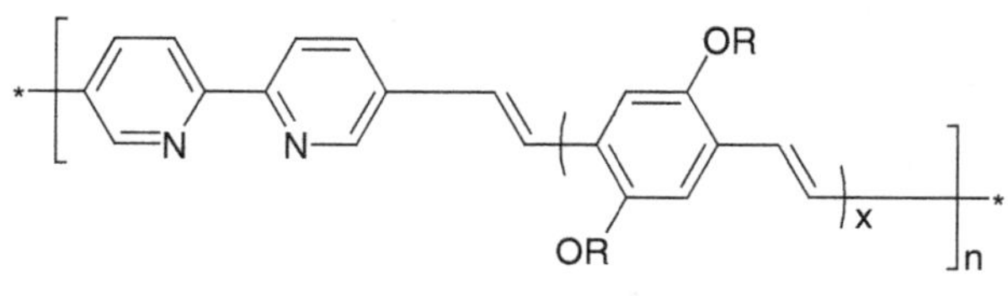

25 (x=1 or 3)

Smith et al. have prepared polymers **26** of this type for sensor applications [63]. Cu^{2+} ions quenched the emission efficiently, but there was partial restoration of the emission in response to excess nitric oxide, which is capable of reducing Cu^{2+}.

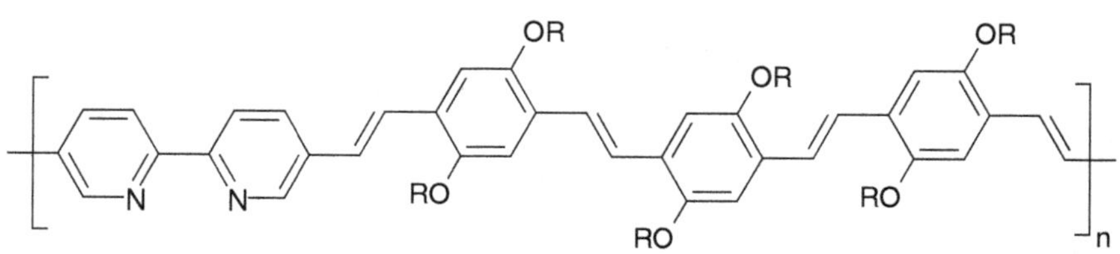

26

In another bipyridine vinylene polymer, Ru chromophores were attached to the bipyridine group before polymerization, **27**, resulting in interesting optical properties. For example, the photoconductivity is increased at the metal complex absorption energy and the polymer displays nonlinear effects such as the photorefractive effect [64]. These results have been compared to related polymers **28** in which the charge on the Ru complex is reduced by using auxiliary diketonate ligands, L_4 [65]. This also lowers the oxidation potential of the complex, leading to higher photoconductive sensitivity of the polymer than **27**. The conductivity also increases with increasing Ru content.

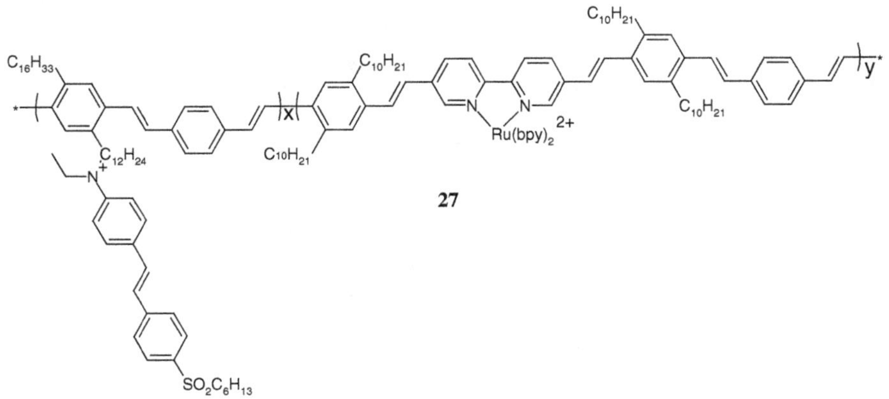

27

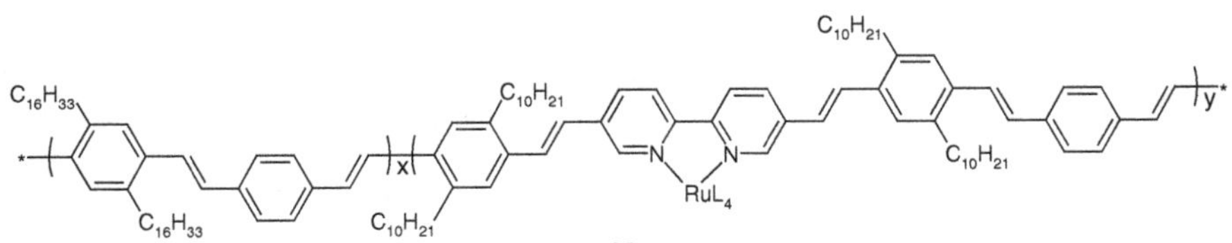

28

Poly(aryl-ethynylene) oligomers **29** ($n = 1$–4) containing bipyridine groups along the backbone were reported by Ziessel and coworkers. The titration of the absorption and fluorescence spectra with Zn^{2+} cations indicates the formation of 1:1, 1:2, 1:3 metal:binding group complexes [66]. Coordination of $Ru(bpy)_2$ groups showed that the photophysical properties were largely those of isolated metal complexes, indicating little communication along the chain [67].

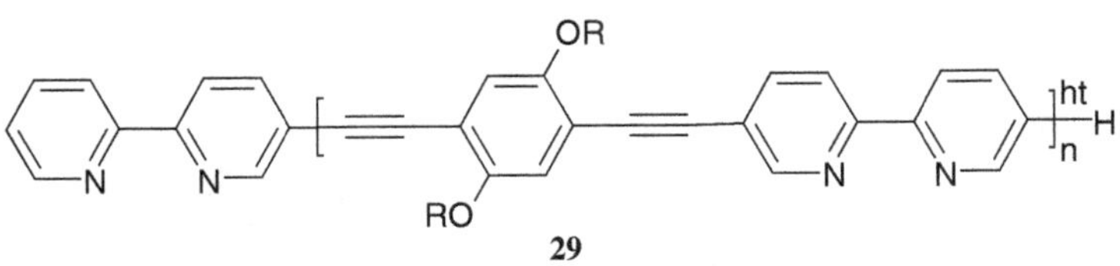

29

Further investigations on the coordination properties of the related fully polymerized material, **30**, were continued by Weder coworkers [68]. Pautzsch and Klemm prepared similar polymers, **31** and **32**, starting from the dibromobipyridine Ru complex that was then coupled to diethynylbenzene by Sonogashira coupling [69]. The polymer bands were red shifted, but the Ru MLCT band was unchanged.

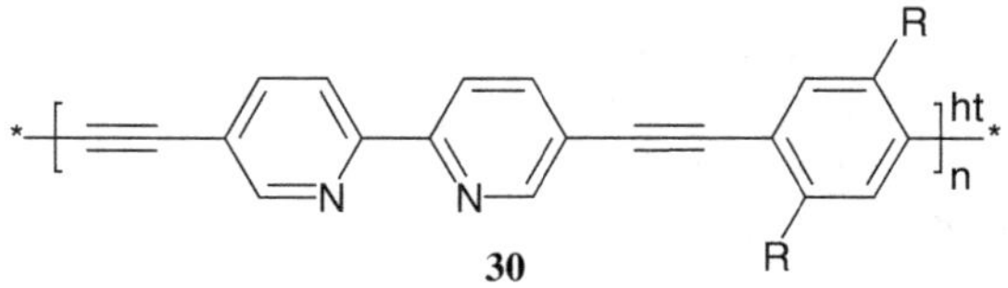

30

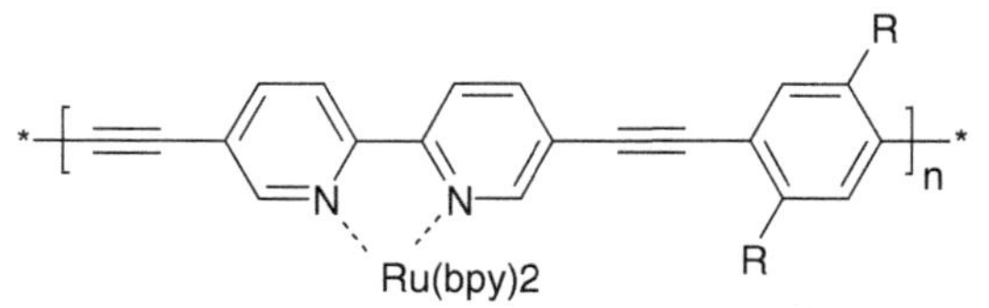

31

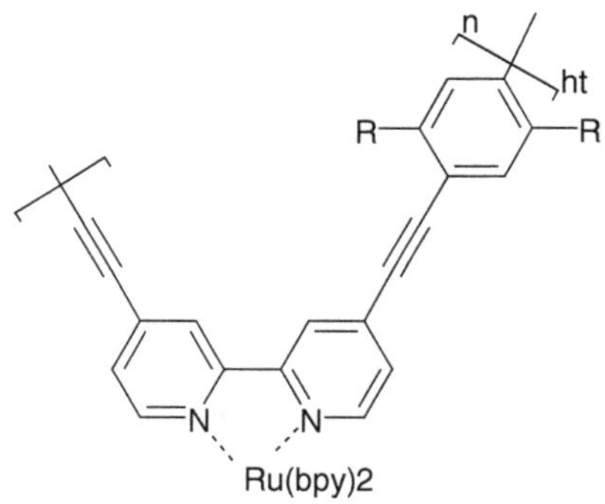

32

Ley and Schanze have also reported on poly(aryl-ethynylene) polymers **33** containing bipyridine groups along the backbone, bound to the Re(I)$(CO)_3$Cl chromophore, to various extents of loading [70]. In addition to the π–π^* polymer backbone absorption, the polymers display an LMCT absorption and fluorescence that is quenched (rather inefficiently) by energy transfer to the Re triplet manifold, which undergoes phosphorescence.

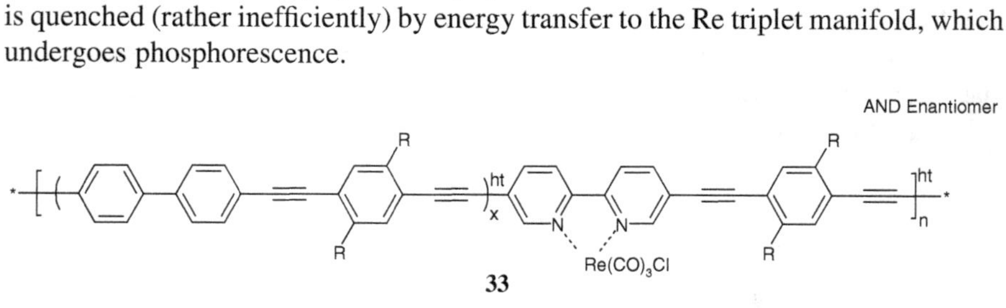

33

Other binding groups have been used, such as dipyridophenazine ligands in polymer **34**, which were found to increase the electron transport rate. LEDs prepared from this material containing Ru at various levels indicated once again energy transfer to the metal complex at higher Ru content [71].

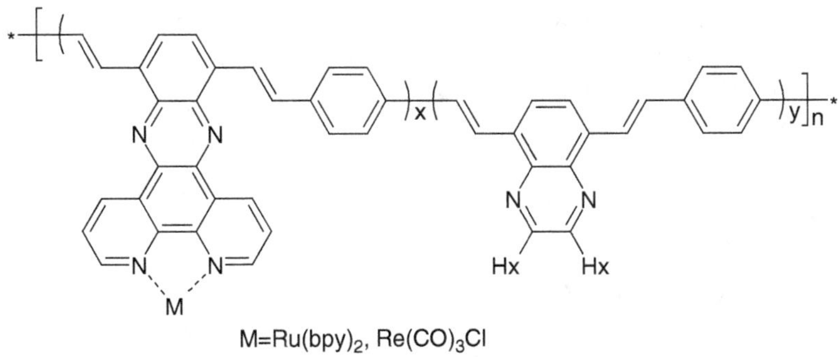

34

Poly(phenylene vinylene)-bridged porphyrin polymers, including **35**, with degrees of polymerization (DPs) of around 12 have been prepared by Wittig reactions [72]. Low energy emission from the porphyrin and higher energy emission from the phenylene vinylene bridges, the latter red shifting and increasing in lifetime and yield as the chain length increased. In the solid state, there was energy transfer to the low energy emission. Polymers were also prepared with alternating metallated (Zn, Cu) and free-base porphyrin groups. These showed energy transfer from the higher energy Zn $\pi\pi^*$ states, probably by the Förster mechanism.

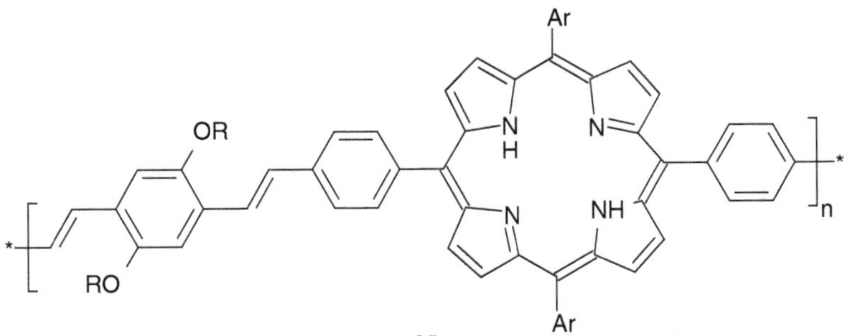

35

$[Ru(terpy)_2]^{2+}$ centers have also been polymerized using the phenylene vinylene bridge [73]. The polymers **36**, when deposited with sulfonated polyaniline in alternating layers, give a light sensitive material, which exhibited a photocurrent of ca. 10 $\mu A\,cm^{-2}$ and a photovoltage of ca. 0.8 V on irradiation.

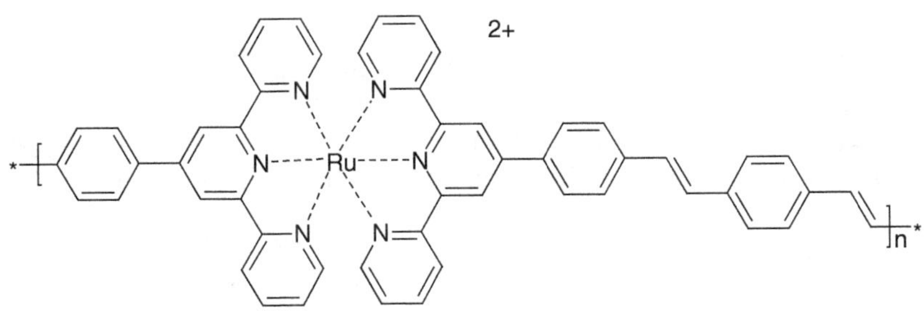

36

The Heck coupling reaction has been shown to be effective in preparing polymers **37** containing rhenium-complexed diimine ligands bridged by phenylene vinylene groups [74]. Photovoltaic devices prepared using this material showed rather low efficiencies.

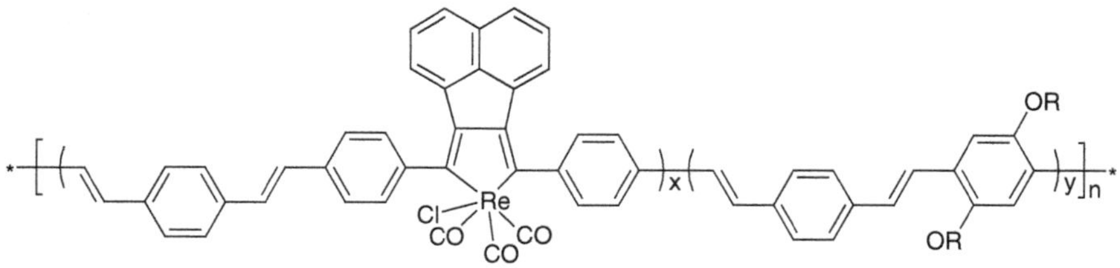

37

Aromatic 1,3,4-oxadiazoles are known to be efficient electron transport materials. When incorporated by Heck coupling into the poly(phenylene) chain in polymers such as **38**, the electron carrier mobilities are indeed increased, while the hole mobilities are unchanged [75]. Light-emitting diodes prepared from this polymer show increasingly metal-dominated emission as the %Ru content is increased.

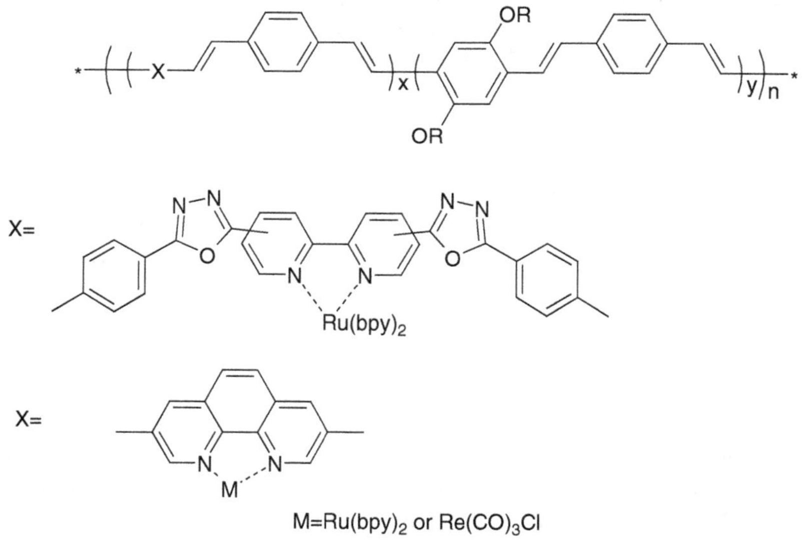

38

Similarly, alternating polymers such as **39** of bipyridine with benzobisthiazole or benzobisoxazole can be prepared by polycondensation of 2,2′-bipyridine-5,5′-dicarboxylic acid with diaminobenzenediols in poly(phosphoric acid) [76]. The polymers can complex to $Ru(bpy)_2$ centers; however, their carrier mobilities were not measured.

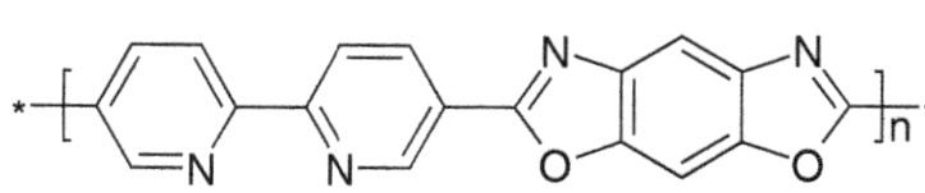

39

Finally, conjugated materials **40** based on poly(phenylene thiophene) and poly (fluorene thiophene) main chain polymers functionalized with pendant trithiocyanato ruthenium terpyridine complexes were synthesized by the Suzuki coupling reaction. Heterojunction photovoltaic cells with the simple structure ITO/polymer/C-60/Al were fabricated. Under simulated AM1.5 solar light illumination, the short circuit currents, open circuit voltages, and power conversion efficiencies of the photovoltaic cells were measured to be 1.53–2.58 mA cm^{-2}, 0.12–0.24 V, and 0.084–0.12%, respectively [77].

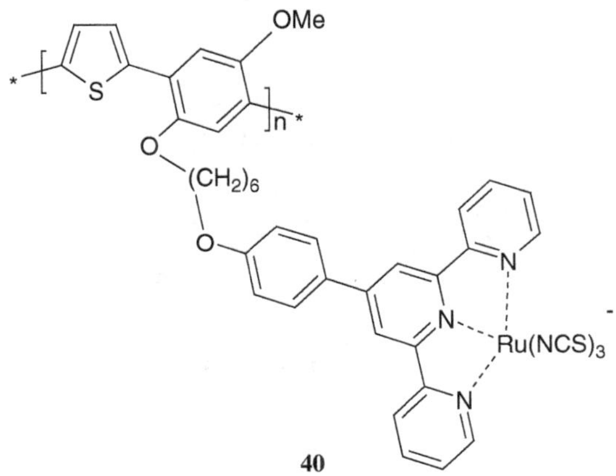

40

7.3.3 Polythiophene-Based Polymers

Polythiophenes with ferrocenes in the main chain **41** have been prepared by electropolymerization of bis-substituted ferrocene monomers, but only for the $n=2$ and $n=3$ monomers [78,79]. The resulting film of the $n=2$ monomer showed a thiophene → Fe(III) LMCT band at 1395 nm. A soluble derivative of the $n=3$ polymer has been prepared. The oxidized form is blue with a thiophene → Fe(III) LMCT band at 590 nm [80]. In general, it has been found that these systems show greater delocalization, the longer the thiophene chain and the better is the matching of the ferrocene and thiophene redox potentials [81,82].

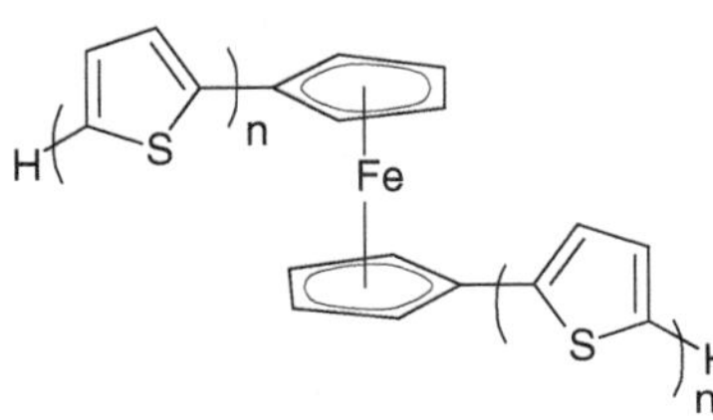

41 (n=1-3)

A similar polymer to **41** above has been prepared with thiophene links between the metal centers **42**, although, in this case, the thiophene was electrochemically coupled together after the terpy complex had been prepared. The red shift in the emission from 498 to 521 nm matched the expected shift due to greater conjugation provided by the bithiophene linkage [83].

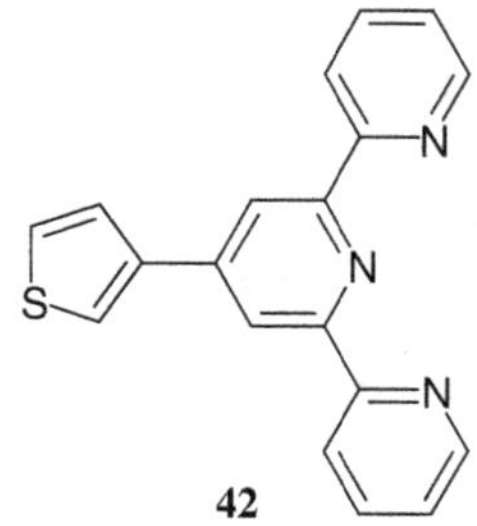

42

Bipyridine ligands **42a** have also been prepared [84] and the resulting electropolymerized films are more conductive when Ru is complexed to the bipyridine [85].

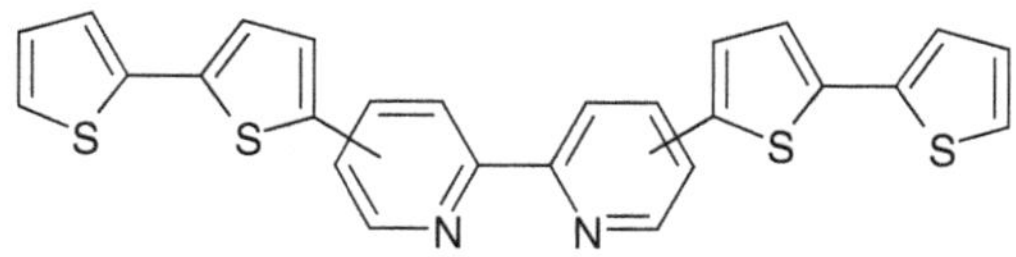

42a

Supramolecular complexes based on these polymers have applications not only in sensing based on the electrochemical response [86,87], but also utilizing changes in the emission properties [88]. A new absorbance band appeared after Zn binding to polymer **43**, and the emission was weaker and red shifted. Interestingly, exposure of the Zn-substituted polymers to alcohol vapor resulted in a restoration of the emission intensity.

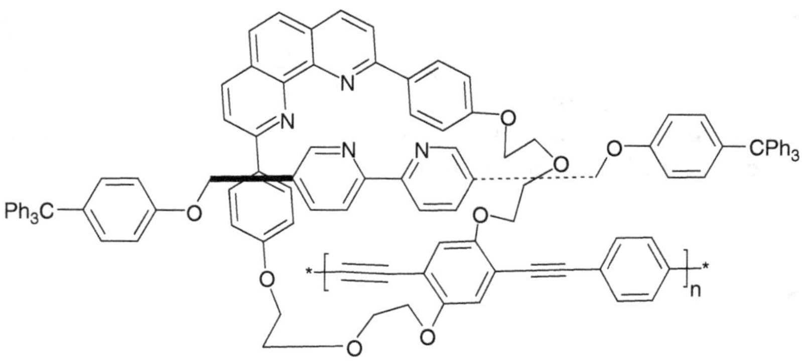

43

An exotic organomolybdenum complex derivative of terthiophene **44** was found to be electropolymerizable to give a thin electrochromic film with two Mo-centered oxidations superimposed upon the thiophene oxidation. The film switched between purple (undoped) and transparent blue (doped). These colors are quite different from poly(terthiophene) itself and are believed to indicate the direct interactions between the complex and the thiophene groups [89]. A sulfide analog has also been studied [90].

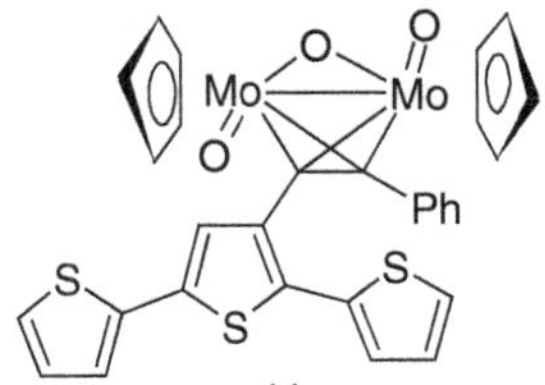

44

7.3.4 Solar Energy Materials and Sensors

The majority of the research into light-emitting polymers has been focused on improving the optical properties for use in organic light-emitting diodes in a display of some sort. A small, but growing, number of research groups have focused their attention on exploiting the properties of these polymers for different applications. In the main, these polymers are excited to produce light; however, these polymers can also work in the reverse, whereby light is used to excite the polymer-forming excitons, which dissociate and the resulting charges produced migrate to the appropriate electrode. In effect, the PLED can also operate as an organic solar cell. Conjugated polymers make good organic solar cells due to not only their excellent charge transport properties, but also their strong absorption in the UV–vis spectrum. To optimize the solar cell, as much light as possible must be harvested and, in particular, the polymer must absorb as much of the solar spectrum as possible. Frequently, copolymers are prepared to lower the bandgap and, therefore, enhance red light absorption as typical conjugated polymers absorb very poorly in the red region of the electromagnetic spectrum. This has been achieved for polyfluorenes by synthesis of an alternating copolymer poly(2,7-(9-(2′-ethylhexyl)-9-hexyl-fluorene)-*alt*-5,5-(4′,7′-di-2-thienyl-2′,1′,3′-benzothiadiazole) (PFDTBT) **45**. This polymer has a maximum absorption of 550 nm, which helps to generate large photocurrents [91].

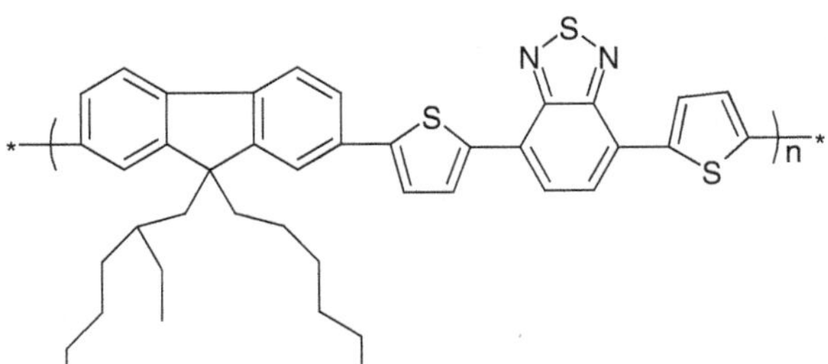

45

Another method of improving these solar cells is to dope the material with a fullerene derivative such as [6]-phenyl-C_{61}-butyric acid methyl ester (PCBM) **46**, which acts as an electron acceptor and improves the charge carrier mobilities of the solar cells. To date, the most efficient polymer solar cell is prepared from poly(3-hexylthiophene) and PCBM and produces an energy conversion efficiency of around 3% [92,93].

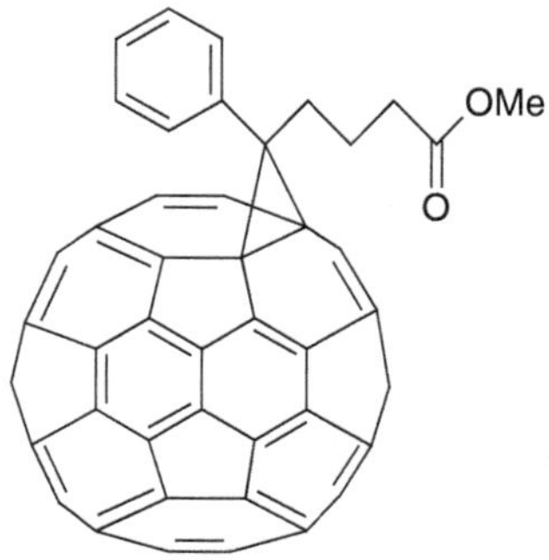

46

A hyperbranched polymer **47** containing Re(bipyridine)$(CO)_3Cl$ chromophores (the $Re(CO)_3Cl^+$ moiety is denoted as [Re]) linked by bipyridineethylene bridges was deposited with a sulfonated polythiophene to yield a solar energy conversion material. This displayed open-circuit voltages of 1.2 V and photocurrents of 27 μA cm^{-2} under solar illumination [94].

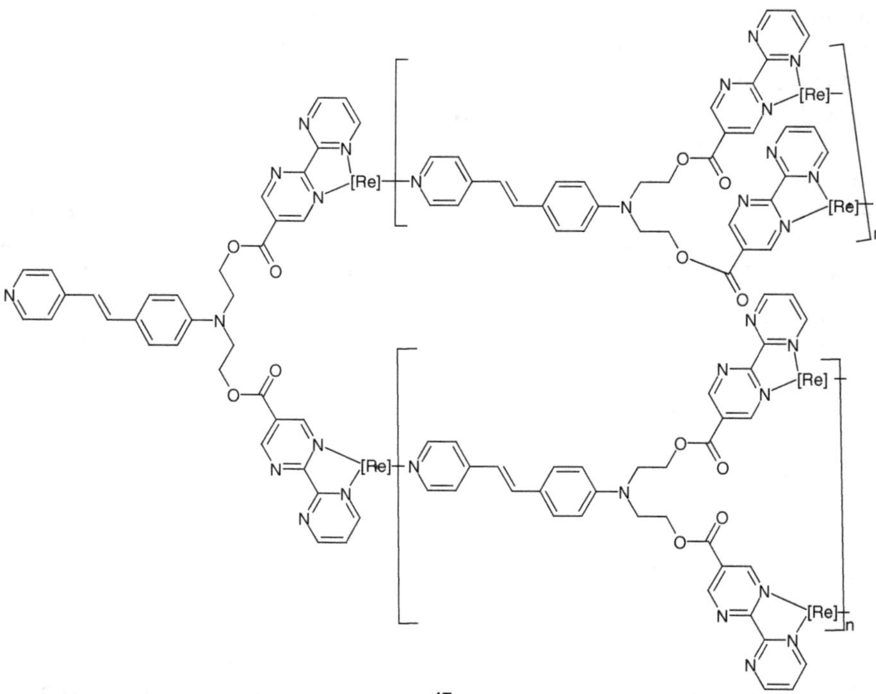

47

Blends of poly(ferrocenylmethylphenyisilane) (PFMPS) **48** and fullerenes as the active layers gave short circuit currents on the order of nanoamperes under white light illumination of ca. 160 mW cm^{-2}, with an open-circuit voltage from ca. 0.3 to 0.45 V [95]. Similar responses were obtained when the fullerenes were covalently attached to the polysilane backbone [96].

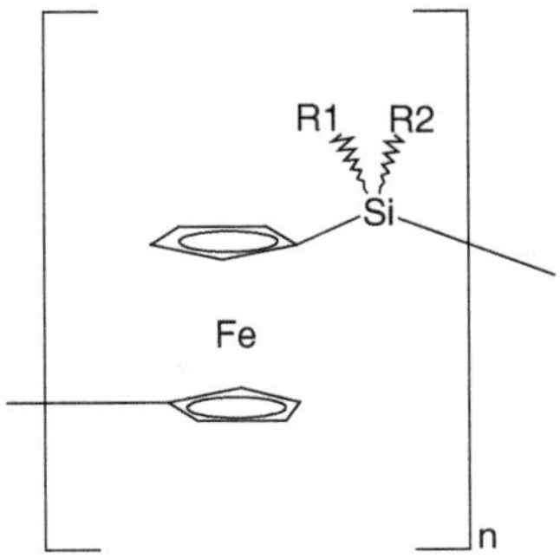

48

7.4 COORDINATION POLYMERS WITH METALS IN THE BACKBONE

The metal–metal interactions in these systems are of particular interest due, in part, to the tremendous effort devoted to understanding the interactions between two metals linked by a bridging ligand in dimers or small oligomers [81].

One of the traditional bridging ligands in coordination chemistry is cyanide. Coordination polymers with interesting photochemical properties have been constructed using the traditional cyanide ligand as a bridge between metals. Thus, ferrocyanide can bridge between Pt(IV)$(NH_3)_4$ centers to form oligomers and polymers **49**. These are believed to photodissociate under light to give ferricyanide and $[Pt(NH_3)_4]^{2+}$ in a two-photon process [97], a reaction that can be applied in the photolithography of the polymers [98].

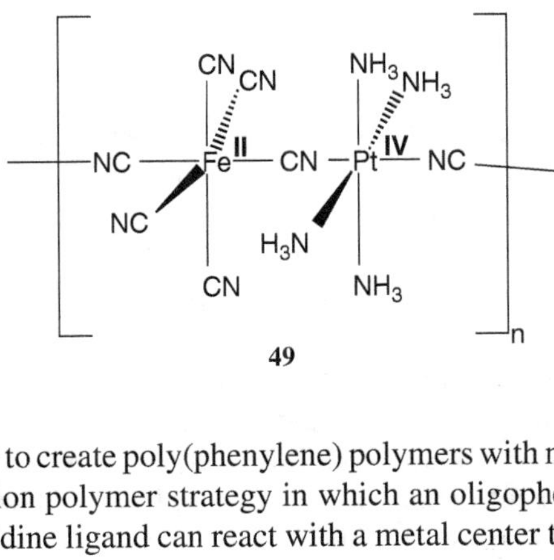

49

An ingenious way to create poly(phenylene) polymers with metals in the backbone is to use a coordination polymer strategy in which an oligophenylene substituted at each end by a terpyridine ligand can react with a metal center to yield a polymer. An iron polymer **50** has been prepared using this method with oligophenylene vinylene bridges [99], as has a ruthenium polymer **51** linked by phenylene vinylene bridges [100]. The ruthenium polymer **52** was made in a slightly different way in two stages, in which the second stage used an unsymmetrical ligand with a phenanthroline at one end and a 2-phenylbipyridine at the other (hence the single positive charge on each Ru) [101]. This polymer has a DP $\geq$ 20 determined by NMR end group analysis. The emission spectroscopy at 77 K of a similar polymer containing chiral Ru centers has been studied, and it was found that the length of the "delocalization box" governs the wavelength of the emission [102].

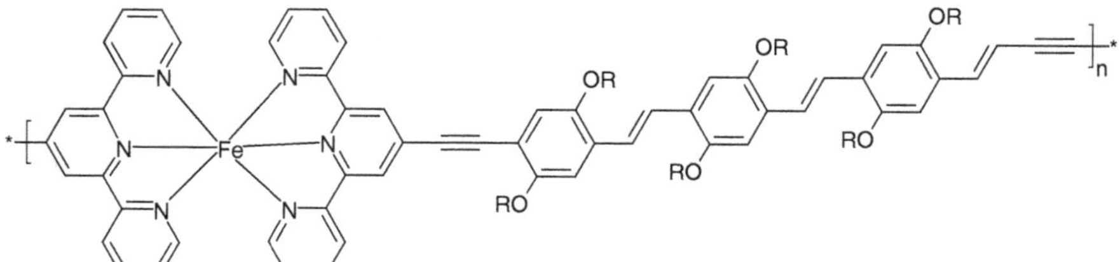

50

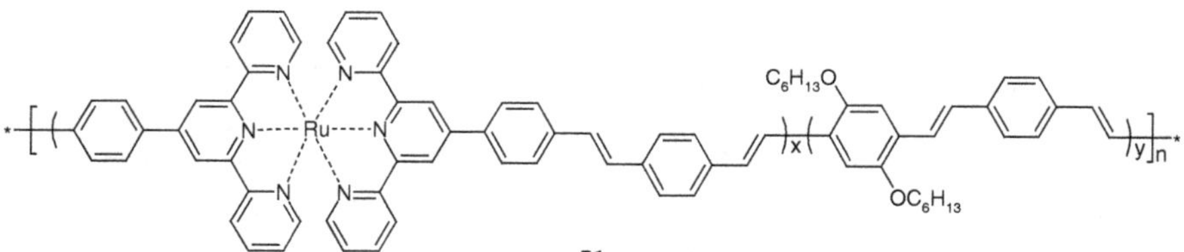

51

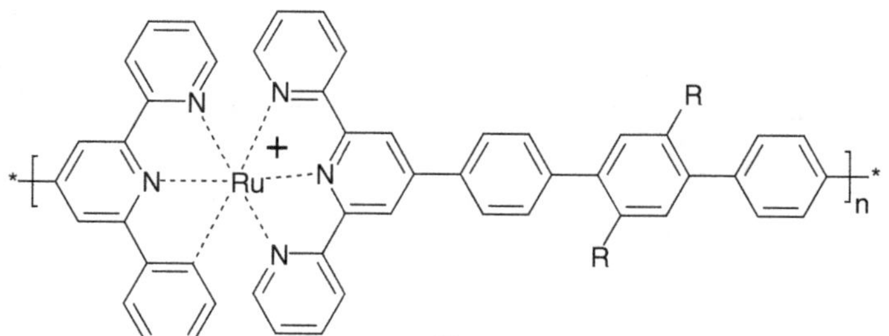

52

Using this strategy, a variety of other linkages between the terpy ligand have been used, including many nonconjugated bridges, and this class of polymer has been reviewed comprehensively [103]. The review also documents how the concept has also been extended to three dimensions to prepare dendrimers.

Platinum(II)-bridged ("polyplatinyne") polymers containing fluorene **53** and carbazole groups have been extensively studied by Wong et al. [104,105]. A wide range of bandgaps can be tailored in these materials according to the conjugation length, and the ISC rate increases as the bandgap increases. Similarly, mercury-bridged fluorene polymers **54** have high ISC rates, enhanced also by the heavy-atom effect on the enhancement of spin–orbit coupling [106]. Fluorene-based polyplatinyne polymers related to those described above containing thiophene groups within the chain appear **55** to be useful as the photoactive material in dye-sensitized solar cells [107–111].

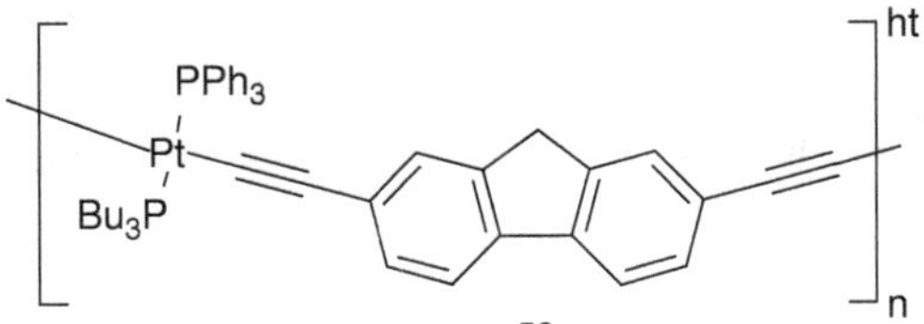

53

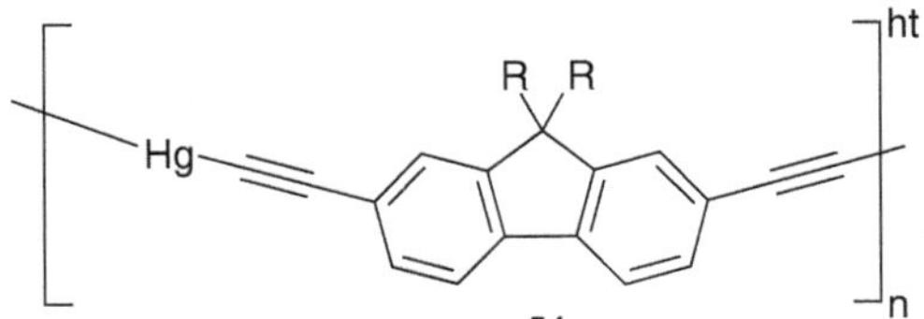

54

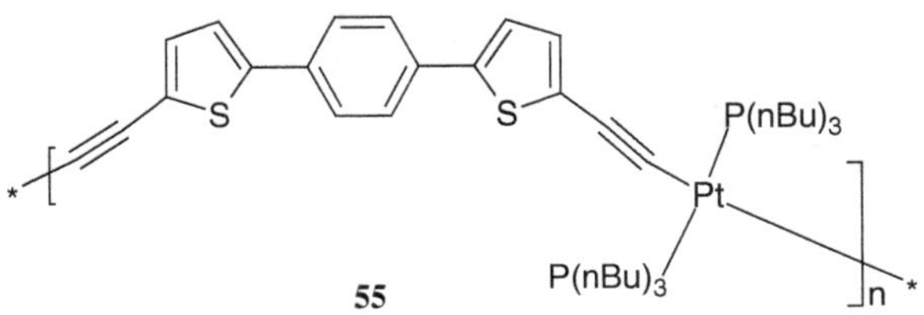

55

Polymer **56**, soluble in DMF or DMSO, can be prepared by the condensation of amino-substituted $[Ru(bpy)_3]^{2+}$ with glyoxal [112]. This polymer showed weaker emission than the free Ru chromophore that was attributed to less efficient intersystem crossing to the ^{3}MLCT state. Similar condensation reactions with diacid anhydrides to produce polyimides have been reported [113]. These polymers showed long-wavelength emission associated with charge-transfer states.

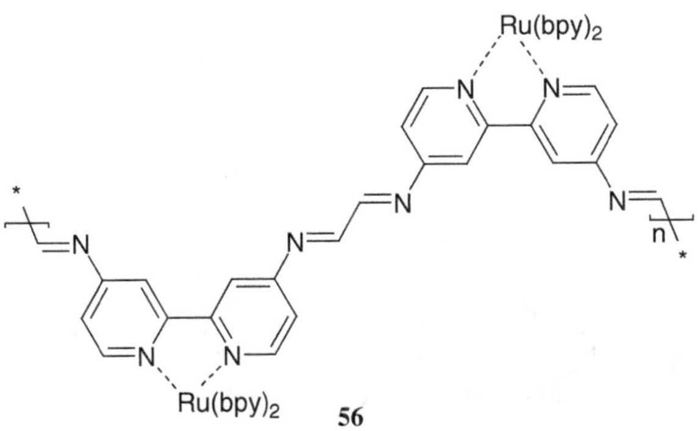

56

Diphosphine-bridged Cu and Ag isocyanide polymeric complexes **57** are luminescent in the solid state [114]. This was proposed to arise from an MLCT excited state (d^{10} metal to both CNR and phosphine π systems). The properties of these polymers have been discussed in relation to the related polymers, $\{M(dmb)_2{}^+\}_n$ (M = Cu, Ag; dmb = 1,8-diisocyano-*p*-menthane). Increasing the number of chromophores in the chain appeared to lead to a distribution of luminescence lifetimes and multiexponential decay. Comparisons were also made with related Pd and Pt polymers, which show monoexponential decay of luminescence, suggesting only weakly interacting chromophores [115].

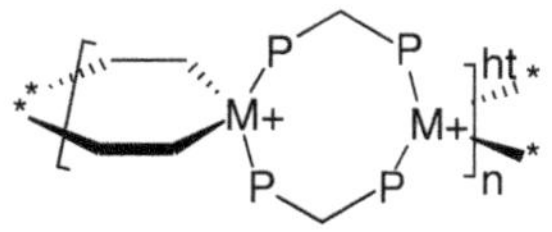

57

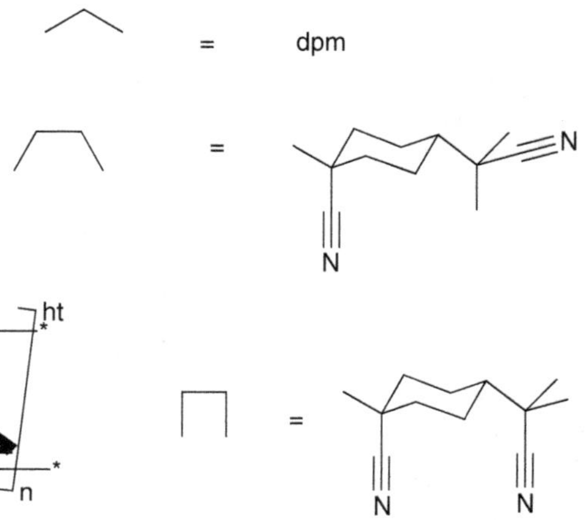

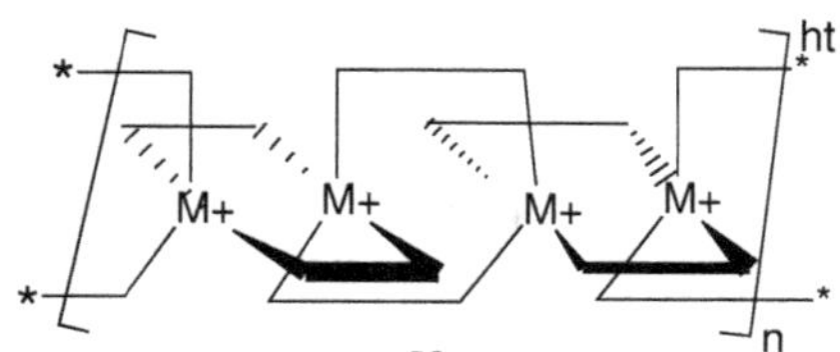

58

Peng and coworkers have prepared polyphenylethynes containing molybdenum polyoxometallates in the main-chain [116] and at the ends of conjugated (**59**) or nonconjugated side-chains [117] by Sonogashira coupling. Fluorescence quenching was strongest in the main-chain polymer and in the side-chain polymer with a conjugated linking unit.

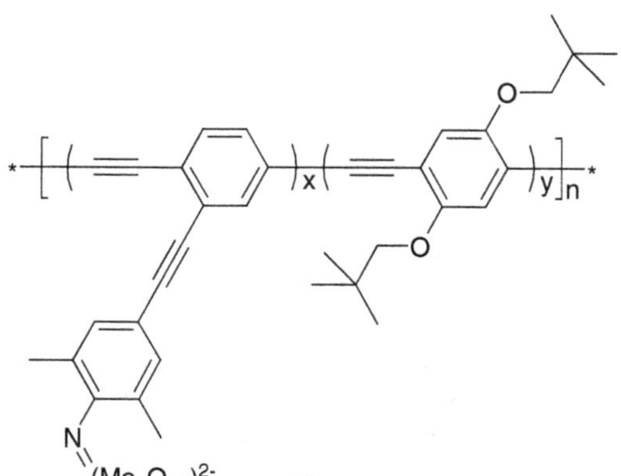

59

There is a large class of metallopolymers based on the linking of the rings in bis (cyclopentadienyl) complexes such as ferrocenes. Conventional condensation polymerization reactions are among those often employed to give nonconjugated polyurethanes, polyureas, polyethers, polyesters, and polyamides. Similarly, dinuclear bis(cyclopentadienyl) complexes with one ring on each metal can be polymerized. Interestingly, the polyamide **60** shown contains a metal–metal bond that can be cleaved photochemically. This property is potentially useful in photolithography [118,119]. Tyler has recently reviewed polymers containing metal–metal bonds [120]. Apart from these polymers, however, the discussion in this chapter is limited to cyclopentadienyl complexes linked directly or by conjugated bridges.

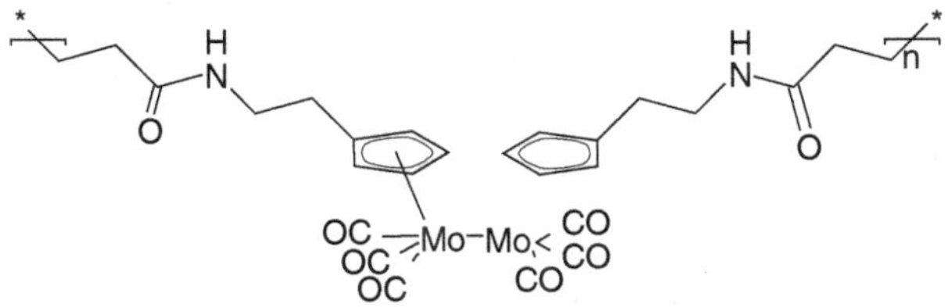

60

The interesting redox chemistry of such conjugated polymeric metallocene systems has been reviewed [121]. For example, oligo(1,11′-dihexylferrocenylenes) **61** show an intervalence charge transfer band (IVCT) on oxidation. For the singly oxidized chains, the IVCT band λ_{max} decreases as the number of ferrocene units increases, as expected. Also, as expected the energy of the IVCT band increases as the oxidation level increases.

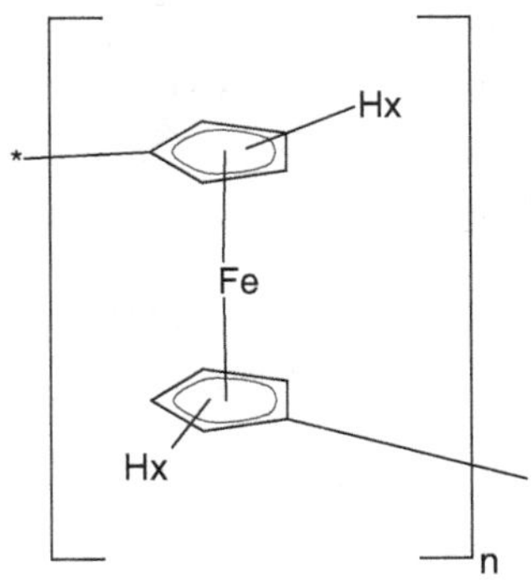

61

Conjugated ruthenocyclopentatriene polymers **62** showed three red shifted absorptions compared to the monomers, whereas a band at 694 nm remained unchanged and was assigned to an LMCT (carbene $\pi \rightarrow$ Ru d) transition [122]. The fact that the reversible Ru reduction wave was unchanged compared to the monomer, and at -1 V falls within the bandgap of the polymer, supports this assignment.

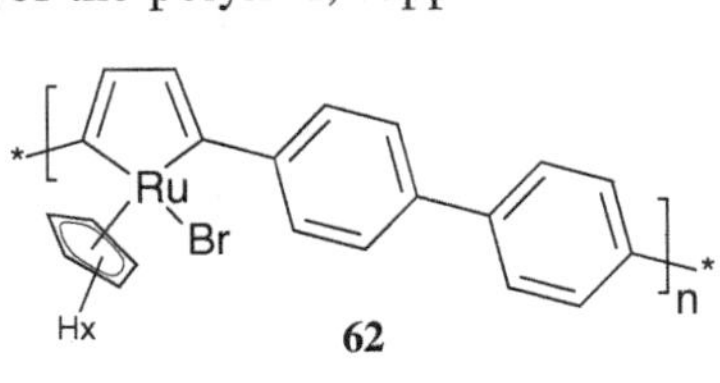

62

7.5 CONCLUSIONS

There is no doubt that the advent of OLEDs and the interest in new solar energy materials has inspired a new wave of activity in the area of metal complex-based polymers and their photochemical behavior. And the ingenuity displayed in designing these new polymers is quite staggering. Perhaps the enthusiasm to evaluate materials for applications has sidelined temporarily the detailed investigation of the photophysics of these materials. Understanding the complexity of these new materials presents a significant challenge to the investigator for some time to come.

REFERENCES

1. J. Ritchie, Ph. D. Thesis, University of St. Andrews, 2005.
2. M.J. Ware, Noble metals for common images in: J.M. Kelly, C.B. McArdle, and M.J.d.F. Maunder (Eds.), *Photochemistry and Polymeric Systems*, Royal Society of Chemistry, Cambridge, 1993.
3. J.H. Alstrum-Acevedo, M.K. Brennaman, and T.J. Meyer, *Inorg. Chem.* **2005**, *44*, 6802.
4. V. Balzani, A. Credi, and M. Venturi, *Chemsuschem* **2008**, *1*, 26.
5. E. Stathatos, R. Lianos, S.M. Zakeeruddin, P. Liska, and M. Gratzel, *Chem. Mater.* **2003**, *15*, 1825.
6. C.K. Chiang, C.R. Fincher, Y.W. Park, A.J. Heeger, H. Shirakawa, E.J. Louis, S.C. Gau, and A.G. Macdiarmid, *Phys. Rev. Lett.* **1977**, *39*, 1098.
7. E. Smela and N. Gadegaard, *Adv. Mater.* **1999**, *11*, 953.
8. V.T. Truong, S.Z. Riddell, and R.F. Muscat, *J. Mater. Sci.* **1998**, *33*, 4971.
9. G. Dufresne, J. Bouchard, M. Belletete, G. Durocher, and M. Leclerc, *Macromolecules* **2000**, *33*, 8252.
10. T.M. Swager and M.J. Marsella, *Adv. Mater.* **1994**, *6*, 595.
11. D.T. McQuade, A.E. Pullen, and T.M. Swager, *Chem. Rev.* **2000**, *100*, 2537.
12. J.H. Burroughes, D.D.C. Bradley, A.R. Brown, R.N. Marks, K. Mackay, R.H. Friend, P.L. Burns, and A.B. Holmes, *Nature* **1990**, *347*, 539.
13. A. Kraft, A.C. Grimsdale, and A.B. Holmes, *Angew. Chem. Int. Ed.* **1998**, *37*, 402.
14. J.A. Barltrop and J.D. Coyle, *Principles of Photochemistry*, Wiley, New York, 1978.
15. S.-H. Hwang and G.R. Newkome, Metallodendrimers and their potential utilitarian applications in: A.S. Abd-El-Aziz and I. Manners (Eds.), *Frontiers in Transition Metal-Containing Polymers*, Wiley, Chichester, 2007.
16. U.S. Schubert, G.R. Newkome, and I. Manners (Eds.), *Metal-Containing and Metallosupramolecular Polymers and Materials*, ACS, Washington, D.C, 2006.
17. A.S. Abd-El-Aziz and I. Manners, *Frontiers in Transition Metal-Containing Polymers*, Wiley, 2007.
18. M.H.V. Huynh, D.M. Dattelbaum, and T.J. Meyer, *Coord. Chem. Rev.* **2005**, *249*, 457.
19. C.N. Fleming, K.A. Maxwell, J.M. DeSimone, T.J. Meyer, and J.M. Papanikolas, *J. Am. Chem. Soc.* **2001**, *123*, 10336.

20. S.L. Mecklenburg, B.M. Peek, J.R. Schoonover, D.G. McCafferty, C.G. Wall, B.W. Erickson, and T.J. Meyer, *J. Am. Chem. Soc.* **1993**, *115*, 5479.
21. X.Y. Wang, R.N. Prabhu, R.H. Schmehl, and M. Weck, *Macromolecules* **2006**, *39*, 3140.
22. X.Y. Wang and M. Weck, *Macromolecules* **2005**, *38*, 7219.
23. S.A. Serron, W.S. Aldridge, C.N. Fleming, R.M. Danell, M.H. Baik, M. Sykora, D.M. Dattelbaum, and T.J. Meyer, *J. Am. Chem. Soc.* **2004**, *126*, 14506.
24. L. Dennany, T.E. Keyes, and R.J. Forster, *Analyst* **2008**, *133*, 753.
25. A.M. Spehar-Deleze, Y. Pellegrin, T.E. Keyes, and R.J. Forster, *Electrochem. Commun.* **2008**, *10*, 984.
26. K.W. Cheng and W.K. Chan, *Langmuir* **2005**, *21*, 5247.
27. Q. Zeng, A. McNally, T.E. Keyes, and R.J. Forster, *Electrochim. Acta* **2008**, *53*, 7033.
28. Q. Zeng, A. McNally, T.E. Keyes, and R.J. Forster, *Electrochem. Commun.* **2008**, *10*, 466.
29. W. Humbs, E. van Veldhoven, H. Zhang, and M. Glasbeek, *Chem. Phys. Lett.* **1999**, *304*, 10.
30. J. Marfurt, W.Y. Zhao, and L. Walder, *J. Chem. Soc.: Chem. Commun.* **1994**, 51.
31. V. Aranyos, J. Hjelm, A. Hagfeldt, and H. Grennberg, *J. Chem. Soc.-Dalton Trans.* **2001**, 1319.
32. V. Aranyos, A. Hagfeldt, H. Grennberg, and E. Figgemeier, *Polyhedron* **2004**, *23*, 589.
33. M.C.E. Bandeira, J.A. Crayston, C.V. Franco, and A. Glidle, *Phys. Chem. Chem. Phys.* **2007**, *9*, 1003.
34. M.C.E. Bandeira, J.A. Crayston, N.S. Concalves, L.K. Noda, A. Glidle, and C.V. Franco, *J. Solid State Electrochem.* **2007**, *11*, 231.
35. S.C. Paulson, S.A. Sapp, and C.M. Elliott, *J. Phys. Chem. B* **2001**, *105*, 8718.
36. K. Takada, G.D. Storrier, F. Pariente, and H.D. Abruna, *J. Phys. Chem. B* **1998**, *102*, 1387.
37. A. Venkatanarayanan, A.M. Spehar-Deleze, L. Dennany, Y. Pellegrin, T.E. Keyes, and R. J. Forster, *Langmuir* **2008**, *24*, 11233.
38. C. Adachi, M.A. Baldo, S.R. Forrest, S. Lamansky, M.E. Thompson, and R.C. Kwong, *Appl. Phys. Lett.* **2001**, *78*, 1622.
39. Y. Kawamura, S. Yanagida, and S.R. Forrest, *J. Appl. Phys.* **2002**, *92*, 87.
40. K.M. Vaeth and C.W. Tang, *J. Appl. Phys.* **2002**, *92*, 3447.
41. C. Adachi, R.C. Kwong, P. Djurovich, V. Adamovich, M.A. Baldo, M.E. Thompson, and S.R. Forrest, *Appl. Phys. Lett.* **2001**, *79*, 2082.
42. T.D. Anthopoulos, J.P.J. Markham, E.B. Namdas, I.D.W. Samuel, S.C. Lo, and P.L. Burn, *Appl. Phys. Lett.* **2003**, *82*, 4824.
43. S.C. Yu, S.J. Hou, and W.K. Chan, *Macromolecules* **1999**, *32*, 5251.
44. C.G. Cameron, T.J. Pittman, and P.G. Pickup, *J. Phys. Chem. B* **2001**, *105*, 8838.
45. C.G. Cameron and P.G. Pickup, *J. Am. Chem. Soc.* **1999**, *121*, 11773.
46. C.G. Cameron and P.G. Pickup, *J. Am. Chem. Soc.* **1999**, *121*, 7710.
47. C.G. Cameron and P.G. Pickup, *Chem. Commun.* **1997**, 303.
48. M.J. MacLachlan, Metal-containing pi-conjugated polymers, in: A.S. Abd-El-Aziz and I. Manners (Eds.), *Frontiers in Transition Metal-Containing Polymers*, Wiley, Chichester, 2007.
49. P.G. Pickup, *J. Mater. Chem.* **1999**, *9*, 1641.

50. Y. Ohmori, M. Uchida, K. Muro, and K. Yoshino, *Jpn. J. Appl. Phys. Part 2: Lett.* **1991**, *30*, L1941.
51. M. Inbasekaran, E. Woo, W.S. Wu, M. Bernius, and L. Wujkowski, *Synth. Met.* **2000**, *111*, 397.
52. T. Yamamoto, A. Morita, Y. Miyazaki, T. Maruyama, H. Wakayama, Z. Zhou, Y. Nakamura, T. Kanbara, S. Sasaki, and K. Kubota, *Macromolecules* **1992**, *25*, 1214.
53. N. Miyaura and A. Suzuki, *Chem. Rev.* **1995**, *95*, 2457.
54. M. Leclerc, *J. Polym. Sci. Part A: Polym. Chem.* **2001**, *39*, 2867.
55. B. Liu, W.L. Yu, J. Pei, S.Y. Liu, Y.H. Lai, and W. Huang, *Macromolecules* **2001**, *34*, 7932.
56. M. Zhang, P. Lu, Y.G. Ma, and J.C. Shen, *J. Phys. Chem. B* **2003**, *107*, 6535.
57. T. Yamamoto, Y. Saitoh, K. Anzai, H. Fukumoto, T. Yasuda, Y. Fujiwara, B.K. Choi, K. Kubota, and T. Miyamae, *Macromolecules* **2003**, *36*, 6722.
58. T. Yasuda and T. Yamamoto, *Macromolecules* **2003**, *36*, 7513.
59. J. Ritchie, A. Ruseckas, P. Andre, C. Munther, M. Van Ryssen, D.E. Vize, J.A. Crayston, and I.D. W. Samuel, *Synth. Met.* **2009**. *159*, 583.
60. Q.D. Ling, E.T. Kang, K.G. Neoh, and W. Huang, *Macromolecules* **2003**, *36*, 6995.
61. T. Yasuda, I. Yamaguchi, and T. Yamamoto, *Adv. Mater.* **2003**, *15*, 293.
62. L.X. Chen, W.J.H. Jager, M.P. Niemczyk, and M.R. Wasielewski, *J. Phys. Chem. A* **1999**, *103*, 4341.
63. R.C. Smith, A.G. Tennyson, A.C. Won, and S.J. Lippard, *Inorg. Chem.* **2006**, *45*, 9367.
64. Z.H. Peng, A.R. Gharavi, and L.P. Yu, *J. Am. Chem. Soc.* **1997**, *119*, 4622.
65. Q. Wang and L.P. Yu, *J. Am. Chem. Soc.* **2000**, *122*, 11806.
66. A. Harriman, S.A. Rostron, A. Khatyr, and R. Ziessel, *Faraday Discuss.* **2006**, *131*, 377.
67. A. Harriman, A. Khatyr, and R. Ziessel, *Res. Chem. Intermed.* **2007**, *33*, 49.
68. A. Kokil, P. Yao, and C. Weder, *Macromolecules* **2005**, *38*, 3800.
69. T. Pautzsch and E. Klemm, *Macromolecules* **2002**, *35*, 1569.
70. K.D. Ley and K.S. Schanze, *Coord. Chem. Rev.* **1998**, *171*, 287.
71. W.K. Chan, P.K. Ng, X. Gong, and S.J. Hou, *J. Mater. Chem.* **1999**, *9*, 2103.
72. B.W. Jiang, S.W. Yang, S.L. Bailey, L.G. Hermans, R.A. Niver, M.A. Bolcar, and W.E. Jones, *Coord. Chem. Rev.* **1998**, *171*, 365.
73. K.Y.K. Man, H.L. Wong, W.K. Chan, C.Y. Kwong, and A.B. Djurisic, *Chem. Mater.* **2004**, *16*, 365.
74. W.K. Chan, C.S. Hui, K.Y.K. Man, K.W. Cheng, H.L. Wong, N.Y. Zhu, and A.B. Djurisic, *Coord. Chem. Rev.* **2005**, *249*, 1351.
75. P.K. Ng, X. Gong, S.H. Chan, L.S.M. Lam, and W.K. Chan, *Chem.-Eur. J.* **2001**, *7*, 4358.
76. S.C. Yu, X. Gong, and W.K. Chan, *Macromolecules* **1998**, *31*, 5639.
77. K.W. Cheng, C.S.C. Mak, W.K. Chan, A.M.C. Ng, and A.B. Djurisic, *J. Polym. Sci. Part A: Polym. Chem.* **2008**, *46*, 1305.
78. Y.B. Zhu and M.O. Wolf, *J. Am. Chem. Soc.* **2000**, *122*, 10121.
79. Y.B. Zhu and M.O. Wolf, *Chem. Mater.* **1999**, *11*, 2995.
80. C.L. Jones and S.J. Higgins, *J. Mater. Chem.* **1999**, *9*, 865.
81. T.L. Stott and M.O. Wolf, *Coord. Chem. Rev.* **2003**, *246*, 89.

82. B.J. Holliday and T.M. Swager, *Chem. Commun.* **2005**, 23.

83. J. Hjelm, E.C. Constable, E. Figgemeier, A. Hagfeldt, R. Handel, C.E. Housecroft, E. Mukhtar, and E. Schofield, *Chem. Commun.* **2002**, 284.

84. S.S. Zhu and T.M. Swager, *Adv. Mater.* **1996**, *8*, 497.

85. S.S. Zhu, R.P. Kingsborough, and T.M. Swager, *J. Mater. Chem.* **1999**, *9*, 2123.

86. S.S. Zhu, P.J. Carroll, and T.M. Swager, *J. Am. Chem. Soc.* **1996**, *118*, 8713.

87. S.S. Zhu and T.M. Swager, *J. Am. Chem. Soc.* **1997**, *119*, 12568.

88. P.H. Kwan, M.J. MacLachlan, and T.M. Swager, *J. Am. Chem. Soc.* **2004**, *126*, 8638.

89. D.H. Kim, D.S. Park, Y.B. Shim, and S.C. Shin, *J. Organ. Chem.* **2000**, *608*, 133.

90. D.H. Kim, J.H. Kim, T.H. Kim, D.M. Kang, Y.H. Kim, Y.B. Shim, and S.C. Shin, *Chem. Mater.* **2003**, *15*, 825.

91. M. Svensson, F.L. Zhang, S.C. Veenstra, W.J.H. Verhees, J.C. Hummelen, J.M. Kroon, O. Inganas, and M.R. Andersson, *Adv. Mater.* **2003**, *15*, 988.

92. F. Padinger, R.S. Rittberger, and N.S. Sariciftci, *Adv. Funct. Mater.* **2003**, *13*, 85.

93. P. Schilinsky, C. Waldauf, and C.J. Brabec, *Appl. Phys. Lett.* **2002**, *81*, 3885.

94. C.W. Tse, K.Y.K. Man, K.W. Cheng, C.S.K. Mak, W.K. Chan, C.T. Yip, Z.T. Liu, and A.B. Djurisic, *Chem.-Eur. J.* **2007**, *13*, 328.

95. P.W. Cyr, E.J.D. Klem, E.H. Sargent, and I. Manners, *Chem. Mater.* **2005**, *17*, 5770.

96. M. Nanjo, P.W. Cyr, K. Liu, E.H. Sargent, and I. Manners, *Adv. Funct. Mater.* **2008**, *18*, 470.

97. C.C. Chang and A.B. Bocarsly, *J. Electroanal. Chem.* **1999**, *470*, 99.

98. A.B. Bocarsly, C.C. Chang, Y. Wu, and E.P. Vicenzi, *J. Chem. Educ.* **1997**, *74*, 663.

99. A. El-Ghayoury, A. Schenning, and E.W. Meijer, *J. Polym. Sci. Part A: Polym. Chem.* **2002**, *40*, 4020.

100. W.K. Chan, X. Gong, and W.Y. Ng, *Appl. Phys. Lett.* **1997**, *71*, 2919.

101. O. Schmelz and M. Rehahn, *E-Polymers* **2002**, 047; http://www.e-polymers.org/journal/papers/rehahn_311002.pdf.

102. J.A. Barron, S. Glazier, S. Bernhard, K. Takada, P.L. Houston, and H.D. Abruna, *Inorg. Chem.* **2003**, *42*, 1448.

103. P.R. Andres and U.S. Schubert, *Adv. Mater.* **2004**, *16*, 1043.

104. W.Y. Wong, G.L. Lu, K.H. Choi, and J.X. Shi, *Macromolecules* **2002**, *35*, 3506.

105. W.Y. Wong, *Coord. Chem. Rev.* **2005**, *249*, 971.

106. W.Y. Wong, L. Liu, S.Y. Poon, K.H. Choi, K.W. Cheah, and J.X. Shi, *Macromolecules* **2004**, *37*, 4496.

107. L. Liu, C.L. Ho, W.Y. Wong, K.Y. Cheung, M.K. Fung, W.T. Lam, A.B. Djurisic, and W.K. Chan, *Adv. Funct. Mater.* **2008**, *18*, 2824.

108. X.Z. Wang, W.Y. Wong, K.Y. Cheung, M.K. Fung, A.B. Djurisic, and W.K. Chan, *Dalton Trans.* **2008**, 5484.

109. W.Y. Wong, X.Z. Wang, H.L. Zhang, K.Y. Cheung, M.K. Fung, A.B. Djurisic, and W.K. Chan, *J. Organomet. Chem.* **2008**, *693*, 3603.

110. W.Y. Wong, X.Z. Wang, Z. He, K.K. Chan, A.B. Djurisic, K.Y. Cheung, C.T. Yip, A.M.C. Ng, Y.Y. Xi, C.S.K. Mak, and W.K. Chan, *J. Am. Chem. Soc.* **2007**, *129*, 14372.

111. W.Y. Wong, X.Z. Wang, Z. He, A.B. Djurisic, C.T. Yip, K.Y. Cheung, H. Wang, C.S.K. Mak, and W.K. Chan, *Nat. Mater.* **2007**, *6*, 521.
112. S.C. Rasmussen, D.W. Thompson, V. Singh, and J.D. Petersen, *Inorg. Chem.* **1996**, *35*, 3449.
113. W.Y. Ng, X. Gong, and W.K. Chan, *Chem. Mater.* **1999**, *11*, 1165.
114. E. Fournier, F. Lebrun, M. Drouin, A. Decken, and P. Harvey, *Inorg. Chem.* **2004**, *43*, 3127.
115. P.D. Harvey and E. Fournier, Luminescence properties of phosphine-isocyanide Cu(I) and Ag(I)-containing oligomers in the solid state, in: U.S. Schubert, G.R. Newkome, and I. Manners (Eds.), Metal-Containing and Metallosupramolecular Polymers and Materials, Washington, D.C. 2006.
116. M. Lu, B.H. Xie, J.H. Kang, F.C. Chen, Y. Yang, and Z.H. Peng, *Chem. Mater.* **2005**, *17*, 402.
117. B.B. Xu, M. Lu, J.H. Kang, D. Wang, J. Brown, and Z.H. Peng, *Chem. Mater.* **2005**, *17*, 2841.
118. G.F. Nieckarz, J.J. Litty, and D.R. Tyler, *J. Organomet. Chem.* **1998**, *554*, 19.
119. G.F. Nieckarz and D.R. Tyler, *Inorg. Chim. Acta* **1996**, *242*, 303.
120. D.R. Tyler, Polymers with metal–metal bonds along their backbones in: A.S. Abd-El-Aziz and I. Manners (Eds.), *Frontiers in Transition Metal-Containing Polymers*, Wiley, Chichester, 2007.
121. H. Nishihara, Redox-based functionalities of multinuclear metal complex systems in: A. S. Abd-El-Aziz and I. Manners (Eds.), *Frontiers in Transition Metal-Containing Polymers*, Wiley, Chichester, 2007.
122. M. Kurashina, M. Murata, T. Watanabe, and H. Nishihara, *J Am. Chem. Soc.* **2003**, *125*, 12420.

8

PHOTOVOLTAIC POLYMER MATERIALS

Hazel Assender and Aaron Barkhouse

8.1 INTRODUCTION

Power consumption and its associated greenhouse gas emissions have reached unprecedented levels and continue to grow. Worldwide installed electrical capacity was 3900 GW in 2005 [1]. This figure is expected to grow to 4200 GW by 2010 and

Photochemistry and Photophysics of Polymer Materials, Edited by Norman S. Allen

will likely reach 6000 GW by 2030 [2]. The majority of this growth is currently being met by thermal generation from coal and natural gas fired power plants. According to the Working Group II Intergovernmental Panel on Climate Change report, the largest increase in greenhouse gas emissions from 1970 to 2004 came from electricity generation [3], and the above estimates suggest that this trend will continue for the foreseeable future unless an economical low-emission alternative to fossil fuel-based generation is found. Photovoltaic (PV) power generation can play an important role in reducing our dependence on fossil fuels for electricity generation, provided its cost can be reduced. For example, the entire electricity demand of the United States, the world's largest energy user, could be met by a 100 km $\times$ 100 km array of 15% efficient PV modules in the deserts of Nevada or Arizona [4]. Alternatively, small-scale PV installations could allow homeowners and businesses to greatly reduce their electricity costs and insulate themselves from rising fuel costs. These are attractive possibilities, but if the goal of widespread PV adoption is to be realized, the cost of producing PV modules must be reduced significantly.

The photovoltaic effect was discovered by Edmund Becquerel in 1839, and the first solid-state photovoltaic device was constructed 40 years later. These early devices were incredibly inefficient, with power conversion efficiencies (the ratio of electrical power produced to incident illumination power) of less than 1%. It was not until 1954 when Chapman, Fuller, and Pearson reported the first silicon solar cell, with an efficiency of 6%, that the possibility of practical photovoltaic power became a reality. This early technology was prohibitively expensive, but developments during the 1950s and 1960s led to the use of silicon photovoltaic technology in applications ranging from satellites to rooftops. Solar power also provided a means of delivering power to remote regions where connection to the electricity grid is not feasible, an application that is playing an important role in developing countries today. Following the 1970's oil crisis, there was a resurgence of interest in photovoltaic technology, leading to a period of intense research and investment into developing a viable alternative to conventional fossil fuel energy sources. Much of the current understanding of the physics of solar cells was developed during this period. In spite of a period in the 1980s and 1990s when public and government support for photovoltaics waned, research continued and today, with assistance from government incentives in a select few states and countries, commercial PV installations are beginning to proliferate.

Today silicon solar cells enjoy the largest share (approximately 90%) of the commercial solar cell market. In the lab, their efficiency has reached 24.7%, which is nearing the theoretical maximum (for a single-bandgap cell with a 1.1 eV bandgap) of 29%, and commercial cells are now about 20% efficient under simulated solar illumination conditions [5]. The main barrier to more widespread adoption is the high cost of modules, which is dominated by the high cost of silicon. Silicon solar cells require high-purity silicon wafers, which undergo expensive high-temperature processing before assembly into the module. These materials and processing costs drive up the price of the end product, and silicon solar modules currently cost just under \$4 USD per watt of peak power (W_p), a factor of 4 higher than conventional (fossil fuel-based) generation technologies [6]. While thin-film cells made from copper indium gallium selenide (CIGS) and amorphous silicon require less material to make and therefore offer a slightly cheaper alternative, at roughly \$3.30/W_p, significant advances are needed to bring the cost of PV modules to a level that is competitive with conventional generation technology. While this is not an easy task, in the absence of international regulation, it may be necessary in order to encourage adoption on the scale necessary to satisfy growing demand and stabilize or reduce anthropogenic CO_2 emissions (i.e., hundreds of GW annually).

8.1.1 Why Organics?

Organic photovoltaics are attractive as an alternative to existing solar energy technology because they combine the potential for good power efficiency with low-cost device processing. While the power conversion efficiency of the state-of-the-art solid-state organic photovoltaic (OPV) cells (between 5% and 7%) is still well behind that of their inorganic counterparts, their theoretical efficiency is the same and the organic semiconducting materials used in OPV devices offer several advantages over conventional (inorganic) semiconductors. OPV materials are lightweight, flexible, and have high absorption coefficients, needing only a few hundred nanometers thickness to absorb most incident light. They are compatible with inexpensive and low-temperature fabrication techniques such as inkjet printing and reel-to-reel processing. This means that OPVs could be deposited cheaply onto flexible lightweight substrates or building materials and could be integrated into virtually any part of an existing structure. Given these significant advantages in the cost, speed, and flexibility of processing and implementation, if the device performance can be improved and stability issues resolved, the significant savings and versatility offered by OPVs may give them a dominant place in the PV market.

8.1.2 Molecular Semiconductors for Solar Cells

To reduce the cost of solar cells, it is desirable to start with raw materials that are cheaper, easier to process, and amenable to low-cost deposition techniques. In particular, it is desirable to use materials that can be deposited from solution or from the vapor phase onto a flexible substrate. This would permit the use of high-throughput fabrication techniques such as reel-to-reel processing and web coating, enabling more rapid production that would greatly lower costs. Molecular semiconductors are one class of materials that meet these processing requirements.

The structures of some molecular semiconductors used in solar cells are shown in Fig. 8.1. While both the small molecules and the polymer backbones appear to consist of alternating single and double bonds in their structure diagrams, in reality, each carbon atom makes σ-bonds to three other atoms via sp^2-p hybridized bonds, leaving one free electron in the p_z orbital that extends above and below the plane of the σ-bonds. These p_z electron orbitals overlap with those of neighboring carbon atoms and form π-bonds, and the electrons occupying these orbitals are therefore delocalized over the entire conjugated network. A molecule with such a delocalized π-bonding network is said to be conjugated. These delocalized π-electron orbitals are the lowest energy orbitals that are occupied in the electronic ground state of the

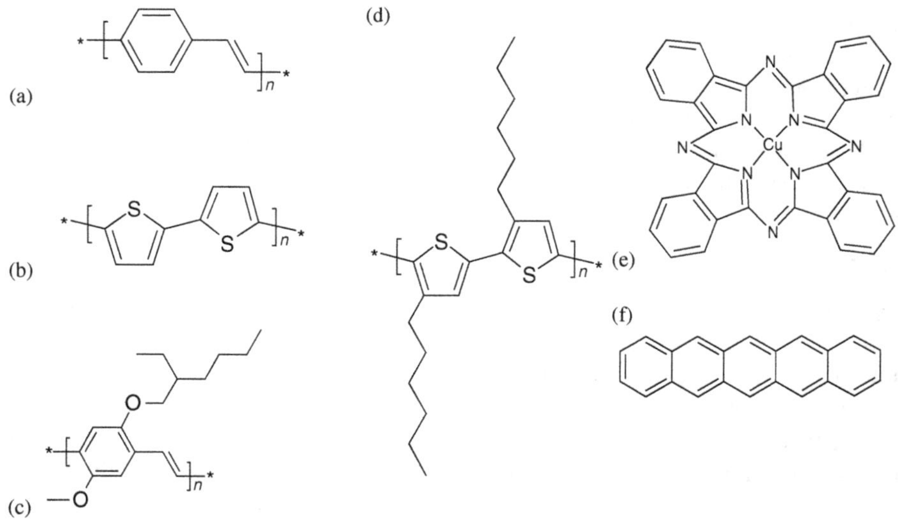

FIGURE 8.1 Molecular structures of some common conjugated polymers and small molecules: (a) poly(*p*-phenylene vinylene), (b) polythiophenes, (c) MEH-PPV, (d) P3HT, (e) copper phthalocyanine, and (f) pentacene.

molecule and are therefore referred to as the highest occupied molecular orbital, or HOMO. The lowest lying excited state (which is unoccupied in the electronic ground state) is the antibonding π^*-orbital and is referred to as the lowest unoccupied molecular orbital, or LUMO.

Because the HOMO–LUMO gap represents the lowest energy photon that can be absorbed by the molecule to create a long-lived excited state and the LUMO is the lowest lying energy level that can accept an electron (e.g., from injection at a contact), the HOMO and LUMO levels dominate the optical and electronic properties of conjugated materials at the low field strengths ($\sim 10^5$ V/cm) and photon energies (a few eV) applicable to solar photovoltaic energy conversion. Since these HOMO–LUMO levels arise from the presence of delocalized π-bonding within the molecule, the extent of conjugation is therefore of central importance in determining the absorption and charge transport properties of a material.

While all the molecules depicted in Fig. 8.1 are molecular semiconductors, it is useful to group them into two smaller categories based on their structure, namely, conjugated polymers (Fig. 8.1a–d) and small molecules (Fig. 8.1e and f). A simple conjugated polymer with its stiff conjugated chain and consequent low solution entropy is generally insoluble and has a degradation temperature below its melting point. The combination of these two properties prevents both solution and melt processing of the material, and so one of two alternative strategies can be deployed. A polymeric precursor molecule can be processed into a device followed by, for example, a heat treatment to release a leaving group from the molecule resulting in the conjugated form of the polymer. Alternatively, and more commonly, flexible side groups are attached to the conjugated backbone to increase the solubility of the molecule. For example, the addition of alkoxy side groups can render a molecule soluble in many common organic solvents, allowing the high-throughput solution processing outlined above. A typical example of this is the modification of the PPV molecule (Fig. 8.1a) to MEH-PPV to increase the solubility (Fig. 8.1c). In the latter case, the flexible side groups on the main, more rigid, conjugated backbone promote the solubilization without interrupting the conjugation and hence conductivity of the molecule. Having said that, the addition of these side groups can modify the solution's processing behavior and hence the resulting polymer microstructure. Also, the addition of the side groups decreases the overall chromophore density and hence possibly the light absorption and the ability for charge to "hop" from one chain to a neighboring one, having implications for charge mobility in a device [7]. Because polymers are extended chains, typically hundreds or thousands of repeat units long, with a high molecular weight distribution, they generally form amorphous, disordered films. This disorder can be reduced by annealing the film above the glass transition temperature of the polymer, encouraging partial crystallization in some regions of the film, with the extent of crystallization depending on the structure and molecular weight of the polymer. This inherent disorder is one of the main factors contributing to the low hole and electron mobilities of most conjugated polymers (10^{-1} to 10^{-7} and 10^{-4} to $10^{-9}\,cm^2/(V\ s)$, respectively, versus 475 and $1500\,cm^2/(V\ s)$ for crystalline silicon) [8]. Another contributing factor to this low mobility is the fact that charges in conjugated polymers generally exist as polarons, or charges accompanied by a distortion of the local lattice structure. These lattice distortions both stabilize the charge carrier and act as scattering sites for other charge carriers. Conjugated polymers also have a low static dielectric constant ($\varepsilon_r \sim 3$), and as a result the primary photoexcitations are not free electrons and holes but bound electron–hole pairs known as excitons, a fact that has significant consequences for device design and performance as will be discussed below. Finally, the bandgap of most conjugated polymers is in the range of 1.9–3.5 eV [18], significantly above that of silicon (1.1 eV) and the ideal gap[1] for a single-bandgap solar cell (1.4 eV) [9]. The development of low-bandgap polymers is being pursued in earnest by many research groups around the world [10–13], and at least one polymer with a 1.4 eV bandgap has already been reported [10].

Small-molecule semiconductors can form more crystalline solids, owing to their smaller size and better molecular packing relative to conjugated polymers [14]. As a result, electron and hole mobilities are generally higher in films of small-molecule semiconductors (up to $6\,cm^2/(V\ s)$ for both electrons in C_{60} and holes in pentacene [15]). Like conjugated polymers, most small-molecule semiconductors are also insoluble, but in this case the addition of alkoxy side groups would inhibit

[1] For a solar cell with a single-bandgap absorber, photons with energy $h\nu < E_g$ are not absorbed. Photons with energy $h\nu > E_g$ are absorbed, but the excitons/free carriers generated quickly relax to the bottom of the conduction band and/or top of the valence band, giving off any energy in excess of the bandgap as heat. This creates a trade-off between losing energy in the form of photons that are not absorbed (large bandgap) and losing excess photon energy ($E = h\nu - E_g$) as heat (small bandgap). For the AM1.5 solar spectrum, the energy lost is minimized (i.e., the efficiency is at a maximum) for an energy gap of approximately 1.4 eV. See Section 8.3.3 for more details.

TABLE 8.1 Optoelectronic Properties of Common Inorganic, Polymer, and Small-Molecule Semiconductors

	Silicon (crystalline)	MEH-PPV	CuPc
E_g (eV)	1.1 (Nelson)	2.1 (Kim)	1.55 (Farag)
α (cm^{-1})	10^4 (Fox)	10^5 (Willekens)	10^6 (Peumans)
μ_h (cm^2/(V s))	475 (Nelson)	10^{-7} (Crone)	10^{-2} to 10^{-7} (Salzman)
μ_e (cm^2/(V s))	1500 (Nelson)	10^{-12} (Crone)	10^{-3}–10^{-4} (Yasuda)

crystallization and degrade the electrical transport properties of the resulting films. For this reason, small-molecule semiconductors are generally deposited from the gas phase. The improved exciton and charge transport properties of small-molecule semiconductors, relative to most conjugated polymers, make them worth investigating as materials for high-performance, low-cost solar cells. Table 8.1 shows the most relevant optoelectronic parameters of silicon, MEH-PPV, and copper phthalocyanine (CuPc) for reference. As shown below, the relative strengths and weaknesses of conjugated polymers and small molecules can be exploited in different photovoltaic device designs.

8.2 PHOTOVOLTAIC DEVICE OPERATION

The basic device design and operation of conventional solar cells are described here to give the reader a familiarity with the current state of the art in commercial solar cell production and to provide a reference point for understanding the operation of organic solar cells (OSCs). The structure, processing, and optoelectronic properties of organic semiconductors suitable for PV devices are then discussed to show their promise for use in low-cost organic solar cells. Finally, the materials and operation of the various types of organic solar cells are described, and the limitations of each outlined to understand its relevance to the rapidly developing field of polymer photovoltaics.

8.2.1 Conventional Solar Cells

Most solar cells in commercial production and use today are p–n junction solar cells. While the p–n junction classification is broad, covering both homojunctions and heterojunctions and many different choices of materials, the basic mechanism underpinning the operation of all p–n junction cells is the same.

In the simplest case, a p–n junction is formed between two layers of the same semiconductor with different types of dopants, creating a p–n homojunction. One semiconductor contains dopants that are acceptor-type (p-type), making it a semiconductor with a Fermi level lower than the intrinsic, or undoped, semiconductor. The other contains donor-type (n-type) dopants, making it a semiconductor with a Fermi level higher than that of the intrinsic material. When the n-type and p-type semiconductor layers are brought into contact, there is a discontinuity in the Fermi level at the interface between the two materials, and charge will flow across this interface until

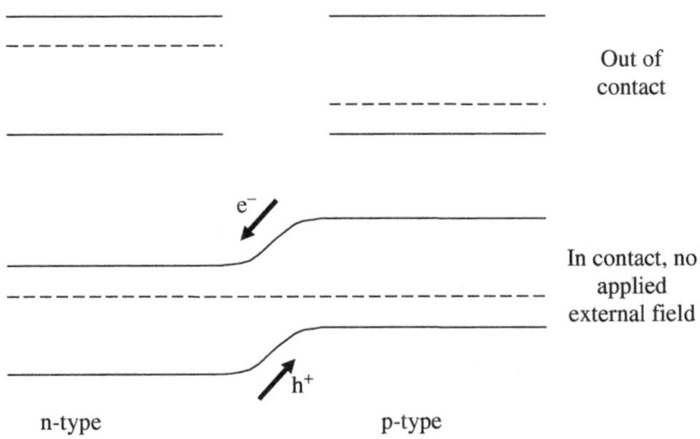

FIGURE 8.2 Schematic drawing of a semiconductor diode at the heart of a PV device. Appropriate doping sets the Fermi level (ε_f) close to the conduction band energy (E_{CB}) in the n-type material and close to the valence band energy (E_{VB}) in the p-type material, which on contact leads to the formation of a local electric field close to the junction directing the flow of any photogenerated charge.

the Fermi levels in the two materials are equal. This generates an excess of electrons in the p-type semiconductor and an excess of holes in the n-type semiconductor. These excess charges are confined by a mutual Coulomb attraction to a thin layer at the interface between the two materials. Thus, in equilibrium, there is no field in the bulk of the n-type and p-type layers, but there exists a thin layer near the interface where an electric field exists that opposes the flow of majority carriers across the interface, but encourages the flow of minority carriers (Fig. 8.2). That is, it prevents electrons from traveling from the n-type layer to the p-type layer, but encourages holes. Similarly, it prevents holes from traveling from the p-type layer to the n-type layer, but encourages electrons. The region of asymmetric charge transport provides the driving force that allows a p–n junction to sustain a voltage under illumination.

In general, the junction region is small compared to the thickness of the device. Thus, under illumination, the majority of photons are absorbed in either the p-type or n-type region of the device. The absorption of a photon results in the generation of a free (i.e., unbound) electron–hole pair. These electrons and holes are free to either (a) recombine with each other (geminate recombination) or other electrons/holes (bimolecular recombination) or (b) diffuse to the p–n interface, where the electric field generated by the p–n junction separates them into a hole in the p-type semiconductor and an electron in the n-type semiconductor. The electrons must then migrate through the n-type layer to the electrode, where they travel through the external circuit and recombine with a hole in the p-type layer at the other electrode, completing the circuit.

8.2.2 Organic Solar Cells

There are four different types of organic or organic/inorganic hybrid solar cells: single-layer, bilayer, bulk heterojunction (BHJ), and dye-sensitized. The basic operation of each device type is described below.

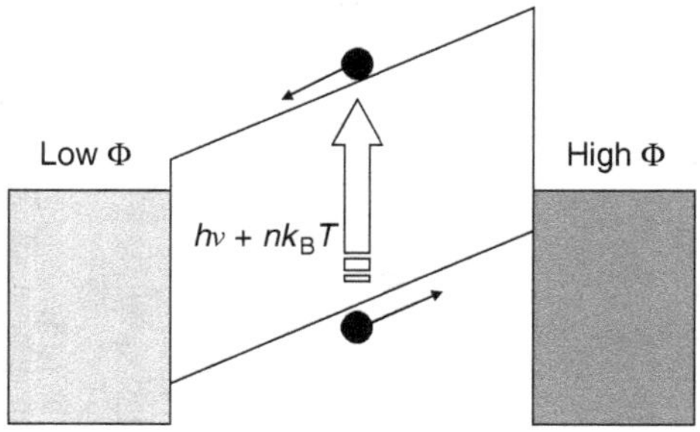

FIGURE 8.3 Schematic drawing of the operation of a single-layer polymer solar cell.

8.2.2.1 ***Single-Layer Polymer Cells*** Single-layer polymer cells (Fig. 8.3) consist of a conjugated polymer layer sandwiched between two metallic electrodes. Metals with different work functions are chosen as the electrode to generate the electric field and asymmetric carrier extraction required for a photovoltaic device. Unlike silicon, the high optical density of conjugated polymers means that only ~100 nm is required to absorb most light at the polymer's absorption maximum. For a typical work function difference of 1 eV for the electrodes, this would give a field on the order of 10^5 V/cm. As mentioned above, due to the low dielectric constant typical of most conjugated polymers ($\varepsilon_r \sim 3$) [16], the primary photoexcitations are not free electrons and holes but bound electron–hole pairs known as excitons. These excitons typically have a binding energy of a few tenths of an eV and a radius of ~1 nm, requiring a field of 10^6 V/cm to dissociate [17]. As a result, very few excitons are dissociated by thermal fluctuations (0.025 eV at room temperature) or the built-in field generated by the contacts, and most simply recombine before being separated. Those free electrons and holes that are generated drift, under the influence of the built-in field provided by the contacts, to opposite electrodes. However, the low hole and electron mobilities in conjugated polymers make the transit times to the electrodes long, so that charge transport and collection must compete with recombination. As a result, single-layer polymer solar cells typically have AM1.5 power conversion efficiencies less than 0.1%.

8.2.2.2 ***Bilayer Cells*** To increase the dissociation of photogenerated excitons, a second semiconductor layer is introduced to the cell. This second semiconductor is chosen such that its conduction band (or LUMO) or valence band (or HOMO) differs from that of the polymer by an amount greater than the exciton binding energy in the polymer. The energetics at the interface between the two semiconductors encourages exciton dissociation into an electron in the material with a lower LUMO level (the electron acceptor) and a hole in the other material (the electron donor). In addition to the improved exciton dissociation, the introduction of a second material provides two significant advantages over single-layer devices. First, the fact that electrons and holes are localized in different materials following exciton dissociation means that each material needs to transport only one type of charge, thereby relaxing the design constraints on the materials. Since conjugated polymers generally have a higher hole

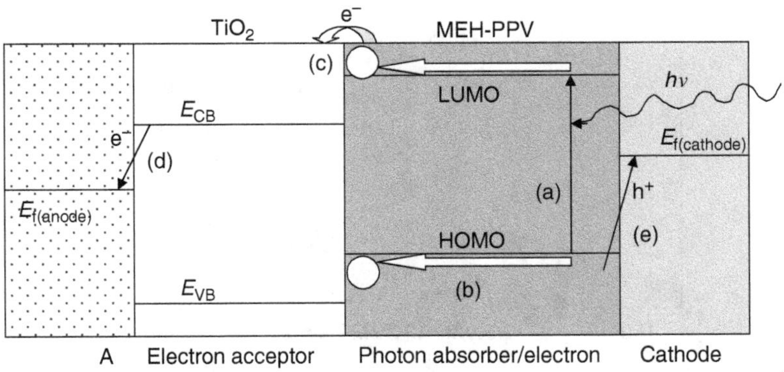

FIGURE 8.4 Schematic drawing of the operation of a bilayer photovoltaic cell (in this example and ITO/TiO_2/MEH-PPV/Au device) showing (a) light absorption, (b) exciton transport, (c) exciton dissociation, and (d) collection of electrons at the anode and (e) holes at the cathode.

mobility than electron mobility, they are generally used as the electron-donating and hole-transporting phase, while a small molecule (e.g., C_{60}) or inorganic semiconductor (e.g., TiO_2) serves as the electron-accepting and electron-transporting phase. Second, since the electrons and holes are already spatially separated upon photogeneration, there is a large concentration gradient of holes in the donor and electrons in the acceptor, which gives rise to a chemical potential that promotes the photovoltaic effect, allowing the open-circuit voltage to exceed the built-in voltage provided by the contacts [17]. Figure 8.4 shows a schematic drawing of the photovoltaic mechanism in bilayer photovoltaic cells.

While the efficiency of molecular (CuPc/C_{60}) bilayer photovoltaic devices has reached 3.6% under one sun (100 mW/cm^2) simulated AM1.5 solar illumination [18], the efficiency of conjugated polymer bilayer cells remains below 1%. There are several reasons for this, including poor spectral match due to the large bandgap (>2 eV) of most conjugated polymers [19], lower charge mobility leading to a buildup of charge at the interface and increased recombination [20], and incomplete exciton harvesting due to the short exciton diffusion length (<10 nm) typical of this class of materials. While the first two problems require materials development, the third issue has been elegantly dealt with by developing a third type of device, known as the bulk heterojunction.

*8.2.2.3 **Bulk Heterojunction Cells*** To overcome the short diffusion length of excitons in conjugated polymers, the donor and acceptor phases can be mixed on a length scale equal to the exciton diffusion length. By distributing the donor–acceptor interface throughout the light absorbing layer in this manner, the fraction of excitons that reach the interface and dissociate is greatly increased. This type of device is known as a bulk heterojunction device. The BHJ device structure, illustrated in Fig. 8.5, contains two bicontinuous phases to maximize the interface area between the phases while allowing a continuous (3D) path of both phases to their respective electrodes to

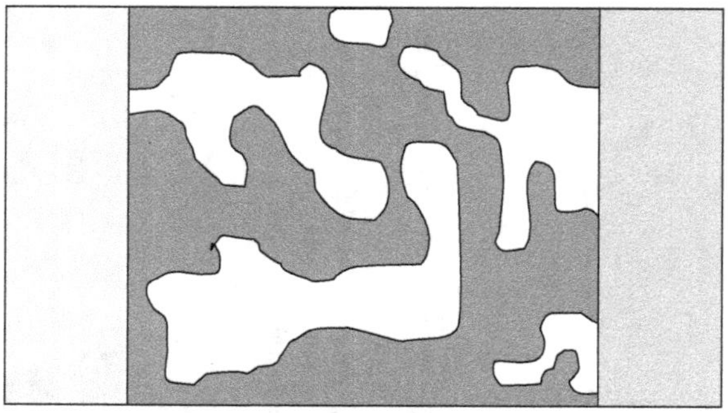

FIGURE 8.5 Schematic drawing of the bulk heterojunction device structure. Note that the scale of phase separation has been exaggerated.

allow charge extraction. While polymer:polymer [21], polymer:nanocrystal [22–26], and small-molecule [27,28] blends have been studied, the most extensively studied system has been the polymer:fullerene BHJ [29–36]. Efficient photoinduced electron transfer and luminescence quenching were first observed in a conjugated polymer: fullerene blend in 1992 [34]. It was later found that even for only 2 wt% PCBM in a polymer:PCBM blend, almost 100% of excitons are dissociated [35]. However, device performance depends not only on light absorption and exciton dissociation, but also on charge separation and transport away from the interface and into the electrodes, and it was found that 80 wt% PCBM is needed to form an efficient electron percolation network and maximize device performance [35].

Bulk heterojunction cells are the most efficient solid-state organic photovoltaic devices reported to date, with one sun AM1.5 power conversion efficiencies of, for example, 6.0% for a P3HT:C_{60}-PCBM blend [37], and recently 5.5% for a low-bandgap polymer/C_{71}-PCBM blend [11]. The performance of BHJ devices is mainly limited by poor spectral match and by recombination of photogenerated charges due to low charge carrier mobility in the blends. For MDMO-PPV:PCBM cells, illumination at the absorption maximum results in nearly 70% of the incident photons being converted into electrons for a 100 nm device [29]. While a 100 nm device is thick enough to absorb the majority of light at the absorption maximum, a thicker layer is required to absorb light toward the edges of the absorption spectrum, but increasing the layer thickness increases transit time and recombination losses, reducing charge collection efficiency and device performance [8].

A variation on the BHJ device structure, known as the ordered bulk heterojunction (Fig. 8.6), has been proposed. By creating a device with two interdigitated layers, the interfacial area could be maximized while still maintaining a direct transport path to the electrodes, minimizing transport losses. Initial attempts at realizing such a structure experimentally have mainly centered around making TiO_2 layers with cylindrical pores (with a diameter in the tens of nanometers) and then infiltrating the polymer layer into the TiO_2 by spin coating and annealing [38]. Success has been limited, however, as the polymer that has infiltrated into the pores seems to have a

FIGURE 8.6 Schematic drawing of the structure of an ordered bulk heterojunction, showing two interdigitated phases sandwiched between the two electrodes.

lower mobility than the bulk of the film, possibly due to steric hindrance from the spatial confinement of the polymer chains resulting in reduced conjugation along the backbone. The recent synthesis of ordered TiO_2 nanotube arrays [39] offers another promising route toward developing ordered BHJs.

As discussed below, the microstructure of the two-phase BHJ system is critical to device performance and needs to be carefully controlled and to be stable under working conditions for the lifetime of the product. The use of block copolymer systems is a very attractive way to be able to form self-assembled nanostructured BHJ systems with thermodynamically stable microstructures. For example, a diblock copolymer consists of molecules in which one end of the chain contains a series of one type of monomer (e.g., an electron acceptor material) and the other end a second type (e.g., an electron donor material). Typical of polymeric systems, the miscibility of the two sections of polymer chain will be low, and the different monomers will tend to phase separate. In the case of the diblock copolymers, however, the coarsening of the phase separation is limited by the dimensions of the molecule as the two phases are held together by the strong covalent bonds at the center of the molecule. The size of the domains of each phase is therefore limited by the size of the "block" within each molecule, as illustrated in Fig. 8.7, and the morphology is determined by the relative volume fractions (lengths) of each block. Therefore, by careful molecular design, a nanometer-scale phase separation with a stable co-continuous morphology can be formed by the natural phase separation process that follows precipitation from a solvent, for example, on thin-film formation. It has even been shown that the morphology of these materials can be influenced by electric, surface, or shear fields, for example, to induce a particular anisotropy such as to align columns of a minority phase in a particular direction. The block copolymer systems are therefore very attractive for their potential application in BHJ polymer photovoltaics. Unfortunately, to date, high-efficiency devices have not been achieved due to lack of block copolymer materials with suitably high electron and hole conduction blocks. Rather than using the block copolymer directly as the semiconducting materials, they can also be exploited as templates for nanopatterned materials such as might be exploited in PV

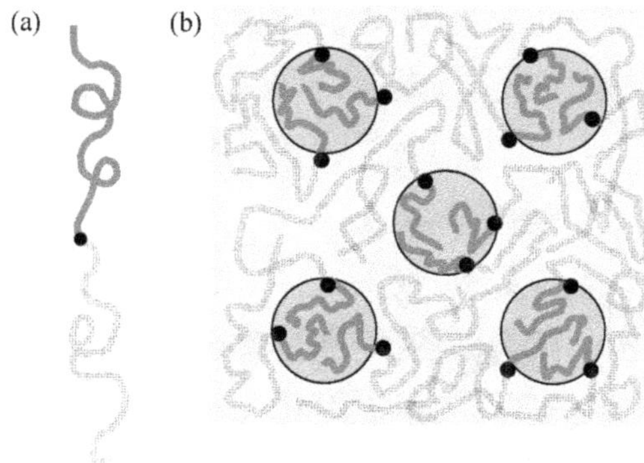

FIGURE 8.7 (a) Schematic representation of a diblock copolymer molecule. The two ends of the molecule each contain repeating units of a single monomer, and the two blocks of different chemistry are covalently bound together. (b) An illustrative example of a phase-separated diblock copolymer system forming, in this case, cylinders of the minority, darker, phase projecting into the page. The relative lengths of each block determine the volume of each phase present. The bonding of the two species within each molecule prevents further coarsening of the microstructure and sets the dimensions of the morphology precisely, leading to an ordered array of the phase-separated structure.

devices. One recent example is that of using electrochemical replication of a nanostructured polymer derived from a diblock copolymer to create a dye-sensitized solar cell [40].

*8.2.2.4 **Dye-Sensitized Cells*** Dye-sensitized solar cells (DSSCs) are slightly more complex than bilayer and bulk heterojunction cells, but as was the case for bilayer cells, the increase in device complexity reduces the number of functions that must be performed by each of the materials. A schematic drawing of a dye-sensitized solar cell is shown in Fig. 8.8. A layer of sintered, interconnected TiO_2 nanoparticles, which serves as the electron transport material (ETM), is coated by a thin layer of light absorbing dye. The remaining pores in the dye-coated TiO_2 layer are then filled with a

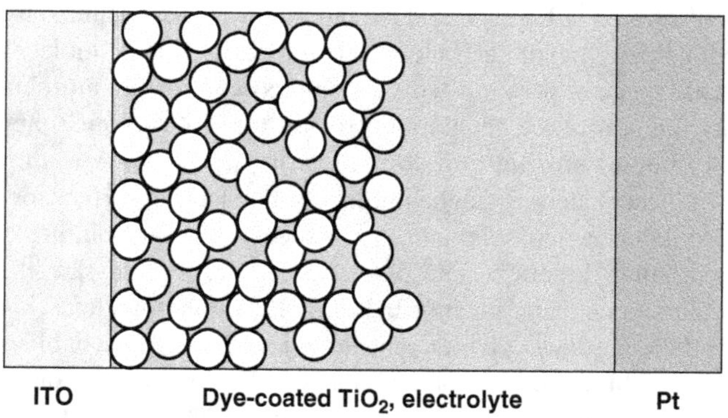

FIGURE 8.8 Schematic drawing of a dye-sensitized solar cell.

hole-transporting material (HTM) or an electrolyte, and the entire TiO_2/dye/HTM assembly is sandwiched between two electrodes. The operation of the cell is similar to that of the BHJ, except that all light absorption occurs in the dye, directly at the interface of the ETM and HTM. This eliminates the need for any exciton migration and allows more flexibility in the choice of light absorber as the absorber no longer needs to transport holes efficiently. Following light absorption, electrons are injected from the dye into the conduction band of the TiO_2. The dye is restored to its ground state by electron transfer from the HTM or electrolyte, and the electron and hole are transferred to opposite electrodes, resulting in electron flow through the external circuit.

While dye-sensitized cells are the most efficient OPV technology to date, with device efficiencies exceeding 10% [41], the use of liquid electrolytes presents problems with device sealing and efficiency. Solid-state HTMs have been studied, but since the sensitizing dyes generally do not conduct charge well, the HTM must be completely infiltrated into the pores and in excellent contact with all the dye molecules, a task that is much more difficult with a solid-state material than with a liquid electrolyte [42]. While DSSCs incorporating solid-state hole-transport materials in place of the liquid electrolyte have thus far shown much lower efficiencies [43], the recent reports in the regions of 4% efficiency under AM1.5 illumination are promising. Quasi-solid-state devices with higher efficiencies (around 6%) are also under development, using low molecular weight organic "gelators."

8.2.3 Routes to Improving OSC Efficiency

It would have become evident from the above discussion that the factors limiting device efficiency are different for each of the device types, although some common limitations exist for OPVs in general. Polymer PV devices (single-layer, bilayer, and bulk heterojunction) would benefit from the development of novel low-bandgap polymers to improve spectral match and photon harvesting, as well as from improved charge mobility, especially for hole transport. For single-layer devices, the main limitation is the inefficient dissociation of excitons. For bilayer devices, the transport of excitons to the dissociating heterointerface is the primary obstacle to improving device performance. For BHJs, the rapid collection of electrons and holes before they can recombine is the main concern. For DSSCs, intimate contact between the solid-state HTM and the dye is of paramount importance.

8.3 POLYMERS IN PHOTOVOLTAICS

There are four main classes of conjugated polymers that have been studied for optoelectronic device applications [44]. They are classified according to the structure of the conjugated backbone and are the poly(*para*-phenylene)s (PPPs), poly(*para*-phenylene vinylene)s (PPVs), polythiophenes (PTs), and polyfluorenes (PFs). Their structures are shown in Fig. 8.9. Of these four classes, the most widely studied and commercially significant have been the PPVs and PFs [45], although the PTs have

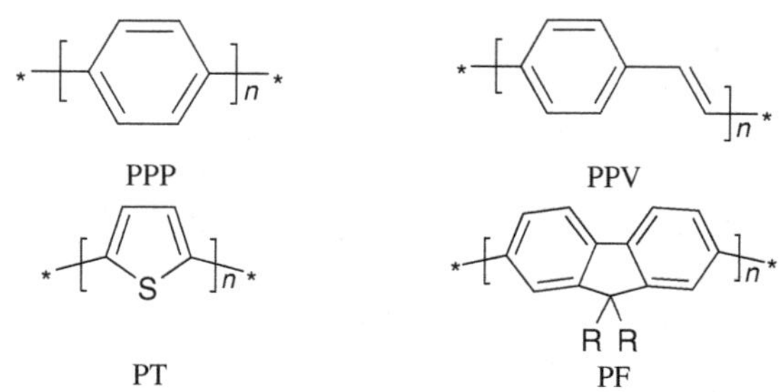

FIGURE 8.9 Linear unsubstituted conjugated polymers: poly(*p*-phenylene), poly(*para*-phenylene vinylene), polythiophenes, and polyfluorenes.

recently garnered much interest due to their high hole mobility and performance in BHJ organic photovoltaic devices [37].

Current OPV research is focused on three key areas, where improvement is needed for commercialization of organic photovoltaics:

- Improving light harvesting
- Improving photogenerated carrier collection:
  - Exciton separation
  - Charge transport
  - Charge injection into electrodes
- Addressing manufacturing and stability issues

All three of these issues can be addressed, to some extent, by understanding and improving the optical, electronic, and stability properties of the photoactive polymer layer in OPV devices.

There are two ways to modify the properties of the polymer in a device. The first is through molecular modification of the polymer. This is generally done during synthesis and involves altering the molecular structure of the polymer either through alterations to the conjugated backbone or through the attachment of tailored side groups to one or more positions on the repeat unit of the polymer. In PPVs, for instance, side groups could be attached either to the phenyl ring or to the vinyl linkage (Fig. 8.10). Both the molecular structure (e.g., size) of such groups and the position at which they are attached can affect device properties, as even small changes in the

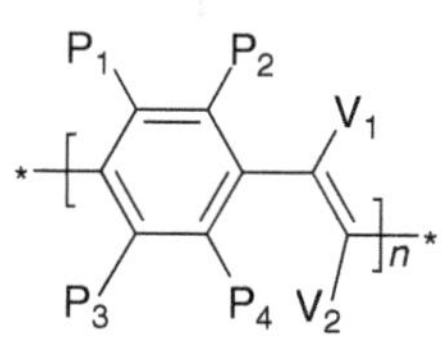

FIGURE 8.10 Substitution positions for PPV, showing the four phenyl positions (P1–P4) and the two vinyl positions (V1 and V2) where side groups may be attached.

conjugated polymers can significantly alter optical and structural properties of polymer films. The second way to modify the polymer involves altering the morphology of the polymer at the macromolecular scale. Significant changes in polymer morphology, and consequently optoelectronic properties and device performance, have been achieved by varying various parameters during film processing or device manufacturing. Changes in the solvent, solution concentration, spin speed, and annealing conditions have all been shown to produce significant changes in polymer morphology and film properties. Changes in the molecular structure of conjugated polymers can also affect morphology, by altering intrachain and interchain interactions, as well as solvent interactions during deposition, leading to changes in the self-assembly characteristics of the polymer.

8.3.1 Molecular Control of Polymer Film and Device Properties

8.3.1.1 Effect of Side Groups on Polymer Solubility While linear conjugated polymers are very useful as optoelectronic devices because of the possibility to tune their optoelectronic properties, almost all such unsubstituted polymers are insoluble. To allow inexpensive solution processing of these materials, it has become common practice to add side groups to the conjugated backbone to improve solubility. The most common form of side group used for this purpose is a long alkyl or alkoxy group, which improves the solubility of many PPV derivatives and thienylene-PPV copolymers in common organic solvents. Increasing the side chain length increases the conformational entropy available to the molecule in solution, hence promoting solubility. Both the size and conformation (e.g., phenyl group versus hexyl group) of added side groups are important for controlling the solubility of conjugated polymers.

8.3.1.2 Effect of Side Groups on the Stability of Conjugated Polymers One of the main obstacles to commercialization of organic PVs is the inherent instability of conjugated polymers. There is currently interest in developing flexible encapsulation materials to prevent oxidative degradation of the polymers. As this could dramatically increase fabrication costs and reduce cell efficiency (e.g., due to potential damage to the cell on application of the encapsulation layers or by loss of transparency), and the encapsulation materials could fail, it is desirable to have inherently stable polymers that are resistant to oxygen and water damage. Since the extended conjugation of the backbone is the source of the semiconducting behavior of these polymers, the inherent stability would ideally be introduced through side groups. This would preserve the conjugation along the backbone, which may otherwise be interrupted by alterations directly to the backbone.

Johansson et al. [46] have reported improved stability of PPV upon addition of phenyl side groups to the phenyl ring in the PPV backbone. The stability was tested by illuminating the polymer in air and measuring the degradation of polymer photoluminescence. They believe that the replacement of a C–H bond by a more stable C–C bond (between the phenyl ring in the backbone and the phenyl side group) is responsible for the improved stability. Cumpston and Jensen have reported improved stability upon addition of phenyl or cyano (CN) groups to the vinyl linkage in a PPV-like polymer [47]. They also observe increased stability when the vinyl linkage is replaced by a triple bond. For MEH-PPV, the phenyl ring appears to be preserved after 9 h of exposure to air under illumination; the C=C vinyl linkage and C–O–C side group signals fade, while more C=O groups are observed. Since the C=C signal fades faster, it appears that the vinyl linkage is the most vulnerable to oxidation, perhaps due to its combination of high electron density and steric exposure to reactive oxygen.

It is apparent that the addition of side groups to either the phenyl ring or the vinyl linkage in PPV-type polymers results in an increase in stability. Both effects may be due to the replacement of reactive C–H bonds with more stable C–C bonds and/or a steric shielding effect of the side groups. While it is encouraging that the addition of side groups can improve both solubility and stability of conjugated polymers, it is also necessary to understand the effect of such substitutions on the optoelectronic properties of the polymers and their films to determine the usefulness of such substitutions to polymer photovoltaic applications.

8.3.1.3 Effect of Side Groups on Optoelectronic Properties of Polymers and Films Most conjugated polymers have a HOMO–LUMO energy gap greater than 2 eV, corresponding to a cutoff wavelength for absorption shorter than 620 nm. This does not compare favorably with the silicon bandgap of 1.1 eV, which is much closer to the 1.4 eV that would give optimal efficiency for the AM1.5 solar spectrum. New materials or material combinations that can extend absorption into the near infrared region would greatly improve light harvesting and device efficiency. Colladet et al. have synthesized a group of low-bandgap polymers for PV applications [48]. Example structures of these thienylene-PPV derivatives are shown in Fig. 8.11. By adding an electron-donating ethylene dioxide side group (Fig. 8.11b) to the thiene group, they lowered the optical bandgap (determined by the onset wavelength for absorption, $E_g = 1242/\lambda_{onset}$) from 1.84 to 1.58 eV. The electron-donating group raises the

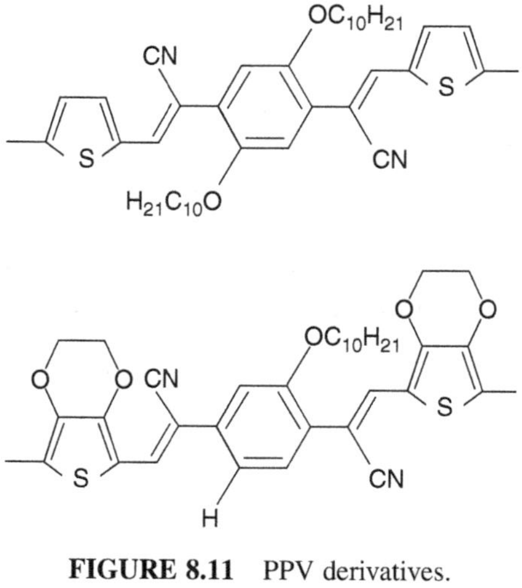

FIGURE 8.11 PPV derivatives.

HOMO of the molecule more than the LUMO (as determined by cyclic voltammetry), thereby decreasing the energy gap. Further addition of a bulky alkyl side chain increased the bandgap by an average of 0.12 eV, attributed to an increase in disorder causing a decrease in conjugation length. It is apparent that the electron affinity of side groups, as well as their position and steric nature, is important in determining their effect on the optoelectronic properties of conjugated polymers.

Developing a clear structure–property relationship for conjugated polymers is complicated by the fact that long polymer chains inevitably have a number of randomly distributed defects leading to a statistical distribution of conjugation lengths [49,50]. To avoid this complication, a number of groups have chosen to study oligomers (molecules with a small number of repeat units, typically less than 50) to try to extract a structure–property relationship in the hope that such knowledge will transfer over to the larger, more complicated polymer system. Hanack et al. studied the effect of cyano (CN) side groups on the optical properties of oligomeric PPV derivatives with three or four phenyl rings [49]. Even a minor variation in the position of cyano groups has a significant impact on the optical properties of the molecule. For instance, moving the cyano group from the inner C atom of the vinyl linkage to the outer C atom of the same vinyl linkage red shifts the UV–visible absorption maximum (in solution) by almost 60 nm and the thin-film PL maximum by over 100 nm. As this work focused on OLED applications, the hexyl groups were introduced to prevent crystallization of the thin films (which is increased by the addition of the cyano groups), since crystallization decreases luminescence quantum yield. Crystallization may also improve charge transport, leading to a further improvement in PV performance. Varying the position of the cyano group on the phenyl rings was also shown to significantly alter the thin-film PL maximum. This is attributed to changes in interchain interactions and may cause changes in the thin-film absorption spectrum as well. For PV applications, both the red shift in absorption maximum and the improved crystallization upon introduction of the cyano groups may be useful for reducing spectral mismatch and improving charge transport, respectively.

Detert and Sugiono studied the effect of adding various side groups to an oligomer of PPV with five benzene rings [50]. Absorption and emission were measured in CH_2Cl_2, and octyl groups are included as side groups on the molecule in all cases to make the compounds soluble. Unfortunately, there are no data presented for the bare, unsubstituted oligomer (i.e., H in all side group positions), so it is difficult to draw concrete conclusions about the effect of any single group on the properties of the molecule. However, it seems clear that the vinyl cyano group red shifts both the absorption and fluorescence maxima by $\sim$20 nm. This shift in absorption agrees with Hanack et al.'s observations [49], while the shift in PL is significantly smaller.[2] The negligible change in the position of the absorption maximum for molecules without the cyano group on the vinyl linkage, despite changes in the location and electron

[2] Note that Hanack et al. measured the PL of thin films, while Detert and Sugiono measured PL in solution. Interchain interactions will be more significant in the thin film, which may cause the magnitude of the spectral shift to change.

affinity of the side groups, indicates that it is possible to tune the electron affinity (at least in solution) without significantly affecting optical properties.

Another way to control interchain interactions in thin films is to introduce bulky side groups that sterically separate polymer chains. This is not desirable for PV applications, where interchain interactions are important for maximizing charge transport, so it is important to understand how interchain interactions can be affected and controlled by the addition of side groups. Jakubiak et al. studied the effect of large dendritic side groups on absorption and luminescence of both isolated (dilute solution) and aggregated (poor solution and solid state) PPV chains [51]. The solution absorption and PL spectra for a first- and second-generation substituted side groups are identical, except for additional absorption in the UV range in the case of the larger side group, possibly due to absorption by the side group. The thin-film absorption spectra are also identical for the two polymers, but the thin-film PL for the smaller substituted molecule is blue shifted slightly further from the solution spectra. The most interesting result might be considered to be the effect of solvent quality. As the solution quality was varied from a good solvent, trichloroethane (100% TCE), to a poor solvent, dimethylsulfoxide (DMSO, 0% TCE), the PLQY of the first-generation (smaller) dendritic side group fell faster. In pure TCE, the PLQY values are equal, while in pure DMSO and thin films the PLQY for the larger substituted group is a factor of 4 higher. As a decrease in PLQY upon going from an isolated state (good solution) to an aggregated state (poor solution or solid state) is known to be caused by an increase in the production of poorly emissive interchain excitations (excimers) [52], one can conclude that the much larger 75-carbon side group in the second-generation material inhibits interchain interactions more effectively than the 24-carbon side group in the first-generation case. Control of these interactions will be important for optimizing charge transport in conjugated polymer films and performance of OPV devices.

Most of the above studies have focused on the optical properties of substituted PPV derivatives, while neglecting or merely speculating at the effect of side groups on the charge transport properties of thin films. Lee et al. studied the hole mobility in fluorene-substituted PPVs [53] and found that attaching electron-donating or electron-withdrawing side groups to the fluorene unit caused a decrease in hole mobility. Attaching a dihexyl fluorene group to the phenyl ring of the PPV backbone (DHF-PPV) resulted in a hole mobility of 2.7×10^{-4} cm^2/(V s), which is among the highest reported to date and two orders of magnitude higher than that of MEH-PPV (1.7×10^{-6} cm^2/(V s)). Attaching a methoxy group to the fluorene group (MDHF-PPV) decreased the mobility to 1.1×10^{-4} cm^2/(V s), while the addition of a cyano group (CNDHF-PPV) decreased the mobility drastically to 4.1×10^{-7} cm^2/(V s), lower than that of MEH-PPV. The level of the HOMO and LUMO remained unchanged, as did the position of the absorption and PL maxima. They suggest that even with a π-electron-rich fluorene side group to facilitate hole transport, the electron-withdrawing (cyano) and electron-donating (methoxy) groups act as trapping centers that inhibit transport. Thus, it appears that while cyano groups can improve the optical properties of conjugated polymers by causing a red shift in absorption, the improvement in OPV device efficiency may be offset by losses in charge transport due to charge trapping effects.

8.3.1.4 ***Summary*** There is a great deal of interest in improving the optoelectronic properties of polymers for use in photovoltaic applications. The charge transport properties of thin films of conjugated polymers depend sensitively on the morphology and macromolecular structure of the films. The morphology and degree of interchain interactions can be controlled during processing, through the choice of solvent, solution concentration, spin speed, and temperature. Morphology and interchain interactions are also sensitive to the molecular structure of the polymers, and this must be taken into account when chemically engineering polymers for photovoltaic applications. The optical and electronic properties of polymers, in both isolated and aggregated states, are also dependent upon the molecular structure of the polymer. The electron affinity, optical absorption spectrum, and degree of interchain interaction (in solution and solid state) can all be tuned by the addition of side groups to the conjugated backbone. When choosing side groups, their location on the conjugated backbone, their electron affinity, and their physical size and steric nature are all important in determining their effects on the optical and electronic properties of polymeric thin films. Keeping all these factors in mind, it may be possible to produce a significant improvement in photovoltaic device performance through careful and calculated engineering of the photoactive polymer structure at the molecular and macromolecular scales.

8.3.2 Polymer Morphology: Effects on Optoelectronic Properties of Films and Devices

8.3.2.1 ***Effects of Solvent on Film Morphology and Film/Device Properties*** There have been a number of reports on the effects of solvent on conjugated polymer conformation, both in solution and in thin films deposited from solution, and the consequent effects on device performance [54–56]. The conformation of MEH-PPV chains in solution varies greatly depending on the choice of solvent [54]. Solvating in chlorobenzene (CB) gives a more extended and open configuration than solvating in tetrahydrofuran (THF), resulting in a red shift in the absorption spectrum for the CB solvated polymers relative to the THF solvated polymers. The red shifts in the absorption and PL spectra are due to longer conjugation lengths in, and greater interchain interactions between, the more open chains in CB solution, while the MEH-PPV chains in THF remained bunched. A red shift in below-gap photoluminescence (i.e., the emission of light with smaller energy than the bandgap for the conjugated polymer) can be observed upon increasing the concentration of the polymer solution. Increasing solution concentration would also increase interchain interaction by decreasing the mean distance between chains, consistent with the observed red shift in below-gap PL at higher concentration. These effects remain when the polymers are deposited as thin films, despite the evaporation of the solvent, indicating a memory effect in the films. The greater interchain interaction in the CB cast films resulted in better charge transport, leading to the conclusion that the optical and electronic properties of MEH-PPV films can be controlled by both the solution concentration and the choice of solvent.

The solvent can also affect the film morphology as observed by scanning force microscopy (SFM) of films cast from different solvents [56]. It is thought that the difference in chain conformation in solution (e.g., where MEH-PPV chains tend to form tightly coiled "spheres" in THF solutions, while adopting a more open configuration in CB, as inferred from the hydrodynamic radii of MEH-PPV chains in the two solvents [54]) is to some degree maintained in the thin films, after spin casting, but we should also bear in mind the possibility that large lumps ($\sim$1 μm across) are simply incompletely dissolved clumps of polymer due to THF being a poorer solvent for the polymer. There can also be an effect of morphology on stability of films attributed to increased oxygen diffusion between closely packed agglomerated spheres of polymer compared with an intertwined chain morphology. We should beware of the possibility that residual solvent would also make a difference to both diffusivity and oxidation. Regardless of the mechanism involved, the fact that the choice of solvent can affect stability has important implications for organic photovoltaic device production, since, as mentioned above, stability is one of the main obstacles to commercialization.

A group at the University of California [57] found that the absorption maximum of MEH-PPV is shifted upon changing the solution concentration or changing the spin speed during film casting. They found that the absorption maximum of MEH-PPV was blue shifted upon increasing the concentration of the polymer solution or decreasing the spin speed during film preparation. They also found that both effects were reproducible in several solvents, including cyclohexanone, THF, chloroform, and *p*-xylene. Both shifts are believed to be due to strong interchain interactions, as discussed above. They believe that a lower spin speed during film deposition results in less stretching of the polymer chains, producing a film in which the polymer chains are more bunched than in a film prepared at a high spin speed. Increasing the concentration of the polymer solution would also result in more interchain interactions during casting and therefore less stretching of the polymer chains. Since an increase in the bunching of individual chains would result in shorter conjugation lengths, intrachain bunching due to insufficient stretching during casting would explain the blue shift in the absorption maximum. However, this effect was lost at high or low concentrations (in this case >1 and <0.4 wt%, respectively), at which composition ranges the absorption spectrum did not shift with spin speed. At high concentrations, the absorption maximum was at low wavelength (497 nm) for all spin speeds, while at low concentrations it was at its highest (510 nm). They conclude that polymer conformation is only spin speed dependent when the cohesive force of the molecules is comparable to the force applied during spinning [57]. When the concentration is too high, the interchain interactions of the highly aggregated polymer chains are too great, and the increase in spin speed (from 2000 to 8000 rpm) is insufficient to stretch the polymer chains, resulting in persistent aggregation and a lower absorption maximum. At lower concentrations, the increased mean distance between polymer chains means that the interchain interactions are insignificant and the polymer chains are easily stretched at all spin speeds measured. This demonstrates that both intrachain aggregation (due to poor choice of solvent) and interchain aggregation (due to high solution concentration) must be accounted for when preparing films for optoelectronic applications. It should be noted that intrachain aggregation will cause both a blue shift in absorption and poor charge transport, while interchain aggregation may actually improve transport due to improved charge transfer between chains. It is also worth noting that, while Nguyen et al.'s observation of a red shift in below-gap PL upon increasing concentration may appear contradictory to the blue shift in absorption reported here, they are actually both consistent with increased interchain aggregation. As later reported by Nguyen et al., the stronger polarization interaction in the aggregate environment produces a red shift in the emission, an effect known as the gas-to-crystal effect.

The University of California group later reported that the performance of MEH-PPV/C_{60} blend solar cells is greatly affected by the choice of solvent [58]. Devices were made with an ITO/PEDOT:PSS/MEH-PPV:C_{60}/Ca geometry, and various blend compositions were studied, from 0% to 12.5% C_{60}.[3] In the absence of C_{60}, devices in which the polymer was spun from a THF solution had a lower short-circuit current than those spun from a xylene solution, with the open-circuit voltage constant at 1.60 V for both solvents. This is consistent with polymer chain bunching and poor interchain transport for MEH-PPV films cast from THF solution, as discussed above. For the MEH-PPV:C_{60} blend, devices made using THF or chloroform (CF) as a solvent had lower short-circuit current and fill factor, as well as a higher open-circuit voltage, than those made using CB, dichlorobenzene (DCB), or xylene. This was the case for all reported blend ratios. This result is consistent with poor mixing of the polymer and C_{60} in the THF and CF, resulting in a device that more closely resembles the plain MEH-PPV device due to phase segregation and poor contact between the polymer and the C_{60}. Kwong et al. [60] obtained similar results for poly(3-hexyl thiophene):TiO_2 nanoparticle blends.

8.3.2.2 Effect of Annealing on Morphology and Device Performance Both Nguyen et al. [55] and Liu et al. [61] have reported an improvement in OLED device performance on annealing of the polymer layer. Nguyen et al. have also taken SFM micrographs of MEH-PPV films before and after annealing and found that there is a decrease in both the size and number of raised features in the film after annealing, which is attributed to an untangling of bunched-up polymer chains upon heating (possibly driven by favorable interchain interactions) [56]. Both groups report that annealing causes a substantial increase in operational current, but that this gain in current does not translate directly into an increase in luminescence efficiency due to an increase in the number of nonemissive interchain species. The source of both the increased current and the increase in nonemissive interchain species is believed to be a relaxation of the polymer chains upon annealing. While the latter is not desirable for LEDs, the luminescence of excited species is not as important for PV applications, which makes annealing an obvious means of improving organic PV device performance. Improvements in PV device performance upon annealing have been reported

[3] It should be noted that this ratio is significantly lower than that usually applied in polymer/C_{60} bulk heterojunction devices, where the C_{60}:polymer ratio is typically 4 : 1 to allow the formation of C_{60} percolation paths for electron collection [59].

for organic/organic blend cells [62] and polymer/dye blends [63]. An increase in current over time for a conjugated polymer/TiO_2 blend, as reported by Grant et al. [64], is believed to be due to an annealing effect at the polymer/metal contact. It is also worth noting that accelerated phase segregation upon annealing of polymer: C_{60} blend cells has been observed [65], raising concerns over the long-term stability of such cells due to the elevated operating temperatures characteristic of PV applications.

8.3.3 Enhancing PV Device Efficiency

We will consider ways in which the efficiency of polymer photovoltaic devices might be improved, taking each step of the PV process in turn.

8.3.3.1 Absorption of Light The overlap of the absorption of the most commonly studied conjugated polymers with the solar spectrum is not all that good—light harvesting in the red to infrared regions of the spectrum in particular is often poor. Synthesis of new polymer materials with a small bandgap (difference in energy between the HOMO and LUMO) will allow these lower energy photons to be absorbed. However, if the absorbing material has a smaller bandgap than the incoming photons, much of the energy from those photons is lost as heat rather than being converted into the desired electricity (Fig. 8.12). For a wide-bandgap material, many of the incoming photons do not have sufficient energy to promote an electron across the gap to make an exciton. Conversely, if the incoming photon has energy larger than the bandgap, electrons can be promoted into a level above the LUMO and/or electrons from below the LUMO can be promoted. This additional energy will typically be lost as heat as the electron and/or hole loses its additional energy before or at the point of charge separation. Such energy loss is of course detrimental to device efficiency.

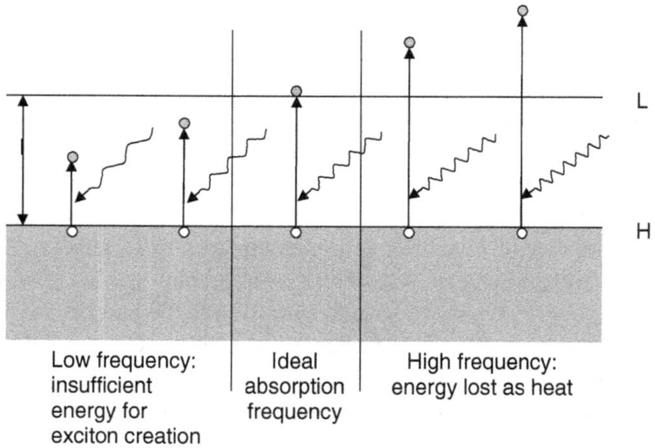

FIGURE 8.12 Schematic illustration of the role of the absorbed frequency on photovoltaic operation. For the purposes of this illustration, it is the electron at the HOMO level that is being promoted in each case.

As a means to tackle this issue, several groups have now reported "tandem" solar cells with two absorbing layers stacked on top of one another, sometimes separated by an electrode material such as TiO_x. The two cells are in series, creating an open-circuit voltage that is the sum of the two cells. Typically, the first cell uses a wide-bandgap polymer to harvest the blue light and being significantly transparent to the lower energy red light that passes through the cell to the back layer where it is absorbed by the lower bandgap material. Thus, two ranges of wavelengths can be harvested, leading to higher efficiency cells, albeit with the greater complexity in device fabrication.

An alternative or additional method under exploration is the incorporation of nanocrystals into the polymer cells. Nanocrystals can be tailored to absorb specific wavelengths of light not only by their chemistry, but also by their size. As the size of the crystal becomes smaller than the Bohr radius of the material, the energy bandgaps widen, changing the color of the light at maximum absorption. The challenge now is to develop very well-controlled nanocrystals (possibly with a range of sizes) well dispersed in the conjugated polymer matrix with a good route to extract charges from excitons generated within the nanocrystals. Developments have included the structuring of the nanocrystals, into rods or tetrapods, for example, to encourage percolation of the nanocrystals allowing them to carry one charge (e.g., the electron) back to the electrode along a continuous path.

Other possibilities for enhancing light harvesting include multiple exciton generation and upconversion nanocrystal systems in which low-energy light can be used to create high-energy excitons, or optical systems, such as waveguides or condensers. Such optical systems have been very effectively used for high-efficiency inorganic systems where a high level of engineering of the device architecture can allow maximum use of the incoming light. Similar approaches with polymer-based systems would need to be compatible with the low-cost/large-area production routes that favor the organic cells.

8.3.3.2 Exciton Diffusion As described above, the exciton formed by photon absorption will typically diffuse to an interface for the charges to be separated. As the exciton is a neutral species, the mechanism is diffusion rather than that of drift under the internal electric field. The lifetime of an exciton is fairly short as they are mostly singlet excitons that can decay radiatively (emitting photons) or nonradiatively (emitting phonons). The excitons lose energy, via a Förster transfer, as they move from higher energy to lower energy (e.g., more conjugated) segments of the polymer. The short lifetime and low mobility lead to short overall diffusion lengths, typically a couple of tens of nanometers. The exciton must reach the interface with the acceptor species with sufficient energy to be transferred into the other phase at the interface.

8.3.3.3 Charge Separation The difference in energy between the polymer LUMO and the electron-accepting material conduction band is usually large (e.g., for MEH-PPV/TiO_2 is it greater than 1 eV) compared to the exciton binding energy, meaning that the exciton dissociation is often very quick and complete. The device's internal electric field must then draw the charges away from the interface to a distance in which the two charges are no longer bound to avoid geminate recombination. Hence, recombination processes form an important limitation to device efficiency and are strongly influenced by the device morphology.

Lithium has been added to the electrolyte of liquid electrolyte DSSCs as early as 1991 [66], as the presence of Li^+ on the surface of TiO_2 can lower the energy of the conduction band by as much as 1 V [67]. This can improve the overlap between the acceptor states in the TiO_2 and the vibrationally relaxed excited states of the dye, improving the quantum yield of charge injection (i.e., the number of injected electrons per absorbed photon). Lloyd et al. [68] found that adding lithium triflate ($LiCF_3SO_3$) to all-organic TIPS-pentacene/C_{60} bilayer devices improved device efficiencies, believed to be because the built-in field provided by the contacts caused Li^+ to drift toward the electrode, while Tf^- anions were driven to the organic/organic interface. This would result in a sharper electric field at the interface, which would sweep holes away from the heterojunction, reducing recombination.

8.3.3.4 Extraction of Charge to the Electrodes The charges are removed to the electrodes under a combination of drift (the influence of the device electric field) and diffusion due to the concentration gradient of charges away from the charge separation interface. To retain device neutrality, of course, both charges need to be removed from the device, and ideally there is a good balance between the mobility of the electron and hole species. Often the holes are carried in a polymeric phase, and the electrons in a polymer, inorganic (e.g., TiO_2), or fullerene phase. In general, the hole mobility is poor by comparison. Keeping the path to the electrodes as short as possible by creation of a thin-film device with an ideal heterojunction morphology reduces the internal resistance of the device, but this needs to be balanced by the requirements for good light absorption (which would favor a thick device) and exciton diffusion to the interface over short lengths (requiring a highly structured interface). Annealing the devices after deposition is often found to be beneficial, due to a combination of coarsening the morphology and ordering the polymer (and, where relevant, a dispersed phase such as fullerene) to improve charge conductivity in the device. The addition of ions to the polymer may also increase the charge mobility; for example, Barkhouse et al. [69] have shown significant improvements in device performance due to increased hole mobility, leading to efficiencies in excess of 1% for bilayer MEH-PPV/TiO_2 devices on the addition of Li salts.

8.3.4 Photovoltaic Device Fabrication

This section seeks to set out a typical fabrication route for lab-scale manufacture and testing of PV devices. There is much interest at present in developing “scale-up processing,” and the focus of effort is based around roll-to-roll printing of polymer-based systems, using solvent-based deposition. There remains significant opportunity for further development of this large-area deposition, which is required for exploitation of the potential of polymer PV materials. Issues will include the patterning and registration of the various layers on what might be an extensible substrate, development of appropriate solution properties for printing technologies, and integration

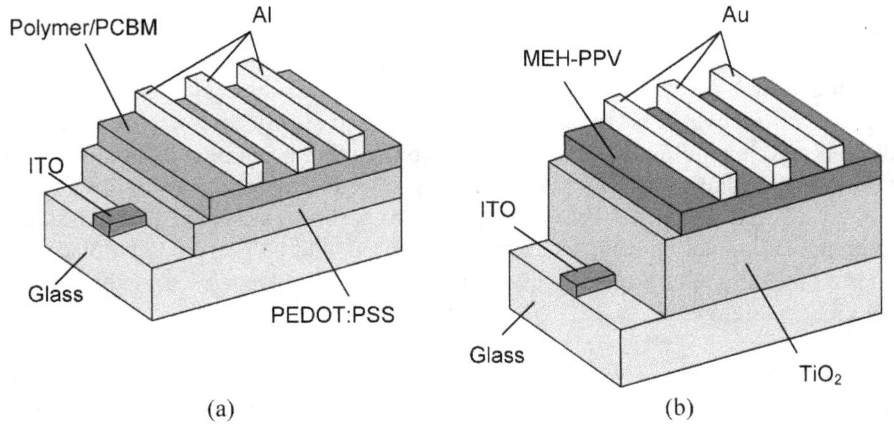

FIGURE 8.13 Schematic diagram of (a) a bulk heterojunction device and (b) an MEH-PPV/TiO_2 bilayer device, as a typical bilayer device.

with other layers such as electrodes, or barrier layers for encapsulation from the atmosphere, some of which may best be deposited in vacuum.

For the lab process, ITO-coated glass is patterned using acid etching to form the transparent bottom electrode. PEDOT:PSS is often applied by spin coating to provide a good electrical contact between the rough surface of the ITO and the polymer deposited on top. For a bulk heterojunction device, the polymer or polymer:inorganic blend layer is spin coated on top. If a TiO_2 layer is used, it is typically applied by doctor blading onto patterned ITO (screen printing or spray pyrolysis can be used especially for larger scale processes), followed by high-temperature sintering before the hole-conducting layer is applied (e.g., by spin coating a polymer solution). Au, Al, or Ca/Al top electrodes are applied by thermal evaporation. A schematic drawing of a complete device is shown in Fig. 8.13. Devices are typically manufactured within a glove box with extremely low levels of water vapor and oxygen present to prevent degradation in properties of the materials.

8.3.5 Photophysical Measurements on Device Materials

8.3.5.1 UV–Visible Absorption Spectrum The absorbance of a material as a function of the wavelength of the incident light can give valuable information about the energy levels within the material. The majority of semiconducting polymers have a bandgap (i.e., HOMO–LUMO gap) greater than 2 eV and typically absorb only in the range of 1.9–3.5 eV [8], corresponding to wavelengths between 650 and 350 nm. Thus, the absorption in the near-ultraviolet–visible region of the electromagnetic spectrum (the UV–visible absorption spectrum) should contain most information about those energy levels in the material that are relevant to photovoltaic device operation.

Light from a source (e.g., tungsten halogen for visible light and deuterium for UV light) is incident on the monochromator before passing through the exit slit. This monochromated light is then split into two beams, one of which passes through the sample under study and the other through a reference sample. The transmitted beams are then incident on silicon photodiode detectors. The difference in the light intensity arriving at these two detectors gives the absorption of the solution/thin-film sample at a given wavelength. The monochromator is rotated slowly to sweep between the desired wavelength limits. UV–visible absorption spectra can be measured for both thin-film and solution samples. For thin films, since the substrate can absorb a small amount of light in the UV–visible region, a bare substrate should be used as the reference. For solution samples, dilute solutions should be prepared and placed in quartz cuvettes, with a quartz cuvette filled with the solvent used in the sample solution as a reference.

8.3.5.2 Photoluminescence Spectrum The plot of the light emitted from a sample as a function of wavelength of the emitted light is referred to as the sample's luminescence spectrum. If the light is emitted following absorption of incident light, the luminescence is referred to as photoluminescence. Like the UV–visible absorption spectrum, the photoluminescence spectrum of a material can give valuable information about the electronic energy levels within the material. In conjunction with the UV–visible absorption spectrum, it can also give some information about energy loss mechanisms within the material, an energy loss showing up as a shift in wavelength between absorbed light and emitted light. This is referred to as a Stokes shift [70].

8.3.5.3 Photoluminescence Quantum Yield The photoluminescence quantum yield (PLQY) of a material is defined by Equation 8.1.

$$\eta = \frac{N_{\mathrm{em}}}{N_{\mathrm{abs}}} \tag{8.1}$$

where N_{abs} and N_{em} are the number of photons absorbed by the sample and subsequently emitted by the sample, respectively. In practice, it is difficult to measure all of the light emitted by or reflected from a sample in all possible directions. It is much more experimentally convenient to measure the light emitted over a solid angle Ω and to use the knowledge of the angular distribution of photon emission from the sample to determine the total number of photons emitted. This is easily done for measurements on samples in solution, as such systems generally have an isotropic angular distribution of emission. For thin-film polymer samples, waveguiding effects and the anisotropic orientation of chromophores can lead to a much more complex angular dependence of emission and reflection from the sample [71]. To avoid measuring the complex angular dependence of emission, an integrating sphere can be used. When light is emitted from a source inside an ideal integrating sphere, the light is redistributed isotropically over the inside surface of the sphere, regardless of the original angular distribution of light. Thus, by measuring the flux of power over a solid angle Ω, the total number of photons emitted, N, can be calculated using Equation 8.2:

$$N = N_{\Omega}\left(\frac{4\pi}{\Omega}\right) \tag{8.2}$$

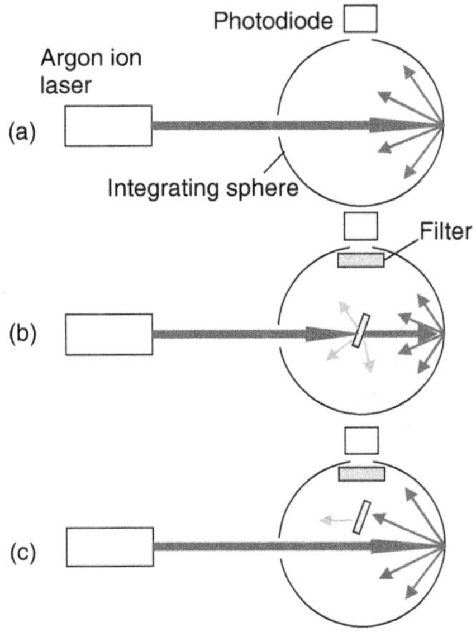

FIGURE 8.14 Schematic diagram of an experimental setup used to measure PLQY. (a) The laser intensity is measured without the filter in place, and then (b) the emission intensity following direct excitation and (c) excitation by scattered light are measured with the filter in place (see text for details).

Figure 8.14 shows the experimental setup used in measuring the PLQY of thin-film samples. Laser light of a suitable wavelength enters the integrating sphere through an aperture in the side of the sphere. The detector is placed at another aperture (e.g., in the top of the sphere), with a (low-pass) filter in front of the detector to prevent scattered laser light from reaching the detector. According to Equation 8.1, we must measure two quantities to calculate the PLQY: the first is the number of photons absorbed by the sample, and the second is the number of photons emitted by the sample following absorption. To determine how much light is absorbed by the sample, we need to know the number of photons incident on the sample. It is more convenient to measure the number of photons incident on the sample multiplied by the solid angle $\Omega/4\pi$, denoted X_{laser}, for reasons that will become clear below. We also need to know the fraction of the incident power that is reflected, R, and the fraction of the incident power that is transmitted, T. X_{laser} is measured by allowing the laser to enter the integrating sphere when it is empty and with the filter removed (Fig. 8.14a). The power arriving at the detector is measured and is converted into the number of photons (per unit time) arriving at the detector using Equation 8.3:

$$X_{\text{laser}} = \frac{P\lambda}{hc} \tag{8.3}$$

R and T are measured outside the sphere using an external power meter. First, the laser beam is shone onto the power meter directly to measure the incident laser power.

The laser beam is then directed onto the sample, and the power meter is positioned to measure the reflected or transmitted power, respectively. The reflected (transmitted) power is then divided by the incident power to obtain R (T). Once R, T, and X_{laser} are known, the number of absorbed photons is given by Equation 8.4:

$$N_{\text{abs}} = (1-R-T)X_{\text{laser}}\left(\frac{4\pi}{\Omega}\right) \tag{8.4}$$

The number of photons emitted following absorption is measured by placing the sample in the integrating sphere and illuminating it with the laser (Fig. 8.14b). The power measured by the detector, P_{sample}, is proportional to the emitted power. The emitted power must be converted into the number of emitted photons using the photoluminescence spectrum of the sample, which has been measured previously. The area under the PL spectrum is set equal to the emitted power (P_{sample}), and the PL curve is then scaled by λ/hc. The area under this scaled curve is equal to the number of photons emitted per unit time, denoted X_{sample}. However, there is one slight complication: the laser light that is transmitted by or reflected from the sample is diffusely reflected off the walls of the sphere, and a portion of this diffuse light will be absorbed by the sample. If this effect was not taken into account, it would lead to an underestimate of the absorbed power and thereby an overestimate of the sample PLQY. As it is difficult to directly measure the amount of diffuse laser light absorbed by the sample, the additional fluorescence created by this additional absorption is measured instead, and this value is subtracted from X_{sample}. The sample is placed in the sphere, but out of the path of the laser (Fig. 8.14c). The laser is incident on the wall of the sphere and is diffusely reflected onto the sample. The fluorescent power generated is measured and is denoted P_{sphere}. X_{sphere} is then calculated in the same way as X_{sample}. Only a fraction of this power is generated during the X_{sample} measurement, as a portion of the laser light has already been absorbed and does not appear as diffuse light. The total number of photons emitted is given by Equation 8.5:

$$X_{\text{emm}} = (X_{\text{sample}}-(R+T)X_{\text{sphere}})\left(\frac{4\pi}{\Omega}\right) \tag{8.5}$$

The PLQY can then be calculated by dividing Equation 8.5 by Equation 8.4, giving Equation 8.6:

$$\eta = \frac{X_{\text{sample}}-(R+T)X_{\text{sphere}}}{(1-R-T)X_{\text{laser}}} \tag{8.6}$$

8.3.5.4 Time-Resolved Photoluminescence In time-resolved photoluminescence (TRL) spectroscopy, the sample is excited by a laser pulse and the sample luminescence is measured as a function of time and wavelength. This can give valuable information about the decay pathways (e.g., radiative versus nonradiative) available to photogenerated excitons, which can help identify the mechanisms limiting device efficiency. A typical experimental apparatus consists of a 400 nm frequency-doubled Ti:sapphire laser with a pulse duration of ~100 fs. The sample is kept under vacuum and excited through a quartz window, while the luminescence (as a function of time and wavelength) is detected with a streak camera. The instrumental response function must be shorter than the lifetime of the sample luminescence. The laser beam is deflected to a different spot on the sample after each measurement to avoid degradation/burning of the illuminated region.

8.3.6 Photovoltaic Device Testing and Characterization

Typically photovoltaic devices are tested by measuring the current–voltage (*I–V*) characteristics in the dark and under white light illumination from a solar simulator with typical illumination power of 100 mW/cm^2. A UV filter might be employed if some components of the device respond unfavorably to high-energy radiation. These measurements help to determine power generation capabilities of the solar cell, as well as to identify whether defects (e.g., shorts and barriers) are present in the device. The wavelength dependence of photocurrent generation can be probed using incident photon conversion efficiency (IPCE) measurements. While slightly more complicated than simple current–voltage measurements, IPCE measurements can help determine whether photon energy-dependent effects such as absorption, exciton transport, or exciton dissociation are limiting device performance.

8.3.6.1 Current–Voltage Measurements The AM1.5 spectrum is the solar radiation that a solar simulator aims to reproduce. The samples are typically illuminated within a test cell that serves two purposes: first, to contain pin contacts to make electrical contact between the bottom (e.g., ITO) and top (e.g., Al) contacts and the outside of the cell, allowing the measurement apparatus to be quickly and securely connected to the device, and for several pixels of device to be tested individually. Second, it serves to isolate the cell from the ambient environment and allow testing under vacuum. Any device constructed under controlled environment should be loaded into the test cell and sealed before being removed from the glove box and positioned in front of the lamp. The cell can then be evacuated to a pressure of, for example, 10^{-5} to 10^{-6} mbar before being illuminated. The test cell is connected to a digital multimeter for *I–V* testing. Care needs to be taken about the area of sample to be illuminated. If, as is often the case, the absorbing species extends between different pixels to be tested (defined by the overlapping area of the two electrodes), it is possible that additional current would be harvested from outside the pixel area if this area is illuminated. This would mean that the current density calculated could be overestimated, as the area of current generation is in fact larger than the nominal area of a pixel.

Figure 8.15 shows a typical *I–V* curve generated. Typical device parameters extracted from these device measurements are the following:

- *Short-Circuit Current Density (J_{SC})*: The short-circuit current is the current produced when there is no load across the device.

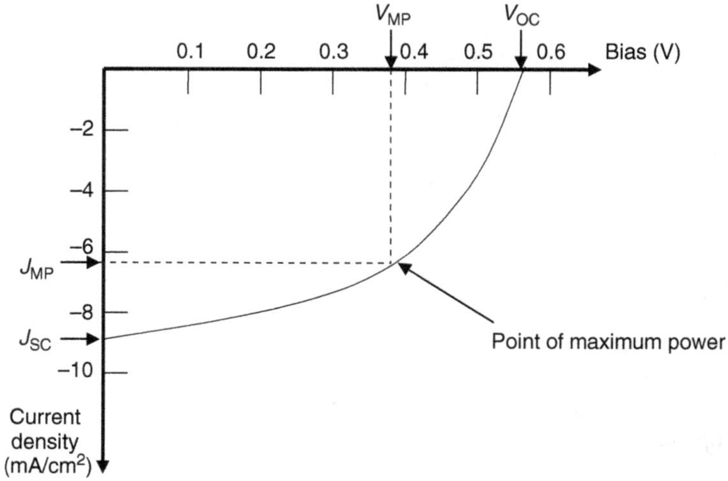

FIGURE 8.15 Typical *I–V* curve measured from a P3HT/fullerene PV device showing rectifying behavior. The key device parameters, V_{OC}, J_{SC}, the maximum power voltage (V_{MP}), and current density (J_{MP}), are indicated.

- *Open-Circuit Voltage* (V_{OC}): The open-circuit voltage is the voltage measured across the device when the current flow is zero (i.e., corresponding to an infinite load).
- *Fill factor (FF)*: The fill factor is a measure of the rectification of the device (essentially its "squareness"). It is defined as the ratio of the maximum power (where the product of the current and the voltage is a maximum) of the device to the product of the short-circuit current and open-circuit voltage.
- *Power Conversion Efficiency* (η_{max}): The power conversion efficiency is the ratio of the power produced by the cell at its maximum power point to the solar power incident on the device.

The quantum efficiency of a device is the ratio of the number of charge carriers collected by the solar cell to the number of photons of a given energy shining on the cell surface. Two types of quantum efficiencies are often quoted: the internal quantum efficiency (IQE) that includes the effects of losses through light transmission and reflection and the external quantum efficiency (EQE) that excludes these effects and instead considers only the fraction of photons that were actually absorbed.

In measuring the *I–V* characteristics, care should be taken about the voltage and illumination history to which the device has been exposed. Sometimes, the short-circuit current or open-circuit voltage can drift slightly under illumination and/or bias, for example, caused by photoconductivity increases in the TiO_2 [72] and/or the buildup of space charge in the polymer [73]. While the appropriate limits of the voltage sweep vary depending on the device under study, generally the limits are in the range of forward bias of at least 1 V (to study field-assisted charge extraction) and reverse bias of at least 1 V beyond the open-circuit voltage of the device (to study reverse-bias injection). The forward bias direction is taken to be the direction that assists photocurrent flow, while reverse bias is the direction that opposed the flow of the photocurrent. Between the short-circuit current ($V_{app} = 0$) and open-circuit voltage ($V_{app} = V_{bi}$) points of the curve, the applied voltage is meant to simulate the effect of a varied external load on device performance and to find the optimum power output of the cell.

I–V curves should be measured in the dark, before and after illumination, both to determine whether any device has developed microscopic shorts (during either device fabrication or testing) resulting in a low shunt resistance and to determine whether the test cell pins are properly contacting the relevant contacts on the device. Poor contact by these pins would result in a large series resistance and mask the behavior of the device under study.

8.3.6.2 Incident Photon Conversion Efficiency The incident photon conversion efficiency is defined as the number of photoelectrons collected per photon incident on the sample and is sometimes referred to as the external quantum efficiency. The efficiency of photocurrent generation as a function of wavelength can provide valuable insight into the mechanism(s) limiting device efficiency. For example, in DSSCs, improved injection from vibrationally relaxed excited states following a downward shift in the TiO_2 conduction band shows up as an increased photocurrent at long wavelengths, where photon energies are insufficient to promote electrons to a vibrationally hot excited state [74]. A comparison of the IPCE spectrum with the absorption spectrum of the component materials in a blend film can also help determine whether both components of a polymer/nanocrystal [75] or polymer/fullerene [36] blend are contributing effectively to the photocurrent. In a bilayer film, comparison of the IPCE and absorption spectra can also indicate which interface serves to dissociate photogenerated excitons [76].

A typical arrangement for IPCE measurements is shown schematically in Fig. 8.16. A light source should be fitted with a matched monochromator to adjust the output wavelength. The output from the monochromator is incident onto a beam splitter. The reflected portion of the beam is incident onto a cosine corrector to minimize the effect of slight angular perturbations of the detector or beam splitter, and then into a spectrometer. The sample is tested under similar vacuum conditions to those in the *I–V* measurement. The output wavelength of the monochromator is swept slowly (e.g., in 10–20 nm increments every 10–30 s) and the photocurrent measured and recorded as a function of wavelength, while the incident power measured by the spectrometer is recorded simultaneously. The photocurrent is converted to the number of electrons generated per unit time, and the number of photons incident on the sample per unit time is calculated using the previously measured transmission/reflection ratio as a function of wavelength. The IPCE is then calculated, as a function of wavelength, as follows:

$$\mathrm{IPCE}(\lambda) = \frac{1.6 \times 10^{19} J(\lambda) hc}{I(\lambda)\lambda} \tag{8.7}$$

A typical IPCE trace is shown in Fig. 8.17

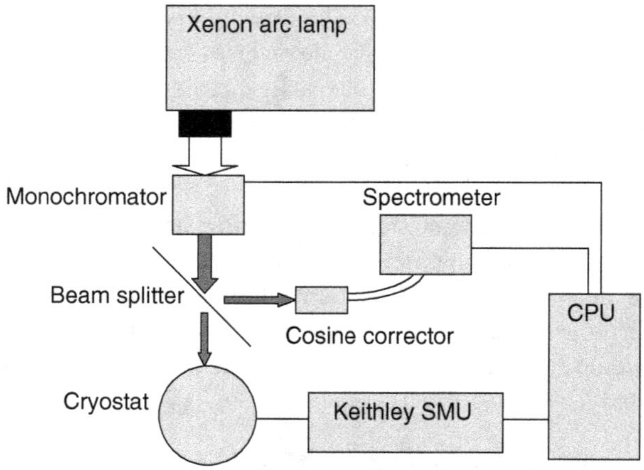

FIGURE 8.16 Schematic drawing of experimental setup used for incident photon conversion efficiency measurements.

8.3.6.3 Mobility Measurements on Hole-Only Diodes Hole-only diodes can be made, for example, by spin coating a layer of hole-conducting polymer material, with a thickness in the range 100 nm–1 μm, onto an ITO-coated glass substrate, and then evaporating Au contacts. *I–V* curves are taken without illumination with the voltage limits set such that the electric field across the polymer is space charge limited (e.g., $\sim 5 \times 10^7$ V/m for MEH-PPV). A field-dependent space charge-limited model can then fit to the data to extract the zero-field hole mobility:

$$J = \frac{9}{8}\varepsilon_r\varepsilon_0\mu(E)\frac{V^2}{L^3} \tag{8.8}$$

$$\mu(E) = \mu_0\, e^{0.89\sqrt{E/E_0}} \tag{8.9}$$

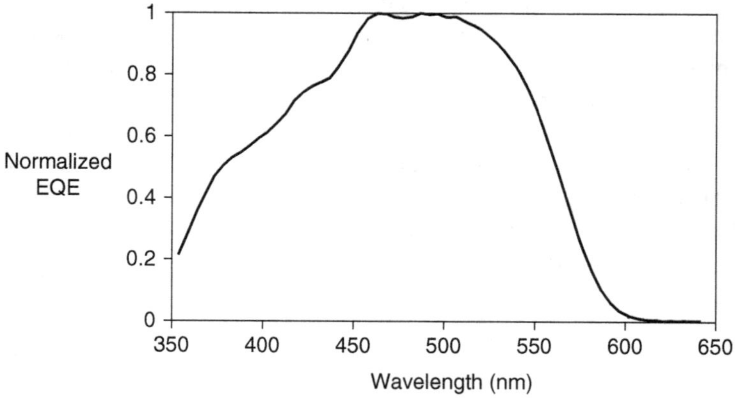

FIGURE 8.17 Typical IPCE trace from an ITO/TiO2/MEH-PPV/Au device.

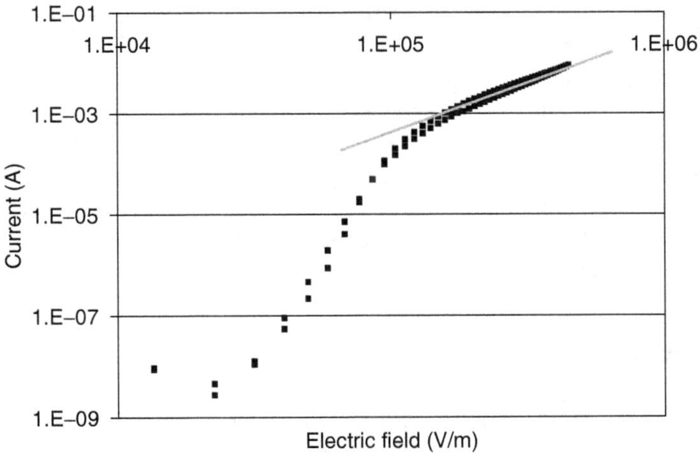

FIGURE 8.18 Typical *I–E* curve for a hole-only diode exhibiting space charge-limited behavior. The gray line has a slope of 2 on the log–log scale and is meant as a guide to the eye.

The zero-field mobility, μ_0, is obtained by performing a linear fit to a plot of ln (J/E^2) versus $E^{1/2}$ in the space charge-limited region defined above. Assuming a suitable value for ε_r, the zero-field hole mobility can be calculated from

$$\mu_0 = \frac{8}{9}\frac{L}{\varepsilon_r\varepsilon_0}\,e^{\zeta} \tag{8.10}$$

where ζ is the intercept of the linear fit with the $\ln(J/E^2)$ axis. Figure 8.18 shows a typical *I–E* plot for an MEH-PPV device with linear fit to the space charge-limited region in which Equation 8.8 is fulfilled.

8.4 CONCLUSION

Given the concerns over global warming and energy supply, photovoltaic materials are now capturing a lot of interest for development and commercialization. Photovoltaics represent the most direct way in which the sun's energy can be captured to generate electrical power. The most established technology in the marketplace is that of Si, but there is considerable interest in flexible PV materials that can be constructed in a semicontinuous roll-to-roll process. Thin-film technologies include inorganic semiconductors on polymer or other flexible substrates, but organic-based devices are also attracting a lot of attention for their prospects of low-cost processing. However, the drive for higher efficiency devices in any material system is an ever present one, and, as the details of the mechanisms of operation of organic devices differ in crucial respects from their inorganic counterparts, it is important to characterize and understand the various steps in the conversion of light into a flow of charge. Improvements in device efficiency have been rapid and it remains to be seen how much further we can go. The state-of-the-art devices make the technology an attractive one to explore commercially, and so now there is considerable growing interest in the processing and stability of devices, and their operation under realistic conditions, alongside the development of new chemistries and device microstructures and geometries. The materials used also need to be suitable for widespread use on grounds of cost (e.g., the current drive to reduce or remove the use of indium in the transparent conducting layer) and environmental impact. It is an exciting time in the development of organic photovoltaics, and it is hoped that such materials can make a significant impact on energy generation in the future.

REFERENCES

1. World Total Electricity Installed Capacity, Energy Information Administration, http// www.eia.doe.gov/pub/international/iealf/table64xls, 2007.
2. World Total Installed Generating Capacity by Region and Country, 2004–2030, Energy Information Administration, http//www.eia.doe.gov/oiaf/ieo/excel/ieoecgtab_1xls, 2007.
3. IPCC, 2007, Summary for policymakers in: M.L. Parryet al. (Eds.), *Climate Change 2007: Impacts, Adaptation and Vulnerability. Contribution of Working Group II to the Fourth Assessment Report of the Intergovernmental Panel on Climate Change*, Cambridge University Press, Cambridge, UK, 2007, pp. 7–22.
4. S.E. Shaheen, D.S. Ginley, and G.E. Jabbour, *MRS Bull.* **2005**, *30*(1), 10–19.
5. M.A. Green, K. Emery, Y. Hisikawa, and W. Warta, *Prog. Photovolt.* **2007**, *15*(5), 375–468.
6. Solar Module Price Highlights: 2009, Solar Buzz, http://www.solarbuzz.com/ Moduleprices.htm, 2009.
7. G.R. Webster, W.J. Mitchell, P.L. Burn, R.K. Thomas, G. Fragneto, J.P.J. Markham, and I. D.W. Samuel, *J. Appl. Phys.* **2002**, *91*, 9066.
8. K.M. Coakley and M.D. McGehee, *Chem. Mater.* **2004**, *16*(23), 4533–4542.
9. J. Nelson, *The Physics of Solar Cells*, Imperial College Press, London, 2003.
10. Z. Zhu, D. Waller, R. Gaudiana, M. Morana, D. Mühlbacher, M. Schorber, and C. Brabec, *Macromolecules* **2007**, *40*(6), 1981–1986.
11. J. Peet, J.Y. Kim, N.E. Coates, W.L. Ma, D. Moses, A.J. Heegar, and G.C. Bazan, *Nat. Mater.* **2007**, *6*, 497–500.
12. W.S. Shin, S.C. Kim, S.J. Lee, H.S. Jeon, M.K. Kim, B.V.K. Naidu, S.H. Jin, J.K. Lee, J. W. Lee, and Y.S. Gal, *J. Polym. Sci. Part A: Polym. Chem.* **2007**, *45*(8), 1394–1402.
13. M.M. Wienk, M.P. Struijk, and R.A.J. Janssen, *Chem. Phys. Lett.* **2006**, *422*(4–6), 488–491.
14. I.H. Campbell and D.L. Smith, *Solid State Phys.: Adv. Res. Appl.* **2001**, *55*, 1–117.
15. T.B. Singh and N.S. Sariciftci, *Annu. Rev. Mater. Res.* **2006**, *36*, 199–230.
16. H.J. Snaith and M. Gratzel, *Appl. Phys. Lett.* **2006**, *89*(26), 262114.
17. B.A. Gregg and M.C. Hanna, *J. Appl. Phys.* **2003**, *93*(6), 3605–3614.
18. J. Xue, S. Uchida, B.P. Rand, and S.R. Forrest, *Appl. Phys. Lett.* **2004**, *84*(16), 3013–3015.
19. S. Gunes, H. Neugebauer, and N.S. Sariciftci, *Chem. Rev.* **2007**, *107*(4), 1324–1338.

20. J. Nelson, J. Kirkpatrick, and P. Ravirajan, *Phys. Rev. B* **2004**, *69*(3), art. no. 035337.
21. R. Pacios and D.D.C. Bradley, *Synth. Met.* **2002**, *127*(1–3), 261–265.
22. Y.S. Liu, L. Wang, D.H. Qin, and Y. Cao, *Chin. Phys. Lett.* **2006**, *23*(12), 3345–3348.
23. Y. Zhou, Y. Li, H. Ahong, J. Hou, Y. Ding, C. Yang, and Y. Li, *Nanotechnology* **2006**, *17*(16), 4041–4047.
24. D.H. Cui, J. Xu, S.Y. Xu, G. Paradee, B.A. Lewis, and M.D. Gerhold, *IEEE Trans. Nanotechnol.* **2006**, *5*(4), 362–367.
25. S. Gunes, H. Neugebauer, N.S. Sariciftci, J. Roiter, M. Kovalenko, G. Pilwein, and W. Heiss, *Adv. Funct. Mater.* **2006**, *16*(8), 1095–1099.
26. D.H. Cui, J. Xu, T. Zhu, G. Paradee, S. Ashok, and M. Gerhold, *Appl. Phys. Lett.* **2006**, *88*(18), 183111–183114.
27. D. Gebeyehu, B. Maenning, J. Drechsel, K. Leo, and M. Pfeiffer, *Sol. Energy Mater. Sol. Cells* **2003**, *79*(1), 81–92.
28. J.G. Xue, B.P. Rand, S. Uchida, and S.R. Forrest, *J. Appl. Phys.* **2005**, *98*(12), 124903.
29. P.W.M. Blom, V.D. Mihailetchi, L.J.A. Koster, and D.E. Markov, *Adv. Mater.* **2007**, *19*(2), 1551.
30. J. Drechsel, B. Männing, D. Gebeyehu, M. Pfeiffer, K. Leo, and H. Hoppe, *Org. Electron.* **2004**, *5*(4), 175–186.
31. V. Dyakonov, *Phys. E: Low-Dimensional Syst. Nanostruct.* **2002**, *14*(1–2), 53–60.
32. A. Gadisa, M. Svensson, M.R. Andersson, and O. Inganäs, *Appl. Phys. Lett.* **2004**, *84*(9), 1609–1611.
33. B.P. Rand, J. Xue, F. Yang, and S.R. Forrest, *Appl. Phys. Lett.* **2005**, *87*(23), 233508.
34. N.S. Sariciftci, L. Smilowitz, A.J. Heegar, and F. Wudl, *Science* **1992**, *258*(5087), 1474–1476.
35. J.K.J. van Duren, X. Yang, J. Loos, C.W.T. Bulle-Lieuwma, A.B. Sieval, J.C. Hummelen, and R.A.J. Janssen, *Adv. Funct. Mater.* **2004**, *14*(5), 425–434.
36. M.M. Wienk, J.M. Kroon, W.J.H. Verhees, J. Knol, J.C. Hummelen, P.A. van Hal, and R.A.J. Janssen, *Angew. Chem. Int. Ed.* **2003**, *42*(29), 3371–3375.
37. K. Kim, J. Liu, M.A.G. Namboothiry, and D.L. Carroll, *Appl. Phys. Lett.* **2007**, *90*(16), 163511–163513.
38. K.M. Coakley and M.D. McGehee, *Appl. Phys. Lett.* **2003**, *83*(16), 3380–3382.
39. C.A. Grimes, *J. Mater. Chem.* **2007**, *17*(15), 1451–1457.
40. E.J.W. Crossland, M. Kamperman, M. Nedelcu, C. Ducati, U. Wiesner, D.-M. Smilgies, G.E.S. Toombes, M.A. Hilmeyer, S. Ludwigs, U. Steiner, and H.J. Snaith, *Nano Lett.* **2009**, *9*, 2807–2812, doi:10.1021/nl803174p.
41. D.B. Kuang, C. Klein, S. Ito, J.E. Moser, R. Humphrey-Baker, N. Evans, F. Duriaux, C. Grätzel, M. Zakeeruddin, and M. Grätzel, *Adv. Mater.* **2007**, *19*(8), 1133–1137.
42. M. Gratzel, *J. Photochem. Photobiol. C: Photochem. Rev.* **2003**, *4*(2), 145–153.
43. G.K.R. Senadeera and J. Fernando, *J. Mater. Sci. Technol.* **2006**, *22*(6), 811–816.
44. T. Ahn, K. Seung-Wan, J. Lee, and H.K. Shim, *Macromolecules* **2002**, *35*(9), 3495–3505.
45. W.J. Mitchell, P.L. Burn, R.K. Thomas, and G. Fragneto, *J. Appl. Phys.* **2004**, *95*(5), 2391–2396.

46. D.M. Johansson, G. Srdanov, G. Yu, M. Theander, O. Ingariäs, and M.R. Andersson, *Macromolecules* **2000**, *33*(7), 2525–2529.
47. B.H. Cumpston and K.F. Jensen, *J. Appl. Polym. Sci.* **1998**, *69*(12), 2451–2458.
48. K. Colladet, M. Nicholas, L. Goris, L. Lutsen, and D. Vanderzande, *Thin Solid Films* **2004**, *451–452*, 7–11.
49. M. Hanack, B. Behnisch, H. Hächi, P. Martinez-Ruiz, and K.H. Schweikart, *Thin Solid Films* **2002**, *417*(1–2), 26–31.
50. H. Detert and E. Sugiono, *Synth. Met.* **2000**, *115*(1–3), 89–92.
51. R. Jakubiak, Z. Bai, and L. Rothber, *Synth. Met.* **2000**, *114*(1), 61–64.
52. B.J. Schwartz, *Annu. Rev. Phys. Chem.* **2003**, *54*, 141–172.
53. S.H. Lee, T. Yasuda, and T. Tsutsui, *J. Appl. Phys.* **2004**, *95*(7), 3825–3827.
54. T.Q. Nguyen, V. Doan, and B.J. Schwartz, *J. Chem. Phys.* **1999**, *110*(8), 4068–4078.
55. T.Q. Nguyen, R.C. Kwong, M.E. Thompson, and B.J. Schwartz, *Appl. Phys. Lett.* **2000**, *76*(17), 2454–2456.
56. T.Q. Nguyen, I.B. Martini, J. Liu, and B.J. Schwartz, *J. Phys. Chem. B,* **2000**, *104*(2), 237–255.
57. Y. Shi, J. Liu, and Y. Yang, *J. Appl. Phys.* **2000**, *87*(9), 4254–4263.
58. J. Liu, Y.J. Shi, and Y. Yang, *Adv. Funct. Mater.* **2001**, *11*(6), 420–424.
59. C. Brabec, V. Dyakonov, J. Parisi, and N.S. Sariciftci, *Organic Photovoltaics: Concepts and Realization*, Springer, Germany, 2003.
60. C.Y. Kwong, A.B. Djurišić, P.C. Chui, K.W. Cheng, and W.K. Chan, *Chem. Phys. Lett.* **2004**, *384*(4–6), 372–375.
61. J. Liu, T.F. Guo, and Y. Yang, *J. Appl. Phys.* **2002**, *91*(3), 1595–1600.
62. J.J. Dittmer, R. Lazzaroni, P. Leclère, P. Moretti, M. Granström, K. Petritsch, E.A. Morseglia, R.H. Friend, J.L. Brédas, H. Rost, and A.B. Holmes, *Sol. Energy Mater. Sol. Cells* **2000**, *61*(1), 53–61.
63. J. Cabanillas-Gonzalez, S. Yeates, and D.D.C. Bradley, *Synth. Met.* **2003**, *139*(3), 637–641.
64. C.D. Grant, A.M. Schwartzberg, G.P. Smestad, J. Kowalik, L.M. Tolbert, and J.Z. Zhang, *Synth. Met.* **2003**, *132*(2), 197–204.
65. X.N. Yang, J.K.J. van Duran, R.A.J. Janssen, M.A.J. Michels, and J. Loos, *Macromolecules* **2004**, *37*(6), 2151–2158.
66. B. Oregan and M. Grätzel, *Nature* **1991**, *353*(6346), 737–740.
67. G. Redmond and D. Fitzmaurice, *J. Phys. Chem.* **1993**, *97*(7), 1426–1430.
68. M.T. Lloyd, A.C. Mayer, A.S. Tayi, A.M. Bowen, T.G. Kasen, D.J. Herman, D.A. Mourey, J.E. Anthony, and G.G. Malliaras, *Org. Elect.* **2006**, *7*(5), 243–248.
69. D.A.R. Barkhouse, M.J. Carey, Z. Xie, K.R. Kirov, B.M. Henry, G.R. Webster, P.L. Burn, and H.E. Assender, *Org. Elect.* in press.
70. M. Fox, *Optical Properties of Solids. Oxford Master Series in Condensed Matter Physics*, Oxford University Press, New York, **2001**.
71. J.C. deMello, H.F. Wittmann, and R.H. Friend, *Adv. Mater.* **1997**, *9*(3), 230–232.
72. Z.B. Xie, B.M. Henry, K.R. Kirov, D.A.R. Barkhouse, V.V. Burlakov, H.E. Smith, C.R.M. Grovenor, H.E. Assender, G.A.D. Briggs, M. Kano, and Y. Tsukahara, *Nanotechnology* **2007**, *18*(14), 145708.

73. M.J. Carey, V.M. Burlakov, B.M. Henry, K.R. Kirov, G.R. Webster, H.E. Assender, G.A.D. Briggs, P.L. Burn, and C.R.M. Grovenor, *Proc. SPIE: Int. Soc. Opt. Eng.* **2004**, *5215*, 32–40.

74. C.A. Kelly, F. Farzad, D.W. Thompson, J.M. Stipkala, and G.J. Meyer, *Langmuir* **1999**, *15* (20),7047–7054.

75. S. Gunes, K.P. Fritz, H. Neugebauer, N.S. Sariciftci, S. Kumar, and G.D. Scholes, *Sol. Energy Mater. Sol. Cells* **2007**, *91*(5), 420–423.

76. A.C. Arango, L.R. Johnson, V.N. Bliznyuk, Z. Schlesinger, S.A. Carter, and H.H. Hörhold, *Adv. Mater.* **2000**, *12*(22), 1689–1692.

9

ORGANIC LIGHT-EMITTING DIODES

SEBASTIEN FORGET, SEBASTIEN CHENAIS, AND ALAIN SIOVE

9.1 INTRODUCTION

Organic electroluminescent devices have been one of the most highlighted research topics in material sciences during the last two decades. As it is the case for organic electronics, organic *opto*electronics has emerged as a major research field not so much because organic semiconductors would surpass inorganic semiconductors (the latter remain unbeatable as far as integration, speed, and reliability are concerned, for instance), but rather, because organic semiconductors are giving to light sources a new class of unprecedented properties among others, the possibility to realize large and ultrathin "flat sheets" of light (and of course, flat-panel displays), the possibility to control the color over all the optical spectrum by chemically modifying the molecule, or the possibility to make light sources on flexible substrates.

Photochemistry and Photophysics of Polymer Materials, Edited by Norman S. Allen

The first observation of organic electroluminescence was reported several decades ago (1953 under pulsed operation [1] and 1963 in CW [2]); however, the efficiencies were too low to lead to any practical device. A major step was made in the late 1980s with the realization of multilayer thin-film devices with efficiencies over 1%. This work by Tang and Vanslyke, in 1987 [3], who succeeded in fabricating double-layer Organic Light Emitting Diodes with a luminance as high as 1000 cd/m^2 at an operating voltage of 10 V, has spurred on the development of multilayer devices built with small organic molecules. A few years later, Burroughes et al. [4] reported electroluminescence from conjugated polymers (e.g., poly(phenylene vinylene) [PPV]) for the first time, boosting polymer chemistry and physics toward the development of numerous organic light-emitting diodes based on conjugated materials, including both low molecular weight compounds and polymers. This has been at the origin of an "exploding" research concerning polymers, guided by the hope of developing plastic, large-area, full-color, ultra-thin, flat-panel displays, rollable electronic newspapers, wall-hanging color television, and house-lighting.

The challenge for the chemist has been to synthesize π-conjugated polymeric materials that fulfilled most of the following requirements: (1) the targeted emission color, (2) a high-quantum yield of fluorescence (or phosphorescence), (3) high carrier mobilities and above all balanced mobilities for electrons and holes (especially important for single-layer devices), (4) a high purity, (5) good thermal and chemical stabilities, and (6) good film-forming ability for spin-coating processing. All these factors are decisive in the electroluminescence performance and durability of the devices. Along these lines, controlled and regiospecific polymerization processes together with defects-free structures and absence of catalytic residues are required, high molecular weight and well-soluble materials are needed, length of π-conjugation and energy bandgap have to be adapted to the desired color light, and good electron affinity is aimed. Among those requirements, electronic factors are obviously more crucial to be attained. Different strategies have been adopted to vary the emission color and the transport properties of π-conjugated polymers. One of them consists in introducing in the macromolecular chains, appropriate electron donor or acceptor substituents shifting the π-electrons cloud toward red or blue emission. Another strategy implies copolymerization pathways incorporating comonomers of suitable electronic properties. The "isolated chromophores" (main-chain segments or side-chains segments as well) and "kinked linkages" approaches have also been adopted to control the extent of π-conjugation and thereby the emission color.

The aim of this chapter is to give a general survey of the most widely studied electroluminescent homopolymers and more specifically those utilizable for the red, green, and blue emission colors. The review is organized as follows. In the first part, some of the main physical processes that take place in organic light-emitting diodes (OLEDs)/polymer light-emitting diodes (PLEDs) as well as the classical architectures used to realize efficient devices are briefly described. Then, the chemist point of view is adopted in the second part by looking over some of the numerous homopolymers used in the PLED field. The consequences of the chemical design of both the macromolecular chain and the substituents attached onto the monomer unit on structure–properties relationships are examined, particularly through the emitted color aspect. Effects of structural regularity and purity on luminescence properties are also presented. Finally, in the last part, some recent results in the increasingly studied field of white light organic emitters are described.

The present overview is done without claiming to be exhaustive, given the tremendous literature covering the field, particularly that is related to copolymers, the synthesis and properties of which are not addressed herein. Thus, the reader might refer to general reviews [5–11], and to the book of Shinar published in 2004 [12] and the one, in 2006, of Müllen [13] and Scherf, both on OLEDs. More specialized reviews on polymer families for light-emitting devices have also been published as, for example, those relative to poly(fluorene)s [14–19], poly(carbazole)s [20–22], and poly(phenylene)s [22–28] as blue light-emitting candidates [10], those concerning poly(phenylene vinylene)s and their derivatives [4,9,25,29–37] utilizable as emitting materials within the color domain spanning from green to red, and those concerning poly(alkylthiophene)s [5,8,38–40] specifically dedicated to red emission.

9.2 POLYMER LIGHT-EMITTING DIODES: PHYSICAL PROPERTIES

Two kinds of organic devices are distinguished, depending of the nature of the organic material used: on the one hand, the "small molecules" (SMOLEDs standing for Small Molecule OLEDs) and on the other hand, PLEDs. As far as physical processes are concerned, most properties are common to both devices; the generic term "OLED" will be used in the following unless otherwise specified. A difference will emerge from the complexity of the devices. Indeed, OLEDs based on small molecules are fabricated by thermal evaporation under high vacuum, so that there is theoretically no limit on the number of deposited layers; this is not the case for PLEDs that are too heavy to be evaporated and those that are thus deposited under the form of films by spin-casting techniques: the number of layers is then limited by the miscibility of the solvents.

The structure of a typical OLED is shown in Fig 9.1. Several organic semiconductors are sandwiched between two electrodes. The anode (most of the time made of indium tin oxide [ITO]) is transparent in the visible range to allow the light escaping, while the cathode is a low-work function metal (Ca, Mg:Ag, and Al). When one is interested in making practical devices, it is clear that a single molecule or polymer cannot gather all the qualities required for building an efficient OLED (that is being at the same time a good light emitter, exhibiting high and balanced mobilities for electrons and holes, and having the adequate energy levels for injection of both types or carriers). As a result, carriers with the highest mobility (e.g., holes) will recombine with opposite carriers at the vicinity of the counter-electrode, resulting in quenching and low quantum efficiency. Multilayer devices are then used to push the recombination zone away from the electrodes, as well as to modify the internal electric fields (in virtue of charge accumulation at the interfaces), and thus, modifying the injection characteristics, and also to have more degrees of freedom in the design by assigning different materials to different roles.

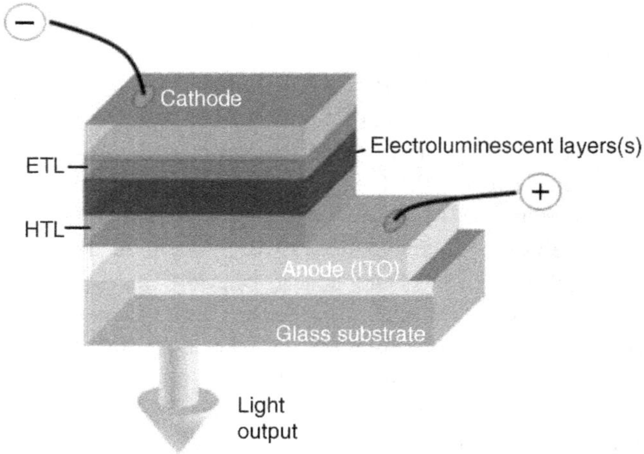

FIGURE 9.1 Schematic of a typical OLED structure.

Thus, one will use some materials referred to as electron/hole transport layer (ETL/HTL), which will be designed to exhibit a high electron/hole mobility, and accordingly electroluminescent material (EML), and electron/hole injection layer (EIL/HIL).

The operating principle of OLEDs is very different from conventional p–n junction LEDs. As will be illustrated below, the nature of charge transport is fundamentally different because it occurs by hopping between localized states rather than by coherent motion within extended bands. Another important difference is that p or n-doping is not mandatory, as opposed to inorganic LEDs. Although doping of organic semiconductors has brought significant progress in the performance of LEDs recently [41], a p–n junction is in general not realized in practical devices. The rectification behavior of an OLED is due to the difference in the work functions of the electrodes used. Let us examine as an example the OLED described in Fig. 9.2 to give a very quick overview of the device operation. Under zero bias, the Fermi levels of the two metals align, causing a built-in voltage to appear across the organic materials equal to the difference between electrodes work functions. Assuming a negligible concentration of free carriers, the potential inside the organics is the one that we should expect in a capacitor, that is, linear with position. The flat-band condition is reached when the applied voltage exceeds the difference in the work functions (a few volts in general), this also corresponds approximately to the threshold of the device. The application of a forward bias above this value causes carriers to be *injected* in the material. Holes will be injected from ITO to the HOMO level of the HIL. The role of this first layer (a classical material used here is copper phtalocyanine [CuPc]) is to lower the energy barrier to facilitate hole injection into the HTL, which exhibits a high hole mobility. On the other side, electrons are injected from the cathode to an ETL, also being luminescent (EML) in this case (the archetypal material being Alq_3). Those layers transport holes and electrons efficiently until they accumulate on both sides of the HTL/ETL interface because of the energetic and mobility barrier.

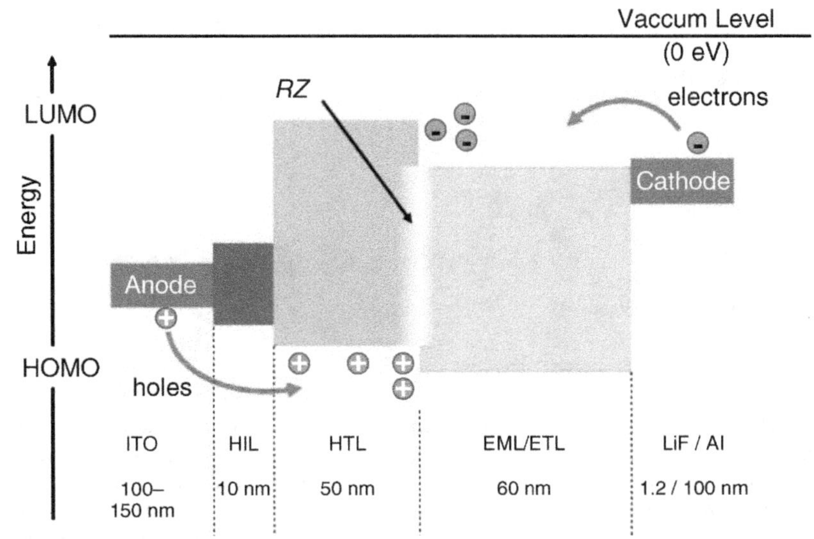

FIGURE 9.2 Energetic diagram of a typical OLED.

The localized high concentration of carriers with opposite charges is favorable for exciton formation in the vicinity of the interface. This site is the so-called recombination zone (RZ). The emitted photons are then the result of the desexcitation of those excitons, with energy equal to the excitonic gap.

The different processes evoked above will be described in detail.

9.2.1 Electronic Transport

9.2.1.1 Injection Mechanisms Consider the injection of one type of carrier (say electrons) in the LUMO of an organic semiconductor. A first approximation of what happens can be obtained by assuming a vacuum-level alignment through the metal/semiconductor interface. The injection barrier is determined by the difference between the electron affinity (or the ionization potential for holes) and the electrode work function. However, this simple picture deviates from reality by two aspects. The first one is the barrier lowering by the external electric field in virtue of the Schottky effect. Indeed, as an electron is extracted from the metal, it interacts with the remaining free electrons and undergoes an "image potential" that adds up to the linear potential of the electric field; the effective barrier with this combined potential is lowered in energy with respect to the bare barrier by a factor $\propto \sqrt{F}$ ($F =$ applied electric field).

The second aspect is the interaction between the first monolayers of organic molecules and the metal that can induce a significant dipole, meaning that the simple rule of vacuum-level alignment is no longer valid [42]. There have been endless debates about the physical nature of injection from a metal to an organic semiconductor, and one can still find numerous contradictory conclusions in textbooks or review articles.

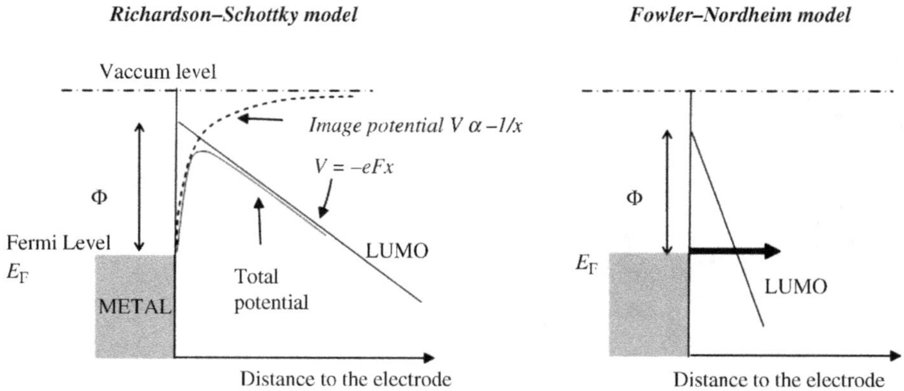

FIGURE 9.3 Schematic presentation of the limit models used to describe injection in organic semiconductors. **Φ** is the barrier for electrons in absence of field, that is, the difference between the metal work function and the electronic affinity of the organic material. Dipoles at the metal/organic interfaces are not considered here, so that vacuum levels are aligned. (*Left*) Richardson–Schottky mechanism. The carriers are injected by thermoionic emission and then drift/diffuse to the top of the total potential (obtained by adding the image potential and the potential imposed by the applied electric field *F*). (*Right*) Fowler–Nordheim mechanism. Injection occurs via tunneling. The image force is neglected.

Two limit models are considered (see Fig. 9.3), the first is the Fowler–Nordheim mechanism, which assumes tunneling through a perfect triangular barrier (the image potential is neglected) and does not consider the localized aspect of the sites. The main characteristic of this model is that the current density is independent of the temperature and scales as a function of the electric field as follows:

$$J_{\text{Fowler-Nordheim}} \propto F^2 \exp\left(-\frac{b}{F}\right)$$

where b is a constant. On the other hand, the Richardson–Schottky model assumes thermoionic (or thermoelectronic) emission. In this model, the current strongly depends on the temperature. The dependence on F will reflect the fact that thermoionic emission will be assisted by the electric field, via the barrier lowering evoked above. Current can be written as:

$$J_{\text{Richardson-Schottky}} \propto T^2 \exp\left(-\frac{\chi-\beta\sqrt{F}}{k_B T}\right)$$

where χ is the bare energetical barrier and β is a constant.

Whereas early publications have explained experimental results within the framework of the Fowler–Nordheim mechanism alone [43], recent publications [44–46] have attributed the injection of carriers to a combination of both mechanisms: at low fields and high temperature, thermoionic emission is considered dominant. On the contrary, for high electric fields (typically >2 MV/cm), injection would essentially occur via tunneling.

In fact, both models fail at explaining quantitatively the values of the currents observed [47,48]. The main reason is that the real nature of charge transport via hopping is ignored. When quantitative models are needed, one will have to turn to more elaborated models that take into account the subtleties of transport in organic media, as those developed by Gartstein and Conwell [49] or by Arkhipov et al. [50]. Let us evoke now some of the aspects of transport in organic semiconductors.

9.2.1.2 Transport in Disordered Organic Semiconductors Polymeric or small molecular materials used in OLEDs form disordered amorphous films with no translational long-range symmetry: band-like models for transport, involving the presence of delocalized states over distances much higher than the typical dimension of a molecule, and are not pertinent in this context. Charge carriers are transported by hopping, characterized by incoherent jumps between isolated molecular sites, which are thought to be the molecules themselves or the conjugated polymer chain segments. The probability for a carrier to jump on a neighboring site will depend on the site energy and distance, and will vary with temperature and electric field. Charge carriers exist in the form of *polarons*: upon adding or removing an electron from a molecule, the molecular skeleton must change because of the readjustment of the individual bond lengths. The neighboring molecules will also be affected and will rearrange to minimize the overall energy. The "polaron" also includes the change in energy due to the electronic polarization of the surrounding sites. The polaron effect, as it further stabilizes the electrical charge on the site, leads to the so-called transport levels, which are below the LUMO and above the HOMO levels, respectively. Furthermore, independent of the polaron concept, Bässler has introduced the idea that hopping occurs between sites whose HOMO and LUMO levels are randomly distributed within a Gaussian density of states (DOS) that reflects the structural disorder [51].

The starting point of all transport models is an expression of the probability W_{ij} for a carrier to jump from site i to site j. Two types of expressions are usually considered: for disordered materials with a weak electron/phonon coupling and at low temperatures, only disorder is considered and the polaronic effects are neglected. In this case the Miller–Abrahams jump rate is used

$$W_{ij} = \upsilon_0 \exp\left(\frac{-2r_{ij}}{b}\right)\exp\left(-\frac{|E_i - E_j| + E_i - E_j}{2k_B T}\right)$$

where υ_0 is the frequency factor, E_i the energy of site i, and b is the localization radius of a charge carrier.

On the contrary, when electron–phonon coupling is strong and at high temperatures, the Marcus theory is preferred [52], given here under its so-called symmetrical jump rate form:

$$W_{ij} = \frac{J^2}{\hbar}\sqrt{\frac{\pi}{2E_p k_B T}}\exp\left(-\frac{E_p}{2k_B T}\right)\exp\left(\frac{E_i - E_j}{2k_B T}\right)$$

where J is the transfer integral between sites and E_p is the polaron binding energy. From this, one can obtain using, for example, the Monte-Carlo [53], Master Equation [54], or the effective medium approach [55], a macroscopic number, the *mobility*, which is defined as the average drift velocity of carriers per unit electric field. The hole mobilities are typically $10^{-7} - 10^{-3}\,\mathrm{cm}^2/(\mathrm{V\,s})$, while the electron mobilities are generally found to be lower by one or two orders of magnitude. Mobilities in disordered materials are lower than in organic crystals (e.g., a hole mobility as high as $20\,\mathrm{cm}^2/(\mathrm{V\,s})$ has been reported in rubrene [56]) and far much lower than in inorganic crystals (for silicon $\mu \sim 10^3\,\mathrm{cm}^2/(\mathrm{V\,s})$).

Many studies have focused on the dependence of mobility on temperature and electric field [53,57,58], and more recently on carrier density [59]. Lets see briefly how these three parameters physically influence the mobility.

The mobility increases with temperature: this behavior is the opposite of what is observed in crystals, and provides a clear proof that conduction occurs through thermally assisted hops. Whereas a simple Arrhenius law for the zero-field mobility $\ln(\mu_0) \propto 1/T$ can be used to fit the data in many systems [57], it has been shown that a dependence of the type $\ln(\mu_0) \propto 1/T^2$ is more adequate [60], although its physical interpretation can hide both disorder or polaron-related effects [55].

The electric field dependence is, in all systems, very well described by a Poole–Frenkel law:

$$\mu(F) = \mu_0 \exp\left(\gamma\sqrt{F}\right)$$

where F is the electric field and μ_0 is the mobility at zero field. Whereas this formula is obtained by assuming the presence of traps whose coulombic potential is lowered by a Shottky effect very similar to the effect occuring at the metal/organic interface, it was soon recognized that the physical origin of this equation is not the presence of traps, but is related to the intrinsic charge transport of disordered materials [61].

At last, it has been reported recently [59,62] that the mobility increases with the number of carriers: indeed, the first injected carriers tend to occupy the lowest-energy sites of the DOS, which are also the sites requiring the highest thermal energy to escape; the following carriers have to hop between higher-lying states, so that the mean activation energy is lowered and mobility increases.

Summarizing the whole physics of transport in a single general expression of the mobility is still an open issue. But for using it in practice, a simple formula as the one proposed by Bässler [53] is very often used, as it describes very well the field and temperature dependence in many materials:

$$\mu(T,F) = \mu_0 \exp\left[-\left(\frac{2\sigma}{3k_B T}\right)^2\right] \exp\left[C_0\sqrt{F}\left(\left(\frac{\sigma}{k_B T}\right)^2 - \Sigma^2\right)\right]$$

where σ is the width of the Gaussian DOS and quantifies energetical disorder, Σ is the same for positional disorder, and C_0 is a constant.

From the knowledge of the mobility, the prediction of $I(V)$ curves for a given device will be strongly dependent on the nature of the injection (is the current injection-limited or transport-limited?), on the number of layers, and so forth. A very simple expression for the current density can be found in the case of a *hole-only* single layer device (e.g., ITO/PPV/Au) in which the current density is *space-charge limited.* In this case, the current density J is given by the Mott–Gurney (or Child) law:

$$J = \frac{9}{8}\varepsilon\mu\frac{V^2}{L^3}$$

where $\varepsilon = \varepsilon_r\varepsilon_0$ is the permittivity, μ the carrier mobility, V the applied bias, L the length of the organic material, and J is the current density.

This law is valid when the current is not limited by injection (low-injection barriers forming ohmic contacts), when the current is high enough to enter the space-charge limiting regime, and not too high to assume that the mobility is still a constant, that is, will not vary significantly with the electric field as exposed above. This is, however, a useful formula as it can be used, for instance, for measuring mobilities.

However, for any more complex (and realistic) structure, one has to examine on a case-by-case basis what happens, and no general $I(V)$ evolution can be predicted. Charge accumulation at the interfaces will play a key role in modifying the electric fields in each portion of the device, which in turn influence the injection characteristics and ultimately the balance of charge carriers [63].

9.2.2 Exciton Formation and Diffusion

Excitons are supposed to be Frenkel excitons rather than Mott–Wannier excitons, which means that they are spatially limited to one excited molecule. An exciton in organic semiconductors is a quite different object than in inorganic crystals, because it looks like just an excited molecule. However, the strong coulomb interaction within the molecule itself between the electron wavefunction and the hole wavefunction causes the exciton binding energy to be quite high (typically ~0.5 eV), and affects the position of the energy levels occupied by the hole and the electron. To summarize, starting from the HOMO and LUMO levels, one should not confuse them with the transport levels, as defined above, also called polaronic levels (shifted downward in energy for the LUMO and upward in energy for the HOMO, which reflect the reorganizing energy of the molecule upon capture of one carrier) and not confuse them either with the excitonic levels, the latter being separated by the so-called optical gaps, which are themselves shifted downward (respectively, upwards) in energy relative to the transport levels of electrons (respectively, holes). The generation rate is governed by a Langevin-type mechanism, which means that the rate of recombination is proportional to the density of holes and electrons. In practical OLEDs, most excitons should recombine at an organic/organic interface, where one expects the carriers to be blocked to enhance the recombination probability. The blockage can be done either in terms of energy (e.g., in Fig. 9.2 both electrons and holes appear to be blocked by HOMO and LUMO offsets at the HTL/ETL interface) and/or by a mobility contrast.

Basic quantum statistics predict that the recombination of uncorrelated electrons and holes forms triplet excitons with a threefold higher probability than singlet excitons. The transition from the triplet excited state to the singlet ground state is then a forbidden transition. Most luminescent polymers or small molecules contain low-Z elements and spin–orbit coupling is weak. One major milestone in the development of OLEDs has been the synthesis of efficient *phosphorescent* materials in which triplet excitons can radiatively decay in virtue of a strong spin–orbit coupling induced by the presence of a heavy element such as iridium or platinum. The archetypal phosphorescent molecule is *fac*-tris(2-phenylpyridine)iridium ($Ir(ppy)_3$): it has been used widely in very efficients OLEDs, as a guest either in small-molecule hosts such as 4,4′-*N*,*N*′-dicarbazole-biphenyl (CBP) [64] or in a polymeric hosts [65].

Another important fact about excitons is their ability to diffuse, once formed at the very close vicinity of the recombination zone. This mechanism not only contributes to "fade out" the recombination zone, but can also be used in a positive way to obtain color mixing (e.g., [66]). Typical orders of magnitude for exciton diffusion lengths are a few nanometers for singlets and much longer, up to 100 nm, for triplet excitons [67].

9.2.3 Optical Properties and Efficiency

The major applications of OLEDs/PLEDs are displays and lightings: the performance of the devices has to take into account the photopic response of the eye, as well as the photometric properties of the organic light source. The corresponding units used in the field are consequently the lumen (measurement of the total luminous flux) and the candela (lumen/steradian, measuring the angle-dependence of the luminous flux) emitted toward the eye, and related to the electrical power injected in the diode, given in lm/W. However, the most pertinent parameter for display applications is the luminance (in cd/m^2), that is, the flux per unit solid angle and per unit apparent area. One great advantage of OLEDs over LEDs is that they can be described with an excellent approximation as Lambertian sources, that is, sources whose luminance does not depend on the viewing angle and position.

In the following paragraphs, the effects that can affect the luminance and the global efficiency of OLEDs will be discussed briefly.

9.2.3.1 Cavity Effects An OLED is basically composed of several organic layers (total thickness around 100–200 nm with a refractive index around 1.7, that is, an optical thickness of the order of magnitude of the visible wavelength (see Fig. 9.2)) sandwiched between two metallic electrodes acting as partially reflecting mirrors. This configuration naturally forms an optical Fabry–Perot microcavity: interference of light between the different interfaces within the devices can modify the angular distribution of the emitted light, the spectrum for a given angle, and the total efficiency. Only optical modes, which satisfy the boundary conditions defined by the cavity may be supported, and the excited molecule can therefore only radiate into these modes. Becker et al. [68] and Burns et al. [69] described the influence of the emissive dipole position (i.e., the recombination zone) relatively to the metallic mirrors. Emission intensity can vary by more than one order of magnitude, depending

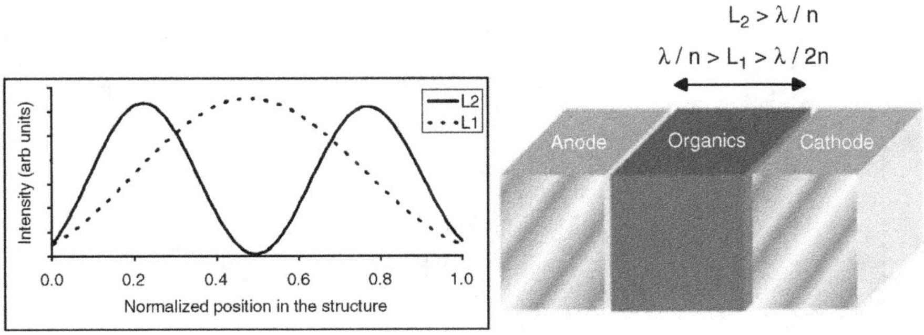

FIGURE 9.4 Illustration of the microcavity effects for different cavity lengths.

on whether the emissive layer was placed at a node or an antinode of the optical cavity (see Fig. 9.4, for an illustration).

Microcavity effects have to be taken into account when defining the OLED structure to ensure that the recombination zone is located at an antinode. Theoretical modeling is also useful to predict the influence of the cavity on the angular distribution of the light. The interference effects can not only be penalizing in terms of luminous uniformity in some cases, but also be used as a strategy to tune the emission color, to obtain spectrally narrow emission or to enhance the light intensity in a given direction.

It is important to remind that the first mode exists only when the total optical thickness between the mirrors is bigger than half the optical wavelength, which corresponds to a physical thickness of around 300 nm for the organic layers. The classical OLEDs are generally thinner, and the cavity effects between the electrodes are consequently absent in most cases. However, the archetypal anode is made of a transparent metal, namely, ITO, with a thickness of more than a hundred of nanometers: the interferences can, therefore, take place between the metallic cathode and the glass substrate/ITO interface (the reflectivity of this interface is given by the index contrast between ITO and glass). The choice of the ITO thickness is then crucial from this point of view when designing an OLED.

9.2.3.2 External Quantum Efficiency Putting the interference effects and the photometric facet aside, the simpler way to determine the efficiency of an OLED is to measure how many photons are emitted outside the device, in the forward direction for each injected electron. This ratio η_{ext} is called "external quantum efficiency."

This is a key issue not only for obvious energy-consumption considerations, but also for its effect on the lifetime of the devices (this last parameter depends strongly on the ability to operate at low voltage to obtain a given luminance). Each process at stake in the OLED during the long way from charge injection to photon emission contributes to decrease this ratio far below unity. A simple way to put this into equation is to decompose the external quantum efficiency as follows:

$$\eta_{ext} = \beta \rho_{st} \eta_{PL} \xi$$

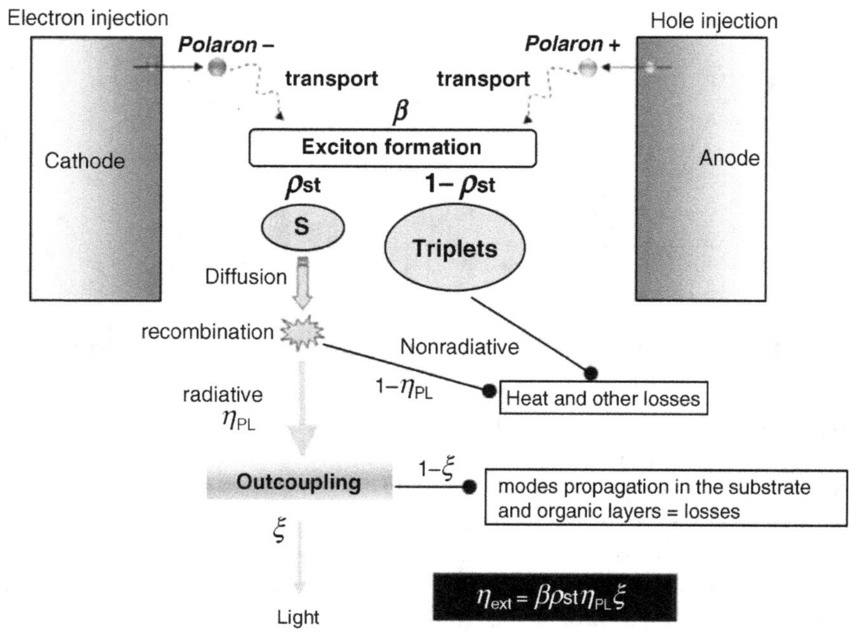

FIGURE 9.5 Schematic description of the parameters involved in the global efficiency of an OLED.

Let us detail each of those parameters (see Fig. 9.5):

- β is a measure of the balance between holes and electrons, and of their recombination probability. It is a very difficult parameter to measure, but some authors argue that it is close to unity in the more efficient OLEDs [70].
- ρ_{st} is the ratio between singlet and triplet excitons. When a pair of polarons recombines, the spin statistics imply that only one exciton over four is a singlet (as parallel spin pairs would produce triplet excitons, whereas antiparallel pairs would recombine equally to singlet and triplet excitons). For fluorescent OLEDs, only singlet excitons decay radiatively, leading to a value around 0.25 for ρ_{st} (although this value was at first experimentally confirmed [71], some recent studies—experiment and theory—reported $\rho_{st} > 0.25$, especially in polymers, but the order of magnitude remains the same [72,73]). As mentioned above, this parameter is not pertinent for electrophosphorescent OLEDs where the triplet exciton decay could be radiative, leading to much higher efficiencies.
- η_{PL} is the photoluminescent quantum yield, that is, the proportion of radiative exciton recombination paths. In many dyes in solution, this ratio is close to 100%, but it strongly drops as the concentration increases and even more in solid-state thin films (this well-known effect is called "concentration quenching," and is due to the creation of nonradiative decay paths—torsional quenching, fission of singlet into triplet excitons, and dissociation of excitons).

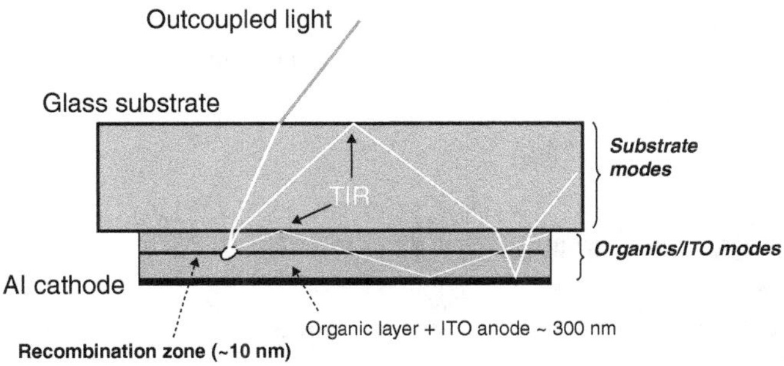

FIGURE 9.6 Outcoupling efficiency for an OLED.

- ξ is the fraction of generated photons that could break away from the device (see Fig. 9.6). A large number of photons are indeed trapped inside the OLEDs because of total internal reflection (TIR) and waveguiding in the organic layers or in the substrate. This effect is much more critical in inorganic LEDs, where the contrast index is higher. In OLEDs, the angle θ (see Fig. 9.6) should be below 1/n to avoid TIR, leading to a solid angle $\Omega = 2\pi(1 - \cos\theta) \approx \pi\theta^2 \approx \pi/n^2$.

  For small molecular weight, OLEDs with randomly oriented dipoles, and taking into account the light reflected by the back cathode (roughly multiplying by a factor of 2), $\xi \approx 1/(2n^2)$ is obtained. With $n = 1.7$, we find ξ around 17%. A more accurate estimation including optical interferences leads to 25% [70]. This factor is closer to 30% for PLEDs where the emissive dipoles are assumed to be aligned in the substrate plane during the spin-coating process. Moreover, numerous studies demonstrated that an important enhancement of this coefficient can be obtained through the use of various schemes, roughly falling in two categories. The first approach is to modify the geometry of the device, for instance, surface texturing [74], microlens array [75], scattering layers [76], additional outcoupling layers [77], shaped substrates [78], or nanopatterned surfaces [79]. Another approach is to engineer the spatial profile of the optical modes to efficiently couple the emission to escaping modes. The use of 1D (Bragg reflectors) [80] and 2D periodic [81] and nonperiodic structures [82] has been reported, even if the former implies strong variations in brightness and color with angle.

A crude estimation of the global efficiency of a fluorescent OLED is then $\eta_{ext} \approx 1 \times 0.25 \times 0.25 \times \eta_{PL} \approx 0.06\eta_{PL}$, leading to a maximal efficiency for very good photoluminescent materials (η_{PL} close to unity) around 5%. From a more optimistic point of view, with adequate outcoupling structures and the use of phosphorescent materials (allowing a nearly 100% internal quantum efficiency by harvesting the triplet excitons), it is likely that more than 20% is a realistic value.[1]

[1] The record-efficiency of OLEDs is based on small molecules. Their characteristics are often unpublished and protected by patents. Novaled® reports on 16% and 19% external quantum efficiencies for red and green phosphorescent OLED, respectively, whereas Universal Display® produced a 20% white OLED.

After this brief overview of the physical concepts at play in OLEDs and PLEDs, the details of the chemicals of PLEDs will be described further, with a classification based on their color emission.

9.3 π-CONJUGATED HOMOPOLYMERS AS ACTIVE ELECTROLUMINESCENT MATERIALS

9.3.1 Wide Bandgap Electroluminescent Homopolymers as Blue Light-Emitting Materials

Special attention has been paid on blue light, as it is difficult to achieve with both inorganic and organic materials. In addition, it can be pointed out that a blue PLED alone, may generate the two other fundamental colours (green and red) by using proper dyes, whereas the reverse cannot be achieved. Poly(fluorene)s and poly (phenylene)s are the most-studied polymer families. Other blue light-emitting polyaromatics such as poly(carbazole)s (2,7 or 3,6-linked) have also been used in PLEDs for their good hole transporting properties.

9.3.1.1 Poly(para-phenylene)s [4,8,21–26] Numerous syntheses of poly(*para*-phenylene) (PPP, **I**) (Fig. 9.7) have been reported, for example, direct oxidation of benzene [83], dehalogenative polycondensation of *para*-dihalobenzene using Ni complexes [84] or electrochemical oxidation [85], and electroreductive methods [86]. All these routes yield insoluble and intractable materials. Nevertheless, oligomers are processable by vacuum deposition [87]. The first single-layer blue light-emitting diode based on PPP sandwiched between ITO and Al electrodes has been described by Leising and coworkers [88] and exhibited an electroluminescence (EL) peak at 485 nm (2.6 eV) and a low external quantum efficiency (η_{ext} ca. 0.01%).

To obtain soluble PPP homopolymers, two main strategies have been used. The first is the so-called precursor route. It consists in starting from soluble materials that are chemically or thermally converted into fully aromatic polymers by elimination of leaving groups. An example is the precursor route utilized by Ballard et al. [89] (Fig. 9.8). Synthesis is based on 5,6-*cis*-dihydroxycyclohexa-1,3-diene as starting material and originating from the bacterial oxidation of benzene by *Pseudomonas*

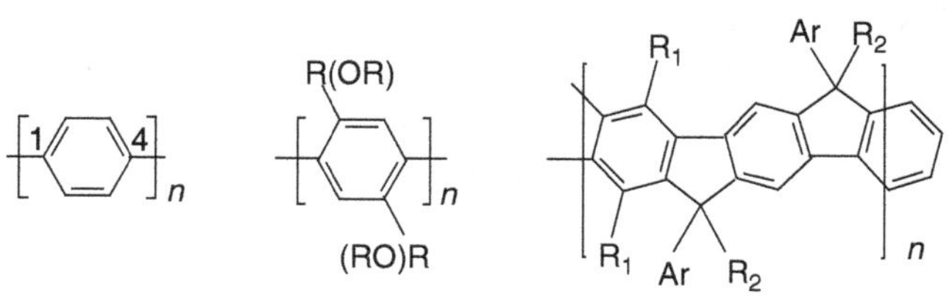

I II III

FIGURE 9.7 Representative poly(phenylene)s.

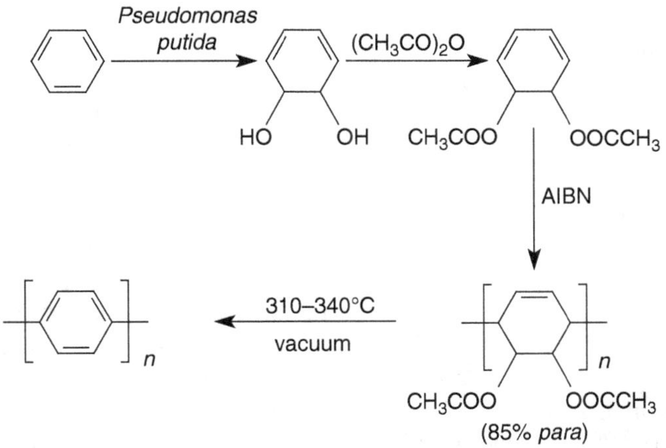

FIGURE 9.8 Ballard et al. precursor route to PPP [89].

putida. Afterwards, cyclohexanediol is esterified and polymerized via radical process initiated by 2,2′-azobisisobutyronitrile (AIBN).

Nevertheless, the resulting PPPs are not structurally perfect, as the aromatization process gives rise to mixed *para* and *ortho* linkages (15% *ortho*) and to interchain side-reactions leading to cross-linking. A nonbacterial synthesis of *cis*-diesters of cyclohexadiene followed by a stereospecific polymerization has been developed by Grubbs and coworkers to prepare a totally *para* precursor polymer [90] (Fig. 9.9).

The stereospecific living polymerization of a trimethylsiloxane derivative of cyclohexanediol has been achieved using bis((η^3-allyl) trifluoroacetylacetonato-nickel(II)(as a catalyst. Although a pure *para*-linked polycyclohexadiene precursor is obtained, pyrolysis of the difunctional cyclohexadiene is accompanied by chain scission and formation of more or less structurally well-defined low molecular weight products.

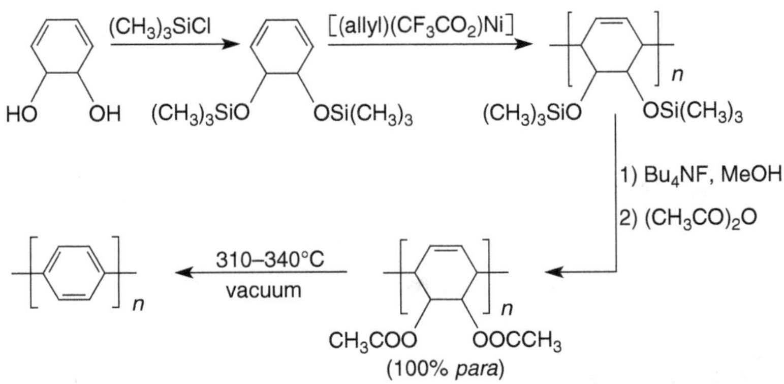

FIGURE 9.9 Grubbs and coworkers precursor route to PPP [90].

The second important strategy is the linkage onto the phenylenic monomer units of various solubilizing side groups, including alkyl, aryl, alkoxy substituents on 2, (5) position(s) as in polymer **II**. A lot of studies have been devoted to these soluble PPPs derivatives (see e.g., Refs. [35,91–95]). Nevertheless, the counterpart of the gain of solubility induced by side groups is the decrease of the π-conjugation by twisting of aromatic nuclei out-of-plane with torsion angles from around 20° in PPP up to 70° in substituted PPPs [96]. As a result, electronic conjugation is reduced and thus, HOMO–LUMO gap is increased (3.0–3.2 eV), resulting in a blue-shifted emission generally accompanied by a decrease of electroluminescence quantum yield [6].

To release steric hindrance, alkoxy groups have been attached onto the aromatic rings as the oxygen atom plays as a planar connection. One of the main synthetic routes to alkoxy-PPPs involves transition-metal-catalyzed polycondensations as, for example, the nickel-catalyzed Yamamoto coupling of dihalo or bismethylate arylenes (Fig. 9.10) [91,95], and the palladium-catalyzed Suzuki cross-coupling of aryldiboronic acids with dibromoarylenes or homocoupling of bromoarylboronic acids [93,94,97,98]. The latter method leads to structurally pure high-molecular weight polymers with a degree of polycondensation of ca. 150 [94,98].

Alkoxy-substituted PPP derivatives has been successfully used as active blue light-emitting materials in PLEDs [23]. High-efficiency single- and double-layer PLEDs based on soluble blue light-emitting poly(2-decyloxy-1,4-phenylene) (DOx-PPP) have been reported by Yang et al. [26,99]. In the double-layer configuration, a hole-transport layer of poly(vinylcarbazole) (PVK) is interposed between ITO and DOx-PPP. Because of the differential solubility of the two polymer materials in aromatic solvents, DOx-PPP can be spin-cast from toluene solution, onto PVK without dissolving this underlying layer. With such a structure, a blue-violet light and good EL performances are obtained, with a $\eta_{ext} = 3\%$ and a luminance of 490 cd/m^2 at 30 V.

In the early 1990s, an approach to obtain PPPs with optimized properties, that is, good solubility and extended electronic conjugation, have been used by and Scherf and Müllen [100,101], through a structure in which adjacent phenylenic units

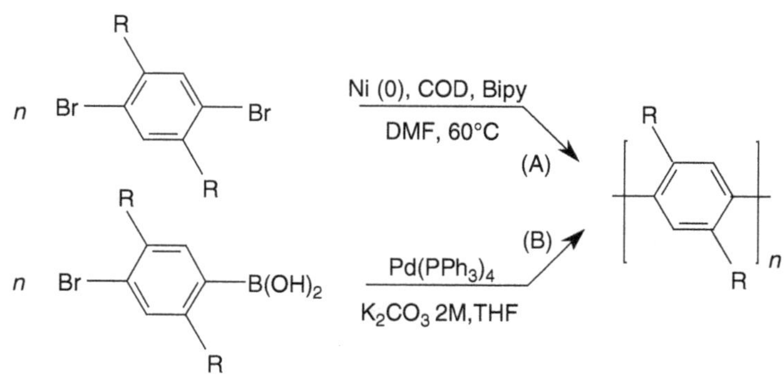

FIGURE 9.10 Syntheses of poly(2,5-substituted-phenylene)s by Yamamoto (A) and Suzuki (B) coupling reactions.

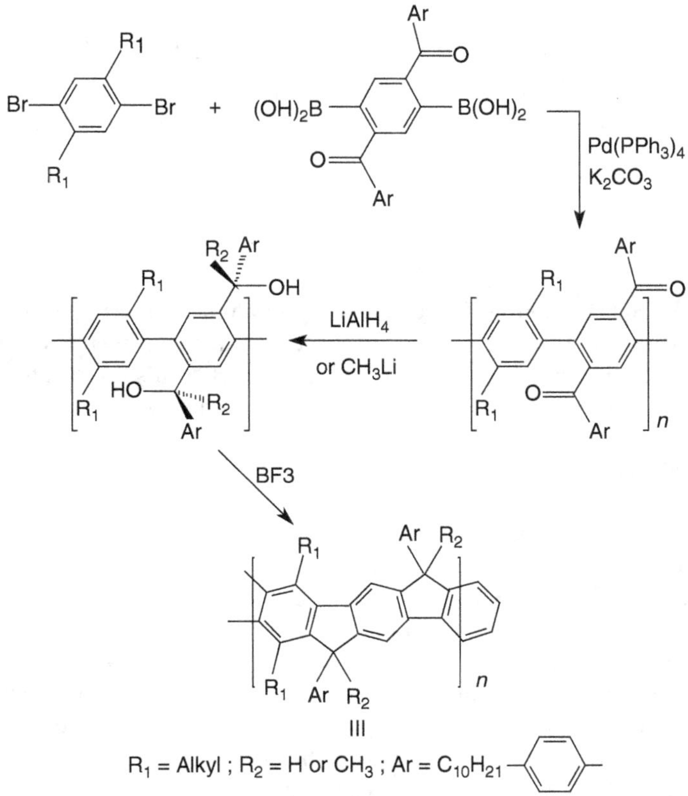

FIGURE 9.11 Synthesis of ladder-type PPPs [59,60].

are forced to remain planar by methylidene bridges. These ladder-type poly(*para*-phenylene)s (LPPP)s **III** have been synthetized via Suzuki coupling methodology (Fig. 9.11). The molecular weights of around 2×10^4 were obtained. Materials show an energy bandgap $E_g = 2.7$ eV and a sharp blue fluorescence in solution (at 450 nm), which is unfortunately broadened and red-shifted in the solid state (450–650 nm), thus spoiling the blue color purity (to green-yellow). The large bathochromic shift has been ascribed to the formation of excimers due to π-stacking of planar segments.

To recover a blue luminescence and to avoid excimer formation, a higher quality ladder polymer was synthetized, so-called m-LPPP, in which a methyl group is linked to the methine bridge, so lowering the interchain aggregation. The polymer exhibits a blue-green emission. A single-layer device utilizing m-LPPP as an active material sandwiched between ITO and Al electrodes shows an external quantum efficiency of 1% [23].

9.3.1.2 Poly(fluorene)s [14–19] Poly(2,7-fluorene)s **IV** constitute another important class of polarylenes that have been developed later than polyphenylenes and applied to blue PLEDs (Fig. 9.12). They possess bridged biphenyl units similar to that

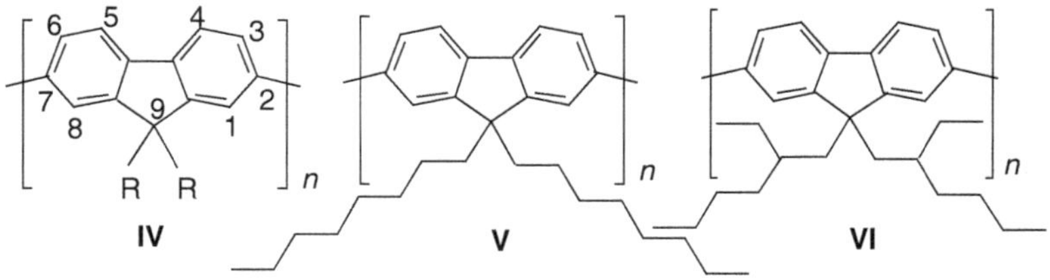

IV V VI

FIGURE 9.12 Some typical poly(9,9-substituted-2,7-fluorene)s.

of LPPP **III**. Substitution at the remote sp^3 C-9 atom does not induce important twisting of the conjugated skeleton due to steric intrachain interactions. Physicochemical properties may be tuned via interactions between substituents at the C-9 position [19]. Indeed, as it has been shown for poly(9,9-dioctyl-2,7-fluorene) **V**, alkyl side-chain interactions lead to the formation of a specific morphology, resulting in their crystallization and inducing a quasiplanar polymer chain structure LPPP-like [17,19].

As for PPP, polyfluorenes have been prepared by direct oxidation of fluorene [102], but low molecular weight materials (M_w up to 5×10^3) were obtained. Moreover, oxidative coupling was accompanied by side-reactions leading to unwanted linkages other than 2,7 and to cross-linking. Important improvements have been achieved with the synthesis of regioregular poly(9,9-dialkyl-2,7-fluorene)s **IV** via, for example, transition-metal-catalyzed Yamamoto [103] or Suzuki polycondensation pathways [14,104]. Pd-catalyzed Suzuki coupling yields relatively high molecular weight polymers (typically 5×10^4) showing high fluorescence quantum efficiencies (50–90%). Photoluminescence of **IV** is characterized by an emission spectrum with a transition of 2.9 eV. Their HOMO and LUMO levels are still the subject of discussion, and different energy values have been reported, that is, within the range 5.6–5.8 and 2.1–2.6 eV, respectively [16,19].

In the solid state, 9-substituted fluorene polymers exhibit a green component emission what might be due to excimer formation originating from the intermolecular interactions between the aromatic units. Alkyl side-chain interactions would be involved in this solid-state packing. These supposed excimers are characterized by a broad emission lying from ca. 480 to 650 nm (with a maximum at ca. 540 nm) located in the green region of the spectrum [10,105]. Nevertheless, formation of excimers still remains questionable, as freshly prepared poly(9,9-dioctyl-2,7-fluorene) **V** films do not exhibit any spectral evidence of low-energy band until a significant oxidation occurs. Occurrence of keto-defects has been proposed to explain the green emission band [17,105–108]. These fluorenone sites are present in pristine 9-monoalkylated fluorene homopolymers, but not in the case of the dialkylated derivative. Formation of hydroperoxide anions at the 9 position of the monomer unit by reaction of fluorenyl carbanions (originating from the Yamamoto synthesis) with atmospheric oxygen during the work-up, followed by a rearrangement to the fluorenone centers, has been proposed to explain the presence of fluorenone centers [17,106] as depicted.

Concerning dialkylated fluorene homopolymers, although they show no detectable trace of low-energy emission peak at 2.3 eV (540 nm), neither in solution nor in the

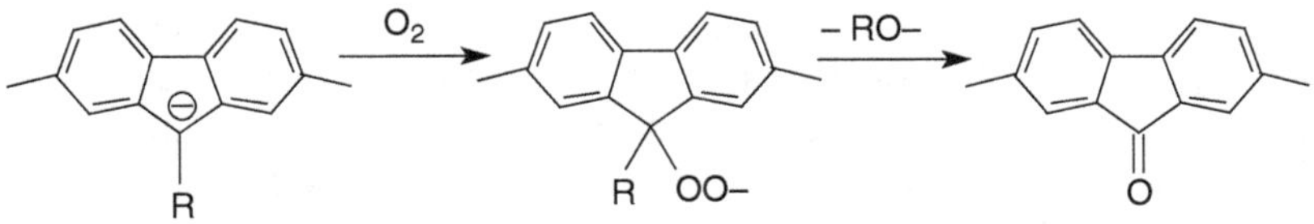

FIGURE 9.13 Proposed reaction mechanism of formation of keto-defects.

solid state, a color change into blue-greenish light appears under application of a bias voltage to PLEDs after 30 min, under ambient atmosphere [17]. This finding demontrates that keto-defects also affect the dialkylated methine bridge of the fluorene moiety and electrooxidation of the monomer units occuring under turn on of operating devices (Fig. 9.13).

The first PLED based on polyfluorenes with ITO and Al as electrodes was reported in 1991 by Ohmori et al. [109]. Later, PLEDs utilizing **V** and a calcium cathode have been developed [16]. These devices were found to show an electroluminescence efficiency of 0.25 cd/A, a power efficiency of 0.04 lm/W, and a maximal luminance of 600 cd/m^2 [107]. Maximal electroluminescence performance of 2 cd/A, 1 lm/W, 3550 cd/m^2 has been obtained with ITO/**V**/Ca devices using high molecular weight **V** fraction [110]. LiF/Ca/Al cathodes have been used to reduce the energy barrier height to electron injection into the conduction band of **V** and an ITO anode recovered by an injecting interlayer of poly(ethylenedioxythiophene)/poly(styrene sulfonic acid) (PEDOT/PSS) to improve hole injection into its valence band [111]. With such devices, luminance of 1600 cd/m^2 at 5 V and a power efficiency of 3 lm/W were attained.

Furthermore, blue emission and polarized electroluminescence have been reported by Grell et al. [112] for PLEDs utilizing liquid crystalline LC homopolyfluorenes as emitters. Indeed, many 9,9′-dialkyl PFs show rich phase morphologies, as thermotropic liquid crystal mesophases, which make them candidates for obtaining oriented active layers [16,17,19,107]. For example, aligned poyfluorenes with branched side groups as poly(9,9-bis(2-ethylhexyl)-2,7-fluorene)s **VI** have been used as blue emitters in PLEDs displaying a luminance of 45 cd/m^2 at 19 V [112].

Thus, although polyfluorenes have been, and still remain, the subject of a considerable interest, they suffer from two main drawbacks for PLEDs applications:

- First, their sensitivity to methine bridge oxidation yielding ketones that induce undesired damage in the optical properties and particularly in light emission.
- Second, relatively weak ability for charge injection and transport.

For these reasons, their "first cousin" polycarbazoles have received more interest.

9.3.1.3 Poly(carbazole)s [20–22] Polycarbazoles contain rigid bridged biphenyl units similar to that of polyfluorenes resulting in materials with a large bandgap adapted to blue emission. The main difference between carbazole and fluorene units is that the former is fully aromatic, the 9 position, that is, the nitrogen atom being involved in the πconjugated sytem. Thus, carbazole units are more stable and the *N*-functionalization provides a means to modify the optoelectronic and semi-conducting properties. The 3 and 6positions are the most reactive of the carbazole ring and can be easily substituted [21,113]. This peculiarity explains why all the studies concerning carbazole-based conjugated polymer materials have been devoted to 3,6-linked *N*-substituted monomer units [21,114–119] until Morin and Leclerc described, in 2001, the first syntheses of polycarbazoles with 2,7-linkages [120]. Another quality of carbazole materials is their remarkable hole-transporting properties. For example, poly(*N*-butyl-3,6-carbazole) shows a hole-drift mobility of about 1.5×10^{-3} cm^2/(V s) [21], which is five times higher than that of polyfluorene.

Poly(3,6-carbazole)s ***VII*** Unlike phenylenes and fluorenes, direct oxidation of carbazoles does not yield polymers, but mostly dimers, that is, 3,3′-bicarbazyls [121] (Fig. 9.14). The stabilization of bicarbazylium radical-cation throughout the π-conjugation extended between the two nitrogen atoms of the oxidized dimer, precludes the follow-up (poly)coupling reaction. However, Siove and Ades have synthesized a high polymer (degree of polymerization of around 130) based on a 3,6-carbazole unit by oxidative polymerization of a hindered

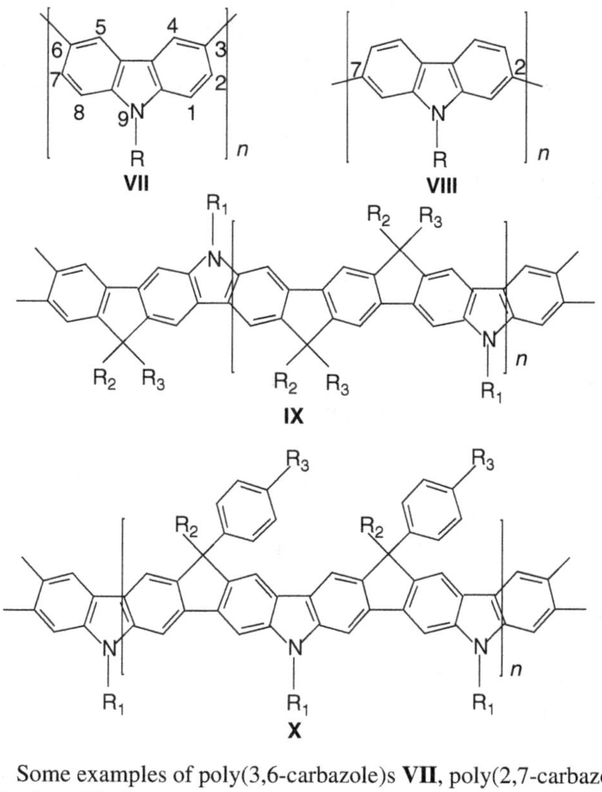

FIGURE 9.14 Some examples of poly(3,6-carbazole)s **VII**, poly(2,7-carbazole)s **VIII**, and ladder-type derivatives **IX**, **X**.

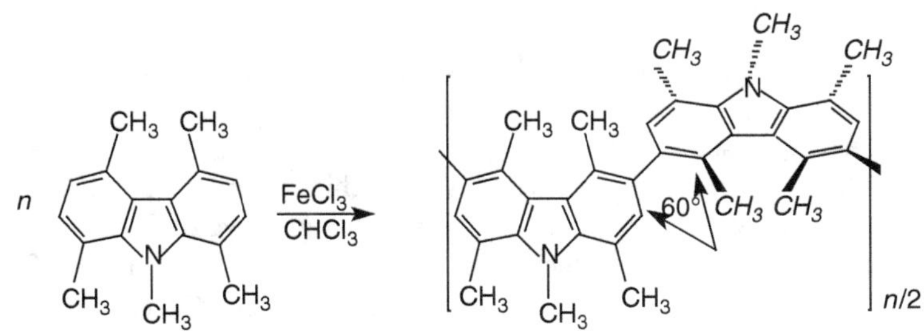

FIGURE 9.15 Oxidative polymerization route to poly(1,4,5,8,9-pentamethylcarbazole).

1,4,5,8,9-pentamethylcarbazole with $FeCl_3$ (Fig. 9.15) [116]. Steric hindrance induces an out-of-plane torsion angle of 60° between adjacent units, in such a way that π-conjugation is prevented, which allows the polycoupling. Deviation from the coplanarity allows this polycarbazole to emit a bright pure violet-blue light peaking at 400 nm, with no excimer formation. Energy bandgap E_g has been evaluated at 3.25 eV [116].

Concurrent to the oxidation route, reductive polymerizations of 3,6-dihalocarbazoles from, for example, Grignard [114], electrochemical [113,115], and chemical nickel-catalyzed [118,119] coupling reactions have been developed (Fig. 9.16). Depending on the nature of the *N*-substituents, more or less soluble polymers with molecular weights lower than 10^4 are commonly obtained. Poly(3,6-carbazole)s **VII** possess a conjugation length limited to dyads (dimeric units) [122]. In such a way, energy bandgap of both polymers and dimers have found to be very

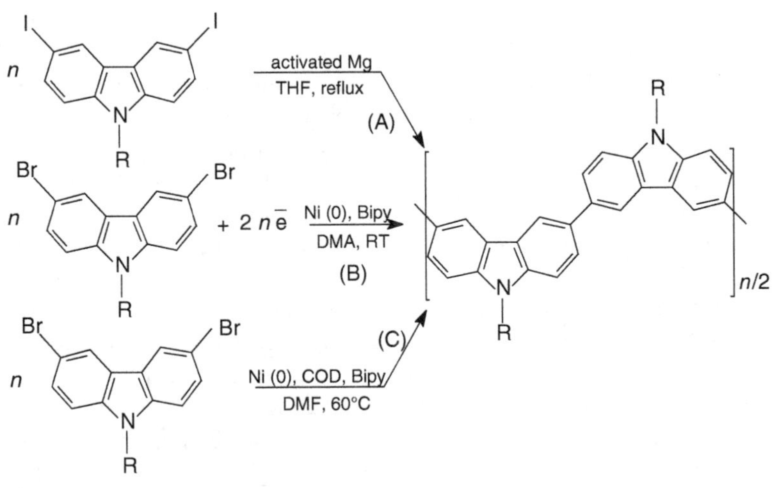

FIGURE 9.16 Syntheses of poly(3,6-carbazole)s by Grignard (A), electroreductive (B), and Yamamoto coupling reactions (C).

close and around 3.1 eV, thus, leading to a blue photoluminescence peaking at 420 nm [123]. Thanks to this feature, both the polymers and dimers may be used as active blue-emitting layers in PLEDs [123] or “small-molecule”-based OLEDs [124,125], respectively.

The first PLEDs based on 3,6-carbazoles was achieved in 1996, by Zuppiroli and coworkers with a poly(*N*-butyl-3,6-carbazole) in a single-layer device with ITO and Al electrodes [123]. Diodes produced blue emission with a low EL performance ($\eta_{ext} = 0.07\%$ and a few cd/m^2 at 15 V). Aiming at increasing the EL performances, five-layers OLEDs based on small molecules were fabricated using carbazole dimers as emitters. A pure blue light-emitting device has been achieved with CIE coordinates $x = 0.158$, $y = 0.169$, $\lambda_{peak} = 456$ nm, a luminance of 1000 cd/m^2, luminance efficiency of 4.7 cd/A at 10 V and a $\eta_{ext} = 3.3\%$ [125].

Poly(2,7-carbazole)s ***VIII*** Poly(2,7-carbazole)s are a relatively young class of conjugated polymers, as their synthesis is far from being as straightforward as that of 3,6-carbazole-based materials. Indeed, both 2 and 7 positions, in the meta-position of the amino group of carbazole, cannot be directly functionalized by standard electrophilic aromatic substitution. Therefore, different strategies involving precursor biphenyl units have been developed to synthesize 2,7-functionalized carbazoles [21]. One of the most convenient methods consists of the Cadogan ring closure of a 2-nitrobiphenyl derivative [21,126] as shown for the synthesis of N-(2′-ethylhexyl)-2,7-dichlorocarbazole and the 2,7-diboronate derivative [127]. **VIII** are then obtained by the standard Yamamoto or Suzuki coupling reactions (Fig. 9.17) [120,127,128]. As these materials have a longer conjugation length (polyphenylenic-like) than the 3,6 derivatives, energy bandgap is a little lower, that is, $E_g = 2.8$ eV, the maximum of the corresponding blue emission in the solid state being at ca. 440 nm [128]. Among the poly(2,7-carbazole)s, poly(N-(2-ethylhexyl)-2,7-carbazolylene) (PEHC) was found to be the best emitter in PLED. Blue electroluminescence ($\lambda_{max} = 452$ nm), that is, luminance up to 1500 cd/m^2 at 10 V can be obtained [21,129].

The preparation of 2,7-functionalized carbazole units has allowed the first synthesis of ladder polymers with carbazole moieties. Using similar strategies than those reported for the synthesis of ladder-type poly(*para*-phenylene) **III**, Scherf and coworkers have synthesized a carbazole analog, LPPPC **IX** [130]. It exhibits a blue fluorescence in solution, centered at 460 nm with a lower energy band at 500 nm. A similar approach has been used by Müllen and coworkers [131], to synthesize carbazole ladder-type polymers **X** having greater carbazole content than **IX**. The synthesis involves the Suzuki cross-coupling of the carbazole 2,7-diboronic acid with a 3,6-diacyl-2,7-dibromocarbazole. The resulting polyketones are transformed into polyalcohols, which lead after ring closure, to the expected ladder polymers **X**. They show a blue-green fluorescence in solution, at around 480 nm with a low-energy emission band at 507 nm. As in the case of **III**, because of their planar nature, emission spectra of these polymers exhibit band in the green region, suggesting that excimers emission dominates.

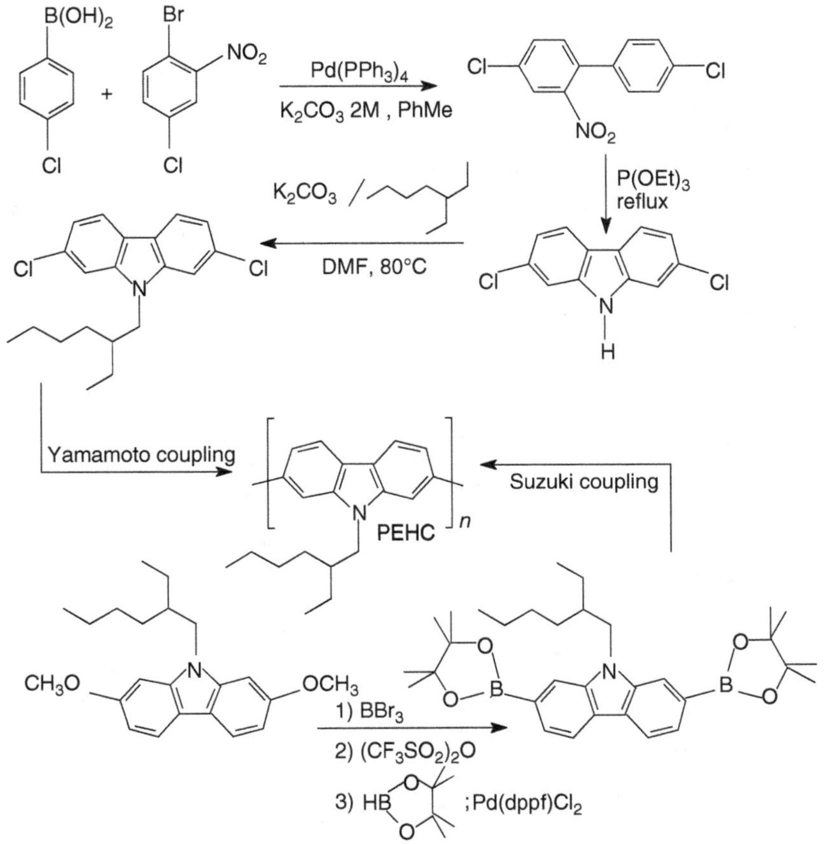

FIGURE 9.17 Syntheses of poly(2,7-carbazole)s by Yamamoto and Suzuki couplings from 2,7-difunctional carbazoles.

9.3.2 Poly(para-phenylene vinylene)s from Green to Red Light-Emitting Materials

The discovery, in 1990, of electroluminescence of poly(*para*-phenylene vinylene) (PPV) by Burroughes et al. at Cambridge [4] has been at the origin of the considerable development of PLEDs (Fig. 9.18). The first PLED based on a PPV emitting single-layer exhibited an EL in the green-yellow region of the spectrum and a low external quantum efficiency of the order of 0.05% [4]. Since, important progress has been made, and PPV and its derivatives have represented the most extensively studied family of electroluminescent polymers.

9.3.2.1 Synthesis Poly(*para*-phenylene vinylene) (PPV, **XI**) is the archetypical fluorescent polymer. It emits a bright green-yellow light with two emission peaks at 520 nm (2.38 eV) and 550 nm (2.25 eV). Direct synthesis of **XI** yields an intractable,

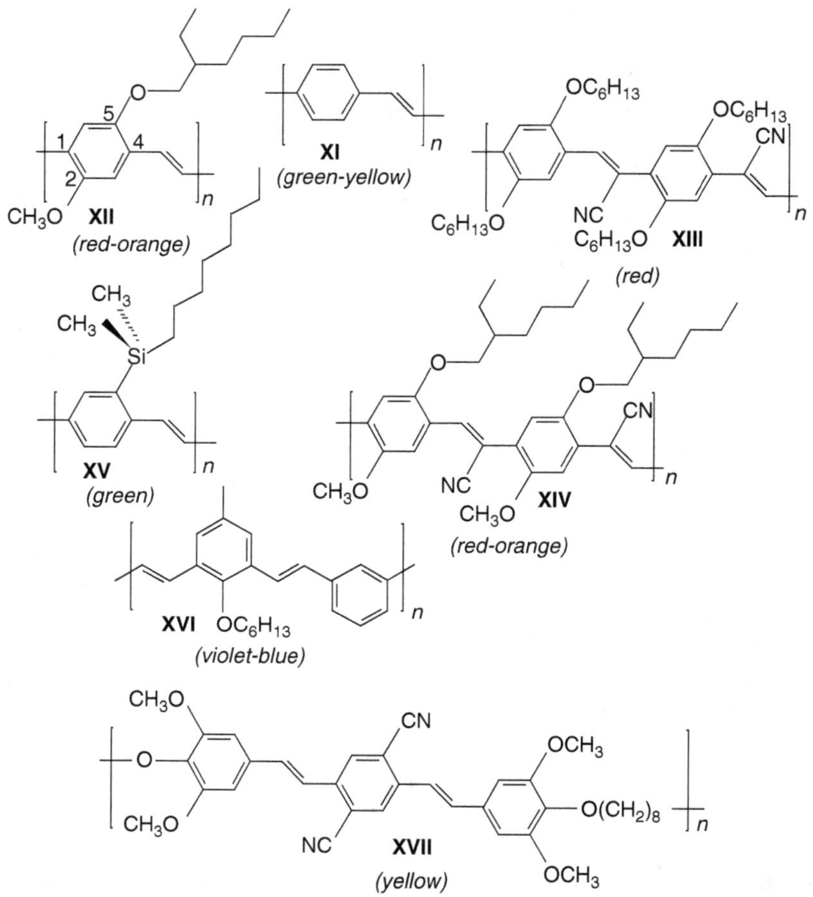

FIGURE 9.18 Examples of PPVs derivatives and their emission color.

infusible, and insoluble material. Hitherto, many efforts have been made toward the synthesis of processable PPV materials from which solution-cast films can be obtained. Two main approaches have been followed: the precursor polymer route and the solubilizing side-group approach. The precursor route to PPV, initiated by Wessling [132] and Wessling and Zimmerman [133] proceeds via a water soluble prepolymer bearing sulfonium chloride salts as depicted for the poly(xylylidene tetrahydrothienylium chloride) [134,135] (Fig. 9.19). Then, the precursor polymer can be cast as thin films, which are thermally converted (200–300 °C) into **XI**, by elimination of leaving groups under vacuum.

Interestingly, methanol is able to displace sulfonium groups giving a more thermally stable prepolymer, soluble in organic medium, thereby allowing processing in milder conditions by the use of more volatile solvents [135,136]. The mechanism of Wessling polymerization has been the subject of controversy concerning the nature of propagating active centers, that is, radicals or

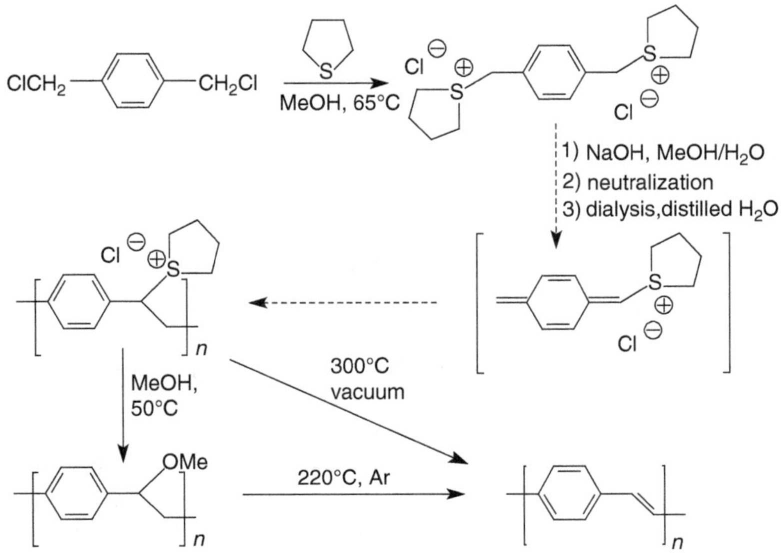

FIGURE 9.19 Wessling–Zimmerman route to PPV.

anions [137,138]. From a practical point of view, the reaction goes better when carried out under an inert atmosphere, thermal conversion into PPV occurring at lower temperatures [139,140]. Indeed, polymerization of sulfonium salts in basic medium generally generates alcohol functions, which are sensitive to oxidation and can be thermally converted into carbonyls. These defects in the π-conjugated macromolecular chain shorten the effective conjugation length [140], act as quenching sites, and have been taken as responsible for the relatively low luminescence quantum efficiency of around 30% found for standard PPV [140,141]. It has been found that thermal elimination to PPV could be lowered to 115 °C by using a bromide or a dodecylbenzenesulfonate (DBS) sulfonium salts instead of a chloride [142–144]. Other precursor routes have been taken for the synthesis of PPV, such as, those using sulfone and sulfoxide monomers, which allow polymerization in organic media and easier purifications of materials [145,146].

The second approach to PPV, is the Gilch synthesis [147], which consists in the introduction of solubilizing side groups, as represented in the case of the most studied dialkoxy derivative: poly[2-methoxy-5-(2′-ethylhexyloxy)-1,4-phenylene vinylene] (MEH-PPV, **XII**) [29,148–150] (Fig. 9.20).

Bis(halomethyl) benzene derivatives undergo a dehydrohalogenation-condensation polymerization induced by a large excess of potassium-*tert*-butoxide in THF solution. Nevertheless, Gilch route needs a fine control of concentration, to avoid cross-linking of polymerization products [151]. Moreover, it has been shown from NMR studies, that PPVs issued from this polymerization process, exhibited ethane, dihaloethane, and ethyne defects due to irregular coupling (3.0–4.4% irregular bonds) that decrease PLEDs performance [152–154]. To avoid drawbacks due to the utilization of an excess of base, a little bit less than the stoichiometric conditions

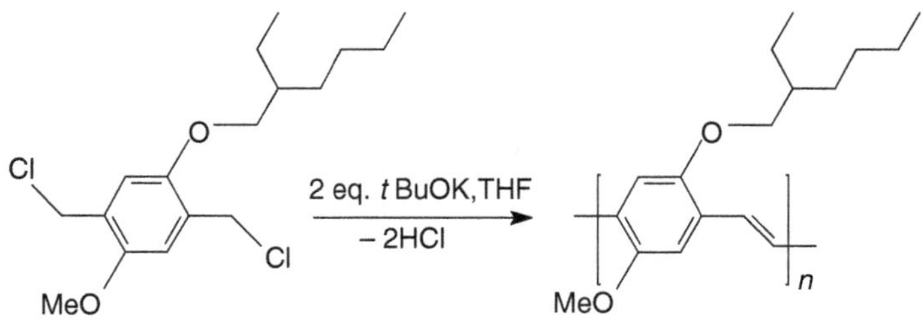

FIGURE 9.20 Gilch synthesis of MEH-PPV **XII**.

in the alkoxide allow obtaining a soluble halo precursor polymer, which can be subsequently thermally converted into 2,5-dialkoxy PPVs [150,155–159].

Other reactions have been used to create the vinylene double bonds of PPVs, such as the Wittig, Horner, and Knoevenagel condensations of terephtaldehydes with bisphosphonium, bisphonates, or bisacetonitriles, respectively [160–162]. These condensation polymerizations lead to polymers having lower molecular weights ($M_w < \sim 10^4$) than those yielded by the Gilch route. Wittig reaction mainly affords *cis*-vinylene units, whereas Horner and Knoevenagel condensations give *trans*-structures. Hence, a large variety of polymer structures are accessible and numerous soluble alkyl-, alkoxy-, and cyano-substituted PPVs have been reported [6,8,9,31,37,163–170]. Transition metal-catalyzed reactions such as Heck and Suzuki coupling or metathesis have been used to synthesize PPVs. The Heck polycondensation reaction involves the palladium-catalyzed coupling of dihalobenzenes with olefins in the presence of ligands as phosphines and of a base as triethylamine [171–174]. Reaction usually leads to high conversions and *trans*-1,2-substituted vinylene moieties accompanied with substantial amount of 1,1-vinylidene structures in the polymer backbone [175,176]. The Suzuki cross-coupling of 1,2-dibromoethylene and 1,4-aryldiboronic acid has been used to prepare PPV. Biaryl moieties (3%) due to abnormal coupling are present as defects in the polymer backbone [177].

Ring-opening metathesis polymerization (ROMP) of substituted bicyclo octadienes or paracyclophane-enes initiated by Grubbs molybdenum, tungsten-based carbenes have been used to prepare PPVs [178–181]. The "living character" of ROMP has been exploited to prepare soluble well-defined precursors, which can be converted into **XI**. Yu and Turner have used ROMP of tetra octyloxy-substituted paracyclophanedienes initiated by reactive ruthenium-based carbenes to prepare monodisperse, soluble yellow fluorescent PPV with an alternating *cis–trans* microstructure and molecular weights as calculated [178] (Fig. 9.21).

9.3.2.2 Structure–Emission Color Relationships Apart from copolymerization (not addressed herein), colour tuning of PPVs can be obtained by varying the nature of the substituents, the stereoregularity of the skeleton, and the π-conjugation length.

Color Tuning by Substituents, Electronic, and Steric Effects The electronic structures of PPV and ring-substituted derivatives have been the subject of several

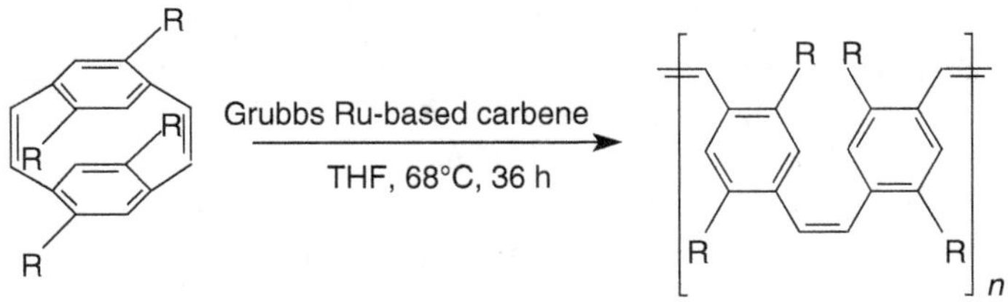

FIGURE 9.21 ROMP of paracyclophanedienes to tetraoctyloxy-substituted PPV [178].

experimental and theoretical studies as, for example, those of Brédas and coworkers [9,25,182–185]. Substitution on the phenylene ring by electron-donating or electron-accepting groups causes modifications of the π-bands and the associated density of states together with the bandgap. Attachment of electron donors, such as alkoxy groups, results in a decrease of the energy bandgap showing up in the red shift of the emission. For example, the 2-methoxy-sustituted PPV is a yellow emitter with PL and EL peaks at 560 nm (2.21 eV) [186], whereas the unsubstituted **XIII** is a green-yellow light-emitting polymer with emission maxima at 520 and 550 nm [4]. For MEH-PPV **XII**, the HOMO–LUMO gap is ca. 2.1 eV and a red-orange emission peaking at ca. 610 nm is obtained [29,148–150]. PLEDs utilizing MEH-PPV were reported for the first time in 1991 by Braun and Heeger [29]. A two-layer electroluminescent device, with an acid-doped polyaniline as a hole-injecting layer onto ITO, **XII** as an emitter and a calcium cathode, exhibits an external quantum efficiency of 2–2.5%, a luminous efficiency of 3–4.5 lm/W, and a luminance of 10,000 cd/m^2 [187]. Nonetheless, degradation and failure of diodes in a single-layer configuration using ITO and Ca as electrodes have been reported by Scott et al. at IBM [188]. It has been shown that the vinylene groups undergo an oxidation reaction, with formation of carbonyls arising from extraction of oxygen atoms from the ITO anode [14,188]. Thus, an oxygen-blocking barrier may be one of the reasons of improved EL performances obtained by diodes, which use a polyaniline layer at the anode interface [149,187]. Another soluble highly efficient related PPV 2,5-substituted by a methoxy group and a 3,7-dimethyloctyloxy side-chain has been synthesized and utilized in a single-layer diode to produce a red-orange electroluminescence ($\lambda_{max} = 610$ nm). The device exhibits an external efficiency as high as 2.1%, a luminous effiency of 3 lm/W, and a luminance of 100 cd/m^2 [33].

Introduction of the electron-withdrawing groups on the PPV skeleton increases the electron affinity (EA), so that the barrier of electron injection is reduced and thereby, the hole–electron balance is improved. The Cambridge group was the first to report the synthesis of the 2,5-dihexyloxy-cyano-substituted PPV **XIII** and that of the MEH-CN-PPV **XIV**, two alkyloxy-disubstituted-PPV with a cyano group onto the vinylene unit [31,166]. The electron-accepting effect of the nitrile group has been shown to increase the binding energy of both the HOMO and the LUMO states while keeping a similar bandgap [183]. Cyclic voltammetry measurements on MEH-CN-PPV **XIV** showed that the reduction onset is shifted about 0.5 V toward the anodic domain in comparison with MEH-PPV **XII**, what is in agreement with the increase of its EA.

HOMO–LUMO gap of CN-PPV **XIII** is about 2.1 eV (590 nm) and two-layer electroluminescent devices made of ITO/PPV (as a hole transporting layer)/CN-PPV (as emitter)/Al or Ca, exhibit a red electroluminescence with a peak at 710 nm and a η_{ext}) of about 1% [31]. The same two-layer configuration devices based on MEH-CN-PPV **XIV** exhibit a red-orange electroluminescence peaking at ca. 600 nm, a $\eta_{ext} = 2.5\%$, a luminous effiency of 2.5 lm/W, and a luminance of 1000 cd/m^2 at 6 V [166,189]. The blue-shifted emission of MEH-CN-PPV in comparison with that of CN-PPV has been ascribed to a steric effect of the branched ethylhexyl side-chain, which induced a slight twisting of the polymer backbone.

Steric effect is revealed in the trialkylsilyl-sustituted PPV series, because the silyl group does not show any electronic effect. Attachment of a trialkylsilyl group onto the phenylene moieties as in DMOS-PPV **XV** widens the HOMO–LUMO gap by out-of-plane torsion of the polymer skeleton. As a consequence, a green luminescence is observed peaking at around 500–530 nm [190–195]. **XV** thin films sandwiched between an ITO anode and an aluminum cathode have been found by Hoger et al. to exhibit a green luminescence with a maximum at 500 nm and a $\eta_{ext} = 0.1\%$ [190].

Color Tuning by π-Conjugation Length Control Two other strategies have been adopted to control the emission color of π-conjugated polymers and particularly PPVs (see e.g., Refs. [6,37]) via disruption of the conjugation length. The first, known as the "kinked linkages" strategy, uses meta-phenylene linkages that create kinks in the structure. The second, known as the "isolated main-chain chromophores" approach, consists in the introduction of nonconjugated segments in the skeleton to confine π-conjugation within short blocks.

meta-Phenylene rings were expected to reduce the conjugation length and offer a convenient route toward the blue light. Unfortunately, the substitution of the 2,5-positions of the phenylene nucleus by alkoxy solubilizing side groups induces a red shift of the energy gap arising from their electron-donating character. To overcome this drawback, alternating copolymers of a *meta*-linked phenylene vinylene and an alkoxy 2,5- or 2,3-dialkoxy-*para*-linked phenylene vinylene have been prepared [196–199]. Liao et al. and Ding et al. [198,199] have reported PLEDs built with poly(*meta*-phenylene vinylene-*alt*-2-methyl-3-alkoxy-*meta*-phenylene vinylene) **XVI** giving a purple-blue emission at 465 nm, a luminance of 15 cd/m^2, and $\eta_{ext} = 0.01\%$.

The "isolated main-chain chromophores" approach consists in the shortening of the effective conjugation length to widen the bandgap and thereby to blue-shift the emission. One way, used by Hay and Klavetter [200] and by Karasz and co-workers [201–203] proceeds by the linkage of a functional phenylene vinylene segment (commonly 3 units) with flexible spacers. These copolymers have been prepared by Wittig or Knoevenagel polycondensations and emit in the blue region [200–203], but the emission spectra are broadened due to the dilution effect of the chromophores by the insulating spacers. Green and yellow emissions have also been produced by copolymers of cyano-containing PPV-based chromophores and a flexible spacer as shown for the yellow light-emitting polymer **XVII** [201].

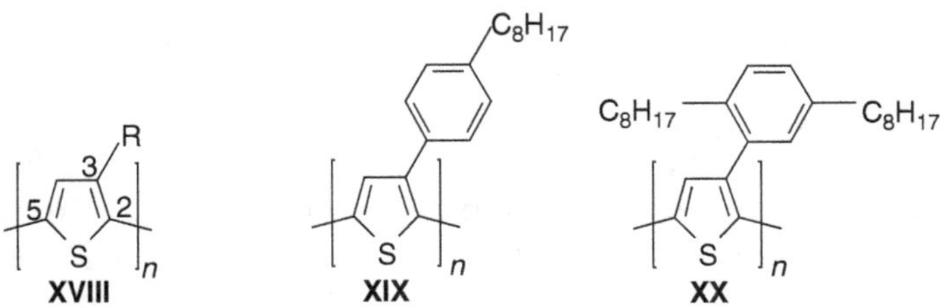

FIGURE 9.22 A few members of the poly(3-alkylthiophene)s family.

9.3.3 Poly(alkylthiophene)s for Red Light Emission

Poly(3-alkylthiophene)s (PATs), which are soluble analogs of the intractable unsubstituted polythiophene, are one of the most widely studied class among the conjugated polyheterocycles due to their unique electronic properties that can be tuned by modification of their structure (Fig. 9.22). Their chemistry and their conducting and electroluminescence properties have been the subject of several well-documented reviews [5,6,8,38–40,204,205]. PATs possess a low bandgap around 2 eV and a HOMO level of 4.6 eV, which make them red-orange emitters (620–640 nm) also well suited for hole injection from ITO. It is noteworthy that all the emission colors have been achieved by controlling the conjugation length from backbone distortion (steric effect) and from the chain regioregularity. Unfortunately, due to their aggregation propensity and the heavy-atom effect of the sulfur on the intersystem crossing of singlet excitons to triplet-states [206–208], nonradiative decay is strongly increased and EL efficiencies are low, in the order of a few tenth of percent [38,141,209–212]. In spite of this handicap, PATs remain interesting candidates for red-light emission, as the latter required low bandgap materials and is difficult to attain with the other great light-emitting polymer families.

PATs **XVIII** have received attention due to their solubility in organic solvents induced by hydrocarbon side-chains, as the first works in 1986 [213–215]. Polymerization of 3-substituted thiophenes may result in three different configurations, that is, head-to-tail (HT), head-to-head (HH), and tail-to-tail (TT) (Fig. 9.23. Different synthetic methods have been carried out that have been extensively reviewed in particular by Roncali [204] and McCullough [205].

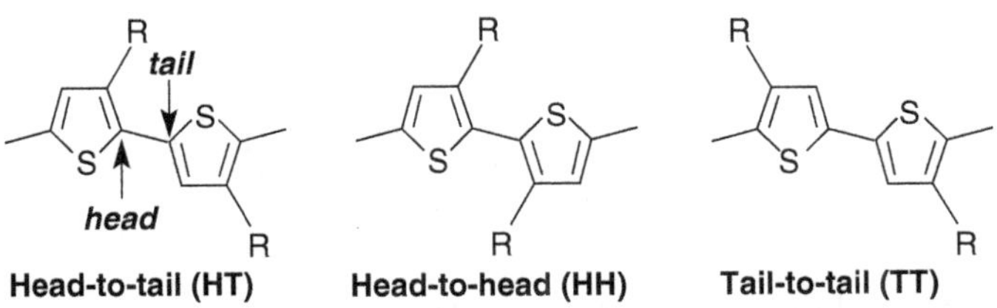

FIGURE 9.23 Three possible regiochemical couplings in PATs.

Oxidative coupling of 3-alkylthiophenes by $FeCl_3$ has been used to prepare PATs [216–220]. Leclerc et al. [217] have shown that this method yields PATs (with alkyl = hexyl, octyl, decyl, dodecyl) with molecular weight in the range 3–8 × 10^4 and an HT content of 80% (Fig. 9.23). Nevertheless, the reproducibility of this method has been questioned, as variable results were obtained while repeating experiments as, for example, molecular weights ranging from 5.4 × 10^4 to 12.2 × 10^4 [218] and levels of iron residues from 0.15% to 9.6% [221]. The latter acts as quenching impurities for excitons and decreases the EL performances of PLEDs [211]. Improvements of the method have been achieved with poly(3-octylthiophene) with $M_w = 7 \times 10^4$, a very low iron content (80 ppm), and an HT proportion of 84% [222,223]. An HT content as high as 94% has been obtained for poly(3-(4-octylphenyl)thiophene) **XIX** by Andersson et al. [224]. The regioselectivity of the oxidative polymerization has been ascribed to both the steric hindrance of the octylphenyl side group and its electronic effect on the 2 position, preventing 2,2′-(TT) coupling. A reaction mechanism involving radical-cationic growing chain-ends has been proposed to explain the regioregular polymerization.

Nickel-catalyzed coupling of 2,5-dihalo-3-alkylthiophenes via (magnesium halides) Grignard species has been used to prepare **XVIII** (**A**) [220,225,226]. Low molecular weights and regiorandom polymers were generally obtained. Highly regioselective routes to PATs using nickel-mediated coupling of a 2-halo-3-alkylthiophene-5-magnesium halide (**B**) or a 5-zinc halide (**C**) have been reported in 1992, by McCullough et al. [227–229] and Rieke and coworkers [230–232], respectively (Figs. 9.24 and 9.25). These methods produce highly regioregular PATs with more than 98% HT links and molecular weights in the range 2–4 × 10^4. Later, a simpler method has been developed by the McCullough group to synthesize 95% HT PATs **XVIII** [233,234], using an 85/15 mixture of 2-bromo-3-alkyl-5-bromomagnesio-thiophene and its 2-bromomagnesio-5-bromo isomer, and a Ni-based catalyst.

The regioregularity of PATs plays an important role on their bandgap, as HH couplings provoke an out-of-plane torsion of the thiophene units due to strong steric repulsion of the alkyl side groups. This twisting destroys conjugation and opposes to intrachain charge mobility [205]. Hadziioannou et al. have shown that HH dyads limit

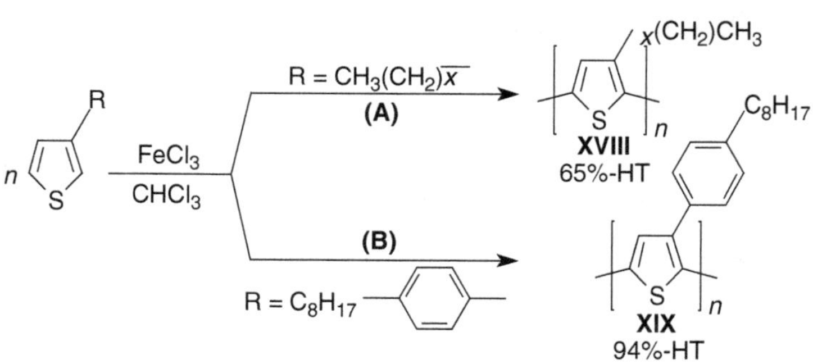

FIGURE 9.24 Oxidation polymerization route to regiorandom (A) and regioregular (B) PATs.

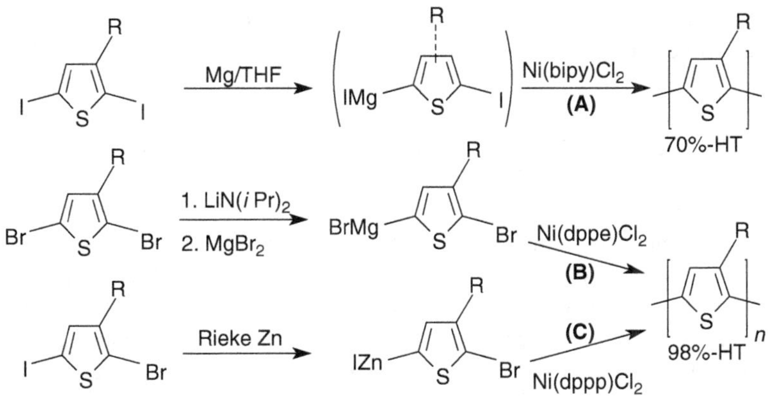

FIGURE 9.25 Transition metal-catalyzed polymerizations to regioirregular (A) and highly regioregular PATs by the McCullough (B) and Rieke (C) routes.

the π-conjugation length [235]. They obtained a fine color control through the variation of the length of coplanar thiophene blocks between HH links and found an energy bandgap varying as the inverse of the thiophene blocks length, as calculated by Brédas et al. [236]. It has been reported that deviation from coplanarity has to remain lower than 30° to achieve a sufficient overlapping of π-orbitals to create a conduction band [237]. Torsion of HH dyads induces an increase of the bandgap with a concomitant blue-shift of the emission spectrum. For example, emission of poly(3-hexylthiophene) (P3hT) films have been found to undergo a red shift from 608 to 670 nm as a result of increasing the percentage of HT dyads from 50% to 80% [238]. A highly regioregular P3hT with 98% HT exhibits a pure-red luminescence peaking at 717 nm [211]. As an unfavorable consequence of increasing the HT regioregular chain planarity is the decrease of photoluminescence efficiency (from 0.8% for 50% HT to 0.2% for 80% HT) arising from chain aggregation in the solid state. Solution to this problem has been found with mono or di-alkylphenyl sustituents as in poly(3-(4-octylphenyl)thiophene) **XIX** and poly(3-(2,5-dioctylphenyl)thiophene) **XX**, in which the coplanarity is preserved and interchain interactions are prevented due to the perpendicular orientation of the phenyl group to that of the thiophene unit. Andersson et al. [40,239] have reported photoluminescence efficiencies of 9% and 24% and red light-emission peaking at 677 and 720 nm for films of **XIX** and **XX**, respectively.

The first PLEDs based on polythiophenes were first reported, in 1991, by Ohmori et al. [240] who utilized PATs **XVIII** (from the oxidation method and with A = $C_{12}H_{25}$, $C_{18}H_{37}$, $C_{22}H_{45}$) in single-layer red-orange electroluminescent devices [240]. The next year, Heeger and coworkers fabricated a diode using poly(3-octylthiophene), showing a red-orange electroluminescence with a low external quantum efficiency ($\eta_{ext} = 0.025\%$) [241]. As more efficient PLEDs have been prepared by the Andersson group utilizing poly(3-(4-octylphenyl))thiophene **XIX** and poly(3-(2,5-dioctylphenyl))thiophene **XX** [40,210,239]. The former shows a red light emission at 670 nm (1.86 eV) and η_{ext} of 0.3% [210]. Diodes built with the latter, produce red light and show η_{ext} of 0.1% and 0.7%, in a single-layer configuration or in a double-layer one with a hole-blocking/electron transporting layer, respectively [40,239].

9.4 MOLECULAR AND MACROMOLECULAR ORGANIC LIGHT-EMITTING SYSTEMS FOR WHITE EMISSION

Lighting accounts for around 20% of the electricity consumed every year in the United States, with almost 40% of that amount consumed by inefficient incandescent lamps. This has generated huge interest in the use of alternative white light sources, and especially organic devices well known for their low-cost manufacturability. During the past few years, organic and polymer light-emitting diodes for white emission (WOLEDs and WPLEDs) have consequently been the subject of a challenging research to improve their efficiency, stability, and color quality (see Fig. 9.26).

To achieve white emission from fluorescent small-molecule based OLEDS, various approaches have been used (see [242] or [243] for a complete review) falling into two categories: down-conversion by phosphors, where a blue OLED is used to excite several green and red phosphors to produce white light and color mixing using multicolor emitters in the same OLED device. The last technique is more efficient as no losses associated with the wavelength conversion processes occur. The color mixing could be achieved by several ways:

- *Multilayer Devices*: The OLED structure consists in several emissive layers (or a single-layer doped with multiple emitters) with different emission colors. By controlling the recombination zone(s) (e.g., through the use of several electron and/or hole blocking layers), it is then possible to balance the different colors

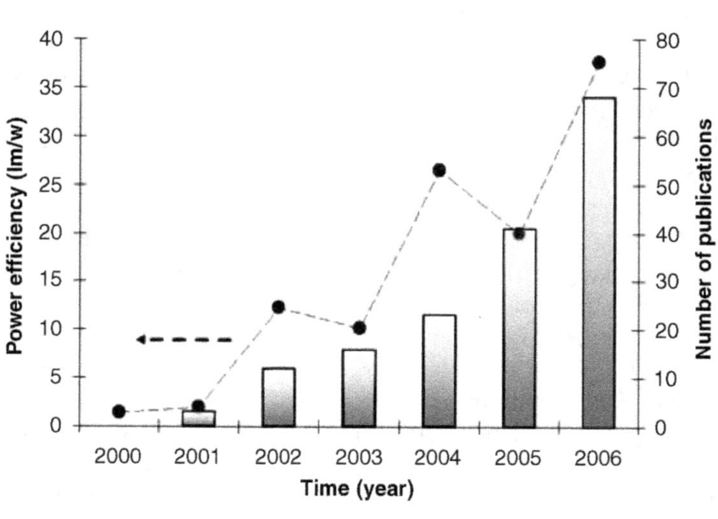

FIGURE 9.26 Evolution of number of journal reports on WOLED together with the power efficiency (based on the references given in [242]).

(red, green, and blue or complementary colors, e.g., blue and yellow) to obtain white light [66,242,244–250]. Among these various devices, numerous multilayer doped-type WOLEDs [244,248] and nondoped ones [66,249,250] have been fabricated. For example, by adjusting both the thickness and position of a ultrathin sensing layer of rubrene as a yellow emitter within the recombination zone of excitons formed at the hole transport/blue emissive interface, a fine chromaticity tuning may be obtained from deep blue to pure yellow via bright white with CIE coordinates (0.32, 0.31) close to the equi-energy white point (0.33, 0.33) [66]. An important breakthrough in the field has been brought by Forrest and coworkers in the early 2000s, with all-phosphor-doped organic light-emitting diodes (PHOLEDs) for white emission (see [242,251] and references therein). As previously emphasized, these PHOLEDs have the potential for internal quantum efficiency of 100% because both the singlet and triplet populations are involved. They use organometallic phosphors based on heavy atoms, such as platinum or iridium, to produce high efficiency red and green phosphorescence (efficient and stable blue phosphorescent dopants are still a bottleneck). For example, eight-layer PHOLEDs consisting of a blue fluorophore, green and red iridium-based phosphors as dopants of an organic conducting host, have been built to produce high-efficiency white emission [251]. In such a device configuration, singlet and triplet excitons are channeled along two different optimized ways minimizing transfer energy losses (see Fig. 9.27). These devices challenge incandescent lamps by exhibiting a η_{ext} of 18.4%, a power efficiency of 24 lm/W at 500 cd/m^2, with CIE coordinates (0.38, 0.40), and a CRI of 85.

- *Exciplex Emission*: A solution to obtain white light while reducing the number of layers and dopants is to use a broadly emitting exciplex (i.e., an excited state formed by two charges located on two dissimilar molecules). Fluorescent exciplex systems exhibit low efficiency [252], but phosphorescent ones lead to broad efficient WOLED [253].
- *Other Approaches*: Other alternative approaches have been reported, such as the SOLED (for stacked OLED) consisting in three OLEDs (red, green, and blue) stacked with four electrodes and three independent voltage sources to control independently the emission of each of the device [254]. It is also possible to employ multimode resonant microcavities [255,256] (see Section 9.1.3.1): several modes overlap the spectrum of an organic material, in such a way that two or three modes are selected to result in white light emission. However, this last approach is very angle-sensitive, limiting its interest for lighting applications.

Similar strategies have been followed to obtain white emission from fluorescent polymer systems and phosphorescent ones as well. However, polymer blends have generally been used to fabricated WPLEDs (as multilayer spin-coated polymer assemblies are difficult to obtain owing to close solubility parameters of materials), such as red, green, blue light-emitting three-polymer systems [257–259] and two-

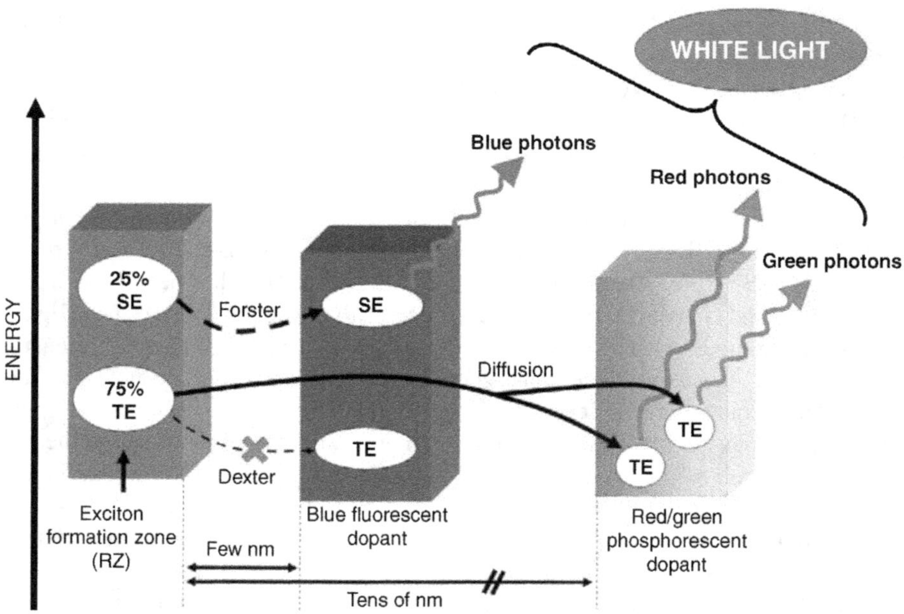

FIGURE 9.27 White fluorescent/phosphorescent OLED principle (after Sun et al. [251]). The high-energy singlet excitons are transferred via short-range Förster process to the radiative singlet excitons of the blue fluorescent emitter. The triplet excitons diffuse over a long distance (tens of nm) and transfer their energy to the lower energy emissive triplet state of the phosphorescent dopants.

polymer blends containing macromolecular materials emitting two complementary colors [260–265]. For example, an efficient WPLED has been reported by Huang et al. by blending polyfluorene **V** as a blue emitter, with a small percent of MEHPPV **XII** as a source of orange-red light. A quantum efficiency η_{ext} of 6% and a power efficiency of 16 lm/W at 110 cd/m^2, with CIE coordinates (0.36, 0.40) were obtained [264]. WPLEDs from a single copolymer with main-chain or/and side-chain segments emitting two complementary colors or the three fundamentals have been also demonstrated [266–268].

To improve quantum efficiency of WPLEDs, phosphorescent polymer-based systems made of a polymer host and organo-Pt or -Ir phosphors have been used [65,269–276]. Due to its high-energy blue-emissive excited state, PVK has been employed as a host for red, green, and blue phosphorescent cyclometalated Ir (III) complexes [65,271]. PHOLEDs containing such blends were found to exhibit a η_{ext} of 2.1%, with CIE coordinates (0.33, 0.42), CRI of 77 [271]. Poly(fluorene)s as **V** or **VI** have also been utilized in PHOLEDs as a blue emissive polymeric material in association with Ir(III) phosphors [272,273] PHOLEDs from a single fluorene-based copolymer by mixing singlet and triplet emission have been reported [275,276]. These copolymers use **V** as a blue light-emitting backbone, short benzothiadiazole main-chain segments as green emitters, and Ir(III)-based complexes side-chain groups as red-light phosphors. Devices display a white light emission with CIE values of (0.34, 0.30), a luminance efficiency of 2.8 cd/A. Multilayer PHOLEDs based on **V** and a red Ir(III)-based phosphorescent emitter have been reported by Gong et al. [273] to exhibit a high-quality white light with a power efficiency of 6 lm/W, a luminance efficiency of 21 cd/A, a CRI value of 92, and stable ideal-white CIE coordinates of (0.33, 0.33).

Although electrophosphorescent polymer devices of the latter-type are attractive candidates for development of low-cost large-area solid-state lighting sources, they are still of a lower quality to that of their small-molecule-based PHOLEDs analogs.

REFERENCES

1. A. Bernanose, M. Comte, and P. Vouaux, *J. Chim. Phys.* (in french) **1953**, *50*, 64–68.
2. M. Pope, H.P. Kallmann, and P. Magnante, *J. Chem. Phys.* **1963**, *38*, 2042–2043.
3. C.W. Tang and S.A. Vanslyke, *Appl. Phys. Lett.* **1987**, *51*(12), 913–915.
4. J.H. Burroughes, et al. *Nature* **1990**, *347*(6293), 539–541.
5. J.R. Skotheim and R.L. Elsenbaumer, (Eds.), *Handbook of Conducting Polymers*, Marcel Dekker, New York, 1998.
6. A. Kraft, A.C. Grimsdale, and A.B. Holmes, *Angew. Chem.-Int. Ed.* **1998**, *37*(4), 402–428.
7. P. Bernier, S. Lefrant, and G. Bidan,(Eds.), *Advances in Synthetic Metals: Twenty Years of Progress and Technology*, Elsevier, Lausanne, 1999.
8. J.L. Reddinger and J.R. Reynolds, Molecular engineering of pi-conjugated polymers in: *Radical Polymerisation Polyeletrolytes* Springer, Berlin, 1999, pp. 57–122.
9. R.H. Friend, et al. *Nature* **1999**, *397*(6715), 121–128.
10. D.Y. Kim, H.N. Cho, and C.Y. Kim, *Prog. Polym. Sci.* **2000**, *25*(8), 1089–1139.
11. U. Mitschke and P. Bauerle, *J. Mater. Chem.* **2000**, *10*(7), 1471–1507.
12. J. Shinar (Ed.), *Organic Light-Emitting Devices*, Springer-Verlag, New York, 2004.
13. S. Müllen,(Ed.), *Organic Light-Emitting Devices*, Wiley-VCH, Weinheim, Germany, 2006.
14. M.T. Bernius, et al. *Adv. Mater.* **2000**, *12*(23), 1737–1750.
15. M. Leclerc, *J. Polym. Sci. Part A: Polym. Chem.* **2001**, *39*(17), 2867–2873.
16. D. Neher, *Macromol. Rapid Commun.* **2001**, *22*(17), 1366–1385.
17. U. Scherf and E.J.W. List, *Adv. Mater.* **2002**, *14*(7), 477–487.
18. P.A. Lane,C in: J. Shinar (Ed.), *Organic Light-Emitting Devices*, Springer Verlag, New York, 2004, pp. 265–301.
19. P. Chen, et al. *Polym. Int.* **2006**, *55*(5), 473–490.
20. J.V. Grazulevicius, et al. *Prog. Polym. Sci.* **2003**, *28*(9), 1297–1353.
21. J.F. Morin, et al. *Macromol. Rapid Commun.* **2005**, *26*(10), 761–778.
22. S. Grigalevicius, *Synth. Met.* **2006**, *156*(1), 1–12.
23. S. Tasch, W. Graupner, G. Leising, and J. Shinar (Eds.), *Organic Light-Emitting Devices*, Springer Verlag, New York, 2004, pp. 205–244.
24. U. Fahnenstich, et al. *Makromol. Chem.-Macromol. Symp.* **1992**, *54–55*, 465–476.

25. M. Remmers, et al. *Macromolecules* **1996**, *29*(23), 7432–7445.
26. Y. Yang, Q. Pei, and A.J. Heeger, *J. Appl. Phys.* **1996**, *79*(2), 934–939.
27. S. Tasch, et al. *Adv. Mater.* **1997**, *9*(1), 33.
28. A.C. Grimsdale and K. Müllen, *Angew. Chem.-Int. Ed.* **2005**, *44*(35), 5592–5629.
29. D. Braun and A.J. Heeger, *Appl. Phys. Lett.* **1991**, *58*(18), 1982–1984.
30. P.L. Burn, et al. *Nature* **1992**, *356*(6364), 47–49.
31. N.C. Greenham, et al. *Nature* **1993**, *365*(6447), 628–630.
32. N.C. Greenham, and R.H. Friend, in: F.S.E.H. Erhenreich (Ed.), *Solid State Physics*, Academic Press, San Diego, CA, 1995, pp. 1–149.
33. J. Salbeck, *Ber. Bunsen-Ges.-Phys. Chem. Chem. Phys.* **1996**, *100*(10), 1667–1677.
34. A. Hilberer, et al. *Macromol. Chem. Phys.* **1997**, *198*(7), 2211–2235.
35. U. Scherf (Ed.), Oligo- and polyarylenes, oligo- and polyarylenevinylenes, in: *Carbon Rich Compounds II*, Springer-Verlag, New York, 1999, pp. 163–222.
36. H.-K. Shim, J.J.-I. in: K.-S. Lee (Ed.), *Polymers for Photonics Applications I. Advances in Polymer Science*, Vol. 158, Springer-Verlag, New York, 2002, pp. 194–243.
37. A.C. Grimsdale and S. Müllen,(Eds.), *Organic Light-Emitting Devices*, Wiley-VCH, Weinheim, 2006, pp. 215–243.
38. I.F. Perepichka, et al. *Adv. Mater.* **2005**, *17*(19), 2281–2305.
39. M. Berggren, et al. *Nature* **1994**, *372*(6505), 444–446.
40. M.R. Andersson, et al. *J. Mater. Chem.* **1999**, *9*(9), 1933–1940.
41. M. Pfeiffer, et al. *Org. Elect.* **2003**, *4*(2–3), 89–103.
42. I.G. Hill, et al. *Appl.Phys. Lett.* **1998**, *73*(5), 662–664.
43. I.D. Parker, *J. Appl. Phys.* **1994**, *75*(3), 1656–1666.
44. P.S. Davids, I.H. Campbell, and D.L. Smith, *J. Appl. Phys.* **1997**, *82*(12), 6319–6325.
45. S. Karg, M. Meier, and W. Riess, *J. Appl. Phys.* **1997**, *82*(4), 1951–1960.
46. B.K. Crone, et al. *Appl. Phys. Lett.* **1998**, *73*(21), 3162–3164.
47. S. Barth, et al. *Synth. Met.* **2000**, *111–112*, 327.
48. P.S. Davids, et al. *Appl. Phys. Lett.* **1996**, *69*(15), 2270–2272.
49. Y.N. Gartstein and E.M. Conwell, *Chem. Phys. Lett.* **1996**, *255*(1–3), 93.
50. V.I. Arkhipov, et al. *J. Appl. Phys.* **1998**, *84*(2), 848–856.
51. H. Bässler, *Phys. Status Solidi B* **1981**, *107*(1), 9–54.
52. R.A. Marcus, *Rev. Mod. Phys.* **1993**, *65*(3), 599.
53. H. Bässler, *Phys. Status Solidi B* **1993**, *175*(1), 15–56.
54. E. Tutis, B. Ivo, and B. Detlef, *Phys. Rev. B (Condens. Matter Mater. Phys.)* **2004**, *70*(16), 161202.
55. I.I. Fishchuk, et al. *Phys. Rev. B* **2003**, *67*(22), 224303.
56. V. Podzorov, et al. *Phys. Rev. Lett.* **2004**, *93*(8), 086602.
57. P.W.M. Blom, M.J.M. de Jong, and M.G. van Munster, *Phys. Rev. B* **1997**, *55*(2), R656.
58. H.C.F. Martens, P.W.M. Blom, and H.F.M. Schoo, *Phys. Rev. B* **2000**, *61*(11), 7489.
59. W.F. Pasveer, et al. *Phys. Rev. Lett.* **2005**, *94*(20), 206601.
60. S. Heun and P.M. Borsenberger, *Chem. Phys.* **1995**, *200*(1–2), 245.
61. W.D. Gill, *J. Appl. Phys.* **1972**, *43*(12), 5033–5040.

62. I.I. Fishchuk, et al. *Phys. Rev. B (Condens. Matter Mater. Phys.)* **2007**, *76*(4), 045210.
63. E. Tutis, D. Berner, and L. Zuppiroli, *J. Appl. Phys.* **2003**, *93*(8), 4594–4602.
64. M.A. Baldo, et al. *Appl. Phys. Lett.* **1999**, *75*(1), 4–6.
65. X.H. Yang, et al. *Adv. Mater.* **2006**, *18*(7), 948.
66. H. Choukri, et al. *Appl. Phys. Lett.* **2006**, *89*(18), 183513.
67. N.C. Giebink, Y. Sun, and S.R. Forrest, *Org. Elect.* **2006**, *7*(5), 375–386.
68. H. Becker, S.E. Burns, and R.H. Friend, *Phys. Rev. B* **1997**, *56*(4), 1893–1905.
69. S.E. Burns, et al. *Adv. Mater.* **1997**, *9*(5), 395.
70. J.S. Kim, et al. *J. Appl. Phys.* **2000**, *88*(2), 1073–1081.
71. M.A. Baldo, et al. *Phys. Rev. B* **1999**, *60*(20), 14422–14428.
72. V. Bulovic, et al. *Phys. Rev. B* **1998**, *58*(7), 3730.
73. M. Wohlgenannt and, Z.V. Vardeny, *J. Phys. Condens. Matter.* **2003**, *15*(3), R83–R107.
74. P.T. Worthing and, W.L. Barnes, *Appl. Phys. Lett.* **2001**, *79*(19), 3035–3037.
75. S. Moller and S.R. Forrest, *J. Appl. Phys.* **2002**, *91*(5), 3324–3327.
76. T. Tsutsui, et al. *Adv. Mater.* **2001**, *13*(15), 1149–1152.
77. Q. Huang, et al. *Appl. Phys. Lett.* **2006**, *88*(11), 113515.
78. M.H. Lu and J.C. Sturm, *J. Appl. Phys.* **2002**, *91*(2), 595–604.
79. Y.J. Lee, et al. *Appl. Phys. Lett.* **2003**, *82*(21), 3779–3781.
80. T. Tsutsui, et al. *Appl. Phys. Lett.* **1994**, *65*(15), 1868–1870.
81. S.H. Fan, et al. *Phys. Rev. Lett.* **1997**, *78*(17), 3294–3297.
82. M. Agrawal and P. Peumans, *Opt. Exp.* **2007**, *15*(15), 9715.
83. P. Kovacic and M.B. Jones, *Chem. Rev.* **1987**, *87*(2), 357–379.
84. T. Yamamoto, Y. Hayashi, and A. Yamamoto, *Bull. Chem. Soc. Jpn.* **1978**, *51*(7), 2091–2097.
85. M. Delamar, et al. *Electrochim. Acta* **1982**, *27*(1), 61–65.
86. J.F. Fauvarque, et al. *Makromol. Chem. Macromol. Chem. Phys.* **1985**, *186*(12), 2415–2425.
87. K. Miyashita and M. Kaneko, *Synth. Met.* **1995**, *68*(2), 161–165.
88. G. Grem, et al. *Adv. Mater.* **1992**, *4*(1), 36–37.
89. D.G.H. Ballard, et al. *Macromolecules* **1988**, *21*(2), 294–304.
90. D.L. Gin, V.P. Conticello, and R.H. Grubbs, *J. Am. Chem. Soc.* **1992**, *114*(8), 3167–3169.
91. T. Yamamoto, *Prog. Polym. Sci.* **1992**, *17*(6), 1153–1205.
92. A.D. Schluter and G. Wegner, *Acta Polym.* **1993**, *44*(2), 59–69.
93. T. Vahlenkamp and G. Wegner, *Macromol. Chem. Phys.* **1994**, *195*(6), 1933–1952.
94. M. Remmers, M. Schulze, and G. Wegner, *Macromol. Rapid Commun.* **1996**, *17*(4), 239–252.
95. V. Percec, et al. *Macromolecules* **1996**, *29*(11), 3727–3735.
96. K.C. Park, et al. *Macromolecules* **1996**, *29*(22), 7149–7154.
97. M. Rehahn, et al. *Polymer* **1989**, *30*(6), 1060–1062.
98. A.D. Schluter, *J. Polym. Sci. Part A: Polym. Chem.* **2001**, *39*(10), 1533–1556.
99. Y. Yang, Q. Pei, and A.J. Heeger, *Synth. Met.* **1996**, *78*(3), 263–267.
100. U. Scherf and K. Müllen, *Macromolecules* **1992**, *25*(13), 3546–3548.

101. U. Scherf and K. Müllen, *Makromol. Chem. Rapid Commun.* **1991**, *12*(8), 489–497.
102. M. Fukuda, K. Sawada, and K. Yoshino, *Jpn. J. Appl. Phys. Part 2: Lett.* **1989**, *28*(8), L1433–L1435.
103. Q.B. Pei and Y. Yang, *J. Am. Chem. Soc.* **1996**, *118*(31), 7416–7417.
104. M. Ranger, D. Rondeau, and M. Leclerc, *Macromolecules* **1997**, *30*(25), 7686–7691.
105. S. Gamerith, et al. *Phys. Status Solidi A: Appl. Res.* **2004**, *201*(6), 1132–1151.
106. E.J.W. List, et al. *Adv. Mater.* **2002**, *14*(5), 374–378.
107. A.W. Grice, et al. *Appl. Phys. Lett.* **1998**, *73*(5), 629–631.
108. M. Gaal, E.J.W. List, and U. Scherf, *Macromolecules* **2003**, *36*(11), 4236–4237.
109. Y. Ohmori, et al. *Jpn. J. Appl. Phys. Part 2: Lett.* **1991**, *30*(11B), L1941–L1943.
110. K.H. Weinfurtner, et al. *Appl. Phys. Lett.* **2000**, *76*(18), 2502–2504.
111. D.D.C. Bradley, et al. *Synth. Met.* **1991**, *43*(1–2), 3135–3141.
112. M. Grell, et al. *Adv. Mater.* **1999**, *11*(8), 671–675.
113. J.F. Ambrose and R.F. Nelson, *J. Electrochem. Soc. Electrochem. Sci. Technol.* **1967**, *115*, 1159.
114. S.A. Jenekhe, S.T. Wellinghoff, and J.F. Reed, *Mol. Cryst. Liq. Cryst.* **1984**, *105*(1–4), 175–189.
115. A. Siove, et al. *Polym. Int.* **1995**, *37*(3), 171–177.
116. A. Siove and D. Ades, *Polymer* **2004**, *45*(12), 4045–4049.
117. A. Siove, et al. *Makromol. Chem. Macromol. Chem. Phys.* **1989**, *190*(6), 1361–1368.
118. A. Iraqi and I. Wataru, *Synth. Met.* **2001**, *119*(1–3), 159–160.
119. J. Ostrauskaite and P. Strohriegl, *Macromol. Chem. Phys.* **2003**, *204*(14), 1713–1718.
120. J.F. Morin and M. Leclerc, *Macromolecules* **2001**, *34*(14), 4680–4682.
121. A. Siove, et al. *J. Chim. Phys. Phys.-Chim. Biol.* **1995**, *92*(4), 787–790.
122. Y. Pelous, et al. *Polym. Commun.* **1990**, *31*(9), 341–342.
123. B. Romero, et al. *Synth. Met.* **1996**, *80*(3), 271–277.
124. D.B. Romero, et al. *Adv. Mater.* **1997**, *9*(15), 1158.
125. A. Fischer, et al. *J. Phys. D: Appl. Phys.* **2006**, *39*(5), 917–922.
126. J.I.G. Cadogan, M. Cameron-Wood, R.K. Mackie, and R.J.G. Searle, *J. Chem. Soc.* **1965**, 4831.
127. G. Zotti, et al. *Macromolecules* **2002**, *35*(6), 2122–2128.
128. J.F. Morin and M. Leclerc, *Macromolecules* **2002**, *35*(22), 8413–8417.
129. J.F. Morin, et al. *Appl. Phys. Lett.* **2002**, *80*(3), 341–343.
130. S.A. Patil, U. Scherf, and A. Kadashchuk, *Adv. Funct. Mater.* **2003**, *13*(8), 609–614.
131. F. Dierschke, A.C. Grimsdale, and K. Müllen, *Macromol. Chem. Phys.* **2004**, *205*(9), 1147–1154.
132. R.A. Wessling, *J. Polym. Sci. Polym. Symp.* **1985**, *72*, 55–66.
133. R.A. Wessling and R.G. Zimmerman, **1968**, USA, US Patent 3 401 152.
134. R.W. Lenz, et al. *J. Polym. Sci. Pt A: Polym. Chem.* **1988**, *26*(12), 3241–3249.
135. P.L. Burn, et al. *J. Chem. Soc. Perkin Trans. 1* **1992** *23*, 3225–3231.
136. D.A. Halliday, et al. *Synth. Met.* **1993**, *55*(2–3), 902–907.
137. L. Hontis, et al. *Polymer* **2001**, *42*(13), 5793–5796.

138. L. Hontis, et al. *Synth. Met.* **2001**, *119*(1–3), 135–136.
139. M. Herold, et al. *Synth. Met.* **1996**, *76*(1–3), 109–112.
140. N.C. Greenham and R.H. Friend,in: J. Shinar (Ed.), *Organic Light-Emitting Devices*, Springer Verlag, New York, 2004.
141. N.C. Greenham, et al. *Chem. Phys. Lett.* **1995**, *241*(1–2), 89–96.
142. R.O. Garay, et al. *Adv. Mater.* **1993**, *5*(7–8), 561–564.
143. A. Marletta, et al. *Adv. Mater.* **2000**, *12*(1), 69.
144. J.C. Carter, et al. *Appl. Phys. Lett.* **1997**, *71*(1), 34–36.
145. F. Louwet, D. Vanderzande, and J. Gelan, *Synth. Met.* **1995**, *69*(1–3), 509–510.
146. L. Lutsen, et al. *Macromolecules* **1999**, *32*(20), 6517–6525.
147. H.G. Gilch and W.L. Wheelwright, *J. Polym. Sci. A1* **1966**, *4*, 1337.
148. A.J. Heeger and D. Braun, *Chem. Abstr.* **1993**, *118*, 157401j, 1992.
149. G. Gustafsson, et al. *Nature* **1992**, *357*(6378), 477–479.
150. G.J. Sarnecki, et al. *Synth. Met.* **1993**, *55*(2–3), 914–917.
151. B.R. Hsieh and W.A. Feld, *Abstr. Papers Am. Chem. Soc.* **1993**, *206*, 62-POLY.
152. H. Becker, et al. *Macromolecules* **1999**, *32*(15), 4925–4932.
153. H. Becker, et al. *Adv. Mater.* **2000**, *12*(1), 42.
154. H. Becker, et al. *Synth. Met.* **2001**, *122*(1), 105–110.
155. F. Wudl, et al. *Abstr. Papers Am. Chem. Soc.* **1993**, *205*, 426-POLY.
156. W.J. Swatos and B. Gordon, *Abstr. Papers Am. Chem. Soc.* **1990**, *199*, 230-POLY.
157. B.R. Hsieh, et al. *Adv. Mater.* **1995**, *7*(1), 36–38.
158. N.N. Barashkov, et al. *Synth. Met.* **1995**, *75*(2), 153–160.
159. H. Trad, M. Majdoub, and J. Davenas, *Mater. Sci. Eng. C: Biomimetic Supramol. Syst.* **2006**, *26*(2–3), 334–339.
160. R.N. McDonald and T.W. Campbell, *J. Am. Chem. Soc.* **1960**, *82*, 4669.
161. H.H. Hörhold and J. Opfermann, *Makromol. Chem.* **1970**, *131*, 105.
162. R.W. Lenz and C.E. Handlovits, *J. Org. Chem.* **1960**, *25*, 813.
163. H.H. Horhold and M. Helbig, *Makromol. Chem. Macromol. Symp.* **1987**, *12*, 229–258.
164. Z. Yang, B. Hu, and F.E. Karasz, *Macromolecules* **1995**, *28*(18), 6151–6154.
165. S. Pfeiffer and H.H. Horhold, *Macromol. Chem. Phys.* **1999**, *200*(8), 1870–1878.
166. D.R. Baigent, et al. *Synth. Met.* **1994**, *67*(1–3), 3–10.
167. S.C. Moratti, et al. *Synth. Met.* **1995**, *71*(1–3), 2117–2120.
168. C. Liedenbaum, et al. *Synth. Met.* **1997**, *91*(1–3), 109–111.
169. K.D. Kim, et al. *Macromolecules* **1998**, *31*(21), 7267–7272.
170. A. Cirpan, et al. *Synth. Met.* **2006**, *156*(2–4), 282–286.
171. R.F. Heck, *Pure Appl. Chem.* **1981**, *53*(12), 2323–2332.
172. A. Greiner and W. Heitz, *Makromol. Chem.-Rapid Commun.* **1988**, *9*(8), 581–588.
173. H.P. Weitzel and K. Müllen, *Makromol. Chem.-Macromol. Chem. Phys.* **1990**, *191*(11), 2837–2851.
174. Z.N. Bao, et al. *Macromolecules* **1993**, *26*(20), 5281–5286.
175. S. Klingelhofer, et al. *Macromol. Chem. Phys.* **1997**, *198*(5), 1511–1530.
176. Y.B. Liu, P.M. Lahti, and F. La, *Polymer* **1998**, *39*(21), 5241–5244.

177. F. Koch and W. Heitz, *Macromol. Chem. Phys.* **1997**, *198*(5), 1531–1544.
178. C.Y. Yu and M.L. Turner, *Angew. Chem.-Int. Ed.* **2006**, *45*(46), 7797–7800.
179. Y.J. Miao and G.C. Bazan, *J. Am. Chem. Soc.* **1994**, *116*(20), 9379–9380.
180. S. Tasch, et al. *Adv. Mater.* **1995**, *7*(11), 903–906.
181. L. Pu, M.W. Wagaman, and R.H. Grubbs, *Macromolecules* **1996**, *29*(4), 1138–1143.
182. M. Fahlman, et al. *Macromolecules* **1995**, *28*(6), 1959–1965.
183. J.L. Brédas and A.J. Heeger, *Chem. Phys. Lett.* **1994**, *217*(5–6), 507–512.
184. J. Cornil, et al. *J. Am. Chem. Soc.* **1998**, *120*(6), 1289–1299.
185. B.S. Chuah, et al. *Synth. Met.* **1999**, *102*(1–3), 935–936.
186. T. Zyung, et al. *Synth. Met.* **1995**, *71*(1–3), 2167–2169.
187. G. Yu, *Synth. Met.* **1996**, *80*(2), 143–150.
188. J.C. Scott, et al. *J. Appl. Phys.* **1996**, *79*(5), 2745–2751.
189. J.J.M. Halls, et al. *Thin Solid Films* **1996**, *276*(1–2), 13–20.
190. S. Hoger, et al. *Chem. Mater.* **1994**, *6*(2), 171–173.
191. D.H. Hwang, et al. *J. Chem. Soc.-Chem. Commun.* **1994**, *21*, 2461–2462.
192. D.H. Hwang, et al. *Chem. Commun.* **1996**, *19*, 2241–2242.
193. S.T. Kim, et al. *Adv. Mater.* **1996**, *8*(12), 979–982.
194. H.Y. Chu, et al. *Synth. Met.* **1999**, *101*(1–3), 216–217.
195. R.E. Martin, et al. *Chem. Commun.* **2000**, *4*, 291–292.
196. T. Ahn, et al. *Macromolecules* **1999**, *32*(10), 3279–3285.
197. F. Cacialli, et al. *Synth. Met.* **2000**, *111*, 155–158.
198. L. Liao, et al. *Macromolecules* **2001**, *34*(21), 7300–7305.
199. L.M. Ding, et al. *Macromolecules* **2003**, *36*(19), 7301–7307.
200. M. Hay and F.L. Klavetter, *J. Am. Chem. Soc.* **1995**, *117*(27), 7112–7118.
201. M.R. Pinto, et al. *Polymer* **2000**, *41*(7), 2603–2611.
202. Z. Yang, I. Sokolik, and F.E. Karasz, *Macromolecules* **1993**, *26*(5), 1188–1190.
203. M. Zheng, et al. *Macromolecules* **2001**, *34*(12), 4124–4129.
204. J. Roncali, *Chem. Rev.* **1997**, *97*(1), 173–205.
205. R.D. McCullough, *Adv. Mater.* **1998**, *10*(2), 93.
206. O. Inganas, et al. *Opt. Mater.* **1998**, *9*(1–4), 104–108.
207. D. Beljonne, et al. *J. Chem. Phys.* **1995**, *102*(5), 2042–2049.
208. D. Beljonne, et al. *J. Phys. Chem. A* **2001**, *105*(15), 3899–3907.
209. J. Pei, et al. *Macromolecules* **2000**, *33*(7), 2462–2471.
210. M.R. Andersson, et al. *Macromolecules* **1995**, *28*(22), 7525–7529.
211. F. Chen, et al. *J. Mater. Chem.* **1996**, *6*(11), 1763–1766.
212. P. Barta, et al. *J. Appl. Phys.* **1998**, *84*(11), 6279–6284.
213. K.Y. Jen, G.G. Miller, and R.L. Elsenbaumer, *J. Chem. Soc.-Chem. Commun.* **1986**, *17*, 1346–1347.
214. M. Sato, S. Tanaka, and K. Kaeriyama, *J. Chem. Soc.-Chem. Commun.* **1986**, *11*, 873–874.
215. S. Hotta, et al. *Macromolecules* **1987**, *20*(1), 212–215.

216. S. Hotta, M. Soga, and N. Sonoda, *Synth. Met.* **1988**, *26*(3), 267–279.
217. M. Leclerc, F.M. Diaz, and G. Wegner, *Makromol. Chem.-Macromol. Chem. Phys.* **1989**, *190*(12), 3105–3116.
218. M. Pomerantz, et al. *Synth. Met.* **1991**, *41*(3), 825–830.
219. V.M. Niemi, et al. *Polymer* **1992**, *33*(7), 1559–1562.
220. H.Y. Mao, B. Xu, and S. Holdcroft, *Macromolecules* **1993**, *26*(5), 1163–1169.
221. M.S.A. Abdou, et al. *Chem. Mater.* **1995**, *7*(4), 631–641.
222. J. Laakso, H. Jarvinen, and B. Skagerberg, *Synth. Met.* **1993**, *55*(2–3), 1204–1208.
223. H. Jarvinen, et al. *Synth. Met.* **1995**, *69*(1–3), 299–300.
224. M.R. Andersson, et al. *Macromolecules* **1994**, *27*(22), 6503–6506.
225. T. Yamamoto, et al. *Macromolecules* **1992**, *25*(4), 1214–1223.
226. H.Y. Mao, and S. Holdcroft, *Macromolecules* **1992**, *25*(2), 554–558.
227. R.D. McCullough, et al. *J. Org. Chem.* **1993**, *58*(4), 904–912.
228. R.D. McCullough, et al. *Synth. Met.* **1995**, *69*(1–3), 279–282.
229. R.D. McCullough and R.D. Lowe, *J. Chem. Soc.-Chem. Commun.* **1992**, *1*, 70–72.
230. T.A. Chen and R.D. Rieke, *Synth. Met.* **1993**, *60*(2), 175–177.
231. T.A. Chen and R.D. Rieke, *J. Am. Chem. Soc.* **1992**, *114*(25), 10087–10088.
232. T.A. Chen, X.M. Wu, and R.D. Rieke, *J. Am. Chem. Soc.* **1995**, *117*(1), 233–244.
233. R.S. Loewe, et al. *Macromolecules* **2001**, *34*(13), 4324–4333.
234. R.S. Loewe, S.M. Khersonsky, and R.D. McCullough, *Adv. Mater.* **1999**, *11*(3), 250.
235. R.E. Gill, et al. *Adv. Mater.* **1994**, *6*(2), 132–135.
236. J.L. Brédas, et al. *J. Am.Chem. Soc.* **1983**, *105*(22), 6555–6559.
237. J.L. Brédas, *J. Chem. Phys.* **1985**, *82*(8), 3808–3811.
238. B. Xu and S. Holdcroft, *Macromolecules* **1993**, *26*(17), 4457–4460.
239. M.R. Andersson, et al. *Synth. Met.* **1997**, *85*(1–3), 1383–1384.
240. Y. Ohmori, et al. *Solid State Commun.* **1991**, *80*(8), 605–608.
241. D. Braun, et al. *J. Appl. Phys.* **1992**, *72*(2), 564–568.
242. B.W. D'Andrade and S.R. Forrest, *Adv. Mater.* **2004**, *16*(18), 1585–1595.
243. A. Misra, et al. *Semicond. Sci. Technol.* **2006**, *21*(7), R35–R47.
244. G. Li and J. Shinar, *Appl. Phys. Lett.* **2003**, *83*(26), 5359–5361.
245. M. Stolka (Ed.), *Organic Light-Emitting Diodes for General Illumination, OIDA Publication*, OIDA, Washington, DC, 2002.
246. A.J. Heeger, *Rev. Mod. Phys.* **2001**, *73*(3), 681–700.
247. A.J. Heeger, *Solid State Commun.* **1998**, *107*(11), 673–679.
248. K.O. Cheon and J. Shinar, *Appl. Phys. Lett.* **2002**, *81*(9), 1738–1740.
249. T. Tsuji, et al. *Appl. Phys. Lett.* **2002**, *81*(18), 3329–3331.
250. W.F. Xie, et al. *J. Phys. D: Appl. Phys.* **2003**, *36*(19), 2331–2334.
251. Y.R. Sun, et al. *Nature* **2006**, *440*(7086), 908–912.
252. J. Feng, et al. *Appl. Phys. Lett.* **2001**, *78*(25), 3947–3949.
253. B.W. D'Andrade, et al. *Adv. Mater.* **2002**, *14*(15), 1032.
254. G. Parthasarathy, G. Gu, and S.R. Forrest, *Adv. Mater.* **1999**, *11*(11), 907.

255. A. Dodabalapur, L.J. Rothberg, and T.M. Miller, *Appl. Phys. Lett.* **1994**, *65*(18), 2308–2310.
256. A. Dodabalapur, et al. *Appl. Phys. Lett.* **1994**, *64*(19), 2486–2488.
257. M. Hamaguchi and K. Yoshino, *Appl. Phys. Lett.* **1996**, *69*(2), 143–145.
258. C.C. Huang, et al. *Appl. Phys. Lett.* **2004**, *84*(7), 1195–1197.
259. Y.H. Xu, et al. *Appl. Phys. Lett.*, **2005**, *86*,(16), 163502.
260. G.K. Ho, et al. *Appl. Phys. Lett.* **2004**, *85*(20), 4576–4578.
261. M. Granstrom and O. Inganas, *Appl. Phys. Lett.* **1996**, *68*(2), 147–149.
262. S. Tasch, et al. *Appl. Phys. Lett.* **1997**, *71*(20), 2883–2885.
263. Y.Z. Wang, et al. *Appl. Phys. Lett.* **1999**, *74*(24), 3613–3615.
264. J.S. Huang, et al. *Adv. Mater.* **2006**, *18*(1), 114–117.
265. S.K. Lee, et al. *J. Polym. Sci., Part A: Polym. Chem.* **2007**, *45*, 1199.
266. S.K. Lee, et al. *Adv. Funct. Mater.* **2005**, *15*(10), 1647–1655.
267. J. Liu, et al. *Adv. Funct. Mater.* **2006**, *16*(7), 957–965.
268. G.L. Tu, et al. *Adv. Funct. Mater.* **2006**, *16*(1), 101–106.
269. T.-F. Guo, et al. *Org. Electron.* **2000**, *1*(1), 15.
270. S. Lamansky, et al. *Org. Electron.* **2001**, *2*(1), 53.
271. Y. Kawamura, S. Yanagida, and S.R. Forrest, *J. Appl. Phys.* **2002**, *92*(1), 87–93.
272. X. Gong, et al. *Adv. Mater.* **2004**, *16*(7), 615.
273. X. Gong, et al. *Adv. Mater.* **2005**, *17*(17), 2053.
274. H.A. Al Attar, et al. *Appl. Phys. Lett.* **2005**, *86*.
275. J.X. Jiang, et al. *Adv. Mater.* **2006**, *18*(13), 1769.
276. H.Y. Zhen, et al. *Macromol. Rapid Commun.* **2006**, *27*,(24), 2095–2100.

10

PHOTOINITIATORS FOR FREE RADICAL POLYMERIZATION REACTIONS

Jean Pierre Fouassier, Xavier Allonas, Jacques Lalevée, and Céline Dietlin

Photochemistry and Photophysics of Polymer Materials, Edited by Norman S. Allen

10.1 INTRODUCTION

Light-induced polymerization reactions are largely encountered in many industrial applications. The basic idea is to readily transform a liquid resin or a soft film into a solid film upon light exposure to form either a coating as developed in the UV curing area (this is the larger part of the radiation curing area that also includes the electron beam curing) or an image as used in the (laser) imaging area. The starting resin is in fact a formulation that consists in an oligomer (the nature of its skeleton governs the final properties of the cured coating), a monomer (to facilitate the handling of the mixture), a photoinitiating system (to start the polymerization reaction), and various additives depending on the applications (formulation agents, stabilizers, pigments, fillers, etc.). The monomer and the oligomer possess more than one (usually two to three, sometimes even four) reactive function so that the polymerization leads to an insoluble polymer network.

UV curing application—using conventional Hg or Xe lamps—is extensively used in various following industrial sectors where new applications are continuously emerging: (i) flooring, packaging, release coatings, powder coatings, wood and medium-density fiber MDF panels, automotive, pipe lining related industries that use coatings, varnishes, and paints for many applications on a large variety of substrates such as wood, plastics, metal, papers, optical fibers, and so on, (ii) adhesive related industries such as laminating, pressure sensitive, hot melt, and so on, (iii) graphic arts related industries such as drying of inks, ink jets, labels, overprint varnishes OPV, protective and decorative coatings, and so on, (iv) dentistry and medicine related industries such as restorative and preventative denture relining, wound dressing, ophtalmic lenses, glasses, artificial eye lens, drug microencapsulation, and so on, and (v) microelectronics related industries such as soldering resists, mask repairs, encapsulants, conductive screen ink, metal conductor layer, and so on. Driving forces lie on (i) performance—high surface quality (chemical, mechanical, gloss, scratch, and abrasion resistance), application versatility, enhanced product durability, and so on, (ii) economy—energy saving, low temperature operation (cold cure), small space requirements, fast cure speed, and so on, and (iii) ecology concerns—nearly no volatile organic compounds VOCs (the very low emissions are due to the fact that the solvent used to adjust the viscosity of the formulation is a reactive monomer), very low extractables, development of waterborne coatings, and so on. UV Curing represents a green technology. In 2006, the production of raw materials was about 200, 000 T with an annual growth rate around 9%.

The imaging technology industries—where lasers are very often used currently—appear in high-tech sectors combining photochemistry, organic and polymer chemistry, physics, optics, electronics such as (i) microelectronics—photoresists for the printed circuits, integrated circuits, very large and ultralarge scale integration circuits, and laser direct imaging (LDI) technology that allows to write complex relief structures for the manufacture of microcircuits or to pattern selective areas in microelectronic packaging, and so on, (ii) graphic arts—manufacture of conventional printing plates, computer-to-plate technology that directly helps to reproduce a document on a printing plate, and so on, (iii) 3D machining (or three-dimensional photopolymerization or stereolithography)—which is giving the possibility to make objects for prototyping applications, (iv) optics—holographic recording and information storage, computer generated and embossed holograms, manufacture of optical elements (diffraction grating, mirrors, lenses, waveguides, array illuminators, and display applications), design of structured materials on the nanoscale size, and so on.

Great efforts are taken at present in (i) the development of the curing equipments—excimer lamps, spot lamps, light emitting diodes, radio frequency excited lamps, visible light sources, laser diodes, and so on, (ii) the continuous proposal of new monomers and oligomers for specific applications and properties, and (iii) the design of new photosensitive systems being able to work in well-defined conditions.

As far as the polymerization reactions are concerned in UV curing and imaging areas, they are mostly based on a radical process. Cationic photopolymerization is noticeably less used. Anionic photopolymerization is rather inexistent. Photolatent base generation technology is expected to be developed in the future.

In this chapter, we will focus on photosensitive systems that are used in free radical photopolymerization reactions. We will give the most exhaustive presentation of the commercially used or potentially interesting systems developed on a laboratory scale together with the characteristics of their excited-state properties. We will also show how modern time resolved laser spectroscopy techniques and quantum mechanical calculations allow to probe the photophysical/photochemical properties as well as the chemical reactivity of a given photoinitiating system.

Many papers appeared in the scientific literature more than two decades ago. Numerous books [1] and reviews [2] have been constantly published. We do not intend to give here neither an exhaustive list of references nor a survey of the patent literature that, however, is very important. The selection of the articles was rather a hard task; pioneer works are cited and the references provided in this chapter will mainly refer to papers published during the last 10 years.

10.2 PROPERTIES OF A PHOTOINITIATING SYSTEM

10.2.1 Production of Radicals

A photoinitiating system (PIS) [1i] consists at least in a photoinitiator I. Very often, a coinitiator coI, a radical scavenger RS or a photosensitizer S can be added. Basically, a PIS must lead to radicals that can initiate the polymerization (see (10.1), for example).

$$\mathrm{I} \xrightarrow{h\nu} \mathrm{R}^\bullet \xrightarrow{\mathrm{M}} \mathrm{RM}^\bullet \longrightarrow \text{Polymer} \tag{10.1}$$

The photoinitiator I is usually an organic molecule (except in few cases where it corresponds to an organo metallic compound). Upon excitation by light, I is promoted from its ground singlet state S_0 to its first excited singlet state S_1 and then converted into its triplet state T_1 via a fast intersystem crossing. In many cases, this transient T_1 state yields the reactive radicals $R^\bullet$ that can attack a monomer molecule M and initiate the polymerization. The free radical production can be affected by cage effect on the primary radical pair formed upon photolysis [3]. Many deactivation routes take place

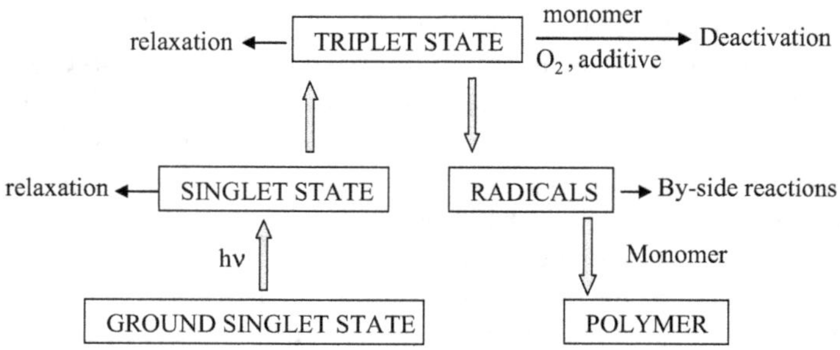

SCHEME 10.1

in the T_1 state: relaxation processes of the spectroscopic states, by-side reactions of the radicals, quenching reactions by oxygen [4], monomer (see, for example, Ref. [5]), additives (UV absorbers UVA, hindered amine light stabiliers HALS [6]), environment (phenols [7–9], micelles [10], etc.). Scheme 10.1 shows the pathways that efficiently lead to the initiation of the polymerization.

The photochemical and photophysical processes that occur in an electronically excited molecule are well described in many books [11]. Radicals in photoinitiators are produced through several following typical processes [1i]:

(i) A photoscission of a C–C, C–S, C–N, and C–P bonds:

$$S_0 \xrightarrow{h\nu} S_1 \longrightarrow T_1 \longrightarrow R^\bullet + R'^\bullet \qquad (10.2)$$

In a general way, an adequate substitution at the R_1 and R_2 moieties of a R_1–R_2 photoinitiator has led to a large variety of compounds. Most cleavable compounds are based on the benzoyl chromophore R_1.

(ii) An hydrogen abstraction reaction between I and coI, which plays the role of a hydrogen donor DH (such as an alcohol, a thiol, etc.); two radicals are formed: one on the donor and another on I (hydroxy radical $K^\bullet$ if I is a ketone).

$$S_0 \xrightarrow{h\nu} S_1 \xrightarrow{DH} T_1 \longrightarrow K^\bullet + D^\bullet \qquad (10.3)$$

(iii) An electron transfer process between I and coI (such as an amine) generating a charge transfer complex (CTC) that is followed by a proton transfer leading to an aminoalkyl radical and a radical on I (e.g., $K^\bullet$)

$$S_0 \xrightarrow{h\nu} S_1 \longrightarrow T_1 \xrightarrow{AH} [CTC] \longrightarrow A^\bullet + K^\bullet \qquad (10.4)$$

The efficiency of PIS is described by two quantum yields: the initiation quantum yield ϕ_i, which represents the number of starting polymer chains per photon absorbed (it depends on the efficiency of the processes involved in the excited states) and the polymerization quantum yield ϕ_m, which is the number of monomer units polymerized per photon absorbed.

The photopolymerization steps can thus be divided into two main parts: the photochemical event that leads to the first monomer radical, the classical chemical propagation and termination processes of the reaction. The rate of a radical polymerization is defined by Equation 10.5, where k_p and k_t are the propagation and termination rate constants and I_{abs} the amount of light absorbed.

$$R_p = k_p/k_t{}^{1/2}[M](I_{abs}\phi_i)^{1/2} \tag{10.5}$$

It is obvious that the rate of polymerization is governed by three types of factors: the experimental conditions (the characteristics of the light source, the wavelength of irradiation, the number of photons emitted, the monomer and initiator concentrations, the presence of additives and oxygen, etc.), the nature of the initiator, and the nature of the monomer.

10.2.2 Absorption of a Photoinitiating System

The spectral absorption range of I is a decisive factor: the wavelength range of the I absorption has to match (see, for example, Fig. 10.1) the spectral emission range of the light source. Polychromatic sources (in general, Hg lamp, Xe lamp, Hg-Xe lamp, and doped Hg lamp) as well as monochromatic sources (lasers) are used. Basic principles in photochemical technologies are presented in Ref. [12].

The molar absorption coefficient ε of the chromophoric group determines the value of the absorbed light intensity I_{abs} (Equation 10.6) by a light beam having an intensity I_0 at a given wavelength. As a consequence, a high absorption is achieved by using molecules having high extinction coefficients and being able to absorb at

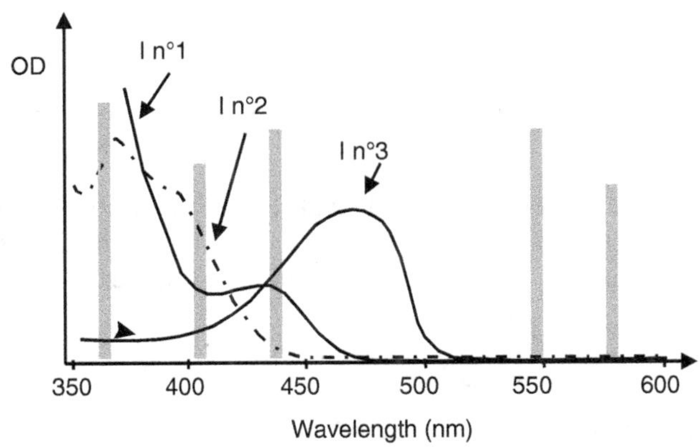

FIGURE 10.1 Typical absorption spectra of photoinitiators that match the emission of a Hg lamp.

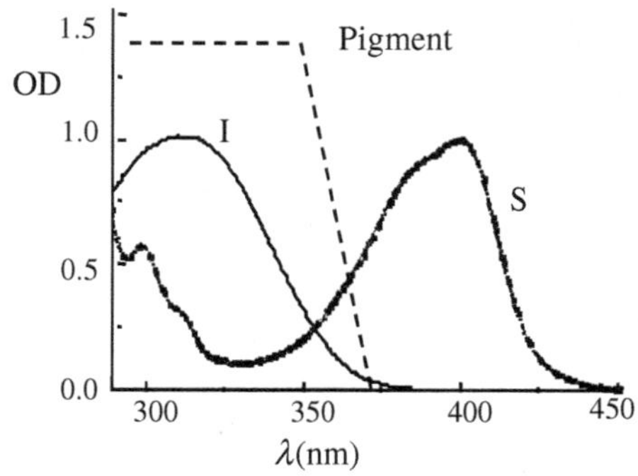

FIGURE 10.2 Role of the pigment on the absorption of the excitation light by a photonitiator I and a photosensitizer S.

the emission lines of the lamp.

$$I_{abs} = I_0[1 - \exp(-2.3\varepsilon lc)] \tag{10.6}$$

The absorption arises from S_0 to S_1 (or S_2 state) transition. It generally involves π, n and π^* molecular orbitals MO, sometimes σ and σ^* MO and d MO in metal organic complexes. The $\pi\pi^*$ transitions are symmetry allowed (high ε) whereas the $n\pi^*$ transitions are symmetry forbidden (low ε).

When pigmented or colored media are used, a spectral window has to be found to excite I. It may happen that the direct excitation of I is impossible. In that case, a photosensitizer S must be added (Fig. 10.2).

The role of S is to absorb the light and to transfer the excess of energy to the photoinitiator through the well-known energy transfer process (for example, (10.7)): the process is efficient only if the energy level of the donor (here designed as ^3S) is higher than that of the acceptor (here designed as ^3I).

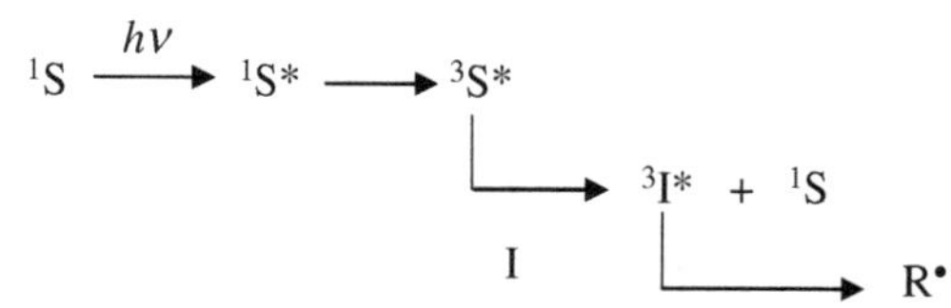

(10.7)

An electron transfer process from S to I (or I to S as a function of the molecules) can be encountered. By extension, it is also depicted as a photosensitization: the efficiency of this process is driven by the free enthalpy change of the reaction.

10.2.3 Choice of a Photoinitiating System

A good photoinitiating system for industrial applications must obviously fulfill a lot of practical requirements: a good solubility or compatibility in the monomer/

oligomer matrix, a good shelf stability, a good acceptance of the marketing considerations (low price, etc.), an excellent safety (absence of odor, toxicity, extractable compounds, etc.), a convenient handling of the formulation, no effect on the final properties of the polymerized material, no formation (or as low as possible) of yellowing photolysis products, and so on. In addition, however, it must exhibit a high reactivity leading to a fast (or ultrafast) polymerization: this is obtained through a well-adapted spectral range absorption, high molar absorption coefficients, an excellent photochemical reactivity, and a high chemical reactivity of the generated radicals.

10.3 AVAILABLE PHOTOINITIATORS: OVERVIEW

In this part, we will present an overview of the radical photoinitiating systems mentioned in the literature. They are classified as one-component (or type I system), two-component (type II systems), and multicomponent photoinitiating systems. According to the chemical structures and the composition of the system, they exhibit a photosensitivity under polychromatic UV lights or UV/visible lights or under laser exposure. Typical absorption spectra of relevant compounds are given in Fig. 10.3.

10.3.1 Main Structures of Radical Photoinitiating

10.3.1.1 One-Component Photoinitiating Systems

Benzoin Ethers Benzoin ethers were the most widely used photoinitiators for a long time [13–15]. The cleavage process in the triplet state (10.8) is very fast compared to that of some related structures ($>10^{10}\,s^{-1}$). The cleavage yield is almost unity for 2,2-dimethoxy—2 phenyl-acetophenone DMPA or 2-methoxy -2 phenyl acetophenone BME. A secondary photochemical cleavage (the possible thermal cleavage is less efficient) can arise in the dimethoxy benzyl radical and leads to a methyl $R^{\bullet}$ radical and an alkyl benzoate.

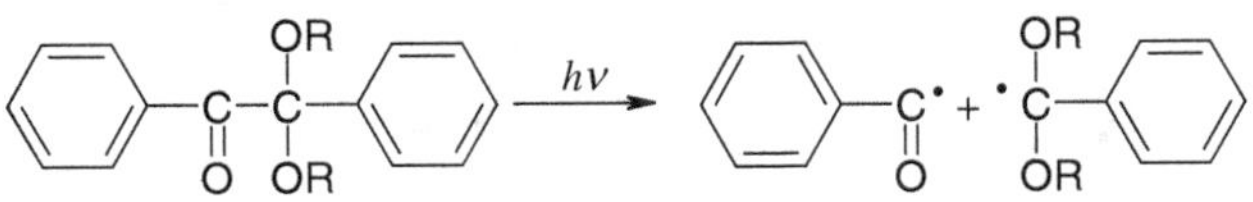

(10.8)

Dialkoxyacetophenones These compounds [16,2i] exhibit a fast cleavage ($\geq 10^{9}\,s^{-1}$) and a quantum yield of polymerization closed to that of DMPA (10.9).

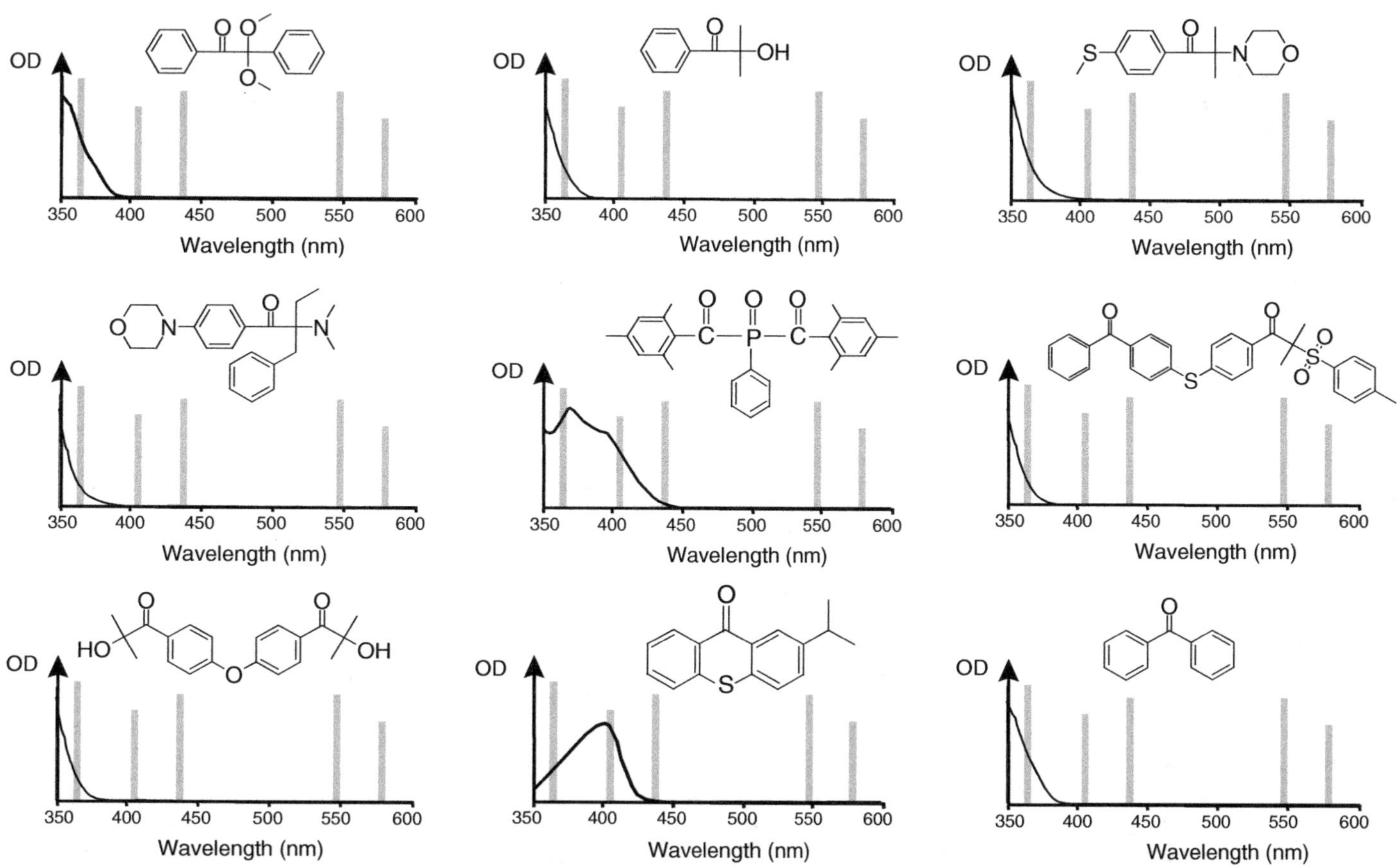

FIGURE 10.3 Typical absorption spectra of usual compounds used as photoinitiators. The characteristic lines of a typical Hg lamp are shown.

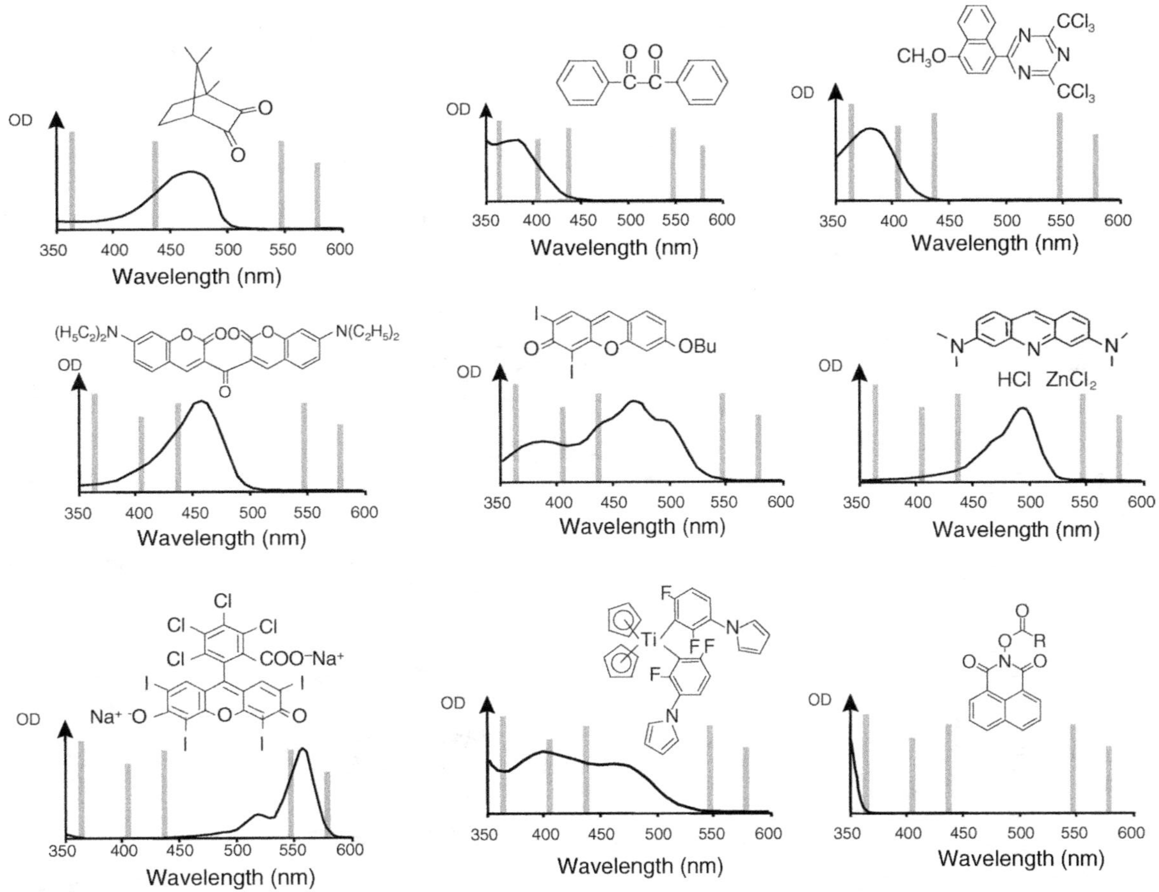

FIGURE 10.3 (Continued)

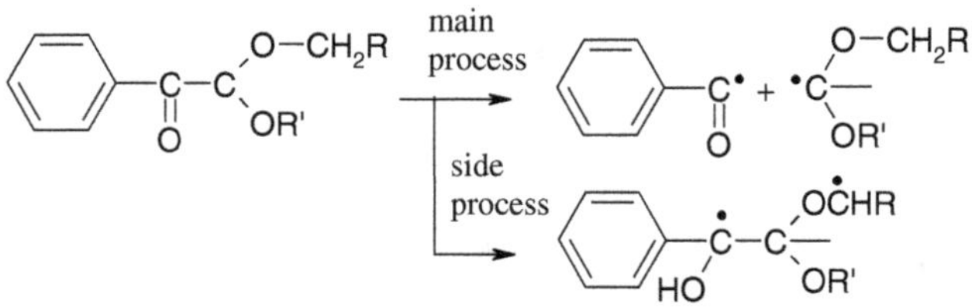

(10.9)

Hydroxy Alkyl Phenyl Ketones The basic structures, 2-hydroxy-2-methyl-1-phenyl-1-propanone HAP and 1-hydroxy-cyclohexyl-1-phenyl ketone HCAP, are also well known [17] and still largely used (10.10) and studied [2i,f,18]. The *para* substitution of the benzoyl group (with an amino, an ether, or a thioether group) drastically changes the electronic transitions but does not significantly affect the reactivity of the derived benzoyl initiating radical.

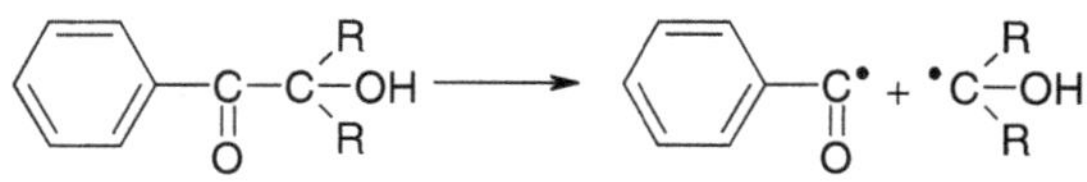

(10.10)

Compounds containing a pyridinoyl, a furane, a thiophene, or a pyrrole chromophore instead of the benzoyl group as well as a new structure containing two furane moieties have been recently proposed [19–22] (10.11)

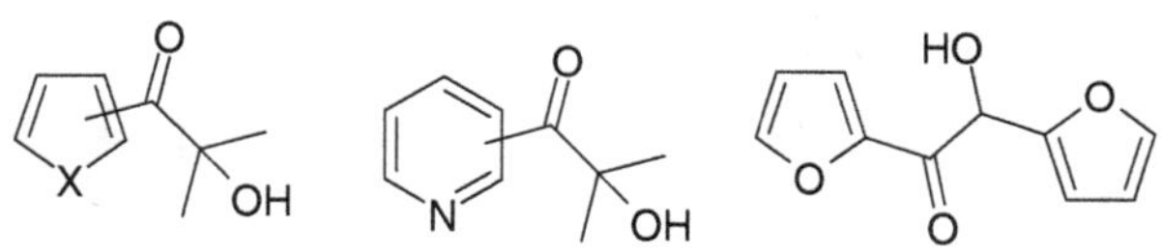

(10.11)

The interest of an introduction of an ynone or an enone as a chromophoric group was also checked [23] (10.12).

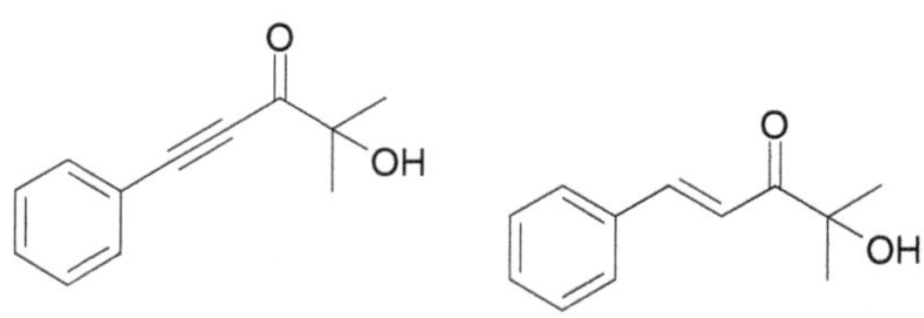

(10.12)

Benzoyl Oxime Esters Few works have been devoted to the photochemistry of these photoinitiators [2f,24–26]. The cleavage of 1-phenyl 2-propanedione-2 (ethoxycarbonyl) oxime PDO occurs at the N–O bond (10.13): a further cleavage process leads to a benzoyl radical, a nitrile and carbon dioxide. The photochemistry of other keto oxime derivatives is rather complex. A suitable substitution on the phenyl ring allows to shift the absorption.

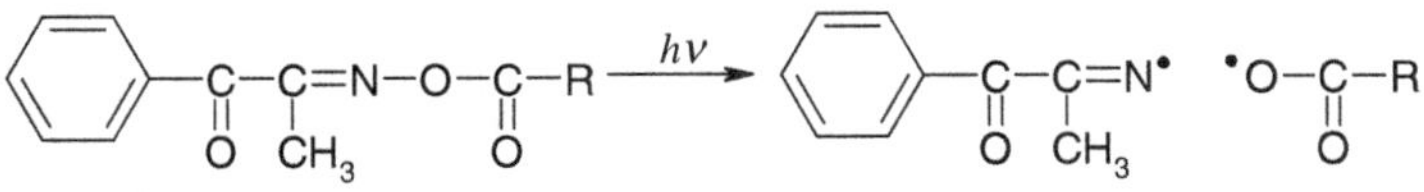

(10.13)

Benzoyl Phosphine Oxides The photochemistry of these efficient photoinitiators has been studied in great details [2f,27–29]. Efficient radicals are produced in the triplet state on a very short timescale. The 2,4,6-trimethyl benzoyl-diphenylphosphine oxide TPO compound was the interesting starting molecule of the series (10.14).

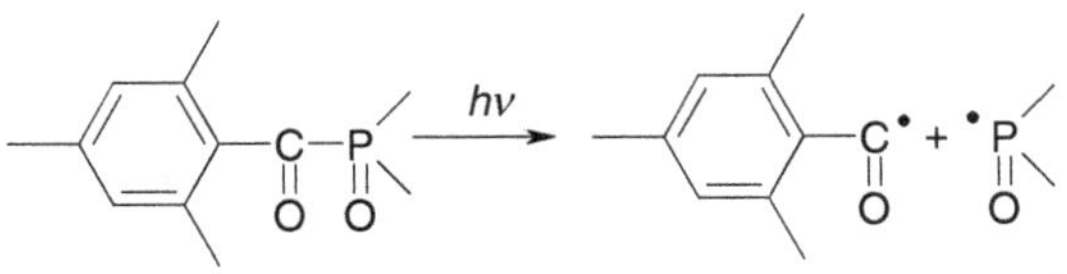

(10.14)

Bis- acyl phoshine derivatives BAPO (for example, (10.15)) containing two substituted benzoyl moieties linked to the P atom were proposed later [1s]. These compounds exhibit an excellent UV-near visible absorption suitable in white-pigmented coatings and an excellent reactivity due to a fast cleavage and a highly efficient phosphinoyl radical. Moreover, the decrease of the ground-state absorption when increasing the UV exposure enables the film to transmit more light so that thick samples can be cured.

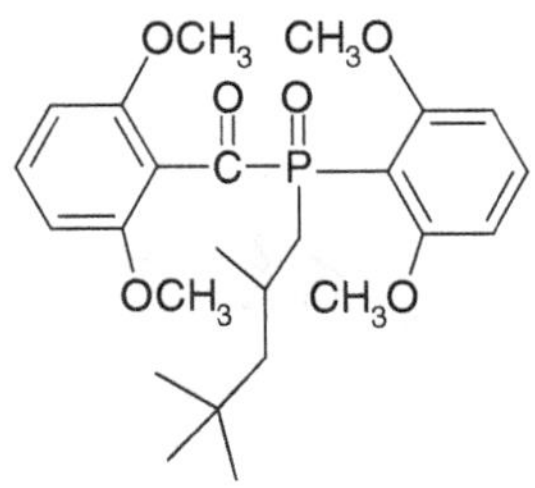

(10.15)

Morpholino- and Amino-Ketones These classes of photoinitiators present a good activity [1s]. The cleavage mechanisms ((10.16) and (10.17)) and the reactivity of the excited states have been fully reported [2i,30]. Excellent representatives of morpholino ketones are: 2-methyl-1-(4-methylthiobenzoyl)-2-morpholino-propan-1-one TMPK (10.16) and 2-methyl-1-(benzoyl)-2-morpholino-propan-1-one MPK. In amino ketones, the introduction of a morpholino moiety on a benzoyl group such as in 2-benzyl-2(diethylamino)-4′-morpholinobutyrophenone MPPK (10.17) red shifts the absorption. In both cases, efficient initiating radicals are generated.

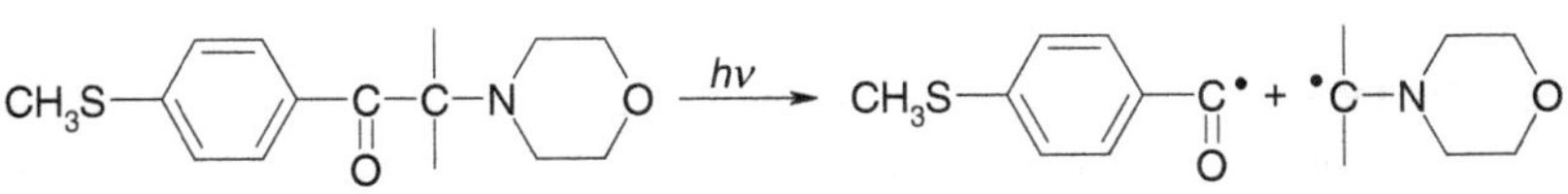

(10.16)

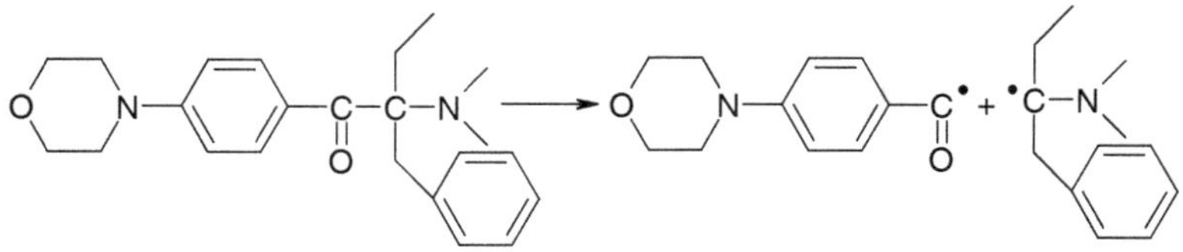

(10.17)

Sulfides The β bond cleavage of phenacyl phenyl sulfides has been recently reviewed [31]. The cleavage of the C–S bond in the benzoylphenyl phenyl sulfide (10.18) [32] occurs in a rather long-lived triplet state. This system, however, works better in the presence of an amine. The role of a sulfide group in a given structure has also been checked [33].

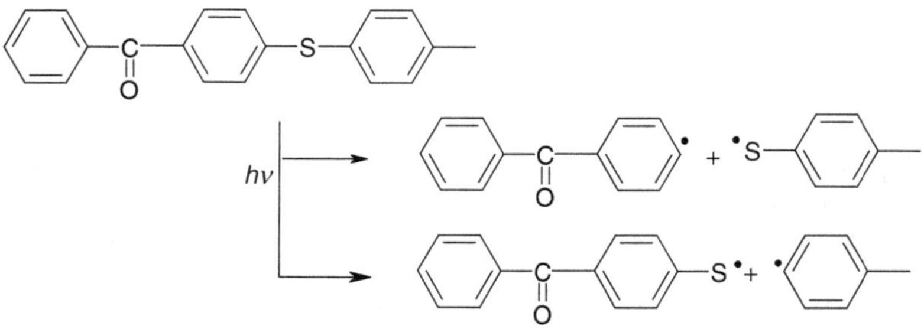

(10.18)

Sulfoxides Ketosulfoxide derivatives (10.19), where R is an alkyl or a phenyl group, exhibit two cleavage processes [34]: one in the triplet state (lifetime: 7.5 ns for R = methyl) and another in the singlet state (lifetime ∼ 3 ns).

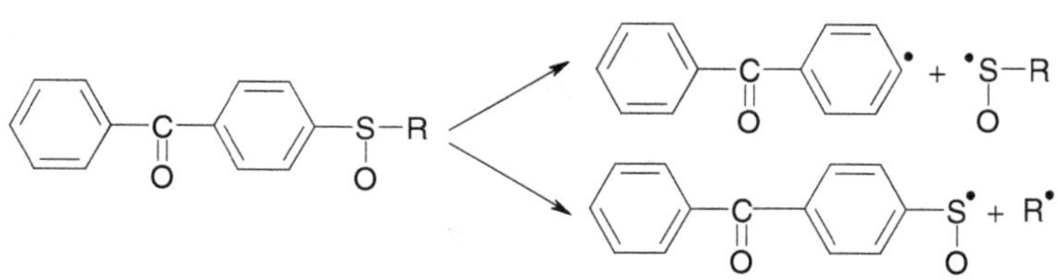

(10.19)

Sulfonyl Ketones Sulfonyl ketones (10.20) undergo a β-cleavage process that occurs in the triplet state [2i,35,36]. A large variety of compounds have been developed [37]. The properties of the benzoyl chromophore are strongly modified by the presence of a sulfoxide, a thiother or a phenyl substituent at the *para* position.

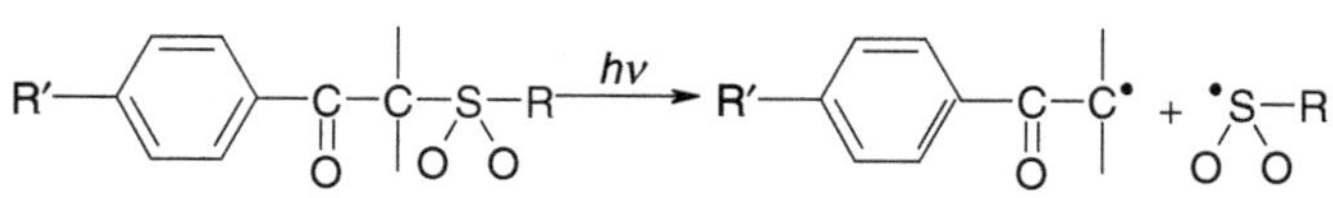

(10.20)

A new and very efficient bifunctional compound BPSK (X = H) [38] exhibiting a better light absorption has been recently developed for the curing of printing inks (10.21). The β-cleavage still occurs at the C–S bond [39].

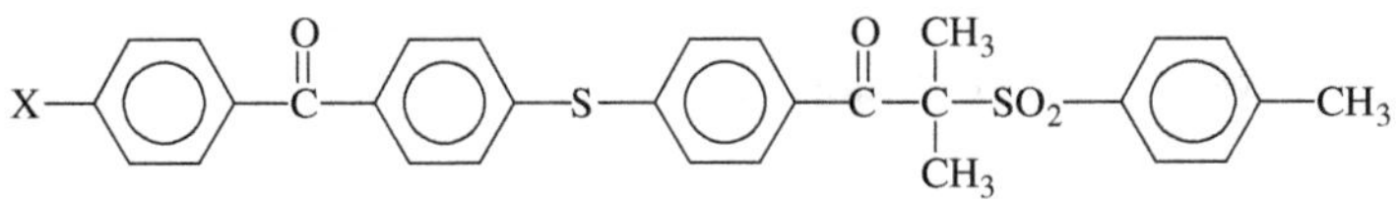

(10.21)

Oxysulfonyl Ketones The triplet cleavage of oxysulfonyl ketones (10.22) occurs at the C–O bond [40,41] but their radical activity is poor [42]. Interestingly, they are able to act as photolatent acid catalyst through the generation of a sulfonic acid under UV light exposure (for example, through a hydrogen abstraction reaction between the oxy sulfonyl radical and a hydrogen donor).

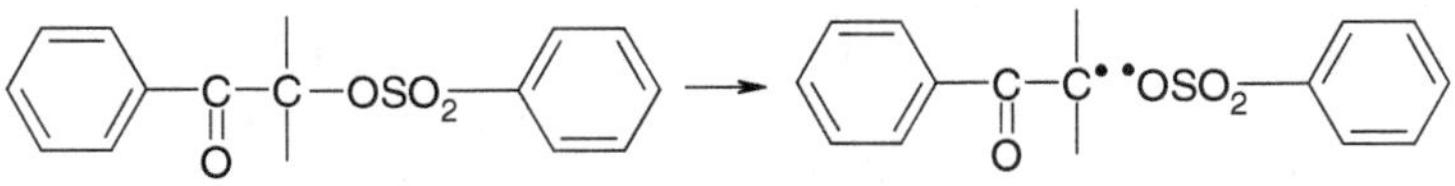

(10.22)

Changing the substitution at the β C atom leads to a more efficient radical photoinitiator (10.23): it allows to favor the α-cleavage process (yielding a benzoyl radical and a second benzyl type radical that liberates an oxysulfonyl radical through a subsequent cleavage).

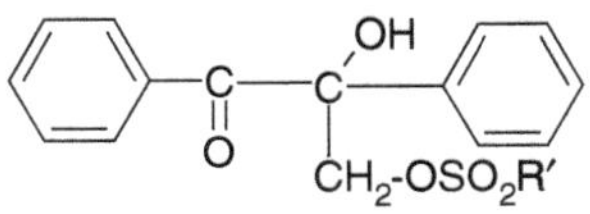

(10.23)

S-Thiobenzoate Derivatives These compounds exhibit an α-cleavage process (10.24) whose efficiency is dependent on the presence of a benzoyl moiety or a benzophenone moiety as the absorbing chromophore [43,44]. Usual benzoyl skeleton-based derivatives work according to a generally efficient cleavage (short triplet state ~2 ns) whereas benzophenone derivatives have a long lifetime triplet state (~500–700 ns).

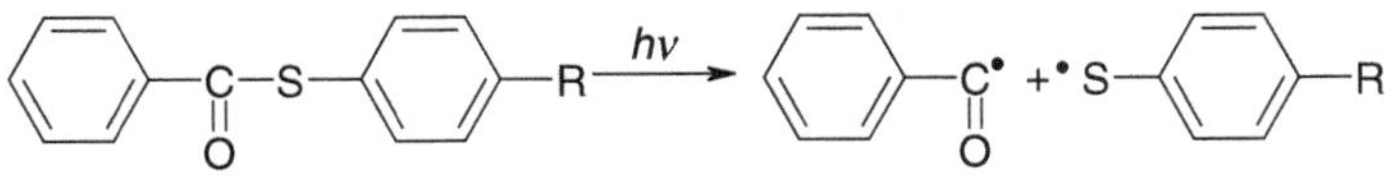

(10.24)

Halogenated Ketones and Trichloromethyl Triazines It is known that halogenated ketones liberate a Cl radical upon light exposure [45]. Substituted tris-(trichloromethyl)-1,3,5-triazines (10.25) generate a chlorine radical and a carbon-centered radical through the cleavage of a C–Cl bond [46,47]; their absorption maxima are located around 400 nm.

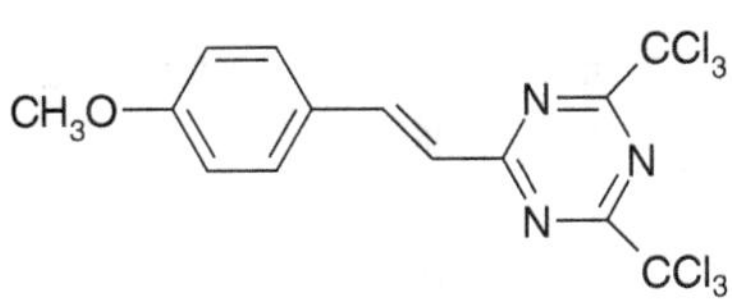

(10.25)

Peroxides A representative compound is the benzoyl peroxide (10.26) that decomposes into two benzoyloxy radicals [48]. However, safety consideration strongly limits the use of such peroxides.

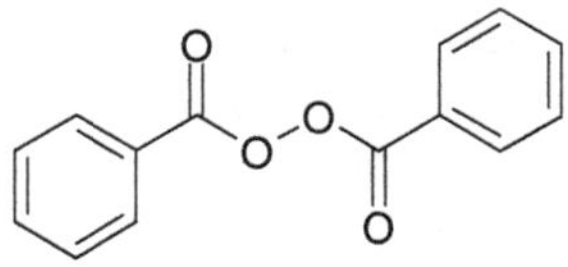

(10.26)

Alkylphenylglyoxylates Methyl phenyl glyoxylate reacts in its triplet state with a monomer bearing an abstractable hydrogen (10.27) and forms an initiating radical on the monomer. A competitive reaction is the formation of a triplet 1,4-biradical that is quenched by oxygen: through a multistep process, this results in the generation of a benzoyl radical, a hydroxyl radical, formaldehyde and carbon dioxide [49].

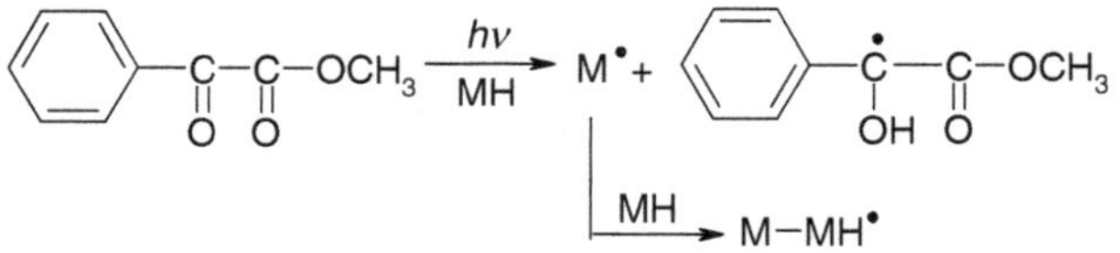

(10.27)

Azides and Aromatic Bis-Azides They possess a N_3 group and decompose into nitrenes [50,51] that may participate in cross-linking reactions through the addition of the nitrene to a double bond (10.28).

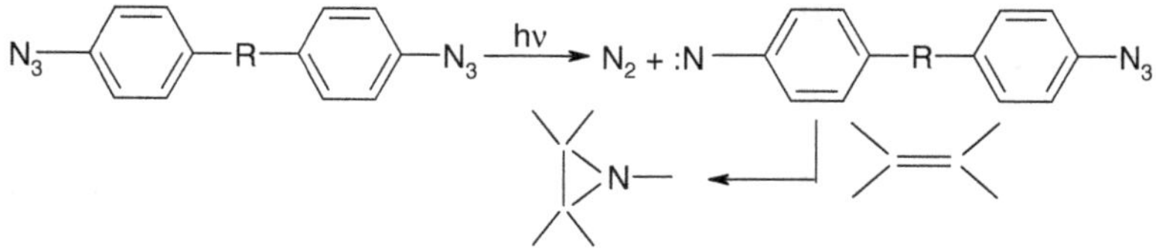

(10.28)

Azo, Disulfide, Disilane, and Diselenide Derivatives Azo compounds [52,53] are easily cleaved: they release nitrogen (10.29).

$$R_1{-}N{=}N{-}R_2 \xrightarrow{h\nu} R_1^{\bullet} + R_2^{\bullet} + N_2 \tag{10.29}$$

Disulfides (such as dithiodiethanol [54]) lead to two thiyl radicals (10.30) that are known to exhibit a low sensitivity to oxygen.

$$R_1{-}S{-}S{-}R_2 \xrightarrow{h\nu} R_1{-}S^{\bullet} + {}^{\bullet}S{-}R_2 \tag{10.30}$$

The Si–Si bond cleavage has been poorly explored. Nevertheless, a new disilane derivative (10.31) has been very recently proposed as a very efficient cleavable photoinitiator [55].

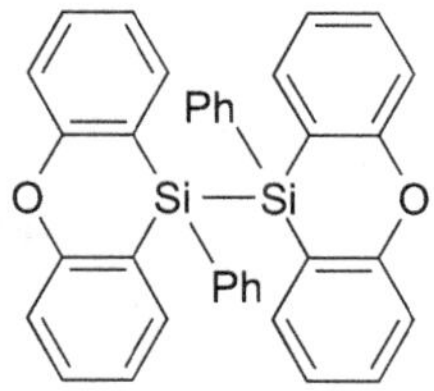

(10.31)

Diphenyldiselenides cleave according to a homolytic way [56].

Peresters In peresters [57–61], a very efficient cleavage of the O–O bond occurs; then, a fast decarboxylation yields an aryl and an alkoxy radicals (10.32).

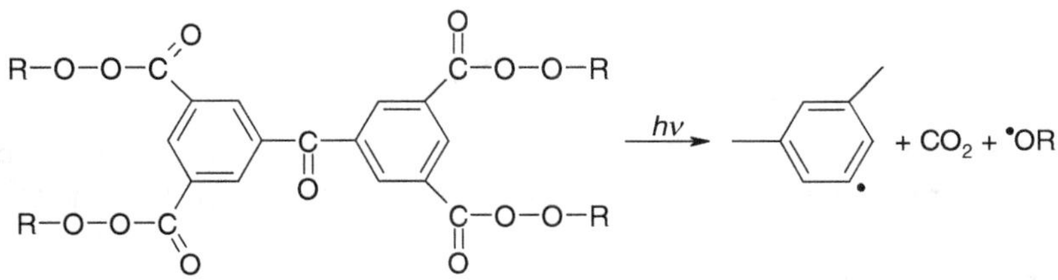

(10.32)

Barton Ester Derivatives Barton ester derivatives are based [62–65] on *O*-acyl derivatives of *N*-hydroxypyridine-2-thione and related compounds such as *O*-acyl-*N*-hydroxy-thiazole-2(3*H*)-thiones (10.33). The excited-state processes lie on a fast singlet-state cleavage leading to thiyl and alkyl radicals. Working on the substitution of the five-membered ring or the nature of the alkyl radical derived from the *O*-acyl moiety might lead to more efficient and more stable compounds [66].

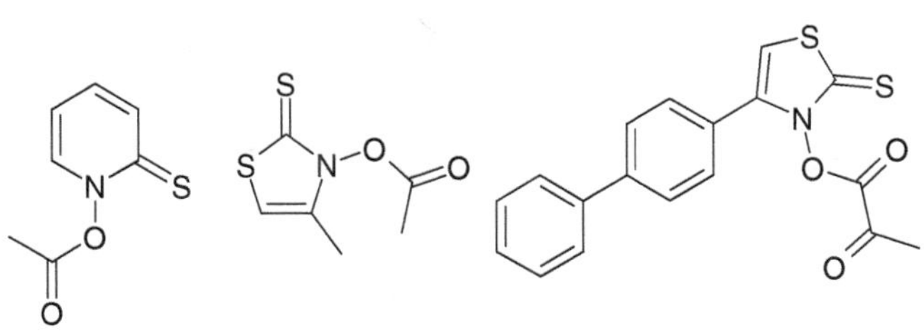

(10.33)

Hydroxamic Acids and Esters The photodissociation of hydroxamic acids and ester derivatives (10.34) was reported for a long time, although their photochemistry remains unclear [67–69]. The decomposition of dibenzoylhydroxylamines is considered as occurring through a N–O bond cleavage; photopolymerization of methylmethacrylate was reported to be effective.

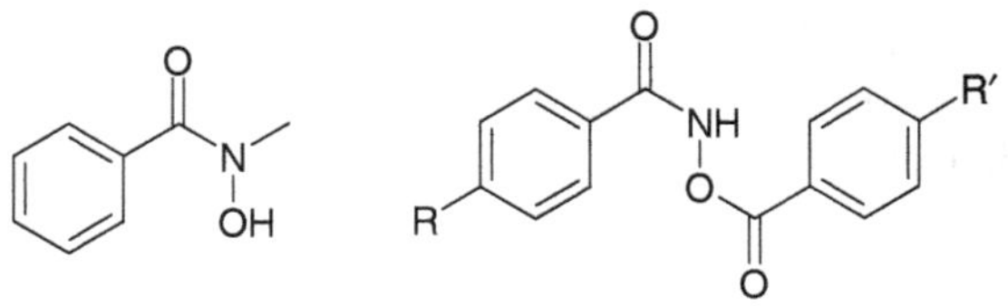

(10.34)

Organoborates Organoborates are composed of a borate anion and a visible light absorbing counter cation (10.35) so that the absorption can be tuned by changing this cation (e.g., cyanine, styrylpyridinium, etc.). After electron transfer and formation of a boranyl radical, an initiating $R^{\bullet}$ radical is generated [70–74].

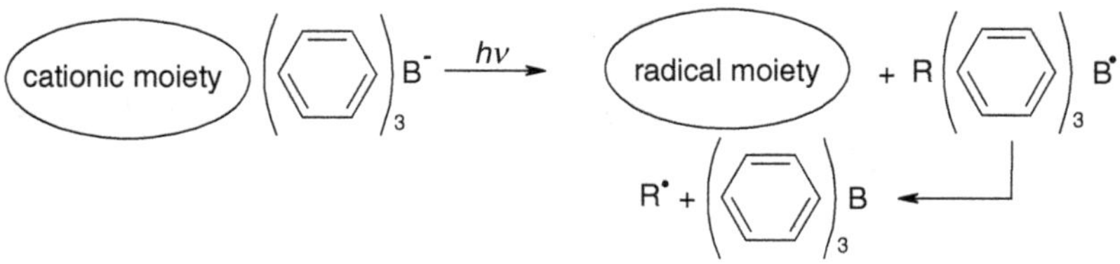

(10.35)

Organometallic Compounds The most widely used compounds [1s,2m] are the titanocene derivatives (10.36), which exhibit an excellent absorption around 500 nm and very efficiently eliminate an aryl and a cyclopentadienyl ligand yielding an unsaturated titanium species and an arylated cyclopentadiene. The Ti structure adds to the CO group of one acrylate unit (may be two) and generates the initiating radicals. Other systems include metal carbonyl (Mn, Fe, Mo, Cr, Os, Re, Ru, etc. [75,76]).

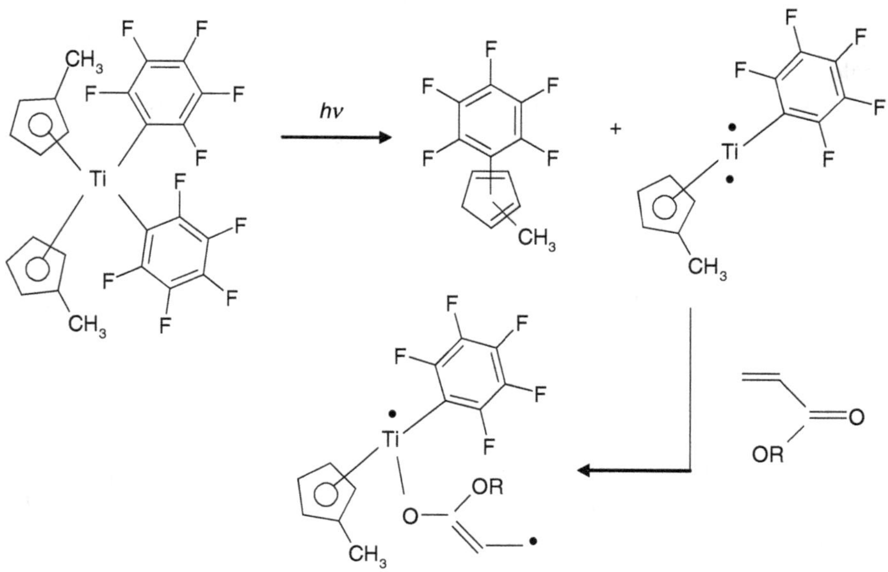

(10.36)

Metal Salts and Metallic Salt Complexes Various metallic cations can be encountered: Co, Fe, Ni, Cu, Cr, and so on [77]. For example, the decomposition of azido-pentaamine cobalt (III) yields an initiating azide radical; the reductive interaction between the chromic acid and a suitable monomer such as acrylamide generates a radical on the monomer (10.37) [78]. Other systems are described in Ref. [79].

$$[Co(NH_3)_5N_3] \xrightarrow{h\nu} N_3^{\bullet} + Co^{2+} + 5\,NH_3$$
$$HCrO_4^- \xrightarrow[M]{h\nu} Cr(V) + M^{\bullet} \qquad (10.37)$$

Light Absorbing Amines Upon light exposure, such an amine (10.38) can be directly excited and then react with an acrylate monomer (see Ref. [2c]). A radical is thus created on the monomer (the monomer acts as an electron acceptor); an alkyl radical is obviously formed on the amine.

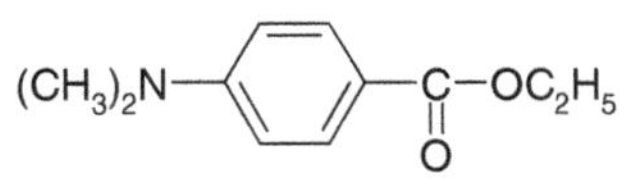

(10.38)

Substituted, Multifunctional, or Polymer Supported Photoinitiators The basic idea is to develop compounds having a better solubility or compatibility in the monomer/oligomer matrix, a good cleavage efficiency and being able to display low volatility, nonyellowing properties and nonreleasing odorous photolysis by-products [80–83]. This has been first achieved using photoinitiators with a long alkyl chain: their reactivity does not significantly differ from that of the unsubstituted parent compounds. A more interesting example relates to a base oligomeric molecule containing the hydroxyl alkyl phenone moiety [84] (10.39) that exhibits interesting properties in UV curing applications.

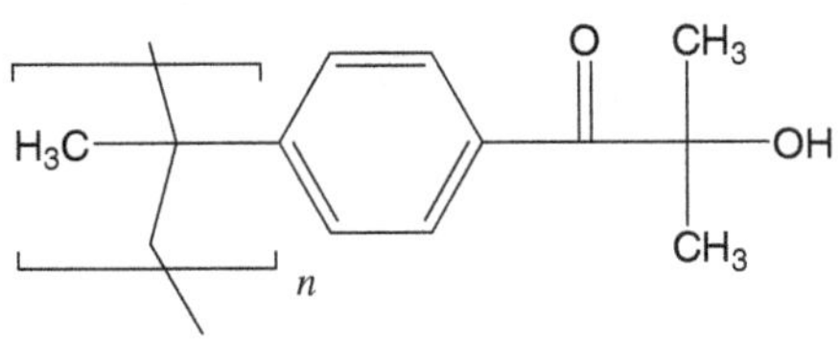

(10.39)

Other examples include polymeric camphorquinones [85] or phosphine oxides [86], monodisperse oligomeric/polymeric compounds [87], and so on.

As reported in a recent review [88], difunctional photoinitiators containing two cleavable moieties exhibit a practical efficiency sometimes better than that of the mono derivatives. In the case of the compound shown in (10.40) (where $X = CH_2$ or O), the photochemical reactivity is almost not affected [89]. A better light absorption is noted: this is due to a large change of the molecular orbitals.

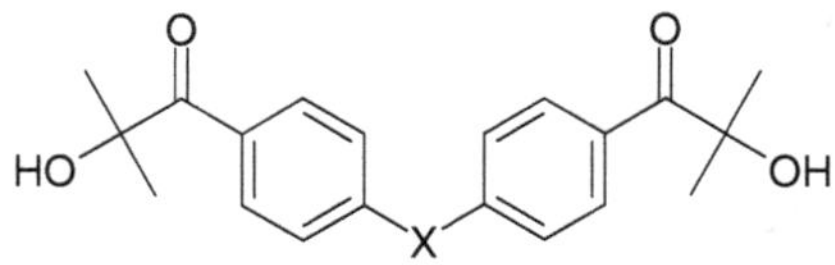

(10.40)

The reactivity of a variety of copolymerizable photoinitiators (where an acrylate group is introduced) has been also checked and discussed in Refs [2i,90,91].

Specific compounds with a long alkyl chain terminated by a fluorinated substituent as well as a mixture of a lithium alkyl sulfonate and a suitable molecule bearing a sulfonic ester behave as surface active photoinitiators [1s] and help to reduce the detrimental effect of the air inhibition in the top layer of the coating.

Water-Soluble Photoinitiators Due to their polar structure, hydrophilic or water-soluble photoinitiators derived from basic structures (hydroxyl alkyl ketones [92] such as the one in (10.41), phosphine oxides [93,94], etc.) or modified compounds (e.g., by carbohydrates [95]) are usable in water-based matrices (water-soluble monomers and aqueous dispersions). Their reactivity has been studied [96,97].

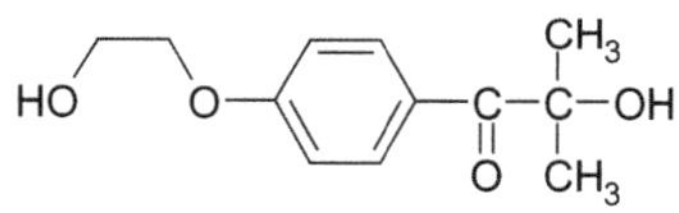

(10.41)

Photoinitiators as Iniferters The availability of iniferters that behave as an initiator, a transfer agent and a terminator is rather limited [98,99]. A renewal of interest is noted for the development of efficient compounds usable in controlled radical photopolymerization reactions [100–104]. Examples of cleavable photoiniferters are shown in (10.42): the particular tetrazole derivative shown here is noticeably attractive due the generation of a tetrazoyl radical that presents a low selectivity and a high efficiency for the addition to acrylate double bonds [105].

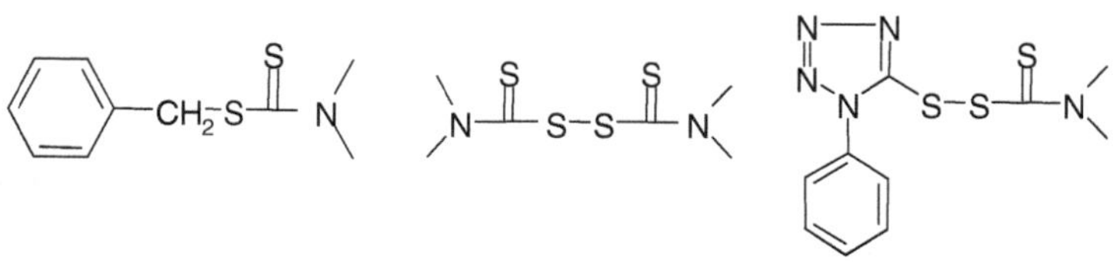

(10.42)

Photoinitiators for Multiphotonic Absorption Multiphoton three-dimensional micro- and nanofabrication is an emerging technology where a laser beam activates the medium by a multiphoton excitation of the photoinitiator. Conventional one-photon UV sensitive cleavable photoinitiators such as bis-acyl phosphine oxides or benzoin ethers can be activated at 700–900 nm (which is thought to lead to the generation of the same usual initiating radicals) but their two-photon absorption cross sections are rather low. New developments are under way [106–111].

Miscellaneous Systems Many systems have been mentioned [1i,m,2i]: acyloxy and acylsilyl phosphine oxides, phosphine sulfides, cyclic compounds, benzoyloxaziridine derivatives, dibenzoylmethane derivatives, triazene and pentaazadiene moiety containing compounds. New developments include benzyl benzoin benzyl ethers [112], dithiocarbamates [113], ketoamides [114], phosphonates [115], bromoacetylpyrene [116], alkylimides [117], aryloxy naphthalene [118], oligosilanes [119], bisphosphine sulfides [120], sulfamic esters of benzoin ethers [121], sulfur [122], or carbohydrate [123] containing compounds.

10.3.1.2 Two-Component Photoinitiating Systems

Usual Ketone-Based Systems Main classes of uncleavable photoinitiators [2i] are based on benzophenones, thioxanthones and camphorquinones (10.43), benzils, α-diketones, anthraquinones, ketocoumarins, and so on.

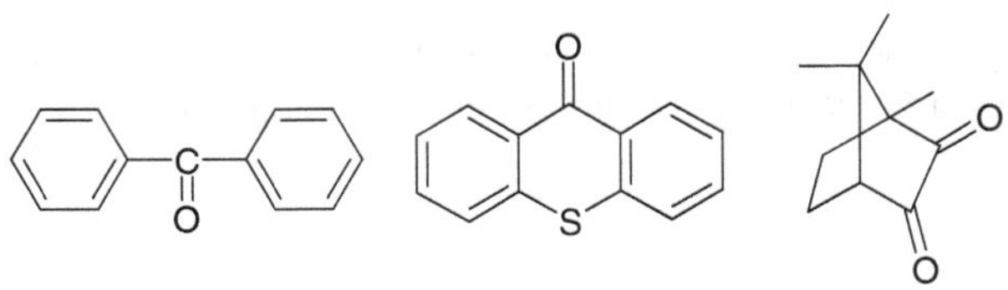

(10.43)

MECHANISM These systems usually work in their triplet state according to a primary electron transfer in the presence of an amine as shown in (10.44).

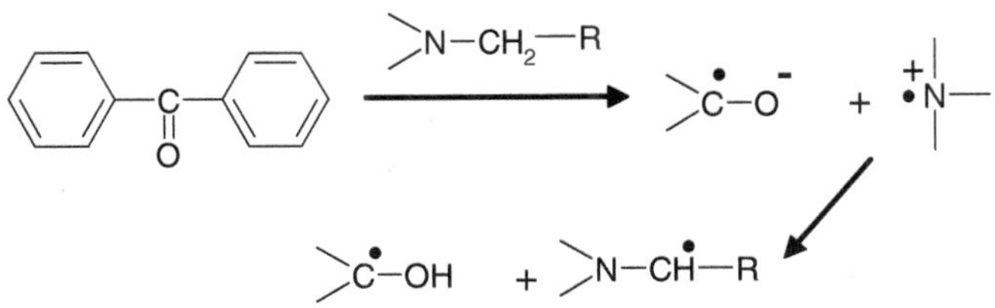

(10.44)

Some examples of widely used amines, in industries, are displayed in (10.45).

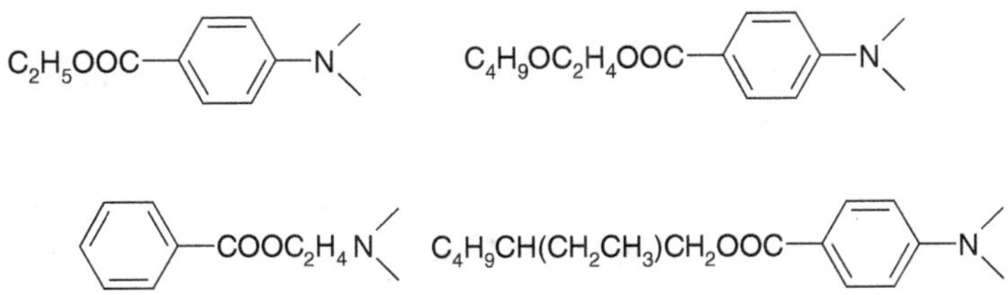

(10.45)

Amine functional acrylates, polymeric amines and copolymerizable tertiary amines which accelerate the cure of UV system have also been proposed (see, for example, (10.46) and Ref. [124]; studied in Ref. [125]).

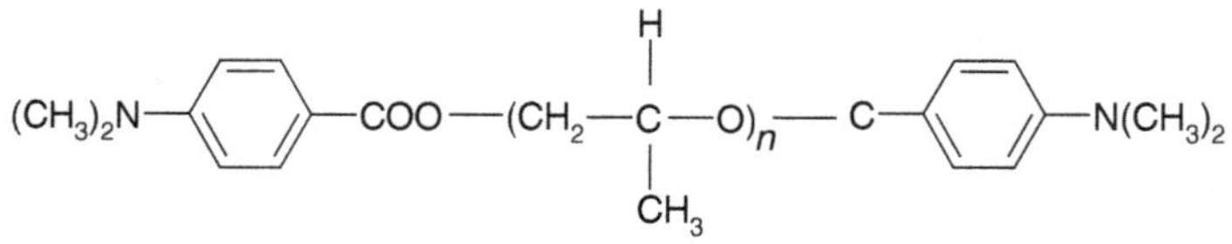
(10.46)

Other interesting coinitiators (10.47) are phenylglycine derivatives (a carbon-centered radical is formed after electron transfer, proton transfer, and decarboxylation) and mercaptans (where a thiyl radical is generated). The role of the mercaptan and the design of newly reactive compounds based on the iniferter structure described above (10.42) have been outlined in Ref. [104].

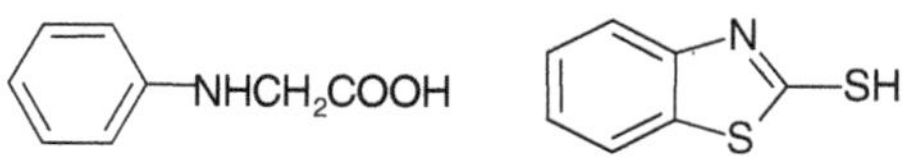
(10.47)

Benzoxazines were shown to be good hydrogen donors [126]. Silanes (for example, (10.48)) were recently proposed as very efficient coinitiators [127,128].

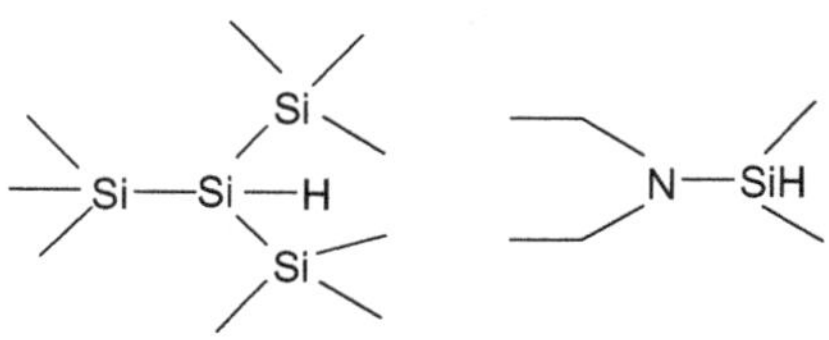
(10.48)

Hindered amines used as light stabilizers are also able to react with ketone triplet states and yield an initiating radical [6].

Benzophenones and thioxanthones can also work through a direct hydrogen abstraction reaction in the presence of H donor such as alcohols or THF (10.49).

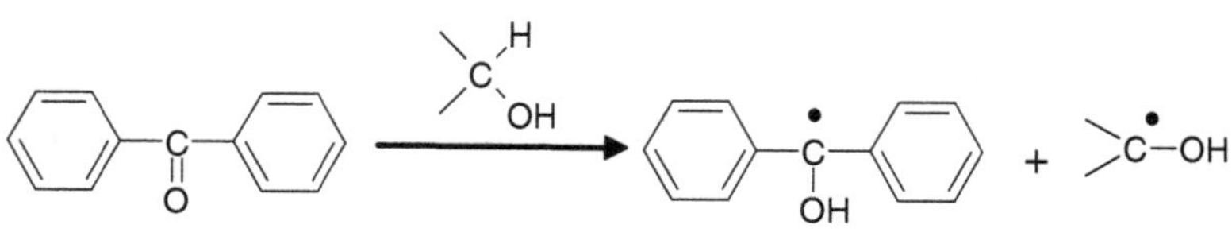
(10.49)

The hydrogen donor can also be a polymer chain possessing a labile hydrogen: the hydrogen abstraction process creates a macromolecular radical that will induce a cross-linking reaction [129].

Some ketones mentioned as type I cleavable systems (such as the benzophenone-ketosulfone structure BPSK (10.21) exhibit a better efficiency in the presence of a coinitiator.

REACTIVITY The photopolymerization activity of many substituted benzophenone derivatives has been checked as a function of the hydrogen donor: amines [130–133], thiols [134–137], aminoacids [138], and phenylglycine [139]. The role of a biphotonic excitation [140], the participation of higher excited states [141,142], and the excited-state dynamics [143] have been also investigated.

Similar works have been carried out with thioxanthones [144–147], anthraquinones [148], camphorquinone [149–152], ketocoumarins [153], xanthones [154], α-di-ketones [1i,155], and so on.

Several other substituted benzophenone systems [1i] such as aminobenzophenones (Mischler's ketone and others), trimethylsilylbenzophenones, alkylthiobenzophenones [156], aryloxy benzophenones [157], maleimidophenoxyl benzophenones [158], polycyclic aromatic ketones (such as tetralone derivatives) were shown to exhibit some interesting properties.

The substitution at the 1,4-position (compared to the usual 2-substitution as in 2-Cl or 2-isopropyl thioxanthone) of the thioxanthone skeleton (10.50) has a larger effect. Interactions with amines are rather complex on the picosecond timescale [159]. Photolysis under steady-state conditions as well as nanosecond laser spectroscopy suggests a possible mechanism for a C–Cl bond breaking.

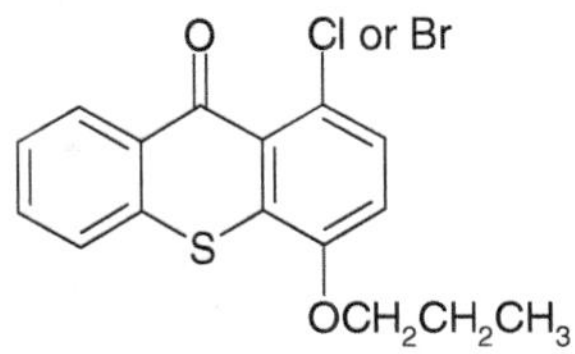

(10.50)

Carefully selected α-diketones allow an excellent efficiency in the near UV-visible region for the photopolymerization of clear thick moulded objects [160]. Excited-state properties of various compounds have been described in Ref. [161].

Amine containing ketone based photoinitiators bearing an amine moiety in the benzophenone or thioxanthone series [162] as well as a new thioxanthone derivative (where the coinitiator such as a thiol [163], an acetic acid derivative [164,165] is grafted) have been developed (10.51); Covalently bonded camphorquinone/amine [166] or amino alkyl ketone/thiol [167] were also recently reported.

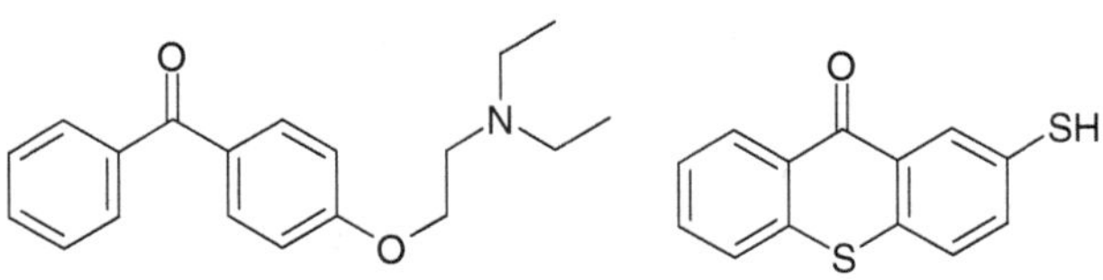

(10.51)

Other following ketone families have been checked: fluorenone, flavone, anthrone, quinone, naphtacenequinone, quinoline [1i], benzodioxinone [168], azidoxanthone and azidofluorene [169], benzils [170], nitroanilines [171–173], and so on.

DERIVATIVES WITH SPECIFIC PROPERTIES As in type I systems, various multifunctional or polymeric compounds are continuously proposed (see Refs [88,163] and references therein): acrylated derivatives of benzophenone (10.52) and thioxanthone, oligomeric phenylglyoxylate (or benzophenone, thioxanthone) [174] (10.53), polymeric systems with a pendant amine moiety and a pendant photoinitiator moiety (benzophenone, thioxanthone, and camphorquinone) [1i,175–181], covalently bonded amine and side-chain ketone moieties [182–184], and so on. Polymeric photoinitiators bearing a thioxanthone and a morpholino ketone moiety [180] lead to an efficient energy transfer process that appears as much higher than that observed for the corresponding low molecular weight structural models mixtures and an improvement of the photoinitiator activity of the acrylate polymerization is thus achieved.

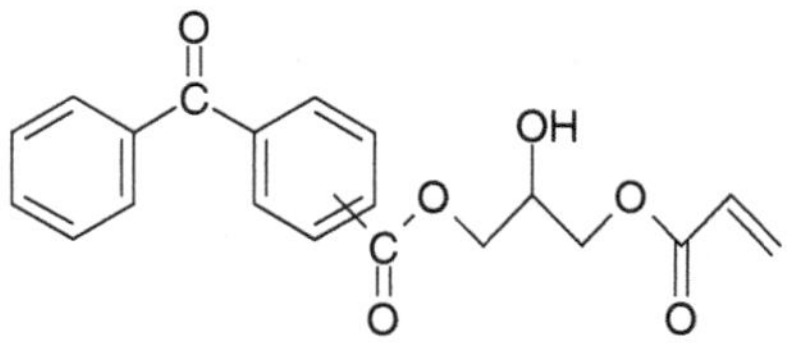

(10.52)

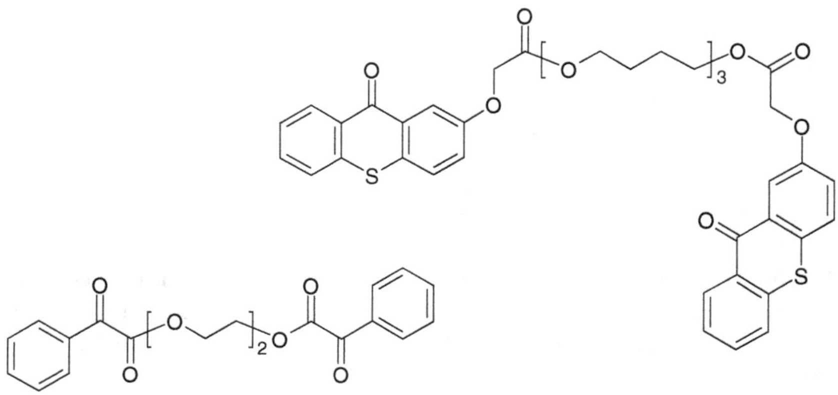

(10.53)

Various structures derived from water-soluble neutral or ionic benzophenones, benzils, thioxanthones (10.54) have been studied for the photopolymerization of water-soluble monomers, monomers in direct and reverse micelles or microemulsions, aqueous dispersions of monomers, water-based coatings [185–188], and so on.

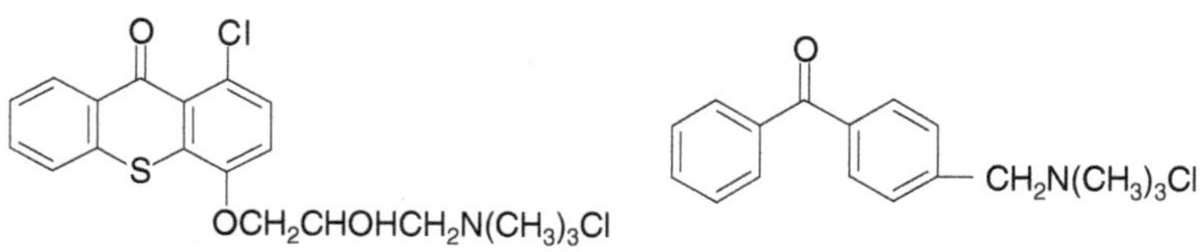

(10.54)

Some particular ketones usable as multiphoton photoinitiators have been synthesized. For example, a compound containing two-electron donating groups linked by a conjugated chain (10.55) is sensitive in the range 800–1000 nm; the initiation mechanism is not fully understood: an electron transfer with the monomer might be the primary process [189]. Newly developed cross-conjugated photoinitiators with bathochromic shifts exhibit one and two photon activity [190].

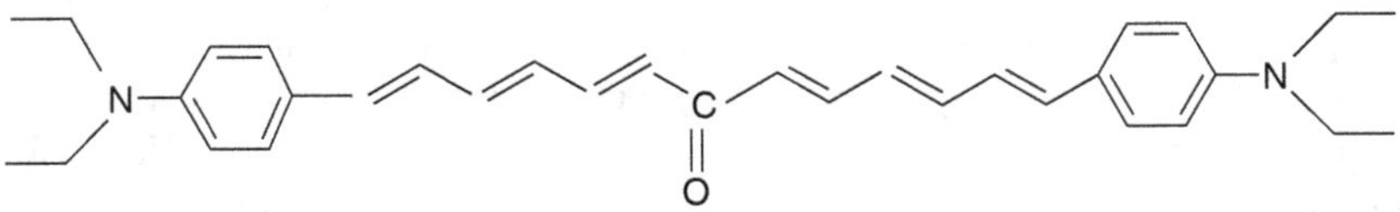

(10.55)

Maleimides The interaction between an excited *N*-substituted maleimide and an amine (10.56) leads to two initiating radicals [191,192].

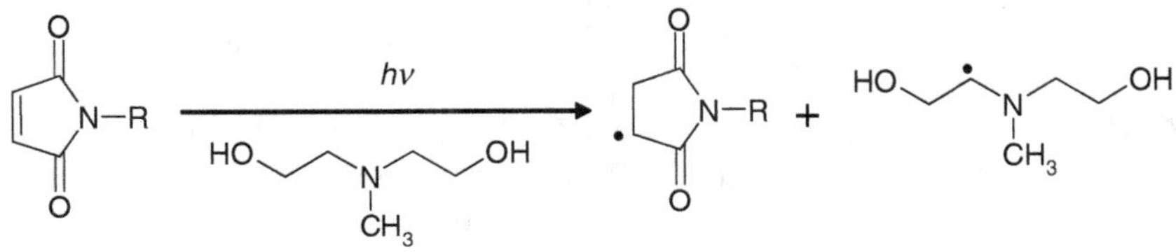

(10.56)

Electron transfer is also observed between a benzophenone and a maleimide, where two radicals are generated.

Donor/Acceptor Systems Upon excitation, a monomer donor (e.g., styrene) undergoes an electron transfer with a monomer acceptor (e.g., maleic anhydride). Then, the donor radical cation and the acceptor radical anion can recombine to form a biradical: a recent review was provided in Ref. [193].

Dye-Based Systems Many compounds can be used [1h,i]. Classical examples are found in the numerous available dyes and colored molecules [1i] such as xanthenic dyes (Rose Bengal, eosin [194,195], etc.), thiazines (methylene blue, etc.), acridines, *N*-methylacridone, phenosafranines, thiopyronines, riboflavines [196,197], phenoxazines [198], pyrromethenes [199], polymethines, fluorones, squarylium [200], julolidine dyes, and so on. (see, for example, Rose Bengal, acriflavine, and acridine orange in (10.57)).

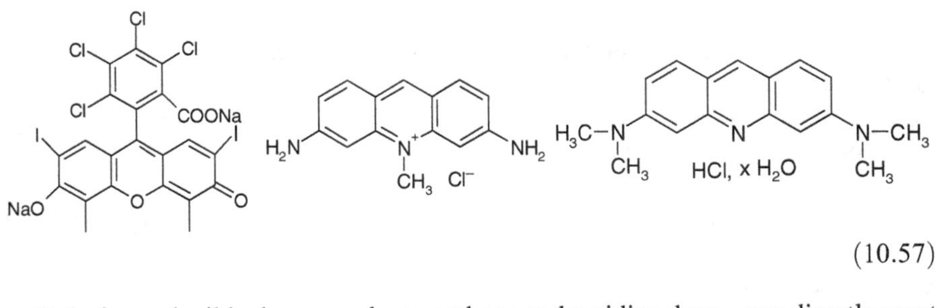

(10.57)

Only few reducible dyes—such as xanthene and acridine dyes—can directly react with electron deficient monomers but the efficiency remains low. As a consequence, dyes behave as type II photoinitiators. For example (10.58), in the case of Rose Bengal RB, a photoreduction of the dye occurs through an electron transfer with electron donors (such as amines, *N*-phenylglycines, benzyltrimethylstannanes, thiols, etc.) that yields a semireduced form on the dye and then a protonated semireduced form: this process is accompanied by the generation of an initiating aminoalkyl radical. Peroxides can also be decomposed in the presence of dyes.

$$^{1,3}RB \xrightarrow{AH} RB^{\bullet -} + AH^{\bullet +} \longrightarrow RBH^{\bullet} + A^{\bullet} \qquad (10.58)$$

The mechanisms involved in various systems have been reported (for example, see Refs [201,202]) as well as the search for efficient systems under visible light exposure usable in various applications [203–205].

Other miscellaneous systems [2c] include phenoxazones, quinolinones, phtalocyanines, benzopyranones, rhodanines, Rose Bengal peroxybenzoate, crystal violet/ benzofuranone derivatives, dimethyl aminostyryl benzothiazolinium iodides, etc.

Bis-Arylimidazole Derivatives Cl-bis-imidazole derivatives HABI—the bis (2,4,5-triphenylimidazole) is named lophine, sometimes, represented by L_2—are largely used in the laser imaging area (10.59) and large efforts are oriented to the design and the study of more efficient derivatives [187,206–209]. Because of the low bond energy between the two imidazoyl moieties (lophyl radical $L^{\bullet}$), these compounds exhibit a very fast cleavage in the S_1 state leading to two lophyl radicals $L^{\cdot}$ that then react with electron/hydrogen donors such as mercaptans (an initiating sulfur-centered radical is formed), *N*-phenyl glycine, and so on. More interestingly, they can be sensitized through a photoinduced electron transfer process (see below).

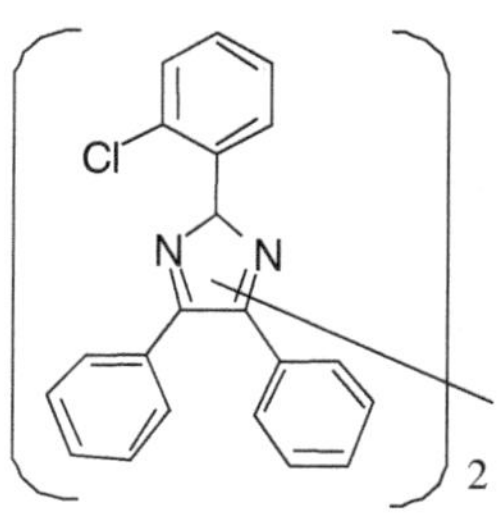

(10.59)

Coumarin Derivatives Coumarins are known as an interesting class of laser dyes. Various structures are able to behave as type II photoinitiators (they work in their first excited singlet state) or to sensitize the photolysis of peresters or bis-aryl imidazoles [210] (see below).

Pyrylium and Thiopyrylium Salts In the presence of additives such as a perester and in conjunction with ketocoumarins and coumarins, pyrylium, and thiopyrylium salts allow to decompose peresters. Addition of a diphenyl iodonium salt or a bromo compound such as CBr_4 to a thiopyrylium salt leads to an electron transfer process resulting in the generation of radicals [211].

Ketone/Ketone-Based Systems As stated above, the photosensitized decomposition of a photoinitiator I in the presence of a photosensitizer S can be described as resulting from the two main processes largely encountered in photochemistry: energy transfer or electron transfer.

Energy transfer is rather seldom encountered [2j] due to the obvious difficulty to have the S triplet-state level higher than that of I. The most typical case of energy transfer is encountered between a thioxanthone derivative and a morpholino ketone derivative (10.60): the excitation moves from the thioxanthone triplet state to the morpholino ketone triplet state. With other thioxanthone derivatives, electron transfer followed by proton transfer with the amine moiety can compete to some extent [212]. The sensitized cleavage in covalently linked thioxanthone derivative/cleavable photoinitiator was recently investigated and discussed [213].

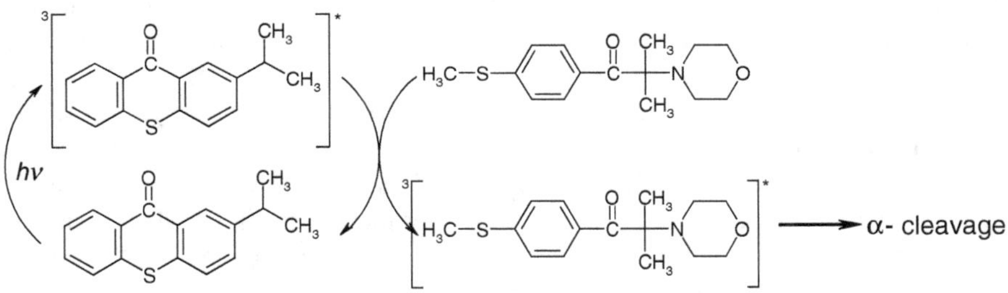

(10.60)

In the formerly proposed mixture of benzophenone and Michler's ketone, an electron transfer obviously arises between the two compounds and yields an aminoalkyl radical.

Organo Metallic Compound/Ketone-Based Systems A nice example of a bond cleavage via an electron transfer reaction was shown in the ruthenium tris bipyridine/morpholinoketone system: the readily formed radical cation on the amine cleaves into an imino cation and a benzoyl radical [214].

Photosensitizer-Linked Photoinitiator or Coinitiator-Based Systems Attempts have been made to incorporate the dye and the energy or the electron donor in the same molecule. An old example was a polymeric system where a morpholino ketone and a thioxanthone derivative were grafted [215]. A first recent example corresponds to a dye linked to a triazine derivative (10.61) where an intramolecular electron transfer yields a radical anion on the triazine moiety that results in the production of a radical [216,217].

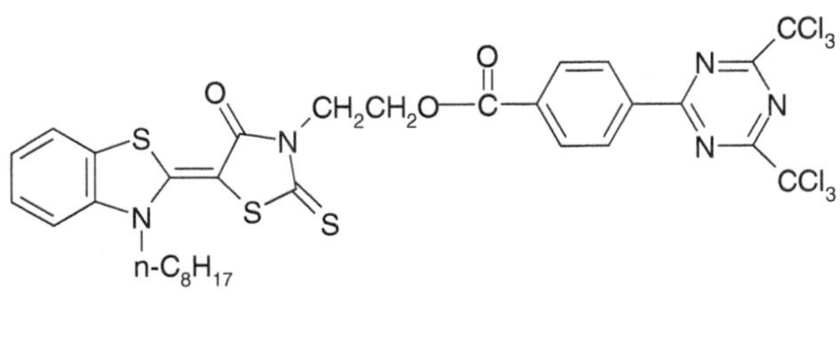

(10.61)

A second example relates to an eosin linked to an *O*-acyloxime (10.62) where radicals are generated according to a complex mechanism [218].

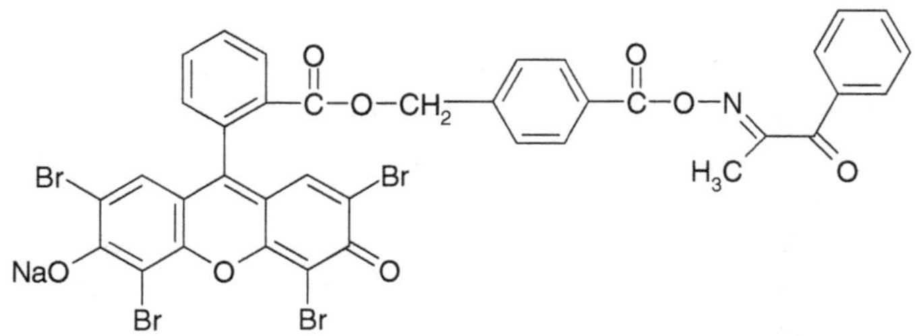

(10.62)

Other examples can be found in xanthene-thiol [219], carbazole-triazine [220], photosensitizer-ketones [221]. In both last cases, the reactivity of the dye–donor linked photoinitiator is higher than that of a physical mixture of the dye and donor moieties.

Photoinitiator/Peroxide or Hydroperoxide-Based Systems In these systems where useful effects occur [1i], oxygen-centered radicals are generated; the mechanism is not clear. In the same way, in the 1-benzoyl-cyclohexanol HAP/benzophenone BP system under air, the radicals that are formed after the α-cleavage of HAP consume oxygen and further allow the generation of a hydroperoxide ROOH whose decomposition is sensitized by BP, and so on.

Miscellaneous Systems Other systems have been described [1i,m]: ferrocene/carbon tetrachloride (a trichloromethyl radical is formed), ferrocene/carbon tetrabromide, metal carbonyl/onium salts (e.g., the [cyclopentadienyl Fe $(CO)_2]_2$/diaryliodonium hexafluorophosphate combination where a phenyl radical is generated), benzene chromium tricarbonyl/halide derivative, and so on.

10.3.1.3 Multicomponent Photoinitiating Systems Many multicomponent photoinitiating systems have been proposed so far in order to improve the efficiency of the radical production: they are very often used in the design of high-speed photopolymers [222–224].

Generally Encountered Mechanism Most of them behave according to the following mechanism (10.63). The C1/C2 combination (where C1 is the absorbing species) is used to produce the initiating radical 1; the concomitantly formed radical 2 is a scavenger of the growing polymer chain. The idea is to eliminate this radical by using an appropriate quencher C3.

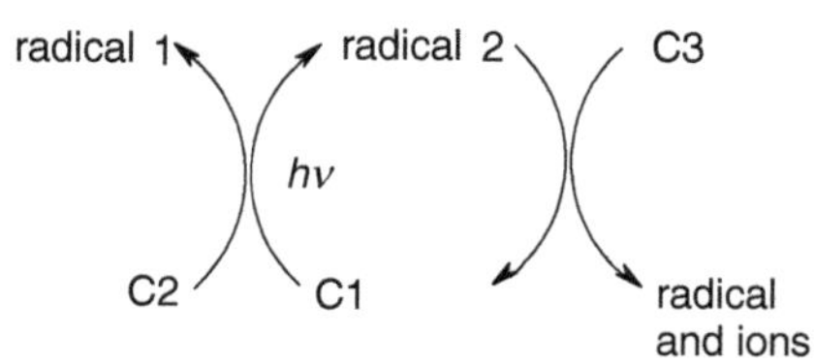

(10.63)

The following systems obey to such a mechanism:

(a) Ketone/amine/bromo compound or onium salt (see, for example, (10.64) and Ref. [225], where C1 is isopropylthioxanthone ITX, C2 an amine AH, and C3 is carbon tetrabromide). Radical 2 of (10.63) is here a ketyl type radical on ITX: the presence of ketyl radicals is known as having a detrimental effect on the rate of polymerization.

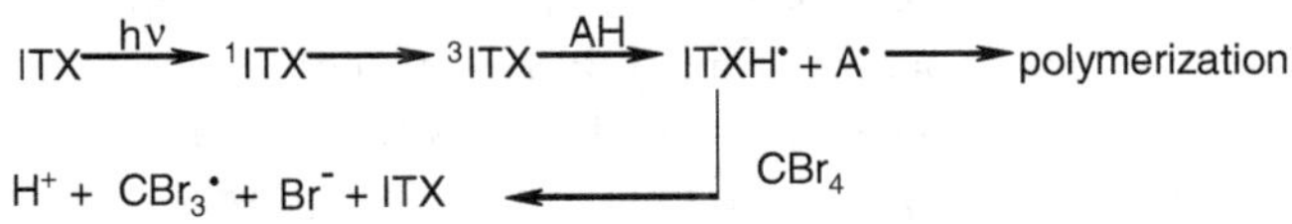

(10.64)

(b) Ketocoumarin/amine/onium salt [226].

(c) Eosin/amine/onium salt, bromo compound, THF or PDO (keto-oxime derivative) [227] as shown in (10.65).

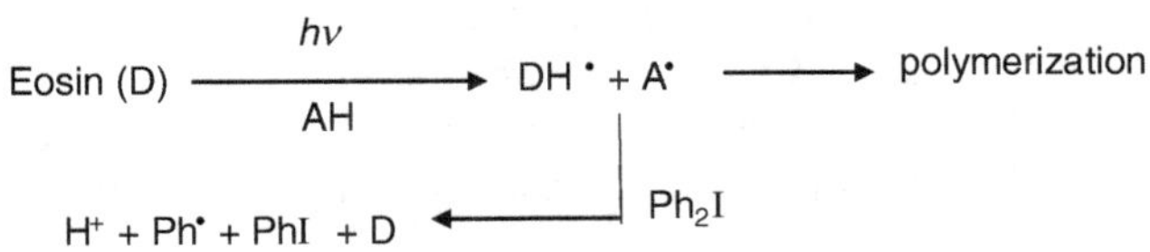

(10.65)

(d) Methylene blue/amine/onium salt [228,229].

(e) Dye (e.g., merocyanine, safranine)/amine/bromo compound or onium salt [230,231].

(f) Thioxanthene dye/amine/onium salt or bromo compound [232].

(g) Acridinium cation/dihydropyridine/onium salt [233].

(h) Ketone/amine/maleimide or maleic anhydride [234] as described in (10.66).

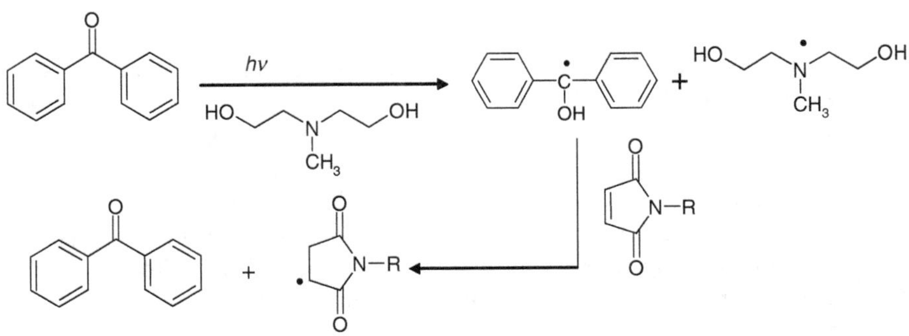

(10.66)

(i) Ketone (isopropylthioxanthone/benzophenone sulfonylketone/amine) [235].

(j) Dye/amine/triazine derivatives [236,237].

Other Mechanisms Other three-component systems involve different mechanisms (see Refs [1i,m] and references therein). For example, in the dye (crystal violet, phenosafranine, methylene blue, thiopyronine)/amine/ketone (acetophenone, benzophenone, thioxanthone, 4,4′-bis-dimethylamino benzophenone) combi-nation, the mechanism is quite complex. In the camphorquinone CQ/amino-benzophenone ABP two-component system, both CQ and ABP absorbs the light; an aminoalkyl radical on ABP and a ketyl radical on CQ are formed.

In the (keto)coumarin/amine/ferrocenium salt system, the ferrocenium salt plays a crucial role that is rather complex. In a three-component photoinitiator system [238,239] consisting of a coumarin, an iron arene complex such as $CpFe^{+}Ar$ and a phenylglycine derivative as an amine, the first step of the photoreaction occurs between the dye and the complex according to an electron process. The amine reacts with the radical (created on the complex) through hydrogen abstraction. Therefore, no detrimental ketyl radicals are formed.

The reaction mechanisms have been explored in the coumarin, ketocoumarin, or titanocene derivative/bis-aryl imidazole Cl-HABI/mercaptobenzoxazole system [240]. For example (10.67), when the sensitizer is the coumarin C1, an electron transfer occurs and the formed Cl-HABI radical anion leads to a lophyl radical that undergoes a hydrogen transfer with the thiol SH (the lophyl radical anion reacts with the thiol radical cation). The C1/SH interaction can also participate in the initiation process. The role of new thiols and disulfides has been recently studied [241].

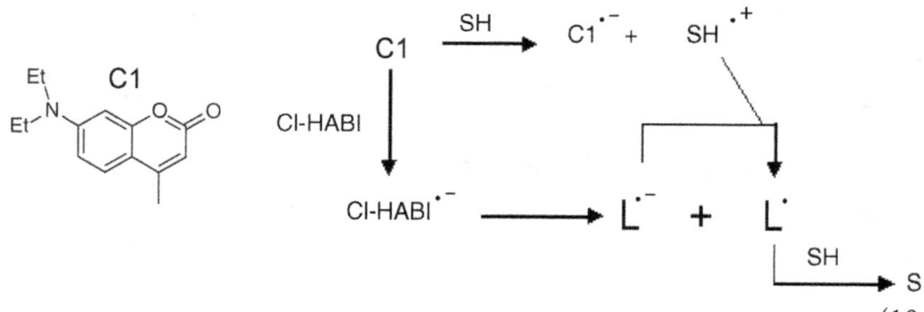

(10.67)

The processes involved in the dye (methylene blue)/amine/cobalt salts system have also been described [242].

The particular behavior of the ketone/silane derivative/additive combination proposed as an efficient photoinitiating system under air (better than the reference ketone/amine/additive system) was shown in Ref. [128].

Finally, four-component systems are also available; the complexity of the mechanisms involved increases. This is exemplified by the dye (Rose Bengal, eosin)/ferrocenium salt/amine/hydroperoxide system [243,244]. The direct photolysis of the dyes and the salt in the presence of oxygen and hydroperoxide has been investigated. In a general way, the primary process is the interaction between the dye and the ferrocenium salt that results in an electron transfer. Then, the reactions of the radical (formed *in situ* on the ferrocenium salt) with the amine, oxygen, and hydroperoxide lead to radicals. Oxygen is thus eliminated. Dark reactions can participate to the polymerization initiation: this allows to explain the body cure of thick samples where the light cannot deeply penetrate. The synthesis, the properties, and the specific role of an intramolecular ion pair complex between Rose Bengal and the ferrocenium salt were very recently outlined [245]. Another example concerns the coumarin or diethyl amino benzophenone/amine/bis-aryl imidazole/onium salt combination. The rather complex mechanism has been recently discussed [246].

10.3.1.4 Other Photoinitiating Systems The photopolymerization of methyl-methacrylate in the presence of nanosized TiO_2 particles has been discussed [247].

Although no photoinitiator is used, the "photopolymerization without light" (a thermal process in suitable systems can form an electronically excited molecule), which was recently reported [248] has to be mentioned. The reaction between an oxalate ester and an hydrogen peroxide produces a high-energy content molecule which decomposes into radicals (similar to those resulting from a direct light excitation) that further add to the monomer: this could be helpful for the polymerization of coatings in the hidden parts of a substrate (remote cure technology).

10.4 REACTIVITY OF PHOTOINITIATORS

This part refers to a general discussion of the photoinitiator reactivity based on both experimental and theoretical considerations. It shows how typical data on the transient states can be obtained and how the driving factors of the processes involved can be understood: cleavage, amine interaction, monomer quenching, oxygen quenching, radical addition, radical quenching, and so on. By using both these information and the polymerization results, it becomes now possible to derive very interesting (photo) chemical reactivity/practical efficiency relationships. The huge progress made in the theoretical approach really improves the interpretation of these relationships compared to what was done in the past.

10.4.1 Excited-State Processes

The overall diagram of evolution of the excited states and reactive intermediates of a photoinitiating system working through its triplet state can be depicted in Scheme 10.2 [249]. Various time resolved laser techniques (absorption spectroscopy in the nanosecond and picosecond timescales), photothermal methods (thermal lens spectrometry and laser-induced photocalorimetry), photoconductivity, laser-induced step scan FTIR vibrational spectroscopy, CIDEP-ESR and CIDNP-NMR) as well as quantum mechanical calculations (performed at high level of theory) provide unique kinetic and thermodynamical data on the processes that govern the overall efficiency of PIS.

A lot of information are currently available, for example, transient absorption spectra, lifetimes of excited states τ, interaction rate constants k, intersystem crossing quantum yield Φ_{isc}, triplet-state energy levels E_T, dissociation quantum yields of cleavable photoinitiators, bond dissociation energies (BDEs) of amines or thiols used

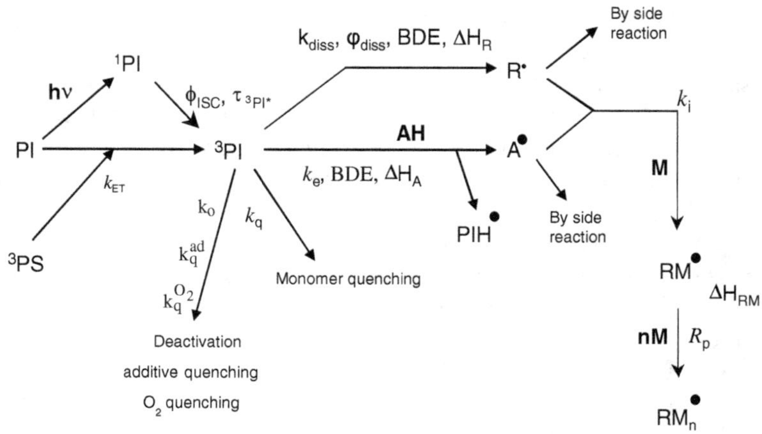

SCHEME 10.2

as coinitiators, formation enthalpies of the initiating radicals ΔH_A and ΔH_R, interaction rate constants between these radicals and the monomer double bonds, addition reaction enthalpy ΔH_{RM}, kinetics of the ions or radical ions generation and recombination, and so on.

Radiative and nonradiative relaxation processes (characterized by a first-order rate constant k) arise from the S_1 and T_1 states according to (10.68) and compete with the photoinitiation process as already shown in Scheme 10.1:

$$\begin{array}{lll} \text{-fluorescence:} & S_1 \rightarrow S_0 + h\nu & k_r \\ \text{-internal conversion:} & S_1 \rightarrow S_0 + kT & k_{IC} \\ \text{-intersystem crossing:} & S_1 \rightarrow T_1 & k_{ST} \\ \text{-phosphorescence:} & T_1 \rightarrow S_0 + h\nu & k_p \\ \text{-intersystem crossing:} & T_1 \rightarrow S_0 + kT & k'_{IC} \end{array} \tag{10.68}$$

The lifetimes of the S_1 and T_1 states are defined by (10.69) and the quantum yields of these different processes can be evaluated as in (10.70).

$$\tau_f = \frac{1}{(k_r + k_{IC} + k_{ST})}; \quad \tau_T = \frac{1}{(k_p + k'_{IC})} \tag{10.69}$$

$$\phi_f = k_r \tau_f; \quad \phi_{ST} = k_{ST} \tau_f \tag{10.70}$$

The following example helps to understand how the kinetics affects the practical efficiency expressed by the rate of polymerization R_p. The dissociation quantum yield in the case of a cleavage process in the absence of oxygen is given by (10.71) and the corresponding initiation quantum yield ϕi connected with R_p is defined by (10.72) where $\sum k$ stands for all the processes involved in the disappearance of the initiating radicals.

$$\phi_{\mathrm{diss}} = \phi_{\mathrm{ST}}\left(\frac{k_\alpha}{k_0 + k_\alpha}\right) \tag{10.71}$$

$$\phi_{\mathrm{i}} = \phi_{\mathrm{ST}}\left(\frac{k_\alpha}{(k_0 + k_\alpha + k_{\mathrm{q}}[\mathrm{M}])}\right)\left(\frac{k_{\mathrm{q}}[\mathrm{M}]}{\sum k}\right) \tag{10.72}$$

The longer the lifetime of the triplet state, the less efficient is the process of the radical generation and the more efficient is the physical quenching by the monomer or oxygen. In addition, this detrimental effect increases upon increasing the monomer concentration and therefore, photoinitiators having long triplet lifetimes will not likely favor the initiation of the polymerization in a fluid concentrated solution of monomer. In bulk monomer/oligomer media where the diffusion rate constants drastically decrease, the monomer quenching is kinetically less efficient.

More than 20 years of research with laser techniques have made possible the proposals of many diagrams of excited-state processes based on kinetic measurements. In the very recent years, a series of works involving laser-induced photocalorimetry and quantum mechanical calculations considerably improves the panel of available data. Today, as mentioned above, such schemes are largely known (see, for example, Refs [1i,m,2i]) for (i) many type I photoinitiators: benzoin ethers, hydroxyalkyl ketones, dialkoxyacetophenones, α-keto-oxime esters, phosphineoxides, sulfonyl ketones, morpholino ketones, and so on, (ii) many type II photoinitiating systems: ketone (benzophenone, thioxanthone, camphorquinone, ketocoumarin, etc.)/amine or thiol, dyes (Rose Bengal, methylene blue, coumarin, etc.)/amine, etc., (iii) some multicomponent photoinitiating systems: dye or ketone/amine/iodonium salt or bromo compound, coumarin or ketone/HABI/amine or thiol, and so on. The role of the structure and the substitution have been largely explored. The particular features observed in multifunctional, copolymerizable, hydrophilic, and water-soluble photoinitiators have also been discussed. Most of these results were obtained in solution. Table 10.1 gathers some typical data.

10.4.2 Analysis of the Primary Processes

Time resolved laser spectroscopy is likely the best method to get kinetic information on transient states. More information can currently be gained from quantum mechanical calculations at relatively high level of theory using the Gaussian Suite of programs (*ab initio* and density functional theory).

10.4.2.1 Kinetics of the Cleavage Process in Type I Photoinitiators Working now on the picosecond timescale (with a pump-probe laser setup) shows the short-lived transient absorptions observed upon light excitation [250]. For example [251], the cleavage process of DMPA (2,2-dimethoxy –2 phenyl-acetophenone) occurs

TABLE 10.1 Examples of Available Data Concerning Usual Photoinitiators: Triplet State Lifetimes τ_T^0, Bond Dissociation Energies BDE, Quantum Yields of Dissociation ϕ_{diss}, Triplet State Energy Levels, Rate Constants of Interaction k_q with Oxygen, Monomer and Hydrogen Donor, Quantum Yields of Intersystem Crossing ϕ_{isc}.

Compound		τ_T^0 (ns)	ϕ_{diss}	BDE (kcal/mol)	E_T (kcal/mol)	$10^{-6}\,k_q$ (l/mol/s) 0_2	$10^{-6}\,k_q$ (l/mol/s) Monomer	$10^{-6}\,k_q$ (l/mol/s) H-donor	ϕ_{isc}
DMPA	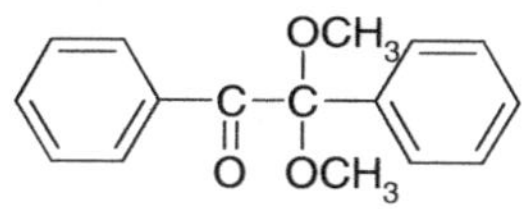	0.25	0.95	50.9	66				
BME	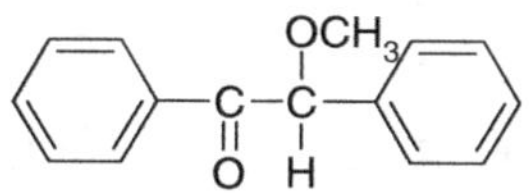		0.5						
HCAP	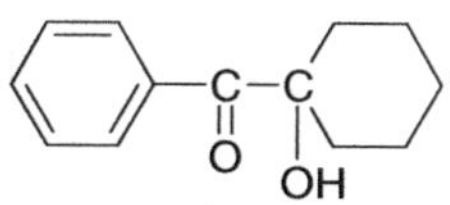	0.38	0.8	63.5	67				
HAP	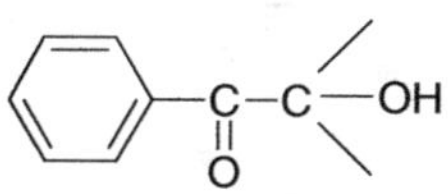	0.45	0.8	65	72		250 (MMA)		
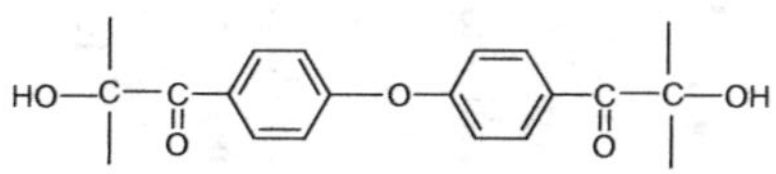		7	1	57.1					

TPMK	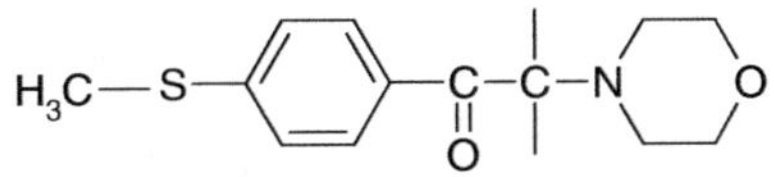	10	0.3	56.8	61	2 (MMA)
MPK	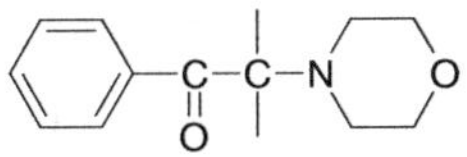	1	0.3	61.9	62	
MPPK	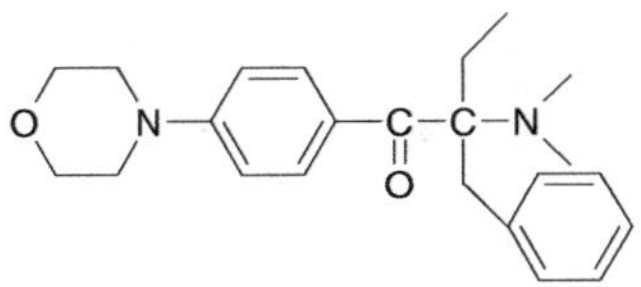	1700	0.9			3.5 (MMA)
TPO	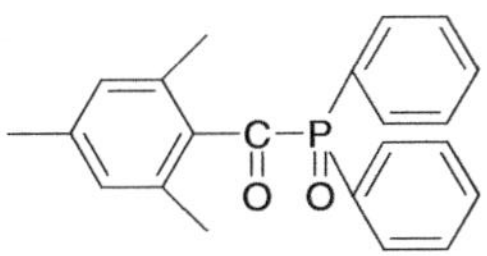	0.12	0.7	57	63	
	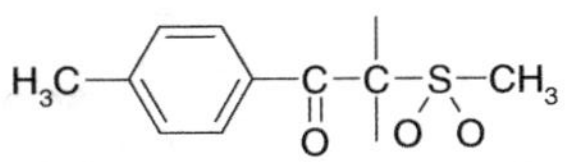	175				4O (MMA)

(continued)

Table 10.1 (*Continued*)

Compound	τ_T^0 (ns)	ϕ_{diss}	BDE (kcal/mol)	E_T (kcal/mol)	10^{-6} k_q (l/mol/s) O_2	10^{-6} k_q (l/mol/s) Monomer	10^{-6} k_q (l/mol/s) H-donor	ϕ_{isc}
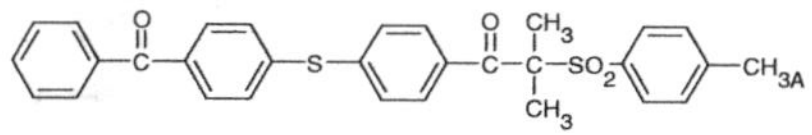	>1000		37.7	66		0.18 (MMA)	390 (MDEA)	0.78
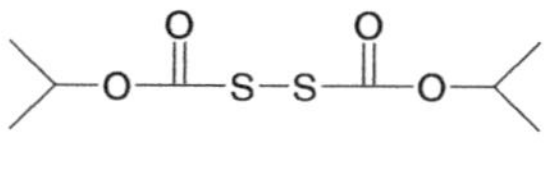		0.38 (S_1)	23.2					
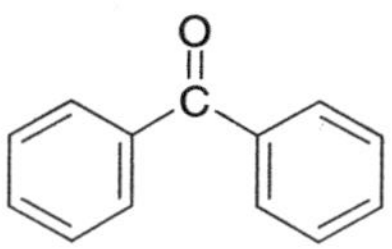				69		66 (MMA) 34 (AN)* 5.4 (VA)* 78 (VE)* 5.5 (MA)*	1300 (MDEA) 4.2 (THF) 1.6(isopropanol)	
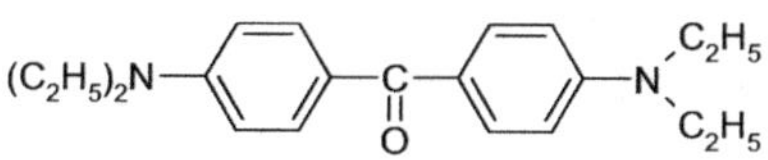	>1000				12500	0.3 (BA)	1500 (NPG)	0.95
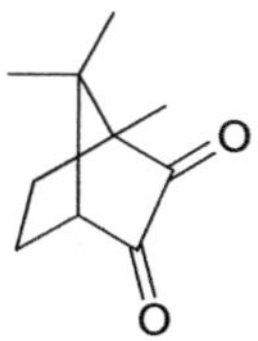	>1000			51.6	200		600 (MDEA)	1

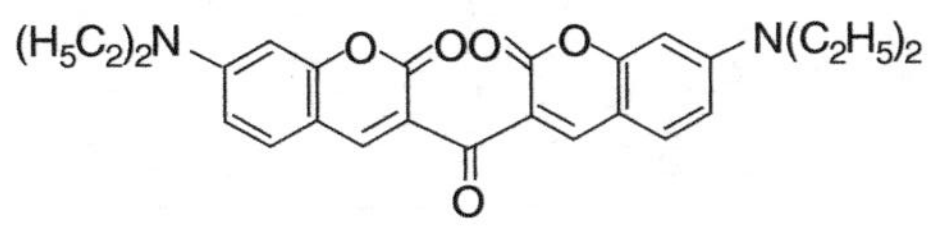	50		950 (MBO)	0.92
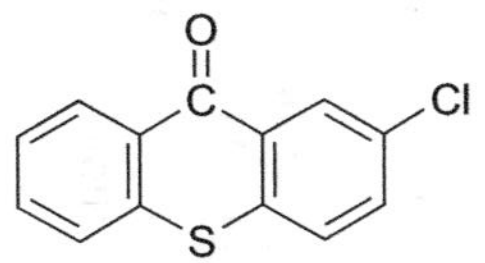		2 (MMA) 0.4 (AN)* 0.02 (VA)* 1.6 (VE)*		0.9
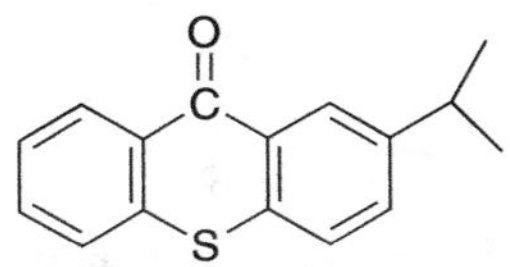		3 (MMA)	2600 (BEA)	

MDEA: methyldiethanolamine, NPG: N-phenylglycine, MBO: mercaptobenzoxale, BEA: alkyl p-dimethylamino benzoate, MA: methylacrylate, MMA: methylmethacrylate, AN: acrylonitrile, VA: vinylacetate, VE: vinylether. See text.

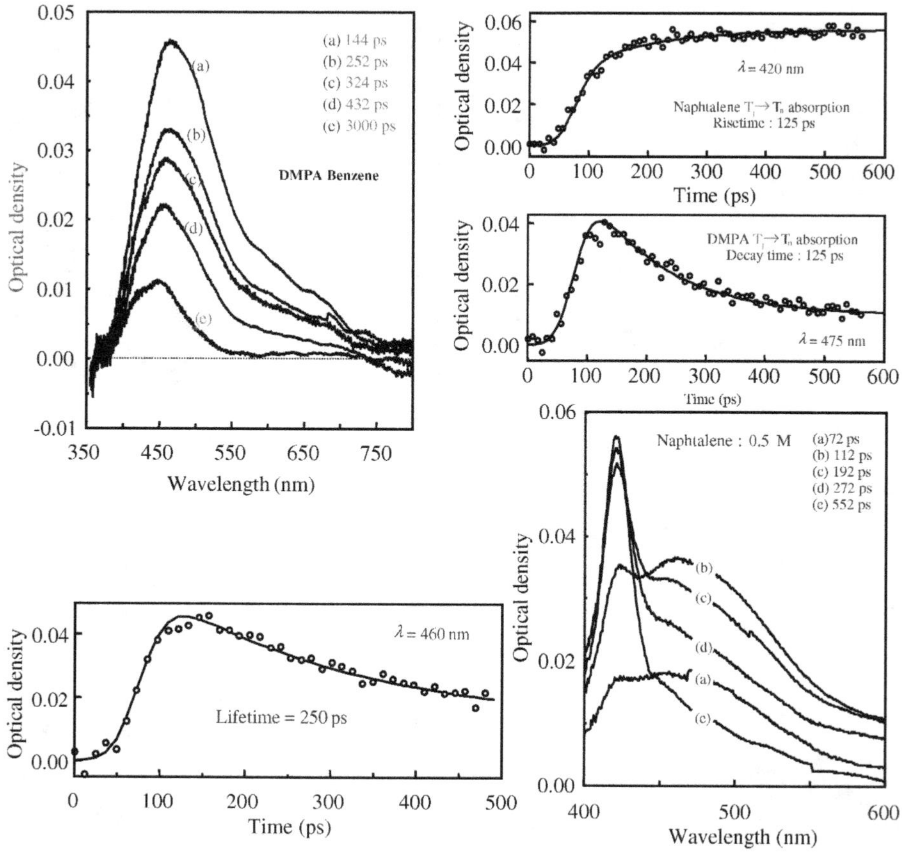

FIGURE 10.4 Cleavage process of DMPA: transient absorption spectra in benzene and benzene/ 0.5 M of naphtalene as a function of time; decay and growing traces of the transients (DMPA triplet state in benzene, naphthalene and DMPA triplet states in benzene/0.5 M of naphthalene. See text.

within 250 ps in benzene (Fig. 10.4). The quenching by a triplet acceptor confirms the assignment for a triplet–triplet absorption: in the presence of naphtalene, a new long-lived absorption due to the naphtalene triplet as well as the DMPA triplet-state quenching are observed. The rise time of the naphtalene triplet state (formed during the energy transfer process from DMPA to naphtalene) corresponds to the triplet-state decay of DMPA. Similar works have been recently done on a large variety of photoinitiators [250].

These experiments can be still completed by working on the nanosecond timescale (using a classical laser flash photolysis device) as formerly done for many years. Longer lived transient states are thus observed and their interaction with other molecules easily followed [1v,2f,i,252–254]. The same holds true for radicals exhibiting an easily detectable optical absorption.

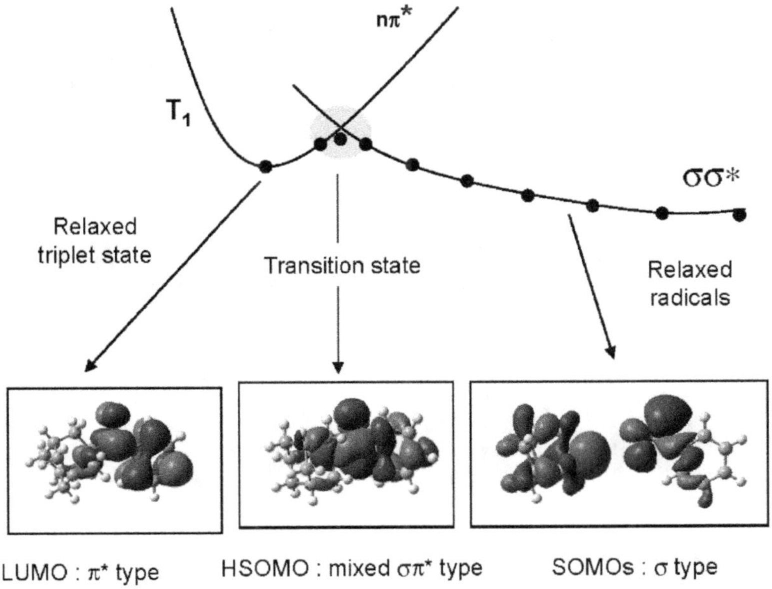

FIGURE 10.5 The cleavage process of HAP. Potential energy surface along the bond being broken and molecular orbitals involved. See text.

10.4.2.2 Theoretical Approach of the Cleavage Process For example, the theoretical investigation of the cleavage process of HAP (2-hydroxy-2-methyl-1-phenyl-1-propanone) has been recently described [255]. The triplet state is calculated to be of nπ* nature with an energy of 75.7 kcal/mol, in very good agreement with the experimental value of 71.4 kcal/mol. This demonstrates the ability of the theoretical methods to accurately predict the photophysical properties of the photoinitiators. By varying the C–C bond distance, the potential energy surface along the bond being broken can be calculated (Fig. 10.5).

After geometry optimization, the transition state is found higher (by 3.5 kcal/mol) than the relaxed triplet state. As a consequence, the cleavage process is thermally activated. The dissociation energy of the C–C bond is computed to be 65.5 kcal/mol. Excitation of HAP quantitatively leads to the triplet state from which the dissociation can occur. The very low energy barrier explains the high value of the reported dissociation quantum yield (0.8). All these properties explain the high efficiency of HAP as a photoinitiator.

The molecular orbitals involved in the different structures (ground state and excited state of the starting molecule, radicals) as well as in the transition state can also be calculated (Fig. 10.5).

This approach was also used for phosphine oxides [256] and has been already extended, in the last few years, to other structures such as other hydroxyalkyl ketones [257], [4-(methylthio) phenyl]-2-morpholinopropanone [258], Barton's esters [66], sulfonyl ketones [250], and so on.

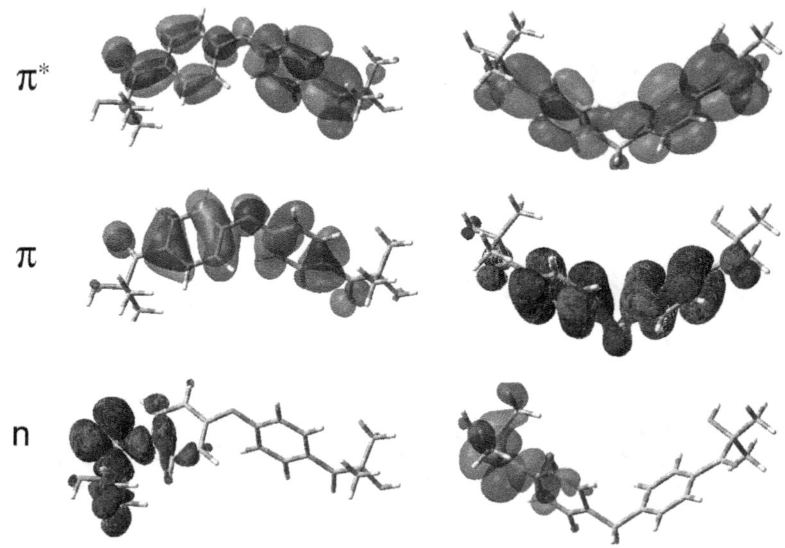

FIGURE 10.6 Calculated main n, π, π* molecular orbitals of two difunctional photoinitiators - dfI1 (left) and dfI2 (right) - at a B3LYP/6-31G* level. See text.

10.4.2.3 Specific Behavior of Difunctional Photoinitiators A fascinating question refers to the description of the MOs in difunctional photoinitiators dfI where two monofunctional photoinitiators are covalently linked: can such a molecule be considered as two independent moieties? The reactivity of two selected dfIs consisting in two HAP moieties linked by either an oxygen atom (dfI1) or a methylene group (dfI2) was very recently discussed [89]. Calculations show that the absorption properties must not be considered in any case as due to the sum of two HAP moieties. The n-orbitals are localized only on one of the two benzoyl groups of the dfI whereas the π-orbitals are partially delocalized over the entire molecule (Fig. 10.6). This could explain why the $\pi\pi^*$-transitions are more affected than the nπ^*-transitions by the coupling of two HAP moieties. The delocalization of the π-orbitals takes place through the nonbinding orbital of the oxygen atom in dfI1 and through a hyperconjugation in dfI2 (the molecule is twisted, which allows the coupling of the two π-systems). No strong modification of the photochemical reactivity of these dfI compared to HAP was observed: this was in agreement with the almost unchanged practical efficiency of the photopolymerization reaction under a monochromatic light exposure. The change in the ground absorption spectra as well as in the values of the molar extinction coefficients leads to a higher amount of absorbed energy and obviously explains the higher rates of polymerization obtained upon exposure to polychromatic lights when using the same w/w photoinitiator concentration.

10.4.2.4 Bond Dissociation Energy and Cleavage Process Bond dissociation energies are theoretically accessible. A common statement suggests that the dissociation efficiency should increase with the BDE decrease. However, the dissocia-

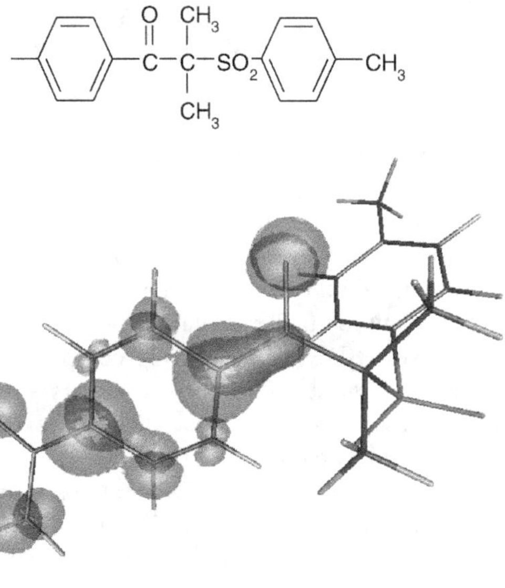

FIGURE 10.7 Ketosulfone derivatives: plot of the triplet spin density of the p-phenyl substituted compound.

tion quantum yields of cleavable photoinitiators do not correlate [259] with the dissociation enthalpy ΔH_{diss} expressed by (10.73) (where E_T is the triplet energy). As a consequence, the enthalpy is not the driving factor for the cleavage process.

$$\Delta H_{diss} = BDE - E_T \tag{10.73}$$

The explanation should be searched elsewhere. After light absorption, the energy should be localized near the bond to dissociate. It seems that the spin density at the $C_{C=O}$ of the cleavable bond has a strong influence. This was recently exemplified in ketosulfone derivatives (Fig. 10.7) where a β-cleavage process occurs in their triplet states. Calculations of the BDEs of the C–S bond effectively show that a bond breaking occurs when the triplet state is higher than the corresponding BDE but the evaluation of the enthalpy change demonstrates that the reaction is strongly exothermic although a slow cleavage process is observed in some cases: actually, the substituent on the benzoyl moiety increases the delocalization of the triplet state. The spin density of two selected derivatives (where the substituent at the *para* position of the benzoyl moiety corresponds to H or a phenyl ring) is 0.01 (phenyl) and 0.54 (H) whereas the corresponding triplet-state lifetimes are 6 and 0.13 μs, respectively: this result accounts for the relative cleavage efficiency [260].

10.4.2.5 Cage Effect The cage effect dynamics after excitation of the photoinitiator has been studied [261]. In fact, the radicals generated upon cleavage or amine interaction constitute a triplet radical pair that readily dissociate into free radicals in nonviscous media. In a viscous medium, the radical pair undergoes intersystem crossing and forms a singlet radical pair where the radicals can react within the cage: this obviously affects the yield in escaping free radicals. The effect of a magnetic field has been discussed [262].

10.4.2.6 Bond Dissociation Energy and Hydrogen Transfer in Type II Photoinitiators In type II systems, the production of the initiating radical depends on the rate constant of the electron transfer process (whose correlations with the donor/acceptor ability have been largely discussed in photochemistry) and the efficiency of the proton transfer. One could expect that the BDE of the hydrogen donor plays a significant role as the more stable radical (which is the easiest to form) should correspond to the lowest BDE of the starting compound. Many works, however, devoted to the BDE of a large set of primary, secondary and tertiary aliphatic amines show that there is no real correlation. One plausible explanation is that the BDE governs a direct hydrogen abstraction rather than a proton transfer. General comments are the following [263]: (i) the α(C–H) BDEs are almost similar (around 90 kcal/mol) for unconstrained amines; (ii) the evolution of N–H and α(C–H) bond strength allows to underline the slight influence of C- and *N*-alkylation on the BDEs; (iii) the effect of the *N*-alkylation on the α(C–H) BDEs is ascribed to a two orbital–three electron interaction; (iv) the C-alkylation has clearly no effect on the α(C–H) BDEs; (v) the stabilization observed in the effect of the *N*-alkylation on the N–H BDEs occurs through C–C hyperconjugation; and (vi) in constrained amines, the role of a through bond stabilization is clearly observed.

The regioselectivity of the hydrogen abstraction reaction between the photoinitiator and an amine remains an outstanding problem that is the subject of debates and discussions. A recent paper [264] based on ESR and laser flash photolysis clearly shows the presence of the aminoalkyl radical and the significant absence of any aminyl radical in primary and secondary amines; such an aminyl radical is, however, readily observed in the same experimental conditions when using tetramethyl piperidine.

10.4.3 Reactivity of Radicals

10.4.3.1 Experimental Analysis Despite the large number of works devoted to the investigation of the radical $R^{\bullet}$ reactivity, the determination of the experimental values of the radical addition rate constants to a monomer double bond remains difficult. Only a small part of these values is based on a direct detection of the produced radical (flash photolysis, ESR). In fact, radicals are very often difficult to observe because of their absorption band below or close to 300 nm. Moreover, the classical overlap of the absorption spectra in the UV (due to the presence of different transient and ground-state species inherent to the production route when using laser flash photolysis) often prevents the direct observation. Indirect methods were tentatively used but they involve a complex set of reactions and are rather difficult to carry out. The same holds true for laser-induced photocalorimetry [265].

Nevertheless, a lot of radicals were already studied in the past years, for example, hydroxyalkyl [266], benzoyl [267], benzyl [268,269], alkylamine [270], thiyl [271], xanthone ketyl [272], benzyl peroxyl [273], phenacetyl [274], sulfydryl [275],

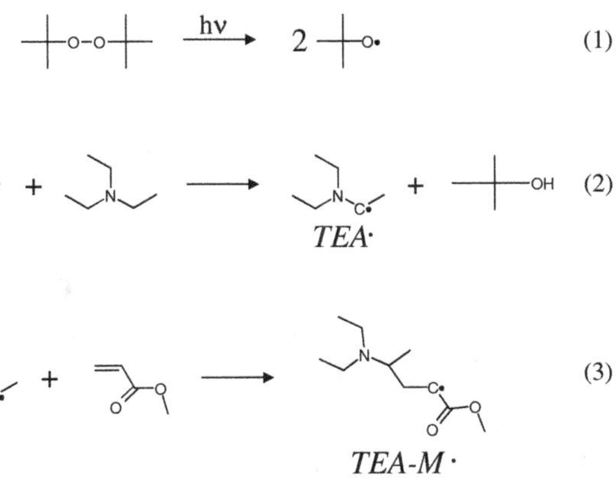

SCHEME 10.3

benzodithiazolyl [276], phenyl selenyl [277], benzoyl peroxyl [278], cumyloxyl [279], diphenylpentadienyl [280], phenoxyl [281], (benzoyl phenyl) diphenyl methyl [282], hydroxyisopropyl [283], phosphinoyl [284], aryloxyl [285–287], AIBN [288], DMPA [289] derived radicals, and so on.

The challenge was still to be able to detect the $R^{\bullet}$ radicals and more interestingly the $RM^{\bullet}$ radicals (where M stands for a given monomer) through a direct optical detection that then will allow an easy access to the addition rate constants of $RM^{\bullet}$ to various compounds. This was recently achieved [290] in the case of acrylate radicals using a reaction that consists in three consecutive steps (Scheme 10.3).

The first step is the generation of a *tert*-butoxyl radical through the photochemical decomposition of *tert*-butylperoxide; the second step corresponds to a α(C–H) hydrogen abstraction reaction from triethylamine as proposed many years ago [291]. The originality of the present procedure lies in the third step that corresponds to the addition of the aminoalkyl radical to a (meth) acrylate double bond. Indeed, upon addition of, for example, methyl acrylate MA or methyl methacrylate MMA, new $RM^{\bullet}$ transients are observed with absorption maxima at 470 and 425 nm for MA and MMA, respectively, as shown in Fig. 10.8.

The direct detection of monomer radicals according to this kind of procedure as well as using other experiments allowed to deeply investigate a large variety of thiyl [292], heterocyclic thiyl [293], lactone alkyl [294], silyl [127], aminoakyl [295], oxyl and peroxyl [296] radicals, and so on. Many addition rate constants of various initiating radicals to monomer double bonds have therefore been determined: they correspond to the usual initiation rate constant k_i. Examples are shown in Table 10.2.

10.4.3.2 Reactivity Toward Monomers According to the Arrhenius's law, the addition rate constant is a function of the activation energy (the "barrier height"). The addition reaction is usually described by a state correlation diagram (see, for example, Ref. [297]) that depicts the reaction profile (expressed as the energy of the compounds involved when going from the reactants to the products) as a function of the reaction coordinate (Scheme 10.4)—it involves the reactant ground-state configuration $R^{\bullet}/^{1}M$,

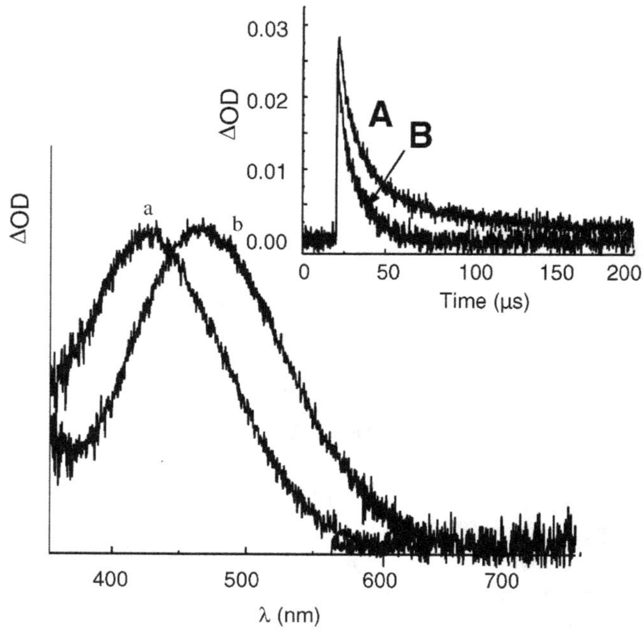

FIGURE 10.8 Transient absorption spectra of the TEA-MMA$^\bullet$ (a) and TEA-MA$^\bullet$ (b) radicals (taken at 1 μs after the laser pulse and normalized at the same maximum absorption). Insert: Decay traces of TEA-MA$^\bullet$ observed at 470 nm for [MA] = 0.2M (A) and 4.0 M (B).

the excited reactant configuration $R^\bullet/^3M$, and two polar charge transfer configurations R^+/M^- and R^-/M^+, these two last configurations can mix with the other two and lead to a change of the crossing region. In addition to possible repulsive steric effects, the exothermicity of the reaction (i.e., the reaction enthalpy ΔH_r) as well as the polar effects strongly affect the barrier height.

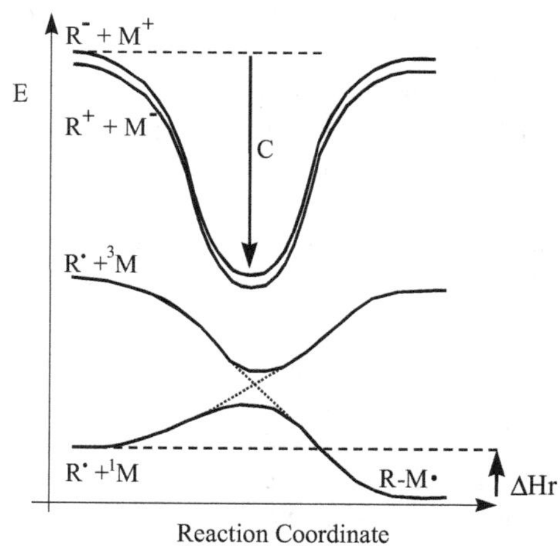

SCHEME 10.4

TABLE 10.2 Addition Rate Constants k_i of Various Radicals to Monomers; Reaction Rate Constants k_r of These Radicals with Oxygen, an Amine, a Stabilizer (HQME: Hydroquinone-Methyl Ether) and a Spin Trap (TEMPO: 2,2,6,6 Tetramethylpiperidine N-Oxyl).

	Addition rate constants k_i (in 10^6 l/mol/s)							Reaction rate constants k_r (10^6 l/mol/s)				Barriers (kJ/mol)			
Initiating radical	MMA	MA	BA	BMA	VA	VE	AN	AH	O_2	HQME	TEMPO	AN	MA	VA	VE
Initiating radical															
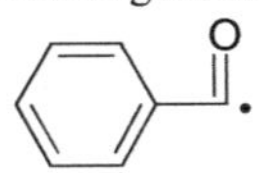	0.09		0.27–0.5									19.5	23.1	34.4	34.8
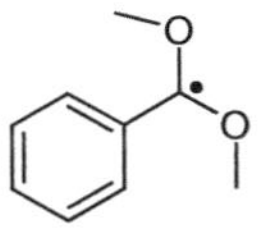		<0.01		<0.5											
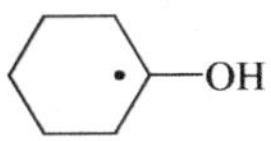			11–25												
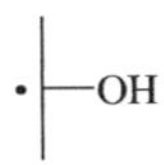	16	35	13									5.6	7.7	29.2	31.4
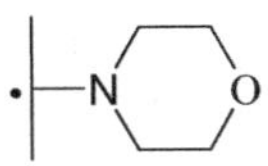		45	29–40												

(continued)

Table 10.2 (*Continued*)

Initiating radical	Addition rate constants k_i (in 10^6 l/mol/s)							Reaction rate constants k_r (10^6 l/mol/s)				Barriers (kJ/mol)			
	MMA	MA	BA	BMA	VA	VE	AN	AH	O_2	HQME	TEMPO	AN	MA	VA	VE
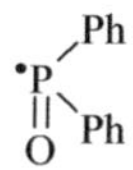	60–110	23–33													
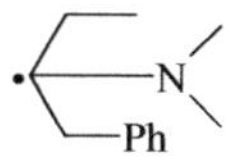		6.1													
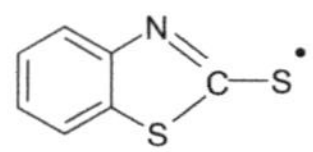		0.63			0.8	1	0.37					22.8	23.3	15.1	13.1
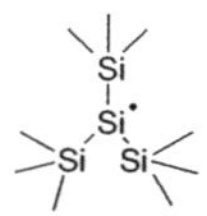		22			1.2	0.2	51			12000		8.6	9.5	10.4	10.2
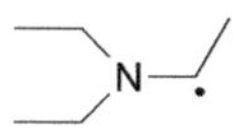	33	41	30	35	<0.01	<0.01	<0.01		3000	<0.5	330		2		
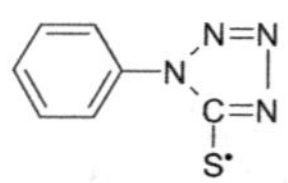		100			200	440	25					11.5	8.7	0.2	0

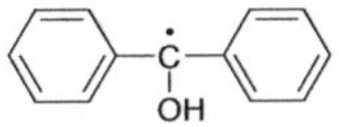		<0.009					
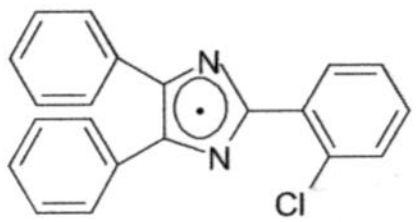				1.2 (NPG)			
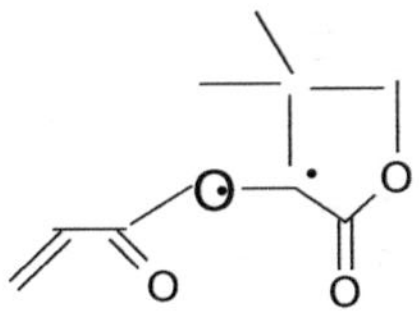		<0.5			<0.5		
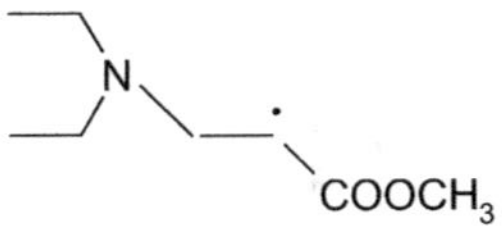		0.02			3000	<0.5	330
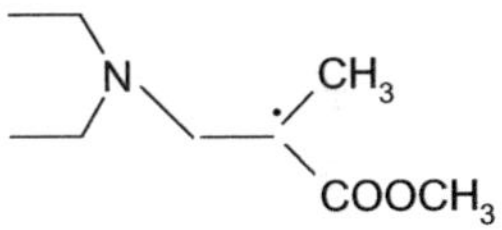	<0.005				1900	<0.5	180
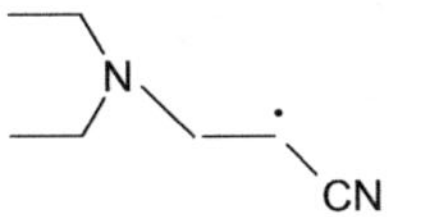			<0.005				

Barriers (in kJ/mol) for the monomer addition. See abbreviations in Table 1. See text.

Examples of computed barrier values E_a are gathered in Table 10.2. The barrier is expressed by (10.74) where E_{enth} is the enthalpy term and ΔE_{pol} the polar energy change.

$$E_a = E_{enth} - \Delta E_{pol} \tag{10.74}$$

The separation of the relative contributions of the enthalpy and polar effect to the activation energies allows to explain the reactivity. For addition reactions that are only governed by the exothermicity, the barrier E_a is usually described by a linear relationship with ΔH_r (Equation 10.75).

$$E_a = a + b\Delta H_r \tag{10.75}$$

The polar effects characterize the interaction between the single occupied molecular orbital SOMO of the radical and the lowest unoccupied molecular LUMO of the double bond. This induces a charge transfer from the radical to the monomer in the transition state TS and change the TS energy. On the opposite, the interaction between the SOMO and the highest occupied molecular HOMO leads to a charge transfer from the monomer to the radical. The electron deficient or electron-rich character of a double bond or a radical is represented by the absolute electronegativity χ and the hardness η, which are connected with the ionization potential (IP) and the electronic affinity (EA) (Equation 10.76). The polar energy change ΔE_{pol} is expressed from the χ and η values of the monomer and the radical (Equation 10.77).

$$\chi = \frac{IP + EA}{2}; \quad \eta = \frac{IP - EA}{2} \tag{10.76}$$

$$\Delta E_{pol} = \frac{(\chi_M - \chi_R)^2}{4(\eta_M - \eta_R)} \tag{10.77}$$

This approach has been applied [295,298–301] to the reactivity of a large series of carbon-centered radicals (including the benzoyl, hydroxyl alkyl, and aminoalkyl radicals) toward various monomer units (acrylate, methacrylate, vinylether, vinylacetate, etc.). Calculating E_a and ΔE_{pol} yields E_{enth}. Linear relationships between E_{enth} and ΔH_r are generally found (Equation 10.78).

$$E_{enth} = C + c\,\Delta H_r \tag{10.78}$$

The E_{enth} term is also expressed by (10.79) where $E_a{}^0$ is the barrier at $\Delta H_r = 0$ in the absence of any polar effect and ΔE_{enth} the contribution of the enthalpy change to the barrier.

$$E_{enth} = E_a^0 - \Delta E_{enth} \tag{10.79}$$

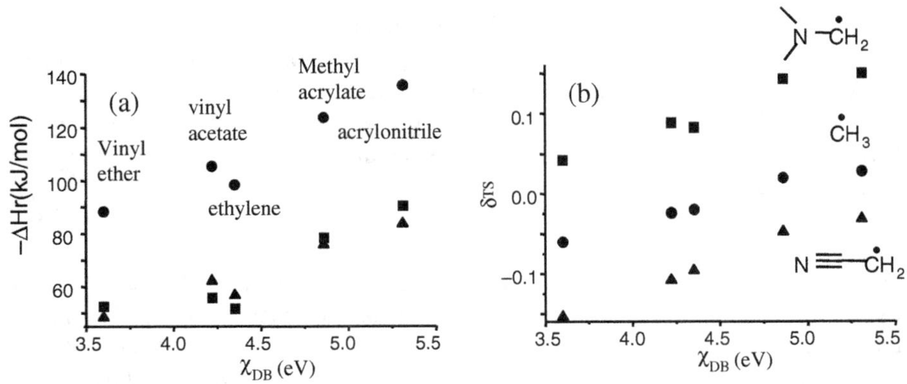

FIGURE 10.9 Addition of initiating radicals $(CH_3)_2$ N $CH_2^{\bullet}$ (square), $CH_3^{\bullet}$ (circle), NC-$CH_2^{\bullet}$ (triangle) to double bonds (vinylether, vinylacetate, ethylene, methylacrylate, acrylonitrile). Evolution of (a) the reaction enthalpy ΔH_r and (b) the amount of charge transfer δ^{TS} vs. the absolute electronegativity of the alkene χ_{DB}.

The enthalpy change ΔE_{enth} is therefore a function of the addition reaction enthalpy ($-c\ \Delta H_r$). The calculated barrier is now expressed by (10.80).

$$E_a = C - \Delta E_{pol} - \Delta E_{enth} \tag{10.80}$$

For example, Fig. 10.9a shows the enthalpy effect as represented by the evolution of the enthalpy (calculated as the difference between the energy of the products and the energy the reactants optimized structures) versus the monomer electronegativity for three particular radicals. Let us consider the addition of the aminoalkyl radical. The enthalpy increases when going from vinyl ether (52.6 kJ/mol) to acrylonitrile (90.3 kJ/mol). On the other side, the barrier decreases from vinyl ether (53.8 kJ/mol) to acrylonitrile (6.3 kJ/mol). As a consequence, it is obvious that the enthalpy partly governs the barrier.

The evolution of the amount of charge transfer δ^{TS} in the transition state reveals the polar effects (Fig. 10.9b)—electron deficient monomers in the presence of nucleophilic radical ($R_2NCH_2^{\bullet}$) increases δ^{TS} whereas electron-rich monomers/electrophilic radical ($NCCH_2^{\bullet}$) couples increase the net charge transfer. The absolute value of δ^{TS} increases from vinyl ether to acrylonitrile for $R_2NCH_2^{\bullet}$ and from acrylonitrile to vinyl ether for $NCCH_2^{\bullet}$. A detailed calculation shows that the polar effect in the addition of the aminoalkyl radical is less important for vinyl ether (1.4 kJ/mol) than for acrylonitrile (17.4 kJ/mol). This outlines the lowering of the barrier by the polar effect. It thus appears that the proposed theoretical treatment adequately describes the reactivity trends and can be confidently used to predict the efficiency evolution for the addition of a given radical to different monomer double bonds.

A similar approach has been used for the addition of other radicals to acrylate monomers such as the well-known benzoyl and hydroxyl isopropyl initiating radicals [300], aminoalkyl radicals [302], thiyl radicals (derived from tetrazole or mercaptobenzoxazole, mercaptobenzimidazole, mercaptobenzothiazole structures) [293], various carbon [299] or other sulfur-centered radicals, silyl radicals [127], and aminyl radicals [264].

Substantial progresses have been shown in the past years as most of the experimental k_i values are now explained on the basis of a theoretical approach. All these works allow to (i) describe the transition state, (ii) explain the role of the enthalpy/polar effects in the addition properties in a very large set of radical/monomer couples, (iii) interpret the large scale of values for the addition rate constants, (iv) study the reversibility of the thiyl radical addition, and (v) design radicals exhibiting both a high reactivity and a low selectivity.

10.4.3.3 Reactivity Toward Oxygen, Hydrogen Donors, and Additives This is exemplified by the following results obtained for the reaction between initiating radicals and propagating radicals and well-known radical polymerization inhibitors such as oxygen, HQME (hydroquinone methylether) or TEMPO (2,2,6,6, tetramethylpiperidine *N*-oxyl radical).

In the case of acrylate radicals [290], the reaction with oxygen is found nearly diffusion controlled (close to $3 \times 10^9 \, M^{-1} \, s^{-1}$). This result is in agreement with both the usual high reactivity of carbon-centered radicals with oxygen and the strong inhibition of radical polymerization reactions under air. The slightly lower reactivity observed for TEA-MMA$^\bullet$ toward oxygen and TEMPO can probably be ascribed to a weak steric hindrance effect in TEA-MMA$^\bullet$ compared to TEA-MA$^\bullet$. The reactivity of (meth)acrylate radicals with TEMPO [302] is particularly worthwhile: the interaction rate constants are close to $3 \times 10^8 \, M^{-1} \, s^{-1}$. These values are also in line with the very efficient reactions observed between other carbon-centered radicals and TEMPO. A very low interaction of HQME with TEA-MA$^\bullet$ and TEA-MMA$^\bullet$ is observed (rate constants lower than $5 \times 10^5 \, M^{-1} \, s^{-1}$).

When considering the aminoalkyl radicals [302], it was clearly found that both their addition to oxygen and their recombination with TEMPO are strongly governed by the reaction exothermicity. A large number of rate constant values were obtained for the aminoalkyl radical/oxygen (0.04–$3 \times 10^9 \, M^{-1} \, s^{-1}$) and aminoalkyl radical/TEMPO interactions (0.002–$5 \times 10^8 \, M^{-1} \, s^{-1}$). In the meantime, the generation of the decarboxylated aminoalkyl radical derived from *N*-phenyl glycine was unambiguously demonstrated [302].

The interaction of the tris(trimethylsilyl)silyl TTMSS$^\bullet$ and triethylsilyl TES$^\bullet$ radicals with HQME [127] exhibits very high rate constants (close to the diffusion rate constant). The Si-H BDEs are 79.1 and 90.1 kcal/mol for TTMSS and TES, respectively. The TES$^\bullet$/HQME interaction thus corresponds to an exothermic hydrogen transfer process. Interestingly, for TTMSS$^\bullet$/HQME, an almost endothermic process is expected since BDE(O–H) $\sim$ BDE(Si-H) but a diffusion controlled reaction is observed. A part of this reactivity is attributed to the participation of polar effects, the electronegativity of these structures (particularly TTMSS) being important. On the opposite, the HQME/TEA$^\bullet$ (triethylamine derived aminoalkyl radical) interaction is very low although the BDE of TEA is high. The polar effects still contribute here, but in the opposite direction, as TEA$^\bullet$ is a nucleophilic radical and TTMSS an electrophilic radical.

10.4.3.4 Reactivity of the Propagating Radicals The direct observation of the acrylate radicals allows to consider the kinetics in solution as a function of the acrylate concentration [303]. At low concentration, the kinetics is described by a second-order law (k_a) that corresponds to the recombination of two propagating radicals (TEA-MA$^•$). At high concentration, the kinetics of the subsequent addition reaction (k_b) of TEA-MA$^•$ to MA is fitted according to a first-order law (pseudo-first-order rate constant: k_b[MA]). These reactions can mimic the propagation k_p and termination k_t rate constants of a polymerization reaction in solution. The latter is nearly diffusion controlled in solution (value close to $10^9\,mol^{-1}\,l\,s^{-1}$). The obtained k_b values appear, at first sight, as close to those determined by pulsed laser polymerization experiments or calculated for low double bond conversion in bulk by the dark polymerization method. Basically, the k_b values therefore correspond to the addition of TEA-M$^•$ to the monomer. According to rather similar calculated barriers for the addition of H-M$^•$, R-M$^•$, R-(M)$_n$-M$^•$ (with $n = 1$–8 [304]) and to the fact that the corresponding preexponential factors should be rather close, the corresponding addition rate constant k_p for the propagating radicals should not significantly differ from k_b. TEA-MA$^•$ could thus be a good representative of the propagating polymeric acrylate radicals. In conclusion, the k_a value likely corresponds to k_t in the early stages of the polymerization reaction and k_b is probably a good acceptable value for k_p in fluid media. These measurements have been extended to acrylonitrile, butyl acrylate, hexane-1,6-diol diacrylate (HDDA), acryloxy-β,β,dimethyl-γ-butyrolactone (known as a highly reactive acrylate monomer), and 2-(2-ethoxy-ethoxy) ethyl acrylate (DEEA).

Other monomers (for example acrylamide, styrene, vinylpyrrolidone, vinylcarbazole, vinyl ether, allyl ether, etc.) can be likely studied by carefully selecting other initiating radicals. Cross-polymerization might also be investigated, for example, the addition of TEA-M$^•$ to vinyl ether since TEA$^•$ does not react with vinyl ether. The results obtained with this procedure can also be extended to the behavior of different acrylate structures in photopolymerization experiments carried out in bulk.

10.4.3.5 Reactivity in Bulk The reactivity in bulk (where the diffusion limits the mobility of the transient species) compared to that in model conditions in solution deserves to be deeply investigated [305]. Photopolymerization experiments clearly show the huge effect of the viscosity of the medium on the relative efficiency of photoinitiators. One has to be careful when extrapolating data from solution to bulk experiments. Apart from practical effects (such as a better compatibility), the change of the efficiency when going from bulk to solution is mainly attributed to the fact that the second-order interaction rate constants are very high in solution but are diffusion controlled in bulk, thereby decreasing the values determined in solution by several orders of magnitude. This is particularly true for the quenching rate constants of triplet states by amines, monomers, oxygen and the initiation rate constants corresponding to the addition of the radicals to the monomer double bonds. The consequence is that the rate constants of 10^9 and $10^6\,M^{-1}\,s^{-1}$ in solution are leveled off to about $10^5\,M^{-1}\,s^{-1}$ in a high viscosity medium so that the apparent higher reactivity in solution could disappear in bulk. For example, the viscosity of multifunctional monomers is 10, 15, 115 cp at 25°C for HDDA (hexane diol diacrylate), TPGDA (tetrapropyleneglycol diacrylate), TMPTA (trimethylol propane triacrylate), respectively. The viscosity of an epoxy acrylate oligomer is 3000 cp at 60°C. For comparison, the viscosity of usual organic solvents is a few cp, that of a toluene/13.7% PMMA (molecular weight: 120,000) is 700 cp. The estimated diffusion rate constant is about 1.1×10^{10}, 8.4×10^6, and $4 \times 10^5\,M^{-1}\,s^{-1}$ in media having a viscosity of 4700 and >1000 cp.

As the two partners have to move before interacting, the viscosity effect on the generation of initiating radicals is important with type II photoinitiators. This is obviously not the case with type I photoinitiators; however, the viscosity has a direct effect on the escape of the radicals from the radical pair formed from the cleavage process: this does not change the rate constant but could modify the dissociation quantum yield. The same also holds true in the radical pair formed in a type II system. The evaluation of this phenomenon shows an increase of the cage effect and suggests that the photoinitiator should not have a constant efficiency during the polymerization. A recent work, however, shows [305] that a quite good correlation exits between the relative efficiency of a set of cleavable photoinitiators in a viscous monomer/oligomer media and their dissociation quantum yield measured in solution (Fig. 10.10). This is due to the fact that (i) the cleavage is very fast (the monomer quenching of the triplet state is inefficient) and (ii) the initiation rate constants are leveled off. The correlation remains valid in less viscous or fluid media providing that the monomer quenching of the photoinitiator triplet state is weak (i.e., there is no change of the dissociation quantum yield) and the radical addition to the monomer is efficient. This is the case for most photoinitiators: they have rather short-lived triplet states and most usual photoinitiating radicals have a similar reactivity toward acrylates and methacrylates (excepted the benzoyl radical). The dependence is more complicated in the case of methacrylates because of the high values for the monomer quenching rate constants.

This approach shows that a deep knowledge of the photophysics and photochemistry of the photoinitiators in solution can thus serve as a good basis for the prediction of the polymerization efficiency in film experiments. Several recent approaches also suggest that a direct investigation on the excited-state processes involved in bulk

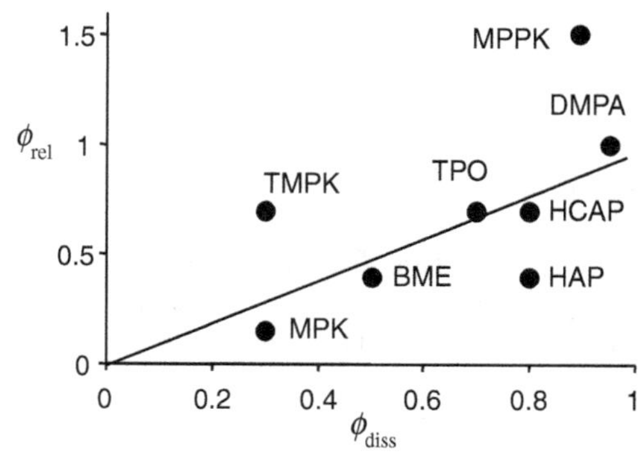

FIGURE 10.10 Plot of the relative efficiency observed for the polymerization of an epoxy acrylate formulation in the presence of various photoinitiators as a function of their dissociation quantum yields. See the formulas of DMPA, BME, HCAP, HAP, TMPK, MPK, MPPK and TPO in Table 1.

media or in high-speed photopolymer coating layers may lead to a deeper insight into the understanding of the practical efficiency [306–308].

10.4.3.6 Reactivity of Radicals in Controlled Photopolymerization Reactions A conventional radical polymerization terminates when the reactive intermediates are destroyed or rendered inactive through bimolecular reactions. The radical generation is irreversible without any possibility to control the final properties of the polymer formed such as the number average molecular weight Mn and the molecular weight distribution (MWD). It is therefore difficult to achieve a highly uniform and well-defined product; the same holds true in specialized applications (where a control is necessary) such as the design of block copolymers or star copolymers, preparation of end functional polymers for grafting, multilayer systems, and so on. In thermal controlled or living radical polymerization [309], the radicals are reversibly generated according to three methodologies largely used in thermal polymerization: atom transfer radical polymerization (ATRP), reversible addition-fragmentation transfer (RAFT), and nitroxide-mediated polymerization (NMP). The use of iniferters is rather limited.

Few things are known in the control of photopolymerization reactions. In a light-induced reaction, a photoiniferter can be used. Both the initiation and the reversible termination are photoinduced. The mechanism of a classical living radical photopolymerization process is recalled in e21.

$$
\begin{aligned}
&\text{R-DC} \xrightarrow{h\nu} \text{R}^{\bullet} + \text{DC}^{\bullet} \\
&\text{R}^{\bullet} + \text{M} \longrightarrow \text{RM}^{\bullet} \\
&\text{RM}^{\bullet} + \text{M} \longrightarrow \text{RM}_{\text{n}}{}^{\bullet} \\
&\text{RM}_{\text{n}}{}^{\bullet} + \text{DC}^{\bullet} \underset{h\nu}{\rightleftarrows} \text{RM}_{\text{n}}\text{-DC}
\end{aligned}
\tag{10.81}
$$

The first reaction corresponds to the cleavage of the photoiniferter structure R-DC to produce an initiating $R^{\bullet}$ and a terminator $DC^{\bullet}$ radicals. The second and third reactions refer to the usual addition of the initiating radical to the first monomer unit and the propagation of the polymerization. The crucial difference with a noncontrolled radical polymerization process lies on the preferred termination of the reaction by the $DC^{\bullet}$ radical: the fourth reaction is photochemically reversible leading to the living character of the polymerization. This reaction as well as the further propagation repeatedly occurs. The living character is achieved providing that the exchange between the reactive and the dormant species is fast in comparison with the propagation. The molecular weight should increase with the conversion, in contrast to the classical radical (photo) polymerizations.

Experimental as well as theoretical works will probably help in the design of high-performance structures. These systems are very often tested with fluid monomers. The synthesis of copolymers (e.g., PMMA-PS) has been achieved in [105]. The insertion of dormant species (R-M_n-DC) in a polymer matrix is worthwhile for many reactions. Applications in viscous monomer/oligomer media might be of interest.

10.5 APPLICATIONS OF PHOTOINITIATORS

Photoinitiators are encountered in a large variety of reactions that will be briefly presented in the following paragraphs. A recent review chapter [2o] gives an insight on these reactions and the properties of the polymerized material observed nowadays in the radiation curing and laser imaging sectors.

10.5.1 Radiation Curing Area

10.5.1.1 Film Photopolymerization of Acrylates The radical polymerization of a formulation based on multifunctional monomers (10.82) and oligomers (10.83) in film is very fast (in the second time range under mercury lamps as in industrial conditions) and sensitive to the presence of oxygen [1a,f,g,i,o,p]. The photopolymerization of thick clear coatings can be achieved if a suitable photoinitiating system is used: the penetration of the light in the depth of the film is favored if the photolysis products do not absorb, that is, if a good bleaching occurs [1s]. The photopolymerization of thick-pigmented coatings appeared as a real challenge because of the strong absorption of the pigment. High-performance suitable photoinitiating systems using a four component system or bis acylphosphine oxides have been proposed: they allow the curing of paints usable on industrial lines [244].

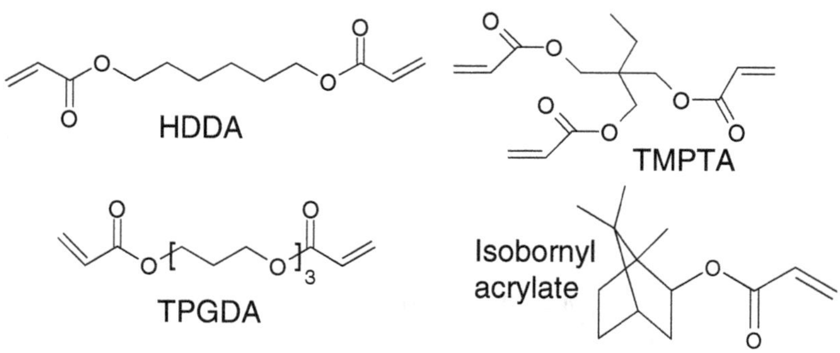

(10.82)

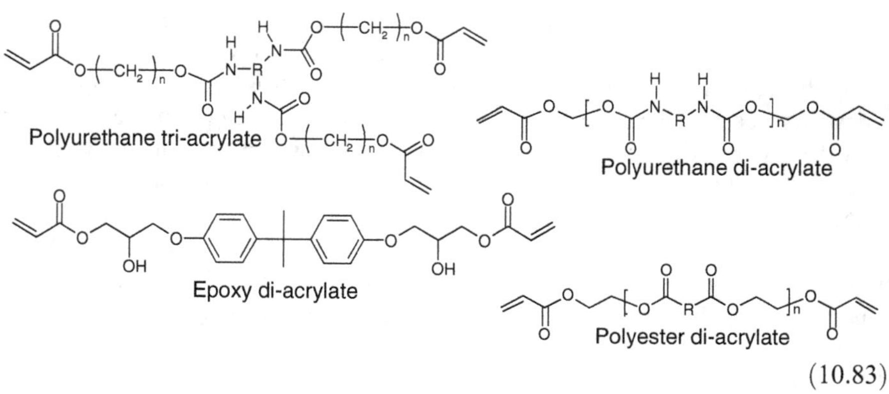

(10.83)

10.5.1.2 Thiol-Ene Photopolymerization The thiol-ene polymerization of suitable systems (10.84) in film is insensitive to oxygen. The reaction refers to the addition of a thiol to a double bond (e.g., vinyl, allyl, acrylate, and methacrylate) [310] and has led in these past years to a new revival of interest [311,312]. Thiol-vinyl ether or -allyl ether polymerization shows some following interesting features: very fast process, low or even no oxygen inhibition effect and formation of highly cross-linked networks with good adhesion, and physical and mechanical properties.

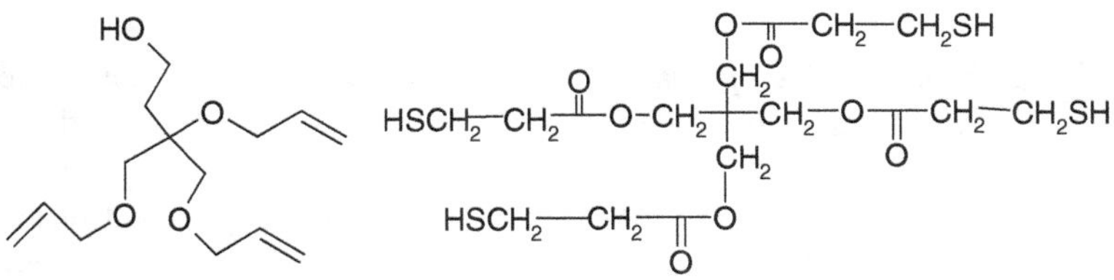

(10.84)

The polymerization reactions in photoinitiator free conditions were recently revisited: this reaction could be helpful if a non-UV-absorbing coating must be obtained because in that case no photoinitiator decomposition products are present in the cured coating. The addition of a UV photoinitiator, however, allows to increase the cure speed.

10.5.1.3 Photopolymerization of Waterborne Light Curable Systems The basic idea is to develop solventless (also designed as waterborne) systems to answer to the pressure of the new environmental regulations [1o]. The water-based formulations are less viscous than the conventional acrylate mixtures. These systems are available as water-soluble media (e.g., urethane acrylates with low molecular weights water-soluble polyether acrylate), emulsions (e.g., ethylene-vinyl acetate- methacrylate or methacrylate- butylacrylate- glycidylmethacrylate), and dispersions (e.g., urethane acrylates, epoxy acrylates or polyester acrylates stabilized with an added surfactant or functionalized with a surfactant moiety as shown in (10.85)). Having, however, good surface properties requires to dry the coating before light exposure. Sunlight curing of pigmented waterborne paints under air using camphorquinone was successfully achieved [313].

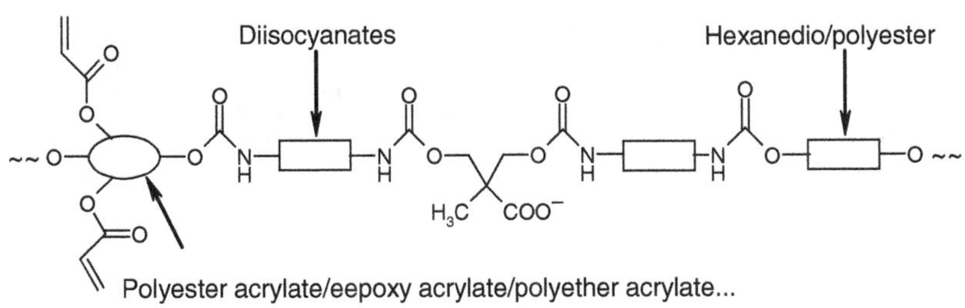

(10.85)

10.5.1.4 Photopolymerization of Powder Formulations Powder coatings are very attractive [314]. The powder is applied on the substrate and exposed to an IR source to get the coalescence of the solid particules; then, the cross-linking reaction is carried out under UV (or visible) light in a very short time and at a temperature about 100–120°C. Largely encountered systems are based on, for example, polyesters (with maleates or fumarates unsaturations) and polyurethane vinyl ethers that copolymerize according to a radical process, amorphous polyesters and functionalized polyesters (with allyl ethers).

10.5.1.5 Charge Transfer Photopolymerization This reaction shows some interesting features [315]: very high cure speed (as that obtained with acrylates) and no or low sensitivity to oxygen. Photoinitiator free formulations can be used. In that case (for example, (10.86)), an acceptor molecule (maleimide) containing an electron poor moiety (absorption around 300–350) and a donor molecule (vinyl ether or an allyl ether) bearing an electron-rich moiety (absorption below 250 nm) lead to the generation of a charge transfer complex. This complex can be either formed in the ground state (and then excited) or generated in the excited state (after reaction between, for example, the donor and the excited acceptor). The polymerization efficiency sharply increases in the presence of a UV or visible radical photoinitiator. The curing of a maleimide-allyl or vinyl ether film formulation can be achieved within 1 s in the presence of a four-component photoinitiating system [204].

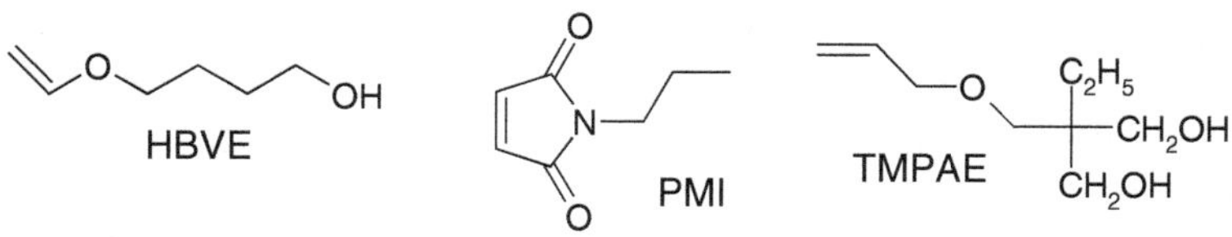

(10.86)

10.5.1.6 Dual Cure Photopolymerization The dual cure technique is a two-step process involving two different types of chemistry [1i]. It usually consists in combining a UV irradiation and a thermal drying process and usually appears as a two-pack material, typically based on a polyisocyanate and an acrylate (10.87). The acrylate is

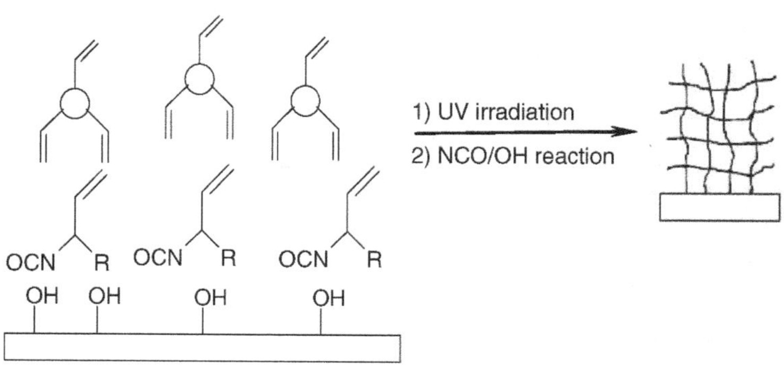

first cured upon light exposure and the subsequent dark reaction of the isocyanates with the hydroxyl groups of the acrylates allows to completely cure the coating.

10.5.1.7 Hybrid Cure Photopolymerization The hybrid cure technique [1i] corresponds to two polymerization mechanisms. For example, a blend of radical M_R and cationic M_C monomers/oligomers (10.87) reacts through a concurrent radical/cationic polymerization mechanism. Cross-linked copolymers combining the properties of the two homopolymer networks and interpenetrating networks exhibiting new properties can be formed. The starting point of the reaction can be driven by a selective and separate excitation of the two photoinitiators depending on the wavelengths used: the possibility of this two-step procedure (thanks to the photochemical initiation process) allows to separately generate (and upon the demand) the two networks.

$$\begin{matrix} M_C \\ M_R \end{matrix} \xrightarrow{h\nu_1} \begin{matrix} -(M_C)_n- \\ M_R \end{matrix} \xrightarrow{h\nu_2} \begin{matrix} -(M_C)_n- \\ -(M_R)_n- \end{matrix} \qquad (10.87)$$

10.5.1.8 Hybrid Sol–Gel Photopolymerization Photosensitive organic–inorganic sol–gel glasses are interesting alternative materials that combine the properties of glasses and polymers [316]. A typical hybrid sol–gel material consists (10.88) in an acrylated silane derivative (which in a first step, leads to a glass-like material via hydrolysis and condensation) and a photoinitiator that allows the free radical photopolymerization of the acrylate. The requirements for the choice of the photoinitiator in such a matrix are crucial: solubility, thermal stability, PH compatibility, polarity, and so on.

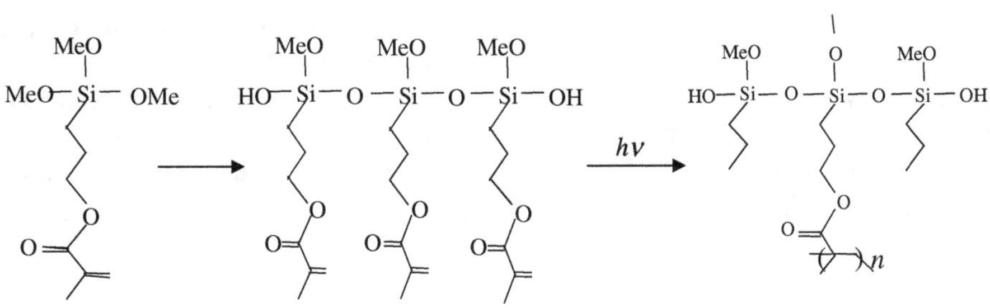

(10.88)

10.5.2 Imaging and Laser Imaging

Phoinitiators usable in graphic arts for the manufacture of negative and positive printing plates and the drying of thin-pigmented printing inks in classical and ink-jet devices must be efficient. Specific properties are very often searched, for example, for applications in the food packaging industry.

In microelectronics, direct laser imaging LDI [317] allows to write complex relief patterns for the manufacture of microcircuits. In the graphic arts area, computer-to-plate (CTP) technology [318] is a quite similar process used to reproduce a document on a printing plate. In both LDI and CTP applications, the photoinitiator has to be obviously sensitive to the laser line. In the past years, the search for suitable systems was strongly driven by the laser industry according to the development of new, powerful and cheaper laser sources: the spectral sensitivity thus moved to 488, 514.5, and 532 nm, then to 351, 364, and 355 nm, and recently to 405 nm.

Three-dimensional machining (or 3D photopolymerization or stereolithography) gives the possibility to make objects, even with complex forms, for prototyping applications. A laser beam is used for the excitation. Creating 3D microscale structures for microelectromechanical, microoptics and microfluidic applications requires to use high peak power laser pulses allowing a multiphoton (typically two photon) of the photoinitiator at the focal point.

In the optics area, the development of photoinitiators for selective laser excitation and the understanding of its efficiency in such particular media have known a real interest, for example, the manufacture of gratings, microlenses or waveguides, the holographic recording, and storage of informations.

10.6 CONCLUSION

This review paper intended to show the state-of-the-art in photoinitiators and photosensitizers of free radical polymerization reactions for the radiation curing and laser imaging technologies. It outlines the huge and continuously emerging development of tailor made systems. This has been made possible, thanks to the experience accumulated during more than 20 years of research, the wide opportunities in organic synthesis, the recent development of more powerful time resolved laser spectroscopy and analytical techniques, the larger use of quantum mechanical calculations and—the last but not the least—the researcher ingenuity. There is no doubt that ever more efficient and well adapted new photoinitiating systems will be proposed in the future[1].

[1]This chapter has been prepared in a view to be ready for the end of 2007. During the last two/three years, many papers have appeared in the literature and outlined the fantastic developments of new photoinitiating systems lying on newly explored radical chemistry and chemical structures (e.g. involving Si, Ge, B, P, Sn atoms, complexes, metal based compounds, macromolecular or bifunctional photoinitiators, multi-component combinations...) or exhibiting a high potential for applications of radical photopolymerization reactions in a lot of experimental conditions (using low viscosity acrylates or various media, incorporating nanoparticles, formulating films for the design of controlled architectures, operating under air, irradiating with visible lights or sunlight, working under low light intensity, elaborating two-photon sensitive systems for the nanotechnology area...). Unfortunately, none of these very recent and promising items could be cited here !

REFERENCES

1. (a) C.G. Roffey, *Photopolymerization of Surface Coatings*, John Wiley & Sons, Inc., New York, **1982**. (b) J.F. Rabek, *Mechanisms of Photophysical and Photochemical Reactions in Polymer: Theory and Practical Applications*, Wiley, New York, **1987**. (c) C.E. Hoyle and J.F. Kinstle (Eds.), *Radiation Curing of Polymeric Materials*, ACS Symposium Series, 417, **1990**. (d) J.P. Fouassier and J.F. Rabek (Eds.), *Lasers in Polymer Science and Technology: Applications*, CRC Press, Boca Raton, **1990**. (e) H. Bottcher, *Technical Applications of Photochemistry*, Deutscher Verlag fur grundstoffindustrie, Leipzig, **1991**. (f) S.P. Pappas, *UV-Curing: Science and Technology,* Tech Mark Corporation, Stamford, 1986, Plenum Press, New York, **1992**. (g) J.P. Fouassier and J.F. Rabek (Eds.), *Radiation Curing in Polymer Science and Technology*, Chapman & Hall, London, **1993**. (h) V. Krongauz and A. Trifunac (Eds.), *Photoresponsive Polymers*, Chapman and Hall, New York, **1994**. (i) J.P. Fouassier, *Photoinitiation, Photopolymerization, Photocuring*, Hanser, Münich, **1995**. (j) N.S. Allen, M. Edge, I.R. Bellobono, and E. Selli (Eds.), *Current Trends in Polymer Photochemistry*, Ellis Horwood, New York, **1995**. (k) A. B. Scranton, A. Bowman, and R.W. Peiffer (Eds.), *Photopolymerization: Fundamentals and Applications*, ACS Symposium Series 673, Washington DC, **1997**. (l) *Chemistry & Technology of UV & EB Formulation for Coatings, Inks and Paints*, K.T. Olldring Ed., John Wiley and Sita, Vol. I–VIII, London, **1997**. (m) J.P. Fouassier, *Photoinitiated Polymerization: Theory and Applications*, Rapra Review Reports, 9, 4, Rapra Tech. Ltd, Shawbury, **1998**. (n) R. Holman (Ed.), *UV and EB Chemistry*, Sita Technology Ltd, London, **1999**. (o) S. Davidson, *Exploring the Science, Technology and Application of UV and EB Curing*, Sita Technology Ltd, London, **1999**. (p) D.C. Neckers, *UV and EB at the Millenium*, Sita Technology Ltd, London, **1999**. (q) J.P. Fouassier (Ed.), *Photosensitive Systems for Photopolymerization Reactions, Trends in Photochemistry and Photobiology*, Vol. 5, Research Trends, Trivandrum, India, **1999**. (r) J.P. Fouassier, *Light Induced Polymerization Reactions, Trends in Photochemistry and Photobiology*, Vol. 7, Research Trends, Trivandrum India, **2001**. (s) K. Dietliker, *A Compilation of Photoinitiators Commercially Available for UV Today*, Sita Technology Ltd, London, **2002**. (t) J.G. Drobny, *Radiation Technology for Polymers*, CRC Press LLC, Boca Raton, **2003**. (u) K.D. Belfied and J.V. Crivello (Eds.), *Photoinitiated Polymerization*, ACS Symposium Series 847, Washington, DC, **2003**. (v) J. P. Fouassier (Ed.), *Photochemistry and UV Curing*, Research Signpost, Trivandrum, India, **2006**. (w) R. Schwalm, *UV Coatings: Basics, Recent Developments and New Applications*, Elsevier, Oxford, UK, **2007**.
2. (a) J.P. Fouassier, *Rec. Res. Dev. Photochem. Photobiol.* **2000**, *4*, 51–74. (b) J.P. Fouassier, *Curr. Trends Polym. Sci.* **1999**. *4*. (c) J.P. Fouassier, *Rec. Res. Dev. Polym. Sci.* **2000**, *4*, 131–145. (d) J.P. Fouassier, X. Allonas, and D. Burget, *Progr. Org. Coat.* **2003**, *47*, 16–36, (e) J. Paczkowski, and D.C. Neckers, *Electron Transfer Chem.* **2001**, *5*, 516–585. (f) W. Schnabel,in: J.P. Fouassier and J.F. Rabek (Eds.), *Lasers in Polymer Science and Technology*, Vol. II, CRC Press, Boca Raton, **2000**. (g) T. Urano, *J. Photopolym. Sci. Tech.* **2003**, *16*(1), 129–156. (h) J.P. Fouassier, X. Allonas, and D. Burget, *Progr. Org. Coat.* **2003**, *47*, 16–36. (i) J.P. Fouassier, D. Ruhlmann, B. Graff, F. Morlet-Savary, and F. Wieder, *Progr. Org. Coat.* **1995**, *25*, 235–271. (j) J.P. Fouassier, D. Ruhlmann, B. Graff, and F. Wieder, *Progr. Org. Coat.* **1995**, *25*, 169–202. (k) J.P. Fouassier, *Progr. Org. Coat.* **1990**, *18*, 229–252.(l) H.J. Timpe, in: J.P. Fouassier (Ed.), *Radiation Curing in Polymer Science and Technology*, Elsevier, Barking, UK, **1993**. (m) A.F. Cunningham and

V. Desobry, in: J.P. Fouassier (Ed.), *Radiation Curing in Polymer Science and Technology*, Elsevier, Barking, UK, **1993**. (n) W.A. Green and A.W. Timms, in: J.P. Fouassier (Ed.), *Radiation Curing in Polymer Science and Technology*, Elsevier, Barking, UK, **1993**. (o) J. P. Fouassier, X. Allonas, and J. Lalevée,in: K. Matyjaszewski, Y. Gnanou, and L. Leibler (Eds.), *Macromolecular Engineering: From Precise Macromolecular Synthesis to Macroscopic Materials Properties and Applications*, Vol. 1, Wiley-VCH, Weinheim, **2007**, pp. 643–672.

3. A.N. Savitsky, H. Paul, and A.I. Shushin, *Helv. Chim. Acta*, **2006**, *89*(10), 2533–2543.
4. K. Sirovatka Padon and A.B. Scranton, *J. Polym. Sci. A Polym. Chem.* **2000**, *38*(18), 3336–3346.
5. V. Lemée, D. Burget, P. Jacques, and J.P. Fouassier, *J. Polym. Sci. A Polym. Chem.* **2000**, *38*(10), 1785–1794.
6. J.P. Fouassier, D. Ruhlmann, and A. Erddalane, *Macromolecules* **1993**, *26*(4), 721–728.
7. M. Dossot, H. Obeid, X. Allonas, P. Jacques, J.P. Fouassier, and A. Merlin, *J. Appl. Polym. Sci.* **2004**, *92*(2), 1154–1164.
8. A. Valdehenito, E.A. Lissi, and M.V. Encinas, *Macromol. Chem. Phys.* **2001**, *202*(12), 2581–2585.
9. I. Morino, M. Wakasa, and H. Hayashi, *Mol. Phys.* **2002**, *100*(9), 1283–1289.
10. I. Capek, *Trends Photochem. Photobiol.* **2001**, *7*, 147–157.
11. (a) R.P. Wayne, *Principles and Applications of Photochemistry*, Oxford Science and Publications, Oxford, **1988**. (b) N.J. Turro, *Modern Molecular Photochemistry*, Benjamin, New York, **1978**, re-edited in 1990. (c) A. Gilbert, J. Baggot, *Essentials of Molecular Photochemistry*, Blackwell Scientific Publications, Oxford, **1991**. (d) P. Suppan, *Chemistry and Light*, The Royal Society of Chemistry, **1994**.
12. A. Braun, M.T. Maurette, and E. Oliveros, *Technologie Photochimique*, Presses Polytechniques Romandes, Lausanne, **1986**.
13. H.A. Gour, C.J. Groenenboom, H.J. Hageman, G.T.M. Hakwoort, P. Osterholl, T. Overeem, R.I. Polman, and S. Van des Werf, *Makromol. Chem.* **1984**, *185*, 1795.
14. P. Jaegermann, F. Lendzian, G. Rist, and K. Mobius, *Chem. Phys. Lett.* **1987**, *140*(6), 615–619.
15. W.U. Palm and H. Dreeskamp, *J. Photochem.* **1990**, *52*(3), 439–450.
16. J.E. Christensen, A.F. Jacobine, and C.J.V. Scaiano, *Radiat. Curing* **1981**, *8*(3), 1.
17. J. Eichler, C.P. Herz, I. Naito, and W. Schnabel, *J. Photochem.* **1980**, *12*(3), 225–234.
18. S. Jockusch, M.S. Landis, B. Freiermuth, and N.J. Turro, *Macromolecules* **2001**, *34*(6), 1619–1626.
19. R. Liskra, *Heterocycles* **2001**, *55*(8), 1475–1486.
20. R. Liskra and D. Herzog, *J. Polym. Sci. A Polym. Chem.* **2004**, *42*(3), 752–764.
21. R. Liskra, B. Seidl, and G. Grabner, *Polym. Preprints* **2004**, *45*(2), 75–76.
22. R. Liska, in: J.P. Fouassier (Ed.), *Photochemistry and UV Curing: New Trends*, Research Signpost, Trivandrum, India, **2006**.
23. R. Liska and B. Seidl, *J. Polym. Sci. A Polym. Chem.* **2005**, *43*(1), 101–111.
24. C.G. Groenenboom, H.J. Hageman, P. Oosterhoff, T. Overeem, and J. Verbeek, *J. Photochem. Photobiol. A Chem.* **1997**, *107*(1–3), 261–269.

25. H.J. Hageman, P. Oosterhoff, and J. Verbeek, *J. Photochem. Photobiol. A Chem.* **1999**, *121* (3), 207–211.
26. R. Mallavia, R. Sastre, and F. Amat-Guerri, *J. Photochem. Photobiol. A Chem.* **1999**, *138*, 207.
27. T. Sumiyoshi, W. Schnabel, and A. Henne, *J. Photochem.* **1986**, *32*(2), 191–201.
28. J.E. Baxter, R.S. Davidson, and H.J. Hageman, *Polymer* **1988**, *29*(9), 1569–1574.
29. S. Jockusch, I.V. Koptyug, P.F. McGarry, G.W. Sluggett, N.J. Turro, and D.M. Watkins, *J. Am. Chem. Soc.* **1997**, *119*(47), 11495–11501.
30. V. Desobry, K. Dietliker, R. Husler, L. Misev, M. Rembold, G. Rist, and W. Rutsch, in: C.E. Hoyle, and J.F. Kinstle (Eds.), *ACS Symposium Serires 247*, Washington, **1989**.
31. M. Yamaji, S. Wakabayashi, and S. Tobita, *Res. on Chem. Intermed.*, **2006**, *32* (8),749–758.
32. J.P. Fouassier and D.J. Lougnot, *Polymer Commun.* **1990**, *31*, 418–421.
33. E. Andrzejewska, *Trends Photochem. Photobiol.* **2001**, *7*, 1–9.
34. J.P. Fouassier, D.J. Lougnot, and L. Avar, *Polymer* **1995**, *36*(26), 5005–5010.
35. J.P. Fouassier, D.J. Lougnot, and J.C. Scaiano, *Chem. Phys. Lett.* **1989**, *160*, 335–340.
36. J.P. Fouassier, V. Lemée, A. Espanet, D. Burget, F. Morlet-Savary, P. Di Battista, and G. Li Bassi, *Eur. Polym. J.* **1997**, *33*(6), 881–896.
37. G. Li Bassi, L. Cadona, and F. Broggi, Radcure 86, Tech. Paper 4–27, SME Ed., Dearborn, Michigan, **1986**.
38. M. Cattaneonti, *Proc. Rad. Tech. 2002*, **2002**, p. 403.
39. X. Allonas, C. Grotzinger, J. Lalevée, J.P. Fouassier, and M. Visconti, *Eur. Polym. J.* **2001**, *37*, 897–906.
40. H.J. Hageman and P. Oosterhoff, *Macromol. Chem. Phys.* **2000**, *201*(14), 1687–1690.
41. J.P. Fouassier and D. Burr, *Macromolecules* **1990**, *23*, 3615–3619.
42. G. Berner, G. Rist, W. Rutsch, and R. Kirchmayer, Radcure Basel, Technical Paper FC85-446, SME, Dearborn, Michigan, **1985**.
43. H. Tomioka, Y. Takimoto, M. Kawabata, M. Harada, J.P. Fouassier, and D. Ruhlmann, *J. Photochem. Photobiol. A Chem.* **1990**, *53*, 359–372.
44. F. Morlet-Savary, J.P. Fouassier, and H. Tomioka, *Polymer* **1992**, *33*(19), 4202–4206.
45. G. Berner, R. Kirchmayer, and G. Rist, *J. Oil. Chem. Assoc.* **1978**, *61*, 105.
46. J.H. Kim and H.L. Kim, *Chem. Phys. Lett.* **2001**, *333*(1,2), 45–50.
47. K. Kawamura, *Chem. Lett.* **2003**, *32*(9), 832–833.
48. M. Buback, M. Kling, S. Schmatz, and J. Schroeder, *Phys. Chem. Chem. Phys.* **2004**, *6*(24), 5441–5455.
49. A.V. Fedorov, E.O. Danilov, M.A.J. Rodgers, and D. C. Neckers, *J. Am. Chem. Soc.* **2001**, *123*(21), 5136–5137.
50. N. Yasuda, S. Yamamoto, Y. Wada, and S. Yanagida, *J. Polym. Sci. A Polym. Chem.* **2001**, *39*(24), 4196–4205.
51. M.-L. Tsao and M.S. Platz, *J. Phys. Chem. A* **2004**, *37*(24), 8984–8991.
52. J. Alvarez, M.V. Encinas, and E.A. Lissi, *Macromol. Chem. Phys.* **1999**, *200*(10), 2411–2415.
53. W. Sander, A. Strehl, and M. Winkler, *Eur. J. Org. Chem.* **2001**, *20*, 3771–3778.

54. S.-R. Baruah and D.K. Kakati, *J. Appl. Polym. Sci.* **2006**, *100*(2), 1601–1606.
55. J. Lalevee, M. El Roz, F. Morlet-Savary, B. Graff, X. Allonas, and J.P. Fouassier, *Macromolecules* **2007**, *40*, 8527–8530 .
56. T.S. Kwon, K. Suzuki, K. Takagi, H. Kunisada, and Y. Yuki, *J. Macromol. Sci. Pure Appl. Chem.* **2001**, *A38*(5–6), 591–604.
57. N.S. Allen, S.J. Hardy, A.F. Jacobine, D.M. Glaser, B. Yang, D. Wolf, F. Catalina, S. Navaratnam, and B.J. Parsons, *J. Appl. Polym. Sci.* **1991**, *42*(5), 1169.
58. D.C. Neckers, I.I. Abu Abdoun, and L. Thijs, *Macromolecules* **1984**, *17*(3), 282–288.
59. B.K. Shah and D.C. Neckers, *J. Amer. Chem. Soc.* **2004**, *126*(6), 1830–1835.
60. B.K. Shah, A. Gusev, M.A.J. Rodgers, and D.C. Neckers, *J. Phys. Chem. A* **2004**, *108*(28), 5926–5931.
61. F. Morlet-Savary, F. Wieder, and J.P. Fouassier, *J. Chem. Soc. Faraday Trans.* **1997**, *93*(22), 3931–3937.
62. D.H.R. Barton and S.Z. Zard, *Pure Appl. Chem.* **1986**, *58*(5), 675–684.
63. B.C. Bales, J.H. Horner, X. Hueng, M. Newcomb, D. Crich, and M.M. Greenberg, *J. Am. Chem. Soc.* **2001**, *123*(16), 3623–3629.
64. M. Newcomb, J.H. Horner, P.O. Whitted, D. Crich, X. Huang, Q. Yao, and H. Zipse, *J. Am. Chem. Soc.* **1999**, *121*(46), 10685–10694.
65. B.M. Aveline, I.E. Kochevar, and R.W. Redmond, *J. Am. Chem. Soc.* **1996**, *118*(42), 10113–10123.
66. C. Dietlin, X. Allonas, F. Morlet-Savary, J.P. Fouassier, M. Visconti, G. Norcini, and S. Romagnano, *J. App. Polym. Sci.* **2008**, *109*, 828–833.
67. M.M. Alam, M. Fujitsuka, and O. Ito, *Rec. Res. Dev. Phys. Chem.* **2000**, *4*, 369.
68. C. Bohne, R. Boch, and J.C. Scaiano, *J. Org. Chem.* **1990**, *55*(19), 5414–5418.
69. M. Newcomb, J.H. Horner, P.O. Whitted, D. Crich, X. Huang, Q. Yao, and H. Zipse, *J. Am. Chem. Soc.* **1999**, *121*(46), 10685–10694.
70. O. Grinevich, P. Serguievski, A.M. Sarker, W. Zhang, A. Mejiritski, and D.C. Neckers, *Macromolecules* **1999**, *32*(2), 328–330.
71. A.M. Sarker, K. Sawabe, B. Strehmel, Y. Kaneko, and D.C. Neckers, *Macromolecules ACS ASAP* **1999**, *32*(16), 5203–5209.
72. J. Kabatc, B. Jedrzejewska, and J. Paczkowski, *J. Polym. Sci. A Polym. Chem.* **2003**, *41*(19), 3017–3026.
73. J. Kabatc, M. Pietrzak, and J. Paczkowski, *J. Chem. Soc. Perkin Trans.* **2002**, *2*(2), 287–295.
74. B. Jedrzejewska, J. Kabatc, M. Pietrzak, and J. Paczkowski, *J. Polym. Sci. A Polym. Chem.* **2002**, *40*(10), 1433–1440.
75. T. Sato, T. Katayose, and M. Seno, *J. Appl. Polym. Sci.* **2000**, *79*(1), 166–175.
76. C.R. Rivarola, S.G. Bertolotti, and C.M. Previtali, *J. Polym. Sci. A Polym. Chem.* **2001**, *39*(24), 4265–4273.
77. M.G. Neumann, C.C. Schmitt, and I.C. Rigoli, *J. Photochem. Photobiol. A Chem.* **2003**, *159*, 145–150.
78. C. Billaud, M. Sarakha, and M. Bolte, *J. Polym. Sci. A Polym. Chem.* **2000**, *38*(21), 3997–4005.

79. J. Jakubiak and J.F. Rabek, *Polimery* **1999**, *44*, 447–461.
80. R. Jantas, T. Wodka, and G. Janowska, *Polimery (Warsaw, Poland)* **2001**, *46*(11–12), 812–816.
81. M. Degirmenci, I. Cianga, and Y. Yagci, *Macromol. Chem. Phys.* **2002**, *203*(10–11), 1279–1284.
82. M. Degirmenci, G. Hizal, and Y. Yagci, *Macromolecules* **2002**, *35*(22), 8265–8270.
83. T. Corrales, F. Catalina, C. Peinado, and N.S. Allen, *J. Photochem. Photobiol. A Chem.* **2003**, *159*, 103–114.
84. J.P. Fouassier, D.J. Lougnot, G. Li Bassi, and C. Nicora, *Polym Commun.* **1989**, *30*, 245–248.
85. L. Angiolini, D. Caretti, and E. Salatelli, *Macromol. Chem. Phys.* **2000**, *201*(18), 2646–2653.
86. V. Castelvetro, M. Molesti, and P. Rolla, *Macromol. Chem. Phys.* **2002**, *203*(10/11), 1486–1496.
87. Z. Liu, M. Weber, N.J. Turro, and B. O'Shaughnessy, *Polym. Prepr. Am. Chem. Soc. Div. Polym. Chem.* **2001**, *42*(2), 771–772.
88. M. Visconti, in: J.P. Fouassier (Ed.), *Photochemistry and UV Curing: New Trends*, Research Signpost, Trivandrum, India, **2006**, p. 153.
89. C. Dietlin, J. Lalevée, X. Allonas, J.P. Fouassier, M. Visconti, G. Li Bassi, and G. Norcini, *J. App. Polym. Sci.* **2008**, *107*, 246–252.
90. W.D. Davies, F.D. Jones, J. Garrett, I. Hutchison, and G. Walton, *Surf. Coat. Int.*, **2000**, *83*(2),72–78.
91. W.D. Davies, F.D. Jones, J. Garrett, I. Hutchinson, and G. Walton, *Surf. Coat. Int. B Coat. Trans.* **2001**, *84*(B3), 213–222.
92. M. Visconti and M. Cattaneo, *Prog. Org. Coat.*, **2000**, *40*(1–4), 243–251.
93. G. Ullrich, B. Ganster, U. Salz, N. Moszner, and R. Liska, *J. Polym. Sci. A Polym. Chem.* **2006**, *44*(5), 1686–1700.
94. J.H. de Groot, K. Dillingham, H. Deuring, H.J. Haitjema, F.J. van Beijma, K. Hodd, and S. Norrby, *Biomacromolecules* **2001**, *2*(4), 1271–1278.
95. R. Liska, *J. Polym. Sci. A Polym. Chem.* **2002**, *40*(10), 1504–1518.
96. J.P. Fouassier, D. Burr, and F. Wieder, *J. Polym. Sci. A Polym. Chem.* **1991**, *29*, 1319–1327.
97. K. Vacek, J. Geimer, D. Beckert, and R. Mehnert, *J. Chem. Soc. Perkin Trans.* **1999**, *2* (11), 2469–2471.
98. T. Otsu and T. Taraki, *Polym. Bull.* **1986**, *16*(4), 277–284.
99. D. Bertin, B. Boutevin, P. Gramain, J.M. Fabre, and C. Montginoul, *Eur. Polym. J.* **1998**, *34*(1), 85–90.
100. A. Ajayaghosh and R. Francis, *J. Am. Chem. Soc.* **1999**, *121*(28), 6599–6606.
101. S.H. Qin, D.Q. Qin, and K.Y. Qiu, *Chinese J. Polym. Sci.* **2001**, *19*(5), 441–445.
102. K. Rathore, K.R. Reddy, N.S. Tomer, S.M. Desai, R.P. Singh, *J. Appl. Polym. Sci.* **2004**, *93*(1),348–355.
103. S.H. Qin and K.Y. Qiu, *Eur. Polym. J.* **2001**, *37*(4), 711–717.
104. J. Lalevée, M. El Roz, X. Allonas, and J.P. Fouassier, *J. Polym. Sci.* **2007**, *45*, 2436–2442.
105. J. Lalevée, X. Allonas, and J.P. Fouassier, *Macromolecules* **2006**, *39*, 8216–8218.
106. Y. Tian, M. Zhang, X. Yu, G. Xu, Y. Ren, J. Yang, J. Wu, X. Zhang, X. Tao, S. Zhang, and M. Jiang, *Chem. Phys. Lett.* **2004**, *388*(4–6), 325–329.

107. Y. Yan, X. Tao, Y. Sun, G. Xu, C. Wang, J. Yang, X. Zhao, Y. Wu, Y. Ren, X. Yu, and M. Jiang, *Mater. Sci. Eng. B* **2004**, *113*(2), 170–174.

108. Y. Yan, X. Tao, Y. Sun, W. Yu, C. Wang, G. Xu, J. Yang, Y. Wu, X. Zhao, and M. Jiang, *J. Mol. Struct.* **2004**, *733*(1–3), 83–87.

109. Y. Boiko, *J. Non linear Opt. Phys. Mater.* **2005**, *14*(1), 79–84.

110. Z.-L. Huang, N. Li, Y.-F. Sun, H.-Z. Wang, H.-C. Song, and Z.-L. Xu, *J. Mol. Struct.* **2003**, *657*(1–3), 343–350.

111. G.Y. Zhou, D. Wang, Y.P. Tian, Z.S. Shao, and M.H. Jiang, *Appl. Phys. B* **2004**, *78*(3–4), 397–400.

112. H. Tomioka, N. Takase, Y. Maeyama, K. Hida, V. Lemée, J.P. Fouassier, and D. Burget, *Res. Chem. Intermed.* **2001**, *27*(1,2), 47–59.

113. P.K. Bhuyan and D.K. Kakati, *J. Appl. Polym. Sci.* **2005**, *98*(5), 2320–2328.

114. C. Ma, M.G. Steinmetz, E.J. Kopatz, and R. Rathore, *J. Org. Chem.* **2005**, *70*(11), 4431–4442.

115. L. Macarie, I. Manoviciu, V. Manoviciu, G. Dehelean, G. Ilia, S. Iliescu, A. Popa, and N. Plesu, *Revista de Chimie (Bucharest, Romania)* **2002**, *53*(7), 568–571.

116. S. Daswal and A. Mishra, *J. Appl. Polym. Sci.* **2006**, *99*(3), 920–926.

117. K. Sakayori, Y. Shibasaki, and M. Ueda, *J. Polym. Sci. A Polym. Chem.* **2005**, *43*(22), 5571–5580.

118. K. Tanaka, R. Akimoto, T. Igarashi, and T. Sakurai, *J. Polym. Sci. A Polym. Chem.* **2006**, *44*(1), 25–31.

119. V.V. Semenov, E.Y. Ladilina, N.F. Cherepennikova, and T.A. Chesnokova, *Russ. J. Appl. Chem.* **2002**, *75*(1), 127–134.

120. R.E. Medsker, A. Sebenik, and H.J. Harwood, *Polym. Bull. (Berlin, Germany)*, **2002**, *48*(1), 17–23.

121. H.J. Hageman, *J. Photochem. Photobiol. A Chem.* **1998**, *117*(3), 235–238.

122. J. Paczkowski, Z. Kucybala, F. Scigalski, and A. Wrzyszczynski, *J. Photochem. Photobiol. A Chem.* **2003**, *159*, 115–125.

123. R. Liska, S. Knaus, H. Gruber, and J. Wendrinsky, *Surf. Coat. Int.* **2000**, *83*(6), 297–303.

124. D. Anderson, Procedings of the Radtech Europe Conference 435–444, Vincentz Network, Hannover, **2005**.

125. N.S. Allen, M.C. Marin, M. Edge, D.W. Davies, J. Garrett, and F. Jones, *Polym. Degrad. Stab.* **2001**, *73*(1), 119–139.

126. M.A. Tasdelen, B. Kiskan, and Y. Yagci, *Macromol. Rapid Commun.* **2006**, *27*(18), 1539–1544.

127. J. Lalevée, X. Allonas, and J.P. Fouassier, *J. Org. Chem.* **2007**, *72*, 6434–6439.

128. J. Lalevée, A. Dirani, M. El Roz, X. Allonas, and J.P. Fouassier, *Macromolecules* **2008**, *41*, 2003–2010.

129. Q. Wu and B. Qu, *Polym. Eng. Sci.* **2001**, *41*(7), 1220–1226.

130. K. Viswanathan, C.E. Hoyle, E.S. Joensson, C. Nason, and K. Lindgren, *Macromolecules* **2002**, *35*(21), 7963–7967.

131. C. Valderas, S. Bertolotti, C.M. Previtali, and M.V. Encinas, *J. Polym. Sci. A Polym. Chem.* **2002**, *40*(16), 2888–2893.

132. J. Paczkowski and D.C. Neckers, *Electron Transf. Chem.* **2001**, *5*, 516–585.

133. J. Kabatc, B. Jedrzejewska, and J. Paczkowski, *Macromol. Mater. Eng.* **2006**, *291*(6), 646–654.
134. E. Andrzejewska, D. Zych-Tomkowiak, M.B. Bogacki, and M. Andrzejewski, *Macromolecules* **2004**, *37*(17), 6346–6354.
135. E. Andrzejewska, D. Zych-Tomkowiak, M. Andrzejewski, G.L. Hug, and B. Marciniak, *Macromolecules* **2006**, *93*(11), 3777–3785.
136. J. Lalevée, X. Allonas, L. Zadoina, and J.P. Fouassier, *J. Polym. Sci. Polym. Chem.* **2007**, *45*, 2494–2502.
137. E. Andrjewska, in: J.P. Fouassier, *Photochemistry and UV Curing: New Trends*, Research Signpost, Trivandrum, India, **2006**, p. 127.
138. F. Scigalski and J. Paczkowski, *J. Appl. Polym. Sci.* **2005**, *97*(1), 358–365.
139. G. Ullrich, P. Burtscher, U. Salz, N. Moszner, and R. Liska, *J. Polym. Sci. A Polym. Chem.* **2006**, *44*(1), 115–125.
140. J.R. Woodward, T.S. Lin, Y. Sakaguchi, and H. Hayashi, *Mol. Phys.* **2002**, *100*(8), 1235–1244.
141. X. Cai, M. Sakamoto, M. Hara, A. Sugimoto, S. Tojo, K. Kawai, M. Endo, M. Fujitsuka, and T. Majima, *Photochem. Photobiol. Sci.* **2003**, *2*(11), 1209–1214.
142. X. Cai, M. Sakamoto, M. Fujitsuka, and T. Majima, *Chem. Eur. J.* **2005**, *11*(22), 6471–6477.
143. D.K. Palit, *Res. Chem. Intermed.* **2005**, *31*(1–3), 205–225.
144. T. Corrales, C. Peinado, F. Catalina, M.G. Neumann, N.S. Allen, A.M. Rufs, and M.V. Encinas, *Polymer* **2000**, *41*(26), 9103–9109.
145. M.V. Encinas, A.M. Rufs, T. Corrales, F. Catalina, C. Peinado, K. Schmith, M.G. Neumann, and N.S. Allen, *Polymer* **2002**, *43*(14), 3909–3913.
146. C. Ley, F. Morlet-Savary, P. Jacques, and J.P. Fouassier, *Chem. Phys.* **2000**, *255*, 335–346.
147. X. Allonas, C. Ley, C. Bibaut, P. Jacques, and J.P. Fouassier, *Chem. Phys. Lett.* **2000**, *322*, 483–490.
148. M. Shah, N.S. Allen, M. Edge, S. Navaratnam, and F. Catalina, *J. Appl. Polym. Sci.* **1996**, *62*(2), 319–340.
149. I. Pyszka, Z. Kucybala, and J. Paczkowski, *Macromol. Chem. Phys.* **2004**, *205*(17), 2371–2375.
150. J. Nie, L.A. Linden, J.F. Rabek, J.P. Fouassier, F. Morlet-Savary, F. Scigalski, A. Wrzyszczynski, and E. Andrzejewska, *Acta Polym.* **1998**, *49*, 145–161.
151. J. Nie, E. Andrzejewska, J.F. Rabek, L.A. Linden, J.P. Fouassier, J. Paczkowski, F. Scigalski, and A. Wrzyszczynski, *Macromol. Chem. Phys.* **1999**, *200*, 1692–1701.
152. X. Allonas, J.P. Fouassier, L. Angiolini, and D. Caretti, *Helv. Chim. Acta* **2001**, *84*, 2577–2588.
153. J.P. Fouassier, S.K. Wu, *J. Appl. Polym. Sci.* **1992**, *44*, 1779–1786.
154. C. Ley, F. Morlet-Savary, J.P. Fouassier, and P. Jacques, *J. Photochem. Photobiol.* **2000**, *137*, 87–92.
155. T. Okutsu, M. Ooyama, H. Hiratsuka, J. Tsuchiya, and K. Obi, *J. Phys. Chem. A* **1999**, *104*(2),288–292.
156. H. Wang, J. Wei, X. Jiang, and J. Yin, *Macromol. Chem. Phys.* **2006**, *207*(12), 1080–1086.

157. R. Nagarajan, J.S. BowersJr., Z. Wu, H. Cui, R. Bao, S. Jonsson, R. McCartney, C.U. PittmanJr., L. Cao, and R. Ran, *Surf. Coat. Int.* **1999**, *82*(7), 344–347.

158. H. Wang, Y. Shi, J. Wei, X. Jiang, and J. Yin, *J. Appl. Polym. Sci.* **2006**, *101*(4), 2347–2354.

159. N.S. Allen, N.G. Salleh, M. Edge, M. Shah, C. Ley, F. Morlet-Savary, J.P. Fouassier, F. Catalina, A. Green, S. Navaratnam, and B.J. Parsons, *Polymer* **1999**, *40*, 4181–4193.

160. C. Grotzinger, D. Burget, J.P. Fouassier, G. Rirchard, O. Primel, and S. Yean, *Procedings of the RadTech USA International Conference*, **2006**.

161. J.P. Malval, C. Dietlin, X. Allonas, and J.P. Fouassier, *J. Photochem. Photobiol.*, in press.

162. T. Corrales, F. Catalina, N.S. Allen, and C. Peinado,in: J.P. Fouassier (Ed.), *Photochemistry and UV Curing: New trends*, Research Signpost, Trivandrum, India, **2006**, p. 31.

163. L. Cokbaglan, N. Arsu, Y. Yagci, S. Jockusch, and N.J. Turro, *Macromolecules* **2003**, *36*(8), 2649–2653.

164. M. Aydin, N. Arsu, and Y. Yagci, *Macromol. Rapid Commun.* **2003**, *24*(12), 718–723.

165. M. Aydin, N. Arsu, Y. Yagci, S. Jockusch, and N.J. Turro, *Macromolecules* **2005**, *38*(10), 4133–4138.

166. G. Ullrich, D. Herzog, R. Liskra, P. Burtscher, and N. Moszner, *J. Polym. Sci. A Polym. Chem.* **2004**, *42*(19), 4948–4963.

167. H. Kura, H. Oka, M. Ohwa, T. Matsumura, A. Kimura, Y. Iwasaki, T. Ohno, M. Matsumura, and H. Murai, *J. Polym. Sci. B Polym. Phys.* **2005**, *43*(13), 1684–1695.

168. M.A. Tasdelen, V. Kumbaraci, N. Talinli, and Y. Yagci, *Polymer* **2006**, *47*(22), 7611–7614.

169. E. Novikova, A. Kolendo, V. Syromyatnikov, L. Avramenko, T. Prot, and K. Golec, *Polimery (Warsaw, Poland)*, **2001**, *46*(6), 406–413.

170. V. Wintgens, J.C. Netto-Ferreira, and J.C. Scaiano, *Photochem. Photobiol. Sci.* **2002**, *1* (3), 184–189.

171. A. Costela, I. Garcia-Moreno, J. Dabrio, and R. Sastre, *J. Polym. Sci. A Polym. Chem.* **1997**, *35*(17), 3801–3812.

172. M.V. Encinas, A.M. Rufs, E. Norambuena, and C. Giannotti, *J. Polym. Sci. A Polym. Chem.* **2000**, *38*(12), 2269–2273.

173. A. Costela, I. Garcia-Moreno, O. Garcia, and R. Sastre, *Chem. Phys. Lett.* **2001**, *347* (1,2,3), 115–120.

174. F. Bertens, Y. Qingjin, and D. Gu, *Procedings of the RadTech Europe* **2005**, *1*, 471–478.

175. T. Corrales, F. Catalina, C. Peinado, N.S. Allen, A.M. Rufs, C. Bueno, M.V. Encinas, *Polymer* **2002**, *43*(17), 4591–4597.

176. X. Jiang, H. Xu, and J. Yin, *Polymer* **2003**, *45*(1), 133–140.

177. X. Jiang and J. Yie, *Macromol. Rapid Commun.* **2004**, *25*(6), 748–752.

178. X. Jiang and J. Yin, *J. Appl. Polym. Sci* **2004**, *94*(6), 2395–2400.

179. J. Wei, H. Wang, X. Jiang, and J. Yin, *Macromol. Chem. Phys.* **2006**, *207*(19), 1752–1763.

180. J.P. Fouassier, D. Ruhlmann, K. Zahouily, L. Angiolini, C. Carlini, and N. Lelli, *Polymer* **1992**, *33*(17), 3569–3573.

181. L. Pouliquen, X. Coqueret, F. Morlet-Savary, and J.P. Fouassier, *Macromolecules* **1995**, *28*(24), 8028–8034.

182. J. Wei, H. Wang, X. Jiang, and J. Yin, *Macromol. Chem. Phys.* **2007**, *208*(3), 287–294.

183. J. Wei, H. Wang, and J. Yin, *J. Polym. Sci. A Polym. Chem.* **2007**, *45*(4), 576–587.

184. H. Wang, J. Wei, X. Jiang, and J. Yin, *Polym. Int.* **2007**, *56*(2), 200–207.

185. D.J. Lougnot and J.P. Fouassier, in: J.P. Fouassier and J.F. Rabek (Eds.), *Lasers in Polymer Science and Technology: Applications*, CRC. Press, Boca Raton, **1990**.

186. D.J. Lougnot, C. Turck, and J.P. Fouassier, *Macromolecules* **1989**, *22*, 108–116.

187. F. Catalina, C. Peinado, M. Blanco, N.S. Allen, T. Corrales, and I. Lukac, *Polymer* **1998**, *39*(18), 4399–4408.

188. T. Corales, F. Catalina, N.S. Allen, and C. Peinado, *J. Photochem. Photobiol. A Chem.* **2005**, *169*(1), 95–100.

189. K. Belfield, *Procedings of the Radtech Europe Conference*, 2, 75–82, Vincentz Network, Hannover, **2005**.

190. C. Heller, N. Pucher, B. Seidl, K.R. Kalinyaprak, G. Ullrich, L. Kuna, V. Satzinger, V. Schmidt, L. Lichtenegger, J. Stampfl, and R. Liska, *J. Polym. Sci. A Polym. Chem.* **2007**.

191. C.K. Nguyen, R.S. Smith, T.B. Cavitt, C.E. Hoyle, S. Jonsson, C.W. Miller, and S.P. Pappas, *Polym. Prepr.* **2001**, *42*(2), 707–708.

192. H. Wang, J. Wei, X. Jiang, and J. Yin, *Polym. Int.* **2006**, *55*(8), 930–937.

193. E.S. Jonsson and C.E. Hoyle,in: J.P. Fouassier (Ed.), *Photochemistry and UV Curing: New Trends*, Research Signpost, Trivandrum, India, 165, **2006**.

194. Y. Bi and D.C. Neckers, *J. Photochem. Photobiol. A Chem.* **1993**, *74*, 221–228.

195. R. Mallavia, F. Amat-Guerri, A. Fimia, and R. Sastre, *Macromolecules* **1994**, *27*(9), 2643–2646.

196. S.G. Bertolotti, C.M. Previtali, A.M. Rufs, and M.V. Encinas, *Macromolecules* **1999**, *32*(9), 2920–2924.

197. M.V. Encinas, A.M. Rufs, S. Bertolotti, and C.M. Previtali, *Macromolecules* **2001**, *34*(9), 2845–2847.

198. L. Villegas, M.V. Encinas, A.M. Rufs, C. Bueno, S. Bertolotti, and C.M. Previtali, *J. Polym. Sci. A Polym. Chem.* **2001**, *39*(23), 4074–4082.

199. O. Garcia, A. Costela, I. Garcia-Moreno, and R. Sastre, *Macromol. Chem. Phys.* **2003**, *204*(18), 2233–2239.

200. H. Yong, W. Zhou, G. Liu, L.M. Zhen, and E. Wang, *J. Photopolym. Sci. Technol.* **2000**, *13*(2), 253–258.

201. M.G. Neumann, C.C. Schmitt, and B.E. Goi, *J. Photochem. Photobiol. A Chem.* **2005**, *174*(3), 239–245.

202. H.J. Timpe, S. Ulrich, C. Decker, and J.P. Fouassier, *Macromolecules* **1993**, *26*, 4560–4566.

203. D. Burget, C. Mallein, and J.P. Fouassier, *Polymer* **2003**, *44*(25), 7671–7678.

204. D. Burget, C. Mallein, and J.P. Fouassier, *Polymer* **2004**, *45*, 6561–6567.

205. F. Mauguière-Guyonnet, D. Burget, and J.P. Fouassier, *Prog. Org. Coat.* **2006**, *57*, 23–32.

206. M. Bendyk, B. Jedrzejewska, J. Paczkowski, and L.-A. Linden, *Polimery (Warsaw, Poland)*, **2002**, *47*(9), 654–656.

207. Y.-T. Shi, J. Yin, M. Kaji, and H. Yori, *Polym. Int.* **2006**, *55*(3), 330–339.

208. Y. Shi, J. Yin, M. Kaji, and H. Yori, *Polym. Eng. Sci.* **2006**, *46*(4), 474–479.

209. X. Allonas, H. Obeid, J.P. Fouassier, M. Kaji, Y. Ichihashi, and Y. Murakami, *J. Photopolym. Sci. Tech.* **2003**, *16*(1), 123–128.

210. X. Allonas, J.P. Fouassier, M. Kaji, M. Miyasaka, and T. Hidaka, *Polymer* **2001**, *42*, 7627–7634.
211. F. Morlet-Savary, J.P. Fouassier, T. Matsumoto, K. Inomata, *Polym. Adv. Technol.* **1994**, *5*, 56–62.
212. G. Rist, A. Borer, K. Dietliker, V. Desobry, J.P. Fouassier, and D. Ruhlmann, *Macromolecules* **1992**, *25*, 4182–4193.
213. K. Dietliker, S. Broillet, B. Hellrung, P. Rzadek, G. Rist, J. Wirz, D. Neshchadin, and G. Gescheidt, *Helv. Chem. Acta* **2006**, *89*, 2211–2225.
214. L.Y. Lee, C. Xi, C. Giannotti, and D.G. Whitten, *J. Am. Chem. Soc.* **1986**, *113*, 2304.
215. L. Angiolini, D. Caretti, C. Carlini, E. Corelli, J.P. Fouassier, and F. Morlet-Savary, *Polymer* **1995**, *36*(21), 4055–4060.
216. K. Kawamura and K. Kato, *Polym. Adv. Technol.* **2004**, *15*(6), 324–328.
217. K. Kawamura, *J. Photochem. Photobiol. A Chem.* **2004**, *162*(2–3), 329–338.
218. D. Burget, J.P. Fouassier, F. Amat-Guerri, R. Mallavia, and R. Sastre, *Acta Polym.* **1999**, *50*, 337–346.
219. F. Scigalski and J. Paczkowski, *Polimery (Warsaw, Poland)*, **2001**, *46*(9), 613–621.
220. K. Kawamura, Y. Aotani, and H. Tomioka, *J. Phys. Chem. B* **2003**, *107*(19), 4579–4586.
221. R. Liska, *J. Polym. Sci. A Polym. Chem.* **2004**, *42*(9), 2285–2301.
222. T. Urano, E. Hino, H. Ito, M. Shimizu, and T. Yamaoka, *Polym. Adv. Technol.* **1998**, *9*(12), 825–830.
223. T. Urano, H. Ito, and T. Yamaoka, *Polym. Adv. Technol.* **1999**, *10*(6), 321–328.
224. M. Harada, M. Kawabata, and Y. Takimoto, *J. Photopolym. Sci. Tech.* **1989**, *2*, 199.
225. J.P. Fouassier, A. Erddalane, F. Morlet-Savary, I. Sumiyoshi, M. Harada, and M. Kawabata, *Macromolecules* **1994**, *27*, 3349–3356.
226. J.P. Fouassier, D. Ruhlmann, Y. Takimoto, M. Harada, and M. Kawabata, *J. Polym. Sci. A Polym. Chem.* **1993**, *31*, 2245–2248.
227. J.P. Fouassier and E. Chesneau, *Macromol. Chem.* **1991**, *192*, 1307–1315.
228. K.S. Padon and A.B. Scranton, *J. Polym. Sci. A Polym. Chem.* **2000**, *38*(18), 3336–3346.
229. K. Kim and A. Scranton, *J. Polym. Sci. A Polym. Chem.* **2004**, *42*(23), 5863–5871.
230. M.L. Gomez, V. Avila, H.A. Montejano, and C.M. Previtali, *Polymer* **2003**, *44*(10), 2875–2881.
231. M.L. Gomez, H.A. Montejano, M. del Valle Bohorquez, and C.M. Previtali, *J. Polym. Sci. A Polym. Chem.* **2004**, *42*(19), 4916–4920.
232. A. Erddalane, J.P. Fouassier, F. Morlet-Savary, and Y. Takimoto, *J. Polym. Sci. A Polym. Chem.* **1996**, *34*, 633–642.
233. H.J. Timpe, S. Ulrich, C. Decker, and J.P. Fouassier, *Eur. Polym. J.* **1994**, *30*(11), 1301–1307.
234. T.B. Cavitt, C.E. Hoyle, V. Kalyanaraman, and S. Jonsson, *Polymer* **2004**, *45*(4), 1119–1123.
235. J.P. Fouassier, X. Allonas, J. Lalevée, and M. Visconti, *J. Polym. Sci. A Polym. Chem.* **2000**, *38*, 4531–4541.
236. C. Grotzinger, D. Burget, P. Jacques, and J.P. Fouassier, *Macromol. Chem. Phys.* **2001**, *202*, 3513–3522.
237. C. Grotzinger, D. Burget, P. Jacques, and J.P. Fouassier, *Polymer* **2003**, *44*, 3671–3677.

238. J.P. Fouassier, F. Morlet-Savary, K. Yamashita, and S. Imahashi, *J. Appl. Polym. Sci.* **1996**, *62*, 1877–1885.

239. J.P. Fouassier, F. Morlet-Savary, K. Yamashita, and S. Imahashi, *Polymer* **1997**, *38*(6), 1415–1421.

240. X. Allonas, J.P. Fouassier, H. Obeid, M. Kaji, and Y. Ichihashi, *J. Photopolym. Sci. Tech.* **2004**, *17*(1), 35–40.

241. J. Lalevée, M. El Roz, F. Morlet-Savary, X. Allonas, and J.P. Fouassier, *Chem. Phys.* **2009**, *210*(5), 311–319.

242. J. Yang and D.C. Neckers, *J. Polym. Sci. A Polym. Chem.* **2004**, *42*(15), 3836–3841.

243. D. Burget and J.P. Fouassier, *J. Chem. Soc. Faraday Trans.* **1998**, *94*(13), 1849–1854.

244. L. Catilaz-Simonin and J.P. Fouassier, *J. Appl. Polym. Sci.* **2001**, *79*, 1911–1923.

245. C. Grotzinger, D. Burget, P. Jacques, and J.P. Fouassier, *J. Appl. Polym. Sci.* **2001**, *81*, 2368–2376.

246. X. Allonas, J.P. Fouassier, M. Kaji, and Y. Murakami, *Photochem. Photobiol. Sci.* **2003**, *2*, 224–229.

247. C. Dong and X. Ni, *J. Macromol. Sci. Pure Appl. Chem.* **2004**, *41*(5), 547–563.

248. A.A. Ermoshkin, D.C. Neckers, and A.V. Fedorov, *Macromolecules* **2006**, *39*, 5669–5674.

249. X. Allonas, J. Lalevée, F. Morlet-Savary, and J.P. Fouassier, *Polymery* **2006**, *51*, 7–8.

250. F. Morlet Savary, X. Allonas, and J.P. Fouassier, in: J.P. Fouassier (Ed.), *Photochemistry and UV Curing: New Trends*, Research Signpost, Trivandrum, India, **2006**, pp. 79–90.

251. X. Allonas, F. Morlet Savary, and J.P. Fouassier, in: J.P. Fouassier (Ed.), *Photochemistry and UV Curing: New Trends*, Research Signpost, Trivandrum, India, 9–16, **2006**.

252. S. Jockusch and N.J. Turro, *J. Am. Chem. Soc.* **1998**, *120*(45), 11773–11777.

253. I.V. Koptyug, N.D. Ghatlia, G.W. Sluggett, N.J. Turro, S. Ganapathy, and W.G. Bentrude, *J. Am. Chem. Soc.* **1995**, *117*(37), 9486–9491.

254. U. Kolczak, G. Rist, K. Dietliker, and J. Wirz, *J. Am. Chem. Soc.* **1996**, *118*(27), 6477–6489.

255. X. Allonas, F. Morlet-Savary, J. Lalevée, and J.P. Fouassier, *Photochem. Photobiol.* **2006**, *82*(1), 88–94.

256. M. Spichty, N.J. Turro, G. Rist, J.L. Birbaum, K. Dietliker, J.P. Wolf, and G. Gescheidt, *J. Photochem. Photobiol. A Chem.* **2001**, *142*(2–3), 209–213.

257. C. Dietlin, X. Allonas, J.P. Fouassier, and A. Defoin, *Photochem. Photobiol. Sci.* **2008**, *7*, 552–557.

258. F. Morlet-Savary, X. Allonas, C. Dietlin, J.P. Malval, and J.P. Fouassier, *J. Photochem.* **2008**, *197*, 342–350.

259. X. Allonas, J. Lalevée, and J.P. Fouassier, *J. Photochem. Photobiol. A Chem.* **2003**, *159* (2), 127–133.

260. X. Allonas, J. Lalevée, and J.P. Fouassier, *J. Photopolym. Sci. Tech.* **2004**, *17*(1), 29–34.

261. I.V. Khudyakov and N.J. Turro, in: J.P. Fouassier (Ed.), *Photochemistry and UV Curing: New Trends*, Research Signpost, Trivandrum, India, **2006**, p. 241.

262. I.V. Khudyakov, N. Arsu, S. Jockusch, and N.J. Turro, *Design. Monom. Polym.* **2003**, *6*(1), 91–101.

263. J. Lalevée, X. Allonas, and J.P. Fouassier, *J. Am. Chem. Soc.* **2002**, *124*(32), 9613–9621.

264. J. Lalevée, D. Gigmes, X. Allonas, and J.P. Fouassier, *Chem. Phys. Lett.* **2007**, *438*, 346–350.

265. X. Allonas, J. Lalevée, J.P. Fouassier, and K.D. Belfield,in: J.V. Crivello (Ed.), *Photoinitiated Polymerization*, ACS Symposium Series 847, **2003**, pp. 140–149.

266. P. Hunt, D.R. Worrall, F. Wilkinson, and S.N. Batchelor, *Photochem. Photobiol. Sci.* **2003**, *2*(5), 518–523.

267. C.S. Colley, D.C. Grills, N.A. Besley, S. Jockusch, P. Matousek, A.W. Parker, N.J. Town, P.M.W. Gill, and M.W. George, *J. Am. Chem. Soc.* **2002**, *124*(50), 14952–14958.

268. T. Lund, P. Christensen, and R. Wilbrandt, *Org. Biomol. Chem.* **2003**, *1*(6), 1020–1025.

269. X.- G. Lei, S. Jockusch, M.F. Ottaviani, and N.J. Turro, *Photochem. Photobiol. Sci.* **2003**, *2*(11), 1095–1100.

270. I. Janovsky, W. Knolle, S. Naumov, and F. Williams, *Chem. Eur. J.* **2004**, *10*(21), 5524–5534.

271. G. Wenska, K. Taras-Goslinska, B. Skalski, G.L. Hug, I. Carmichael, and B. Marciniak, *J. Org. Chem.* **2005**, *70*(3), 982–988.

272. M. Sakamoto, X. Cai, M. Hara, M. Fujistuka, and T. Majima, *J. Phys. Chem. A* **2005**, *109*(11), 2452–2458.

273. G. El Dib, A. Chakir, E. Roth, J. Brion, and D. Daumont, *J. Phys. Chem. A* **2006**, *100*(25), 7848–7857.

274. X. Zhang and W.M. Nau, *J. Phys. Org. Chem.* **2000**, *13*(10), 634–639.

275. T.N. Das, R.E. Huie, P. Neta, and S. Padmaja, *J. Phys. Chem. A* **1999**, *103*(27), 5221–5226.

276. I.V. Vlasyuk, V.A. Bagryansky, N.P. Gritsan, Y.N. Molin, M.A. Yu, Y.V. Gatilov, V.V. Shcherbukhin, and A.V. Zibarek, *Phys. Chem. Chem. Phys.* **2001**, *3*(3), 409–415.

277. I.P. Beletskaya, A.S. Sigeev, V.A. Kuzmin, A.S. Tatikolov, and L. Hevesi, *J. Chem. Soc. Perkin Trans.* **2000**, *2*(1), 107–109.

278. M. Hoshino, R. Konishi, H. Seto, H. Seki, H. Sonoki, T. Yokoyama, and H. Shimamori, *Res. Chem. Intermed.* **2001**, *27*(1–2), 189–204.

279. U. Pischel and W.M. Nau, *J. Am. Chem. Soc.* **2001**, *123*(40), 9727–9737.

280. M.A. Miranda, E. Font-Sanchis, J. Perez-Prieto, and J.C. Scaiano, *J. Org. Chem.* **2002**, *67*(17), 6131–6135.

281. R. Dabestani, I.N. Ivanov, P.F. Britt, M.E. Sigman, and A.C.III. Buchanan, *Preprints of Symposia, American Chemical Society, Division of Fuel Chemistry*, **2002**, *47*(1), 390–392.

282. V.V. Jarikov, A.V. Nikolaitchik, and D.C. Neckers, *J. Phys. Chem.* **2000**, *104*(21), 5131–5140.

283. S. Jockusch and N.J. Turro, *J. Am. Chem. Soc.* **1999**, *121*(16), 3921–3925.

284. I. Gatlik, P. Rzadek, G. Gescheidt, G. Rist, B. Hellrung, J. Wirz, K. Dietliker, G. Hug, M. Kunz, and J.P. Wolf, *J. Am. Chem. Soc.* **1999**, *121*(36), 8332–8336.

285. J.M. Fede, S. Jockusch, N. Lin, R.A. Moss, and N.J. Turro, *Org. Lett.* **2003**, *5*(26), 5027–5030.

286. M. Weber, I.V. Khudyakov, and N.J. Turro, *J. Phys. Chem.* **2002**, *106*(10), 1938–1945.

287. N.J. Turro, X.-G. Lei, S. Jockusch, W. Li, Z. Liu, L. Abrams, and M.F. Ottaviani, *J. Org. Chem.* **2002**, *67*(8), 2606–2618.

288. M. Terazima, Y. Nogami, and T. Tominaga, *Chem. Phys. Lett.* **2000**, *332*(5,6), 503–507.

289. A.L. Konkin, H.K. Roth, M. Schroedner, G.A. Nazmutdinova, A.V. Aganov, T. Ida, and R.R. Garipov, *Chem. Phys.* **2003**, *287*(3), 377–389.

290. J. Lalevée, X. Allonas, and J.P. Fouassier, *Chem. Phys. Lett.* **2005**, *415*(4–6), 287–290.
291. J.C. Scaiano *J. Phys. Chem.* **1981**, *85*, 285–292.
292. J. Lalevée, X. Allonas, and J.P. Fouassier, *J. Org. Chem.* **2006**, *71*, 9723–9727.
293. J. Lalevée, X. Allonas, and J.P. Fouassier, *J. Phys. Chem.* **2006**, *110*, 11605–11612.
294. J. Lalevée, X. Allonas, and J.P. Fouassier, *Chem. Phys. Lett.* **2006**, *429*, 282.
295. J. Lalevée, X. Allonas, S. Genet, and J.P. Fouassier, *J. Am. Chem. Soc.* **2003**, *125*(31), 9377–9380.
296. J. Lalevée, X. Allonas, and J.P. Fouassier, *Chem. Phys. Lett.* **2007**, *445*, 62–67.
297. H. Fischer and L. Radom, *Angew. Chem. Int. Ed.* **2001**, *40*, 1340–1349.
298. J. Lalevée, X. Allonas, and J.P. Fouassier, *J. Phys. Chem. A* **2004**, *108*(19), 4326–4334.
299. J. Lalevée, X. Allonas, and J.P. Fouassier, *J. Org. Chem.* **2005**, *70*(3), 814–819.
300. J. Lalevée, X. Allonas, and J.P. Fouassier, *Macromolecules* **2005**, *38*(10), 4521–4524.
301. J. Lalevée, X. Allonas, J.P. Fouassier, D. Rinaldi, R. Lopez, and J.L. Rivail, *Chem. Phys. Lett.* **2005**, *415*(4–6), 202–205.
302. J. Lalevée, X. Allonas, B. Graff, and J.P. Fouassier, *J. Phys. Chem. A* in press.
303. J. Lalevée, X. Allonas, and J.P. Fouassier, *J. Polym. Sci. A Polym. Chem.* **2006**, *44*(11), 3577–3587.
304. J. Lalevée, X. Allonas, and J.P. Fouassier, *Phys. Lett.* **2008**, *466*, 227–230.
305. J. Lalevée, X. Allonas, S. Jradi, and J.P. Fouassier, *Macromolecules* **2006**, *39*(5), 1872–1879.
306. T. Urano, Y. Tsurutani, M. Ishikawa, and H. Itoh, *J. Photopolym. Sci. Technol.* **2000**, *13* (1), 83–88.
307. T. Urano, M. Ishikawa, and H. Itoh, *Imaging Sci. J.* **1999**, *47*(3), 121–125.
308. T. Hatano, K. Fukui, T. Karatsu, A. Kitamura, and T. Urano, *J. Photopolym. Sci. Technol.* **2000**, *13*(5), 697–701.
309. D. Greszta, D. Mardare, and K. Matyjaszewski, *Macromolecules*, **1994**, *27*, 638–644.
310. A.F. Jacobine,in: J.P. Fouassier (Ed.), *Radiation Curing in Polymer Science and Technology*, Elsevier, Barking, **1993**.
311. C.E. Hoyle, *Proc. RadTech 2002, Indianapolis 674*, **2002**.
312. T.Y. Lee, T.M. Roper, E.S. Jonsson, I. Kuyakov, K. Viswanathan, C. Nason, C. A. Guymon, and C.E. Hoyle, *Polymer* **2003**, *44*(10), 2859–2865.
313. C. Bibaut-Renaud, D. Burget, J.P. Fouassier, C.G. Varelas, J. Thomatos, G. Tsagaropoulos, L.O. Ryrfors, and O.J. Karsonn, *J. Polym. Sci. A Polym. Chem.* **2002**, *40*(18), 3171–3181.
314. M. Manea, K. Ogemark, and L.S. Svensson,in: J.P. Fouassier (Ed.), *Photochemistry and UV Curing: New Trends*, Research Signpost, Trivandrum, India, **2006**, pp. 445–460.
315. S. Jonsson, P.E. Sundell, M. Shimose, J. Owens, and C.E. Hoyle, *Polym. Mat. Sci. Eng.* **1995**, *72*, 470–472.
316. O. Soppera and C. Croutxe Barghorn, *J. Polym. Sci. A Polym. Chem.* **2003**, *41*, 831–838.
317. Y. Ichihashi and M. Kaji, *J. Photopolymer Sci. Tech.* **2004**, *17*, 135–140.
318. S. Suzuki, T. Urano, K. Ito, T. Murayama, I. Hotta, S. Takahara, and T. Yamaoka, *J. Photopolym. Sci. Tech.* **2004**, *17*, 125–129.

11

PHOTOINITIATED CATIONIC POLYMERIZATION: REACTIVITY AND MECHANISTIC ASPECTS

MUHAMMET U. KAHVECI, ALI GORKEM YILMAZ, AND YUSUF YAGCI

11.1 INTRODUCTION

Photoinitiated polymerization is typically a process that transforms a monomer into polymer by a chain reaction initiated by reactive species (free radicals or ions), which are generated from photosensitive compounds, namely, photoinitiators and/or

Photochemistry and Photophysics of Polymer Materials, Edited by Norman S. Allen

$$\mathrm{PI} \xrightarrow{h\nu} \mathrm{H^+} + \mathrm{R^{+\cdot}} + \text{Other products} \quad (11.1)$$

$$\mathrm{R^{+\cdot}} + \mathrm{M} \longrightarrow \text{Polymer} \quad (11.2)$$

$$\mathrm{H^+} + \mathrm{M} \longrightarrow \text{Polymer} \quad (11.3)$$

SCHEME 11.1 General scheme of photoinitiated cationic polymerization.

photosensitizers, by ultraviolet–visible (UV–Vis) light irradiation. In recent decades, it has become powerful industrial process widely used in various applications including coatings, inks, adhesives, varnishes, electronics, photolithography, and dyes because of its excellent advantages [1–3]. It offers high rate of polymerization at ambient temperatures, lower energy cost, and solvent-free formulation, thus elimination of air and water pollution [2,3]. It also devotes temporal and spatial control of the polymerization when high initiation rate is reached [4].

Much effort has been devoted to free radical systems [5,6] mainly due to the availability of a wide range of photoinitiators and the great reactivity of acrylate-based monomers. Although the most popular industrial applications are based on the photoinitiated free radical polymerization there are some drawbacks associated with this type of polymerization, such as the inhibition effect of oxygen and postcure limitations, which may affect the properties of the final product. Several advantages of the photoinitiated cationic polymerization (PCP) over the photoinitiated free radical polymerization have also been reported [2,3,7]. Cationic photopolymerization overcomes volatile emissions, limitations due to molecular oxygen inhibition, toxicity, and problems associated with high viscosity. Furthermore, once initiated, cationically polymerizable monomers such as vinyl ethers (VEs) and epoxides undergo dark-polymerization in which they slowly polymerize without radiation.

General scheme for photoinduced cationic polymerization is depicted in Scheme 11.1. A photosensitive compound, namely, photoinitiator (PI), absorbs incident light and undergoes decomposition leading to production of initiating species. Active species, namely, a radical cation ($R^{+\cdot}$) in turn, react with cationic polymerizable monomers (M), and yield polymer (Scheme 11.1).

This chapter reports recent progress in the PCP, and covers various aspects of the process including initiating systems, photoinitiators, monomers, and applications. Special emphasize is devoted to kinetic and mechanistic details.

11.2 PHOTOINITIATING SYSTEMS FOR PHOTOINITIATED CATIONIC POLYMERIZATION

Since the most significant element of PCP is the cationic photoinitiators, their synthesis and initiation mechanism is one of the most important research areas for polymer science. A compound can be said to be a useful photoinitiator if it has high absorption of light in the UV–Vis region and high quantum yield that can be defined as the number of initiating species formed per photon absorbed. Additionally, the reactivity of the initiating species is an important issue for an efficient photoinitiator.

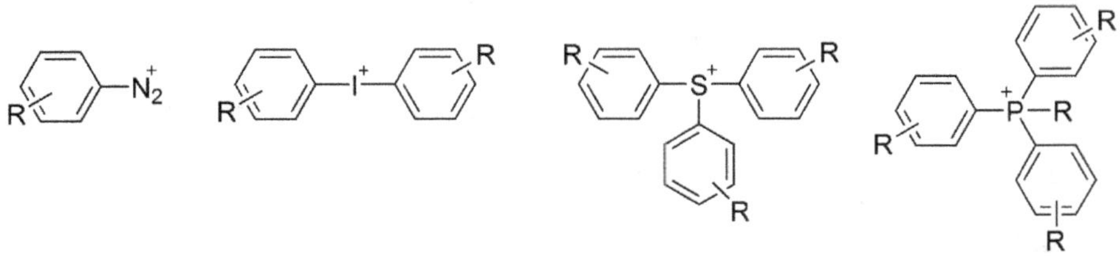

CHART 11.1

11.2.1 Onium Salts

Onium salts are the most widely used cationic photoinitiators. They contain chromophopric groups as the light sensitive body with heteroatoms as cationic centers in the structure. As counterions, mostly inorganic metal complex anions are used [8]. In recent years, onium salts with highly nucleophilic counterions such as Cl^-, Br^-, and I^- have also been used in conjunction with Lewis acids [9–12].

So far, the most frequently used onium salts are aryldiazonium, diaryl iodonium, triarylsulfonium, and tetra alkyl phosphonium salts with nonnucleophillic counter ion (Chart 11.1).

Recently, other onium salts such as *N*-alkoxy pyridinium [13], allylic onium [14,15], trialkyl phenacyl ammonium [16], and dialkyl phenacyl sulfonium salts [17,18] of the following structures (Chart 11.2) are shown to be convenient for producing the initiating species for cationic polymerization.

11.2.1.1 Direct Photolysis Generally, these onium salts generate initiating species upon irradiation at appropriate wavelengths. The mechanism usually referred to as direct photoinitiation [8], and represented in Scheme 11.1. The cationic polymerization of suitable monomers is initiated by both radical cation and/or protonic acid that are generated photochemically upon photolysis of cationic photoinitiators. Most photoinitiators, used in cationic photopolymerization, mainly absorbs light between 225 and 350 nm. For practical applications, however, they are expected to absorb light at quite longer wavelengths. Several attempts have been described to overcome this problem. Three modes of indirect initiation are possible depending on the role played by the additives in the initiation of the polymerization (see below). We will describe below the mechanism of direct photolysis for the individual onium salts.

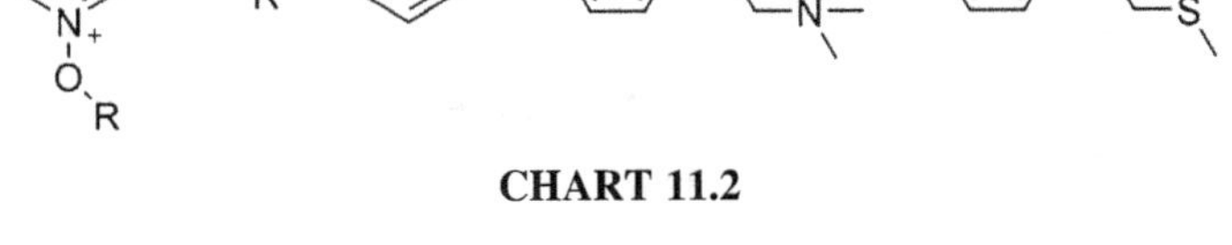

CHART 11.2

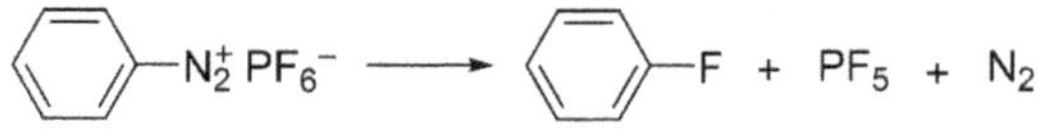 (11.4)

$$PF_5 + ROH \longrightarrow H^+ROPF_5^- \qquad (11.5)$$

SCHEME 11.2

Aryldiazonium Salts Being easily obtained starting from the aniline derivates, these salts produce Lewis acids upon irradiation that can initiate polymerization itself or react with a hydrogen donor compound in the reaction mixture to yield Brønsted acid that is capable of initiating appropriate monomers.

Despite having high quantum yields changing in the range 0.3 and 0.6 [19], aryldiazonium salts suffer from their thermal instability that limits their practical applications as a long time storage is quiet impossible. Furthermore, evolution of nitrogen gas during polymerization causes bubbling to yield porous materials (Scheme 11.2).

Diaryliodonium Salts Diaryliodonium salts (Chart 11.3) are the most frequently used halonium salts as they are easy to obtain and quite reactive [20–22]. The nucleophillic halogen counterion must be replaced by a nonnucleophillic anion in order to prevent the termination of cationic polymerization. As they generally have low spectral sensitivity, an electrophillic substitution reaction can be applied on the aromatic rings to possess electron-donating species that can move absorption bands to lower energies. Alternatively, some special additives can be used to carry out polymerization at longer wavelengths.

Photolysis of diaryliodonium salts take place either through homolytic or through heterolytic cleavage of the halogen-aryl bond to form species which react with a hydrogen donor compound to yield a Brønsted acid that initiates polymerization (Scheme 11.3).

Notably, the electron-donating subtituents on the aromatic structures not only shifts absorption bands to longer wavelengths but also favors photolysis of diaryliodonium salts to afford higher polymerization rates.

Sulfonium Salts Triaryl sulfonium salts (TPSs) are generally produced by the method of Pitt [23]—a Friedel–Crafts condensation of aromatic hydrocarbons with

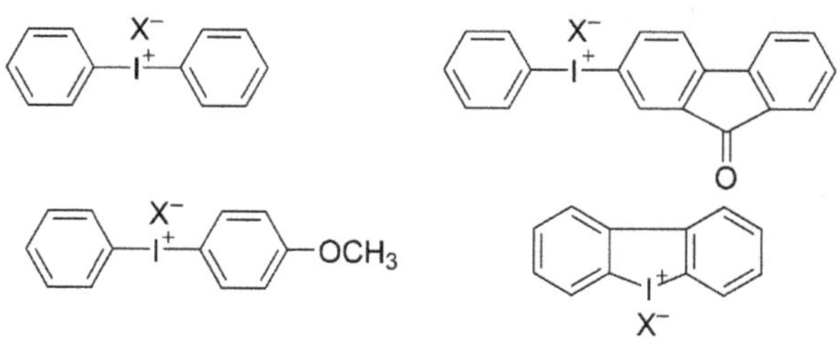

X^-: nonnucleophilic counterion

CHART 11.3

$$Ar_2I^+ MtX_n^- \xrightarrow{h\nu} [Ar_2I^+ MtX_n^-]^* \longrightarrow \left[\begin{matrix} Ar^{\cdot} + ArI^{\cdot +} MtX_n^- \\ Ar^+ MtX_n^- + ArI \end{matrix}\right] \quad (11.6)$$

$$\left[\begin{matrix} Ar^{\cdot} + ArI^{\cdot +} MtX_n^- \\ Ar^+ MtX_n^- + ArI \end{matrix}\right] + R{-}H \longrightarrow H^+ MtX_n^- \quad (11.7)$$

$$H^+ MtX_n^- + M \longrightarrow M^+ MtX_n^- \xrightarrow{M} \text{Polymer} \quad (11.8)$$

RH: solvent or monomer
MtX_n^-: PF_6^-, SbF_6^-, AsF_6^-, etc.

SCHEME 11.3

sulfur dichloride, followed by chlorination and further condensation. Various alkylaryl sulfonium salts may be synthesized by an alkylation of mercaptobenzene [24]. Sulfonium salts that have been the most common utilized for cationic polymerizations are shown in Chart 11.4 [25–27].

The photolysis mechanism is similar with the diaryliodonium salts. When irradiated in appropriate wavelengths, TPSs undergo either a homolytic or a heterolytic cleavage followed by a proton release after some additional steps, which are summarized in Scheme 11.4.

In some special cases, a Brønsted acid does not need to be the only initiating species. If heterolytical cleavage of one of the alkyl groups results with a stable carbocation, polymerization can possibly be initiated by this intermediate structure (Scheme 11.5).

Recently, Endo and coworkers have developed novel sulfonium type initiators that can initiate polymerization either upon irradiation or upon thermal treatment. In addition, these photoinitiators are shown to be functional for both cationic and radical polymerizations. This dual activity is particularly important in hybrid curing systems for coatings and adhesions [28]. Scheme 11.6 contains the general steps for the synthesis of these photoinitiators from sulfides.

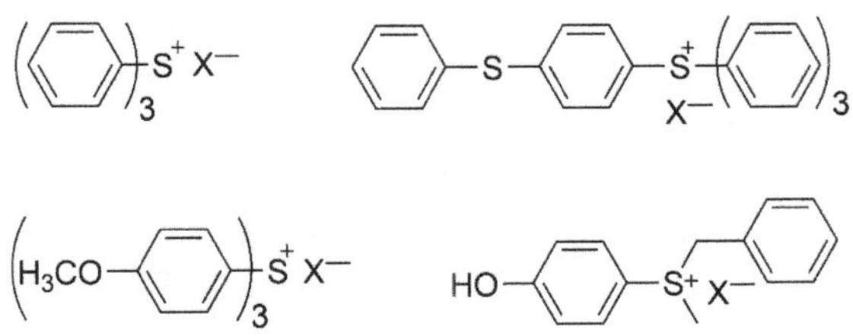

X^-: nonnucleophilic counterion

CHART 11.4

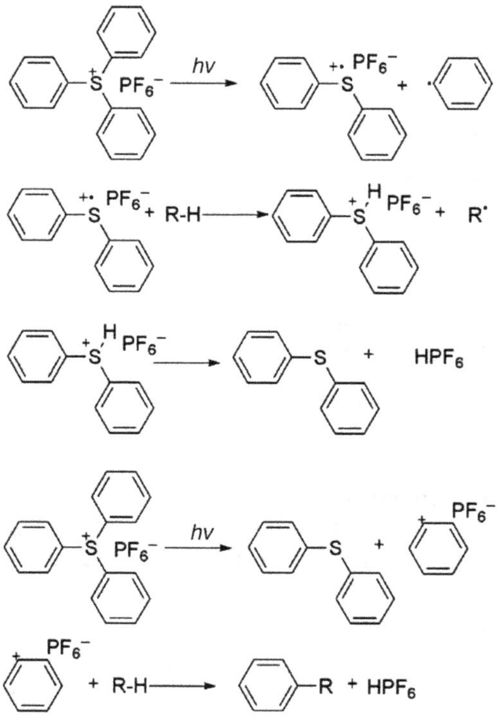

(11.9)

(11.10)

(11.11)

(11.12)

(11.13)

SCHEME 11.4

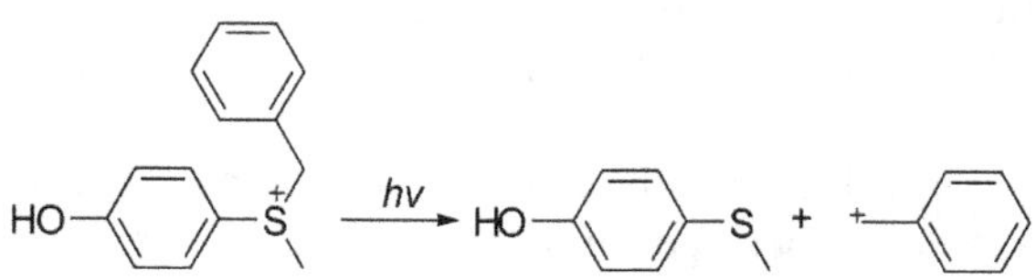

(11.14)

SCHEME 11.5

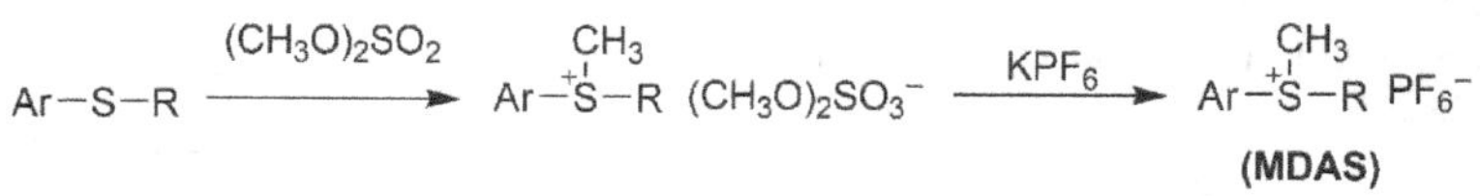

(11.15)

	Ar	R
MDAS-1a	Phenyl	2-Indanyl
MDAS-1b	Phenyl	1-Ethoxycarbonethyl
MDAS-1c	Phenyl	2-Phenyletheyl
MDAS-1d	Phenyl	2-Phenylpropyl
MDAS-1e	Phenyl	2-Methyl-2-phenylpropyl
MDAS-1f	Phenyl	2-(4-Methoxyphenyl)ethyl
MDAS-1g	Phenyl	1-(4-Methoxyphenyl)-2-propyl
MDAS-2a	2-Naphthyl	2-Indanyl
MDAS-2b	2-Naphthyl	1-Ethoxycarbonethyl
MDAS-2c	2-Naphthyl	Methyl

SCHEME 11.6

The real time Fourier transform infrared (FT-IR) studies demonstrated the activity order as, **MDAS-2b**, **MDAS-2a** > **MDAS-1f**, **MDAS-1g** > **MDAS-1a**, **MDAS-1e** > **MDAS-1d**, **MDAS-1c** ≥ **MDAS-1b**. In particular, sulfonium salts having naphthyl groups, **MDAS-2b** and **MDAS-2a**, showed higher photoactivity than simple triphenylsulfonium hexafluorophosphate. **MDAS-2c** showed nearly the same photoactivity as di-*tert*-butylphenyliodonium hexafluorophosphate [28]. This can be attributed to the stability of the intermediates formed during the reaction steps. For example, initiators bearing naphthalene are found to be very reactive, since naphthyl cations, formed upon irradiation are very stable as a result of the resonance contributors. Three probable mechanisms for the photolytic decomposition are shown in Scheme 11.7.

Photoinitiation is believed to take place through a heterolytical and/or a homolytical cleavage (Equations 11.16 and 11.19). Initiators bearing ester groups, **MDAS-1b** and **MDAS-2b**, are more prone to cleave heterolytically and release a proton that eventually initiates cationic polymerization as shown in Equations 11.17 and 11.18. The salts that do not bear ester groups, on the other hand, can both initiate radical and/or cationic polymerization. It is believed that **A** and **C** are the initiating species for the radicalic polymerization whereas **B** is for the cationic polymerization.

Recently, Crivello and coworkers have developed synthesis of a new initiating system called *S,S*-dialkyl-*S*-(3,5-dimethyl-4-hydroxyphenyl)-sulfonium salts (4HPS) [29–31] with absorption maxima in the middle UV region. More recently, the same group also reported a facile synthesis of its isomeric counterparts, *S,S*-dialkyl-*S*-(3,5-dimethyl-2-hydroxyphenyl) sulfonium salts (2HPS) (Chart 11.5).

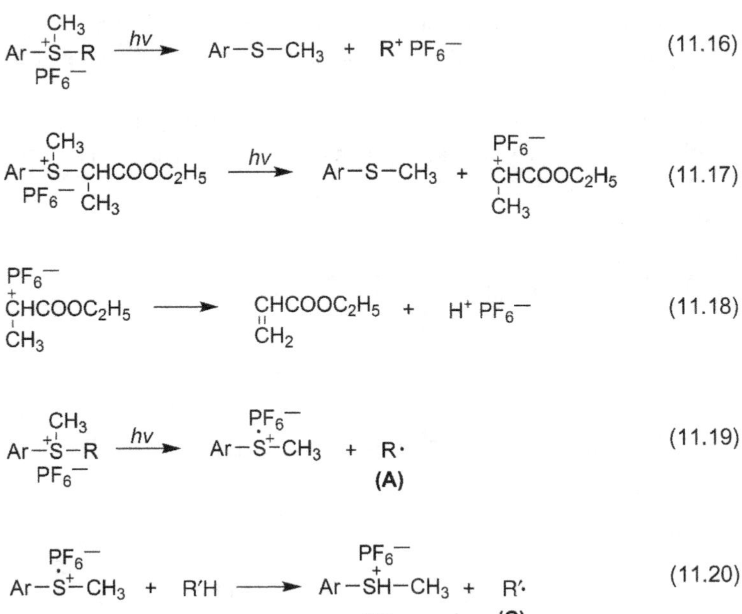

SCHEME 11.7

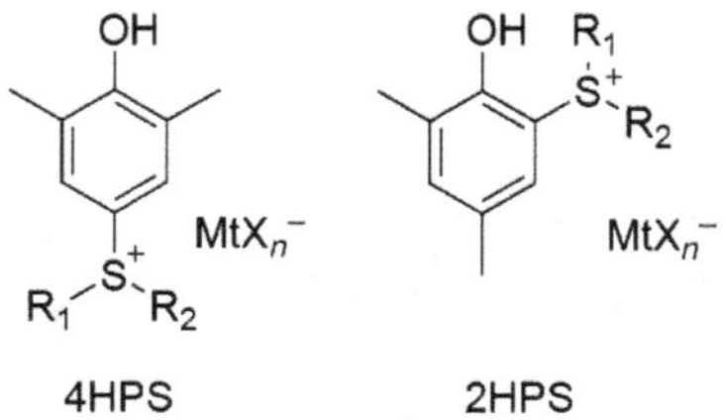

CHART 11.5

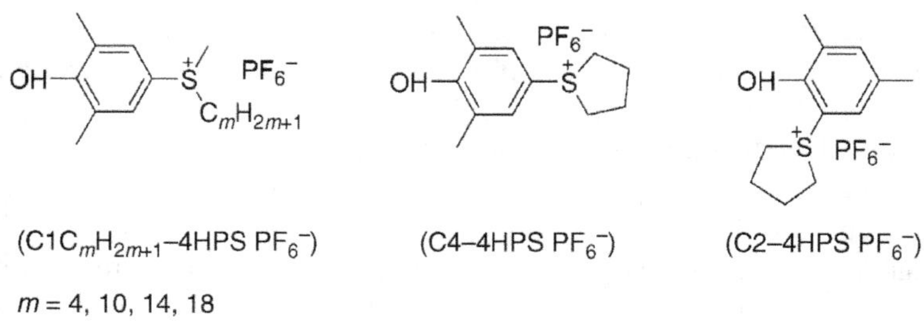

CHART 11.6

Being poorly soluble in common monomers used in photocationic curing, these sulfonium compounds (Chart 11.6) attract little attention. However, deriving the structure with possessing different alkyl groups, improved solubility characteristics can be achieved.

Upon irradiation, appropriate monomers can be polymerized when used in conjunction with 2HPS and 4HPS salts, according to the following mechanism (Scheme 11.8).

As can be seen, the attack of the nucleophillic oxygen of the ylide compound to the growing carbocation chain terminates polymerization. A steric hindrance around

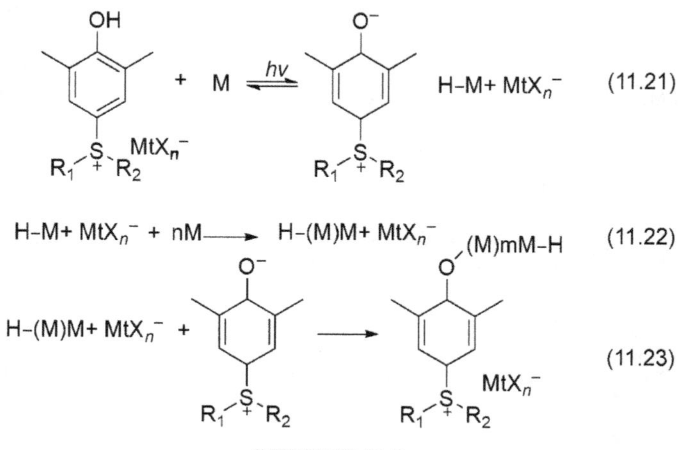

SCHEME 11.8

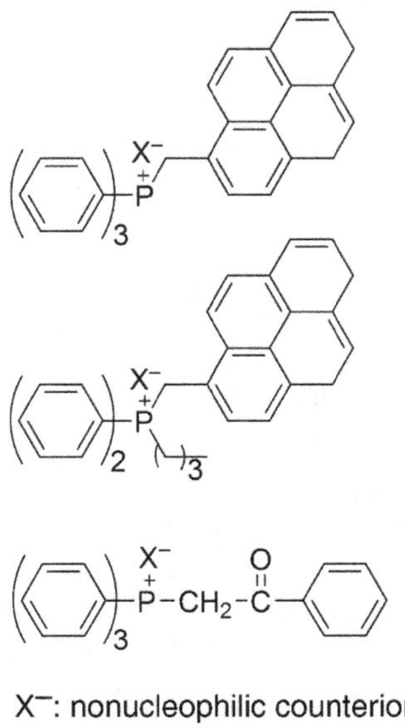

CHART 11.7

the attacking nucleophillic oxygen can thus slow down termination process that can be accounted for higher efficiency of 2HPS rather than 4HPS isomers. Notably, these salts are demonstrated to be of high photosensitivity ($\Phi = 0{,}12$–$0{,}31$) and good thermal stability.

Phosphonium Salts The preparation of phosphonium salts is based on the reaction of chloromethylated or bromomethylated aryl compounds with the corresponding phosphines [32–36]. As illustrated in Chart 11.7, phosphonium salts with absorptions acceptable for direct photolysis have been synthesized.

Benzyl or pyrenylmethyl groups containing phosphonium salts produce the respective carbon centered cations after a heterolytic bond rupture according to Equation 11.24 (Scheme 11.9) [37–39]. These cations are assumed to be the initiating species in cationic polymerization.

Because of releasing very stable cations upon irradiation, phosphonium salts containing pyrenylmethyl groups are excellent initiators for photopolymerization of convenient monomers such as epoxides and vinyl monomers [35,39].

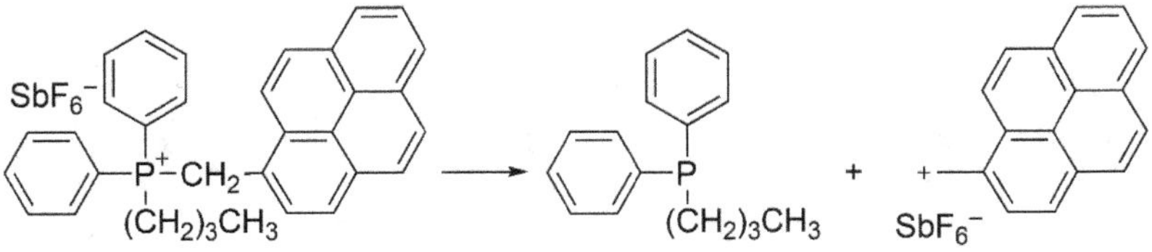

(11.24)

SCHEME 11.9

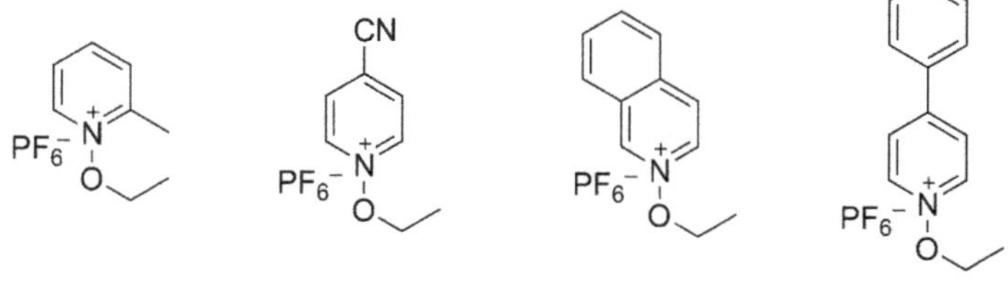

CHART 11.8

N-Alkoxy Pyridinium Salts *N*-Alkoxy Pyridinium salts are obtained with relatively high yields by a reaction of pyridine *N*-oxides with a triethyloxonium salt in methylene chloride or chloroform [13]. Quinolinium salts can also be prepared from the corresponding *N*-oxides [40]. In both cases, an anion exchange is not necessary since the triethyl oxonium salt is available with nonnucleophilic counter anions. The most frequently used photoinitiators of this type are shown in Chart 11.8. The spectral response of these salts is in 260–310 nm range [13].

When irradiated in suitable wavelengths, these salts readily initiate polymerization of appropriate monomers according to the following mechanism as exemplified for the case of *N*-ethoxy-2-methylpyridinium hexafluorophosphate ($EMP^+ PF_6^-$) in Scheme 11.10.

Notably, some reactive monomers such as isobutylvinylether and *N*-vinylcarbazol are observed to polymerize even in dark when used in conjunction with *N*-ethoxy-4 cyanopyridinium (EPP) and *N*-ethoxyisoquinolinium (EIQ) salts. An electron transfer from the monomer to these initiators can be proposed as an explanation for the observed reactivity in the absence of light (Scheme 11.11).

Phenacyl Sulfonium Salts Being thermally rather stable and highly photoresponsive, phenacyl sulfonium salts are significantly attractive for photoinduced cationic polymerization [41,42]. Despite being easily obtained, they suffer from their poor

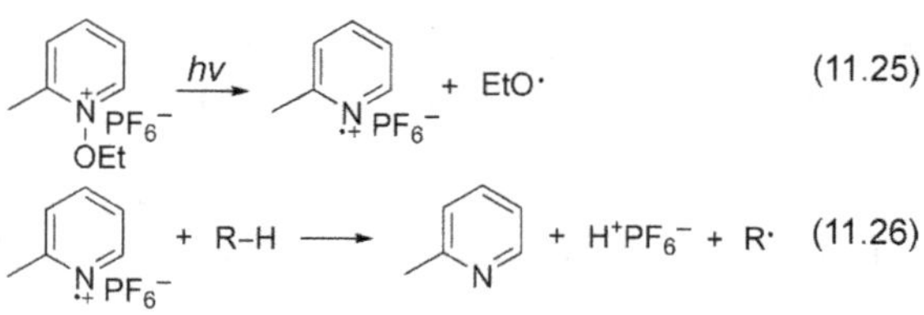

SCHEME 11.10

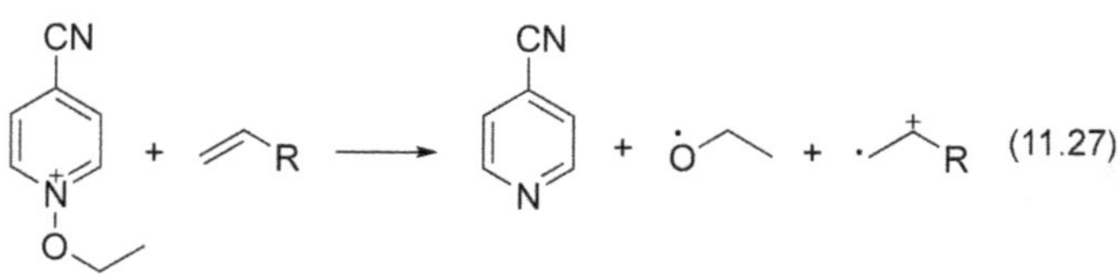

SCHEME 11.11

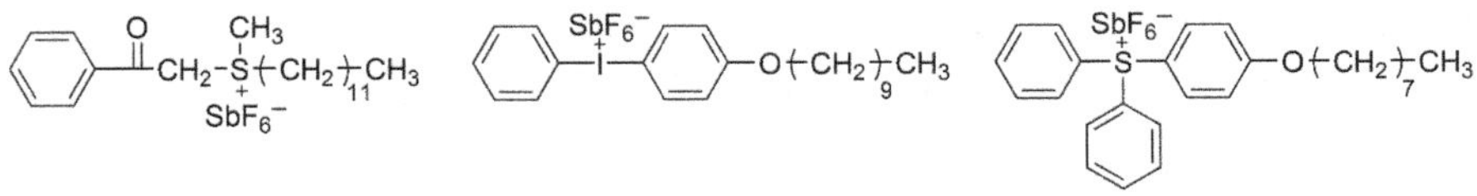

(DPS-12) (IOC-10) (SOC-8)

CHART 11.9

solubility in common monomers that make them unpreferable. However, novel phenacyl sulfonium salts were synthesized via deriving the structure with alkyl substituents to improve solubility properties [17]. Similar strategy was also followed to improve the solubility of iodonium and sulfonium salts. Chart 11.9 shows the chemical structures of the salts having good solubility. The photolysis mechanism of phenacyl sulfonium salts is reversible and shown in Scheme 11.12.

Since the photolysis of the salt is reversible, the formation of the proton is scavenged by the ylide to form the starting photoinitiator when irradiation ceases [43]. During irradiation, polymerization takes place via protonation of a monomer followed by sequential monomer addition. Notably, a termination via addition of a ylide to the growing cationic chain end is feasible. The overall polymerization process is summarized in Scheme 11.13.

Notably, phenacyl sulfonium salts are very stable even in very reactive monomers in dark. For example, a 3% solution of a these photoinitiators in the highly reactive biscycloaliphatic epoxy monomer showed no tendency toward spontaneous polymerization even after 3 months storage in the dark at room temperature.

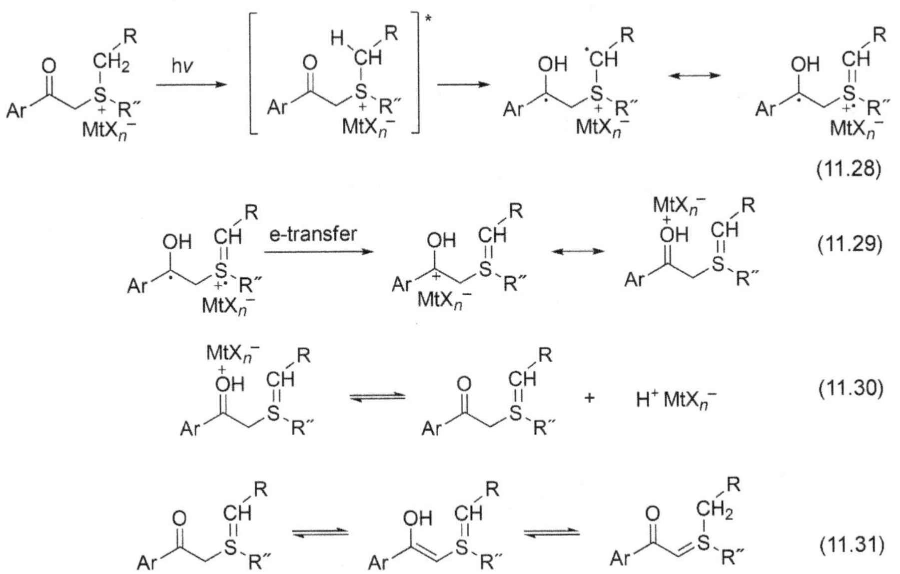

(11.28)

(11.29)

(11.30)

(11.31)

SCHEME 11.12

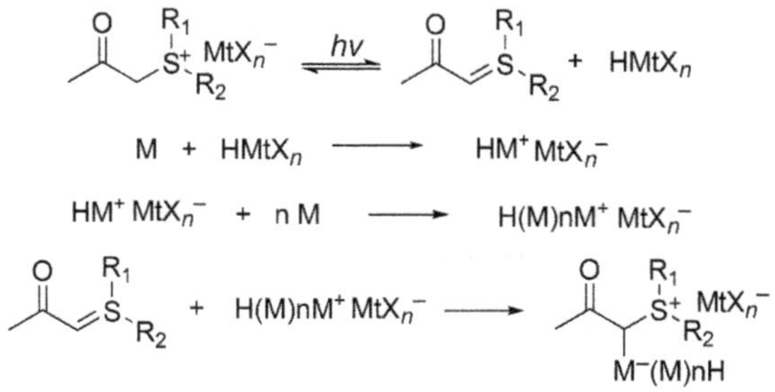

(11.32)
(11.33)
(11.34)
(11.35)

SCHEME 11.13

Phenacylammonium Salts Phenacyl anilinium salts have light absorption 300–350 nm that make them preferable for photoinduced cationic polymerization. Upon irradiation, these salts undergo either a heterolytic cleavage or a homolytic cleavage followed by an electron transfer to yield a cation intermediate that in turn can initiate polymerization of appropriate monomers [16]. Scheme 11.14 shows the mechanism in detail. As can clearly be seen, the photolysis of phenacylammonium salts are irreversible and different than their sulfonium analogs [43].

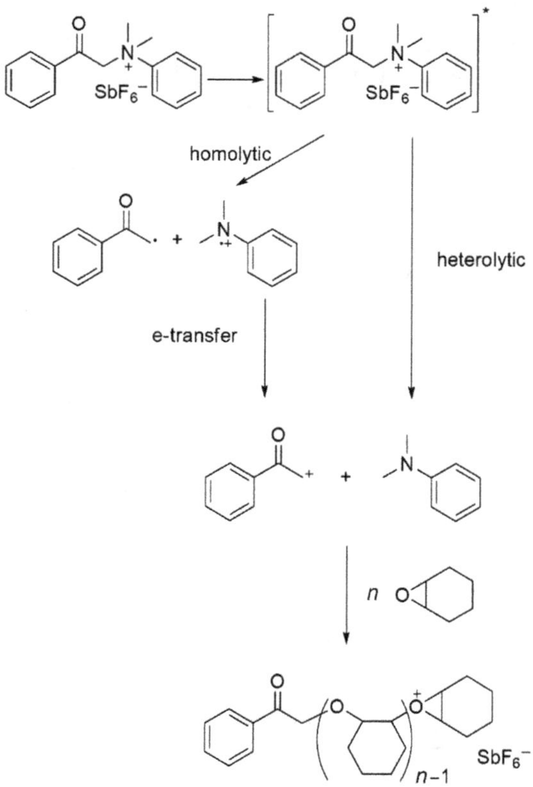

(11.36)
(11.37)
(11.38)

SCHEME 11.14

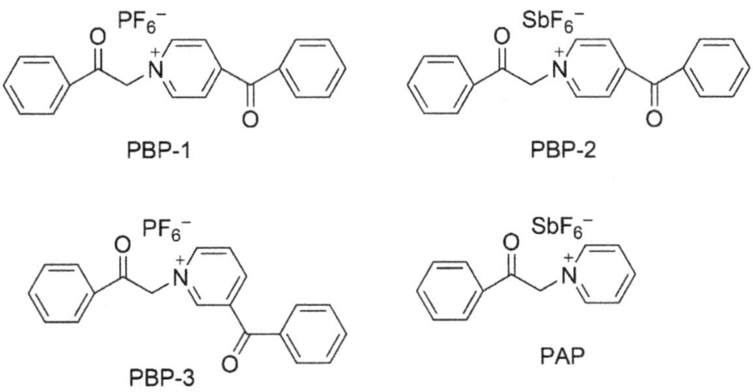

CHART 11.10

Recently, photolysis of a structurally similar salts having additional chromophoric groups was investigated. The structure of so-called phenacyl benzoylpyridinium salts, (PBP-1-3) [44], and together with the unsubstituted model compound (PAP) are shown in Chart 11.10.

Interestingly, these salts are stable and colorless in the solid form while turn to pink color in organic solvents. Detailed spectral investigations reveal that they undergo keto-enol tautomerization in solution, facilitated with a little energy just by solution mixing and that the enol form is responsible for absorption in visible region (Fig. 11.1).

However, it was demonstrated that only the keto form is responsible for the initiation as no polymerization take place in the wavelengths where the enol forms absorb the light. Photoinduced decomposition pathway follow a similar mechanism as their anilinium analogs.

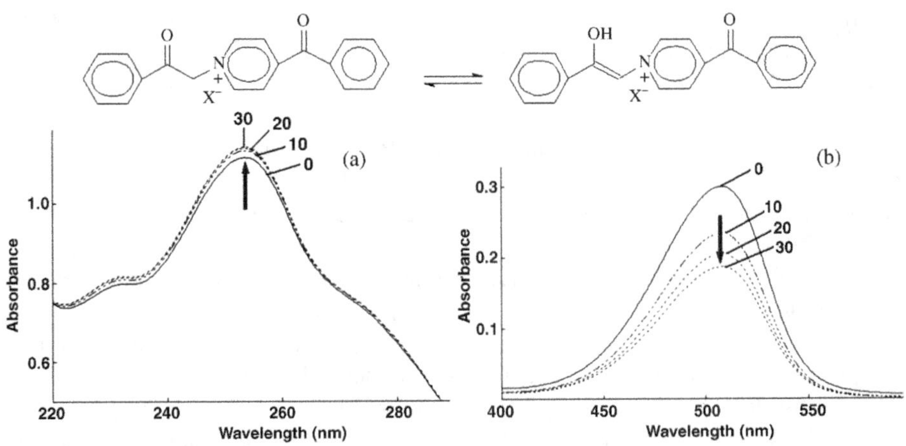

FIGURE 11.1 Typical UV-spectral changes of salt **PBP-1** (see Chart 11.10) on irradiation at $\lambda > 507$ nm under nitrogen in CH_3CN.

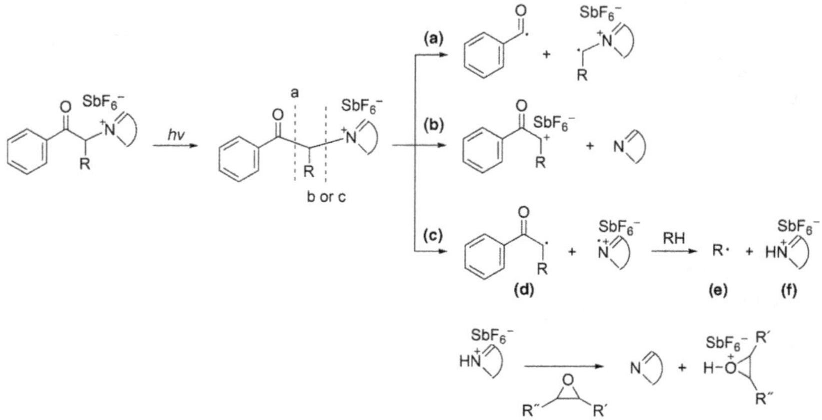

SCHEME 11.15

Endo and coworkers have developed novel phenacyl ammonium salts that are applicable to both cationic and free radical polymerizations for hybrid curing systems consisting of epoxides and acrylates [45]. Moreover, the salts are capable of initiating polymerization both upon heat treatment and irradiation. It was also found out that the ammonium salts worked effectively as photocationic initiators to polymerize a difunctional epoxy monomer (CY179). Scheme 11.15 summarizes the possible mechanisms demonstrating the direct initiation steps of polymerization. Path (a) is thought not to be responsible for cationic photopolymerization whereas path (b) and (c) are assumed to take action simultaneously. In path (b), a heterolytic cleavage of the carbon nitrogen bond occurs while in path (c) a homolytic cleavage followed by a hydrogen abstraction from any hydrogen donor in the reaction media takes place. The active species of radical polymerization seems to be (d) and/or (e), and that of cationic polymerization seems to be (f). Photosensitization and free radical promoted acceleration of this salt has also been investigated in the same study.

N-Methyl-2-Alkylthiobenzothiazolium Salts Recently, another kind of thermal cationic initiator for epoxy resins was developed by Endo and coworkers called *N*-methyl-2-alkylthiobenzothiazolium salts starting from the 2-mercaptobenzothiazoles Scheme 11.16 [46].

These salts are demonstrated to have dual function that can initiate polymerization through either cationic and radicallic pathways. Their photoinitiation activity has been demonstrated by curing of a difunctional epoxy monomer. The following types of pathways were proposed for the initiation (Scheme 11.17).

It is observed that, the reactivity of the photoinitiators is in the following order: **TBTA-3, TBTA-1** > **TBTA-2** > **TBTA-4**. This reactivity order is presumably due to the stability of the intermediates formed in the reaction pathway. The more stable the intermediates formed during photolysis, the more prone the photoinitiator to initiate

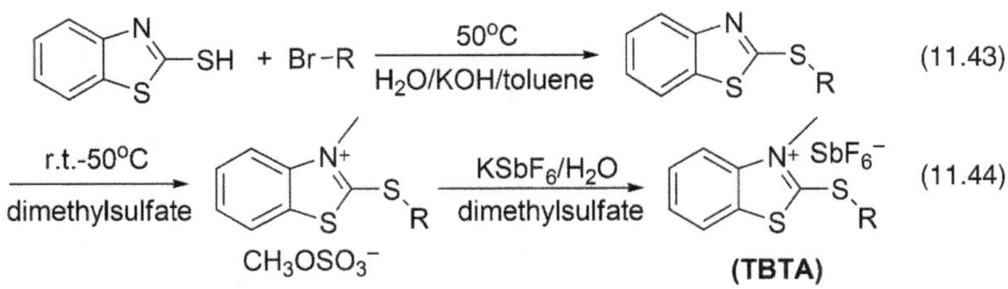

TBTA salts	R–
TBTA-1	$PhCH_2$–
TBTA-2	4-$(NO_2)C_6H_4CH_2$–
TBTA-3	$C_2H_5OCO(Me)CH$–
TBTA-4	Me

SCHEME 11.16

cationic polymerization. It is also notable that the salts **TBTA-3** and **TBTA-1** are more reactive than triphenylsulfonium salts. Another point to emphasize is that, although they do not bear carbonyl groups in the structure as in the case of *N*-phenacylpyridinium salts, these initiators are shown to be still reactive.

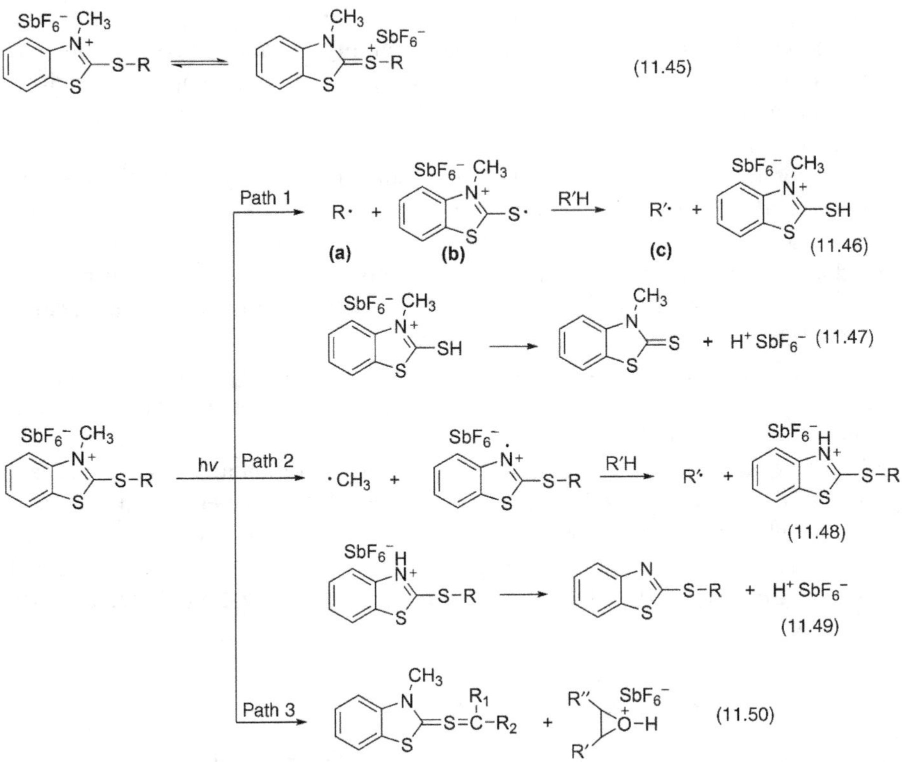

SCHEME 11.17

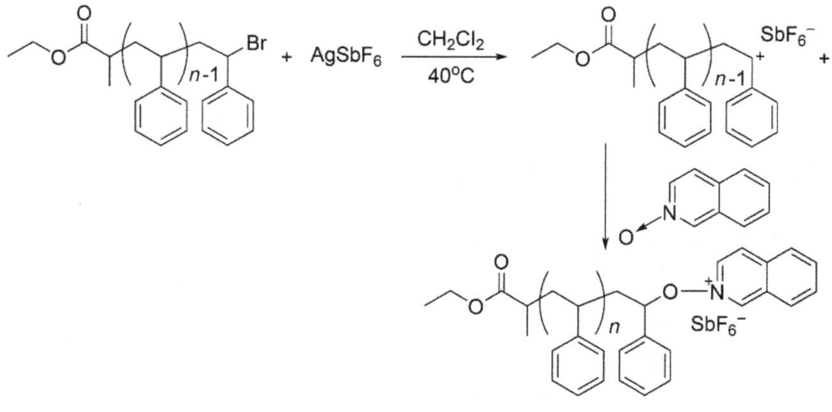

(11.51)

(11.52)

SCHEME 11.18

For practical applications, it was shown that it is possible to photosensitize these salts with novel photosensitizers such as 2,4-dimethylthioxanthone (DMTX) and 2-ethyl-9,10-dimethoxyanthracene (EDMA). The photosensitizer extent the polymerization wavelength to as long as 380 nm [46].

Polymer-Bound Onium Salts Onium salts may be incorporated into polymers. These polymeric initiators often show good miscibility with monomers and polymers generated in the course of the polymerization. Furthermore, a lower order of toxicity is found owing to the inability of high molecular weight polymers to be absorbed by biological systems.

Incorporating iodonium and sulfonium salts to the polymer backbone was achieved and used as macrophotoinitiators [47,48]. Easy to incorporate, *N*-oxides also found applications in this area [49–51]. In a recent study, Yagci and coworkers introduced *N*-oxide functions to the chain end of polystyrene via substitution reaction of halogen terminated polystyrenes obtained by atom transfer radical polymerization (ATRP) (Scheme 11.18).

Although being cationic photoinitiators, these polymers were used to form block copolymer via a free radical mechanism when irradiated in the presence of methylmethacrylate (MMA) [52].

In another study, the same group attached 4-phenyl-pyridinium *N*-oxide as pendant groups to the polychloromethyl styrene and its block copolymer with polystyrene via nucleophillic substitution with chloride as leaving group. The obtained macrophotoinitiators were used to obtain graft copolymers by grafting from method of MMA (Scheme 11.19, path a), and hydroxyl functionalization (Scheme 11.19, path b) [53].

11.2.2 Other Initiating Systems

11.2.2.1 Iron Arene Complexes-Based Photoinitiators Iron arene complexes or ferrocenium salts are attractive photoinitiators for cationic polymerization of

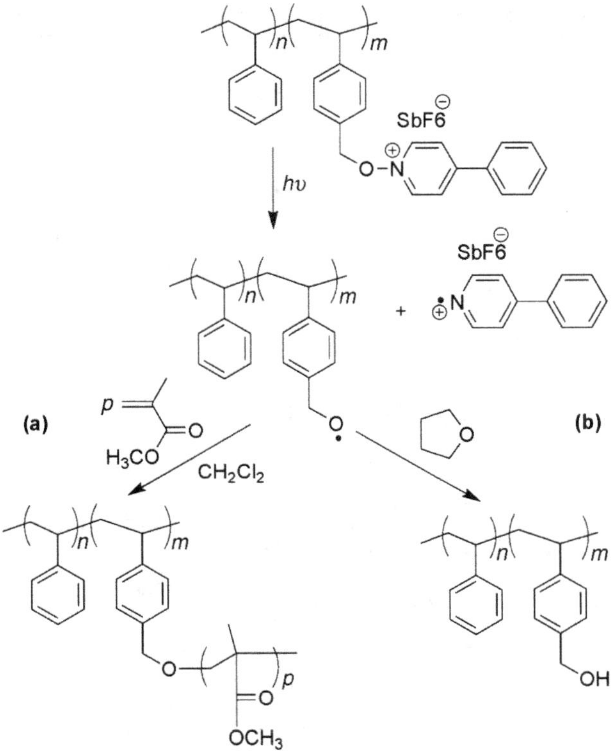

(11.53)

(11.54)

SCHEME 11.19

epoxides [54] since they are readily prepared from nonexpensive starting materials and their photolysis is extremely efficient in the UV and visible regions [55]. Iron arene complexes are usually prepared from ferrocene by exchange of one cyclopentadienyl moiety for a neutral aromatic species in the presence of a Lewis acid, followed by metathesis with a salt possessing an appropriate counter anion (Scheme 11.20) [56–58]. In addition, their spectral responses are localized in the middle region of UV spectrum that can be extended easily to visible light regions by exchange of their ligands (Table 11.1).

Upon irradiation, ferrocenium salts lose their arene ligands leading to generation of iron-based Lewis acids that coordinate with epoxide monomers. One of the monomers undergoes ring opening followed by addition of a new monomer. Arene

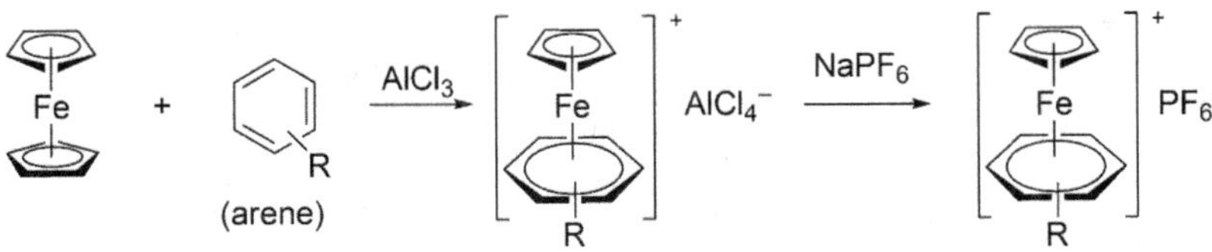

(11.55)

SCHEME 11.20

TABLE 11.1 Spectral Responses of Various Iron Arene Complexes-Based Photoinitiators.

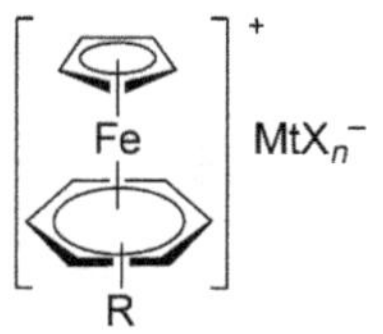	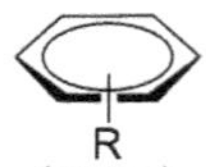 (arene)	MtX_n^-	λ_{max} (nm) (ε_{max} (L/mol/cm))	Reference
Cyclopentadienyl-Fe-cymene hexafluorophosphate (I-261)	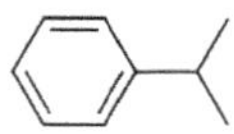	PF_6^-	249 (1.24×10^4); 297 (3.01×10^3); 375 (0.75×10^2)	[59]
Cyclopentadienyl-Fe-carbazole hexafluorophosphate (CFC)	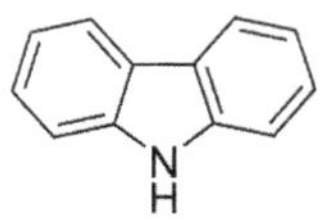	PF_6^-	219 (2.41×10^4); 259 (2.62×10^4); 347 (2.17×10^3); 421 (2.67×10^2)	[59]
Cyclopentadienyl-Fe-*n*-ethyl CFE	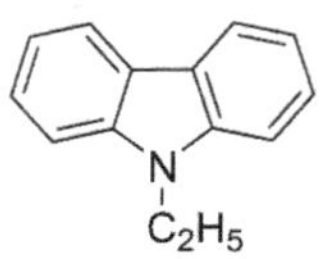	PF_6^-	219 (2.37×10^4); 259 (2.87×10^4); 352 (2.32×10^3); 429 (1.87×10^2)	[59]
Cyclopentadienyl-Fe-aminonaphthalene hexafluorophosphate (CFN)	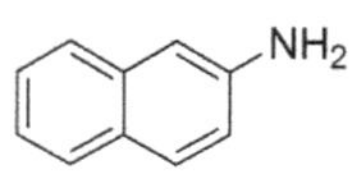	PF_6^-	213 (2.16×10^4); 242 (2.01×10^4); 376 (2.38×10^3); 546 (9.67×10^2)	[59]
Cyclopentadien-Fe-anisol hexafluorophosphate (CFA-PF_6)	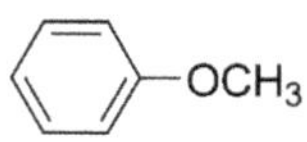	PF_6^-	242 (1.47×10^4); 399 (1.36×10^2); 466 (72)	[60]
Cyclopentadien-Fe-anisol tetrafluoroborate (CFA-BF_4)	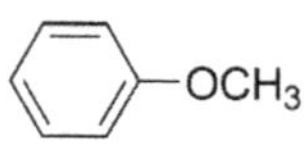	BF_4^-	242 (1.66×10^4); 399 (1.27×10^2); 466 (75)	[60]
Cyclopentadien-Fe-1,4-diethoxybenzene hexafluorophosphate (CFDE)	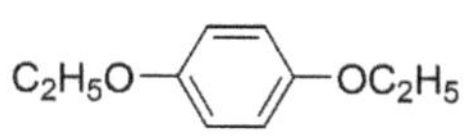	PF_6^-	224 (1.68×10^4); 409 (1.46×10^2); 465 (59)	[60]
Cyclopentadien-Fe-1,4-diphenylether hexafluorophosphate (CFD-PF_6)	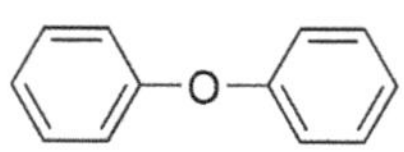	PF_6^-	243 (1.94×10^4); 397 (1.40×10^2); 462 (75)	[60]
Cyclopentadien-Fe-1,4-diphenylether tetrafluoroborate (CFD-BF_4)	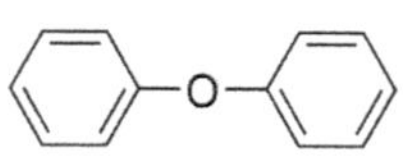	BF_4^-	243 (1.49×10^4); 397 (1.36×10^2); 462 (78)	[60]
Cyclopentadien-Fe-1,4-aniline hexafluorophosphate (CFAn)	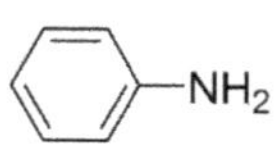	PF_6^-	212 (1.76×10^4); 246 (1.61×10^4); 301 (1.44×10^3); 409 (3.30×10^2)	[60]

TABLE 11.1 (*Continued*)

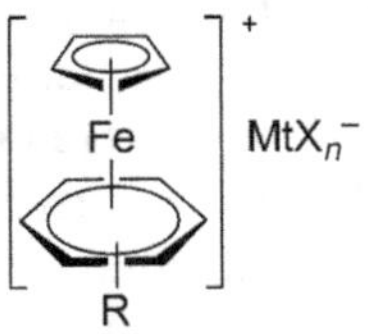	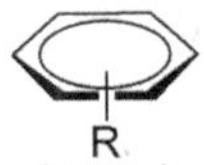 (arene)	MtX_n^-	λ_{max} (nm) (ε_{max} (L/mol/cm))	Reference
Cyclopentadien-Fe-l,4-*p*-methylaniline hexafluorophosphate (CFMA)	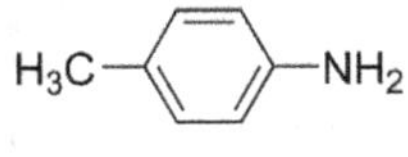	PF_6^-	213 (2.07×10^4); 246 (1.82×10^4); 303 (1.82×10^4); 405 (3.09×10^2)	[60]

complex initiators are convenient for polymerization of only epoxides, which are able to coordinate with iron. Vinyl monomers cannot be polymerized with this type of initiators. The photoinitiated cationic polymerization of epoxides by iron complexes is indicated in Scheme 11.21.

Iron arene complexes have been successfully applied in photopolymerization and UV curing of various epoxides [59–64]. Multifunctional solid novolac resins were also polymerized successfully by using iron arene complexes. The irradiation of these formulations gives a latent image; the cross-linking itself is started only after heating for 3 min at 100°C. The fact that thermal curing can be applied long time after the irradiation has also been utilized in dark curing adhesives for opaque substrates.

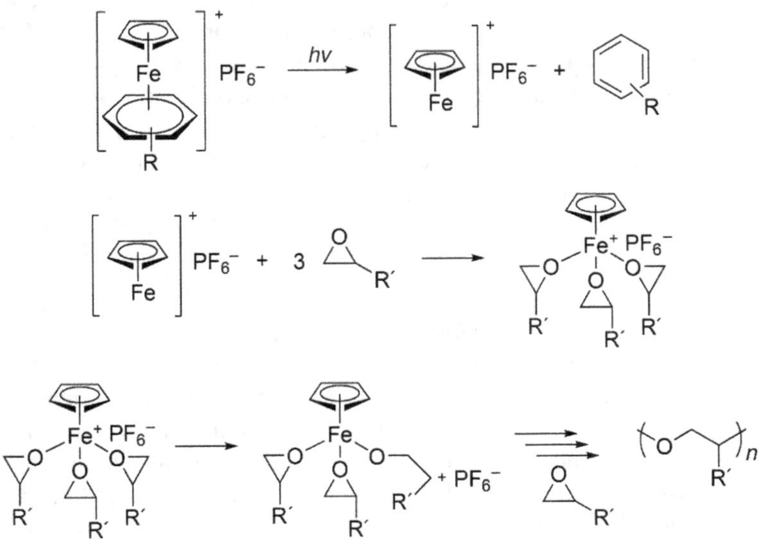

SCHEME 11.21

11.2.2.2 Nonsalt Photointiators for Cationic Polymerization As stated several times in the text, onium salts are the most commonly used photoinitiators in cationic photopolymerization. However, their disadvantages including limited spectral response characterized under 300 nm, poor solubility in organic solvents and the emerging of undesirable and toxic materials such as arsenic or antimony force development of nonionic or nonsalt photoinitiators. This type of initiators does not contain metal atoms or show any ionic character that is the main property of conventional cationic photoinitiators. High solubility in organic solvents, elimination of metal atoms (As, Sb, etc.) and absorption characteristics well suited for deep-UV exposure are some advantages of such photoinitiators.

Nonsalt photoinitiators are widely used in cationic photopolymerization based cross-linking, curing and lithography applications due to their great advantages. Upon photolysis, photochemically active nitrobenzyl esters [65,66], sulfonyl ketones [67], phenacyl sulfones, and phenyl disulfones [68] generate strong organic acids. In addition to sulfur containing photoinitiators, selenide [69] and organosilane [70–72] based photoinitiators were employed to initiate cationic photopolymerization.

Sulfonic (-SO_3H) and Sulfinic (-SO_2H) Acid-Based Nonsalt Photoinitiators Photosensitive tosylate esters of nitrobenzyl [65] and benzoin [73], sulfonyl ketones [67,68], and diphenyl disulfones [68] can be employed in cationic photopolymerization, especially in photolithography and chemical amplification processes. The advantages of these photoinitiators are the ease of synthesis, the absorption wavelengths suitable for deep-UV curing and the high yield of acid formation [65].

Upon their photolysis, corresponding sulfonic or sulfinic acid formed and initiates cationic polymerization. Although each type of photoinitiators has different photolysis process the active species for cationic polymerization are same for each compound. All compounds generating organic acids upon photolysis are summarized in Table 11.2. The mechanism of acid release by *o*-nitrobenzyl tosylate is illustrated in Scheme 11.22. The mechanism involves the insertion of nitro group oxygen into the benzyl C–H bond. The quantum yields of nitrobenzyl ester vary from 0.02 to 0.18 depending on substituent on the benzene ring and the wavelength of the laser pulse [65].

Organosilane-Based Nonsalt Photoinitiators Silanol are also capable of initiating of cationic polymerization of cyclohexene oxide in the presence of aluminum(III) complexes of β-diketones, β-ketoesters or *ortho*-carbonylphenolate upon UV irradiation or moderate heating (40°C) [70,72]. The initiation mechanism is demonstrated on the example of the triphenylsilanol/tris(ethyl acetoacetate) aluminum system in Scheme 11.23. The release of protons is a thermal reaction. The release of Brønsted acid is strongly accelerated by UV light due to the formation of light sensitive intermediates, when the ligands are 1,3-diketones [71].

TABLE 11.2 Sulfonic and Sulfinic Acid-Based Photoinitiators and Corresponding Initiating Species Generated upon Photolysis.

Photoinitiator	Initiating Species	Reference(s)
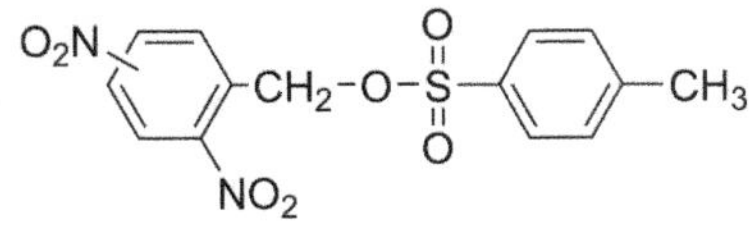	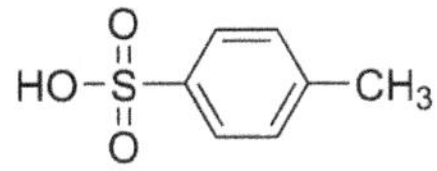	[65,74]
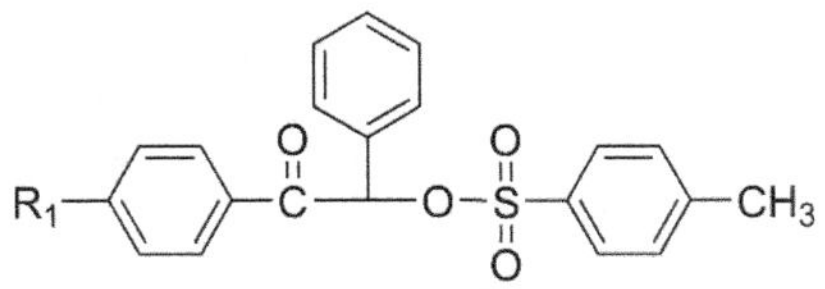	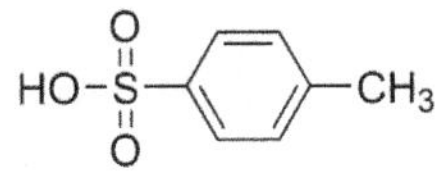	[75]
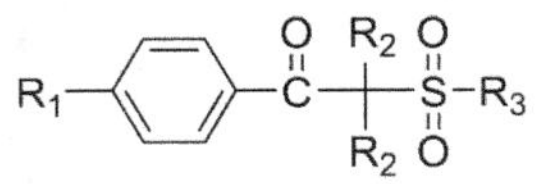	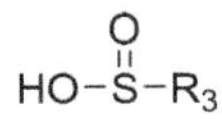	[67,68]
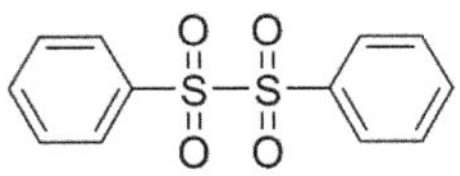	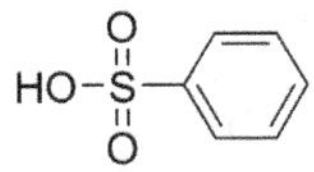	[68]
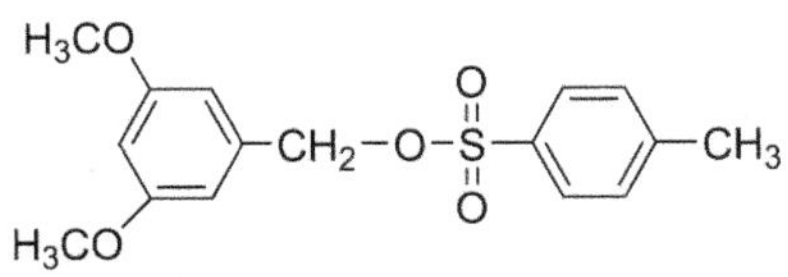	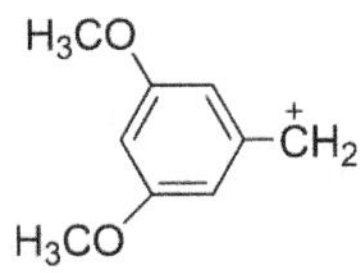	[76]

Selenide-Based Nonsalt Photoinitiator In more recent years, diphenyldiselenide (DPDS) as nonsalt initiator was used for photosensitized cationic polymerization of *N*-vinyl carbazole (NVC) [69]. Diselenide compounds in the presence of aromatic nitriles are well-known photosensitization system for *in situ* generation of electrophilic

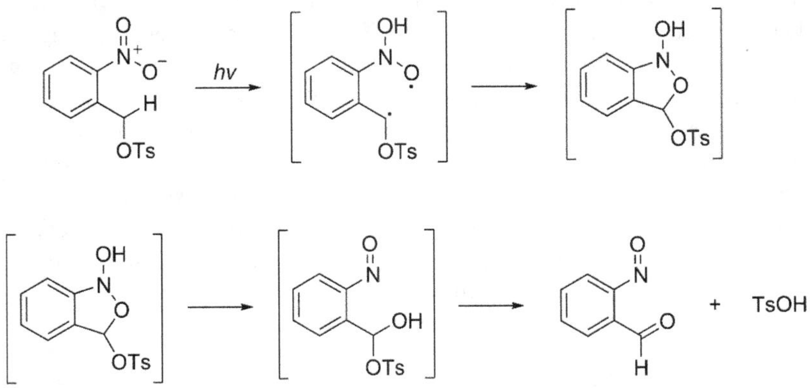

(11.59)

(11.60)

SCHEME 11.22

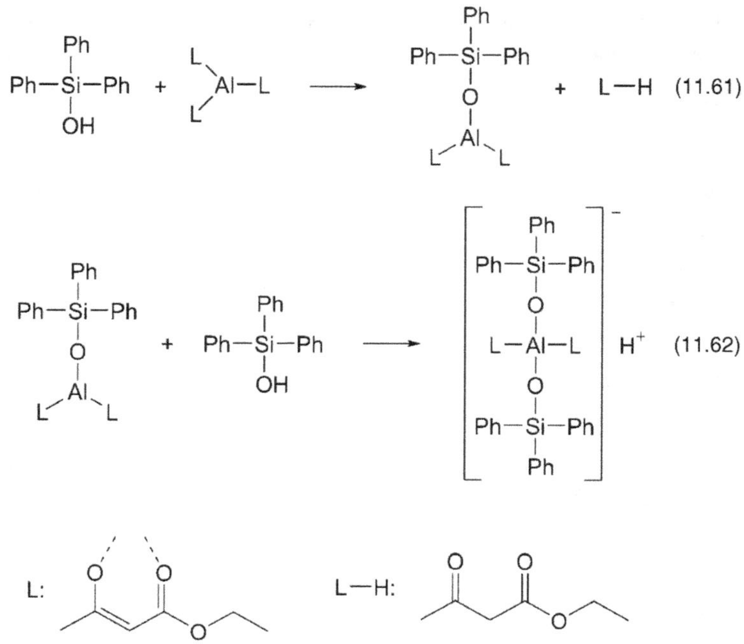

SCHEME 11.23

selenium species. The proposed mechanism is for the photosensitized cationic polymerization of NVC with DPDS and 1,4-dicyanonaphthalene as a representative example is demonstrated in Scheme 11.24. Irradiation of these compounds results in single-electron transfer between the excited singlet state sensitizer and the ground state of DPDS, and, in turn, the formation of selenium radical cation as actual initiating species.

11.2.2.3 Indirect Photolysis Without absorption of the incident photon energy, photochemical processes cannot occur. Medium- and high-pressure mercury lamps that are frequently used as light sources provide emissions at 313 and 366 nm. If daylight is to be used for curing a coating formula, light absorption at wavelengths longer than 400 nm is highly desired. Rather than introducing electron-donating substituents to the structure as mentioned before, some electron-rich compounds such as trimethoxybenzene or hexamethylbenzene can be added to polymerization mixture to form charge transfer complexes (CTCs) with initiators in the electronic ground state that have absorptions at longer wavelengths. Moreover, some special additives can be used in conjunction with photoinitiators to carry out polymerization at longer wavelengths. Notably, in general, the additives are the light absorbing species here. Provided the systems thus obtained do initiate cationic polymerizations, the initiation can be explained through one of the following mechanisms:

(1) *Classical Energy Transfer.* The electronic excitation energy is transferred from the excited additive (sensitizer) to the onium salt initiator producing the

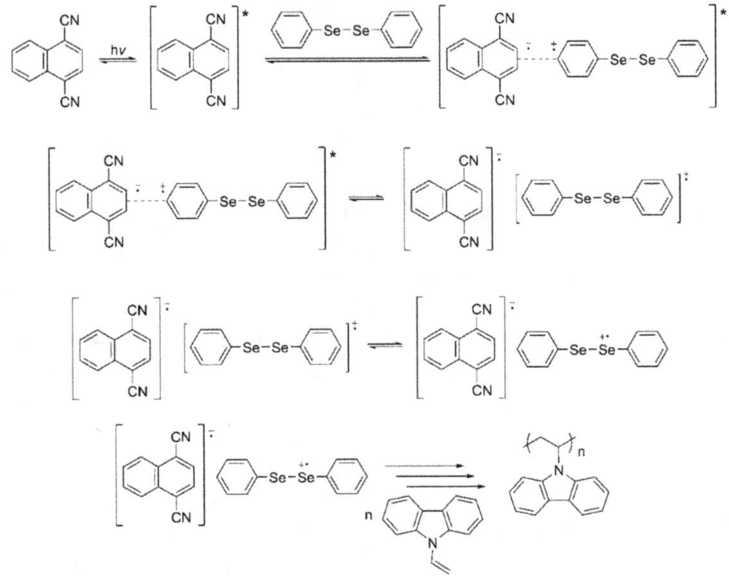

(11.63)

(11.64)

SCHEME 11.24

excited state of the latter. The route of onium salt decomposition often differs from that observed for direct photolysis of the onium salt [77–81].

(2) *Free Radical Promoted.* Many photolytically formed radicals can be oxidized by onium salts. The cations thus generated are used as initiating species for cationic polymerizations [82–84].

Polymerizations initiated via addition-fragmentation reactions can also be classified as an initiation process involving radicalic species. The principle of this class of reactions consists in the reaction of a photolytically formed radical with an allyl-onium salt generating a radical cation intermediate. These reactive species undergo a fragmentation giving rise to the formation of initiating cations.

(3) *Electron Transfer Via Exciplexes.* Sensitizers such as anthracene, perylene, or phenothiazone form exciplexes with onium salts. Being formed in the consequence of light absorption by the sensitizer, these energy-rich complexes consist of nonexcited onium salt and electronically excited sensitizer molecules. In the complexation state, electron transfer to the onium salt is observed, giving rise to positively charged sensitizer species [85,86].

Notably, above-described initiation methods do not involve the electronic excitation of the onium salt. Consequently, the initiation mechanisms are entirely different from that found for direct photolysis of onium salts [87–90].

Sensitization by Classical Energy Transfer This mechanism involves the electronic excitation of the ground state of the sensitizer, a molecule possessing a suitable

$$S \rightarrow S^* \quad (11.65)$$

$$S^* + I \rightarrow S + I^* \quad (11.66)$$

SCHEME 11.25

absorption spectrum, to its excited state. Energy may be transferred from the excited sensitizer (S*) to the onium salt (I) by either resonance excitation or exchange energy transfer (Scheme 11.25). Depending on the two components involved, the energy transfer may proceed either in the excited singlet or in the triplet state.

In consequence of the transfer process, the sensitizer returns to its ground-state and excited onium salt species (I*) are formed. The further reactions may also differ from those, taking place when the onium salt is excited by direct absorption of light. This conclusion has been drawn on the bases of product analyses [91–93]. An obvious explanation for this difference is the spin multiplicity: in the below discussed sensitized excitations triplet states of the onium salts are populated. In contrast to this, through direct irradiation of the onium salt, electrons are excited primary to the singlet state. A sufficient energy transfer requires the excitation energy of the sensitizer $E^*(S)$ to be at least as large as the excitation energy of the photoinitiator $E^*(I)$. The photopolymerization with most onium salts can be sensitized by commonly used photosensitizers, such as acetophenone or naphthalene. However, in many cases, this reaction does not proceed via energy transfer, since most onium salts are capable of oxidizing these sensitizers in an exciplex formed between sensitizer and onium salt.

Diphenyl iodonium salts are shown to take action of energy transfer with suitable additives such as *m*-trifluoromethyl acetophenone [94]. However, energy transfer using TPS salts are shown to be impossible because of unfavorable thermodynamic conditions. Energy transfer sensitization did not turn out to be technically useful, although being a possible pathway in starting the decomposition of onium salts. The reason is that the high triplet energies required allow only the use of sensitizers absorbing at wavelengths below 350 nm. Other multicomponent initiating systems (see below) show a more practical spectral response.

Free Radical Promoted Cationic Polymerization Many photochemically formed radicals can be oxidized by onium salts leading to generation of cations that are considered as initiating species for cationic polymerization according to the following mechanism as shown in Scheme 11.26 [82,84]. In this mechanism, only VE type monomers reacts with the radical (R) generated photochemically from photoinitiator and give rises a new radical (R-M) that is also oxidized by the diaryliodonium salt, whereas cyclic monomers do not undergo such oxidation process. Polymerization of cyclic monomers is initiated by only primary radicals; on the other hand, polymerization of VE monomers is initiated by both primary and secondary radicals.

This process is usually termed as the free radical promoted cationic polymerization. This so-called free radical promoted cationic polymerization is an excellent and fairly flexible type of indirect initiation of cationic polymerization.

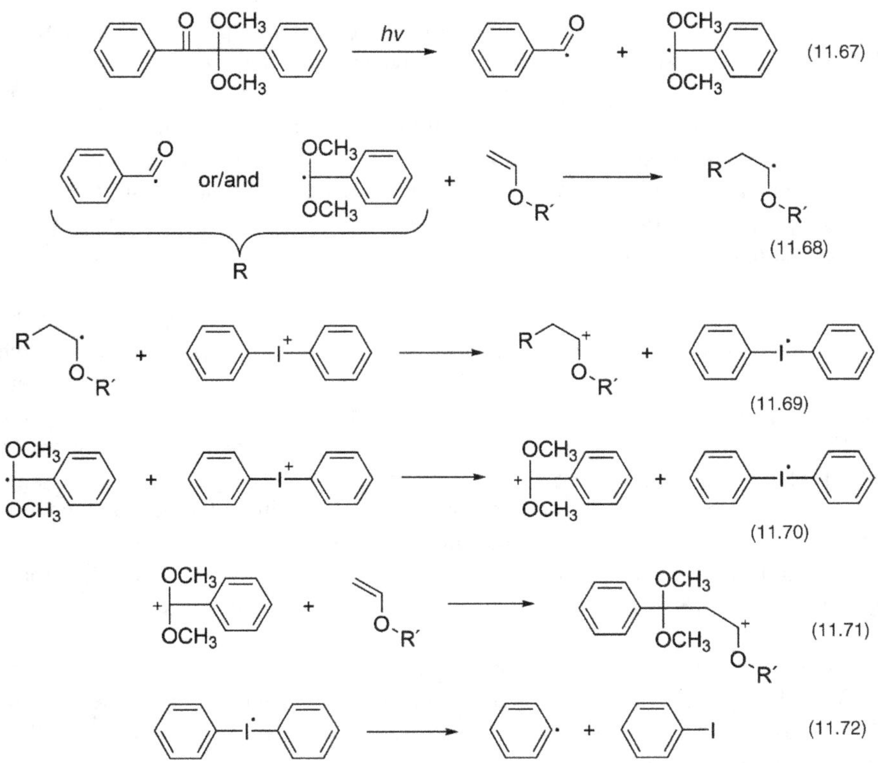

SCHEME 11.26

In recent years, a new method has been developed for cationic photopolymerization applications, especially high resolution requiring processes. This method is known as "two-photon absorption" and regarded as simultaneous absorption of two photons by a molecule. The key for successful two-photon absorption is employment of pulsed lasers that provide high energy required for the absorption process. The main difference between single-absorption and two-absorption is rate of light absorption that is directly proportional to the incident intensity for former, and proportional to the square of the incident intensity for latter [95–97]. Two-photon absorption processes provide the means to activate chemical and physical processes with high spatial resolution due to the nonlinearity of light absorption. Therefore, two-photon absorption processes are widely used for a variety of applications, such as photopolymerization [95–97], three-dimensional (3D) microfabrication [98–101], optical data storage [102], imaging [103], and the controlled release [104] of biological systems. In these reported applications, simultaneous two-photon absorption is utilized, where two photons of lower energy are absorbed by the chromophore simultaneously to populate the excited state.

In addition, nonlinearity of light absorption can also be achieved by stepwise two-photon absorption. In this process, the first photon is absorbed to generate a long-lived excited-state or reaction product [105]. In the second step, a photon of the same or different wavelength is absorbed by the long-lived excited-state or reaction product to generate the final photoproduct. Likewise to simultaneous two-photon absorption, stepwise two-photon absorption occurs with high spatial resolution.

Benzodioxinone and naphtadioxinone were used in stepwise two-photon absorption process in the presence of diphenyliodonium hexafluorophosphate to initiate cationic polymerization of CHO [105]. In fact, this process can be considered as a typical example of two-photon absorption approach in free radical promoted cationic polymerization. Upon absorption of light, benzodioxinone decomposes according to Scheme 11.27. If benzodioxinone or naphtodioxinone absorbs the first photon the molecule releases the benzophenone that then absorbs the second photon to reduce the iodonium salt yielding in protonic acids for initiation of cationic polymerization. According to experimental results, epoxides can readily be polymerized by the photolysis of benzodioxinones in the presence of suitable oxidants such as iodonium salts [105]. (It is obvious that vinyl ethers may also be polymerized with same system.) The process is based on two-photon absorption in which benzophenone is released by the photolysis of benzodioxinone. Subsequent absorption of the benzophenone thus formed eventually produces protonic acids capable of initiating cationic polymerization by successive hydrogen abstraction and redox processes.

Sensitization Via Exciplexes The use of photosensitizers is critical to the success of cationic photopolymerizations in many applications in which photopolymerizations are employed as it accelerates the rates of reactions and requires less energy as they provide polymerizations in longer wavelengths [106].

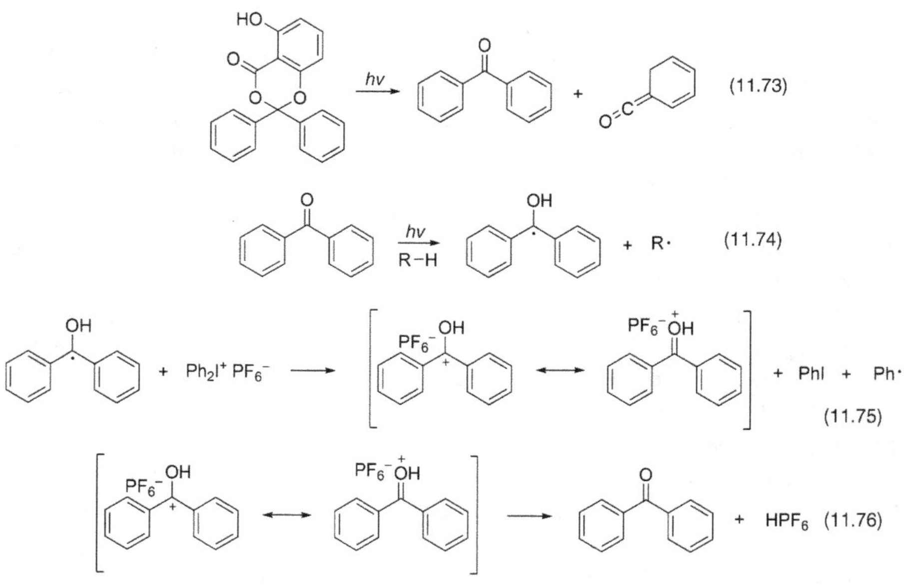

SCHEME 11.27

$$S^* + On^+ \longrightarrow [S^* \ldots On^+] \longrightarrow S^{\bullet +} + On^{\bullet} \quad (11.77)$$

$$S^{\bullet +} \xrightarrow{\text{monomer}} \text{polymer}; \qquad S^{\bullet +} \xrightarrow{\text{R-H}} \text{S-R} + H^+ \quad (11.78)$$

$$H^+ + \text{monomer} \longrightarrow \text{polymer} \quad (11.79)$$

Counter ion is omitted

SCHEME 11.28

Electron-rich polyaromatic compounds such as anthracene, pyrene, and perylene [107] are suitable as photosensitizers as they give redox reactions with DPI salts through exciplex to finally yield the initiating species for photoinduced cationic polymerizations. Scheme 11.28 demonstrates the mechanism of a polymerization followed via exciplex formation through the excited sensitizer with the ground-state onium salt.

However, because that they are poorly soluble in many monomers and are toxic, photosensitizers received little attention. In addition, because of their high vapor pressure, polyaromatic compounds can be lost from thin coatings during polymerization. For these reasons, functionalizing these electron-rich compounds in a way to improve solubility and less toxicity without affecting their absorption and photosensitizing characteristics can be a convenient solution (Chart 11.11).

By attaching formyl groups [106] and reducing them to hydroxymethyl moieties, benzylic alcohol functions can be formed [108], which enhances the polymerization rates when used with epoxides in Brønsted-acid-catalyzed ring-opening cationic polymerizations. As termed by Penczek and Kubisa, the polymerization follows a mechanism called "the activated monomer mechanism" (Scheme 11.29) [109–112].

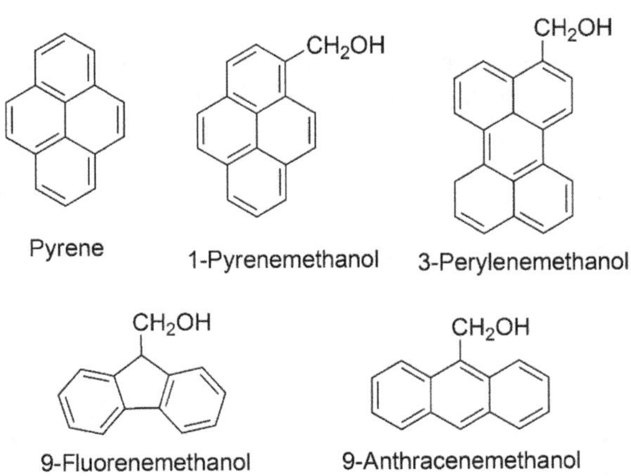

CHART 11.11

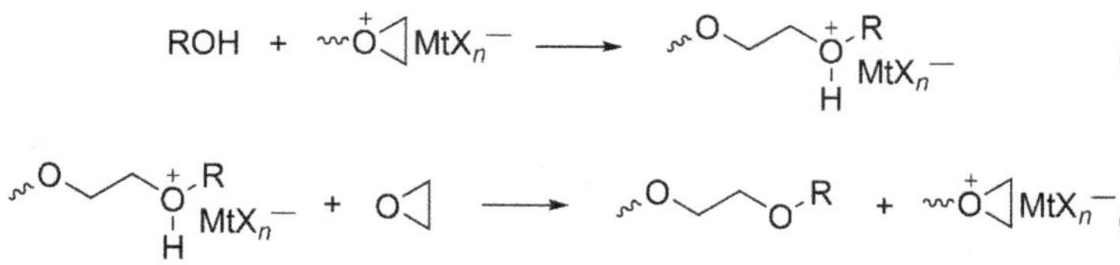

(11.80)

(11.81)

SCHEME 11.29

It is interesting to note that these compounds are linked to the polymer chain as terminal groups, thus decreasing its toxicity since migration rates of polymers in the formulation is incomparatively low than their low molecular weight analogs.

In the recent decay, there is a considerable interest in making green photopolymerizations using photosensitizers. One way to obtain nontoxic polymers through sensitization is to copolymerize compounds that can behave either as a photosensitizer or monomer with different monomers [113]. Another way is to polymerize these monomeric photosensitizers and afterward subject them to sensitize the polymerization of convenient monomers. In both ways, nontoxic and odorless polymers can be obtained after polymerization. Compounds introducing phenothiazine moiety and their polymeric analogs were found the display high efficiency in PCP of vinyl ethers and epoxides.

Recently, Yagci and coworkers investigated the initiation mechanism of a new type of cationic photoinitiator, namely, *N*-phenacyl-*N*,*N*-dimethylanilinium hexafluoroantimonate (PDA^+ SbF_6^-) which initiates the polymerization of appropriate monomers [116]. The proposed mechanism includes irreversible fragmentation of the absorbent salt to yield the initiating species either via a heterolytic cleavage or via a homolytic cleavage followed by subsequent electron transfer between the preformed species; still, forming the same cation that initiates cationic polymerization (Scheme 11.30).

This salt can also be activated by various photosensitizer [114–116] such as perylene anthracene and phenothiazine. The mechanism involves again an electron transfer from the excited sensitizer to the ground-state phenacyl salt (Scheme 11.31).

Notably, thioxanthone and benzophenone do not undergo this redox reaction due to the unfavorable thermodynamic conditions.

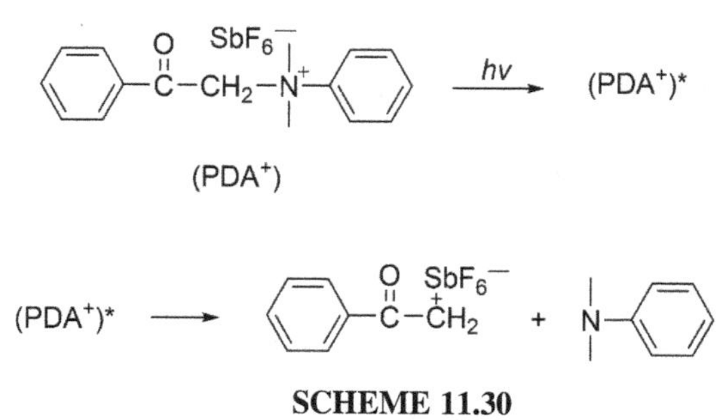

(11.82)

(11.83)

SCHEME 11.30

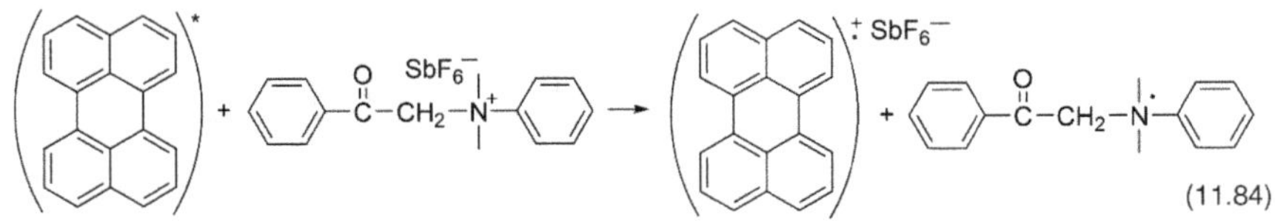

(11.84)

SCHEME 11.31

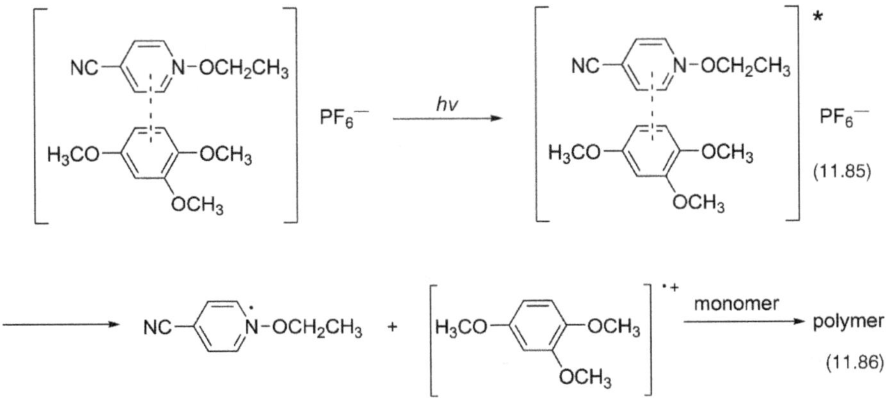

(11.85)

(11.86)

SCHEME 11.32

Ground-State Charge-Transfer Complexes Although it is not considered to be a general method for the indirect initiation, certain salts can undergo decomposition upon irradiation in their appropriate charge transfer complexes. For example, pyridinium salts are capable of forming ground-state CTCs with electron-rich donors such as methyl- and methoxy-substituted benzene [85] Notably, these complexes absorb at relatively high wavelengths, where the components are virtually transparent. For example, the complex formed between *N*-ethoxy-4-cyano pyridinium hexafluorophosphate and 1,2,4-trimethoxybenzene possesses an absorption maximum at 420 nm.

The following mechanism shows the action of a CTC as photoinitiator (Scheme 11.32). Since polymerization takes place even in the presence of a proton scavenger like 2,6-di-*tert*-butylpyridine, an initiation through Brønsted acid formation can totally be excluded.

11.3 MONOMERS

The types of general monomers that may undergo photoinitiated cationic polymerization are vinyl and alkoxy vinyl monomers, heterocyclic monomers involving sulfur, oxygen, and nitrogen atoms in their rings. Cationically polymerizable monomers and their corresponding polymer [117] are summarized in Fig. 11.2. In UV-curing applications, difunctional epoxide and alkyl vinyl ether monomers are

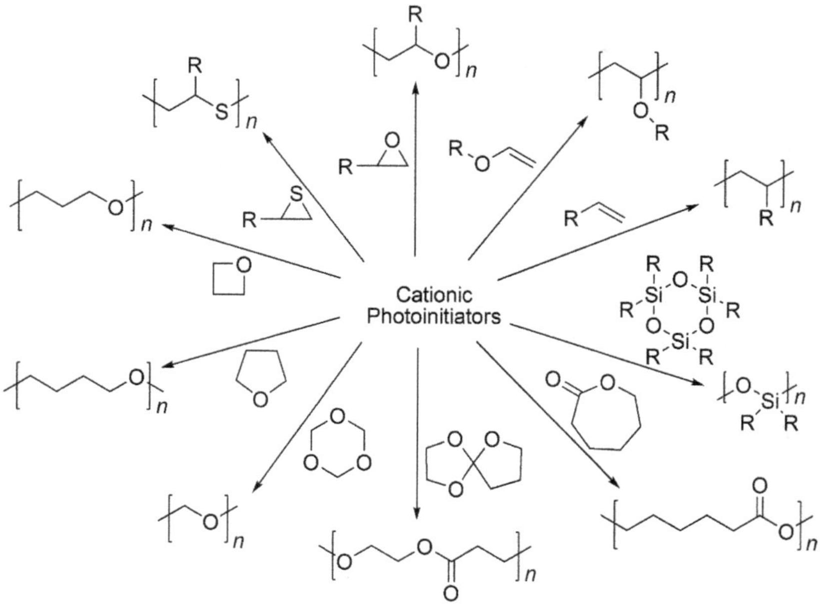

FIGURE 11.2 Photoinitiated cationic polymerization of various monomers.

usually employed. The polymerization mechanisms are outlined for the main monomer groups employed in cationic polymerizations, namely, vinyl ethers and epoxides in Scheme 11.33a,b. As described previously, initiating species, usually protonic acids, are formed upon photolysis of photoinitiators, and these species react with monomer. Depending on type of monomer, they undergo addition to double bond or ring opening.

Although wide variety of monomers is available, demands from many industrial applications promote the design of new monomers with high polymerization rate and improved features. For this reason, several works in recent decades have been conducted to development of new monomers, especially epoxides, with high reactivity. Furthermore, some works have been subjected to synthesis of monomers carrying functional groups allowing for design of the special polymer since nature of the monomers affects the physical and mechanical properties of the resulted polymer. In summary, newly developed monomers containing a wide variety of functional groups may lead to a high monomer conversion, acceleration in polymerization as well as production of polymers with improved properties.

11.3.1 Structurally Different New Monomers

The structure of the monomers has a great influence on both reactivity and the physical and mechanical properties of the corresponding products. Although a large variety of the monomers are now available, newly developed monomers carrying different functional groups may lead to a high conversion and improved properties.

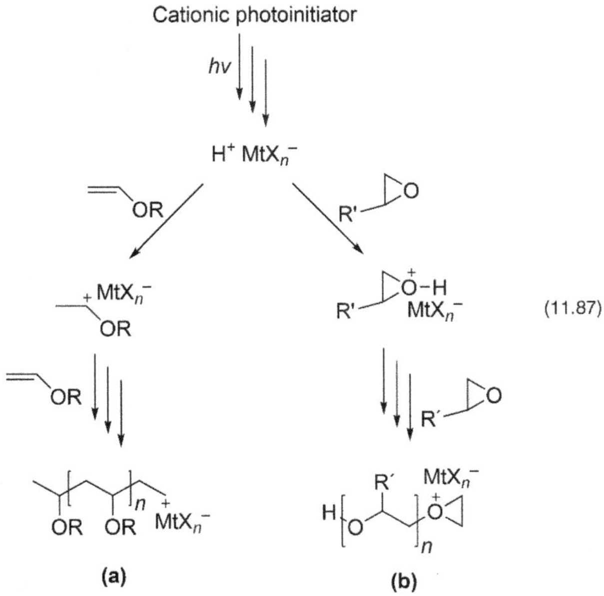

SCHEME 11.33 General mechanism for photoinitiated cationic polymerization of alkyl vinyl ethers (a) and epoxides (b).

11.3.1.1 Photoinitiated Cationic Polymerization of Unusual Monomers Recently, novel monomers and macromonomers (Chart 11.12), namely, benzoxazines, monothiocarbonates, thiophene, and epoxy and vinyl ether functional polymers were reported to undergo photoinitiated cationic polymerization [2,118]. Among them, benzoxazine monomers deserves a special attention as they yield thermosets with several excellent characteristics including heat resistance; good flame retardance; stable dielectric constants; low water absorption and perfect dimensional stability. This type of thermosets is also obtained by thermally activated ring-opening polymerization at elevated temperatures without catalysts [19].

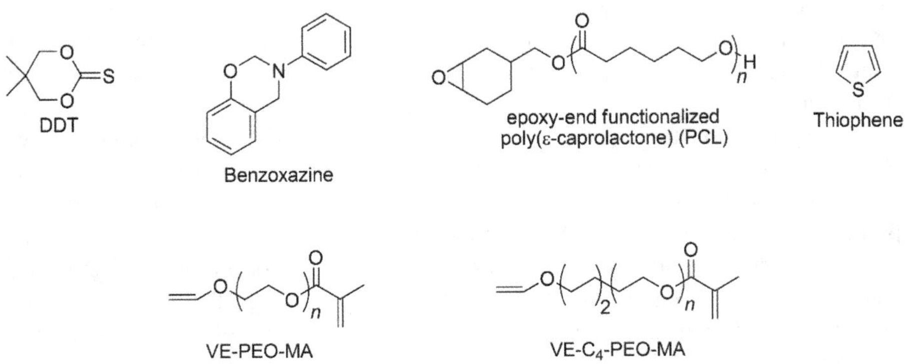

CHART 11.12

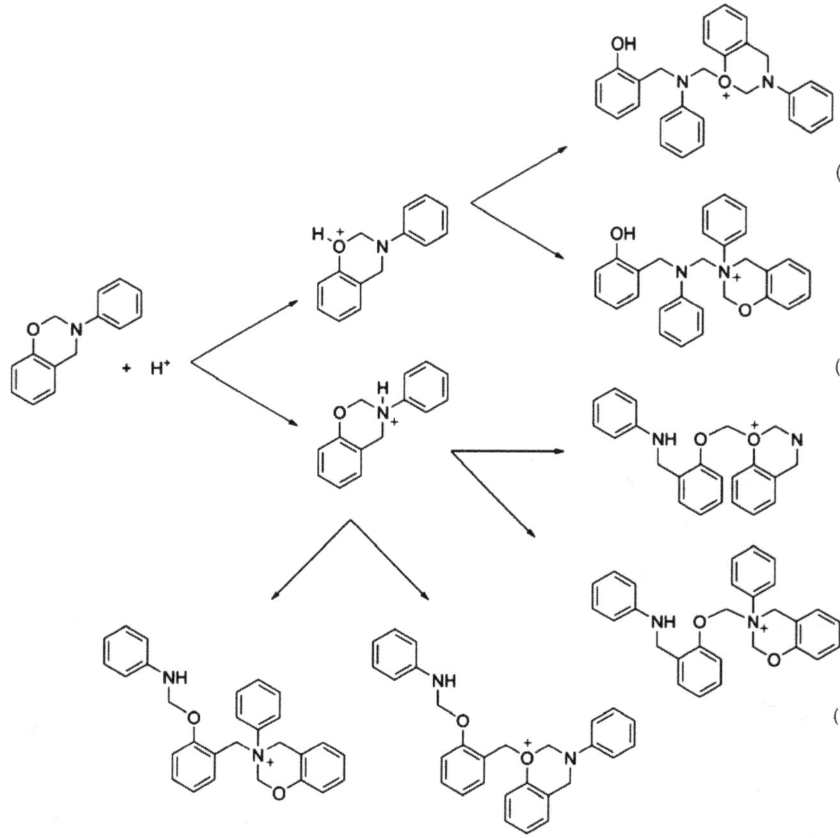

(11.88)

(11.89)

(11.90)

SCHEME 11.34

Photoinitiated cationic polymerization of benzoxazines by onium salts was investigated mechanistically. It was postulated that the first step involves the addition of photocemically generated proton (or carbocation) either to oxygen or to nitrogen atom. Then, the polymerization proceeds via two different routes leading to the formation of different structures (Scheme 11.34) [2,118].

Monothiocarbonates such as 5,5-dimethyl-1,3-dioxane-2-thione reportedly polymerize by using cationic photoinitiators [2]. It was shown that onium salt based direct, and free radical promoted and photosensitization via exciplexes are practicable photoinitiation systems for such cyclic monomers.

An interesting photopolymerization route was reported for thiophene monomers. Detailed laser flash photolysis studies revealed that although the cationic species are involved during the polymerization, the actual mechanism is step-growth (Scheme 11.35).

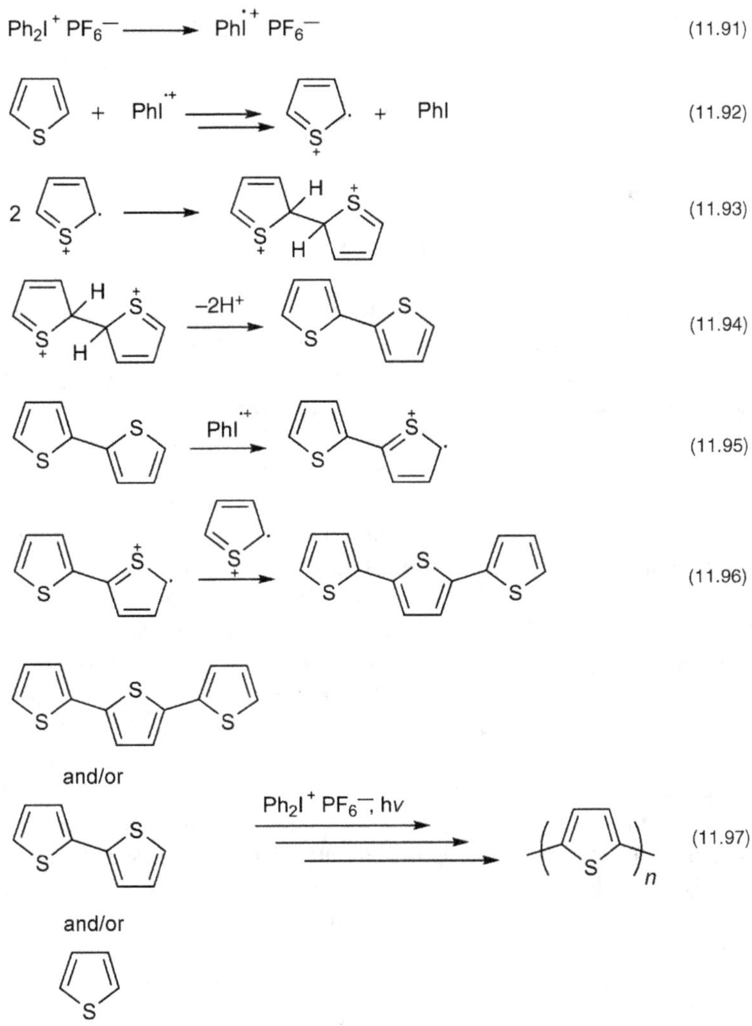

SCHEME 11.35

The solution containing thiophene and diphenyl iodonium salt becomes dark and films are accumulated on the surface of the reaction tube when the tube is exposed to radiation. A conducting freestanding film is essentially obtained.

11.3.1.2 Monomers with Different Functionalities Introduction of other functionalities into monomer structure is a versatile approach to increase the reactivity, provide wavelength tunability, and improve the properties of the networks formed. Monomers attached to preformed polymers [2] or equipped with photosentizers [119], hydroxyl groups [108–110], and groups polymerizable by other modes [2,120,121] were readily prepared. For example, epoxy end functionalized poly (ε-caprolactone)

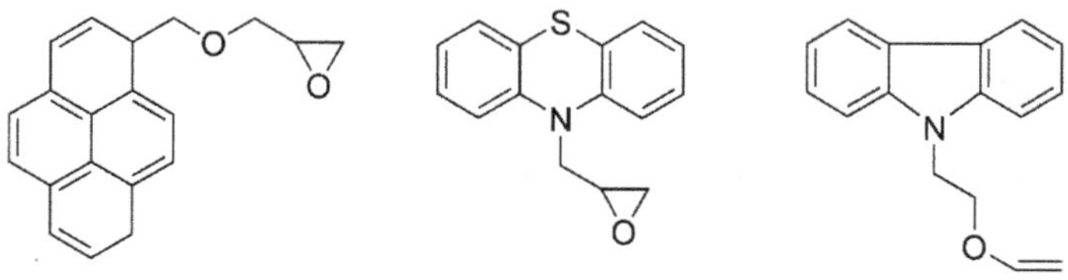

(PEP) (PTEP) (NVCVE)

CHART 11.13

(PCL) macromonomer was photochemically polymerized [2]. It was demonstrated that all the initiation modes described above for onium salts are efficient in the polymerization of such macromonomer with high conversion (>99%) [2]. Also, PCL macromonomer allows generation of poly(cyclohexene oxide)-*g*-poly(ε-caprolactone) graft copolymer with random sequences of PCL on the side chain, when CHO is used as a comonomer. Furthermore, novel hybrid poly(ethylene oxide) (PEO) macromonomers (VE-PEO-MA and VE-C_4-PEO-MA) carrying various vinyl end-groups were designed to obtain network or gelled polymer materials with soft and hydrophilic domains or surfaces by controlled syntheses [2].

In order to decrease the migration of the unreacted sensistizer residues in the cured products, Crivello's group designed photosensitizers carrying catonically polymerizable functional groups [119]. These types of compounds act as a photosensitzer as well as monomer in copolymerization with vinyl ether or epoxy. The sensitizer is bound to the polymer chain, and cannot be released. Therefore, toxicity caused by photosensitizers will be prevented. A pyrene derivative with epoxy function (**PEP**), a phenothiazine derivative with epoxy function (**PTEP**), and a *N*-vinyl carbazole derivative with vinyl ether function (**NVCVE**) were prepared and employed in an onium salt initiated cationic photopolymerizations of epoxides and vinyl ethers (Chart 11.13) [119]. Furthermore, these compounds and their polymers accelerate onium salt induced cationic photopolymerizations.

Compounds containing hydroxyl group accelerate photoinduced cationic polymerization of epoxides as explained previously (*vide ante*) [109,110,122]. Accordingly, 1-pyrenemethanol was designed as an accelerator as well as photosensitizer [119].

11.3.1.3 Expandable Monomers Several attempts have been made to design systems that undergo little or no volume changes upon polymerization since many important technologies, including dental materials, composites without strain stress, stereolithography, and many imaging applications do not tolerate volume shrinkage during polymerization [123–126].

Polymerization kinetics and ring-opening mechanisms of 2-methylene-7-phenyl-1,4,6,9-tetraoxa-spiro [4.4] nonane (MPN) were investigated to develop new spiro orthocarbonate (SOC) monomers, which produce low shrinkage and expanding polymers [123]. The onium salt-induced double ring-opening photopolymerization of this kind of monomers may lead to expansion, which is crucial for dentistry applications. The thermodynamic calculations that were performed to analyze possible ring-opening mechanisms indicate that α-attack was more probable than

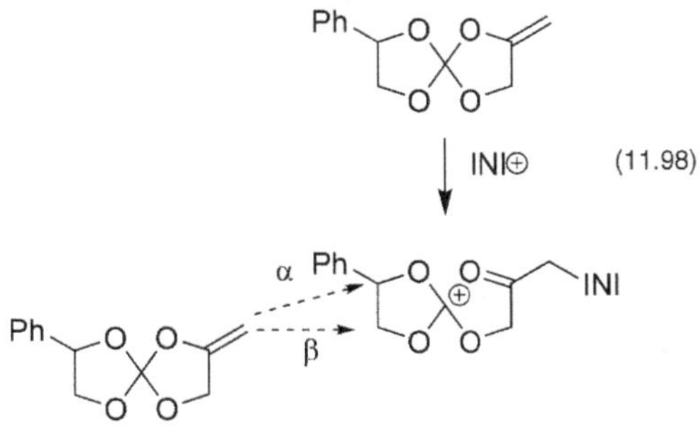

SCHEME 11.36

β-attack as shown in Scheme 11.33. Therefore, polymerization is postulated to occur from the α-attack. Further analyses, concerning the kinetic aspects have been made to investigate the effect of polymerization temperature and the possibility of free radical promoted photoinitiated cationic polymerization. Results exhibit that the polymerization rate was increased with increasing temperature and DMPA concentration. Presence of free radical photoinitiator (DMPA) causes acceleration in the rate of polymerization of MPN as in the case of polymerizations of epoxide and vinyl ether monomers (Scheme 11.36).

Cross-linking of another expanding monomer, 1,5,7,11-tetraoxaspiro [5.5] undecane (**TOSU**), with bisphenol A diglycidyl ether (**BADGE**) and with bifunctional oxetane (**BOXT**) was investigated by Holder [124] and Endo [125], respectively (Chart 11.14). The resulted networks are nonshrinking cross-linked materials.

There are several reports on homopolymerization of TOSU and its derivatives. Although the obtained homopolymers are expanded, they do not have resistance over the adverse effect of some solvents and are not rigid enough for intended applications particularly in dentistry. On the other hand, homopolymerization of difunctional monomers results in shrinking cross-linked networks. Incorporation of difunctional monomers such as BADGE [124] and BOXT [125] with TOSU provides cross-linking as well as solutions to problems associated with volume shrinkage, lack of rigidity and solvent resistance.

The mechanistical investigation of onium salt induced photocross-linking of TOSU with BADGE was performed by matrix-assisted laser desorption/ionization

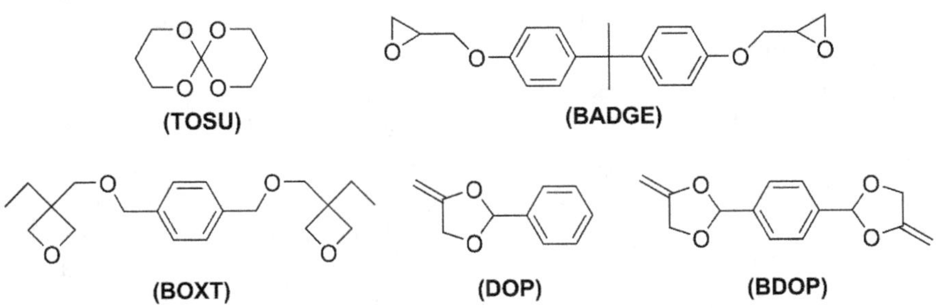

CHART 11.14

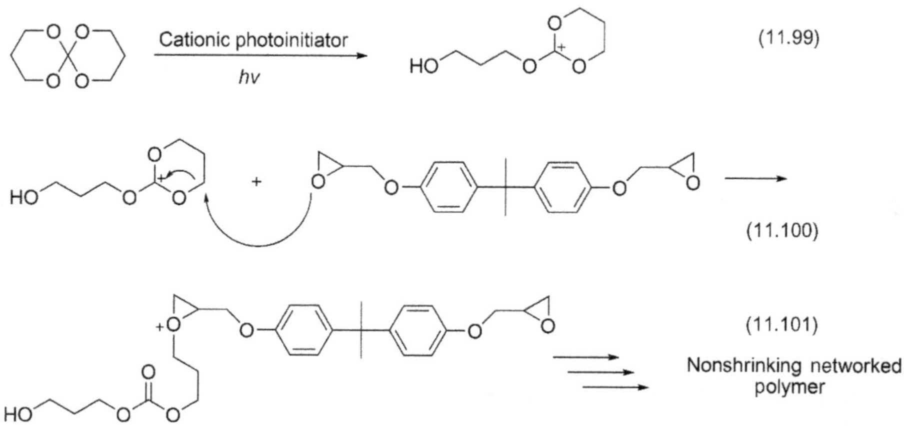

SCHEME 11.37

time-of-flight mass spectrometry (MALDI-TOF MS) analysis. The results reveal that these monomers were polymerized in a sequential manner, that is, first TOSU is polymerized and propagating ends initiates the polymerization of BADGE (Scheme 11.37). Interestingly, while BADGE is not polymerized alone, independent polymerization of TOSU yields only oligomers. During the course of the polymerization, polymeric chains gain or lose cyclic carbonate structure. A similar mechanism was proposed for the thermal cross-linking of TOSU with another bifunctional oxatane monomer BOXT [125]. Obviously, the photochemical cross-linking of this monomer couple seems to be feasible.

Photolysis of appropriate combinations of the mono- and bifunctional monomers, **DOP** and **BDOP** in the presence of diaryliodonium or triarylsulfonium salts yields expanded polymer networks [126]. The resulting cross-linked polymers have suitable T_g values for desired applications.

Several phosphorus-containing vinyl ether and 1-propenyl ether monomers were prepared, and their photoinitiated cationic polymerizations were carried out [127]. The synthesis involves regioselective reaction of glycidyl vinyl ether or 1-propenyl glycidyl ether with diaryl phosphonates in the presence of catalysts [127,128]. The photochemically obtained corresponding polymers exhibited flame retarding properties [127] due to the phosphorous present in the structure. Such materials are considered to be alternative for highly toxic halogenated formulations in UV-curing applications [127,129].

Substituted vinylcyclopropanes would also be intended to be polymerized with volume expansion. Photoinitiated cationic polymerization of 1-cyclopropyl-l-phenyl ethylene, 1-cyclopropyl-l-(*p*-ethoxyphenyl) ethylene and 1-cyclopropyl-l-(*p*-fluorophenyl) ethylene was investigated [130]. Depending on the type of initiator used, various polymer structures were obtained with shrinkage about 11% that is exceptionally high for ring-opening polymerizations. The observed high shrinkage is mainly due to the small ring of cyclopropane. Also, polymer structures and polymerization mechanisms were investigated in detail.

11.4 ACCELERATION OF PHOTOINITIATED CATIONIC POLYMERIZATION

The most important limitation related to commercial applications is relatively low reactivity of epoxide-based resins, commonly used in coatings, compared to acrylate based systems. Therefore, most works have been effort to accelerate ring-opening polymerization of epoxides. There are four different strategies to speed up the polymerization:

(1) Addition of additives (i.e., alcohols) that switch the polymerization mechanism (i.e., activated monomer mechanism) [108–110,119,122,131,132].
(2) Addition of plasticizing agents that dilute polymerization medium and increase diffusion rate of reactants [132,133].
(3) Addition of compounds (i.e., benzyl ethers) that produce relatively stable radicals and cations, and cause chain induced decomposition of onium salts [108–110,122,134].
(4) Employment of hybrid monomers consisting of epoxide and a functional group (usually vinyl ether) with high reactivity instead of conventional epoxide monomers [120,121,135,136].

11.4.1 Acceleration of Photoinitiated Cationic Polymerization by Using Additives

Photoinitiated cationic ring-opening polymerization of epoxy monomers initiated by onium salts are obviously accelerated by addition of benzyl alcohols [108]. The best way to understand how benzyl alcohols are effective in the polymerization is determination of acceleration factor (AF). AF is the ratio of the rate of ring-opening polymerization epoxy in the presence of the benzyl alcohol (Rp_a) to the rate of that in the absence of the alcohol (Rp). Notably, $[M_0]_a$ is the initial concentration of the monomer in the presence of the alcohol, and $[M_0]_a$ is in the absence [108].

$$AF = \frac{Rp_a/[M_0]_a}{Rp/[M_0]} \qquad (11.102)$$

If the addition of alcohols (or additives) causes deceleration, this factor is termed deceleration factor (DF) [121].

The acceleration is expected to occur through simultaneously and/or independently occurring "activated monomer" and "hydrogen abstraction followed by oxidation" processes (Scheme 11.35).

According to the activated monomer mechanism, when the polymerization is performed in the presence of alcohol, an attack from the hydroxyl group to the α-carbon of the oxonium ion is possible due to the high nucleophilicty of this functional group. If this process is predominant, a polymer chain is formed resulting from the successive addition of protonated monomer to the terminal hydroxyl of the growing chain (see Equations 11.81 and 11.82 in Scheme 11.29).

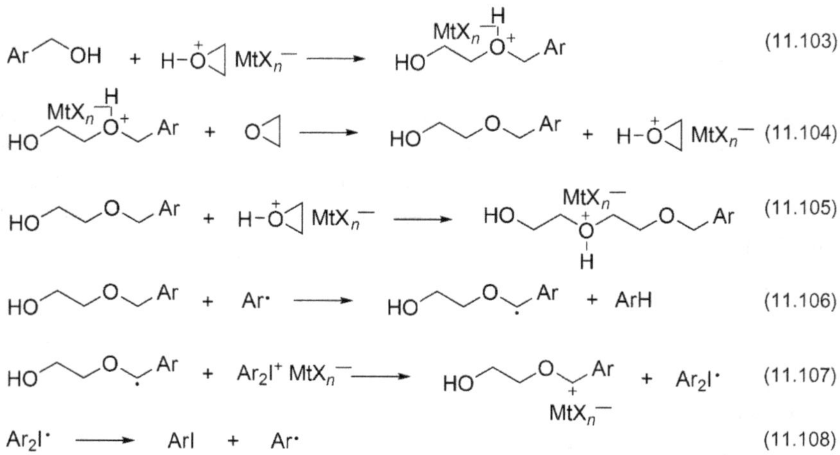

SCHEME 11.38

The second acceleration effect is based on the radical promotion and mainly depends on nature of the substituents on the benzene ring of the benzyl alcohol. In this case, benzyl ether groups containing hydrogen atoms, which are labile to abstraction, are attached to ends of the polymer chains (see Equation 11.106 in Scheme 11.38). In the subsequent stage, benzyl ether hydrogens are abstracted by aryl radicals produced as a result of the photolysis of the onium salt, and benzyl ether radicals are formed (Equation 11.107). These more stable radicals, in turn, are oxidized by the onium salt to corresponding carbocations, which are also stable and able to initiate a new polymerization chain more rapidly (Equation 11.108), as described by Ledwith [82]. This process also results in increase in the quantum yield of the photolysis of the onium salt.

These two mechanisms were thoroughly investigated in photoinitiated cationic polymerization of various epoxide monomers by Crivello and Ortiz [108]. The polymerizations were kinetically monitored via real time FT-IR. The results of kinetic studies indicate that rate of photopolymerization of 4-vinylcyclohexene dioxide (VCHDO) initiated by onium salt (i.e., IOC-10) enhances with the increasing concentration of benzyl alcohol. Notably, structure of the benzyl alcohol also affects rate of the polymerization. Electron-donating groups substituted on *para*-position of the benzene ring such as α-methyl benzyl alcohol and 4-methoxybenzyl alcohol have a positive effect on acceleration since benzyl ether radicals or carbocations formed during polymerization are stabilized by resonance and inductive effect, whereas benzyl alcohols with an electron-withdrawing substituent on their 4-position such as 4-nitrobenzyl alcohol decreases the rate of polymerization. The effect of piperonyl alcohol, which contains highly electron-rich and activating methylene-dioxy group on the benzene ring, was found to be as accelerating as 3,4-dimethoxybenzyl alcohol.

According to AF values, the rate of the polymerization decreases in the order of 3,4-dimethoxybenzyl alcohol > piperonyl alcohol > 4-methoxylbenzyl alcohol >

3,5-dimethoxybenzyl alcohol > benzyl alcohol > 4-chlorobenzyl alcohol > 4-nitrobenzyl alcohol. In conclusion, benzyl alcohol and its derivatives containing electron-donating subtituents are effective in acceleration of onium salt initiated cationic photopolymerization of epoxide monomers.

In another study, a hydroxyl vinyl ether monomer (butandiol vinyl ether) (HBVE) was used as additive in UV curing of both a difunctional and a monofunctional epoxy monomers [132]. It has been shown that in the first case, the additive can act as plasticizing agent whereas in the second case, the hydroxyl containing vinyl ether behaves like a difunctional monomer and therefore allows the curing of a monofunctional epoxy monomer. Kinetic results demonstrate that the addition of HBVE causes an acceleration of the polymerization and increases the monomer conversion due to participation of the activated monomer mechanism. Furthermore, monofunctional epoxy monomers such as cyclohexene oxide can be cured photochemically by using onium salt photoinitiators in the absence of conventional cross-linker. Spectroscopic and thermal studies together with control experiments revealed that cross-linking occurs by the participation of the additive in both cationic and activated monomer polymerizations. Moreover, the process causes changes in the properties of the cured networks that were confirmed by T_g measurements.

As mentioned before, addition of plasticizing agents, which decrease the viscosity, into polymerization medium increase the rate of polymerization. This effect was observed in photoinitiated cationic polymerization of epoxides in the presence of epoxidized soybean oil [133].

11.4.2 Acceleration of Photoinitiated Cationic Polymerization by Using Monomers with High Rectivity

Increase in the polymerization rate can also be achieved by designing and synthesizing new monomers. Therefore, in the recent decades, demands on rapidly polymerizable cationic monomers trigger many groups to design new monomers with high reactivity. The following section is devoted to the several strategies to accelerate the polymerization rate by monomer design.

11.4.2.1 Monomers Bearing Functional Groups Epoxides bearing various functional groups including hydroxy groups (**EPO-1–EPO-3**) and benzyl ether (**EPBE** and **EPMBE**) were designed by Crivello's group to accelerate the cationic polymerization of epoxides as described in Scheme 11.29. Epoxide monomers with hydroxyl groups act as both monomer and chain transfer agents. In accordance with their functions, such structures were termed as "monofers" [134]. Some examples of monofers are depicted in Chart 11.15.

Crivello's group followed either or both of two strategies that described for the additives in acceleration of photoinitiated cationic polymerization of epoxide monomers. These are stabilization of free radicals and cations by resonance and inductive effect, and the activated monomer mechanism. Comparative studies of novel monomers with conventional monomers show that newly designed monomers given in

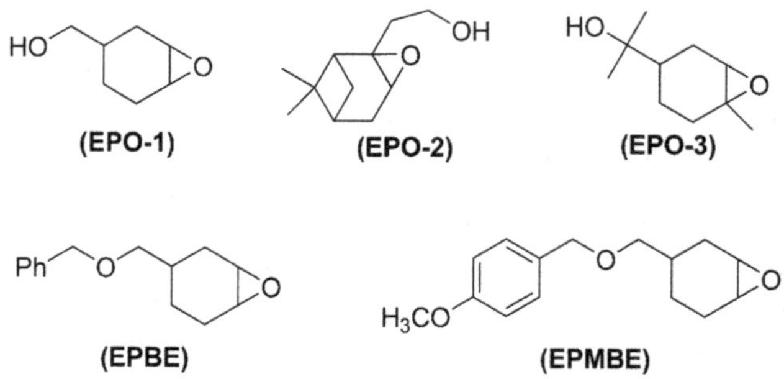

CHART 11.15

Chart 11.15 remarkably enhance the photoinitiated ring-opening polymerization [108,119,131]. The overall polymerization mechanism is outlined in Scheme 11.39. Accordingly, aryl radicals, formed during the photolysis of diaryliodonium salts (Scheme 11.3) abstracts hydrogen from benzyl ether to yield a radical site in the structure. Afterward, the free radical is readily oxidized by the onium salt to yield the corresponding resonance stabilized cation, which is capable of initiating cationic polymerization. This also increases the amount of onium salts consumed (Scheme 11.35), which in turn provides higher quantum yield of the photolysis.

11.4.2.2 Hybrid Monomers Another excellent way to enhance the polymerization rate of epoxides is to design and use hybrid monomers carrying additional functional group with intrinsically higher reactivity. Several works have been reported on this subject [120,121,135]. These studies are mainly related to the investigation of the relationship between epoxides and vinyl ethers when simultaneously polymerized. Studies emphasizing on designing new hybrid monomers and their behavior have also been reported. These studies revealed three main conclusions. First, when only a hybrid monomer or both epoxide and vinyl ether are subjected to photoinitiated cationic polymerization, while epoxide polymerization is accelerated, there is a significant decrease in the rate of vinyl ether polymerization [120,121,135]. In addition, the induction periods of both polymerization modes are shortened and the overall conversions are increased.

Monomers possessing 1-propenyl ether groups are reported to be the most reactive class of monomers toward cationic species. Therefore, a series of hybrid epoxide monomers with 1-propenyl ether group was designed and their behavior in cationic photopolymerization initiated by a diaryliodonium salt was investigated [120]. The

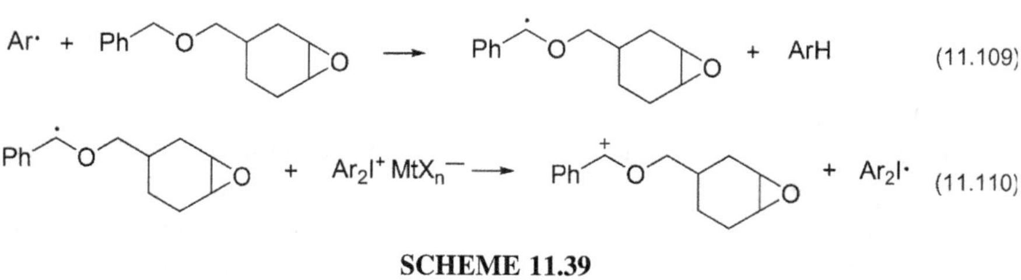

SCHEME 11.39

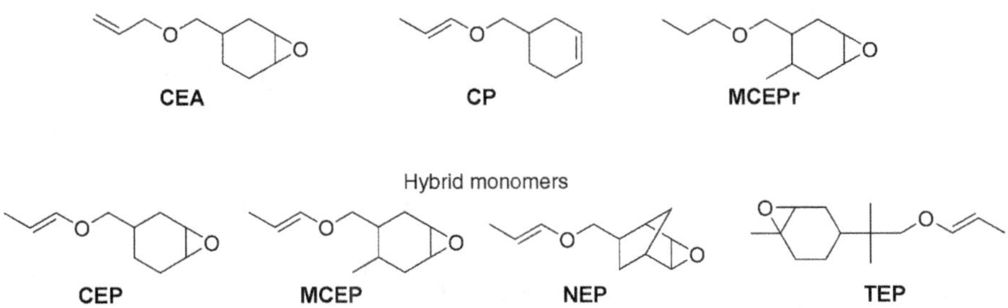

CHART 11.16

structures of the selected hybrid monomers are presented in Chart 11.16. Kinetic investigations associated with independent polymerizations of an epoxide monomer, **CEA**, and a 1-propenyl ether monomer, **CP**, show that rate of the polymerization of 1-propenyl ether group is remarkably higher than that of the epoxide group. However, the report on the simultaneous polymerization of these two monomers with equal concentrations indicates that polymerization of the epoxide is accelerated whereas the rate of polymerization of the 1-propenyl vinyl ether is usually reduced. Similar behavior was observed in polymerizations of monomer couples involving *n*-butyl glycidyl ether (BGE)/*n*-butyl vinyl ether (BVE), phenyl glycidyl ether (PGE)/BVE, *p*-cresyl glycidyl ether (CGE)/BVE, 1,2-epoxydodecane (V12)/BVE, MCEPr/CP, and 3,4-epoxycyclohexylmethyl 3′,4′-epoxycyclohexanecarboxylate (CY179)/BE [121,135]. Furthermore, the kinetic results indicated that the induction periods of both epoxide and vinyl ether polymerizations were shortened since aryl radical attacks the double bond of vinyl group and forms relatively stable radicals and cations leading to amplification of initiating species.

The low viscosity of the reaction media due to ether linkages in the backbone is possibly one of the reasons why polymerizations of hybrid monomers result with high conversions.

According to mechanistic studies, Crivello and coworkers proposed a mechanism (Scheme 11.40), which explains the higher rates of polymerizations of hybrid

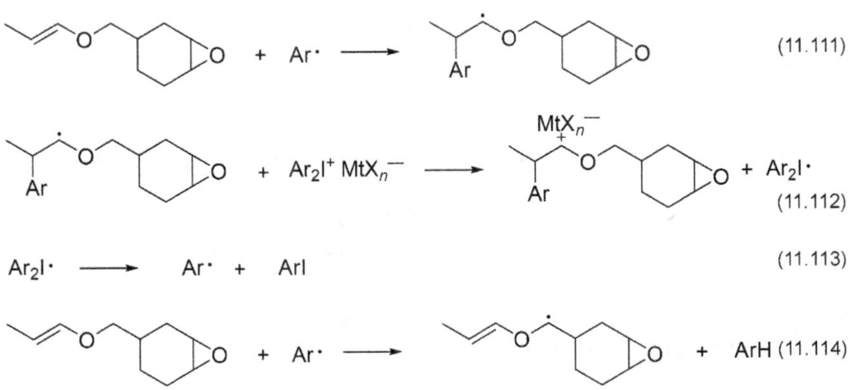

SCHEME 11.40

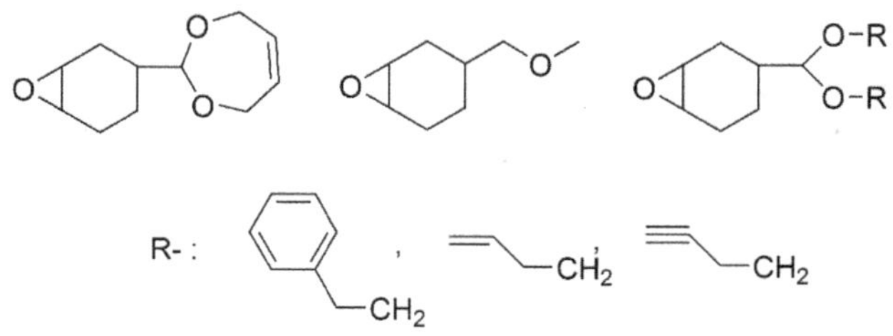

CHART 11.17

monomers. The radical cations formed from the photolysis of onium salts initiates the polymerization of epoxy monomers in the usual manner. However, the acceleration effect is mainly attributed to the participation of aryl radical in various additional reactions. These radicals interact with hybrid monomers in two ways: (i) addition to the double bond of 1-propenyl ether group (Equation 11.112) and (ii) hydrogen abstraction (Equation 11.115). The resulted radicals are eventually oxidized by the onium salt to generate a carbocation that is capable of initiating the polymerization (Equation 11.113). Therefore, these two processes result in acceleration of polymerization of epoxides. On the other hand, the polymerization of 1-propenyl ether group is suppressed due to such competing reactions.

Furthermore, novel epoxide monomers bearing benzyl, allyl, and propargyl acetal and ether groups that can stabilize free radicals through inductive and resonance interactions have been synthesized (Chart 11.17) [136]. Kinetic studies via real-time infrared spectroscopy were carried out for the investigation of the onium salt initiated cationic photopolymerization of the monomers containing these functional groups. As intended, the monomers undergo photoinitiated cationic polymerization at high rates. It is proposed that the above-described radical-induced reactions and conventional protonic acid-catalyzed initiation are responsible for the acceleration.

11.5 PHOTOINITIATED LIVING CATIONIC POLYMERIZATION

The synthesis of well-defined macromolecules with structural control can only be achieved by the controlled/living polymerization methods. In addition to the well established anionic [137] and recently developed radical processes [138], living cationic polymerization [139] of appropriate monomers has also been reported.

A key to the success of the living cationic polymerization of vinyl ethers is the stabilization of the unstable carbocations via suitable nucleophilic counterion. There are two ways to stabilize the carbocations: (1) generation of suitable nucleophilic counterion resulted from the initiator and the catalyst, and (2) addition of nucleophilic agents to the polymerization media. In the first way, Bronsted acids such as hydrogen iodide are employed as the initiators, while Lewis acids such as zinc iodide are employed as the catalysts (Scheme 11.41) [140–143].

Recently, photoinduced living cationic polymerization of isobutyl vinyl ether (IBVE) in the presence of various diphenyliodonium salts and zinc halides was

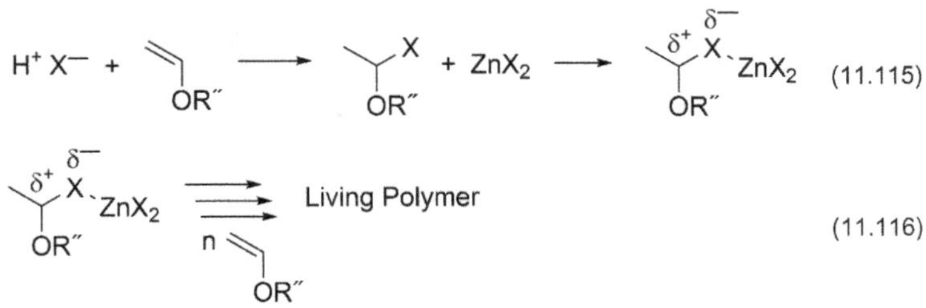

X: Cl, Br, I   R′H: monomer or solvent

SCHEME 11.41

reported by Mah and coworkers [9–12]. Photochemically generated protonic acid reacts with IBVE to form the corresponding monomer adduct. Terminal carbon-halide bond of the adduct is activated by the coordinating effect of zinc halide (Equation 11.116). This activation leads to generation of suitable nucleophilic counterion by stabilizing the growing carbocation (Equation 11.117). Thus, chain-breaking processes are prevented and living cationic polymerization of IBVE proceeds. This approach was successfully employed also in free radical promoted and sensitized systems. The main advantage of these systems is the use of onium salts with highly neucleophlic counter ions such as bromide and chloride. It is well known that the onium compounds with nonnucluophilic counter ion of especially SbF_6^- and AsF_6^-, exhibit high toxicity and low cost efficiency because of their central heavy metals. Duan et al. and Gebel reported that long-term exposure of these metal salts to human skin led to increased incidences of various cancers [144,145]. Furthermore, the preparation of such initiators from their corresponding onium salts with halides requires additional steps, that is, counter anion exchange, thus it this makes them expensive [8,21].

In the second way, addition of nucleophilic agents to the polymerization media has been used to suppress side reactions. For example, when the polymerization is performed in the presence of dimethyl sulfide, the polymers obtained display characteristics of living polymerization such as narrow polydispersity and well-defined chain end functionality due to suppression of side reactions [146]. In this case, the polymerization reaches to an equilibrium between alkoxycarbenium ion (growing chain) and trialkylsulfonium species (Scheme 11.42) [146]. Indeed, the latter reaction is predominant; thus, chain transfer and termination reactions are prevented to lead to living polymerization.

Similar to the strategy employed in vinyl ether polymerization, addition of sulfides was considered as a way to achieve living cationic polymerization of epoxides. The

SCHEME 11.42

(11.118)

(Counterions are omitted)

SCHEME 11.43

effect of additional sulfides on onium salt induced photoinitiated cationic ring-opening polymerization of epoxides was reported by Crivello [147]. Polymerization profiles of the monomers such as cyclohexene oxide and 1, 2-epoxyhexane were monitored by optical pyrometry and real time FT-IR spectroscopy. Dialkyl sulfides inhibit the polymerization rate as a result of its nucleophilic nature, which competes with monomers to attack to the growing chain end cation, and even overwhelms it. On the other hand, if sulfides with moderate basicity such as diaryl sulfides and thianthrene are used, the polymerization is retarded. The polymerization rates and monomer conversions are decreased with increasing amount of aryl sulfides. Scheme 11.43 demonstrates the attack of the sulfide group on the growing oxonium ion instead of the epoxide monomer. Thus, nucleophillic nature of the sulfide determines the role of the additive in the polymerization. Retardation of the ring-opening polymerization of epoxides, may be employed in order to prevent chain transfer and termination reactions and living conditions can be thus be attained.

11.6 UV CURING BY PHOTOINITIATED CATIONIC POLYMERIZATION

UV-radiation curing is defined as transformation of liquid multifunctional monomers or oligomers into solid and insoluble cross-linked polymers using UV or laser radiation. In recent decades, it has become a powerful industrial method widely used in various applications including coatings, inks, adhesives, varnishes, electronics, photolithography, and dyes on due to its great advantages [1–3]. It offers high rate of curing at ambient temperatures, lower energy cost, and solvent-free formulations for curing. Thus, air and water pollution is eliminated [2,3]. It also devotes temporal and spatial control of the curing when high initiation rate is reached [4,148].

UV-radiation curing can be achieved by both cationic and free-radical photopolymerization modes. Although there is a wide range of monomers for the free radical UV curing, cationically polymerizable UV-curing formulations has been becoming more applicable due to its attractive advantages [2,3,7]. It overcomes volatile emissions, limitations due to molecular oxygen inhibition, toxicity, and problems associated with high viscosity. Furthermore, after curing, the unreacted monomers in the formulation can eventually undergo dark-curing processes in which they slowly polymerize.

It should be noted that a large amount of investigations associated with UV curing has been carried out in the recent decade. These investigations mainly concerned kinetic and mechanistic aspects of the curing processes as well as development of monomers (oligomers), photoinitiators, and photosensitizers. However, this section will focus on the recent developments on the new systems particularly those involved controlled functionalization of the polymers and oligomers.

One of the main important problems encountered in UV-curing application is the low reactivity of epoxy based monomers and resins. Thus, significant effords have been made to provide a solution by designing formulations with enhanced polymerization rates. For example, epoxidized soybean is used as a reactive additive in the formulations of various epoxy resins. Epoxidized soybean oil accelerates the photoinduced cross-linking of multifunctional epoxy resins or monomers by decreasing the viscosity and, in turn, increasing the diffusion rate of reactants [133]. Furthermore, the cross-linked polymers containing soybean oil moieties are tighter compared to networks without such moieties, which make them have resistance toward chemicals.

Epoxy resins based on bisphenol A are widely used in industrial application to achieve surface coating. Although they have several elegant properties including good adhesion, high mechanical, thermal, and electrical properties, they are very sensitive to UV irradiation due to its bisphenol A moiety. In a recent work, Penczek and coworkers prepared UV-resistant epoxypolyesters in two steps [149]. Firstly, they synthesized a series of solid polyester with a side chain containing a double bond, which can undergo epoxidation, by using 1,2,3,6-tetrahydrophtalic anhydride (THPA) and various difunctional alcohols (neopentyl glycol, 1,4-butanediol, ethylene glycol, and 4-cyclohexene-1,2-dimethanol. To increase cross-linking density, dicyclopentadiene was added to terminal carboxylic groups of the polyesters. Afterward, the obtained polyesters were epoxidized, and their films containing 0.5% of triarylsulfonium type photoinitiator were prepared and cured. The overall process is summarized in Scheme 11.44. Results of mechanical studies show that the cured films based on ester groups had fine properties meeting with the required standards.

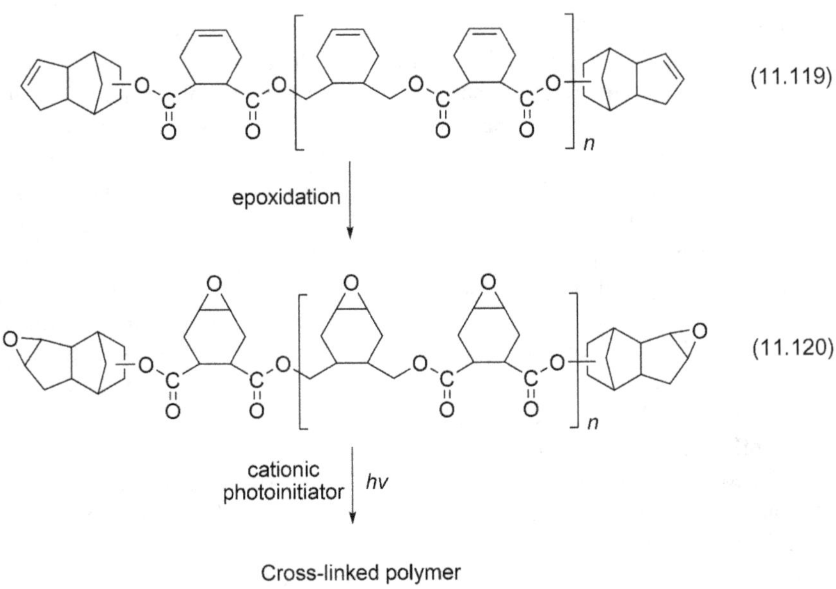

SCHEME 11.44

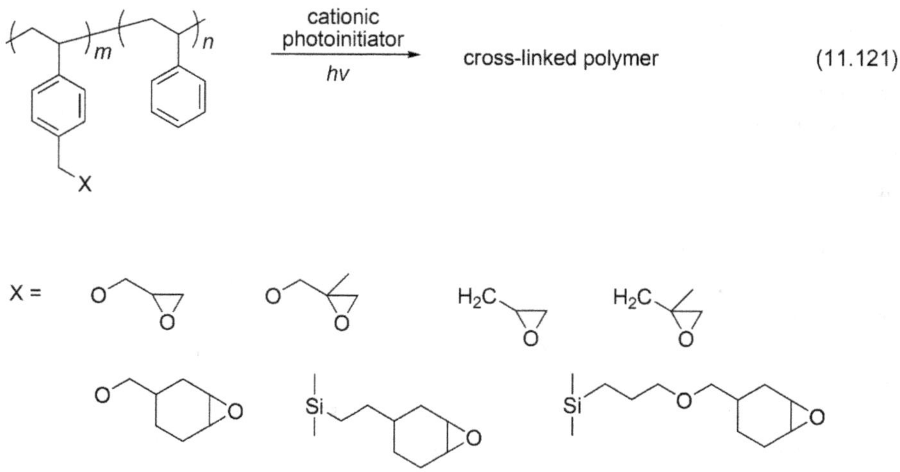

SCHEME 11.45

In another work, improvement of polymerization reactivity of epoxides was attempted [148]. Styrene monomers were modified with epoxide functions that are inert toward radicallic species, and then polymerized in the presence of azobisisobutyronitrile (AIBN). The polymers, ultimately, were cross-linked by sulfonium salt induced photopolymerization of the epoxide groups (Scheme 11.45) [148]. Kinetic results indicate that polymers containing silicon in their pendant group has the highest cross-linking reactivity.

Hybrid monomer strategy can also be used in UV-curing or cross-linking applications. Hydroxyl-terminated poly(3,4-epoxy-1-butene) (polyEPB) is an excellent polymer synthesized from a hybrid monomer consisting of epoxide and vinyl groups [150]. Kinetic studies indicated that two distinct mechanisms cause the acceleration of the polymerization; activated monomer mechanism and free radical induced chain decomposition of the onium salt. These two mechanisms, which are related to two important features of the polyEPB, are discussed in before. In the first mechanism, terminal hydroxyl groups participate in ring-opening polymerization of epoxides by means of the activated monomer mechanism (Scheme 11.29). In the second mechanism, the secondary and tertiary α-allylic ether protons of polyEPBs are readily abstracted by aryl radicals leading to production of relatively stable radicals that can induce decomposition of the onium salt photoinitiators (Scheme 11.46).

As discussed in a previous section, a number of studies have been conducted to increase the rate of cationic polymerization of epoxides. In curing applications, polymerization should be rapid enough for high output of production. In a recent work, the effect of addition of tetraethylene glycol (TEG) or polyEPB on the rate of photoinitiated cationic polymerization of CY179, limonene dioxide (LDO), and 1,2,7,8-diepoxyoctane (DEO) has been investigated [150]. These hydroxyl containing additives were shown to obviously accelerate the polymerization, increase the total epoxide conversion and decrease the induction period.

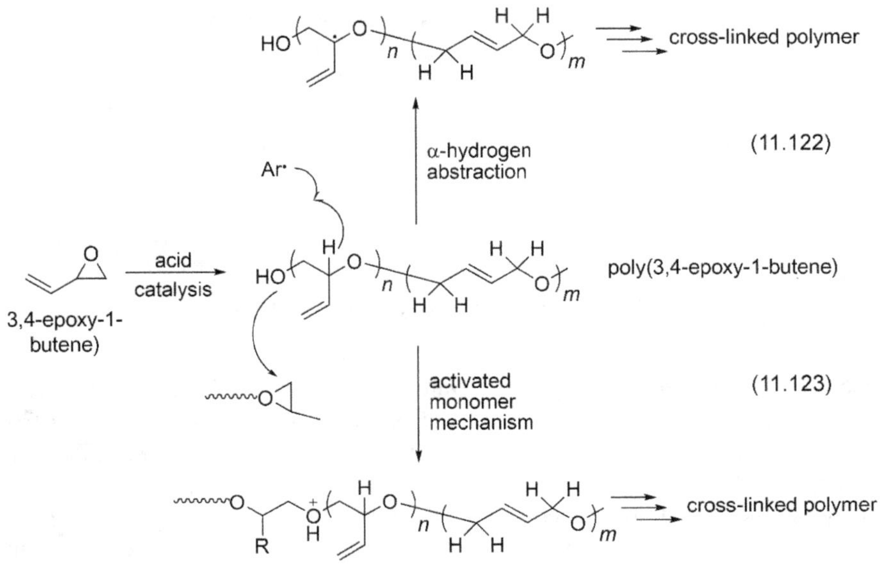

(11.122)

(11.123)

SCHEME 11.46

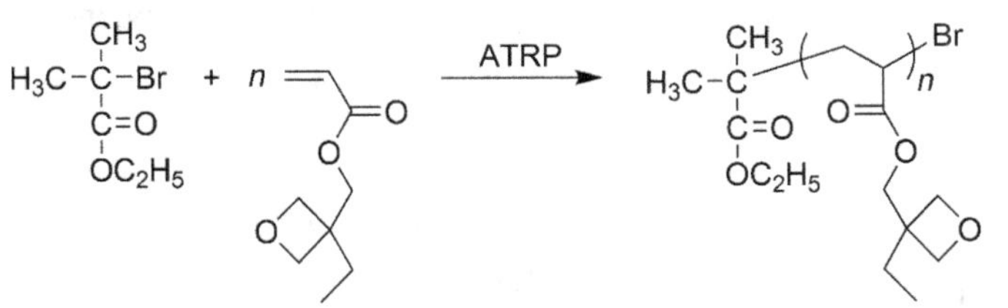

(11.124)

SCHEME 11.47

More recently, Schubert and coworkers has made tailor made macromonomers bearing oxetane moiety which is cross-linkable in the presence of cationic species [151]. Recently developed ATRP has been recognized as a useful method to polymerize monomers bearing different functionalities such as epoxides in the side chain that could find potential applications in coatings and adhesives. Scheme 11.47 illustrates the preperation of polymers with oxetanes as pendant groups that are preserved through ATRP. The resulting polymer is notably cross-linkable when small amount of cationic initiator is supplied.

11.7 PHOTOINITIATED CATIONIC FRONTAL POLYMERIZATION

In recent decades, production of uniform polymers and cross-linked networks has been achieved by frontal polymerization (FP). FP is a method in which a monomer is converted into a polymer via a localized reaction zone that propagates through the

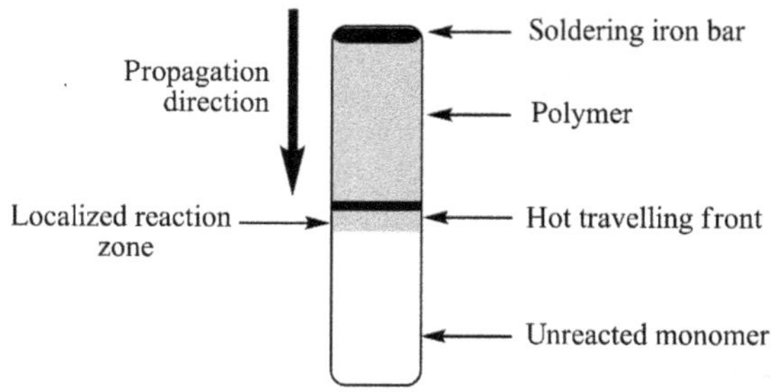

FIGURE 11.3 Schematic representation of frontal polymerization.

coupling of thermal diffusion and temperature-dependent reaction rates (Fig. 11.3). FP was first reported by two Russian scientists, Chechilo and Enikolopyan, for free radical polymerization in 1972 [152]. Pojman and coworkers have studied FP and its applications [153–156].

Once induced, additional energy is not necessary for propagation of polymerization due to highly exothermic nature of the polymerization. Therefore, many polymers can be quickly synthesized by using FP characterized by the localized reaction zone and the fast increasing temperature.

Figure 11.4 is an example for temperature profile of a FP. This figure illustrates how tricaprylmethylammonium persulfate initiated polymerization of tri(ethylene glycol)

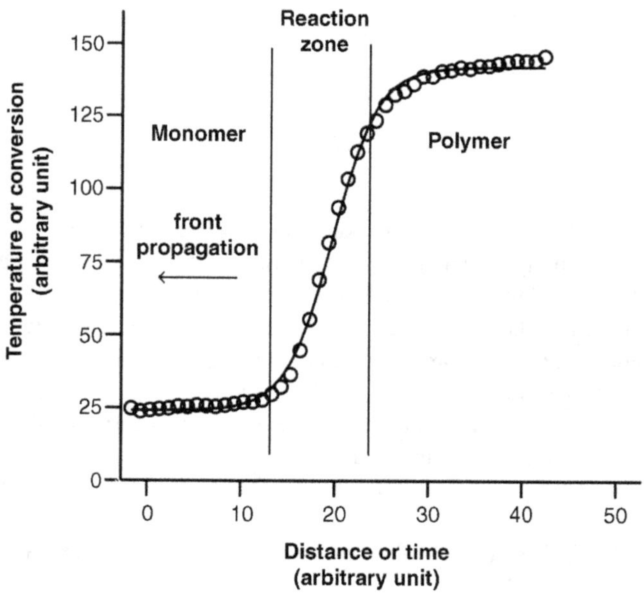

FIGURE 11.4 Temperature profile of the descending front in tri (ethylene glycol) dimethacrylate (TGDMA) and tricaprylmethylammonium persulfate as an initiator. Front velocity was 11.0 mm/min.

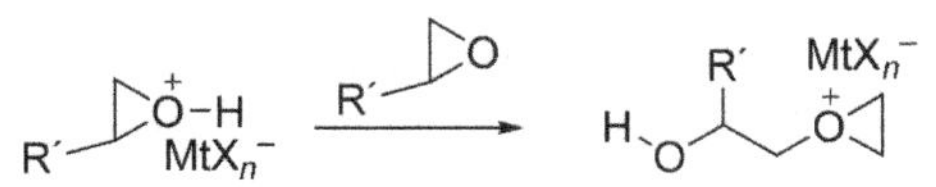

(11.125)

SCHEME 11.48

dimethacrylate proceeds [155]. Some potential applications of FPs can be summarized as the preparation of materials with functionally graded properties, thermosets with phase-separated liquid-crystalline domains, interpenetrating network polymers and blends, and the conservation and restoration of art objects [157].

In the literature, there are many studies on free radical FP. However, the corresponding cationic mode has been scarcely investigated [158]. More recently, photoinitiated cationic FPs of various oxirane and oxetanes has been reported [157]. This type of monomers show remarkable induction period, and polymerization occurs as a rapid front propagating throughout the entire reactants when an irradiated monomer sample is subjected to localized heat. Conversion profile of photopolymerization of 3-ethyl-3-[(phenoxy)-methyl] oxetane initiated by a diaryliodonium salt is also similar to the curve depicted in Fig. 11.4. An induction period is followed by the initiation stage. However, after standing for a certain period, if the sample is heated slightly, polymerization will occur in a very fast fashion. This fact implies that in ring-opening polymerizations, there is an energy requirement to overcome the activation energy necessary for the reaction to proceed (see Equation 11.126 in Scheme 11.48). However, once induced, the polymerization proceeds under autoacceleration conditions, and the monomer is converted into polymer within a very short period of time.

Results of studies on the FP of oxirane and oxetanes demonstrated that some monofunctional and difunctional 3,3-disubstituted oxetane, arylalkyl, and alkyl glycidyl ether monomers display frontal polymerization characteristics [157].

In conclusion, as demonstrated in this chapter, there has been a significant research effort on various aspects of photoiniated cationic polymerization in the past decade. These investigations mainly concerned with the development of specific photoinitiators with high quantum yield, adequate absorption characteristics, good solubility properties, and high reactivity toward some special monomer systems. With the possibilities of combining these needs at least partially, expectations are still running in this area. In addition, new inventions on the practical applications, as well as developments in traditional ones will strongly be demanded. As so, we can conclude that photoinitiated cationic polymerization occupies a significant research area and expects efforts that can bring out new discoveries for the benefit of human life.

REFERENCES

1. C. Decker, Kinetic study and new applications of UV radiation curing. *Macromol. Rapid Commun.* **2002**, *23*(18), 1067–1093.
2. Y. Yagci, Photoinitiated cationic polymerization of unconventional monomers. *Macromol. Symp.* **2006**, *240*(1), 93–101.

3. J.D. Cho and J.W. Hong, Curing kinetics of UV-initiated cationic photopolymerization of divinyl ether photosensitized by thioxanthone. *J. Appl. Polym. Sci.* **2005**, *97*(3), 1345–1351.
4. C. Decker, Light-induced crosslinking polymerization. *Polym. Int.* **2002**, *51*(11), 1141–1150.
5. J.P. Fouassier, *Photoinitiation, Photopolymerization and Photocuring: Fundamentals and Applications*. Hanser Publishers, New York, 1995.
6. Y. Yagci and M.K. Mishra, *Handbook of Radical Vinyl Polymerization*. Marcel Dekker, Inc., New York, 1998.
7. M. Sangermano, et al., Cationic photoinitiated copolymerization of 1-propenyl-vinyl ether systems. *Eur. Polym. J.* **2002**, *38*(4), 655–659.
8. Y. Yagci and I. Reetz, Externally stimulated initiator systems for cationic polymerization. *Prog. Polym. Sci.* **1998**, *23*(8), 1485–1538.
9. S. Kwon, H. Chun, and S. Mah, Photo-induced living cationic polymerization of isobutyl vinyl ether in the presence of various combinations of halides of diphenyliodonium and zinc salts in methylene chloride. *Fibers Polymers*, **2004**, *5*(4), 253–258.
10. S. Kwon, et al., Living cationic polymerization of isobutyl vinyl ether (II): Photoinduced living cationic polymerization in a mixed solvent of toluene and diethyl ether. *J. Appl. Polym. Sci.* **2006**, *101*(6), 3581–3586.
11. M.U. Kahveci, M.A. Tasdelen, and Y. Yagci, Photochemically initiated free radical promoted living cationic polymerization of isobutyl vinyl ether. *Polymer* **2007**, *48*(8), 2199–2202.
12. M.U. Kahveci, M.A. Tasdelen, and Y. Yagci, Photo-induced cross-linking of divinyl ethers by using diphenyliodonium salts with highly nucleophilic counter anions in the presence of zinc halides. *Macromol. Rapid Commun.* **2008**, *29*(3), 202–206.
13. Y. Yagci, A. Kornowski, and W. Schnabel, *N*-Alkoxy-pyridinium and *N*-alkoxy-quinolinium salts as initiators for cationic photopolymerizations. *J. Polym. Sci. A Polym. Chem.* **1992**, *30*(9), 1987–1991.
14. S. Denizligil, Y. Yagci, and C. McArdle, Photochemically and thermally-induced radical promoted cationic polymerization using an allylic sulfonium salt. *Polymer* **1995**, *36*(16), 3093–3098.
15. S. Denizligil, et al., Photosensitized cationic polymerization using allyl sulfonium salt. *Macromol. Chem. Phys.* **1996**, *197*(4), 1233–1240.
16. F. Kasapoglu, et al., Photoinitiated cationic polymerization using a novel phenacyl anilinium salt. *Polymer* **2002**, *43*(8), 2575–2579.
17. J.V. Crivello and S.Q. Kong, Synthesis and characterization of second-generation dialkylphenacylsulfonium salt photoinitiators. *Macromolecules* **2000**, *33*(3), 825–832.
18. J.V. Crivello and S.Q. Kong, Long-wavelength-absorbing dialkylphenacylsulfonium salt photoinitiators: synthesis and photoinduced cationic polymerization. *J. Polym. Sci. A Polym. Chem.* **2000**, *38*(9), 1433–1442.
19. G. Smets, A. Aerts, and J. Vanerum, Photochemical initiation of cationic polymerization and its kinetics. *Polym. J.* **1980**, *12*(9), 539–547.
20. J.V. Crivello and J.H.W. Lam, New photo-initiators for cationic polymerization. *J. Polym. Sci. C Polym. Symp.* **1976**, (56), 383–395.
21. J.V. Crivello and J.H.W. Lam, Diaryliodonium salts: new class of photo-initiators for cationic polymerization. *Macromolecules* **1977**, *10*(6), 1307–1315.

22. J.P. Fouassier, D. Burr, and J.V. Crivello, Photochemistry and photopolymerization activity of diaryliodonium salts. *J. Macromol. Sci. Pure Appl. Chem.* **1994**, *A31*(6), 677–701.
23. H.M. Pitt and Stauffer Chemical Co., **1955**, US Patent No. 2807647.
24. F. Hamazu, et al., Novel benzyl sulfonium salt having an aromatic group on sulfur atom as a latent thermal initiator. *J. Polym. Sci. A Polym. Chem.* **1991**, *29*(11), 1675–1680.
25. J.V. Crivello, in: K. Dietliker (Ed.), *Chemistry and Technology of UV and EB Formulation for Coatings, Inks and Paints*, Vol. 3, SITA Technology, London, 1991.
26. J.V. Crivello and J.H.W. Lam, Complex triarylsulfonium salt photoinitiators.1. The identification, characterization, and syntheses of a New class of triarylsulfonium salt photoinitiators. *J. Polym. Sci. A Polym. Chem.* **1980**, *18*(8), 2677–2695.
27. F. Hamazu, et al., Photopolymerization of epoxide by benzyl(*para*-hydroxyphenyl) methylsulfonium salts as novel photolatent initiators. *Makromol. Chem. Rapid Commun.* **1992**, *13*(4), 203–206.
28. E. Takahashi, F. Sanda, and T. Endo, Photocationic and radical polymerizations of epoxides and acrylates by novel sulfonium salts. *J. Polym. Sci. A Polym. Chem.* **2003**, *41* (23), 3816–3827.
29. J.V. Crivello and J.H.W. Lam, Photoinitiated cationic polymerization by dialkyl-4-hydroxyphenylsulfonium salts. *J. Polym. Sci. A Polym. Chem.* **1980**, *18*(3), 1021–1034.
30. J.V. Crivello and J. Ahn, Synthesis and characterization of second-generation *S,S*-dialkyl-*S*-(dimethylhydroxyphenyl)sulfonium salt photoinitiators. *J. Polym. Sci. A Polym. Chem.* **2003**, *41*(16), 2556–2569.
31. J.V. Crivello and J. Ahn, Photoinduced and thermally induced cationic polymerizations using *S,S*-dialkyl-*S*-(3,5 dimethylhydroxyphenyl)sulfonium salts. *J. Polym. Sci. A Polym. Chem.* **2003**, *41*(16), 2570–2587.
32. A. Kikkawa, T. Takata, and T. Endo, Cationic polymerization of vinyl monomers with latent initiators. 3. 1-(Para-methoxybenzyl)tetrahydrothiophenium hexafluoroantimonate as a new potent cationic initiator. *Makromol. Chem. Macromol. Chem. Phys.* **1991**, *192*(3), 655–662.
33. S.N. Gupta, L. Thijs, and D.C. Neckers, Divinylbenzophenone and poly(divinylbenzophenone): on the synthesis of rigid polymeric photosensitizers. *Macromolecules* **1980**, *13*(5), 1037–1041.
34. I.I. Abu-Abdoun and A. Aale, Cationic photopolymerization of cyclohexene oxide. *Eur. Polym. J.* **1992**, *28*(1), 73–78.
35. T. Takata, K. Takuma, and T. Endo, Photoinitiated cationic polymerization of epoxide with phosphonium salts as novel photolatent initiators. *Makromol. Chem. Rapid Commun.* **1993**, *14*(3), 203–206.
36. I.I. Abu-abdoun and A. Aale, Cationic photopolymerization of styrene by phosphonium and arsonium salts. *J. Macromol. Sci. Pure Appl. Chem.* **1993**, *A30*, 327–336.
37. D.C. Neckers and I.I. Abu-abdoun, Para-para'-bis((triphenylphosphonio)methyl) benzophenone salts as photoinitiators of free-radical and cationic polymerization. *Macromolecules* **1984**, *17*(12), 2468–2473.
38. K. Takuma, T. Takata, and T. Endo, Cationic polymerization of epoxide with benzyl phosphonium salts as the latent thermal initiator. *Macromolecules* **1993**, *26*(4), 862–863.
39. K. Takuma, T. Takata, and T. Endo, Latent cationic initiator: Photoinitiated polymerization of epoxides and vinyl monomers with phosphonium salts. *J. Photopolym. Sci. Technol.* **1993**, *6*, 67–74.

40. C. Reichard, Notiz zur darstellung von *n*-athoxy-pyridium-und-chinoliniumsalzen. *Chemische Berichte Recueil* **1966**, *99*(5), 1769.

41. J.V. Crivello and J.H.W. Lam, Photoinitiated cationic polymerization by dialkylphenacylsulfonium salts. *J. Polym. Sci. A Polym. Chem.* **1979**, *17*(9), 2877–2892.

42. J.V. Crivello and J.L. Lee, Structural and mechanistic studies on the photolysis of dialkylphenacylsulfonium salt cationic photoinitiators. *Macromolecules* **1983**, *16*(6), 864–870.

43. J.V. Crivello and S.Q. Kong, Photoinduced and thermally induced cationic polymerizations using dialkylphenacylsulfonium salts. *Macromolecules* **2000**, *33*(3), 833–842.

44. N. Yonet, N. Bicak, and Y. Yagci, Photoinitiated cationic polymerization of cyclohexene oxide by using phenacyl benzoylpyridinium salts. *Macromolecules* **2006**, *39*(8), 2736–2738.

45. E. Takahashi, F. Sanda, and T. Endo, *Photocationic and radical polymerizations by novel N-phenacylammonium salts*. Journal of Applied Polymer Science, **2004**. *91*(6): p. 3470–3476.

46. E. Takahashi, et al., Novel *N*-methylbenzothiazolium salts as hardeners for epoxy and acrylate monomers. *J. Polym. Sci. A Polym. Chem.* **2003**, *41*(23), 3828–3837.

47. J.V. Crivello and J.L. Lee, The synthesis and characterization of polymer-bound diaryliodonium salts and their use in photo and thermally initiated cationic polymerization. *Polym. Bull.* **1986**, *16*(4), 243–248.

48. H.S. Chao and General Electric Company, **1987**, European Patent Application No. 310882.

49. G. Hizal, Y. Yagci, and W. Schnabel, *N*-Alkoxy pyridinium ion terminated polytetrahydrofurans: synthesis and their use in photoinitiated block copolymerization. *Polymer* **1994**, *35*(20), 4443–4448.

50. G. Hizal, A. Sarman, and Y. Yagci, Synthesis of hydroxy-terminated polytetrahydrofuran by photoinduced process. *Polym. Bull.* **1995**, *35*(5), 567–573.

51. S. Denizligil, A. Baskan, and Y. Yagci, Bifunctional polytetrahydrofuran initiator for sequential photochemical and thermal initiation. *Macromol. Rapid Commun.* **1995**, *16* (5), 387–391.

52. Y.Y. Durmaz, G. Yilmaz, Y. Yagci, and N-Alkoxy pyridinium ion terminated polystyrenes: a facile route to photoinduced block copolymerization. *J. Polym. Sci. A Polym. Chem.* **2007**, *45*(3), 423–428.

53. Y.Y. Durmaz, G. Yimaz, and Y. Yagci, Polymers with side chain *N*-alkoxy pyridinium ions as precursors for photoinduced grafting and modification processes. *Macromol. Chem. Phys.* **2007**, *208*(16), 1737–1743.

54. K. Meier, N. Buhler, H. Zweifel, G. Berner and F. Lohse, **1982**, European Patent Application No. 093915.

55. V. Jakubek and A.J. Lees, Quantitative wavelength-dependent photochemistry of the [CpFe(eta(6)-ipb)]PF6 (ipb = isopropylbenzene) photoinitiator. *Inorg. Chem.* **2000**, *39*(25), 5779–5786.

56. J.F. Helling and W.A. Hendrickson, Pi-cyclohexadienyliron complexes bearing exocyclic double-bonds. *J. Organomet. Chem.* **1977**, *141*(1), 99–105.

57. A.N. Nesmeyanov, N.A. Volkenau, and I.N. Bolesova, The interaction of ferrocene and its derivatives with aromatic compounds. *Tetrahedron Lett.* **1963**, *25*, 1725–1729.

58. A. Roloff, K. Meier, and M. Riediker, Synthetic and metal organic-photochemistry in industry. *Pure Appl. Chem.* **1986**, *58*(9), 1267–1272.

59. T. Wang, et al., Several ferrocenium salts as efficient photoinitiators and thermal initiators for cationic epoxy polymerization. *J. Photochem. Photobiol. A Chem.* **2007**, *187*(2–3), 389–394.
60. T. Wang, P.Y. Wan, and L.J. Ma, Synthesis and characterization of alkoxy and phenoxy-substituted ferrocenium salt cationic photoinitiators. *Chinese J. Chem. Eng.* **2006**, *14*(6), 806–809.
61. T. Wang, B.S. Li, and L.X. Zhang, Carbazole-bound ferrocenium salt as an efficient cationic photoinitiator for epoxy polymerization. *Polym. Int.* **2005**, *54*(9), 1251–1255.
62. T. Wang and Z.H. Wang, Cationic photopolymerization of epoxy systems initiated by cyclopentadien-iron-biphenyl hexafluorophosphate ([Cp-Fe-biphenyl] + PF6-). *Polym. Bull.* **2005**, *53*(5–6), 323–331.
63. V. Andruleviciute, R. Lazauskaite, and J.V. Grazulevicius, Synthesis and photopolymerization of new triphenylamine-based oxiranes. *Des. Monom. Polym.* **2007**, *10*(2), 105–118.
64. V. Andruleviciute, et al., Synthesis and cationic polymerization of oxyranyl-functionalized triphenylamine. *Eur. Polym. J.* **2006**, *42*(5), 1069–1074.
65. F.M. Houlihan, et al., Nitrobenzyl ester chemistry for polymer processes involving chemical amplification. *Macromolecules* **1988**, *21*(7), 2001–2006.
66. T.X. Neenan, et al., Photochemistry and thermochemistry of select 2,6-dinitrobenzyl esters in polymer matrices: studies pertaining to chemical amplification and imaging. *Macromolecules* **1990**, *23*(1), 145–150.
67. D. Ruhlmann and J.P. Fouassier, Structure–property relationships in photoinitiators of polymerization. 8. Sulfonyl ketone derivatives. *Eur. Polym. J.* **1993**, *29*(8), 1079–1088.
68. M. Tsunooka, S. Tanaka, and M. Tanaka, Photocrosslinking of poly(2,3-epoxypropyl methacrylate) films with organic sulfur-compounds. *Makromol. Chem. Rapid Commun.* **1983**, *4*(8), 539–541.
69. M.K. Gupta and R.P. Singh, Diphenyldiselenide as novel non-salt photoinitiator for photosensitized cationic polymerization of *N*-vinyl carbazole. *Macromol. Symp.* **2006**, *240*, 186–193.
70. S. Hayase, et al., Polymerization of cyclohexene oxide with al(acac)3-silanol catalyst. *J. Polym. Sci. A Polym. Chem.* **1981**, *19*(9), 2185–2194.
71. S. Hayase, et al., Photopolymerization of epoxides: a new type of photopolymerization with photodecomposable silyl ether as coinitiator. *Macromolecules* **1985**, *18*(10), 1799–1804.
72. S. Hayase, et al., Polymerization of cyclohexene oxide with aluminum complex silanol catalysts. 5. A catalytic activity dependence on the aluminum chelate structure. *J. Polym. Sci. A Polym. Chem.* **1982**, *20*(11), 3155–3165.
73. H. Rudolph, H.J. Rosenkranz, and H.G. Heine, *Appl. Polym. Symp.* **1975**, *26*, 157.
74. E. Reichmanis, B.C. Smith, and R. Gooden, Ortho-nitrobenzyl photochemistry: solution vs solid-state behavior. *J. Polym. Sci. A Polym. Chem.* **1985**, *23*(1), 1–8.
75. J.P. Fouassier, and D. Burr, Triplet-state reactivity of alpha-sulfonyloxy ketones used as polymerization photoinitiators. *Macromolecules* **1990**, *23*(15), 3615–3619.
76. M.G. Mikhael, A.B. Padias, and H.K. Hall, Synthesis and investigation of *m,m′*-dimethoxybenzyl esters as novel organic soluble cationic photoinitiators. *Macromolecules* **1995**, *28*(17), 5951–5954.

77. J.V. Crivello and M. Sangermano, Visible and long-wavelength photoinitiated cationic polymerization. *J. Polym. Sci. A Polym. Chem.* **2001**, *39*(3), 343–356.
78. J.V. Crivello, and U. Bulut, Indian turmeric and its use in cationic photopolymerizations. *Macromol. Symp.* **2006**, *240*, 1–11.
79. Y. Yagci, I. Lukac, and W. Schnabel, Photosensitized cationic polymerization using *n*-ethoxy-2-methylpyridinium hexafluorophosphate. *Polymer* **1993**, *34*(6), 1130–1133.
80. D. Dossow, et al., Photosensitized cationic polymerization of cyclohexene oxide: a mechanistic study concerning the use of pyridinium-type salts. *Polymer* **1996**, *37*(13), 2821–2826.
81. J.V. Crivello and J.L. Lee, Photosensitized cationic polymerizations using dialkylphenacylsulfonium and dialkyl(4-hydroxyphenyl)sulfonium salt photoinitiators. *Macromolecules* **1981**, *14*(5), 1141–1147.
82. F.A.M. Abdulrasoul, A. Ledwith, and Y. Yagci, Photo-chemical and thermal cationic polymerizations. *Polymer* **1978**, *19*(10), 1219–1222.
83. Y. Yagci and W. Schnabel, New aspects on the photoinitiated free-radical promoted cationic polymerization. *Makromol. Chem. Macromol. Symp.* **1992**, *60*, 133–143.
84. Y. Yagci and A. Ledwith, Mechanistic and kinetic-studies on the photoinitiated polymerization of tetrahydrofuran. *J. Polym. Sci. A Polym. Chem.* **1988**, *26*(7), 1911–1918.
85. G. Hizal, Y. Yagci, and W. Schnabel, Charge-transfer complexes of pyridinium ions and methyl-substituted and methoxy-substituted benzenes as photoinitiators for the cationic polymerization of cyclohexene oxide and related-compounds. *Polymer* **1994**, *35*(11), 2428–2431.
86. G. Hizal, S.E. Emiroglu, and Y. Yagci, Photoinitiated radical polymerization using charge transfer complex of *N*-ethoxy-*p*-cyano-pyridinium salt and 1,2,4-trimethoxybenzene. *Polym. Int.* **1998**, *47*(4), 391–392.
87. I. Reetz, V. Bacak, and Y. Yagci, Addition-fragmentation reactions for cationic polymerization using a novel allyloxy-picolinium salt. *Polym. Int.* **1997**, *43*(1), 27–32.
88. V. Bacak, et al., Addition-fragmentation type initiation of cationic polymerization using allyloxy-pyridinium salts. *Polym. Int.* **1998**, *47*(3), 345–350.
89. Y. Yagci and I. Reetz, Addition: fragmentation type initiators for cationic polymerization. *Macromol. Symp.* **1998**, *132*, 153–164.
90. Y. Yagci and I. Reetz, Addition-fragmentation reactions in polymer chemistry. *React. Funct. Polym.* **1999**, *42*(3), 255–264.
91. J.L. Dektar and N.P. Hacker, Photochemistry of diaryliodonium salts. *J. Org. Chem.* **1990**, *55*(2), 639–647.
92. J.L. Dektar and N.P. Hacker, Comparison of the photochemistry of diarylchloronium, diarylbromonium, and diaryliodonium salts. *J. Org. Chem.* **1991**, *56*(5), 1838–1844.
93. J.L. Dektar and N.P. Hacker, Triphenylsulfonium salt photochemistry: new evidence for triplet excited-state reactions. *J. Org. Chem.* **1988**, *53*(8), 1833–1835.
94. R.J. Devoe, et al., Transient intermediates in the photolysis of iodonium cations. *Can. J. Chem. Revue Canadienne De Chimie* **1987**, *65*(10). 2342–2349.
95. H. B. Sun and S. Kawata, Two-photon photopolymerization and 3D lithographic microfabrication, in: A. Abe, A.-C. Albertsson, R. Duncan et al. (Eds.), *NMR, 3D Analysis, Photopolymerization*. pp. 169–273, Springer Berlin, Heidelberg, 2004.

96. K.D. Belfield and K.J. Schafer, Two-photon photoinitiated polymerization, in: K.D. Belfield and J. V. Crivello (Eds.), *Photoinitiated Polymerization, ACS Symposium Series 847*, pp. 464–481, American Chemical Society, 2003.

97. K.D. Belfield, et al., Near-IR two-photon photoinitiated polymerization using a fluorone/amine initiating system. *J. Am. Chem. Soc.* **2000**, *122*(6), 1217–1218.

98. S. Maruo, O. Nakamura, and S. Kawata, Three-dimensional microfabrication with two-photon-absorbed photopolymerization. *Opt. Lett.* **1997**, *22*(2), 132–134.

99. S. Maruo and K. Ikuta, *Proc. Soc. Photo Opt. Instrument. Eng.* **2000**, *3937*, 106–112.

100. S. Kawata, et al., Finer features for functional microdevices: micromachines can be created with higher resolution using two-photon absorption. *Nature* **2001**, *412*(6848), 697–698.

101. B.H. Cumpston, et al., Two-photon polymerization initiators for three-dimensional optical data storage and microfabrication. *Nature* **1999**, *398*(6722), 51–54.

102. H.E. Pudavar, et al., High-density three-dimensional optical data storage in a stacked compact disk format with two-photon writing and single photon readout. *Appl. Phys. Lett.* **1999**, *74*(9), 1338–1340.

103. W. Denk, 2-Photon scanning photochemical microscopy: mapping ligand-gated ion-channel distributions. *Proc. Natl. Acad. Sci. USA* **1994**, *91*(14), 6629–6633.

104. R.M. Williams and D.W. Piston, W.W. Webb, 2-Photon molecular-excitation provides intrinsic 3-dimensional resolution for laser-based microscopy and microphotochemistry. *FASEB J.* **1994**, *8*(11), 804–813.

105. M.A. Tasdelen, et al., Photoacid generation by stepwise two-photon absorption: photoinitiated cationic polymerization of cyclohexene oxide by using benzodioxinone in the presence of iodonium salt. *Macromolecules* **2008**, *41*(2), 295–297.

106. Y.J. Hua, F.M. Jiang, and J.V. Crivello, Photosensitized onium-salt-induced cationic polymerization with hydroxymethylated polynuclear aromatic hydrocarbons. *Chem. Mater.* **2002**, *14*(5), 2369–2377.

107. J.V. Crivello and J.H.W. Lam, Dye-sensitized photoinitiated cationic polymerization – system–perylene-triarylsulfonium salts. *J. Polym. Sci. A Polym. Chem.* **1979**, *17*(4), 1059–1065.

108. J.V. Crivello and R.A. Ortiz, Benzyl alcohols as accelerators in the photoinitiated cationic polymerization of epoxide monomers. *J Polym. Sci. A Polym. Chem.* **2002**, *40*(14), 2298–2309.

109. K. Brzezinska, et al., Activated monomer mechanism in cationic polymerization: ethylene-oxide, formulation of mechanism. *Makromol. Chem. Rapid Commun.* **1986**, *7*(1), 1–4.

110. S. Penczek, P. Kubisa, and R. Szymanski, Activated monomer propagation in cationic polymerizations. *Makromol. Chem. Macromol. Symp.* **1986**, *3*, 203–220.

111. M. Wojtania, P. Kubisa, and S. Penczek, Polymerization of propylene-oxide by activated monomer mechanism: suppression of macrocyclics formation. *Makromol. Chem. Macromol. Symp.* **1986**, *6*, 201–206.

112. P. Kubisa, Activated monomer mechanism in the cationic polymerization of cyclic ethers. *Makromol. Chem. Macromol. Symp.* **1988**, *13*(4), 203–210.

113. Z. Gomurashvili, Y.J. Hua, and J.V. Crivello, Monomeric and polymeric carbazole photosensitizers for photoinitiated cationic polymerization. *Macromol. Chem. Phys.* **2001**, *202*(11), 2133–2141.

114. Y. Yagci and W. Schnabel, Direct and sensitized photoinitiated cationic polymerization using pyridinium salts. *Macromol. Symp.* **1994**, *85*, 115–127.

115. G. Manivannan and J.P. Fouassier, Primary processes in the photosensitized polymerization of cationic monomers. *J. Polym. Sci. A Polym. Chem.* **1991**, *29*(8), 1113–1124.

116. F. Kasapoglu and Y. Yagci, Photo sensitized cationic polymerization of cyclohexene oxide using a phenacylanilinium salt. *Macromol. Rapid Commun* **2002**, *23*(9), 567–570.

117. J.V. Crivello and K.K. Dietliker, Photoinitiators for free radical and cationic polymerisation, in: G. Bradley, (Ed.), *Chemistry and Technology of UV and EB Formulation for Coatings, Inks and Paints*, 2nd edition, Vol. 3, SITA Technology Ltd., London, 1998.

118. F. Kasapoglu, et al., Photoinitiated cationic polymerization of monofunctional benzoxazine. *J. Polym. Sci. A Polym. Chem.* **2003**, *41*(21), 3320–3328.

119. J.V. Crivello, et al., Advances in the design of photoinitiators, photo-sensitizers and monomers for photoinitiated cationic polymerization. *Macromol. Symp.* **2004**, *215*, 165–177.

120. S.K. Rajaraman, W.A. Mowers, and J.V. Crivello, Novel hybrid monomers bearing cycloaliphatic epoxy and 1-propenyl ether groups. *Macromol.* **1999**, *32*(1), 36–47.

121. J.V. Crivello, et al., Free radical accelerated cationic polymerizations. *Macromol. Symp.* **2000**, *157*, 109–119.

122. P. Kubisa, Hyperbranched polyethers by ring-opening polymerization: contribution of activated monomer mechanism. *J. Polym. Sci. A Polym. Chem.* **2003**, *41*(4), 457–468.

123. J.H. Ge, M. Trujillo-Lemon, and J.W. Stansbury, A mechanistic and kinetic study of the photoinitiated cationic double ring-opening polymerization of 2-methylene-7-phenyl-1,4,6,9-tetraoxa-spiro[4.4]nonane. *Macromolecules* **2006**, *39*(26), 8968–8976.

124. M.D. Miller, et al., Matrix-assisted laser desorption/ionization time-of-flight analysis of the copolymerization reaction of an expanding monomer with a diepoxide. *J. Polym. Sci. A Polym. Chem.* **2005**, *43*(23), 5962–5970.

125. D. Nagai, et al., Non-shrinking networked materials from the cross-linking copolymerization of spiroorthocarbonate with bifunctional oxetane. *Macromol. Rapid Commun.* **2006**, *27*(12), 921–925.

126. K.D. Belfield and F.B. Abdelrazzaq, Photoinitiated cationic cross-linking of 4-methylene-2-phenyl-1,3-dioxolane with 2,2′-(1,4-Phenylene)bis(4-methylene-1,3-dioxolane). *Macromolecules* **1997**, *30*(22), 6985–6988.

127. S. Minegishi, et al., Novel photocurable monomers: the synthesis of difunctional vinyl ethers with a phosphonate group and difunctional 1-propenyl ethers with a phosphonate group and their photoinitiated cationic polymerization. *J. Polym. Sci. A Polym. Chem.* **2005**, *43*(14), 3105–3115.

128. T. Nishikubo, et al., Soluble polymer-supported catalysts containing pendant quaternary onium salt residues for regioselective addition reaction of epoxy compound with active ester. *Macromolecules* **1994**, *27*(25), 7240–7247.

129. K. Kampke-Thiel, et al., Isolation, characterization, and toxicological aspects of volatile organophosphorus compounds from the combustion of flame-retarded epoxy resins with phosphonate substructures. *Chem. Eur. J.* **1998**, *4*(8), 1581–1586.

130. A.A. Al-Doaiss, et al., Photoinitiated cationic polymerization of substituted vinylcyclopropanes. *Macromol. Chem. Phys.* **2005**, *206*(23), 2348–2353.

131. V. James and S.L. Crivello, Photoinitiated cationic polymerization of epoxy alcohol monomers. *J. Polym. Sci. A Polym. Chem.* **2000**, *38*(3), 389–401.

132. M. Sangermano, M.A. Tasdelen, and Y. Yagci, Photoinitiated curing of mono- and bifunctional Epoxides by combination of active chain end and activated monomer cationic polymerization methods. *J. Polym. Sci. A Polym. Chem.* **2007**, *45*(21), 4914–4920.

133. C. Decker, T.T.N. Viet, and H.P. Thi, Photoinitiated cationic polymerization of epoxides. *Polym. Int.* **2001**, *50*(9), 986–997.

134. E.J. Goethals, et al., Synthesis of polymer networks containing degradable polyacetal segments. *Abstr. Pap. Am. Chem. Soc.* **1993**, *205*, 457-POLY.

135. S.K. Rajaraman, W.A. Mowers, and J.V. Crivello, Interaction of epoxy and vinyl ethers during photoinitiated cationic polymerization. *J. Polym. Sci. A Polym. Chem.* **1999**, *37*(21), 4007–4018.

136. J.V. Crivello and R.A. Ortiz, Design and synthesis of highly reactive photopolymerizable epoxy monomers. *J. Polym. Sci. A Polym. Chem.* **2001**, *39*(14), 2385–2395.

137. K. Matyjaszewski and A.H.E. Muller, 50 years of living polymerization. *Prog. Polym. Sci.* **2006**. *31*(12): p. 1039–1040.

138. W.A. Braunecker and K. Matyjaszewski, Controlled/living radical polymerization: Features, developments, and perspectives. *Prog. Polym. Sci.* **2007**, *32*(1), 93–146.

139. R. Faust and J.P. Kennedy, Living carbocationic polymerization. 4. Living polymerization of isobutylene. *J. Polym. Sci. A Polym. Chem.* **1987**, *25*(7), 1847–1869.

140. M. Kamigaito, M. Sawamoto, and T. Higashimura, Living cationic polymerization of vinyl ethers by electophile lewis acid initiating systems. 6. Living cationic polymerization of isobutyl vinyl ether by RCOOH/lewis acid initiating systems: effects of carboxylate ions and lewis acid activators. *Macromolecules* **1991**, *24*(14), 3988–3992.

141. M. Miyamoto, M. Sawamoto, and T. Higashimura, Living polymerization of isobutyl vinyl ether with hydrogen iodide/iodine initiating system. *Macromolecules* **1984**, *17*(3), 265–268.

142. M. Kamigaito, M. Sawamoto, and T. Higashimura, Living cationic polymerization of isobutyl vinyl ether by protonic acid/zinc halide initiating systems: evidence for the halogen exchange with zinc halide in the growing species. *Macromolecules* **1992**, *25*(10), 2587–2591.

143. M. Kamigaito, et al., Living cationic polymerization of isobutyl vinyl ether by benzoic acid derivatives/zinc chloride initiating systems: slow interconversion between dormant and activated growing species. *Macromolecules* **1992**, *25*(24), 6400–6406.

144. T. Gebel, Arsenic and antimony: comparative approach on mechanistic toxicology. *Chem. Biol. Interact.* **1997**, *107*(3), 131–144.

145. G.L. Duan, et al., Characterization of arsenate reductase in the extract of roots and fronds of Chinese brake fern, an arsenic hyperaccumulator. *Plant Physiol.* **2005**, *138*(1), 461–469.

146. V. Percec and D. Tomazos, Liquid-crystalline copoly(vinylether)s containing 4(4′)-methoxy-4′(4)-hydroxy-alpha-methylstilbene constitutional isomers as side groups. *Polym. Bull.* **1987**, *18*(3), 239–246.

147. B. Falk, M.R. Zonca, and J.V. Crivello, Modification of photoinitiated cationic epoxide polymerizations by sulfides. *J. Polym. Sci. A Polym. Chem.* **2005**, *43*(12), 2504–2519.

148. T. Oyama, et al., Photo-crosslinking of polystyrenes having pendant epoxy groups. *React. Funct. Polym.* **2001**, *49*(2), 99–116.

149. P. Penczek, Z.B. Tomaszewski, and A. Bankowska, Epoxypolyesters as film-forming materials. *Macromol. Symp.* **2002**, *187*(1), 243–248.

150. M. Sangermano, S.N. Falling, and J.V. Crivello, Photoinitiated cationic polymerization of epoxy monomers in the presence of poly(3,4-epoxy-1-butene). *J. Macromol. Sci. Pure Appl. Chem.* **2002**, *A39*(11), 1279–1294.

151. N.K. Singha, B. de Ruiter, and U.S. Schubert, Atom transfer radical polymerization of 3-ethyl-3-(acryloyloxy)methyloxetane. *Macromolecules* **2005**, *38*(9), 3596–3600.

152. N.M. Chechilo, N.S. Enikolopyan, and R.Y. Khvilivitskii, Phenomenon of polymerization reaction spreading. *Dokl. Akad. Nauk SSSR* **1972**, *204*(5), 1180.

153. T. Hu, et al., Frontal free-radical copolymerization of urethane-acrylates. *J. Polym. Sci. A Polym. Chem.* **2006**, *44*(9), 3018–3024.

154. Y. Chekanov, et al., Frontal curing of epoxy resins: comparison of mechanical and thermal properties to batch-cured materials. *J. Appl. Polym. Sci.* **1997**, *66*(6), 1209–1216.

155. Y.A. Chekanov and J.A. Pojman, Preparation of functionally gradient materials via frontal polymerization. *J. Appl. Polym. Sci.* **2000**, *78*(13), 2398–2404.

156. J.A. Pojman, G. Curtis, and V.M. Ilyashenko, Frontal polymerization in solution. *J. Am. Chem. Soc.* **1996**, *118*(15), 3783–3784.

157. J.V. Crivello, R. Falk, and M.R. Zonca, Photoinduced cationic ring-opening frontal polymerizations of oxetanes and oxiranes. *J. Polym. Sci. A Polym. Chemi.* **2004**. *42*(7), 1630–1646.

158. J.A. Pojman, et al., Binary frontal polymerization: a new method to produce simultaneous interpenetrating polymer networks (SINs). *J. Polym. Sci. A Polym. Chem.* **1997**, *35*(2), 227–230.

12

PHOTOIMAGING AND LITHOGRAPHIC PROCESSES IN POLYMERS

MARIUS GABRIEL IVAN AND JUAN CESAR (TITO) SCAIANO

Computers have had a tremendous influence in our society, and they are now present in our everyday life. Almost any electronic equipment uses processors or memory to perform duties, from printers to automobiles, game consoles to medicinal equipments, digital photo cameras, cellular phones, or even children toys. The performance

Photochemistry and Photophysics of Polymer Materials, Edited by Norman S. Allen

of the computer chips has increased dramatically over the past few decades, at the same time with the reduction in size of the integrated circuits, roughly following Moore's law [1]. Gordon Moore, one of Intel's cofounders, predicted in a paper in 1965 that the size of the features inside a chip should shrink by 50% every 18 months in order for the microelectronics industry to stay competitive [2]. And indeed, for more than three decades the trend was to continuously decrease the size, from tens of micrometers to tens of nanometers nowadays, with the target of reaching molecular sizes over the next decade [1,3,4]. At the same time with the decrease in feature size the performance increased and the price fell.

Nanolithography is a technology used to manufacture computer chips, dynamic random access (DRAM) memories, optical sensors for digital cameras (CCD, CMOS image sensors), diode arrays, microelectro-mechanical systems (MEMS), photonics, transistors for LCD and OLED displays, silicon substrates for photovoltaic cells, in an ever-growing list. Lithographic printing is done on various semiconducting substrates, usually silicon, and it consists of transferring the pattern of a mask onto a film of resist, and further to the substrate using various types of radiation. Lithographic techniques may be classified depending on the type of radiation used: photolithography, X-ray lithography, ion beam lithography, electron beam lithography, and extreme ultraviolet lithography (EUV) [3–9]. The most widely employed technology by the microelectronics industry to manufacture computer chips is photolithography or optical lithography [1,3–5]. New lithography techniques are continually developed, the so-called soft lithographic methods being a serious alternative to photolithography: nanoimprint lithography (NIL), dip-pen lithography, step-and-flash imprint lithography (SFIL), screen printing, and so forth. Because the overwhelming majority of ICs are obtained through photolithography, this chapter deals with photochemistry of materials used in near-UV (250–400 nm), deep UV (DUV, 190–250 nm), vacuum UV (VUV, <190 nm), and extreme UV (EUV, 13.4 nm) photolithography.

In optical lithography, the pattern of a mask is transferred to a film of photoresist coated on top of a metal or oxide film deposited on top of a silicon wafer, resulting in the formation of a latent image (Fig. 12.1). The exposed regions suffer photochemically induced changes, which alter their solubility compared with the unexposed regions. These transformations usually occur at elevated temperatures (90–110 °C), during baking on a hot plate, a process called post-exposure baking (PEB). During the development step, the latent image is revealed employing a proper developer, usually an aqueous base solution; the exposed regions may be removed, obtaining a positive image; when the unexposed regions are removed by the developer, the image obtained is a negative image. The next step is etching, when the exposed areas of the metal, oxide, or silicon wafer are chemically etched (wet etch or dry etch under plasma), while the areas under the photoresist film are protected against etching, followed by stripping when the remaining photoresist is completely removed. The result is a silicon wafer whose surface contains lines and spaces on the nanometer scale. Further, the spaces are filled with a conductor, usually copper, obtaining the complimentary metal-oxide semiconductor (CMOS).

The photoresist must contain an optically active compound, which upon absorption of a photon undergoes photolysis, affecting the solubility of the exposed

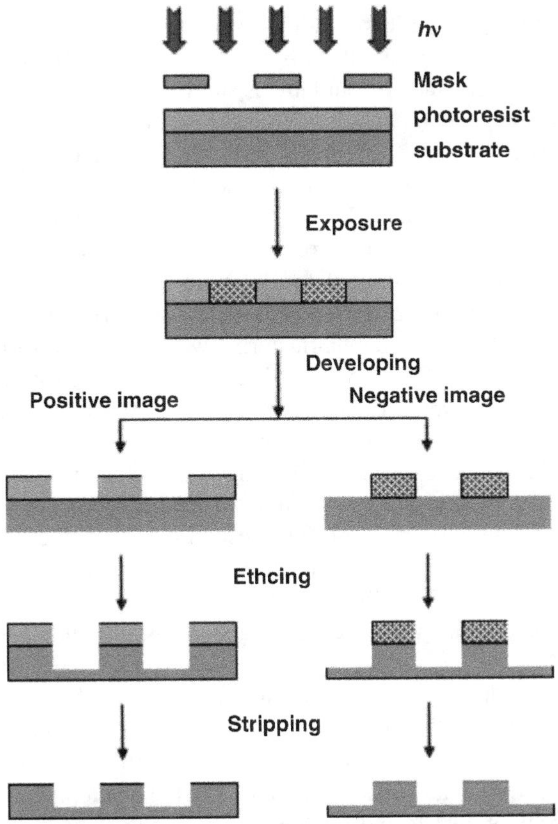

FIGURE 12.1 Photolithographic steps: exposure to light from a lamp or laser, developing with an aqueous base, etching (with HF or plasma), and stripping or removing of remaining photoresist, revealing alternating lines and spaces.

areas compared with the unexposed ones. As the feature size continues to shrink, photoresists must have certain stringent properties, including ability to form defect-free thin films, high contrast, high resolution and sensitivity, adhesion to the substrate, thermal stability, proper absorbance at the exposure wavelength, and slow diffusion. Modern resist films deposited on top of the wafers have thickness of a few hundred of nanometers, thus etch resistance is another important property. The films must withstand harsh etching conditions, therefore, bilayer and multilayer resists have been developed.

The exposure of the coated wafers is done through a complicated optical system. The size of the features is given by the Rayleigh equation, and is directly proportional to the exposure wavelength (λ) and inversely proportional to the numerical aperture (NA), while k_1 is a constant that depends on the equipment used (with values between 0.4 and 1.0)

$$R = k_1\lambda/\mathrm{NA}$$

From Rayleigh's equation, it is clear the size of the features may be decreased by decreasing the exposure wavelength or by increasing the numerical aperture. Both these strategies have been tackled, and this is shown in the decrease of the exposure wavelength over time, accompanied by an increase in NA. NA is a property of the optics used to guide the light from the source onto the resist film.

The first photoresists were designed during the 1970s for exposure to radiation from an Hg lamp (G-line, 436 nm and I-line, 365 nm). I-line resists are still used today in manufacturing the transistors that control the pixels in LCDs and OLED displays. The industry moved toward deep ultraviolet (DUV) during the 1990s, and used excimer lasers as sources of radiation at 248 nm (KrF) and 193 nm (ArF). Following 193 nm, the next expected wavelength used in manufacturing was 157 nm. Absorbance of organic and inorganic materials at 157 nm is very high, forcing the photoresists manufacturer to find solutions to produce more transparent polymers. At the same time, as oxygen absorbs at 157 nm, the exposure should be done under vacuum or under nitrogen. One important obstacle toward exposure at 157 nm was the birefringence of CaF_2, the material used to manufacture the optical parts. All these aspects determined industry to give up 157 nm around 2003, and focus on resolution enhancement techniques that allow extension of photolithography at 193 nm to even smaller features. The sizes currently obtained with 193 nm immersion lithography are 45 nm and 32 nm, with 22 nm half pitch seen as the next potential target for 193 nm immersion coupled with double exposure.

According to IC industry's roadmap, the next generation lithography will probably use exposure at 13.4 nm, in the extreme ultraviolet (EUV), for 22 nm feature size. Due to the huge absorbance of any material at this wavelength, the optics will consist of reflective mirrors, without any lenses involved. The big challenges at 13.4 nm are the new materials for photoresists and the low output of the EUV light sources.

The photoresist is sensitive to the incident radiation and it undergoes (photo) chemical transformations. Photoresists are complex formulations consisting of organic solvent, polymer, photoactive compound (PAC), base, and other chemicals that confer it desired properties. In one-component resists, the polymer is the photoactive compound, whereas in two- or multicomponent resists the photoactive compound undergoes photochemical transformation resulting in new species that interact with the radiation inert polymer, triggering transformations that alter the solubility in the exposed areas.

At the beginning of 1990s, X-ray and E-beam lithography techniques were announced by different groups and authors as the candidates to manufacture IC starting around 1998, but these predictions did not become reality [4,9]. Improvements in optics and resists performance saw the introduction of 248 nm lithography in the second part of the 1990s, and later of 193 nm around the year 2000. Thus, the use of E-beam and X-rays as incident radiation was delayed, according to the forecasts, until around 2010. However, at the present time, the wavelength of exposure is 193 nm, with predictions that it will be used further for 22 nm node [10]. Thus X-ray and E-beam are still not being used for mass production, but for niche applications such as manufacturing of photomasks and templates for imprint lithography. It appears that the technology of choice for 22 nm node will be EUV, but skeptics suggest that this is a costly proposition, and it could be a single node technology, something that the IC industry is not pleased with. Advances in optics, coupled with immersion lithography in high refractive indices liquids, double exposure, or double patterning are the resolution enhancement techniques that will probably allow 22 nm lithography using 193 nm as exposure wavelength [10].

The IC industry used a resist consisting of novolak and diazonaphthoquinone (DNQ) during the 1970s and 1980s for computer chips and DRAM manufacturing. As the feature size became smaller, so did the wavelengths used for exposure. However, in DUV (248 and 193 nm) the number of incident photons per unit area of exposed films is smaller than at 436 and 365 nm, so resists a few orders of magnitude more sensitive than DNQ/novolak resist were needed. The breakthrough came in 1982, when Ito, Willson, and Fréchet proposed a new photoresist, which made use of the chemical amplification (CA) concept [1,6,9,11].

12.1 PHOTOLITHOGRAPHY WITH NONAMPLIFIED RESISTS

During the 1970s and 1980s, the common exposure wavelengths were 436 and 365 nm using the G- and I-lines from an Hg lamp [4]. Both negative- and positive-tone photoresists were employed.

Most of the ICs were obtained making use of a resist formulation discovered by Süss, namely, diazonaphthoquinone as PAC and novolak resin (Chart 12.1) as polymer matrix [12]. The resolution achieved with this resist formulation was smaller than 500 nm [13]. Novolak is obtained through a polycondensation reaction between formaldehyde and cresols [4,14]. The novolak resin is photochemically inert at 436 and 365 nm, and is easily soluble in basic developers due to its phenolic OH groups, but upon addition of naphthoquinone the dissolution rate decreases dramatically [15,16].

The inhibition of dissolution in novolak is particularly interesting. Films of PMMA, for example, containing 10% dissolution inhibitor show a dissolution rate only 10–20% lower than the films without inhibitors. However, addition of 10% DNQ to novolak films slows down the dissolution rate by approximately 20 times [15,17]. After the novolak films containing DNQ are exposed to UV radiation, their solubility increases by a few orders of magnitude, the dissolution rate becoming even higher than in the case of novolak films without any dissolution inhibitor. The explanation of this phenomenon was sought in the photochemistry of DNQ, but no clear picture was offered until relatively recently [15,16]. When exposed to UV radiation, upon absorption of photons diazonaphthoquinone eliminates N_2 forming a reactive carbene biradical that undergoes Wolff rearrangement to form a ketene (Scheme 12.1). The ketene reacts with water traces from the photoresist film, forming indenecarboxylic acid, soluble in basic developers (Scheme 12.1). Although novolak-based

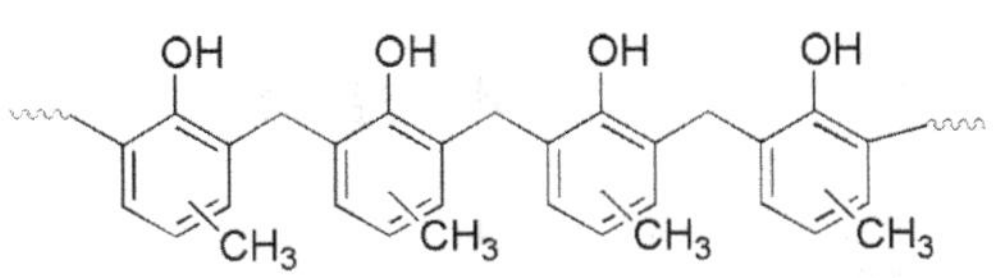

CHART 12.1 The chemical structure of novolak (it may be a mixture of *o*-, *m*-, or *p*-cresols).

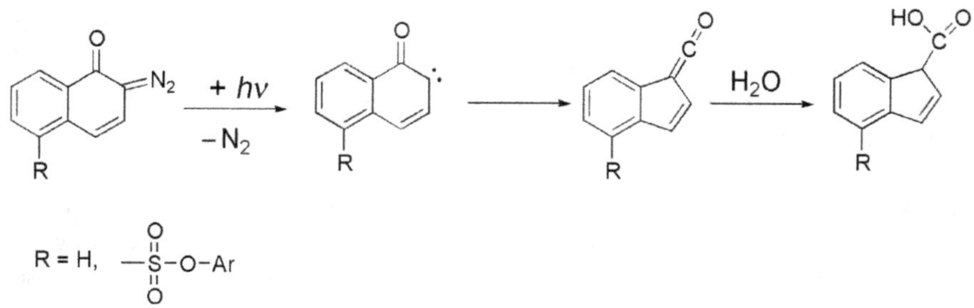

SCHEME 12.1 Photochemical reactions of diazonaphtoquinone derivatives: elimination of molecular nitrogen with formation of a carbene, followed by Wolff rearrangement to give a ketene that reacts with water traces to form a carboxylic acid.

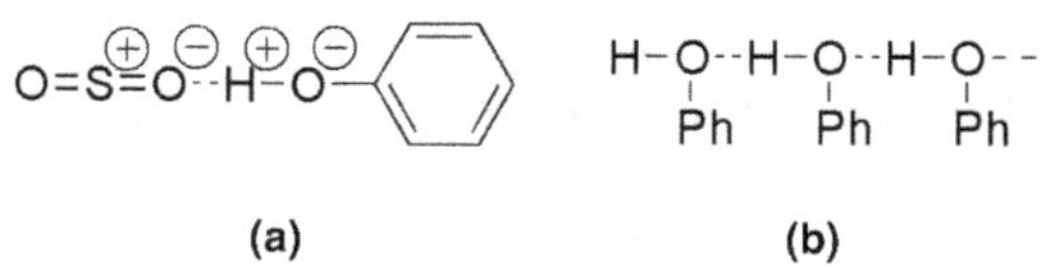

SCHEME 12.2 Polarization induced by $-SO_2-$ group on the OH group from the phenol (a), and the formation of the phenolic string (b).

nonamplified resists are not used in processors and memory manufacturing anymore, they are still used on a large scale in lithographic applications in other fields such as micro-electromechanical systems (MEMS) and flexible panels production. This positive photoresist was largely used for the manufacturing of 1–16 Mb DRAM [6]. Most of the IC manufacturers used multicomponent resists instead of one-component resists, with the DNQ/novolak as the workhorse for more than two decades.

The most often encountered DNQ derivative is a sulfonic ester derivative, with the SO_2 group having a strong influence on the dissolution inhibition [15]. The H bond between $-SO_2-$ group and the phenolic OH polarizes the last one, making the O atom a stronger H acceptor (Scheme 12.2a), which leads in the end to the formation of a second H bond with another phenol (Scheme 12.2b) [15]. Thus, the sulfone group triggers the formation of a chain of hydrogen-bonded phenols, called the "phenol string" by Reiser et al. [15,16]. The Wolff rearrangement is a highly exothermic reaction, with an enthalpy of −65 kcal/mol [15,16]. The heat released during this step may locally increase the temperature up to 200 °C, causing the H bonds formed between phenol strings to break. Breaking of the "phenol strings" coupled with formation of the carboxylic acid, formed as a result of DNQ photolysis, determines the tremendous increase in solubility in the exposed areas.

12.2 CHEMICALLY AMPLIFIED PHOTORESISTS

As the wavelengths used decreased, so did the number of incident photons per unit area. At the same time, the absorbance of the novolac/diazonaphthoquinone photoresist was too high at the 254 nm Hg and the 248 nm KrF excimer laser lines. It became clear that there was a need for a drastic increase in resists sensitivity, eventually by a few orders of magnitude. This increase was possible due to the development of a concept called chemical amplification (CA), developed by Ito, Fréchet and Willson at the beginning of 1980s [1,4,6,8,11]. The idea was simple and efficient: through absorption of one photon create a chemical species that would generate a chain reaction. Thus, one photon would trigger hundreds of reactions, giving an increase in quantum yield of at least two to three orders of magnitude compared with 0.7–0.8 quantum yield of the nonchemically amplified DNQ/novolak positive resists. Theoretically any (photo)chemical reaction that induces a change in polarity or solubility is desirable. Thus, depolymerization, deprotection, rearrangement, polymerization, and cross-linking were potential mechanisms considered for chemical amplification. By far, the most common mechanism in CA resists employed in IC manufacturing for computer chips and DRAM memories is the deprotection of a pendant esteric group, *tert*-butoxycarbonyloxy (*t*-Boc), attached to a polymer chain (Scheme 12.3).

The most widely employed photoactive compound that triggers the chain of chemical reactions is the photoacid generator (PAG), a molecule that upon absorption of a photon undergoes photolysis, producing Brönsted acid and other photoproducts. The acid catalyzes a chemical reaction and it is regenerated at the end of the reaction (Scheme 12.3) [1,4,6,11]. The PAG may be an ionic organic salt or a nonionic organic compound, vide infra.

A key parameter that characterizes any photochemical reaction is the quantum yield. Quantum yield is given by the following formula

$$\Phi = \frac{\text{Number of photochemical events}}{\text{Number of absorbed photons}}$$

In the case of CA photoresists, the quantum yield is given by the number of molecules of acid formed divided by the number of the absorbed photons. Another

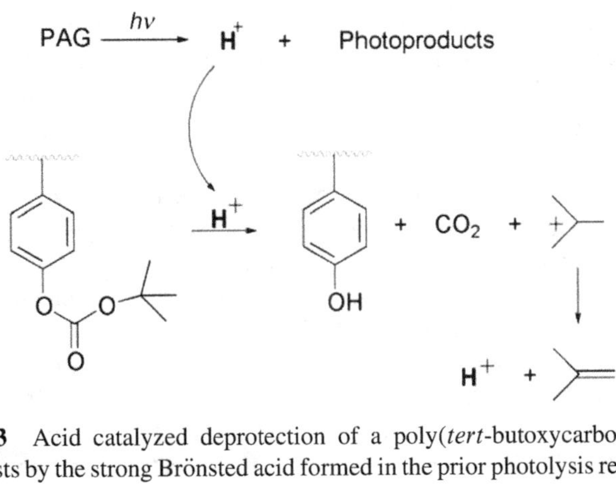

SCHEME 12.3 Acid catalyzed deprotection of a poly(*tert*-butoxycarbonyloxy styrene) PTBOCST resists by the strong Brönsted acid formed in the prior photolysis reaction of a PAG.

parameter critical for CA resists is the catalytic chain length (CCL), which is given by the number of deprotected *t*-Boc groups per proton generated. Literature reports suggest that a molecule of acid is capable of catalyzing from tens to hundreds of *t*-Boc deprotection reactions, for example, 800–1000 [18,19].

A compound must have certain properties to act as an effective PAG, such as formation of strong acid upon photolysis, reasonable absorbance at the exposure wavelength, solubility, thermal stability, long shelf life, adequate diffusion through the polymer matrix, high quantum yield, and so forth [1,4]. The efficiency of acid generation and the CCL are influenced by the chemical structure of the PAG, the counterion, the molecular volume, and the polarity of the polymer [20,21]. The PAG is present in small amounts in resist formulations (usually <5% weight), and consequently their contribution to the overall absorbance of the resist at the exposure wavelength is low. This was an advantage for the photoresist producers, because it allowed them to design formulations for 193 nm lithography without having to develop PAGs more transparent than those used at 248 nm. However, for photolithography at 157 nm and even EUV, more transparent PAGs would be needed.

McKean et al. developed a method that allows quantification of the CCL using infrared (IR) analysis. Films containing *t*-butoxycarbonate phenolic resin, poly(4-*t*-butoxycarbonyloxystyrene, PTBOCST), and 1.0 wt % triphenylsulfonium hexafluoroantimonate (as PAG) were spin coated on top of silicon wafers, then exposed to radiation at 254 nm originating from a Hg lamp. They developed the films and quantified the amount of acid by titration. They monitored the intensity of the carbonyl (C=O) signal before and after postbaking the films, thus obtaining the number of deprotected carbonyl groups. The CCL was calculated as the ratios between the number of deprotected esteric groups and the number of H^+ obtained photochemically. The value of CCL was between 800 and 1000, with an average diffusion of 50 Å [19].

PAGs were employed by Crivello as photoinitiators in cationic polymerization during the 1970s, and their photochemical reaction mechanisms were thoroughly studied and reported [22]. The mechanism of photolysis is influenced not only by the structure of the PAG, but also by the polymer matrix as well. Polymers may act as sensitizers, absorbing photons and releasing the excess energy through energy or electron transfer to the PAG molecules, thus increasing the efficiency of the photoresist. Polymers used for photolithography at 248 nm were comprised mostly of derivatized polyphenols, bearing *t*-Boc pendant groups, and they were acting as sensitizers for certain PAGs. However, at 193 nm the polyphenol-based polymers were too opaque, and they were replaced with more transparent polyacrylic-type polymers. Polyacrylates were not as good sensitizers for the PAGs, although they were also involved in some photochemical reactions.

The most widely employed solubility switch is the deprotection of a *t*-Boc protected polymer [1]. In choosing a PAG to incorporate within a photoresist, one must take into account the solubility of the PAG in the solvent, the miscibility of the PAG and the polymer, possible influence of the polymer on the PAG's photochemistry (e.g., sensitization), diffusion of the acid within the film, the strength of the acid formed, and the activation energy (E_a) for the deprotection reaction.

12.3 IONIC PHOTOACID GENERATORS AND THEIR PHOTOCHEMISTRY

Since the discovery of photochemically initiated cationic polymerization by Crivello et al., new photoacid generators have been designed and reported. There exists a wide variety of PAGs, and they may be divided into two big categories: ionic and nonionic. By far the two most widely employed types of PAGs are sulfonium and iodonium salts (Chart 12.2). The behavior of a PAG is determined by both its ions. The cation is the photochemically active part, it determines the absorption of the PAG, its quantum yield, and whether or not the molecule may be photosensitized [22]. The anion determines the strength of the photochemically formed acid, it helps stabilize the acid through ion pairing, and it plays an important role in diffusion throughout the polymer matrix [22].

Triphenylsulfonium (TPS) salts with various counterions have been widely employed as PAG in CA photoresists, and their photochemistry has been extensively studied [23–25]. Typical counterions for TPS are Br^-, $CF_3SO_3^-$ (triflate), $CF_3(CF_2)_3SO_3^-$, BF_4^-, PF_6^-, and SbF_6^-.

The study of the primary photoproducts of TPS in solution lead to the conclusion that there are two pathways for direct photolysis in the case of TPS salts, both originating from an excited singlet state: homolytic and heterolytic [24]. Following the homolytic cleavage, diphenylsulfinyl radical cation and phenyl radical are formed

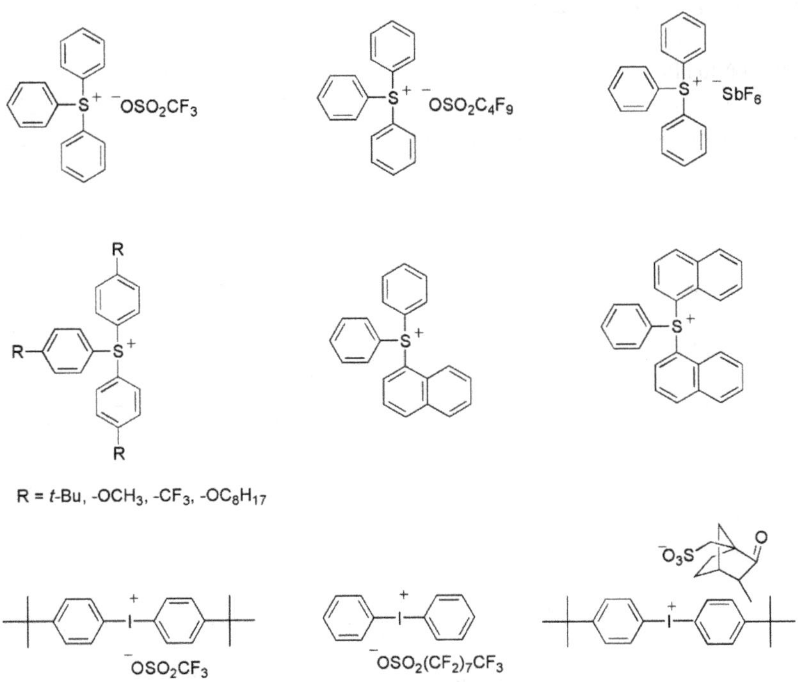

CHART 12.2 Structures of some ionic PAGs.

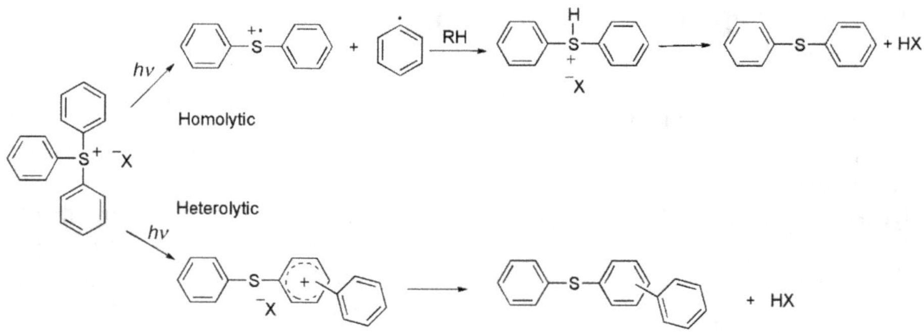

SCHEME 12.4 Photochemistry of triphenylsulfonium triflate: homolytic and heterolytic photolysis following absorption of photons.

in the solvent cage; the two radicals may either recombine to form the starting material or escape the solvent cage and form diphenyl sulfide and acid as main photoproducts (Scheme 12.4). The heterolytic cleavage initially leads to the formation of diphenyl sulfide and phenyl cation. The phenyl cation may attack the diphenyl sulfide and form the starting material, or it electrophilically attacks one of the rings from the diphenyl sulfide, leading to the formation of 2-, 3-, or 4-(phenylthio)biphenyl and acid [24].

The photosensitization of sulfonium salts leads to formation of sulfonium triplet, followed by homolytic cleavage with formation of a triplet radical pair of phenyl radical and diphenylsulphinyl radical cation. The triplet pair reacts only after escaping the solvent cage, and it does not lead to formation of (phenylthio)biphenyl recombination products.

Dektar and Hacker report on the photochemistry of diaryliodonium salts in the case of direct photolysis and in the case of energy transfer from a photosensitizer [26]. The photoproducts and the reaction mechanisms are different in the two situations (Scheme 12.5).

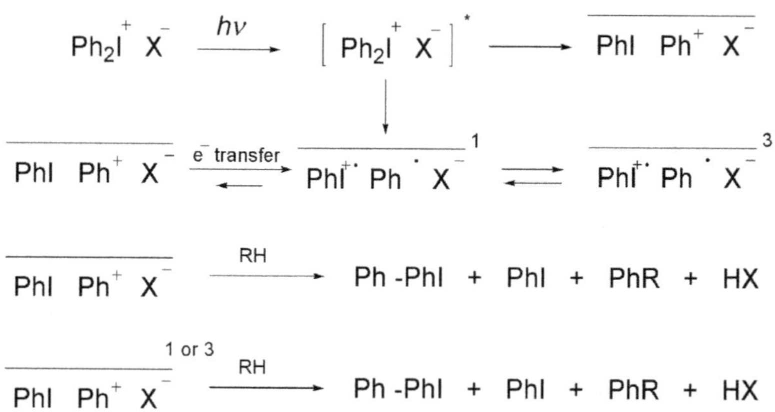

SCHEME 12.5 Proposed photochemistry of diaryliodonium salts in the case of direct photolysis.

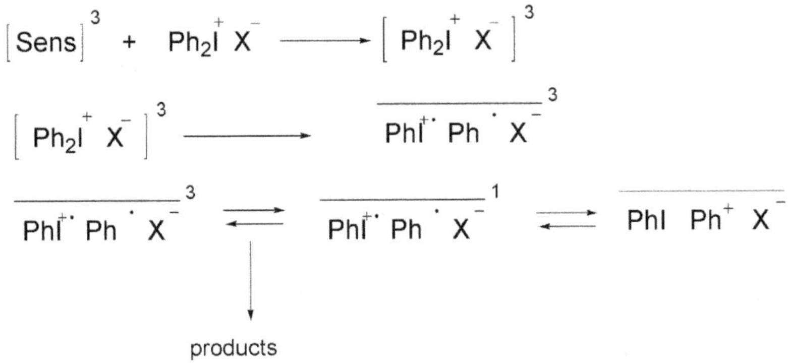

SCHEME 12.6 Photolysis of diaryliodonium salts—sensitized energy transfer.

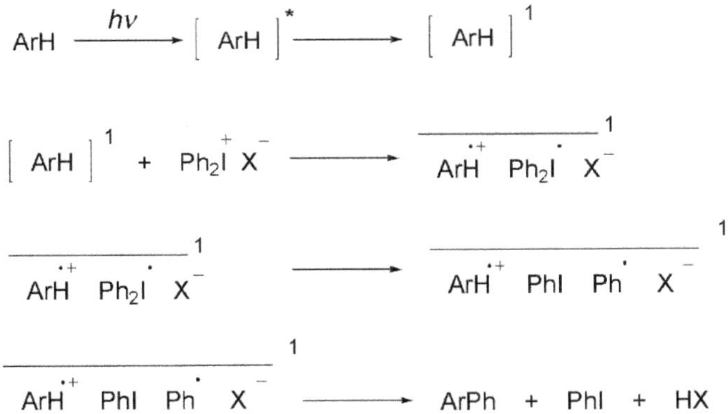

SCHEME 12.7 Photolysis of diaryliodonium salts—electron transfer.

Photosensitization of diaryliodonium salts may occur through energy transfer or electron transfer. The triplet sensitizer transfers energy to the diaryliodonium molecule, which undergoes homolytic cleavage from its triplet state, resulting in a radical pair within the solvent cage (Scheme 12.6).

Photolysis through electron transfer (sensitization) occurs when another molecule absorbs photons, goes into excited state, and then transfers its electron from the LUMO orbital to the PAG molecule (Scheme 12.7).

12.4 NONIONIC PAGs AND THEIR PHOTOCHEMISTRY

Nonionic PAGs have been also used, and their usage may pick up again once the immersion lithography will be implemented. An important issue in immersion lithography is the leakage of ions from the resist film into the liquid, thus causing damage to the optics. However, using nonionic PAGs the leakage may be considerably reduced. The chemical structures of some of the commonly employed nonionic PAGs are presented in Chart 12.3.

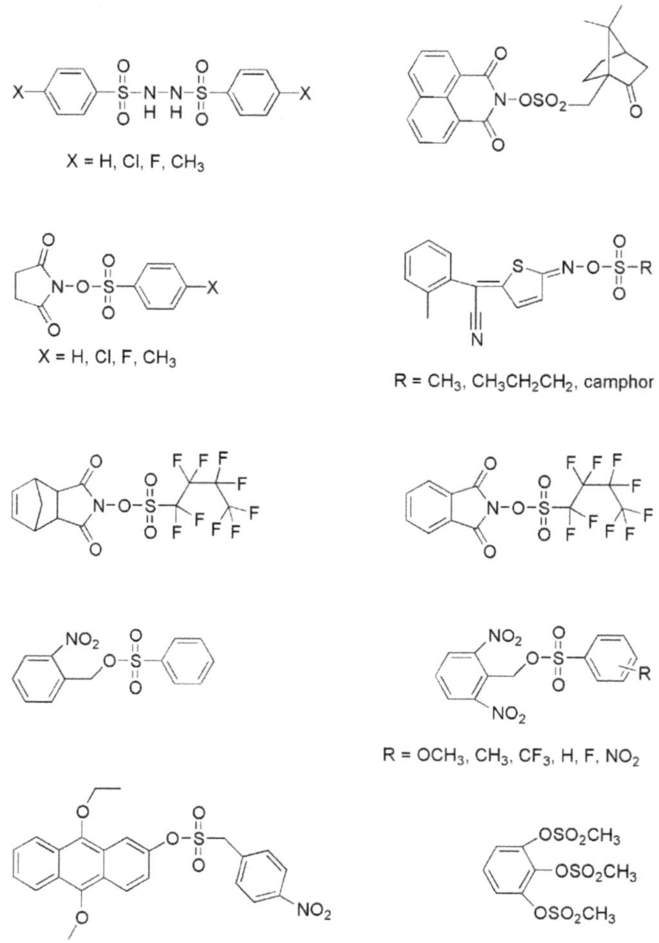

CHART 12.3 Structures of some nonionic PAGs.

Coenjarts et al. and Ortica et al. reported the mechanism of photolysis for nonionic PAGs irradiated with a 266-nm laser. Using laser flash photolysis, they elucidated the mechanism of photolysis and acid formation following direct irradiation as well as sensitization with Xanthone for the first two PAGs on the left column of Chart 12.3 [27,28]. Following absorption of a photon during direct excitation, the next step is homolytic cleavage, yielding arylsulfonyl radical and a nitrogen-centered radical. The arylsulfonyl radical may react with molecular oxygen forming peroxysulfonyl radical, or it may undergo desulfonation, leading to formation of an aryl radical, which further may form peroxide in reaction with molecular oxygen. The nitrogen centered radical is involved in other reactions, and it is the source of the acid.

12.5 ACID DETECTION, QUANTIFICATION, AND CATALYTIC CHAIN LENGTH

For CA photoresists, there are two important parameters that are taken into account: the CCL and the quantum yield of acid generation (Φ). The amount of acid produced during exposure is of paramount importance, because the acid will deprotect the *t*-Boc groups, which act as solubility switch. As one molecule of acid deprotects hundreds of protective groups (CCL), one could easily calculate how much acid should theoretically be needed in the resist film, to have complete deprotection of the polymer in the exposed regions, and avoid diffusion of the acid in unexposed areas, an unwanted process. The quantum yield of acid generation is an important parameter, which gives the overall performance of a PAG, and it is an important criterion for PAG selection. As previously mentioned in this chapter, the quantum yield is given by the ratio between the number of moles of acid and the number of photons absorbed by the PAG.

An important physical process in CA photoresists is the diffusion of the acid through the polymer film. Diffusion ensures that H^+ molecules reach *t*-Boc groups, however, too much diffusion causes deprotection in the unexposed areas, an unwanted process that is becoming increasingly important as feature size decreases. Studies have been performed with the goal of understanding the acid diffusion within films of photoresists during PEB. Most experimental techniques involve optical spectroscopy and fluorescence microscopy, but other techniques were employed as well [29–32]. The reported values describe a diffusion sphere of the H^+ with a diameter varying between 27 and 50 Å, depending on the system studied, the PEB time, and temperature [18,19,33].

Although it may seem a trivial procedure, it is actually very difficult to determine the exact number of photons absorbed by the PAG, because there are other resist components that may absorb in the same region as the PAG, and more importantly, some of the photoproducts obtained through PAG photolysis absorb in the same region as the parent molecule. The second parameter needed to calculate the quantum yield is the number of moles of acid produced. There have been numerous reports on experimental procedures employed to detect and quantify the amount of acid produced.

The detection and quantification of the acid formed have been done in two ways: *in situ* within the photoresist film, and in solution, the exposed film being developed with an appropriate solvent, and the acid titrated with a base in the presence of an acid–base indicator whose spectroscopic properties were easily monitored [18,34–39].

For the detection of acid in solution, a variety of pH sensors that change their absorption/emission properties were employed. In Scheme 12.8, we present a selection of acid sensors employed both in solution and in films [18,34,37,39,40].

In solution, the results are given by the titration curves, as in a typical analytical chemistry titration. The disadvantage of detection and quantification of acid in solution is the small volume of film irradiated. Considering that the typical thickness of the films is a few hundred of nanometers, there is a need for many exposed films to be dissolved in solution. Further, this approach, requiring extraction can only make an "end-point" measurement.

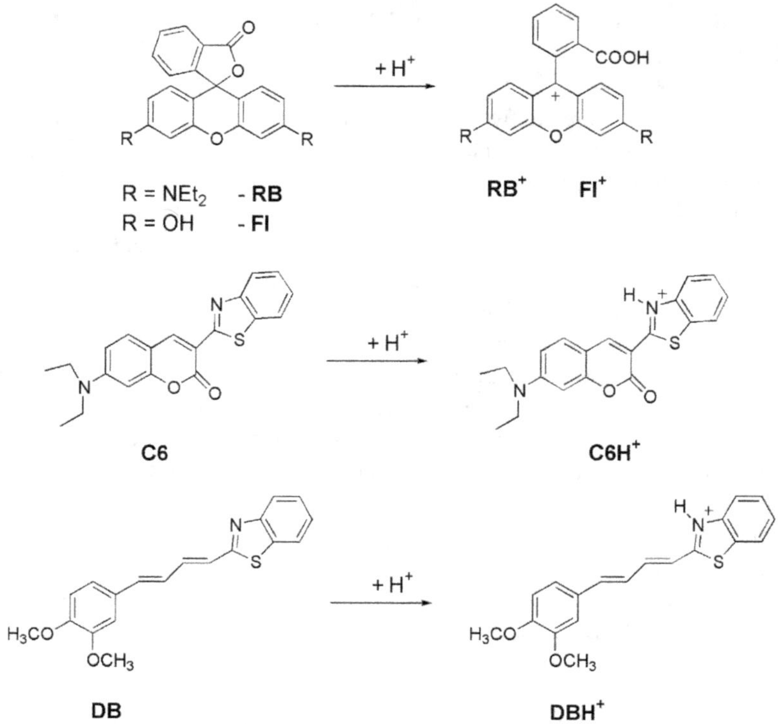

SCHEME 12.8 Chemical reactions involving the acid produced by PAG photolysis and the fluorescent sensors in solution or in films; rhodamine B (RB) and fluorescein (Fl) undergo opening of the lactone ring upon protonation, accompanied by a large bathochromic shift; Coumarin 6 (C6) and 2-[4-(3,4-dimethoxyphenyl)-1,3-butadienyl]benzothiazole (DB) are protonated at the N atom in the thiazole ring, causing a significant red shift.

A more convenient way of detecting and quantifying the acid formed is *in situ*, using fluorescent pH sensors that clearly mark the areas where the acid was formed [18,36,38–42]. These sensors are introduced into the photoresist that is spin coated, then following exposure they image the areas where acid is formed. Fluorescence spectroscopy and microscopy are then employed to capture the latent image and to determine the amount of acid formed. Examples of typical fluorescence images are shown in Fig. 12.2, where the images presented show the formation of acid upon exposure of a fluorinated resist exposed to laser radiation at 157 nm. Small amounts of Coumarin 6, a suitable pH sensor used in many studies [18,34,36–39], were introduced in the resist, then the film exposed through copper mask having hexagonal holes. Using the right set of excitation filters, in Fig. 12.2A, the dark areas are the exposed regions, while the green areas represent the unexposed regions where Coumarin 6 emits. Upon exposure to 80 mJ/cm^2 at 157 nm, the PAG present in film was photolysed and protonated the pH sensor Coumarin 6, see Scheme 12.8, leading to a change in its emission properties, the monoprotonated form emitting in the red region (Fig. 12.2B). In Fig. 12.2B, the

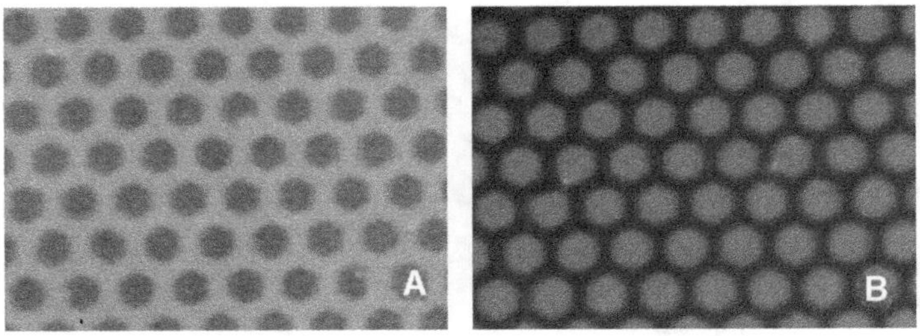

FIGURE 12.2 Fluorescence images of neutral Coumarin 6 (A, green) and protonated Coumarin 6, C6H$^+$ (B, red) recorded with the fluorescence microscope using excitation wavelength 485 ± 12 nm for neutral Coumarin 6 (A), and 546 ± 12 nm for protonated Coumarin 6 (B). The photoresist contained Coumarin 6 as pH sensor, a fluorinated polymer, and PAG, and it was exposed to 80.5 mJ/cm^2 at 157 nm. The green and red colors are the real colors corresponding to those observed under the microscope.

excitation wavelength from the microscope was selected such that only C6H$^+$ would be excited. This is easily observable because the unexposed areas, which emitted in green are now completely dark. Thus, with proper excitation wavelengths it was possible to discriminate the exposed (protonated) and the unexposed (unprotonated) regions on the same film.

The example presented in Fig. 12.2 shows how the acid formation is determined *in situ* by employing a pH sensor that changes its spectroscopic properties upon protonation. Coumarin derivatives bearing a *t*-Boc protecting group dramatically shift their absorption and emission maxima upon deprotection catalyzed by acid [18,40,42]. For example, Frenette et al. developed a prefluorescent sensor, a Coumarin 4 derivative, which mimics the *t*-Boc polymer [40]. The hydroxyl functionality in 7-hydroxy-coumarins was protected with *t*-Boc groups that quench the fluorescence of the molecule. Upon interaction with H$^+$, the *t*-Boc group is deprotected, releasing 7-hydroxycoumarin, a strongly fluorescent molecule. Thus, it is possible to distinguish *in situ*, using fluorescence microscopy, the exposed and unexposed regions of a photoresist, by incorporating the prefluorescent molecule (Fig. 12.3). The two images were obtained employing two derivatives of C4-*t*Boc,

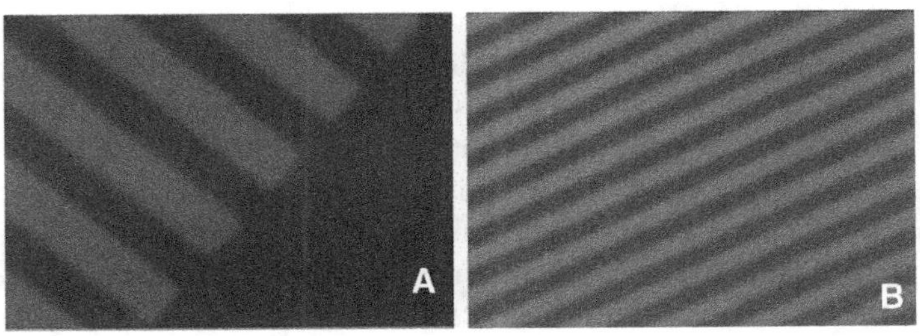

FIGURE 12.3 Fluorescence image of a PMMA film containing the *t*-Boc protected prefluorescent sensor in a 1.43 μm film thickness; the mask had a spacing of 55 μm. The blue regions in (A) and the green regions in (B) are the exposed regions and are the true colors observed with the fluorescence microscope.

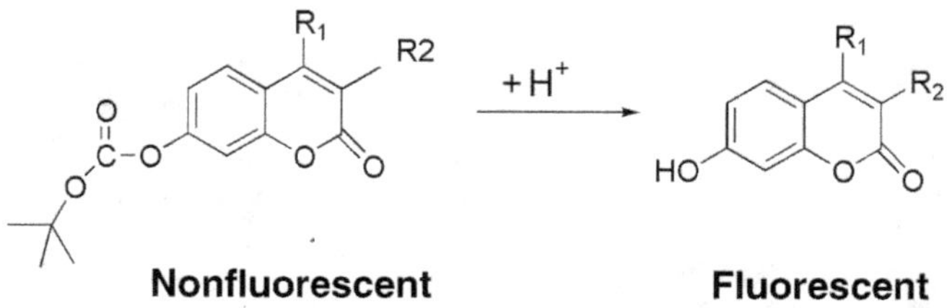

SCHEME 12.9 The deprotection reaction of Coumarin 4-based prefluorescent probes bearing *t*-Boc protective groups. Upon acidolysis, the nonfluorescent sensor becomes highly fluorescent; $R_1 = H$, Me, $R_2 = H$, benzothiazole.

whose deprotection reaction is represented in Scheme 12.9. The prefluorescent probe added to the PMMA film, in Fig. 12.3A, had $R_1 = Me$, $R_2 = H$, while in Fig. 12.3B, $R_1 = H$, and $R_2 =$ benzothiazole.

This particular sensor, C4-*t*Boc, together with Coumarin 6, were used in the development of a method that allows quantification of the CCL employing fluorescence spectroscopy of pH sensors added to the film [18]. The *t*-Boc protected Coumarin 4 gave the number of *t*-Boc deprotected groups in a film, while Coumarin 6 was used as pH sensor doing *in situ* titrations, which gave the number of protons produced. The ratio of *t*-Boc deprotected groups and the number of H^+ gave the CCL.

It is a common practice in microelectronics industry to add to resists formulations small amounts of base, whose role is to stop diffusion of acid in unexposed areas by reacting with the photogenerated acid. Bases added to photoresists in known amounts are also used to quantify the amount of acid produced. Pawlowski and Nealy found that bases added to CA photoresists have a more complex influence on the resist performance than just stoichiometric neutralization of the photogenerated acid. They also influence the dissolution, acting as dissolution inhibitors or promoters [43].

Adding known amounts of base, makes it is possible to quantify the formed acid by doing *in situ* titrations, and monitoring the emission or absorption of pH sensors present in the film.

12.6 POLYMERS FOR DEEP-UV, VACUUM-UV, AND EXTREME-UV PHOTOLITHOGRAPHY

The second most present component in a photoresist, in terms of mass, is the polymer. During postapplication baking (PAB) the solvent evaporates, leaving the polymer as the component with the highest contribution to the film's mass and volume. The polymers employed in modern photoresists are photochemically inert, and must possess a series of properties that makes them suitable for CA photoresists. The glass transition temperature (T_g) is a parameter that has a tremendous influence on the diffusion of H^+ throughout the film. Postexposure baking (PEB) is usually done at ~90–100 °C, requiring the synthesis of polymers with T_g higher than this temperature. To avoid contamination with airborne bases, which would interact with the photogenerated proton and cause surface defects in resist films after development (e.g., T-shaped defects), chemists synthesized photoresists with a low activation energy (E_a)

for the deprotection reaction. Thus, immediately after photolysis of the PAG, and before PEB, the acid would catalyze deprotection reactions, and the airborne bases would have less time to react with the acid [1,44]. However, too low a T_g would negatively impact the resist film, causing flow problems during the PEB step.

There are properties that make a polymer suitable for photolithography in general, but there are requirements for specific exposure wavelengths (248, 193, and 157 nm). For each one of these wavelengths, there are an impressive number of polymers and copolymers developed, although only a small fraction of them made it into resist formulations employed in mass production [1,4].

12.7 POLYMERS FOR 248 nm PHOTOLITHOGRAPHY

The DNQ/novolak photoresists used for 436 and 365 nm optical lithography did not meet the requirements for photolithography at 248 nm, and the CA photoresists were employed at this wavelength. Besides the low quantum yield, an important aspect was the high absorbance of novolak at 248 nm. The photoresists for 248 nm were based on polyhydroxystyrene (PHOST) and modified PHOST, which met the requirements for this exposure wavelength.

The first polymer to be used in a CA photoresist at 248 nm was poly(*tert*-butoxycarbonyloxystyrene) (PTBOCST) [1,4,6,9,44]. The chemical reaction—the deprotection of the *t*-Boc groups catalyzed by H^+—acts like a solubility switch, and is presented in Scheme 12.3. The resist formulation containing PTBOCST was the first one to be used in DUV photolithography under 248 nm exposure for the manufacturing of 1 Mb DRAMs [4]. To obtain a good solubility switch, it is enough to have approximately 10–30% of the hydroxyl groups protected with *t*-Boc [1]. Considering the importance of PTBOCST and PHOST for the microelectronics industry, a variety of methods to obtain poly(4-hydroxystyrene) were published by different groups, for example, living radical, anionic, and cationic polymerization [1,45,46]. The monomers were either 4-hydroxystyrene or derivatives of it. There has been an abundance of reports on polymers developed for photolithography at 248 nm by researchers from industry and academia alike. Due to the space constraints, only a few generic structures of polymers are displayed in Chart 12.4, but the reader is directed toward other sources on polymers used in photolithography [1,4,6,9,44].

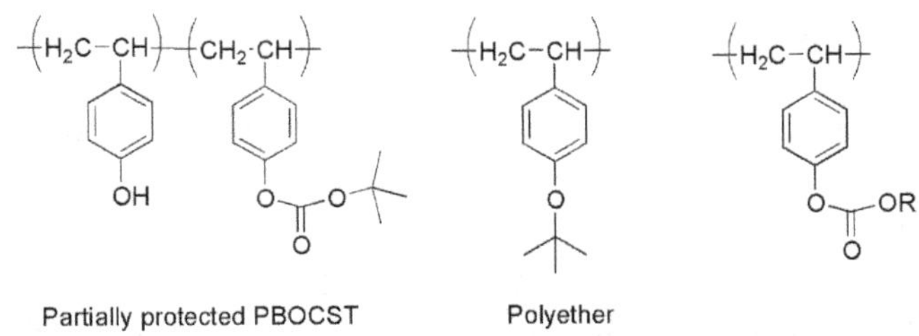

CHART 12.4 Typical structures of (co)polymers employed in resist formulations at 248 nm (esters and ethers bearing different substituents).

Copolymers of 4-hydroxystyrene with various other monomers were synthesized as well. Because the polymer must withstand tough etching conditions, must be transparent at the exposure wavelength, have good solubility, film forming abilities, and good adhesion to the substrate, quite often copolymers of 4-hydroxystyrene with other monomers were developed.

The acid generation efficiency is influenced by the exposure wavelength as well as the polymer matrix. Literature reports show differences in acid formation efficiency at 193 and 248 nm and in phenolic and acrylate-based polymers [47]. In addition to the direct photolysis of the PAG, the phenolic polymers may sensitize the PAG through electron transfer from the polymer matrix to the PAG molecule, unlike the polymethacrylates in which only direct photolysis of the PAG is responsible for acid formation.

12.8 POLYMERS FOR 193 nm PHOTOLITHOGRAPHY

The race toward better computer processors and DRAM chips has fuelled the research for new polymers to be used at 193 nm exposure wavelength. The most important obstacle against using PTBOCSTs as resist at 193 nm was their high opacity, corroborated with the decreased number of incident photons per unit area produced by the ArF lasers compared with the KrF lasers at 248 nm [1,48]. The photoresists used at 193 nm are based on the same principle of CA, and the polymers bear the same *t*-Boc group as solubility switch. However, to avoid excessive absorbance, the polymer backbone, consisting of poly(hydroxystyrene) had to be replaced with an aliphatic polymer that still contained pendant *t*-Boc groups that upon interaction with the photogenerated proton would form groups such as –OH and –COOH that lead to solubilization in aqueous base. Polymethacrylates were developed for photolithography at 193 nm, and a few characteristic structures are displayed in Chart 12.5. Besides methacrylate monomers, norbornenes were found to add to the polymer's transparency at 193 nm, and usually the polymers used for 193 nm are *co-*, *ter-* and

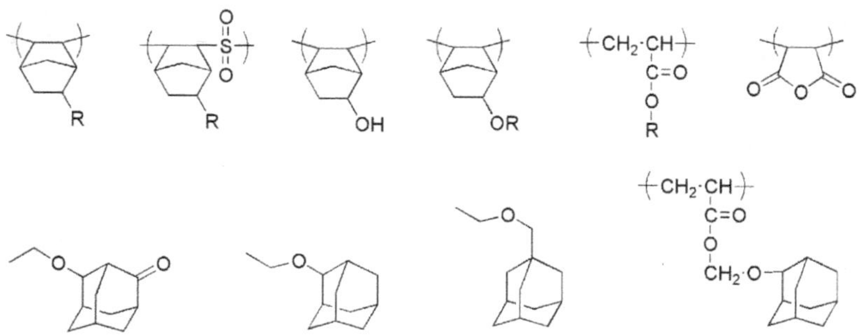

CHART 12.5 Structures of monomers and polymers employed in photolithography in 193 nm optical lithography. Most polymers are *co-*, *ter-*, and *tetra*-polymers of norbornenes bearing various substituents, *tert*-butoxyacrylates, adamantane, and maleimide.

tetra-polymers, each monomer bringing a series of properties desirable for the photoresist.

A difficult obstacle in IC clean rooms is the presence of organic bases in the air. Bases are used in formulations of photoresists and as developers. Organic bases, such as N-methylpyrrolidone (NMP), react with the photochemically formed acid on top of the photoresist films, forming, in the case of positive CA resists, thin layers insoluble in developer. As a consequence, the air in the clean rooms is filtered through activated carbon filters that retain much of the air-contained bases. Other solutions adopted by the industry were the use of photoresist films with smaller free volume, or application of a protective coating that prevents the base from reacting with the acid. To help prevent defects, such as T-shaped skins induced by airborne contamination, researchers developed low E_a (energy of activation) photoresists. The low E_a favors deprotection of *t*-Boc group at room temperature, allowing more acid to be involved in the wanted reaction of deprotection, and not be consumed by airborne bases [49].

Photolithography using exposure at 193 nm is the current technology of choice used by IC manufacturers. Resolution enhancement techniques (RET) allowed engineers to push the limits for which 193 nm is used, from 120 nm node down to 45 and 32 nm nowadays, and most likely for 22 nm as well. Double exposure, immersion lithography in a liquid with high refractive index, and optical proximity correction (OPC) have been employed so far [50–52]. Immersion lithography in water, which has a refractive index of 1.44, is the current technology employed [52]. The main question that scientists and researchers ask themselves concerns the wavelength of the exposure radiation for 22 nm node. So far, EUV seems to be the best-positioned candidate.

12.9 POLYMERS FOR 157 nm OPTICAL LITHOGRAPHY

Photolithography at 157 nm was seen as the next technology to replace 193 nm lithography. There has been a huge volume of research devoted to the synthesis and development of new materials for 157 nm. The greatest obstacle was the high absorbance of organic compounds at this wavelength. The high absorbance determined the usage of thin resist films that had to withstand the harsh processing conditions. Because oxygen absorbs at 157 nm, the exposure has to be done under vacuum or under a nitrogen atmosphere. Another major barrier toward adopting 157 nm as exposure wavelength at industrial scale was the birefringence of CaF_2, the transparent material preferred for manufacturing the expensive optical parts [53,54]. All these obstacles and the high cost associated with overcoming them determined the industry to continue using 193 nm as exposure wavelength, and to implement RET that allow the use of 193 nm for even smaller feature sizes.

The characteristic of polymers developed for 157 nm lithography is the high fluorine content, which considerably increases the transparency at the exposure wavelength [55–63]. Usually, these polymers are *co*,- *ter*-, or *tetra*-polymers, containing monomers that bring specific properties to the material. For example, tetrafluoroethylene brings transparency, *t*-Boc protected groups act as solubility switch, while norbornene bring etch resistance (Chart 12.6). Besides fluorine-containing polymers,

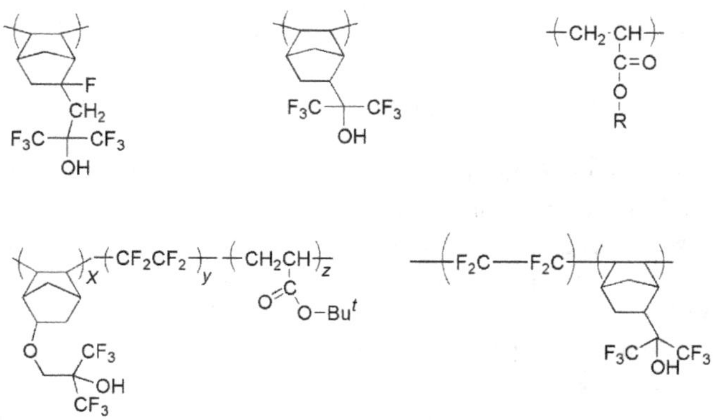

CHART 12.6 Structures of polymers employed for 157 nm optical lithography: fluorinated monomers and norbornene-based polymers were found to work excellent in photoresists for 157 nm optical lithography.

polysilanes, polysiloxanes, and polysilsesquioxanes were considered as useful polymers for photoresists at 157 nm due to their low absorbance and resistance to etching conditions [1,64–66].

The absorbance of a fluorine containing monomer at 157 nm is influenced by the position of the F atom in the molecule. For norbornene, for example, one of the commonly encountered monomers in polymers designed for 157 nm lithography, the absorbance varies greatly depending on the position of the substituting F atom. When two F atoms are introduced in a molecule, the absorption is reduced even more [62].

Although lithography with 157 nm has not been implemented at an industrial scale, some research is still being carried on, and the photochemistry at this wavelength is being studied. Possible applications include laser ablation of polymeric materials, art restoration, and photochemical treatment of polymeric and semiconductor surfaces for chemical modification. The knowledge and the materials developed for 157 nm lithography enable the IC industry now to obtain 32 nm feature sizes employing radiation at 193 nm.

12.10 MATERIALS FOR NEXT GENERATION EXTREME-UV LITHOGRAPHY

According to Rayleigh's law, the feature size (R) decreases as the exposure wavelength decreases. The next exposure wavelength expected to be introduced at industrial scale in IC industry is 13.4 nm in the EUV region of the optical spectrum [7,52,67]. Recent years have witnessed the development of light sources, optics, materials, and equipment for EUV lithography, with promising results [7,67,68]. Design, development, and synthesis of suitable polymers and photoacid generators for EUV resists have seen some progress. The resist for EUV lithography must posses most of the characteristics for other wavelengths, with emphasis on high sensitivity due to the low output of the radiation sources, and the high resolution as the projected feature size will be 22 nm and probably below.

The need for lowline edge roughness (LER) limits the use of high molecular weight polymers, which have a high molecular volume. Besides polymers for photoresist formulation, another type of materials of interest are the molecular glass photoresists, in which the polymer is replaced with monomolecular compounds having a high molecular weight [69–71]. A direct consequence of the low output of the light source in EUV is the necessity to use photoresists with increased sensitivity as compared with 193 nm and 248 nm exposure. Thus, chemically amplified photoresists seem to be a good choice for this wavelength. However, the small feature size combined with acid diffusion may be an inconvenience, because acid diffusion in unexposed areas increases the LER. Nonchemically amplified resists are considered as candidates for EUV [72], as well as organic–inorganic hybrid polymers containing Si in their main chain or in pendant groups [73].

It has been found that at 13.4 nm the absorbance is mainly due to the atoms present in the molecule rather than the type of bonds (σ or π) among atoms [52]. Atoms of Si, C, H, and B have a low photoabsorption cross-section, while F, which considerably reduces the absorbance in polymers employed for 157 nm photoresists, has an absorption cross-section more than four times that of C [52]. Oxygen also increases the absorbance of materials used for EUV lithography and is avoided, making most of the current resist formulations unusable at 13.4 nm. Outgassing is an important obstacle for current photoresist employed in the EUV region [74]. Polystyrenes bearing Si-containing pendant groups have been synthesized and shown to work well in EUV resists, decreasing the amount of outgassed products [73,75]. Poly(hydroxystyrene)-based resists have been shown to have a lower absorption than other resists [76]. Chemical structures of typical Si-containing polymers for EUV are displayed in Chart 12.7.

Outgassing leads to deposition of outgassed molecules on the reflective optics, and as such, low outgassing resists are a priority.

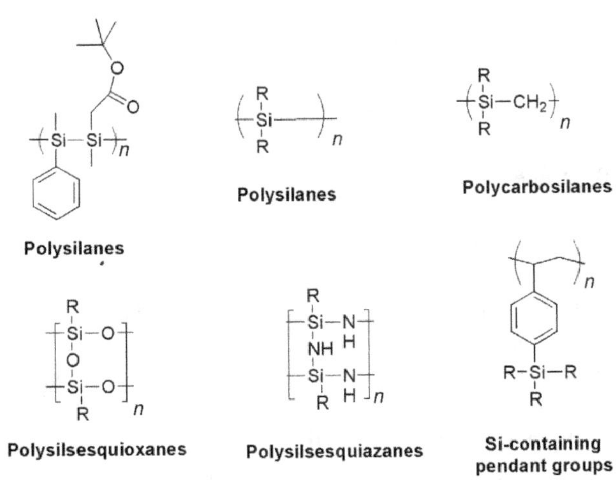

CHART 12.7 Chemical structures of organic and hybrid inorganic–organic polymers employed in EUV lithography.

12.11 PHOTOCHEMICAL REACTIONS INVOLVING POLYMERS IN PHOTORESIST FILMS

Although polymers included in resist formulations are supposed to be transparent at the exposure wavelength, they still absorb at short wavelengths such as 254, 248, 193, and 157 nm, and especially at 13.4 nm, and undergo photolysis reactions [35,77–84]. The photolysis reactions cause main-chain scission, pendant groups scission, accompanied by outgassing when small molecules are released to the environment, and photo cross-linking. Photolysis of *t*-Boc protective group may occur as well [77]. Besides polymers, PAGs and solvent traces left during PEB are also source of outgassing [74,85–87]. Outgassed molecules deposit on the surrounding objects, which is an unwanted process especially when very expensive optical parts are affected. Thus, researchers studied photochemical reactions involving polymers and suggested improvements for polymers used in optical lithography.

UV–Vis radiation may also be used to modify physicochemical properties of polymer surfaces through ablation. For ablation to occur, the polymer has to absorb at the exposure wavelength or have a dopant molecule present in the polymer matrix [88]. A dopant is a molecule that absorbs at the exposure wavelength and releases the energy of the absorbed photons to the polymer as heat, causing the temperature on local environment to rise dramatically up to 500K or more. Ablation itself is a wide topic, and for a comprehensive review on polymer ablation, the reader is directed to the paper of Lippert et al. [89]. Ablation and etching may be photothermal or photochemical, or a combination of both. It has been speculated that UV and Vis laser radiation cause mainly photochemical reactions, while laser in the near infrared (NIR) and IR cause mainly photothermal reactions; however, the ablation mechanism depends on the incident dose, the rate at which it was delivered, and also on the polymer structure [89–97]. Ablation occurs at much higher energy doses than those used in photolithography; thus, only photochemical reactions representative for optical lithography will be introduced here.

Just like in small molecules, the chromophores present in polymer molecules absorb photons and are promoted into electronic excited states, from which they can relax by emitting light (fluorescence and phosphorescence), by releasing the energy as heat into the surrounding environment (internal conversion), or through bond breaking [64,98–100]. Depending on the chromophore and the energy of the absorbed photon, the following electronic transition may occur: π–π^*, σ–σ^*, and n–π^*. π–π^* transition is characteristic to aromatic and unsaturated aliphatic polymers, σ–σ^* transition accompanies photon absorption by saturated aliphatic polymers, while n–π^* transition is characteristic, for example, to carbonyl containing chromophores [99–101]. Formation of photoproducts is influenced not only by the chemical structure of the absorbing molecule, but also by the environment, such as polar/nonpolar, viscosity of the solvent, presence/absence of molecular oxygen, temperature, magnetic field, and so forth [99,100].

The typical C–C bond dissociation energy range between 80 and 100 kcal/mol, while the energy of one mol of photons at 157, 193, and 248 nm are 182, 147, and 115 kcal/mol, respectively. It is expected from these values that the absorbed photons at these wavelengths could lead to photolysis of C–C bonds, particularly at 157 nm, a very energetic radiation that may cause chemical and topological changes of polymers even at relatively low incident doses [35]. Radiation from VUV causes scission of most chemical bonds, such as C–C, C–O, C–S, C–F, and C–H, which have a highbond dissociation energy [78,79,81,83]. However, bond dissociation is just one of the possible mechanisms to release the energy of the absorbed photon, and other relaxation pathways compete with bond breaking.

Scientists have used a wide arsenal of analytical techniques to monitor chemical and physical transformations of polymers following exposure to laser radiation, among which UV–Vis absorption, nuclear magnetic resonance (NMR) spectroscopy, electron spin resonance (ESR) spectroscopy for detection of free radicals, GC/MS analysis, FTIR for detection of various functional groups and bonds, X-ray photoelectron spectroscopy (XPS) for the chemical composition of surfaces, optical, and fluorescence microscopy, atomic force microscopy (AFM) for surface topography, quartz crystal microbalance (QCM) for *in situ* mass loss measurements, and so forth.

The most common primary photoproducts of photochemical reaction are radicals and radical ions [99,100]. Primary photoproducts may recombine to form the starting material, or they may escape the solvent cage and form other chemical compounds through hydrogen abstraction, rearrangement, addition to unsaturated bonds, combination with other radicals or molecular oxygen. Recombination and additions of radicals may lead to formation of double bonds and cross-linking. Photocross-linking decreases the solubility of the polymer in the developer, counterbalancing, for example, the effect of *t*-Boc deprotection in CA photoresists, which is supposed to increase solubility in the exposed areas of the polymer film.

12.12 PHOTOCHEMISTRY OF PMMA

Poly(methyl methacrylate), PMMA, is one of the most studied polymers because it may be included in electron-beam resists, as well as in 193 nm photoresists when bearing a *t*-Boc protected ester group [1,4,84]. In e-beam resist PMMA undergoes main-chain scission, with formation of lower M_w fragments. When exposed to DUV and VUV radiation, the carbonyl group causes Norrish type 1 and Norrish type 2 photoreactions to occur, with cleavage of the C–C bonds in the α and β positions relative to the C=O group, see Scheme 12.10 [77,83,84]. Norrish type 1 reactions cause homolytic cleavage, leading to ester elimination and formation of carbon radicals centered on the main polymer chain, which in a subsequent step may lead to cross-linking. The ester group may lose CO_2 in a subsequent step. Norrish type 2 reactions cause fragmentation of the main chain, with formation of polymer fragments with lower molecular weight. Methyl and *tert*-butyl radicals were detected for PMMA and poly(*t*-butylacrylate), respectively, following exposure to 157 nm [77]. Methyl radicals were detected with ESR spectroscopy when PMMA was exposed to 254 nm radiation at 77 K [84]. The presence of the methyl is due to the pendant $COOCH_3$ fragment, and it does not originate from the main chain [84]. When poly

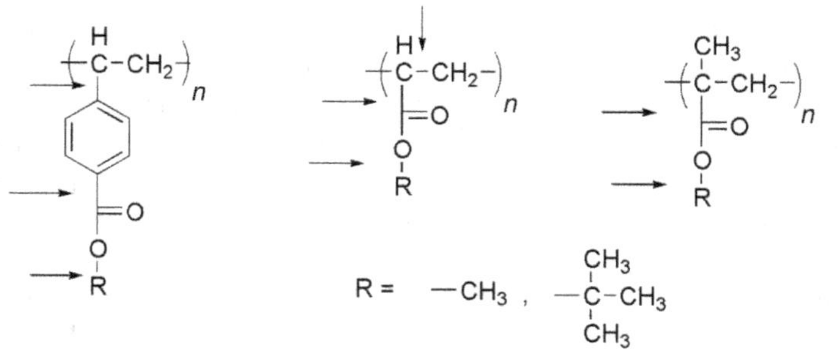

SCHEME 12.10 Photochemical-induced scission that may occur in (meth)acrylate and polystyrene derivative polymers during exposure to DUV and VUV radiation: Norrish type 1, Norrish type 2, ester elimination, main-chain scission, and so forth.

(methacrylic acid) was exposed to 254 nm, methyl radicals were not detected, suggesting clearly that it originates from the $^{\bullet}COOCH_3$ fragment [84]. In the presence of molecular oxygen, the radicals are readily quenched, forming peroxyl radicals.

Outgassing from poly(*t*-butylacrylate) during exposure at 157 nm has been studied, and among the outgassed products isobutene, isobutene, acetone, isobutanal, and isobutanol were detected [77].

Poly(hydroxystyrene) undergoes photocross-linking when exposed to both 157 and 248 nm, while PMMA undergoes chain scission during exposure to the radiation at the same wavelengths [77].

12.13 PHOTOCHEMISTRY OF FLUORINATED MATERIALS

Fluorinated organic polymers are more transparent at 157 nm than other polymers employed for photolithography at 193 and 248 nm [1]. However, chemical and topological transformations have been observed in fluorinated materials exposed to 157 nm [35]. Literature studies report on photodarkening and photodegradation of fluorinated materials exposed to 157 nm. Model fluorocarbons and fluorinated ethers with various amounts of hydrogen in molecule undergo photochemical reactions with formation of different products depending on the presence or absence of oxygen [79]. The high-energy photons at 157 nm can cause ejection of an electron in a process called photoelectron ejection. Formation of HF has been reported in literature, an important finding considering the etching properties of this acid on glass and the optical parts contained within the exposure systems [35,79]. The identity of the acidic species released upon exposure of a polymer with the structure displayed in Chart 12.8 was revealed with a fluorescent sensor embedded in the polymers film. The fluorescence of the sensor is "turned on" in the presence of HF, making an easy monitoring of this species with fluorescence microscopy and spectroscopy [35].

The use of fluorescent sensors at 248, 193, and 157 nm is possible due to their ability to withstand absorption of many photons, up to ten, during a laser pulse, and not be degraded [88]. This is due to excitation of the sensor into high-excited states,

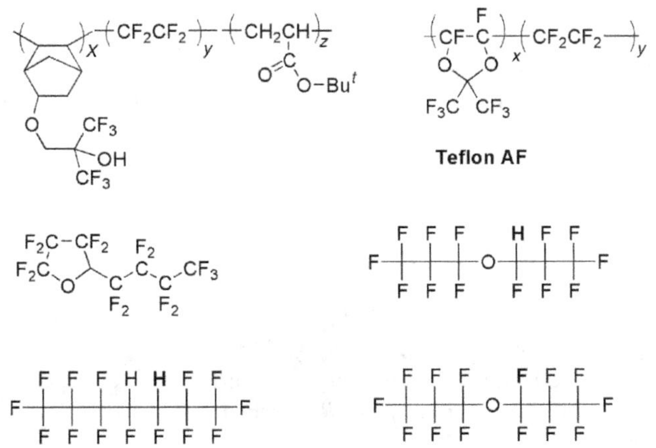

CHART 12.8 Chemical structures of exemplar monomolecular compounds and fluorinated polymers employed for studying photochemistry at 157 nm; C–O and C–H bonds act as chromophores at 157 nm.

followed by rapid decay to S_1, from which reabsorption may occur [88]. The molecule relaxes through internal conversion from S_n to S_1, and then through fluorescence to ground state S_0, emitting photons at longer wavelengths. Intersystem crossing (ISC) from S_1 to the triplet state T_1 is possible sometimes, followed by relaxation to S_0 through emission of a photon, a process called phosphorescence, or through radiationless processes.

The higher the H content in partially fluorinated polymers, the higher the absorbance at 157 nm. In fact, the main chromophore at this wavelength is the C–H bond. Electron ejection was found to occur from partially fluorinated compounds. Exposure of Teflon AF (Chart 12.8) to 157 nm causes C–C and C–O bond cleavage, resulting in formation of carbonyl containing free radicals. Hydrogen trapping leads to formation of an acid fluoride, monitored with FTIR at $\sim$1890 cm^{-1}; hydrolysis of the acid fluoride with water traces leads to formation of carboxylic acid, detected with FTIR at $\sim$1780 cm^{-1}, and with ^{19}F and ^{1}H NMR [79,81]. The double bonds -CF=CF- and >CF=CF_2 were detected with FTIR as well, monitoring the characteristic peaks at 1720 and 1735 cm^{-1}, respectively [78]. ESR analysis of Teflon AF irradiated at 157 nm reveals formation of radicals by main-chain scission, as well as cleavage of C–O bonds in the perfluorodioxole ring, yielding a variety of volatile and non–volatile photochemical products [80].

12.14 PHOTOCHEMISTRY OF Si-CONTAINING POLYMERS

Polysilanes, polysilsesquioxanes, and polysiloxanes have found a wide use in photoresist formulations due to their relatively high transparency, film forming abilities, low outgassing, and resistance to etching [1,64,66].

ACKNOWLEDGMENTS

The authors thank NSERC (Canada) and Rohm and Haas Electronic Materials, whose support over the years stimulated their interest and research in photolithography.

REFERENCES

1. H. Ito, *Adv. Polym. Sci.* **2005**, *172*, 37–245.
2. G. Moore, *Electronics*. **1965**, *38*, 114–117.
3. G.M. Wallraff and W.D. Hinsberg, *Chem. Rev.* **1999**, *99*, 1801–1821.
4. C.G. Willson, Introduction to Microlithography in: L.F. Thompson, C.G. Willson, and M.J. Bowden (Eds.), *ACS Professional Reference Book*, American Chemical Society, Washington, DC, 1994.
5. K. Ronse, *C. R. Phys.* **2006**, *7*, 844–857.
6. H. Ito, *IBM J. Res. Dev.* **2000**, *44*, 119–130.
7. K. Kemp and S. Wurm, *C. R. Phys.* **2006**, *7*, 875–886.
8. S.A. MacDonald, C.G. Willson, and J.M.J. Frechet, *Acc. Chem. Res.* **1994**, *27*, 151–158.
9. E. Reichmanis, F.M. Houlihan, O. Nalamasu, and T.X. Neenan, *Chem. Mater.* **1991**, *3*, 394–407.
10. A. Hand, *Semicond. Int.* **2007**, *30*, 34–38.
11. H. Ito, C.G. Willson, and J.M.J. Frechet, *Symp. VLSI Technol.* **1982**, *260*, 262.
12. O. Sus, *Justus Liebigs Ann. Chem.* **1945**, *557*, 237.
13. E. Gipstein, A.C. Ouano, and T. Tompkins, *J. Electrochem. Soc.: Solid-State Sci. Technol.* **1982**, *129*, 201–205.
14. L.E. Bogan Jr., *Macromolecules* **1992**, *25*, 1966–1969.
15. A. Reiser, J.-P. Huang, X. He, T.-F. Yeh, S. Jha, H.-Y. Shih, M.S. Kim, Y.K. Han, and K. Yan, *Eur. Polym. J.* **2002**, *38*, 619–629.
16. A. Reiser, H.-Y. Shih, T.-F. Yeh, and J.-P. Huang, *Angew. Chem. Int. Ed. Engl.* **1996**, *35*, 2428–2440.
17. K. Ueberreiter and F. Asmussen, *J. Polym. Sci.* **1962**, *57*, 187.
18. M. Frenette, M.G. Ivan, and J.C. Scaiano, *Can. J. Chem.* **2005**, *83*, 869–874.
19. D.R. McKean, U. Schaedeli, and S.A. MacDonald, *J. Polym. Sci. Part A: Polym. Chem.* **1989**, *27*, 3927–3935.
20. G. Pohlers, Y. Suzuki, N. Chan, and J.F. Cameron, *Proc. SPIE* **2002**, *4690*, 178–190.
21. S.L. Ablaza, J.F. Cameron, G. Xu, and W. Yueh, *J. Vac. Sci. Technol. B* **2000**, *18*, 2543–2550.
22. J.V. Crivello, *J. Polym. Sci. Part A: Polym. Chem.* **1999**, *37*, 4241–4254.
23. J.W. Knapczyk, W.E. McEwen, and *J. Org. Chem.* **1970**, *35*, 2539–2543.
24. J.L. Dektar and N.P. Hacker, *J. Am. Chem. Soc.* **1990**, *112*, 6004–6015.
25. J.L. Dektar and N.P. Hacker, *J. Org. Chem.* **1988**, *53*, 1833–1835.
26. J.L. Dektar and N.P. Hacker, *J. Org. Chem.* **1990**, *55*, 639–647.
27. C. Coenjarts, F. Ortica, J.F. Cameron, G. Pohlers, A. Zampini, D. Desilets, H. Liu, and J.C. Scaiano, *Chem. Mater.* **2001**, *13*, 2305–2312.

77. T.H. Fedynyshyn, R.R. Kunz, R.F. Sinta, R.B. Goodman, and S.P. Doran, *J. Vac. Sci. Technol., B* **2000**, *18*, 3332–3339.
78. V.N. Vasilets, I. Hirata, H. Iwata, and Y. Ikada, *J. Polym. Sci. A: Polym. Chem.* **1998**, *36*, 2215–2222.
79. K. Lee, S. Jockusch, N.J. Turro, R.H. French, R.C. Wheland, M.F. Lemon, A.M. Braun, T. Widerschpan, D.A. Dixon, J. Li, M. Ivan, and P. Zimmerman, *J. Am. Chem. Soc.* **2005**, *127*, 8320–8327.
80. I. Blakey, G.A. George, D.J.T. Hill, H. Liu, F. Rasoul, L. Rintoul, P. Zimmerman, and K. Whittaker, *Macromolecules* **2007**, *40*, 8954–8961.
81. C.M. Garza and T. Bierschenk, *Proc. SPIE* **2004**, *5377*, 527–535.
82. G. Peng, D. Yang, J. Liu, and S. He, *J. Appl. Polym. Sci. 90*, **2003**, 115–121.
83. A.C. Fozza, J.E. Klemberg-Sapieha, and M.R. Wertheimer, *Plasmas Polym.* **1999**, *4*, 183–206.
84. H. Hiraoka, *IBM J. Res. Dev.* **1977**, *21*, 121–130.
85. M. Shirai, T. Shinozuka, M. Tsunooka, and T. Itani, *Jpn. J. Appl. Phys.* **2003**, *42*, 3900–3904.
86. S. Irie, T. Hagiwara, K. Fujii, Y. Itakura, Y. Kawasa, K. Egawa, I. Uchino, A. Sumitani, and T. Itani, *J. Vac. Sci. Technol., B* **2004**, *22*, 3513–3517.
87. Y. Matsui, S. Umeda, S. Matsui, S. Seki, S. Tagawa, S. Ishikawa, and T. Itani, *Proc. SPIE* **2003**, *5039*, 121–128.
88. H. Fujiwara, H. Fukumura, and H. Masuhara, *J. Phys. Chem.* **1995**, *99*, 11844–11853.
89. T. Lippert and J.T. Dickinson, *Chem. Rev.* **2003**, *103*, 453–485.
90. R. Srinivasan, B. Braren, and R.W. Dreyfus, *J. Appl. Phys.* **1987**, *61*, 372–376.
91. R. Srinivasan, and V. Mayne-Banton, *Appl. Phys. Lett.* **1982**, *41*, 576–578.
92. S.V. Babu, G.C. D'Couto, and F.D. Egitto, *J. Appl. Phys.* **1992**, *72*, 692–698.
93. A. Costela, I. Garcia-Moreno, F. Florido, J.M. Figuera, R. Sastre, S.M. Hooker, J.S. Cashmore, and C.E. Webb, *J. Appl. Phys.* **1995**, *77*, 2343–2350.
94. J.H. Brannon, J.R. Lankard, A.I. Baise, F. Burns, and J. Kaufman, *J. Appl. Phys.* **1985**, *58*, 2036–2043.
95. Y. Kawamura, K. Toyoda, and S. Namba, *J. Appl. Phys.* **1982**, *53*.
96. S. Lazare and V. Granier, *J. Appl. Phys.* **1988**, *63*, 2110–2115.
97. A. Athanassiou, E. Andreou, D. Fragouli, D. Anglos, S. Georgiou, and C. Fotakis, *J. Photochem. Photobiol. A: Chem.* **2001**, *145*, 229–236.
98. V.Y. Shlyapintokh, *Photochemical Conversion and Stabilization of Polymers*, Hanser Publishers, Munich, Germany, 1984.
99. N.J. Turro, *Modern Molecular Photochemistry*, University Science Books, Sausalito, CA, 1991.
100. A. Gilbert, J. Baggot, and P.J. Wagner, *Essentials of Molecular Photochemistry*, CRC Press, Boca Raton, FL, 1991.
101. B. Valeur, *Molecular Fluorescence—Principles and Applications*, Wiley-VCH, Weinheim, 2002.

13

PHOTOGRAFTING OF POLYMERIC MATERIALS

ALI EKREM MUFTUOGLU, MEHMET ATILLA TASDELEN, AND YUSUF YAGCI

13.1 INTRODUCTION

Covalent attachment of polymer chains with different chemical natures onto a substrate, the so-called grafting, has emerged as an elegant technique in achieving targeted physical and mechanical properties. Graft chains can be positioned either along a polymer backbone resulting in new macromolecular structures or along surfaces in 2D, in which case, providing the desired physical properties on the surfaces and interphases without altering the properties of the bulk.

Synthetic strategies associated with grafting employ various modes of polymerizations, for example, living/controlled techniques, thermally initiated or photoinitiated polymerizations. Among these, photochemical technique offers a number of advantages. First of all, it can be conducted at room temperature. This is quite remarkable when low manufacturing costs are concerned for mass production, as well

Photochemistry and Photophysics of Polymer Materials, Edited by Norman S. Allen

as in cases where functional groups are intolerant to high temperatures. In addition, on account of the selective absorption of various chromophoric functionalities built at targeted positions in the macromolecule, reactive sites can be generated as desired. Finally, such systems allow lithographic patterning of surfaces, in which thermally initiated systems would normally fail.

To date, photoinitiating systems have been inadequate in polymerization control. Apart from quasi-living "iniferter" polymerization, coined by Otsu et. al. [1,2], controlled polymerizations have exploited thermal activation in establishing dynamic equilibrium between active and dormant chains. However, in a number of publications, photoinitiation has successfully served as a means of achieving macromolecular architectures. In particular, graft copolymers have been prepared upon irradiating polymer backbones bearing side-chain photoactive moieties. In this manner, copolymers with almost quantitative graft yields were obtained [3,4].

In principle, one can get access to graft structures by the following well-known strategies: (1) "grafting from" (2) "grafting onto", and (3) "grafting through". In the first one, graft chains are grown out of a polymer backbone. Here, the backbone acts as a side-functional macroinitiator in the polymerization of a certain monomer. In the second method, preformed polymer chains carrying antagonist groups at one end are chemically linked to the backbone. The last one involves copolymerizing macromonomers. That is, the backbone is synthesized through the polymerizable moieties attached to termini of polymeric chains, simultaneously yielding the graft copolymer.

This chapter is intended to give an overview of the synthetic photografting approaches and thereby reveals the fundamental aspects in the photografting of polymeric materials.

13.1.1 Photoinitiation

Before proceeding with the photoinduced grafting methods, a brief introduction to photoinitiation is needed for the realization of photografting. However, the reader is encouraged to refer to comprehensive reviews of photoinitiated polymerization [5–8].

Two types of photoinitiation, the so-called "Norrish type I" and "Norrish type II," are known. In the former, two initiating radicals are generated by α-cleavage when the initiator is exposed to UV light. Benzoin derivatives serve efficiently for this purpose (Scheme 13.1). Such chromophores anchored to the polymer trunk, in chain or side chain, will afford block and graft copolymers, respectively. Here, homopolymer formation is an intrinsic outcome of this kind of initiation that originates from the low molar mass initiator fragment.

In the latter, the chromophore itself does not initiate the polymerization. Instead, a coinitiator should be present in the medium as the source of initiating radicals. The chromophore, also termed as sensitizer, absorbs UV radiation and is excited to a triplet state. This stage is accompanied by abstraction of labile hydrogen from the coinitiator to form two radical species. The radical formed on the sensitizer is usually unreactive and does not take part in initiation. (Scheme 13.2) This way homopolymer formation can be prevented. The most common sensitizers are aromatic ketones such as benzophenones with maximum absorbance at 254 nm. Others include thioxanthones

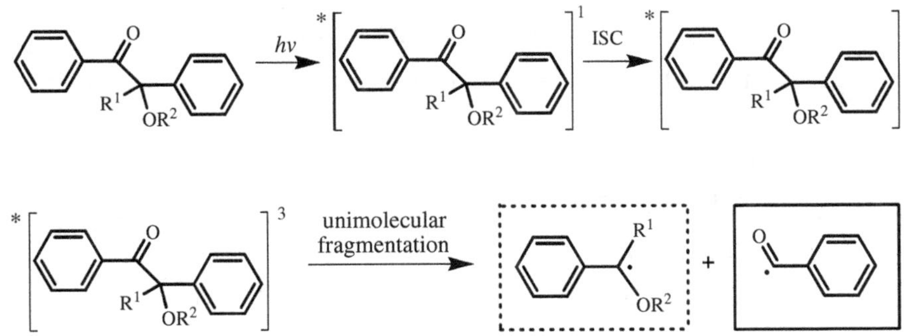

SCHEME 13.1 α-Cleavage type photoinitiation.

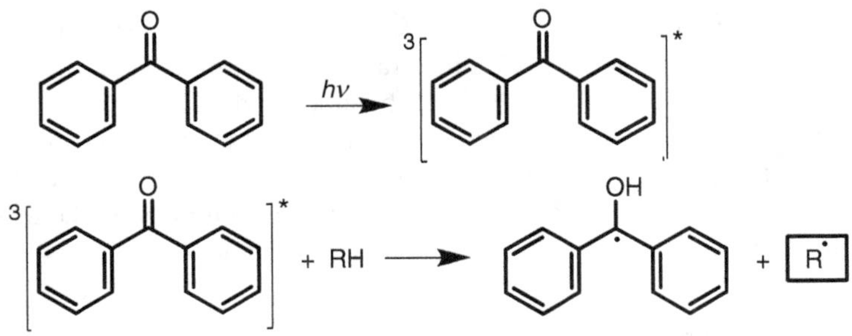

SCHEME 13.2 Hydrogen abstraction type photoinitiation.

and dyes having absorbances in UV–visible region. It is noteworthy that aliphatic ketones, such as butanone, pentan-2-one, pentan-3-one, and heptan-3-one, have been found to be potential sensitizers for photografting when they are in suitable ketone/water/alcohol mixed solvents [9].

Another interesting case is the self-initiation of some acrylic and styrenic monomers upon UV irradiation. Since this process is quite slow, a photoinitiator is normally required. Recent findings on the self-initiation of maleic anhydride, [10] styrene, [11] and a number of acrylic monomers [12] have demonstrated that photopolymerization and photografting could possibly be achieved without using photoinitiators or sensitizers.

Cationic photoinitiators such as onium salts yield cationic species, which are capable of initiating cationic polymerization, upon irradiation. Radical cations and Brønsted acids are generated via irreversible photofragmentation of diaryl iodonium, triaryl sulphonium, and *N*-alkoxy pyridinium salts [13,14] as depicted in Scheme 13.3.

Photoinduced decomposition of the cationic photoinitiator can also be achieved by electron transfer reaction between photoexcited sensitizer and onium salt [15,16] (Scheme 13.4).

The oxidation of photochemically formed radicals by onium salts is another way of producing reactive cations indirectly [17] (Scheme 13.5).

$$On^{+} \xrightarrow{h\nu} On^{+\bullet};\ H^{+}$$

SCHEME 13.3 Photofragmentation of onium salts.

$$(PS)^* + On^+ \xrightarrow{h\nu} PS^{+\bullet} + On^{\bullet}$$

SCHEME 13.4 Electron transfer from photoexcited sensitizers to onium salts.

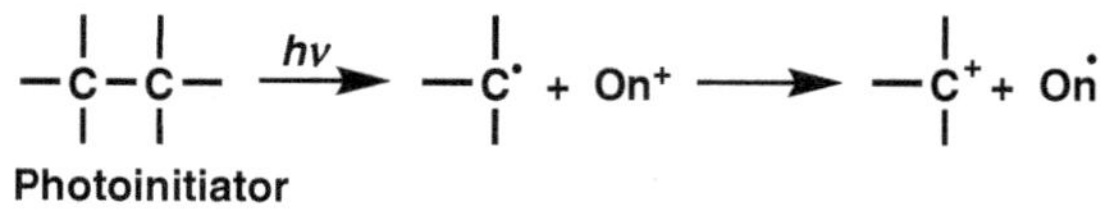

SCHEME 13.5 Oxidation of photogenerated radicals by onium salts.

At this stage, it should be pointed out that a distinction be made between photochemical initiation and irradiation techniques. The former relies on absorption of light in the UV–visible region for either α-cleavage or H-abstraction-type initiation. In the latter, high-energy radiation, such as gamma radiation, is utilized in the generation of radicals on the polymers. UV light can still be employed. However, since no chromophoric groups are involved, the radicals formed are randomly generated on the polymer and site-specific grafting is not possible. UV-assisted insensitized grafting is used in the preparation of brushes and therefore will be mentioned in the text, while irradiation technique is out of the scope and will be excluded.

13.1.2 Surface Grafting

For many years, a great deal of research activity has been carried out in an effort to design macromolecular architectures. Various controlled polymerizations have been applied in making block, graft, star, and hyperbranched polymer structures. Thus, chemically different segments have been combined in a macromolecule with predetermined microstructure, imparting enhanced properties to the material. Indeed, many amazing discoveries have been made in material science in due course. Efforts to prepare complex macromolecules have been directed to a more contemporary issue, namely, surface modification, owing to the need to control the interfacial properties of films and the compatibility of blends [18]. Surfaces can suitably be modified via grafting multiple polymer chains to possess unique properties, such as hydrophilicity, biocompatibility, dyeability, adhesion, antifogging, wear resistance, and printability.

It is not surprising that a vast majority of the literature on surface grafting involves photografting, since, owing to the aforementioned advantages, the method successfully meets the needs of many industrial applications.

13.2 PHOTOINDUCED GRAFTING METHODS

Part of grafting strategies can also be used in association with photoinduced polymerization. As previously stated, "grafting from" methodology consists of polymerization of monomers initiated from active sites distributed along a polymer backbone or a surface. Photoinitiated polymerizations can well be adapted to this

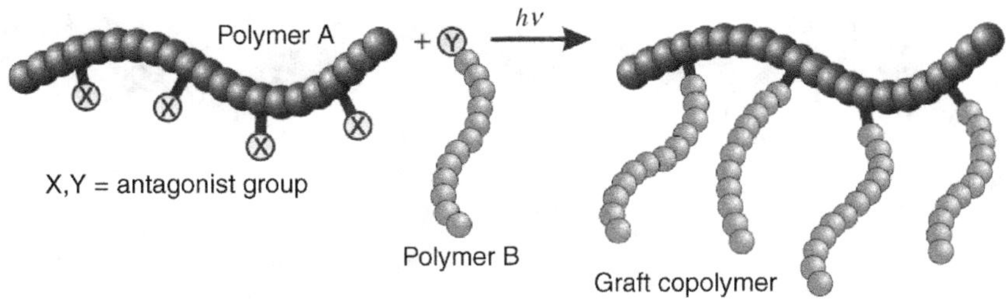

SCHEME 13.6 Synthesis of graft copolymers by photoinduced grafting onto technique.

approach. Both type I and type II initiations may be used effectively. Most of the reported studies are based on this technique.

To the best of our knowledge, no literature work has been reported describing the use of photoinduced "grafting onto" technique, which seems almost impossible to achieve. "Grafting Onto" is best defined as the attachment of end-functional polymeric chains, namely, telechelics, onto specific sites on polymer backbones or surfaces via a chemical reaction (Scheme 13.6).

On the other hand, polymer films can be immobilized onto surfaces quite easily via photochemical ways [19–25]. In principle, this is different from grafting polymer chains to backbones or surfaces since graft chains are linked to substrates through their end-chain functionalities (Scheme 13.6), while film immobilization involves side-chain linkages in the covalent attachment of polymer chains onto surfaces (Scheme 13.7). In a way, this polymer surface coating process is a kind of "grafting onto" as well. Polymer is cast onto a substrate bearing surface-bound benzophenone monolayers and irradiated. Hydrogen atoms on polymers are abstracted by the excited benzophenone resulting radicals on the polymer skeleton besides ketyl radicals (Scheme 13.2). Chains are attached covalently to the surface by coupling of these radicals. Extraction of unbound polymer chains by suitable solvents yields surface-grafted films [24].

"Grafting through" is a common way of preparing synthetic graft copolymers through polymerization of macromonomers, where the mode of initiation is simply of

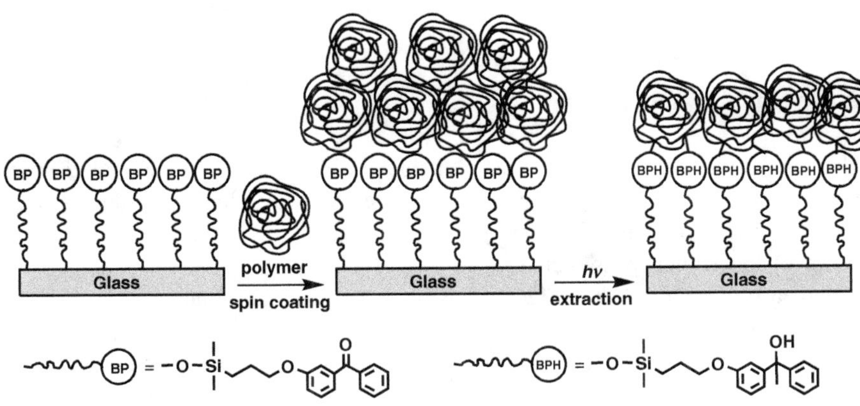

SCHEME 13.7 Schematic representation of polymer surface coating process by photoinduced grafting onto technique.

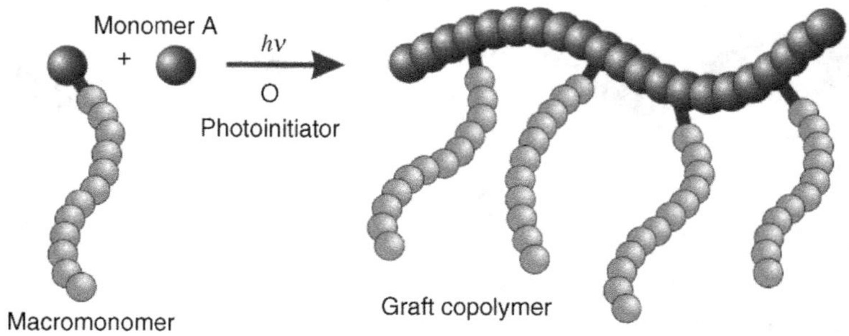

SCHEME 13.8 Synthesis of graft copolymers by photoinduced grafting through technique.

low practical value (Scheme 13.8). Thermally initiated controlled polymerizations are more complementary in "grafting through" strategy than photoinitiated ones. Although a few studies have been reported by several authors [26–28], this technique will not be detailed herein.

13.2.1 Homogeneous Grafting

When the polymeric substance to be grafted has been dissolved in a solution, each and every chain in the sample carries the graft chains after grafting reaction has taken place. Therefore, the substrate is said to be homogeneously grafted. However, this is not the case in surface grafting, wherein the graft chains are merely held on substrate surfaces. As previously noted, usually a modification at the surface molecules only has been intended as to retain the properties in the bulk. Depending on the extent of penetration of the solvent through the substrate surface, the depth of grafting can be adjusted. Surface grafting in thin films or porous membranes sometimes results in homogeneously grafted layers.

Below are provided some examples of photoinduced graft polymerizations in solution, chiefly to elucidate the synthetic approach employing type I and type II initiation. Subsequent pages are devoted to a number of selected surface grafting procedures.

13.2.1.1 Free Radical Systems

α-Cleavage Initiation Initiation can be achieved by the radicals generated upon photolysis of type I initiator moieties at a specific wavelength, as depicted in Scheme 13.9. In this case, homopolymer formation is also expected due to the presence of low molar mass radicals.

In the early work of Norrish, [29] it has been demonstrated that upon irradiation poly(methyl vinyl ketone) in conjunction with acrylonitrile, vinyl acetate, or methyl methacrylate in dioxane solution went through α-cleavage and yielded graft and homopolymers of the corresponding monomers according to the mechanism outlined in Scheme 13.10. The presence of acetaldehyde and methane in the polymerization medium supported the proposed cleavage mechanism. The initial

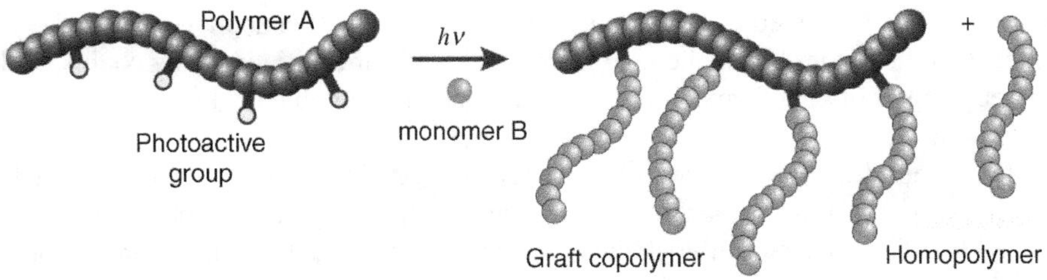

SCHEME 13.9 Homogeneous synthesis of graft copolymers by α-cleavage type photoinitiation.

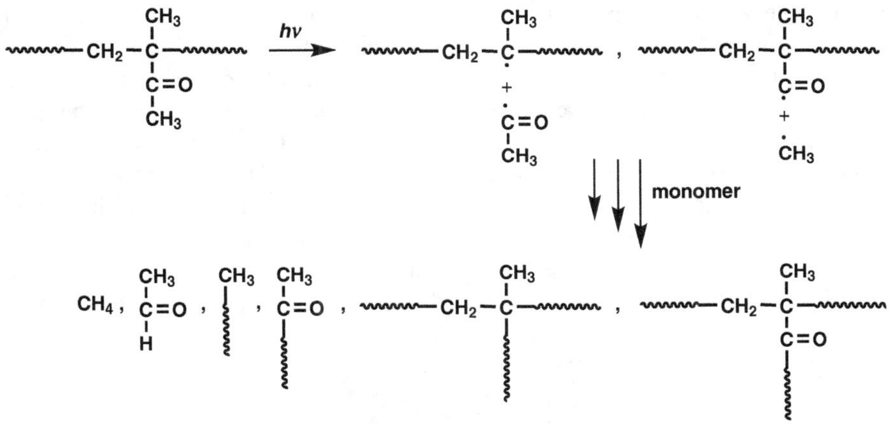

SCHEME 13.10 Photoscission and self-initiation mechanism of poly(phenyl vinyl ketone) in the presence of various monomers.

quantum yield $\Phi(\alpha)$ for photoscission of poly(phenyl vinyl ketone) has been reported elsewhere to be 0.3 [30].

Hydrogen Abstraction Initiation Use is made of type II initiation if the precursor polymer is open to H-abstraction from the side chain by an excited photosensitizer (Scheme 13.11). For instance, tertiary amines are capable of donating hydrogen

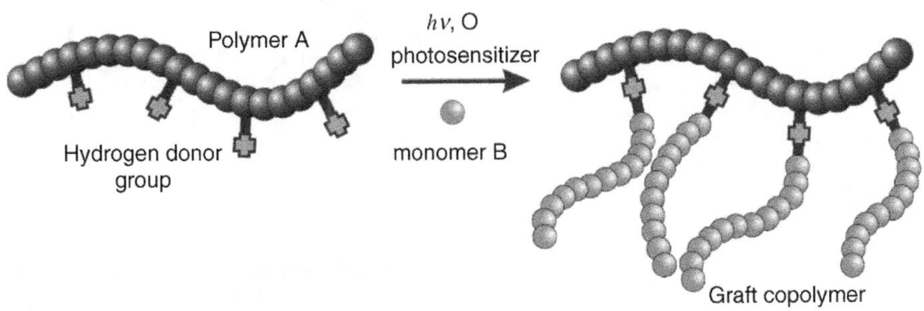

SCHEME 13.11 Homogeneous synthesis of graft copolymers by hydrogen abstraction type photoinitiation.

radicals when irradiated by a UV source in the presence of benzophenone. The ketyl radical formed from benzophenone is unreactive toward initiating new chains, and thereby homopolymer formation is avoided. This is a profitable feature of this kind of initiation.

In a recent study, we have described the preparation of methyl methacrylate brushes on a poly(*N,N*-dimethyl-4-vinylphenethylamine-block-styrene) copolymer backbone [31]. The precursor diblock copolymer was synthesized through living anionic polymerization, in which one of the blocks held single *N,N*-dimethyl amino moiety in each repeating unit. Upon exposure to UV light in the presence of benzophenone, carbon-centered radicals generated via H-abstraction from *N,N*-dimethyl amino groups by the triplet benzophenone initiated the polymerization, the block–graft copolymer eventually being formed (Scheme 13.12). Formation of block–graft copolymer was evidenced by GPC and ^{1}H NMR analyses. Simultaneous formation of cross-linked polymers have also been observed to a greater extent, which was primarily attributed to the participation of unsaturated chain ends, formed by the disproportionation reaction, in further radical attack. It has somewhat been ascribed to macroradical attack by other growing species in the chain termination as well, since methyl methacrylate polymerization is known to be terminated fairly by chain combination.

It is also possible to anchor the sensitizer one way or the other to the trunk polymer. Actually, macrophotoinitiators are frequently used in curing applications and present

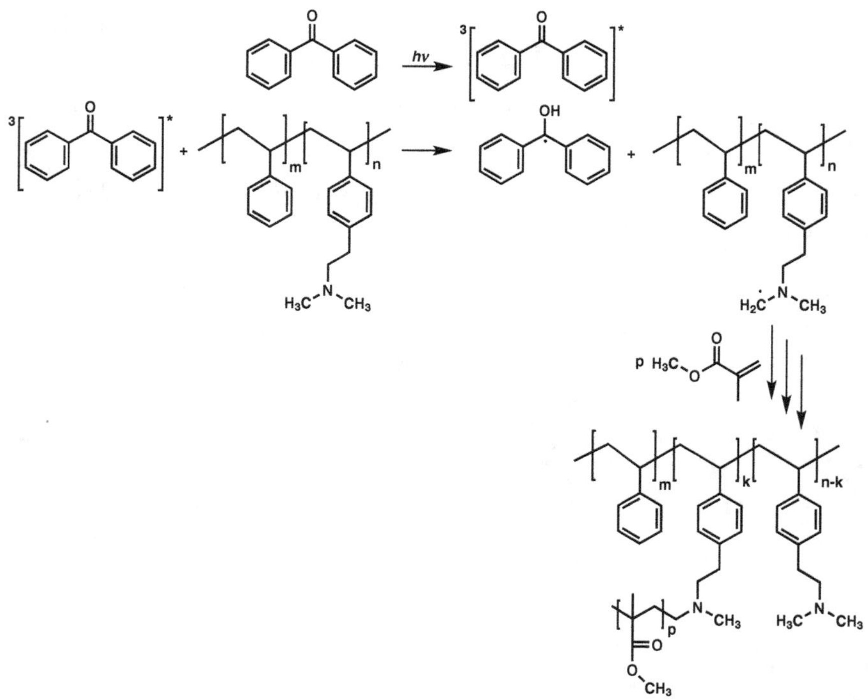

SCHEME 13.12 Photoinduced synthesis of block-graft copolymers using polymeric hydrogen donor and benzophenone.

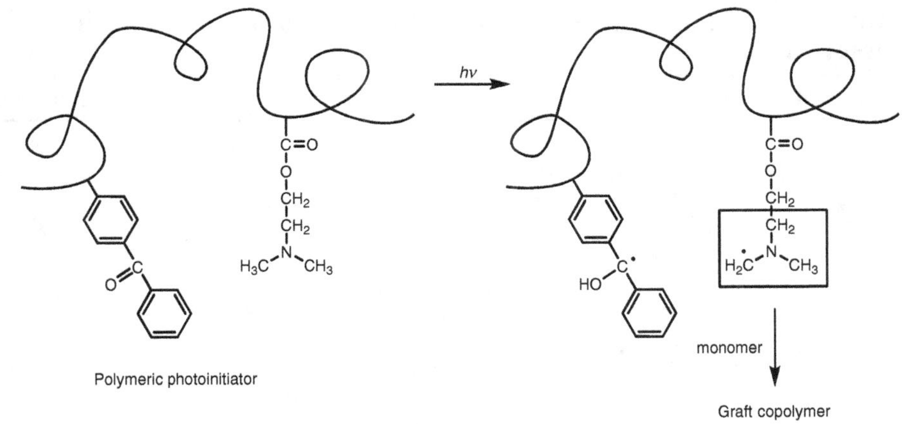

SCHEME 13.13 Synthesis of graft copolymers by one-component type II macrophotoinitiators.

such advantages as non-yellowing, low-odor, low contaminant release, and so on in processing over the low molecular analogues. Additional advantages originate, in the case of photografting, [32] from their macromolecular orientation giving rise to intramolecular reactions that would generate more free radicals and promote grafting while the termination rate of active species is reduced by the low mobility as well as caging of the chains (Scheme 13.13). A series of studies have been carried out by Yin and coworkers in which several sensitizers and coinitiator amines (see Table 13.1) have been incorporated into the same polymer chain and photoinitiation efficiencies assessed for a variety of monomers.

Photoiniferter This technique, developed by Otsu et al., employs dithiocarbamate groups as the iniferter for the controlled growth of polymer chains. An iniferter is a species acting as an initiator, chain transfer agent, and terminator. Upon UV irradiation, a pair of radicals is formed through bond rupture, the mechanism of which is shown in Scheme 13.14. The alkyl radical initiates polymerization, while the dithiocarbamate radical with much less reactivity takes part in the reversible activation and deactivation of the propagating chain end. As long as irradiation is continued, the equilibrium between active and dormant chains prevails and addition of a number of monomers is allowed each time the chain is active in a quasi-living manner. Depending on the location of dithiocarbamate groups in the backbone, block and graft copolymers have been prepared. Table 13.2 lists a variety of photoiniferters used to graft polymerize monomers to specific surfaces.

13.2.1.2 Cationic Systems Photoinduced cationic polymerization can be initiated directly or indirectly upon exposure to UV light in the presence of onium or pyridinium salts with suitable nucleophilic counterions [33]. These initiating systems have been employed in synthesizing block and graft copolymers [34]. For example, side-chain benzoin-containing polymers were used as promoters to yield poly(benzoin acrylate-*g*-cyclohexene oxide) copolymers [35] by using *N*-ethoxy-2-methyl-pyridinium

TABLE 13.1 Various In-Chain or Side-Chain Type II Photoinitiators for Graft Polymerization

Polymeric Photoinitiator	Monomer	References
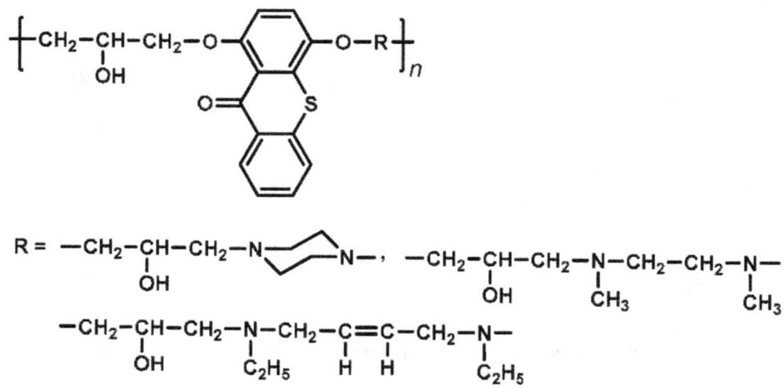	MMA	[36,37]
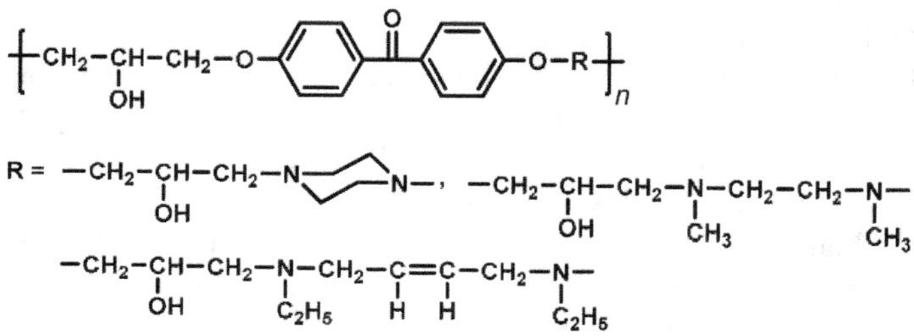	MMA	[38]
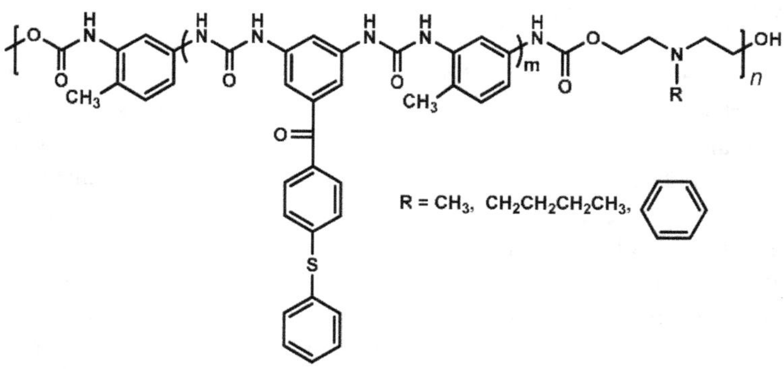	EA	[39]
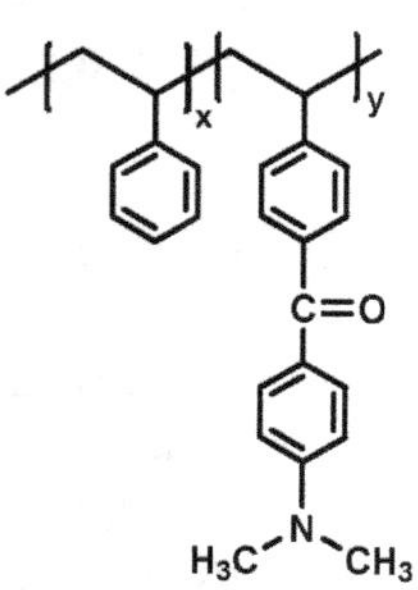	MMA	[32]

Monomers: MMA, methyl methacrylate; EA, ethyl acrylate.

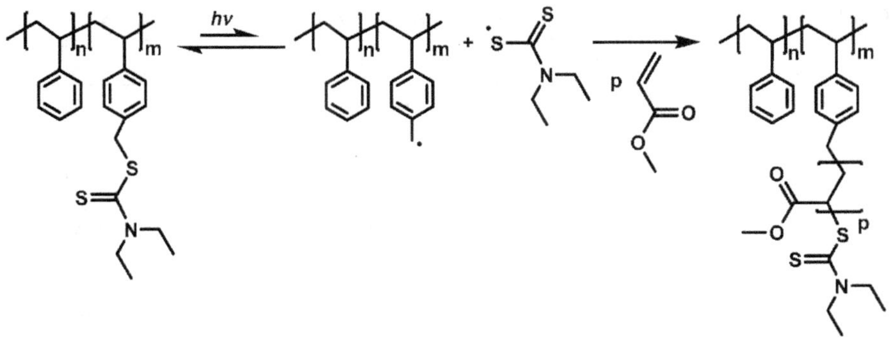

SCHEME 13.14 Photograft polymerization of vinyl monomers by photoiniferter mechanism.

hexafluorophosphate ($EMP^+PF_6^-$) as depicted in Scheme 13.15. Electron donor polymeric benzyl ester radicals are readily oxidized by pyridinium salts to corresponding carbocations that give rise to formation of graft and cross-linked polymers.

A novel well-defined macromonomer of epoxy end-functionalized poly(ε-caprolactone) (PCL) was synthesized and its reactivity in photoinitiated cationic polymerization was examined [28]. PCL macromonomer as the comonomer allowed a rather simple incorporation of PCL side chains into poly(cyclohexene oxide) (PCHO) backbone. This way PCHO-*g*-PCL copolymer with random sequences of the structure shown in Scheme 13.16 is formed.

13.2.2 Surface Photografting

Most of the known procedures for surface grafting are based on "grafting from" technique. The surfaces to be grafted include a wide range of commodity polymers with different forms such as films, membranes, sheets, fibers, foams, granules, besides inorganic surfaces such as gold, silicon, and so on. Some of these surfaces are referred to as unable-to-be-irradiated surfaces due to their shapes or manufacturing specifications [40,41]. Experimental conditions can be selected according to the type of substrate and monomer. Regarding such technical requirements, immersion grafting and vapor-phase grafting procedures may be employed and will be mentioned in the following discussion.

13.2.2.1 Surface Photografting Methods

Mutual Irradiation Technique Mutual irradiation is a direct method in which the substrate, the photoinitiator, and the monomer are simultaneously exposed to UV irradiation. This technique is unable to eliminate the homopolymer formation since a great many radicals caused by irradiation of the monomer will be present in the medium. Only a limited number of works have been reported concerning surface grafting via UV irradiation without utilizing any photoinitiators or sensitizers.

TABLE 13.2 Photoiniferter Polymerization in the Preparation of Surface Brushes

Surface	Photoiniferter	Monomer	References
Glass	DTCS	DMAAm	[42]
		NIPAAm	[42,43]
	XDT	Thiol–DVE3, thiol–VE5015	[44]
		TEGDA,TEGDMA	[44]
	HEMA-E-In	HEMA, NaMAA	[45]
		PEGMA	[45–47]
		MMA, OctMA	[46,47]
	MHEDC, HEDC	DMAAm	[48]
	DC	CMS, DMAAm	[49,50]
		CMS, DMAEMA	[49]
		NIPAAm	[51]
		AA, DMAAm	[52]
Glass, aluminum	CPDTC	MMA, St	[53]
Gold	DTCA	NIPAAm	[54]
Silica	DC	St	[55]
		MMA	[56]
		PEGMA, MAA	[57]
Polyurethane	DC	PEGMA	[58–62]
		OctA, TFEA, flMA, tBA	[60]
		DMAAm	[62]
		HEMA/AA	[63]
		HEMA/DEAEA	[63]
Polystyrene	DC	AAm	[50,64]
		DMAAm, St	[50,65]
		DMAPAAm	[50,65,66]
		MAm,MA, EA, EMA, VP,	[50]
		HEMA, DMAEMA,VA	[50]
		NaAA	[66]
		MAA	[65]
		PEGMA	[67]
		DMAAm/St, AA, BMA, MMA	[68]
	PhSeMSt		[69]
Polystyrene beads	DC	NIPAAm, AA	[70–72]
	XT	AAm, AA, MAA, MMA	[73]
PET	DC	DMAPAAm, HEMA,	[74]
		NaMAA, SMAK	[74]
		DMAAm,	[74,75]
		St	[75]
Poly(phenylene vinylene)	DC	MA, tBA	[76]
Gelatin	DC	NIPAM	[77,78]
Cellulose	DC	EDMA, MAA	[79]

TABLE 13.2 (*Continued*)

Surface	Photoiniferter	Monomer	References
Poly(IB-*co*-CMS)	DC	MMA	[80]
Poly(epichlorohydrin)	DC	MMA, St	[81,82]
Polysilsesquioxane	DC	DMAAm	[83]
Heparin	DC	NIPAM	[84]

Photoiniferters: DTCA, dithiodiundecane-11,1-diylbis[4([(diethylamino)carbonothioyl] thioethyl)phenyl] carbamate; DC, *N,N*-diethyldithiocarbamate; DTCS, *N*-(dithiocarboxy)sarcosine; XDT, *p*-xylene bis(*N,N*-diethyl dithiocarbamate); HEMA-E-In, (methacryloyl ethylenedioxycarbonyl) benzyl *N,N*-diethyldithiocarbamate; CPDTC, 2-cyanoprop-2-yl *N,N'*-dimethyldithiocarbamate; MHEDC, *N*-methyl-*N*-hydroxyethyldithiocarbamate; HEDC, *N,N*-bis(hydroxyethyl) dithiocarbamate; PhSeMSt, *p*-phenylselenomethylstyrene; XT, xanthate.
Monomers: NIPAAm, *N*-isopropylacrylamide; MAA, methacrylic acid; MMA, methyl methacrylate; St, styrene; DMAAm, *N,N*-dimethylacrylamide; PEGMA, poly(ethylene glycol) methacrylate; HEMA, 2-hydroxyethyl methacrylate; DVE3, triethylene glycol divinyl ether; VE5015, Vectomer 5015 vinyl ether; TEGDA, triethylene glycol diacrylate, TEGDMA, tetraethylene glycol dimethacrylate, thiol, pentaerythritol tetra(3-mercaptopropionate); NaMMA, sodium methacrylic acid; OctMA, *n*-octyl methacrylate; AA, acrylic acid; OctA, *n*-octyl acrylate; TFEA, trifluoroethyl acrylate; flMA: fluorescein monoacrylate; tBA, *tert*-butyl acrylate; EDMA, ethylene glycol dimethacrylate; NaAA, sodium acrylic acid; DMAPAAm, *N*-[3-(dimethylamino)propyl]acrylamide; AAm, acrylamide; CMS, chloromethyl styrene; DMAEMA, 2-dimethylaminoethyl methacrylate; MAm, methacrylamide; MA, methyl acrylate; EA, methyl acrylate; EMA, ethyl methacrylate; VA, vinyl acetate; VP, *N*-vinyl-2-pyrrolidone; BMA, butyl methacrylate; SMAK, 3-sulfopropyl methacrylate potassium salt.

Because, in most cases, the radicals generated on the surface are inadequate to afford graft chains with quantitative yields, the addition of a photosensitizer is essential. Once again, the reader's attention is directed to the fact that high-energy radiations such as γ-rays are not regarded as photochemical means.

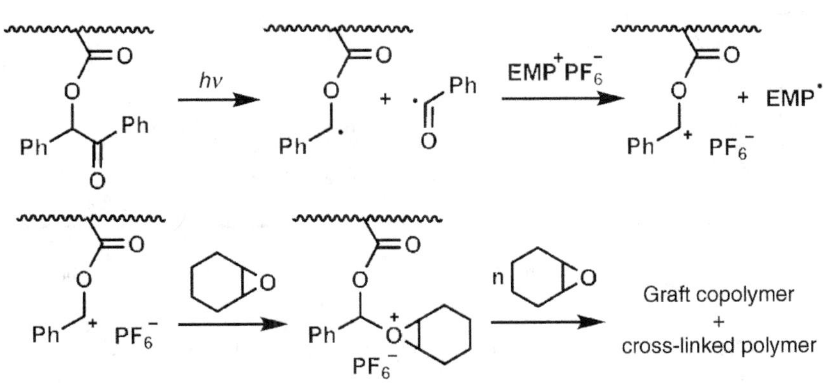

SCHEME 13.15 Synthesis of graft copolymer by indirect photoinduced cationic polymerization

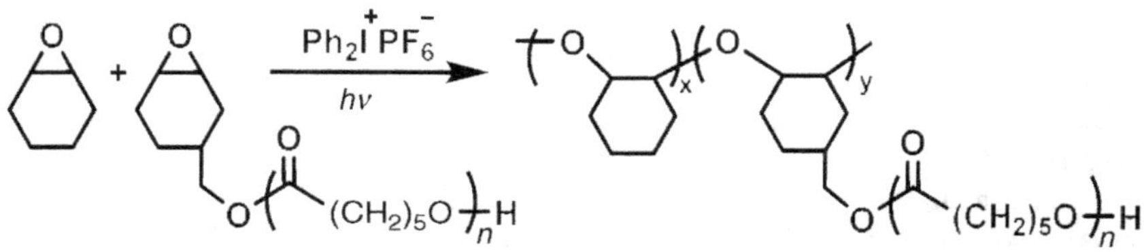

SCHEME 13.16 Synthesis of graft copolymer by direct photoinduced cationic polymerization.

In the work of Uchida et al., using the mutual (simultaneous) irradiation method in the absence of a photosensitizer, polyacrylamide has been grafted onto the surface of a poly(ethylene terephthalate) (PET) film so as to enhance the low water wettability of PET [85]. The PET film immersed in a 10 wt% deaerated solution of acrylamide in water has turned out to be very hydrophilic upon UV irradiation. A pretreatment with benzyl alcohol was carried out to increase the monomer diffusion into the PET matrix, also increasing the grafting efficiency.

Polyethylene (PE) is surface modified via attachment of hydrophilic polymer chains in an attempt to improve wettability, adhesion, lubrication, biocompatibility, and so on. [86,87] In a typical work, methacrylic acid (MAA), and acrylic acid (AA) have been photochemically grafted on high-density polyethylene (HDPE) films using benzophenone as a sensitizer and a variety of solvents [88]. Benzophenone was either dissolved in monomer solution or precoated on PE film. The extent of grafting was maximized as benzophenone coated film was used with the choice of 40%, by volume, acetone/water solution as shown in Fig. 13.1. The authors explained this phenomenon in terms of self-initiation observed as aqueous acetone solution absorbs UV radiation and produces free radicals according to Norrish type I and type II mechanisms (Scheme 13.17). When a hydrogen atom is abstracted from the surface, grafting takes place. Other radicals can form homopolymers or terminate the propagating chains.

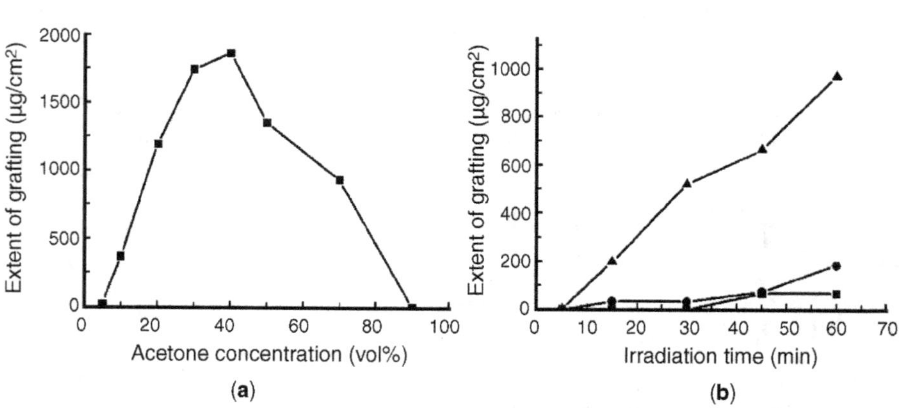

FIGURE 13.1 (a) Change in the of grafting of MAA on PE films with the acetone concentration after 30 min of irradiation (left), (b) grafting of MAA onto PE films: (■) 1 M MAA in acetone with no BP, (•) 1 M MAA in acetone with 1% BP, and (▲) 1 M MAA in acetone with BP precoated on PE (right). Reproduced with permission from Ref. [88], Wiley.

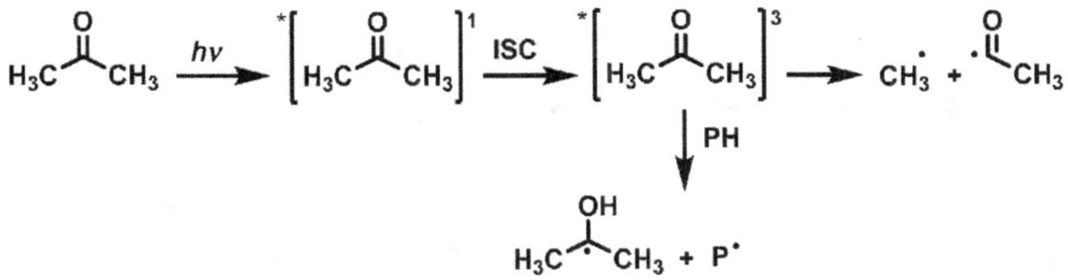

SCHEME 13.17 Photodecomposition and self-initiation mechanism of acetone.

Probably, at higher acetone concentrations, radical recombination and termination are more dominant and hence extent of grafting has been reduced (Fig. 13.1a). The increasing trend, seen in Fig. 13.1b, for the grafting with benzophenone precoated on PE can be attributed to lower degree of quenching by the excited acetone since benzophenone is mostly concentrated at the surface and not in bulk.

Bowman and coworkers have developed a novel sequential UV-induced living graft polymerization method consisting of two steps [89]. In the first step, a surface initiator has been prepared from polymeric substrate and benzophenone upon UV irradiation in the absence of monomer. That is, the semipinacol radicals produced by hydrogen abstraction from the surface polymer recombine with the surface radicals. In the second step, monomer introduced to the polymerization vessel has been polymerized through a living mechanism mediated by the surface-bound semipinacol radicals (Scheme 13.18). The amount of grafting obtained by this method has been determined to be fourfolds that of the one-step grafting method.

Frechet and coworkers [90] compared the efficiency of this two-step approach with that of single-step irradiation in the presence or absence of sensitizers for producing surface grafting of acrylamide, 2-hydroxyethyl methacrylate, vinyl pyrrolidinone, and poly(ethylene glycol) methacrylate onto a poly(butyl methacrylate-*co*-ethylene dimethacrylate) porous monolith with the purpose of reducing protein adsorption. The initiator-free and two-step sequential photografting techniques have turned out to be more efficient than the sensitized single-step photografting for the system of interest.

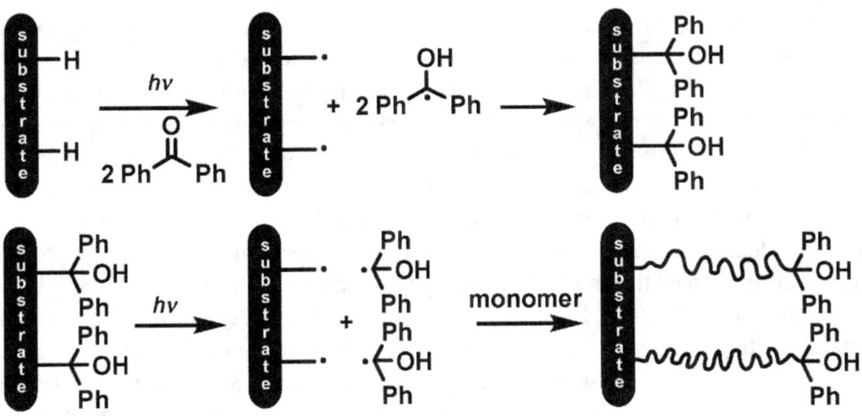

SCHEME 13.18 Schematic representation of photoinduced living graft polymerization.

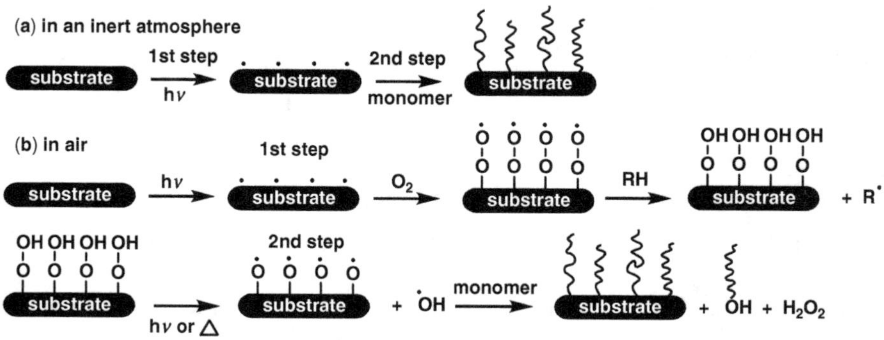

SCHEME 13.19 Pre-irradiation photografting technique: (a) in an inert atmosphere (b) in air.

Preirradiation Technique Another procedure commonly consulted is the preirradiation technique. In this one, the substrate is preirradiated in the presence of air or in an inert atmosphere, followed by addition of the monomer in bulk, in solution or in vapor phase. Oxidation of the surface radicals in air generates surface-bound peroxide groups that are subsequently heated or UV irradiated to generate peroxy radicals. In the other case, the radicals formed at the surface of the substrate in an inert atmosphere or in vacuum survive for a certain period and the introduction of the monomer into the second step affords graft copolymer (Scheme 13.19). The advantage of this method is that the homopolymer formation is prevented (Scheme 13.19a). However, some homopolymer chains may still be formed from hydroxyl radicals as seen in Scheme 13.19b. In addition, films pretreated to possess surface peroxide groups can be stored for a long time before proceeding with grafting. On the other hand, peroxy radicals may initiate oxidative degradation, possibly leading to flaking of the polymer substrate. In any case, specific stabilizers have proved to be successful in partly inhibiting degradation.

In the work of Uyama et al., [91] films of nylon 6, polypropylene (PP), and ethylene vinyl acetate copolymers have been photografted with acrylamide or dimethylacrylamide to make the hydrated surfaces slippery. Here, the preirradiations were either carried out in air with a low-pressure mercury lamp, allowing for the formation of peroxy radicals at the surface or in a deaerated vessel for about 30 min. In the second step, the films bearing the peroxy groups were immersed in an aqueous monomer solution and heated to 50°C after degassing. Other films, preirradiated in the lack of oxygen, were placed in monomer solution containing a small amount of riboflavin, with a simultaneous UV irradiation to yield grafted film. The purpose of irradiation in the second step was to excite riboflavin dye to consume the oxygen, avoiding degassing of the monomer solution. This would presumably prevent inhibition of propagating radicals by oxygen. Grafting has transformed the hydrophobic and nonlubricating films into hydrophilic and slippery surfaces. The coefficient of friction (μ) against a glass plate in water has been measured using a tensile testing machine. Grafting with acrylamide decreased the coefficient of friction (μ) from 30 $\mu g/cm^2$ to less than 0.05 $\mu g/cm^2$, demonstrating a sharp increase in the surface lubricity.

From another point of view, surface photografting techniques can be categorized as liquid-phase (immersion) photografting and vapor-phase photografting. The applicability of these approaches can be extended to modification of unable-to-be-irradiated surfaces, lamination of films, patterning, membrane modification, and so on.

Immersion Photografting As the name implies, the surface to be grafted is immersed in a solution containing sensitizer and monomer or alternatively solutions of each sequentially. The examples given thus far also fall into this category, regardless of the number of steps grafting has been conducted.

Vapor-Phase Photografting Being pretreated with benzophenone, the substrate, in contact with vaporized monomer, is exposed to UV light to achieve grafting [92–94]. Coating benzophenone onto surfaces may be performed by casting the benzophenone solution onto the surface followed by drying [95]. Another way is to expose the substrate to benzophenone vapor in conjunction with vaporized monomer in a solvent-free atmosphere [96]. The advantages of vapor-phase grafting include efficient use of monomer, less homopolymer formation, and easy isolation of the grafted substrate from the polymerization medium. Sometimes, the to-be-grafted surface is dipped in an acetone solution containing around 0.5 wt% poly(vinyl acetate) besides sensitizer [97,98]. The main role of poly(vinyl acetate) addition is to provide homogeneous distribution of the sensitizer on the sample surface.

13.2.2.2 Miscellaneous Applied Surface Grafts

Self-Assembled Monolayers A variety of surface brushes have been prepared bearing self-assembled monolayers (SAMs) as shown in Table 13.3. These include type I and type II photoinitiating surface assemblies from which chains have been elongated, eventually reaching a brush regime. That is when the chains begin to react collectively to environmental stimuli such as changes in pH, temperature, and so on. This unique feature of brushes is the basis of building nano devices for use in high-tech applications.

Among interphases possessing different photoinitiator moieties on the periphery, presented in Table 13.3, dithiocarbamate derivatives surfaces allow a controlled growth of brushes following a "grafting from" pathway. de Boer et al. modified the silicon wafers using organosilane-terminated iniferters so that self-assembled monolayers have been achieved on the surfaces [99]. Upon exposure to UV light, polymerization of styrene and methyl methacrylate has been carried out sequentially, initiated from these monolayered assemblies (Scheme 13.20). Photoiniferter route has allowed for the good control of the grafted layer thickness up to about 100 nm. Presence of the polymer layer on the substrates has been evidenced via both direct and indirect methods such as scanning electron microscopy (SEM), transmission electron microscopy (TEM), contact angles, IR, and X-ray photoelectron spectroscopy (XPS).

Membranes Photografting is a fashionable method in the modification of polymer membranes. [100] Polymeric membranes play an important role in various membrane

TABLE 13.3 Self-Assembled Photoinitiator Monolayers in the Preparation of Brushes

Photoinitiator	Monomer	References
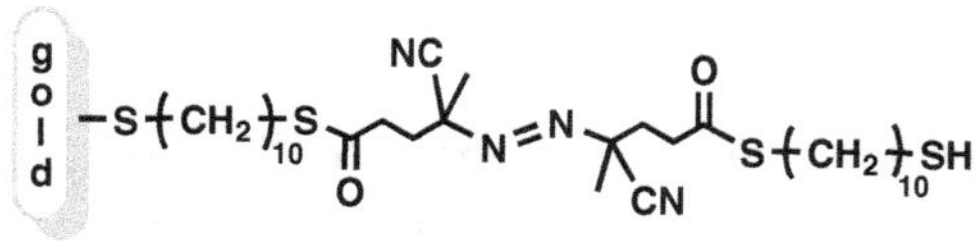	St MMA AN NIPAAm MAA	[101–104] [101] [105] [106] [106]
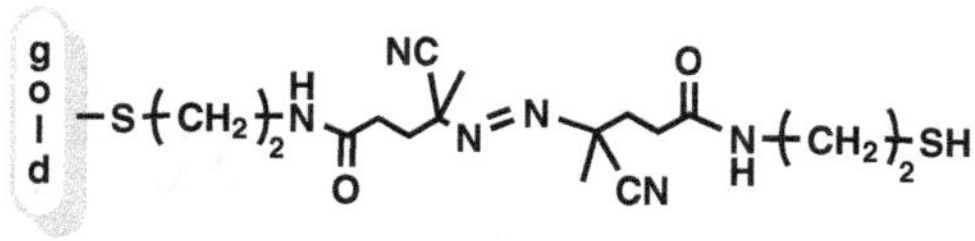	DMAAm MBAA	[107]
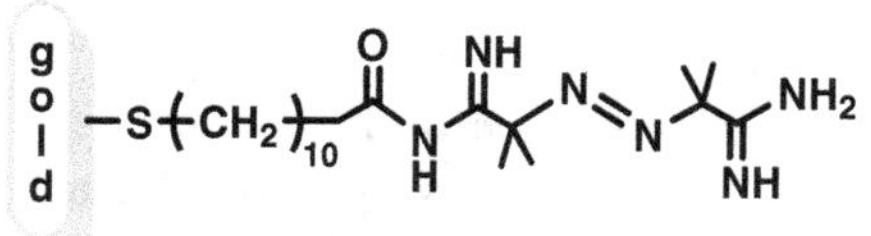	MAA TRIM	[108]
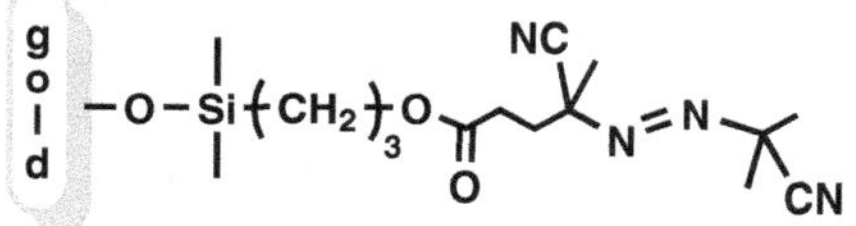	St	[109]
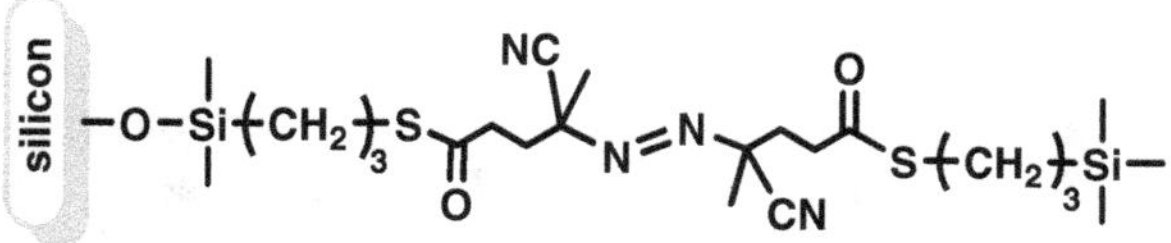	St NDMAA PyMMA AnMMA	[110] [110] [110] [110]
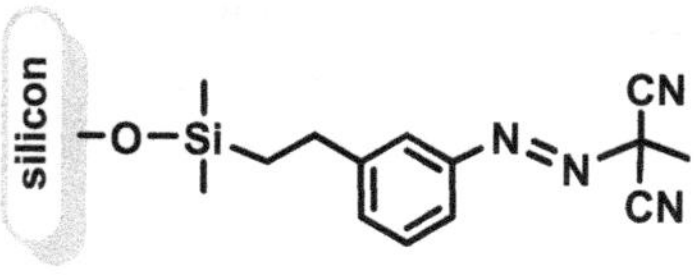	St DMAAm PyMMA AnMMA	[109,110] [110] [110] [110]
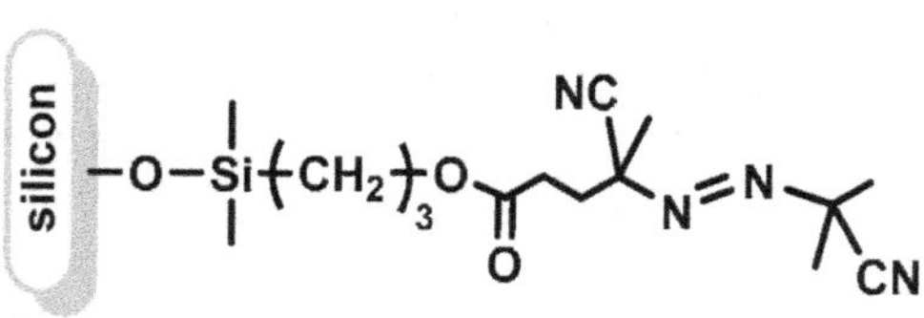	St DMAAm PyMMA AnMMA	[109–111] [110] [110] [110]
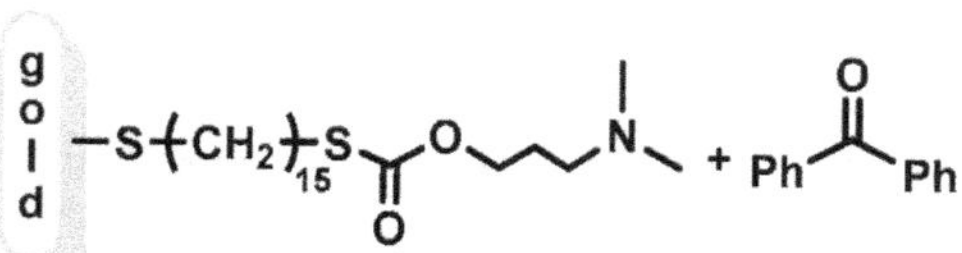	St	[102]

TABLE 13.3 (*Continued*)

Photoinitiator	Monomer	References
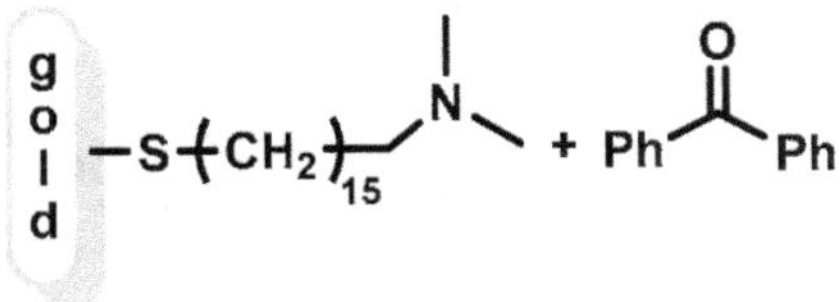	St	[102]
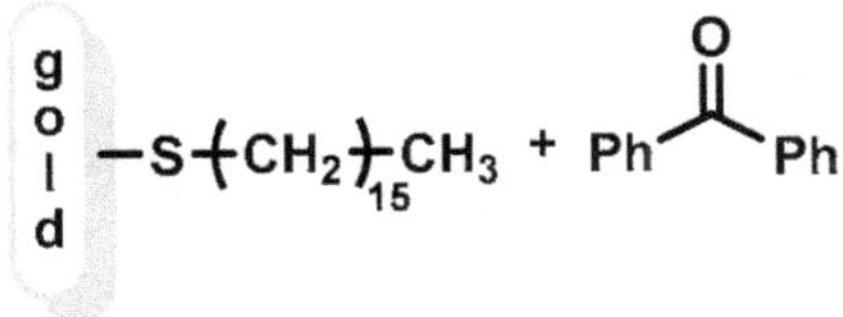	MBAA	[112]
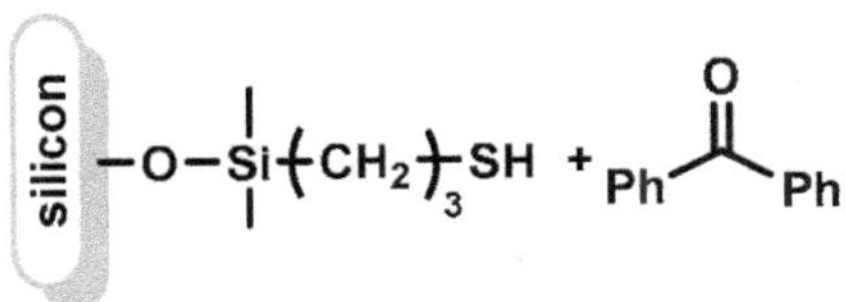	ODVE ODA	[113]
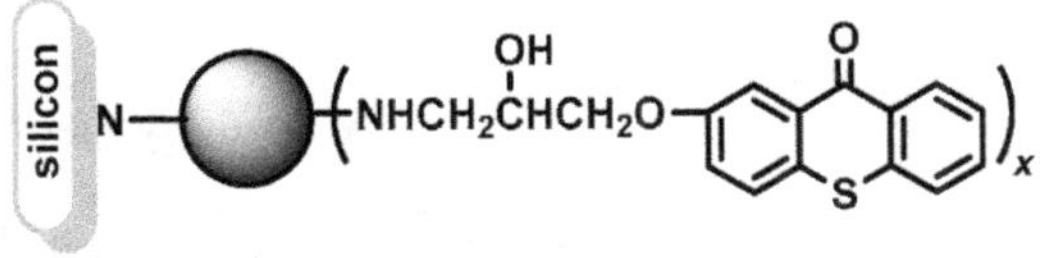	MMA	[114]
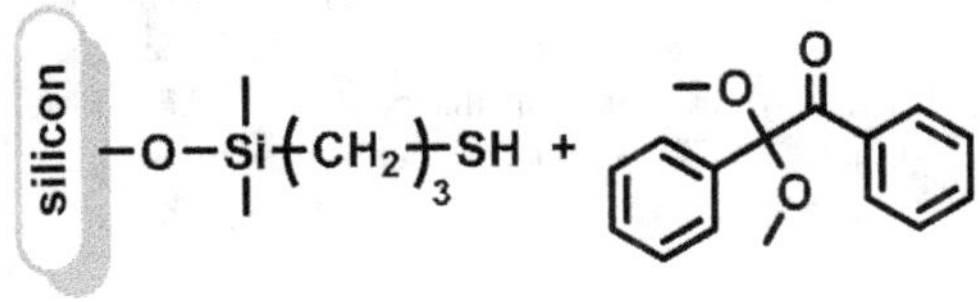	HDT TEGDVE	[115]
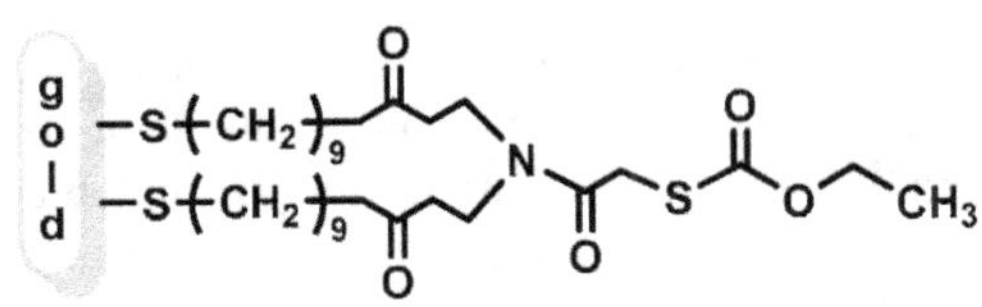	MAA	[116]
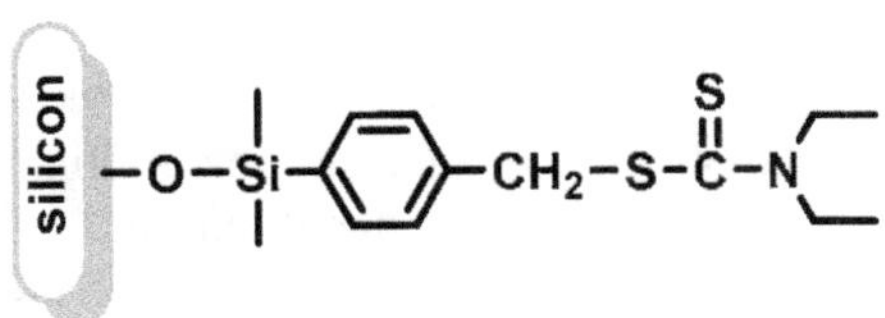	MAA NIPAAm MMA St	[117] [117,118] [99,119–121] [99,122]

(*continued*)

TABLE 13.3 (*Continued*)

Photoinitiator	Monomer	References
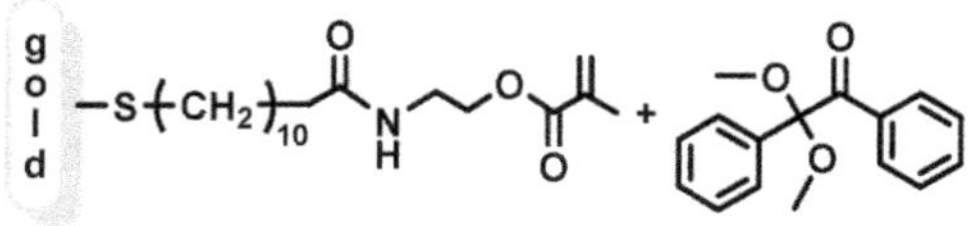	HEMA	[123]
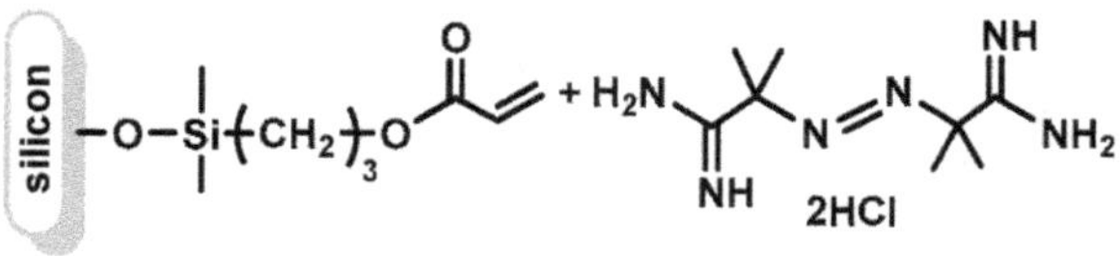	AAm	[124]
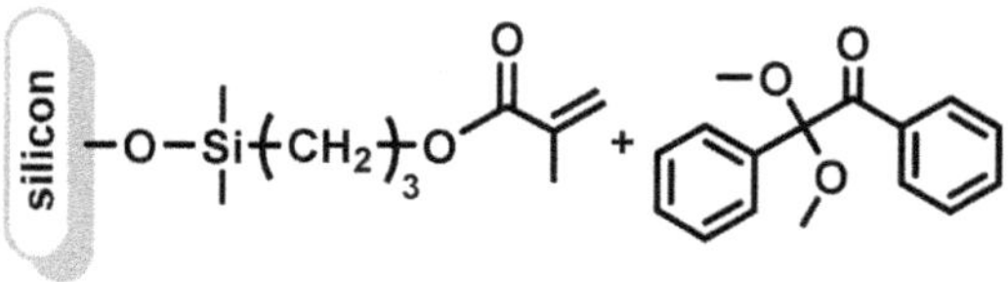	PEG-DA PEG-MA PEG-TA	[125]
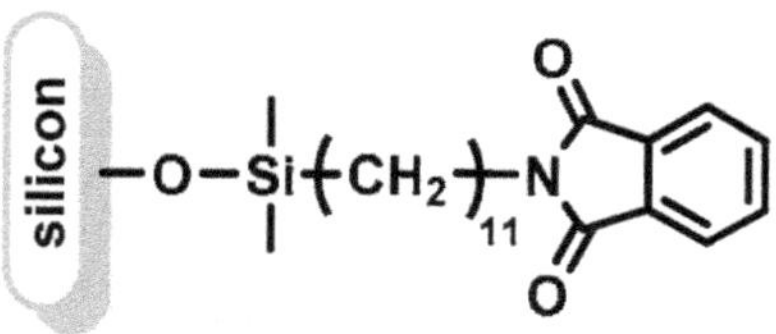	Carbohydrates	[126]

Monomer: St, styrene; MMA, methyl methacrylate; AN, acrylonitrile; NIPAAm, *N*-isopropylacrylamide; MAA, methacrylic acid; NDEAMA, 2-diethylaminoethyl methacrylate; MBAA, *N,N'*-methylene bisacrylamide; TRIM, trimethylolpropane trimethacrylate; ODVE, octadecylvinyl ether; ODA, octadecyl acrylate; DMAAm, *N,N*-dimethylacrylamide; PyMMA, 1-pyrenylmethyl methacrylate; AnMMA, 9-anthracenylmethyl methacrylate; HDT, 1,6-hexane dithiol; TEGDVE, triethyleneglycol divinyl ether; HEMA, hydroxyethyl methacrylate; AAm, acrylamide; PEG-DA, poly(ethylene glycol) diacrylate; PEG-TA, poly(ethylene glycol) tetraacrylate.

separation processes, namely, gas separation, reverse osmosis, pervaporation, ultrafiltration, nanofiltration, microfiltration, as well as in applications such as biomaterials, catalysis, proton conductivity, and so on. Desirable mechanical and separation features can be imparted to hydrocarbon polymer membranes by grafting certain monomers, considering whether they operate in aqueous or nonaqueous conditions.

A novel route to the preparation of proton-conducting membranes has been proposed by Chen et al. [127] The method consists of UV-induced photografting of styrene into polytetrafluoroethylene (PTFE) films and subsequent sulfonation (Scheme 13.21).

It was found that graft chains penetrated through the films and only 7% grafting improved proton conductivity providing better mechanical properties that resemble

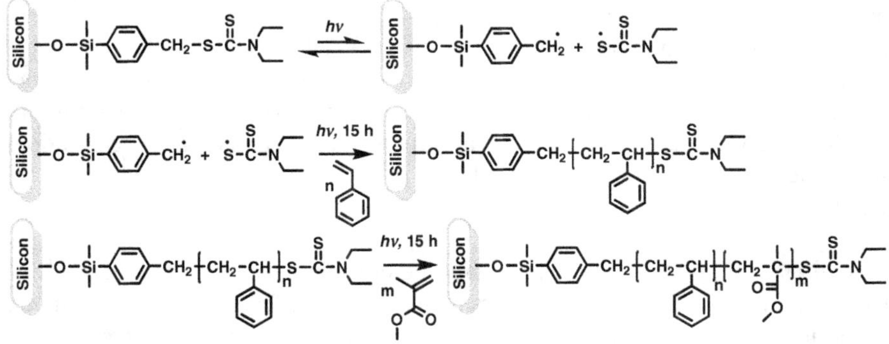

SCHEME 13.20 Self-assembled photoiniferter monolayers in the preparation of brushes.

those of Nafion membrane. A follow-up study [128] includes SEM images of the surface and cross section morphologies, indicating that completely sulfonated graft chains were effectively constructed to form the network for proton transport.

Same authors [129] compared the efficiencies of modified poly(tetrafluoroethylene-*co*-ethyle) films, obtained by vapor- and liquid-phase photografting of styrene, in fuel cell applications. They have found proton conductivities as high as 0.065 and 0.087 S/cm for vapor- and liquid-phase grafts, respectively.

A process comprising photografting poly(acrylic acid) onto polysulfone (PS) ultrafiltration membranes has been studied. [130] The effect of graft parameters such as irradiation time, wavelengths, and monomer concentration on the membrane performance has been examined. Use of *N,N′*-methylene bisacrylamide as a cross-linking agent has resulted in a more stabilized grafting, in which no photoinitiator and no reactant purification are needed. The water softening measurements of the membranes were obtained in dead-end mode and at a 95% recovery. The results demonstrated the feasibility of photografting in the preparation of nanofiltration membranes.

Recently, Ulbricht and coworkers [131] have reported the preparation of low-fouling ultrafiltration membranes by simultaneous photograft copolymerization of hydrophilic poly(ethylene glycol) methacrylate onto a polyethersulfone (PES) membrane. A broad characterization using flux measurement and sieving curve

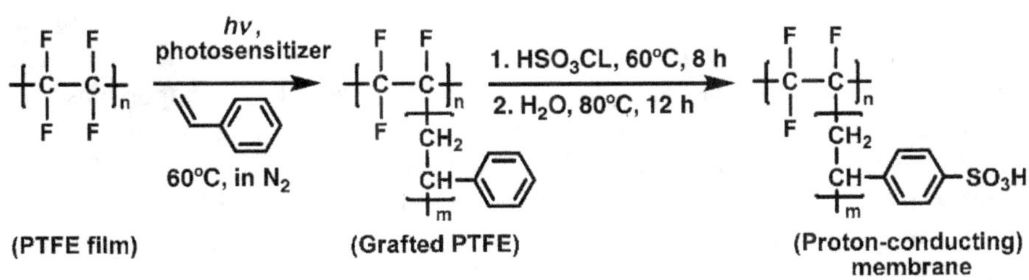

SCHEME 13.21 Preparation of proton-conducting membranes via photografting method.

analysis, FTIR-ATR spectroscopy, contact angle, and zeta potential measurement has been carried out. Membrane performance was assessed utilizing model solutions of sugarcane juice polysaccharides and the protein BSA. All modified membranes showed more resistance to fouling and higher rejection than unmodified ones. Surface functionalization was best achieved with relatively high monomer concentrations (40 g/L) and moderate irradiation periods (1.5–3 min). Some other examples from the literature are presented in Table 13.4.

Lamination of Films A bulk surface grafting procedure has been devised by Yang and Ranby, [132–136] consisting of irradiating a layer of the monomer and initiator placed between two polymer films, as shown in Fig. 13.2. Solution containing the monomer and photoinitiator is introduced to the film surface with a microsyringe as a thin layer having a thickness of 2–5 pm. The films are sandwiched between two quartz plates. The assembled unit is irradiated by UV light from the topside at constant temperature. Photolamination is achieved along with the photografting process. Robust laminates can be prepared from a number of polymer films, for example, polyolefins, polyamides (PA), polyesters, and so on as long as the upper films are transparent to the far UV radiation.

The mechanism of photolamination has been proposed by the authors [133,136,137] as illustrated in Scheme 13.22. It is based on an initial primary grafting reaction followed by a secondary growth period, filling the spaces with branched chains. In most cases, addition of a small amount of cross-linking agents increases the strength of laminates by forming a network between the substrate films.

Lamination by photografting has been employed in obtaining barrier materials. For instance, acrylic acid graft polymerized between two low-density polyethylene (LDPE) films decreases the penetration of oxygen dramatically. Poly(vinyl alcohol) (PVA) interior films photolaminated by acrylic acid polymerization to LDPE films serve as a barrier material for oxygen and water. Such laminates have potential use in packaging industry [137].

Unable-To-Be-Irradiated Surfaces Present techniques employed to modify surfaces via photochemical ways are well established especially with substrates having compact shapes and planar surfaces or with transparent materials that allow the passage of UV light. However, using current techniques, surfaces with opaque or complex shapes are impossible to be modified photochemically, which necessitates the development of new approaches. Recently, Zhang et al. have proposed a novel one-step photografting method to modify the unable-to-be-irradiated surfaces [40,41]. It is well known that sensitizers such as benzophenone have relatively long lifetimes of triplet states. Hence, the sensitizer in vapor phase has been irradiated in one place, absorbing UV light and transforming to a triplet state. Then, it has been transferred to a dark place in which the substrate to be grafted has been located. Here, the excited benzophenone molecules abstract hydrogen atoms from the surface producing radical sites. Benzopinacol-type compounds are produced through combination of the ketyl and alkyl radicals. These compounds could be used as dormant groups in grafting polymerization when monomer is available, preferably

TABLE 13.4 Modification of Polymer Membranes via Photografting Method

Membrane	Monomers	PI	References
Poly(ethylene terephthalate)	NIPAAm	BP	[138,139]
	4VP	BP	[140]
	AA	BP	[40,141]
	NIPAAm	–	[142]
Poly(tetrafluoro ethylene)	St	XT	[127,128]
	HEMA	BP	[143]
	HEA	BP	[143]
	DMAEMA	–	[144,145]
	MMA	–	[145]
Polysulfone	NaSS	HBP	[146]
	NaSS	–	[147,148]
	AC	–	[147]
	AA	–	[130,149–151]
Polyethersulfone	PEGMA	–	[131,152]
	SPE	–	[152]
	AMPS	BP	[153]
	qDMAEMA	BP	[153]
Polypropylene	AA	BP	[40,154–156]
	NVP and MAn	BP	[157]
	NVP and BE	BP	[158]
	AMPS	BP	[159]
	PEGMA	BP	[154]
	DMAEMA	BP	[154,155]
	MAA	BP	[155]
Polyethylene	GMA	BP	[160,161]
	GMA	XT	[162]
	DMAEMA	BP	[163]
	NIPAAm	XT	[164]
	MAA	XT	[164]
	AA	BP	[156]
Low-density polyethylene	St	BP	[40]
	MMA	BP	[40]
	AAm	BP	[40]
	St and MAn	BP	[165]
	NIPAAm	XT	[166,167]
	MAA	XT	[168,169]
	MAA	BP	[155]
	AA	BP	[40,155]
	AA	ITX	[40]
	DMAEMA	BP	[155]
High-density polyethylene	MAA	BP	[155]
	AA	BP	[155]
	DMAEMA	BP	[155]
	NIPAAm	XT	[167]

(continued)

TABLE 13.4 (*Continued*)

Membrane	Monomers	PI	References
Poly(vinylidene fluoride) (PVDF)	AA	BP	[154]
	PEGMA	BP	[154]
	DMAEMA	BP	[154]
Polyacrylonitrile (PAN)	AA	DT	[170]
	qDMAEMA	–	[171]
	NaSS	–	[171,172]
Polyurethane (PU)	AAm	–	[173]
	HEE	–	[173]
	MAA	–	[173]
Polyamide	DEAAm	BP	[174]
Polystyrene (PSt)	NIPAAm	–	[142]
Poly(vinyl chloride) (PVC)	AA	BP	[156]
Poly-L-lactide (PLLA)	HEMA	–	[175]
	MAA	–	[175]
	AAm	–	[175]
Cellulose (C)	NIPAAm	–	[176]
	MAA	DT	[106]
Cellulose acetate (CA)	AA	BP	[154]
	PEGMA	BP	[154]
	DMAEMA	BP	[154]
Polypropylene-*g*-poly(acrylic acid) (PP-*g*-PAA)	NIPAAm	–	[177]
Styrene–butadiene–styrene rubber (SBS)	DMAEMA	–	[178]
	4VP	BEE	[136]
Poly(tetrafluoroethylene-*co*-ethylene) (PETFE)	St	XT	[129,179]
Ethylene vinyl alcohol copolymer (PE-*co*-PVA)	NIPAAm	XT	[167]

Photoinitiators (*PI*): benzophenone, HBP, 4-hydroxybenzophenone; BEE, benzoin ethyl ether; ITX, 2-isopropylthioxanthone; DT, dithiocarbamate.
Monomers: 4VP, 4-vinylpyridine; NIPAAm, *N*-isopropylacrylamide; AA, acrylic acid; PEGMA, poly (ethylene glycol) methacrylate; SPE, *N,N*-dimethyl-*N*-(2-methacryloyloxyethyl-*N*-(3-sulfopropyl)ammonium betaine; AMPS, 2-acrylamido-2-methyl-1-propanesulfonic acid; qDMAEMA, quaternary 2-dimethylaminoethyl methacrylate; St, styrene; HEMA, 2-hydroxyethyl methacrylate; HEA, 2-hydroxyethyl acrylate; DMAEMA, 2-dimethylaminoethyl methacrylate; MAA, methacrylic acid; NaSS: sodium *p*-styrene sulfonate; AC, [(2-acryloyloxy)ethyl]trimethyl ammonium chloride; GMA, glycidyl methacrylate; NVP, *N*-vinylpyrrolidone; MAn, maleic anhydride; BVE: *n*-butyl vinyl ether; AAm, acrylamide; DEAAm, *N,N*-diethylacrylamide; DMAAm, *N,N*-dimethylacrylamide; MMA, methyl methacrylate.

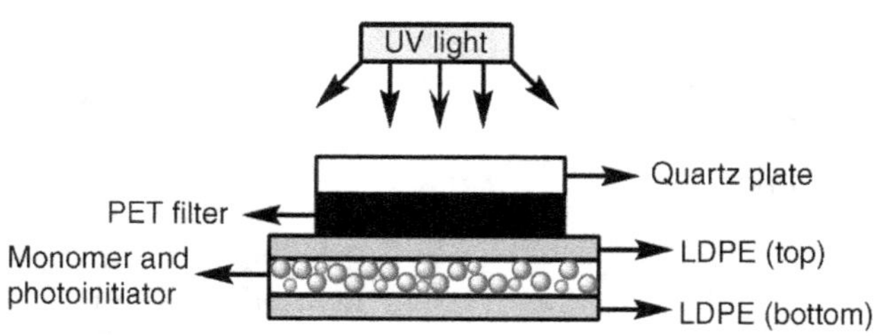

FIGURE 13.2 Photolamination of two LDPE films with a reactive layer of monomer and initiator.

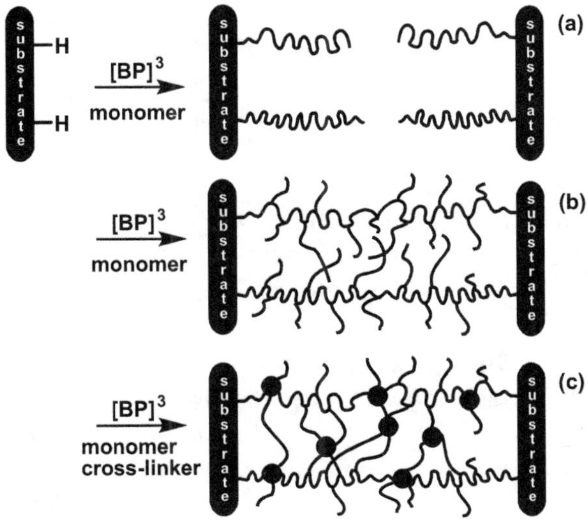

SCHEME 13.22 Schematic mechanism of photolamination by films: (a) primary grafting (b) secondary grafting (c) cross-linking by addition of a cross-linker.

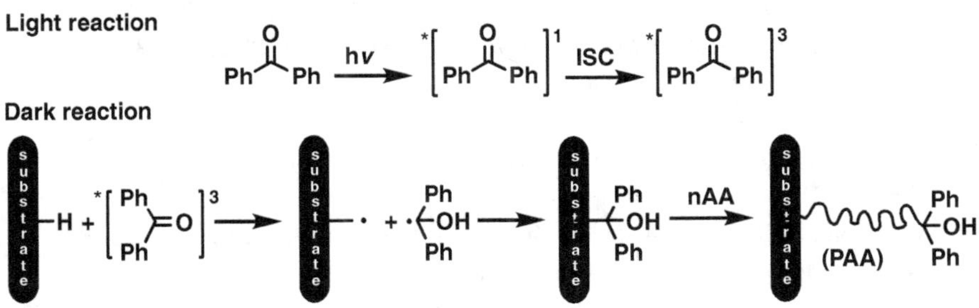

SCHEME 13.23 Mechanism of photografting reaction unable-to-be-irradiated surfaces.

in vapor phase (Scheme 13.23). Grafting was confirmed by UV–vis, ATR-IR, SEM, XPS, and water contact angle measurements.

13.3 CONCLUSION

We have tried to demonstrate all the major photografting approaches regarding homogeneous and surface grafting types. Photografting has appeared to be a versatile tool to properly modify polymeric materials and surfaces, providing desired properties while retaining most of the inherent characteristics. However, there still exist some problems in controlling various factors such as grafting efficiency, homopolymer formation, absorbance selectivity, and so on. Hopefully, new developments to overcome such problems will lead to an improved facility of this technique in commercial applications.

REFERENCES

1. T. Otsu, *J. Polym. Sci. Part A: Polym. Chem.* **2000**, *38*, 2121.
2. T. Otsu, M. Yoshida, *Makromol. Chem. Rapid Commun.* **1982**, *3*, 127.
3. A. Kuriyama, T. Otsu, *Polym. J.* **1984**, *16*, 511.
4. T. Doi, A. Matsumoto, T. Otsu, *J. Polym. Sci. Part A: Polym. Chem.* **1994**, *32*, 2911.
5. M. Kaur, A. K. Srivastava, *J. Macromol. Sci. Part C: Polym. Rev.* **2002**, *C42*, 481.
6. B. M. Monroe, G. C. Weed, *Chem. Rev.* **1993**, *93*, 435.
7. E. Andrzejewska, *Prog. Polym. Sci.* **2001**, *26*, 605.
8. G. Oster, N. L. Yang, *Chem. Rev.* **1968**, *68*, 125.
9. H. L. Wang, H. R. Brown, *Macromol. Rapid Commun.* **2004**, *25*, 1257.
10. J. P. Deng, W. T. Yang, *J. Polym. Sci. Part A: Polym. Chem.* **2001**, *39*, 3246.
11. J. P. Deng, W. T. Yang, B. Ranby, *Macromol. Rapid Commun.* **2001**, *22*, 535.
12. H. L. Wang, H. R. Brown, *Macromol. Rapid Commun.* **2004**, *25*, 1095.
13. Y. Yagci, A. Kornowski, W. Schnabel, *J. Polym. Sci. Part A: Polym. Chem.* **1992**, *30*, 1987.
14. J. V. Crivello, *Adv. Polym. Sci.* **1984**, *62*, 1.
15. Y. Yagci, I. Lukac, W. Schnabel, *Polymer* **1993**, *34*, 1130.
16. J. V. Crivello, J. H. W. Lam, *J. Polym. Sci. Part A: Polym. Chem.* **1979**, *17*, 1059.
17. Y. Yagci, W. Schnabel, *Makromol. Chem. Macromol. Symp.* **1992**, *60*, 133.
18. D. J. Dyer, *Adv. Polym. Sci.* **2006**, *197*, 47.
19. H. Ziani-Cherif, K. Imachi, T. Matsuda, *Macromolecules* **1999**, *32*, 3438.
20. K. M. Defife, K. M. Hagen, D. L. Clapper, J. M. Anderson, *J. Biomater. Sci. Polym. Ed.* **1999**, *10*, 1063.
21. V. T. Bhat, N. R. James, A. Jayakrishnan, *Polym. Int.* **2008**, *57*, 124.
22. M. A. Bartlett, M. D. Yan, *Adv. Mater.* **2001**, *13*, 1449.
23. O. Prucker, C. A. Naumann, J. Ruhe, W. Knoll, C. W. Frank, *J. Am. Chem. Soc.* **1999**, *121*, 8766.
24. N. Adden, A. Hoffmann, G. Gross, H. Windhagen, F. Thorey, H. Menzel, *J. Biomater. Sci. Polym. Ed.* **2007**, *18*, 303.
25. O. Helmecke, A. Hirsch, P. Behrens, H. Menzel, *Colloid. Polym. Sci.* **2008**, *286*, 225.
26. O. Iguerb, P. Bertrand, *Surf. Interface Anal.* **2008**, *40*, 386.
27. I. Capek, *Eur. Polym. J.* **2000**, *36*, 255.
28. M. Degirmenci, O. Izgin, Y. Yagci, *J. Polym. Sci. Part A: Polym. Chem.* **2004**, *42*, 3365.
29. J. E. Guillet, R. G. W. Norrish, *Proc. R. Soc. London Ser. A* **1955**, *233*, 172.
30. I. Naito, W. Schnabel, *Polym. J.* **1984**, *16*, 81.
31. A. E. Muftuoglu, Y. Yagci, K. Se, *Turk. J. Chem.* **2004**, *28*, 469.
32. J. L. Mateo, P. Bosch, F. Catalina, R. Sastre, *J. Polym. Sci. Part A: Polym. Chem.* **1993**, *31*, 153.
33. Y. Yagci, I. Reetz, *Prog. Polym. Sci.* **1998**, *23*, 1485.
34. Y. Yagci, W. Schnabel, *Prog. Polym. Sci.* **1990**, *15*, 551.
35. A. Onen, Y. Yagci, *Eur. Polym. J.* **1992**, *28*, 721.

36. X. S. Jiang, J. Yin, *Macromol. Rapid Commun.* **2004**, *25*, 748.
37. X. S. Jiang, H. Yin, *Polymer* **2004**, *45*, 5057.
38. X. S. Jiang, X. W. Luo, J. Yin, *J. Photoch. Photobio. A* **2005**, *174*, 165.
39. J. Wei, H. Y. Wang, X. S. Jiang, J. Yin, *Macromol. Chem. Phys.* **2006**, *207*, 1752.
40. Z. D. Zhang, L. B. Kong, J. P. Deng, P. Yang, W. T. Yang, *J. Appl. Polym. Sci.* **2006**, *101*, 2269.
41. Z. D. Zhang, L. B. Kong, J. P. Deng, H. Y. Luo, W. T. Yang, *J. Appl. Polym. Sci.* **2007**, *103*, 118.
42. T. Matsuda, S. Ohya, *Langmuir* **2005**, *21*, 9660.
43. A. Idiris, S. Kidoaki, K. Usui, T. Maki, H. Suzuki, M. Ito, M. Aoki, Y. Hayashizaki, T. Matsuda, *Biomacromolecules* **2005**, *6*, 2776.
44. S. K. Reddy, R. P. Sebra, K. S. Anseth, C. N. Bowman, *J. Polym. Sci. Part A: Polym. Chem.* **2005**, *43*, 2134.
45. N. Luo, A. T. Metters, J. B. Hutchison, C. N. Bowman, K. S. Anseth, *Macromolecules* **2003**, *36*, 6739.
46. N. Luo, J. B. Hutchison, K. S. Anseth, C. N. Bowman, *J. Polym. Sci. Part A: Polym. Chem.* **2002**, *40*, 1885.
47. N. Luo, J. B. Hutchison, K. S. Anseth, C. N. Bowman, *Macromolecules* **2002**, *35*, 2487.
48. T. Matsuda, M. Kaneko, S. R. Ge, *Biomaterials* **2003**, *24*, 4507.
49. Y. Nakayama, M. Sudo, K. Uchida, T. Matsuda, *Langmuir* **2002**, *18*, 2601.
50. H. J. Lee, Y. Nakayama, T. Matsuda, *Macromolecules* **1999**, *32*, 6989.
51. S. Kidoaki, S. Ohya, Y. Nakayama, T. Matsuda, *Langmuir* **2001**, *17*, 2402.
52. S. Kidoaki, Y. Nakayama, T. Matsuda, *Langmuir* **2001**, *17*, 1080.
53. M. Yasutake, Y. Andou, S. Hiki, H. Nishida, T. Endo, *Macromol. Chem. Phys.* **2004**, *205*, 492.
54. E. M. Benetti, S. Zapotoczny, J. Vancso, *Adv. Mater.* **2007**, *19*, 268.
55. P. Liu, Z. X. Su, *J. Photoch. Photobio. A* **2004**, *167*, 237.
56. J. Bai, K. Y. Qiu, Y. Wei, *Polym. Int.* **2003**, *52*, 853.
57. B. Ruckert, A. J. Hall, B. Sellergren, *J. Mat. Chem.* **2002**, *12*, 2275.
58. R. P. Sebra, A. M. Kasko, K. S. Anseth, C. N. Bowman, *Sens. Actuators B* **2006**, *119*, 127.
59. R. P. Sebra, K. S. Masters, C. Y. Cheung, C. N. Bowman, K. S. Anseth, *Anal. Chem.* **2006**, *78*, 3144.
60. R. P. Sebra, K. S. Anseth, C. N. Bowman, *J. Polym. Sci. Part A: Polym. Chem.* **2006**, *44*, 1404.
61. R. P. Sebra, K. S. Masters, C. N. Bowman, K. S. Anseth, *Langmuir* **2005**, *21*, 10907.
62. H. J. Lee, T. Matsuda, *J. Biomed. Mater. Res.* **1999**, *47*, 564.
63. H. M. Simms, C. M. Brotherton, B. T. Good, R. H. Davis, K. S. Anseth, C. N. Bowman, *Lab Chip* **2005**, *5*, 151.
64. Y. Nakayama, T. Matsuda, *Macromolecules* **1999**, *32*, 5405.
65. Y. Nakayama, T. Matsuda, *Macromolecules* **1996**, *29*, 8622.
66. Y. Nakayama, J. M. Anderson, T. Matsuda, *J. Biomed. Mater. Res.* **2000**, *53*, 584.
67. Y. K. Joung, J. H. Choi, J. W. Bae, K. D. Park, *Acta Biomater.* **2008**, *4*, 960.
68. Y. Nakayama, T. Matsuda, *Langmuir* **1999**, *15*, 5560.

69. T. S. Kwon, S. Kondo, K. Takagi, H. Kunisada, Y. Yuki, *Polym. J.* **1999**, *31*, 483.
70. S. Tsuji, H. Kawaguchi, *Langmuir* **2004**, *20*, 2449.
71. S. Tsuji, H. Kawaguchi, *Macromolecules* **2006**, *39*, 4338.
72. H. Kawaguchi, Y. Isono, S. Tsuji, *Macromol. Symp.* **2002**, *179*, 75.
73. A. Ajayaghosh, S. Das, *J. Appl. Polym. Sci.* **1992**, *45*, 1617.
74. J. Higashi, Y. Nakayama, R. E. Marchant, T. Matsuda, *Langmuir* **1999**, *15*, 2080.
75. T. Matsuda, J. Nagase, A. Ghoda, Y. Hirano, S. Kidoaki, Y. Nakayama, *Biomaterials* **2003**, *24*, 4517.
76. K. Nagesh, S. Ramakrishnan, *Synth. Met.* **2005**, *155*, 320.
77. S. Ohya, T. Matsuda, *J. Biomater. Sci. Polym. Ed.* **2005**, *16*, 809.
78. N. Morikawa, T. Matsuda, *J. Biomater. Sci. Polym. Ed.* **2002**, *13*, 167.
79. K. Hattori, M. Hiwatari, C. Iiyama, Y. Yoshimi, F. Kohori, K. Sakai, S. A. Piletsky, *J. Membr. Sci.* **2004**, *233*, 169.
80. Y. H. Ma, G. Y. Wu, W. T. Yang, *J. Macromol. Sci. A. Pure Appl. Chem.* **2003**, *A40*, 1147.
81. K. Al-Kaabi, A. J. van Reenen, *J. Appl. Polym. Sci.* **2008**, *108*, 2528.
82. I. Cakmak, H. Baykara, B. Set, *J. Appl. Polym. Sci.* **2008**, *107*, 1604.
83. O. Moriya, S. Yamamoto, T. Kumon, T. Kageyama, A. Kimura, T. Sugizaki, *Chem. Lett.* **2004**, *33*, 224.
84. T. Magoshi, H. Ziani-Cherif, S. Ohya, Y. Nakayama, T. Matsuda, *Langmuir* **2002**, *18*, 4862.
85. E. Uchida, Y. Uyama, Y. Ikada, *J. Polym. Sci. Part A: Polym. Chem.* **1989**, *27*, 527.
86. Y. Uyama, K. Kato, Y. Ikada, *Adv. Polym. Sci.* **1998**, *137*, 1.
87. K. Kato, E. Uchida, E. T. Kang, Y. Uyama, Y. Ikada, *Prog. Polym. Sci.* **2003**, *28*, 209.
88. H. L. Wang, H. R. Brown, *J. Polym. Sci. Part A: Polym. Chem.* **2004**, *42*, 253.
89. H. M. Ma, R. H. Davis, C. N. Bowman, *Macromolecules* **2000**, *33*, 331.
90. T. B. Stachowiak, F. Svec, J. M. J. Frechet, *Chem. Mater.* **2006**, *18*, 5950.
91. Y. Uyama, H. Tadokoro, Y. Ikada, *J. Appl. Polym. Sci.* **1990**, *39*, 489.
92. B. Ranby, *Polym. Eng. Sci.* **1998**, *38*, 1229.
93. K. Allmer, A. Hult, B. Ranby, *J. Polym. Sci. Part A: Polym. Chem.* **1988**, *26*, 2099.
94. B. Ranby, *Makromol. Chem. Macromol. Symp.* **1992**, *63*, 55.
95. Y. Ogiwara, M. Kanda, M. Takumi, H. Kubota, *J. Polym. Sci. Polym. Lett.* **1981**, *19*, 457.
96. A. Wirsen, H. Sun, A. C. Albertsson, *Biomacromolecules* **2005**, *6*, 2697.
97. T. Amornsakchai, H. Kubota, *J. Appl. Polym. Sci.* **1998**, *70*, 465.
98. H. Kubota, Y. Hata, *J. Appl. Polym. Sci.* **1990**, *41*, 689.
99. B. de Boer, H. K. Simon, M. P. L. Werts, E. W. van der Vegte, G. Hadziioannou, *Macromolecules* **2000**, *33*, 349.
100. M. Ulbricht, *Polymer* **2006**, *47*, 2217.
101. J. X. Feng, R. T. Haasch, D. J. Dyer, *Macromolecules* **2004**, *37*, 9525.
102. D. J. Dyer, J. X. Feng, R. Schmidt, V. N. Wong, T. F. Zhao, Y. Yagci, *Macromolecules* **2004**, *37*, 7072.
103. R. Schmidt, T. F. Zhao, J. B. Green, A. J. Dyer, *Langmuir* **2002**, *18*, 1281.

104. R. Paul, R. Schmidt, J. X. Feng, D. J. Dyer, *J. Polym. Sci. Part A: Polym. Chem.* **2002**, *40*, 3284.

105. R. Paul, R. Schmidt, D. J. Dyer, *Langmuir* **2002**, *18*, 8719.

106. M. Kaholek, W. K. Lee, J. X. Feng, B. LaMattina, D. J. Dyer, S. Zauscher, *Chem. Mater.* **2006**, *18*, 3660.

107. M. Lotierzo, O. Y. F. Henry, S. Piletsky, I. Tothill, D. Cullen, M. Kania, B. Hock, A. P. F. Turner, *Biosens. Bioelectron.* **2004**, *20*, 145.

108. T. Piacham, A. Josell, H. Arwin, V. Prachayasittikul, L. Ye, *Anal. Chim. Acta* **2005**, *536*, 191.

109. O. Prucker, J. Habicht, I. J. Park, J. Ruhe, *Mater. Sci. Eng. C* **1999**, *8–9*, 291.

110. X. H. Chen, L. M. Tolbert, C. L. Henderson, D. W. Hess, J. Ruhe, *J. Vac. Sci. Technol. B* **2001**, *19*, 2013.

111. O. Prucker, M. Schimmel, G. Tovar, W. Knoll, J. Ruhe, *Adv. Mater.* **1998**, *10*, 1073.

112. H. Yang, D. Lazos, M. Ulbricht, *J. Appl. Polym. Sci.* **2005**, *97*, 158.

113. E. Besson, A. M. Gue, J. Sudor, H. Korri-Youssoufi, N. Jaffrezic, J. Tardy, *Langmuir* **2006**, *22*, 8346.

114. X. S. Jiang, J. Yin, *Chem. Commun.* **2005**, 4927.

115. V. S. Khire, D. S. W. Benoit, K. S. Anseth, C. N. Bowman, *J. Polym. Sci. Part A: Polym. Chem.* **2006**, *44*, 7027.

116. M. Niwa, M. Date, N. Higashi, *Macromolecules* **1996**, *29*, 3681.

117. S. B. Rahane, J. A. Floyd, A. T. Metters, S. M. Kilbey, *Adv. Funct. Mater.* **2008**, *18*, 1232.

118. L. Liang, P. C. Rieke, G. E. Fryxell, J. Liu, M. H. Engehard, K. L. Alford, *J. Phys. Chem. B* **2000**, *104*, 11667.

119. B. P. Harris, J. K. Kutty, E. W. Fritz, C. K. Webb, K. J. L. Burg, A. T. Metters, *Langmuir* **2006**, *22*, 4467.

120. B. P. Harris, A. T. Metters, *Macromolecules* **2006**, *39*, 2764.

121. S. B. Rahane, S. M. Kilbey, A. T. Metters, *Macromolecules* **2005**, *38*, 8202.

122. S. B. Rahane, A. T. Metters, S. M. Kilbey, *Macromolecules* **2006**, *39*, 8987.

123. I. S. Lokuge, P. W. Bohn, *Langmuir* **2005**, *21*, 1979.

124. B. J. Kirby, A. R. Wheeler, R. N. Zare, J. A. Fruetel, T. J. Shepodd, *Lab Chip* **2003**, *3*, 5.

125. A. Revzin, R. J. Russell, V. K. Yadavalli, W. G. Koh, C. Deister, D. D. Hile, M. B. Mellott, M. V. Pishko, *Langmuir* **2001**, *17*, 5440.

126. G. T. Carroll, D. N. Wang, N. J. Turro, J. T. Koberstein, *Langmuir* **2006**, *22*, 2899.

127. J. H. Chen, M. Asano, Y. Maekawa, T. Sakamura, H. Kubota, M. Yoshida, *Electrochem. Solid-State Lett.* **2006**, *9*, G326.

128. M. Asano, J. Chen, Y. Maekawa, T. Sakamura, H. Kubota, M. Yoshida, *J. Polym. Sci. Part A: Polym. Chem.* **2007**, *45*, 2624.

129. J. H. Chen, M. Asano, Y. Maekawa, T. Sakamura, H. Kubota, M. Yoshidaa, *Electrochem. Solid-State Lett.* **2006**, *9*, G184.

130. S. Bequet, T. Abenoza, P. Aptel, J. M. Espenan, J. C. Remigy, A. Ricard, *Desalination* **2000**, *131*, 299.

131. H. Susanto, M. Balakrishnan, M. Ulbricht, *J. Membr. Sci.* **2007**, *288*, 157.

132. B. Ranby, W. T. Yang, O. Tretinnikov, *Nucl. Instrum. Methods B* **1999**, *151*, 301.
133. B. Ranby, *Int. J. Adhes. Adhes.* **1999**, *19*, 337.
134. W. T. Yang, B. Ranby, *J. Appl. Polym. Sci.* **1996**, *62*, 533.
135. W. T. Yang, B. Ranby, *J. Appl. Polym. Sci.* **1996**, *62*, 545.
136. W. T. Yang, B. Ranby, *J. Appl. Polym. Sci.* **1997**, *63*, 1723.
137. B. Ranby, *Mater. Res. Innovations* **1998**, *2*, 64.
138. C. Geismann, A. Yaroshchuk, M. Ulbricht, *Langmuir* **2007**, *23*, 76.
139. B. Yang, W. T. Yang, *J. Membr. Sci.* **2003**, *218*, 247.
140. B. Yang, W. T. Yang, *J. Membr. Sci.* **2005**, *258*, 133.
141. B. Yang, W. T. Yang, *J. Macromol. Sci. A. Pure Appl. Chem.* **2003**, *A40*, 309.
142. P. S. Curti, M. R. de Moura, W. Veiga, E. Radovanovic, A. F. Rubira, E. C. Muniz, *Appl. Surf. Sci.* **2005**, *245*, 223.
143. K. Yamada, Y. Iizawa, J. Yamada, M. Hirata, *J. Appl. Polym. Sci.* **2006**, *102*, 4886.
144. K. Yamada, T. Gondo, M. Hirata, *J. Appl. Polym. Sci.* **2001**, *81*, 1595.
145. K. Yamada, T. Ebihara, T. Gondo, K. Sakasegawa, M. Hirata, *J. Appl. Polym. Sci.* **1996**, *61*, 1899.
146. A. Akbari, S. Desclaux, J. C. Rouch, J. C. Remigy, *J. Membr. Sci.* **2007**, *297*, 243.
147. A. Akbari, S. Desclaux, J. C. Rouch, P. Aptel, J. C. Remigy, *J. Membr. Sci.* **2006**, *286*, 342.
148. A. Akbari, S. Desclaux, J. C. Remigy, P. Aptel, *Desalination* **2002**, *149*, 101.
149. S. Bequet, J. C. Remigy, J. C. Rouch, J. M. Espenan, M. Clifton, P. Aptel, *Desalination* **2002**, *144*, 9.
150. T. Goma-Bilongo, A. Akbari, M. J. Clifton, J. C. Remigy, *J. Membr. Sci.* **2006**, *278*, 308.
151. J. K. Shim, Y. B. Lee, Y. M. Lee, *J. Appl. Polym. Sci.* **1999**, *74*, 75.
152. H. Susanto, M. Ulbricht, *Langmuir* **2007**, *23*, 7818.
153. N. Hilal, L. Al-Khatib, B. P. Atkin, V. Kochkodan, N. Potapchenko, *Desalination* **2003**, *158*, 65.
154. H. M. Ma, R. H. Davis, C. N. Bowman, *Polymer* **2001**, *42*, 8333.
155. K. Yamada, J. Kimura, M. Hirata, *J. Appl. Polym. Sci.* **2003**, *87*, 2244.
156. V. Costamagna, D. Wunderlin, M. Larranaga, I. Mondragon, M. Strumia, *J. Appl. Polym. Sci.* **2006**, *102*, 2254.
157. C. M. Xing, J. P. Deng, W. T. Yang, *Macromol. Chem. Phys.* **2005**, *206*, 1106.
158. C. M. Xing, J. P. Deng, W. T. Yang, *Polym. J.* **2003**, *35*, 613.
159. S. A. Piletsky, H. Matuschewski, U. Schedler, A. Wilpert, E. V. Piletska, T. A. Thiele, M. Ulbricht, *Macromolecules* **2000**, *33*, 3092.
160. K. Yamada, Y. Saitoh, Y. Haga, K. Matsuda, M. Hirata, *J. Appl. Polym. Sci.* **2006**, *102*, 5965.
161. K. Yamada, R. Nagano, M. Hirata, *J. Appl. Polym. Sci.* **2006**, *99*, 1895.
162. G. S. Irwan, S. Kuroda, H. Kubota, T. Kondo, *J. Appl. Polym. Sci.* **2004**, *93*, 994.
163. K. Yamada, T. Taki, K. Sato, M. Hirata, *J. Appl. Polym. Sci.* **2003**, *89*, 2535.
164. T. Peng, Y. L. Cheng, *Polymer* **2001**, *42*, 2091.
165. H. P. Deng, W. T. Yang, *Eur. Polym. J.* **2005**, *41*, 2685.

166. G. S. Irwan, S. Kuroda, H. Kubota, T. Kondo, *J. Appl. Polym. Sci.* **2003**, *87*, 458.
167. H. Kubota, N. Nagaoka, R. Katakai, M. Yoshida, H. Omichi, Y. Hata, *J. Appl. Polym. Sci.* **1994**, *51*, 925.
168. T. Kondo, H. Kubota, R. Katakai, *Eur. Polym. J.* **1998**, *34*, 1099.
169. L. Q. Zhao, G. S. Irwan, T. Kondo, H. Kubota, *Eur. Polym. J.* **2000**, *36*, 1591.
170. H. Y. Wang, T. Kobayashi, N. Fujii, *J. Chem. Technol. Biotechnol.* **1997**, *70*, 355.
171. T. Kobayashi, K. Kumagai, Y. Nosaka, H. Miyama, N. Fujii, H. Tanzawa, *J. Appl. Polym. Sci.* **1991**, *43*, 1037.
172. T. Kobayashi, N. Fujii, *J. Appl. Polym. Sci.* **1992**, *45*, 1897.
173. J. J. Guan, C. Y. Gao, L. X. Feng, J. C. Shen, *J. Appl. Polym. Sci.* **2000**, *77*, 2505.
174. G. G. Wu, Y. P. Li, M. Han, X. X. Liu, *J. Membr. Sci.* **2006**, *283*, 13.
175. Z. W. Ma, C. Y. Gao, J. Juan, J. Ji, Y. H. Gong, J. C. Shen, *J. Appl. Polym. Sci.* **2002**, *85*, 2163.
176. H. Kubota, N. Shiobara, *React. Funct. Polym.* **1998**, *37*, 219.
177. J. M. Yang, H. T. Lin, *J. Membr. Sci.* **2004**, *243*, 1.
178. J. M. Yang, C. P. C. Chian, K. Y. Hsu, *J. Membr. Sci.* **1999**, *153*, 175.
179. J. H. Chen, M. Asano, Y. Maekawa, T. Sakamura, H. Kubota, M. Yoshida, *J. Membr. Sci.* **2006**, *283*, 373.

14

PHOTOABLATION OF POLYMER MATERIALS

LUKAS URECH AND THOMAS LIPPERT

Photochemistry and Photophysics of Polymer Materials, Edited by Norman S. Allen

14.1 INTRODUCTION

Laser ablation of polymers was first reported by Srinivasan and Mayne-Banton [1] and Kawamura et al. [2] in 1982. Since then, numerous reviews on laser ablation of a large variety of polymers and the different proposed ablation mechanisms have been published [3–11]. There is still an ongoing discussion about the ablation mechanisms, for example, whether it is dominated by photothermal or photochemical processes.

Since its discovery, laser polymer processing has become an important field of applied and fundamental research. The research can be separated into two fields, the investigation of the ablation mechanism and its modeling and the application of laser ablation to produce novel materials. Laser ablation is used as an analytical tool in matrix-assisted laser desorption/ionization (MALDI) [12,13] and laser-induced breakdown spectroscopy (LIBS) [14] or as preparative tool for pulsed laser deposition (PLD) of synthetic polymers [15,16] and of inorganic films [17,18].

The industrial applications for polymers in laser ablation can be divided into two main groups: in applications where a structure is produced in the polymer and into a second group, where the ablation products are of specific interest.

Structuring of polymers today is industrially used for the production of nozzles for inkjet printers [19] and to prepare via holes in multichip modules through polyimide (PI) by IBM [20], as well as for many other applications, for example, fabrication of microoptical devices [21] and microfluidic channels [22–25].

Examples for the second group are polymers as fuel in the micro laser plasma thruster (μLPT), PLD of polymers, matrix-assisted pulsed laser evaporation (MAPLE), which is a deposition technique that can be used to deposit highly uniform thin films [26], or laser-induced forward transfer (LIFT) [27–29], which can be used to produce microstructures by transferring an irradiated area of a target film to an acceptor substrate. The polymer can be the transferred material, or just functions as driving force in the transfer.

14.1.1 Fundamental Issues of Laser Ablation

For an understanding of polymer ablation the main ablation parameters have to be explained and their definition have to be discussed in detail. Also, the most frequently proposed ablation mechanisms and models will be discussed.

14.1.1.1 Ablation Parameters The main parameters that describe polymer ablation are the ablation rate, $d(F)$, and the ablation threshold fluence, F_{th}, which is defined as the minimum fluence where the onset of ablation can be observed. A third important parameter is the effective absorption coefficient, α_{eff}, which yields information on the mechanisms that take place in the ablation process when compared to the linear absorption coefficient, α_{lin}, that is measured on thin unirradiated polymer films.

The ablation process is often described by the following equation [30,31]:

$$d(F) = \frac{1}{\alpha_{eff}} \ln\left(\frac{F}{F_{th}}\right) \tag{14.1}$$

Also, the method how the ablation parameters are acquired can have a pronounced influence on the results. The ablation rate can be defined either as the depth of the ablation crater after one pulse at a given fluence, or as the slope of a linear fit of a plot of the ablation depth versus the pulse number for a given fluence. Very different ablation rates can result from the two different measurement methods. This is especially the case for materials where ablation does not start with the first pulse, but after multiple pulses, or if the ablation crater depth after one pulse is too small to be measured. The process that occurs if ablation does not start with the first laser pulse is called *incubation*. It is related to physical or chemical modifications of the material by the first few laser pulses, which often results in an increase of the absorption at the irradiation wavelength [32,33], for example, the formation of double bonds in poly (methylmethacrylate) (PMMA). Incubation is normally observed only for polymers with low absorption coefficients at the irradiation wavelength.

The method applied to measure the depth of the ablated area or the removed mass can also have an influence on the ablation parameters. If profilometric measurements (optical interferometry, mechanical stylus [34], or atomic force microscopy [35]) are used to calculate the ablation rate, a sharp ablation threshold can be defined. This is also supported by reflectivity [36] and acoustic measurements [37]. In mass loss measurements, such as mass spectrometry or with a quartz crystal microbalance (QCM), the so-called *Arrhenius tail* [38] has been observed for certain conditions. The Arrhenius tail describes a region in the very low fluence range, where a linear increase of detected ablation products is observed, which is followed by a much faster increase, that coincides with removal rates of the profilometric measurements [39].

Even if these different approaches are taken into account, it is often the case that the ablation rate cannot be defined with a single set of parameters. Therefore, one set of parameters has to be defined for each fluence range in which different processes dominate the ablation process and thereby influence the ablation rate. In Fig. 14.1, the dependence of the ablation rate on the irradiation fluence is illustrated as a generic

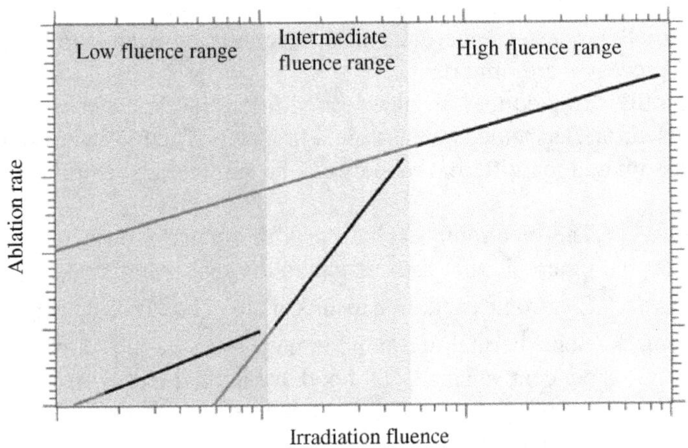

FIGURE 14.1 Schematic plot of the three fluence ranges that are typically observed for polymers. The three ranges are indicated with different shades of gray.

scheme, which is typical for most polymers. The intersection of the gray extensions of the schematic ablation rates (black lines) with the x-axis of the ablation rate versus irradiation fluence plot is the threshold fluence and varies for each fluence range. Also, a different effective absorption coefficient can be defined for each region.

Three fluence ranges are visible, which can be characterized as follows:

Low Fluence Range:

- From this fluence range, the ablation threshold fluence is normally defined.
- Incubation can be observed at these low fluences.

Intermediate Fluence Range:

- Increase of the slope of the ablation rate that is caused by a more efficient decomposition of the polymer. Energy that has been gained from an exothermic decomposition of the polymer can also increase the ablation rate.

High Fluence Range:

- The incident laser light is screened by solid, liquid, and gaseous ablation products and the laser produced plasma. This leads to similar ablation rates for many polymers [5] at high fluences.

It is therefore of great importance to consider not only the values for the different ablation parameters, but also information about the technique of analysis and for which fluence range they are valid. An interpolation to values beyond the measurement range is also not advisable, as not all three ranges have to be present for all polymers and irradiation conditions.

14.1.2 Ablation Mechanisms

Even after 25 years of research in the field of laser polymer ablation, there is still an ongoing discussion about the ablation mechanisms, for example, whether in addition to photothermal processes, photochemical reactions or even photophysical and mechanical processes are important.

It is generally accepted that for nanosecond laser pulses, the energy of the laser photons is used for electronic excitation in a first step. The following steps are still under discussion and the different models can be summarized as followed:

Photothermal: The electronic excitation is thermalized on a picosecond (ps) timescale that then results in thermal bond breaking [40–44].

Photochemical: Electronic excitation results in direct bond breaking [5,30,45–47].

Photophysical: Both thermal and nonthermal processes play a role. Two independent channels of bond breaking [48,49] or different bond breaking energies for ground-state and electronically excited-state chromophores are applied [50,51] in this model. It is most adequate for short laser pulses in the picosecond and femtosecond ranges [52].

Another way to distinguish the different models is by separating them into *surface* and *volume* models. In the volume models, the different processes that eventually result in ablation take place within the bulk of the material. In the surface models, the processes that are responsible for the material removal take place within a few monolayers below the surface. The different models can be described as follows:

Photochemical Surface Models: Valid for long pulses and/or higher irradiation fluences [53].

Thermal Surface Models: These models are mainly developed for metal ablation and do not consider the sharp ablation threshold, but can describe the occurrence of an *Arrhenius tail* [43,44,49,54].

Photochemical Volume Models: These models describe a sharp ablation threshold and a logarithmic increase of the ablation crater depth with the number of laser pulses, but the *Arrhenius tail* is not accounted for [3,5,30,45,46]. A linear dependence can be described with models that consider the motion of the ablation front, but ignore the screen effects caused by the plasma plume.

Thermal Volume Models: These models are often oversimplified by neglecting the movement of the solid–gas interface and therefore result in too high temperatures [38,42]

Volume Photothermal Model: In this model by Arnold and Bityurin [55], a thermal surface model and a photochemical volume model have been combined. In this model, it is assumed that photothermal bond breaking takes place within the bulk polymer. When a density of broken bonds reaches a critical value, ablation begins. This model can account for sharp ablation thresholds and Arrhenius tails.

A new coarse-grained chemical reaction model (CGCRM) has been proposed by Garrison and coworkers [56,57]. In this model a kinetic Monte Carlos approache that includes a probabilistic element is used to predict when reactions occur. It is thereby possible to avoid the use of a chemically correct interaction potential. The CGCRM uses known chemical reactions along with their probabilities and exothermicities for a specific material to estimate the effect of chemical reactions on the ablation process.

A new coarse grained molecular dynamics model was developed to study the role of thermal, mechanical and chemical reactions in the onset of the ablation process of PMMA [58]. In this model, the laser energy is absorbed in different ways, i.e. pure heating and Norrish type I and II reactions. Mechanical stresses and pressure are dominant for very short pulses in the stress confinement regime and can initiate ablation by a mechanical breakdown of the polymer in the case of pure heating. For longer pulse lengths, the ejection process is mainly thermally activated. This can be well described with thermal models based on thermally activated bond breaking processes. The presence of small molecules and gaseous products can not be accounted for by a purely thermal mechanism. A modeling of the photoablation channels requires a two-step ablation model, which incorporates the effect of photolysis of the polymer and the creation of new species, that is then followed by a thermally activated removal step.

The breathing sphere model was enhanced by Garrison and coworkers [59–62] to allow the photons to break a bond in the molecule and to describe subsequent abstraction and recombination reactions. The model was initially applied to chlorobenzene, where good correlation with experimental data was found.

The new concept that arises from these calculations is the difference in the temporal and spatial deposition of the available energy in photochemical and photothermal mechanisms. This concept provides the foundation to make specific comparisons with experiment and to explain experimental results as summarized below:

- It was found that photochemical reactions release additional energy into the irradiated sample and decrease the average cohesive energy and thereby decrease the value of the ablation threshold. The yield of emitted fragments becomes significant only above the ablation threshold.
- The presence of a shockwave with a high initial velocity, large clusters in the plume, and high velocities of particles in the plume are explained by the fast rise in energy deposition in time scales from 20 to 150 ps.
- The chemical reactions that take place above the surface after the laser pulse on longer timescales explain the higher background density in the plume with photochemical ablation than observed for photothermal ablation.
- The presence of a combination of a thermal mechanism below the ablation threshold and a volume ejection mechanism above the threshold explains why volatile products such as HCl and the matrix are only observed below threshold and all products are observed above threshold.
- The absence of heat deposited below $\sim$1.5 times the penetration depth may help to explain the cold etching process in far UV photoablation as used commercially in the corrective eye surgery, LASIK.

The different models include many material parameters and several of these parameters are obtained from fitting of experimental data and have to be adjusted to fit each polymer [9,63]. It is worth mentioning in this context that polyimide is probably the most studied polymer in laser ablation and is also the material for which most ablation models are applied, but great care has to be taken for which type of polyimide the data have been obtained. Polyimide describes a whole group of polymer that can range from soluble polymers to insoluble films and even photosensitive polymers with very different properties [10]. Even products with the same name, such as Kapton™, are not one polymer, but there are also many different types of Kapton that are defined with additional letters, for example, HN.

In general, it can be said that polymers that show a photochemical ablation behavior at the irradiation wavelength would be preferable for structuring, as the damage of the surrounding material due to a thermal processes is minimized and less carbonization is observed. A conversion of the polymer into gaseous products is also of advantage, as no or only minor amounts of ablation products are redeposited on the structured surface and additional cleaning procedures may not be necessary.

14.2 CLASSIFICATION OF POLYMERS USED FOR LASER ABLATION

Polymers can be classified in many different ways, such as their ablation behavior, the mechanisms that take place during ablation, and their suitability for selected applications.

Classification by Decomposition Behavior: The decomposition mechanism is a reasonable way to classify polymers. They can either depolymerize upon irradiation, for example, poly(methylmethacrylate), or decompose into fragments such as polyimides or polycarbonates. This method of classification is closely related to the synthesis of polymers. Polymers that are produced by radical polymerization from monomers, which contain double bonds, are likely to depolymerize into monomers, while polymers that have been formed by reactions like polycondensation will not depolymerize into monomers upon irradiation, but will be decomposed into different fragments. The second group cannot be used to produce films with the same structure or molecular weight as the original material with methods such as PLD.

Classification by the Absorber: Polymers can have intrinsic absorption at the irradiation wavelength or a dopant can be used to induce the necessary absorption. The dopant can be implemented either into the polymer matrix on a molecular level or as absorber particles in the nanometer to micrometer range. The absorbing species can also be included into the polymer backbone or side chains to increase the absorption and thereby forms a copolymer with intrinsic absorption.

By Their Availability: Many *commercially available polymers* such as poly (methylmethacrylate), poly(vinylchloride), Poly(ethylene terephthalate) (PET), and so on have been extensively studied in recent years [64], but they have several characteristics that make them unsuitable for high-quality structuring. The most prominent are low sensitivity, carbonization upon irradiation, and redeposition of ablation products on the polymer surface [65]. To improve the ablation properties for specific applications, polymers or polymer-absorber systems were designed to fit special requirements. The most important criteria for these *doped and designed polymers* for structuring are as follows:

- High absorption coefficients ($\geq$20,000 cm^{-1}) at the irradiation wavelength.
- Exothermic decomposition at well-defined positions of the polymer backbone.
- Decomposition of the polymer into gaseous products that do not contaminate the polymer surface [66,67].

14.3 LASER SOURCES

The most frequently used lasers for polymer ablation are *excimer (excited dimer)* lasers [68,69]. Their main advantages are the emission in the UV range, where most polymers reveal a high absorption, a relatively homogeneous beam profile, and the possibility to use a mask to image a small section of the laser beam on the sample. Common excimer laser wavelengths are 157 nm (F_2), 193 nm (ArF), 222 nm (KrCl), 248 nm (KrF), 308 nm (XeCl), and 351 nm (XeF) [70]. A disadvantage of excimer lasers is the limited lifetime of the gas used as laser medium that has to be exchanged quite often and thereby contributes to the high cost of the excimer laser photons. The gas has to be exchanged more often for the shorter wavelengths.

An alternative to the excimer lasers are frequency multiplied solid-state laser. The main disadvantage of these lasers is their strong coherence that makes it difficult to use a mask to cut a part of the laser beam without having diffraction patterns on the irradiated surface.

14.4 COMMERCIALLY AVAILABLE POLYMERS

14.4.1 Poly(methylmethacrylate)

A widely studied commercially available polymer is poly(methylmethacrylate) (Fig. 14.2). It is a polymer that can be completely depolymerized by heating above the ceiling temperature (T_C). It is possible only to achieve 100% monomer as product by laser irradiation with a CO_2 laser ($\lambda \approx 9.6$ or 10.6 μm) [71]. About 1% monomer can be detected in the ablation products after irradiation with 248 nm laser light, and about 18% monomer can be produced with 193 nm [71,72].

For irradiation with 308 nm, PMMA is not suitable, as it has a very low absorption coefficient at this wavelength. Structures produced with 308 nm are of poor quality [73], while for irradiation with shorter wavelengths, high-quality structures can be obtained. Ablation with 193 nm is most suitable, as no incubation is observed [74–77].

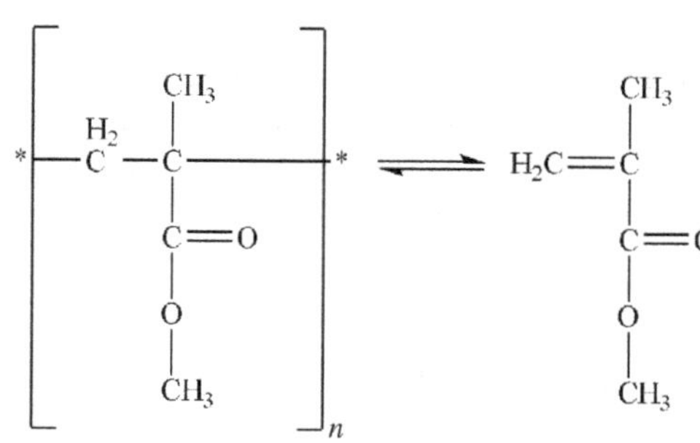

FIGURE 14.2 Chemical structure of poly(methylmethacrylate) and its monomer (right).

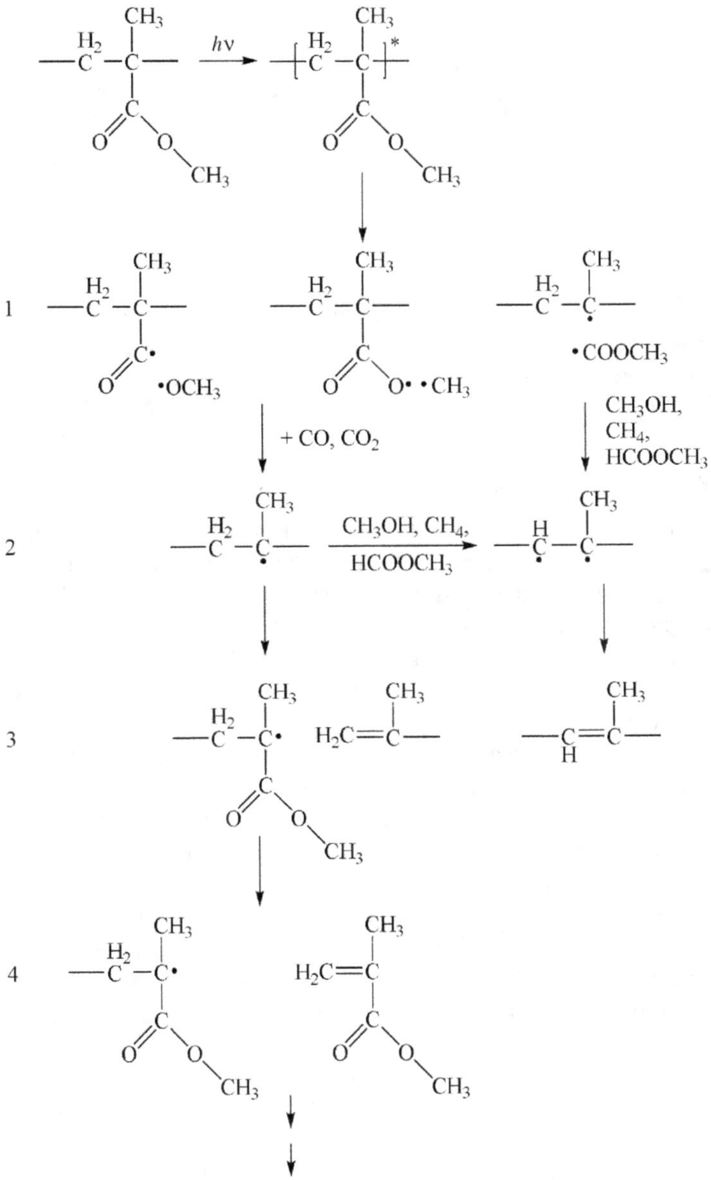

FIGURE 14.3 Proposed decomposition pathway of PMMA after irradiation with UV laser.

In Fig. 14.3, the decomposition pathway of PMMA upon irradiation with UV laser light is shown [78] as it has been defined from products detected with time-of-flight mass spectroscopy (TOF-MS). The first three steps can be associated with incubation, whereas in step 4 depolymerization takes place. Incubation, as observed for PMMA, is often present in polymers that show weak absorption at the irradiation wavelength

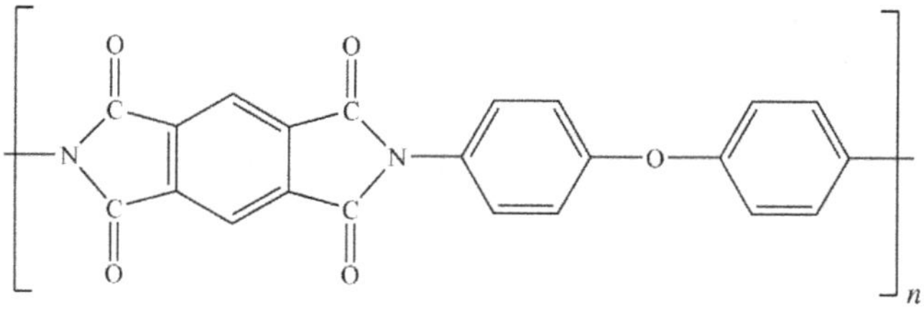

FIGURE 14.4 Chemical structure of polyimide (Kapton HN).

($\alpha_{lin} < 8000\,cm^{-1}$). It has been shown by Stuke [33] and Srinivasan [32] that during incubation a gradual increase in the absorption can be observed until ablation starts and that material is removed as vapor during incubation, even if no etching is observed.

14.4.2 Polyimide (KAPTON)

Polyimide or its most common type Kapton HN (chemical structure shown in Fig. 14.4) is the most studied polymer, as it can be ablated with all common excimer laser wavelengths and *pulse lengths* (τ). Most ablation models are based on Kapton, as its material properties are well characterized.

In nanosecond shadowgraphy measurements, a shockwave was visible for all excimer wavelengths, but the composition of the ablation plume varies with the irradiation wavelengths [79]. At 248 nm almost no solid ablation products were observed and the material removal was confined in time duration to the laser pulse. With increasing wavelengths, the amount of solid material increases. The temporal limitation of the ablation process to the laser pulse was confirmed with interferometry measurements [80]. For irradiation with 351 nm, pronounced surface swelling was observed. The swelling was then followed by the ejection of solid ablation products and a prolonged material removal. The change in the composition of the ablation plume and the duration of the ablation process has been assigned to a shift in the ablation mechanism, that is, photochemical at 248 nm to photothermal at 351 nm. No Arrhenius tail was observed with QCM measurements for 193 nm laser irradiation, which also implies a photochemical ablation process at low irradiation wavelengths. This is also in good correlation with the linear absorption coefficient of the polymer, which shows a strong decrease from 300,000 cm^{-1} at 193 nm to about 10,000 at 351 nm (see Fig. 14.5).

Different methods, such as gas-phase FTIR [81], gas chromatography–mass spectrometry [82] and quadrupole MS [83], and emission spectroscopy [72,84], have been used to analyze the ablation products. With mass spectroscopy, fragments with masses that correspond to the following species have been detected: C_2H_2, HCN, CO, CO_2, C_4H_2, C_6H_2, C_6H_2–CN, C_6H_5–CNO, and a species at mass 153 nm that can be assigned to a cyanonaphthalene structure. With emission spectroscopy, mainly low mass species such as C_2 and CN are detected.

In industry, polyimides are widely used as dielectric in microelectronics for multichip modules [85,86] and printed circuit boards [87], as substrate for inkjet

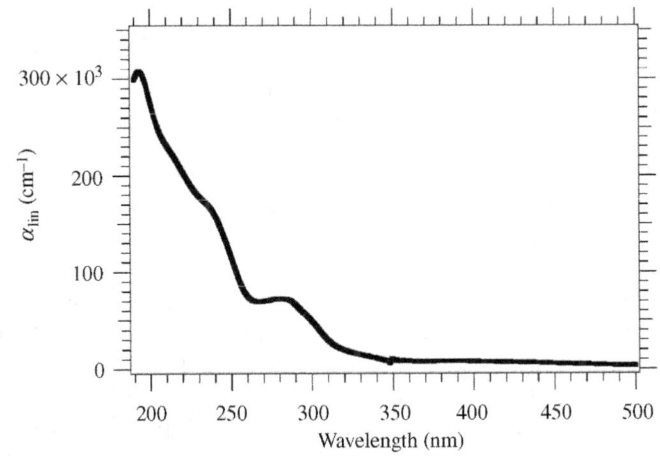

FIGURE 14.5 Linear absorption coefficient of a thin polyimide (similar to Kapton) measured from thin films.

printer nozzles [19], and as precursor for graphite materials [88]. One important feature of the polyimides is their characteristic carbonization upon laser irradiation. The carbonized material can be used to produce conduction areas on the nonconduction polymer [89,90].

14.4.3 Other Polymers

14.4.3.1 Poly(ethylene terephthalate) Poly(ethylene terephthalate) (chemical structure shown in Fig. 14.6) has been studied for most UV wavelengths (308 nm [91], 266 nm [92], 248 nm, 193 nm [93,94], and 157 nm [95]). At 157 nm, a low ablation threshold and mainly gaseous ablation products are observed, which suggests an ablation process that is dominated by photochemical reactions.

With different irradiation wavelengths, the photothermal part of the mechanism is changed and the surface crystallinity can be lowered (irradiation at 248 and 308 nm [96]) or increased with shorter wavelengths 193 nm [97].

14.4.3.2 Fluoropolymers The different fluoropolymers, poly(tetrafluoroethylene) (PTFE, Teflon, Fig. 14.7), ethylene-tetrafluoroethylene copolymers, and poly (vinylidene fluoride) (PVDF), are commercially very important polymers. The most

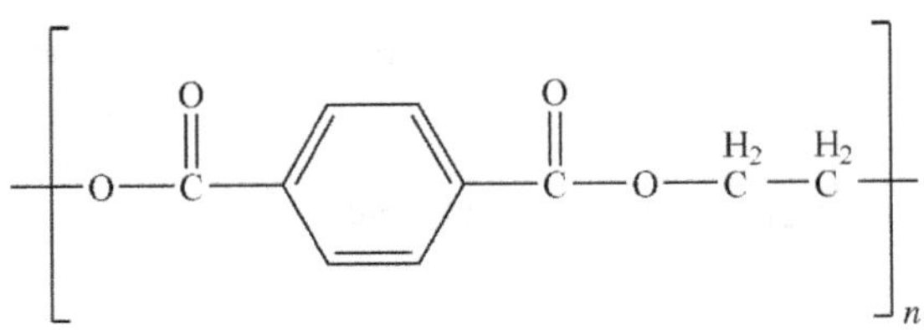

FIGURE 14.6 Chemical structure of poly(ethylene terephthalate).

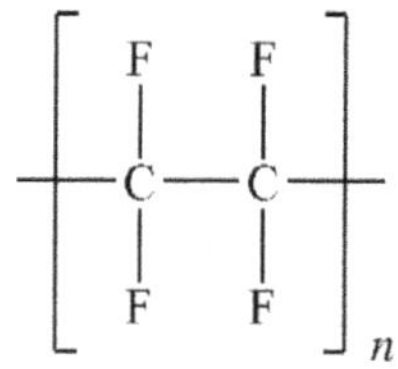

FIGURE 14.7 Chemical structure of poly(tertrafluoroethylene).

important features of PTFE are its chemical inertness, thermal stability, and hydrophobicity. It is better suited to be ablated with femtosecond laser irradiation [98], as it absorbs only weakly at wavelengths longer than 193 nm.

The ablation rate of PTFE at a constant fluence increases for irradiation between 193 and 308 nm and reaches a maximum value of 60 μm/pulse before it starts decreasing with longer wavelengths [99]. The ablation process could also be explained by applying "simple" photothermal or photochemical models.

With laser treatment in reactive atmosphere (e.g., hydrazine [100,101]) or under various liquids (e.g., $B(CH_3)_3$ [102,103]), the F–C bonds on the surface can be broken, CH_3 radicals or amino groups can be incorporated, allowing to apply a coating after laser treatment. This surface modification takes place without detectable ablation. By irradiating a PTFE foil with a F_2 laser (157 nm) in N_2 atmosphere, the surface roughness can be lowered and the transparency can be improved, without changing the film stability [104].

14.4.4 Polymer Ablation with Ultrashort Pulses

Picosecond laser pulses in the UV range do not result in better ablation behavior than nanosecond laser pulses. This is different for doped polymers. Experiments with doped PMMA (an IR-absorber, i.e., IR-165 for ablation with near-IR laser and diazomeldrum's acid (DMA) for ablation with UV lasers) with nanosecond and picosecond laser irradiation in the UV (266 nm) and near-IR (1064 nm) range have shown that, in the IR, neat features could be produced with picosecond laser irradiation, while nanosecond irradiation only results in rough surface features [105]. This corresponds well with the different behavior of the two absorbers. With IR-165 the polymer is matrix is heated by a fast vibrational relaxation and multiphonon uppumping [106]. This leads to a higher temperature jump for the picosecond irradiation, which causes ablation, while for nanosecond pulses only lower temperatures are reached.

In the UV, DMA ablation is attributed to cyclic multiphoton absorption [107] and only swelling was observed for the picosecond pulses, whereas the polymer could be ablated with nanosecond pulses.

Several studies to determine the ablation mechanisms for picosecond laser ablation were focused on spectroscopy (coherent anti-Stokes Raman scattering (CARS), absorption, and ultrafast imaging) [108–113]. It has been shown that pulses in the picosecond range produce fast temperature jumps and solid-state shockwaves that are not observed for longer pulses. This results in pressure jumps when the film is heated faster than the hydrodynamic volume relaxation time. This pressure is then released by a rarefaction wave. The pressure produced in thin films can reach ~0.5 GPa and is generated with more or less equal amounts from shock and thermochemical decomposition. In the picosecond range, a "shock-assisted photothermal ablation" takes place.

Laser ablation with laser pulses in the femtosecond (fs) range yields unique advantages, that is, negligible heat affected zone, lower ablation threshold fluence, plasma shielding is not an issue, and the possibility to structure materials that are transparent at the irradiation wavelength.

High-quality structures can be produced in PMMA with 160 fs laser pulses at 308 nm [114], where the polymer could only be "damaged" with nanosecond pulses at the same wavelength. An ablation threshold for PMMA at 248 nm that was five times lower for 300 fs than for nanosecond pulses [115] and the structures were of better quality. Ablation of PTFE with femtosecond pulses leads to high-quality structures [98]. In the IR range, multiphoton absorption is assumed to be dominant for the ablation of PMMA and PTFE.

Near-IR solid-state lasers (e.g., Ti:sapphire) with chirped pulse amplification produce laser light with high brightness and very short pulses around 800 nm [116]. 150 fs laser pulse experiments on PI, polycarbonates (PC), PET, and PMMA have shown an increase in the single pulse ablation threshold from 1 J/cm for PI to 2.6 J/cm for PMMA. This corresponds well with the optical bandgaps of these polymers and indicates a multiphoton process. Incubation effects were observed for all polymers, but are more pronounced for PMMA, PC, and PET than for PI and PTFE, which are more stable [117–120]. Clear signs of molt redeposition of material can be observed for all polymers, except PI, which is not surprising, as it decomposes and does not melt.

14.5 DOPED POLYMERS

14.5.1 Motivation

One approach to improve the ablation behavior of commercially available polymers is to increase the absorption at the irradiation wavelength by adding an additional absorber. The ablation of doped polymers was reviewed in 1997 by Lippert et al. [121] and the polymers and the ablation mechanism were classified according to the absorption properties of the absorber-polymer system. The properties changed from systems, where only the dopant is absorbed, to systems, where absorption occurs only in the polymer. It was suggested that ablation results from a mixture of processes that originate from the polymer and the dopant. The properties of the dopant result in processes that can dominate the ablation mechanisms.

An important factor is whether the dopant is decomposing or not. A photolabile dopant that decomposes into gaseous products leads to pronounced surface swelling at low irradiation fluences, while this behavior is much less pronounced for "photostable" dopants. Thermoelastic stress can also be induced in the polymer below the ablation threshold fluence by localized heating and thermal expansion of the polymer. This stress is then released in acoustic waves and thermal conduction into the surrounding material. The resulting transient and quasi-static thermoelastic stresses can lead to material damage and even material ejection. At high fluences, very high ablation rates [122] can be achieved, but with the drawback of pronounced surface melting. In the case of photostable dopants, less surface swelling, lower ablation rates and structures with higher quality are observed.

For all doped systems, it has to be considered that the amount of dopant is limited (typically $\leq$10 wt%) and that polymer properties such as T_g may change (to lower values).

14.5.2 Doped PMMA to Investigate the Ablation Mechanism

A higher ablation threshold fluence was found for PMMA doped with photosensitive organic compounds (iodonaphthalene and iodophenanthrene) with a higher molecular weight for irradiation with 248 nm laser light. This can be explained by the larger number of bonds that have to be broken and by the higher pressure produced by gaseous ablation products that is necessary to remove the longer fragments [123]. Also, higher surface temperatures were determined for PMMA with higher molecular weight for irradiation with 248 and 193 nm. These results can be described by a bulk photothermal model, in which a critical concentration of monomer and oligomer has to be reached before ablation occurs. This critical concentration is reached at higher temperatures for PMMA with higher molecular weight [124]. The viscosity of the polymer with different M_w in the irradiated area differs less than expected from the literature values. A strong dependence of the polymer viscosity on the molecular weight has been suggested, but the measured data reveal similar values all molecular weights. This can be explained by the higher temperatures that were observed for the higher molecular weights.

14.5.3 Doped PMMA for Structuring

Different dopants were added to PMMA to investigate the ablation mechanism during UV irradiation. The dopants that were used ranged from polyaromatic compounds to compounds that contained photochemically active groups [121].

Groups of dopants that were tested contain the triazene group (–N=N–N), as they are photochemically well studied [125–127] and also release a large amount of nitrogen when they are photochemically decomposed. Pronounced swelling has been observed by SEM analysis of the ablation craters at low irradiation fluence (Fig. 14.8, top), which is caused by gaseous products produced by the decomposition of the photolabile dopants. It has been suggested that the released nitrogen and other gaseous ablation products act as carriers for larger ablation fragments.

With increasing fluence and dopant concentration, high ablation rates of up to 80 μm can be achieved, but pronounced signs of surface melting are always visible [122] (Fig. 14.8, bottom), which is an indication for the presence of a

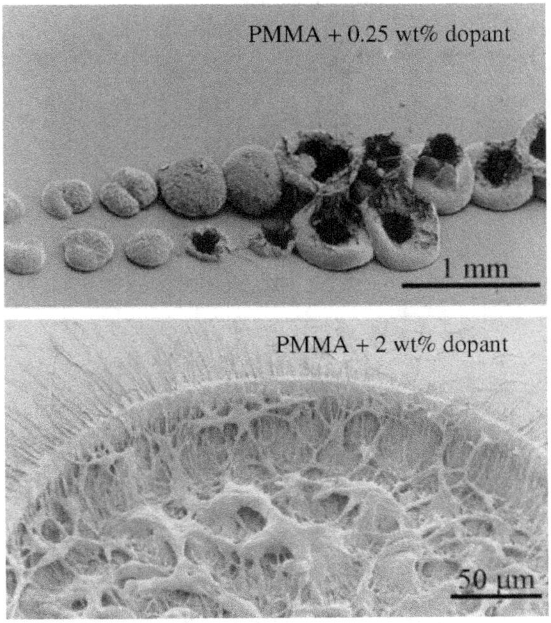

FIGURE 14.8 Irradiated PMMA with different dopant concentration after irradiation with 308 nm. In the top image, PMMA with 0.25 wt% triazene was used for ablation with one laser pulse per position. The irradiation fluence increases from left to right. Pronounced swelling and bubble formation are visible. In the bottom image, PMMA with 2 wt% is shown after two laser pulses.

photothermal mechanism. A possible reason for these pronounced thermal effects could be that the maximum amount of dopant that can be added to the polymers is $\approx$10%, which limits the achievable temperature (energy/volume).

14.5.4 Doped Polymers as Fuel for Laser Plasma Thrusters

Another quite different application utilizes dopants to induce absorption for near-IR irradiation from diode lasers. The plasma created by laser ablation of the doped polymer acts as a microthruster for small satellites. The operating principle and setup are described in detail elsewhere [128,129].

The main demand for a polymer as fuel for the plasma thruster is its exothermic decomposition. This chemically stored energy can be utilized to release higher energies in the form of thrust than the energy deposited in the polymer by the laser [130]. As absorbers for the near-IR wavelengths, carbon nanoparticles (+C) or IR-dyes (+IR) were used. As fuel polymer, glycidyl azide polymer (GAP) was studied (Fig. 14.9). GAP was originally developed as binder for solid propellant rockets [131,132], but it meets the demands for the LPT, such as easy handling, exothermic decomposition (decomposition enthalpy of −2050 J/g), and a well-defined decomposition temperature (249°C).

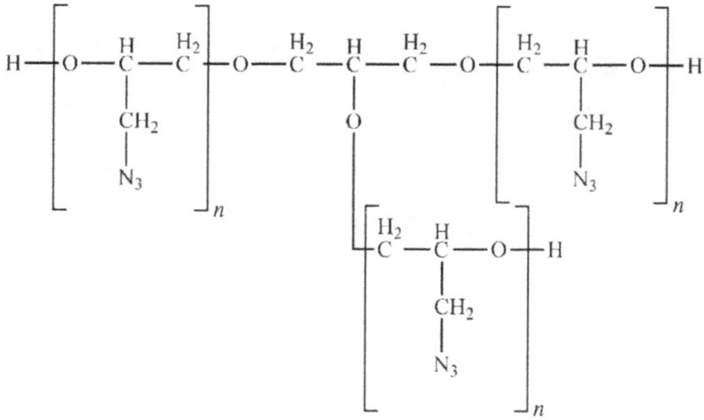

FIGURE 14.9 Chemical structure of GAP polyol.

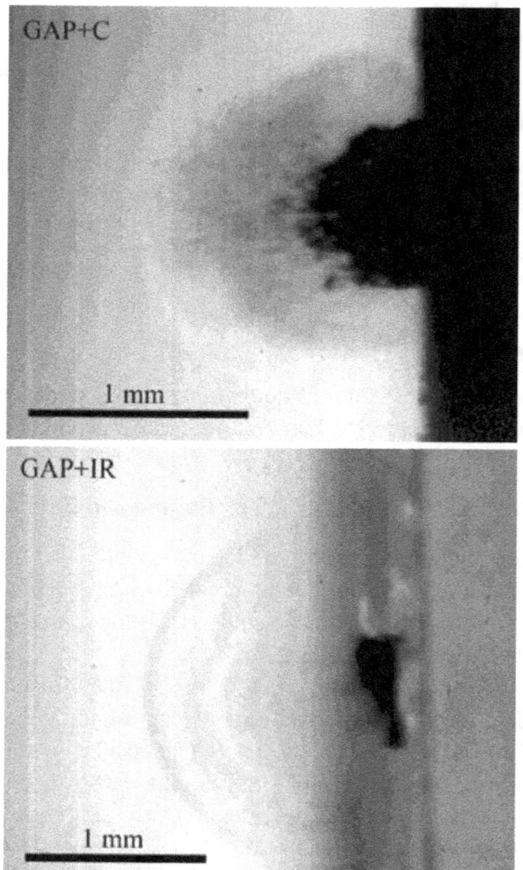

FIGURE 14.10 Shadowgraphy images of carbon (left) and IR-dye (right) doped GAP. The image was taken 1 μs after the laser pulse.

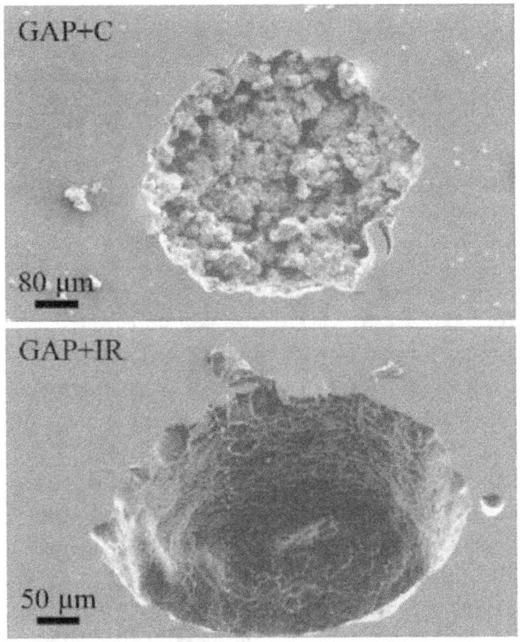

FIGURE 14.11 SEM images of the ablation spots of GAP + C (top) and GAP + IR (bottom).

The different dopants for GAP had only a small influence on the ablation properties, such as ablation rate and threshold fluence [133]. The most pronounced differences are observed in the ejected fragments detected in the shadowgraphy measurements and the ablation crater appearance. In the nanosecond-shadowgraphy image (Fig. 14.10), large fragments of solid and liquid ablation products are observed for GAP + C. In the ablation plume of GAP + IR, it seems that the ablated material is transformed completely into gaseous products. This is a desired effect, as more energy is gained by decomposing the polymer completely.

An SEM investigation of the ablation crater confirmed these results by showing an ablation crater with steep, smooth walls for GAP + IR, whereas the crater of GAP + C is quite rough, with deep holes and a very uneven bottom (Fig. 14.11).

14.6 DESIGNED POLYMERS

14.6.1 Triazene Polymers

New polymers have been developed to further improve the quality of the ablation process. One approach was to incorporate the triazene unit into the polymer backbone. A unique feature of these triazene polymers (TP, chemical structure shown in Fig. 14.12) is the possibility to adjust the absorption maximum by varying the "X"-component in Fig. 14.12 [134]. The absorption maximum of such triazene

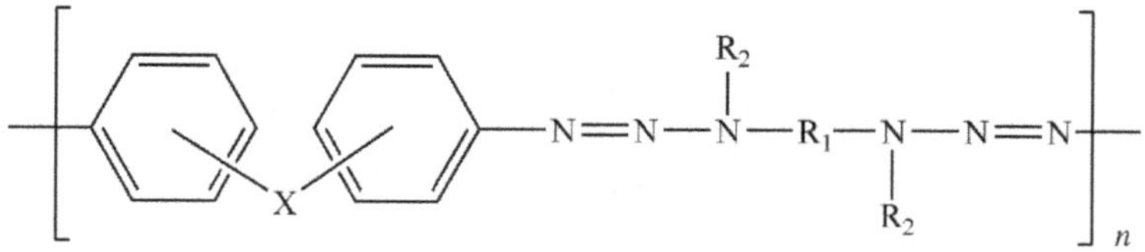

FIGURE 14.12 Chemical structure of the triazene polymers.

polymers can be tuned from 290 to 360 nm and maximum linear absorption coefficients of up to 200,000 cm^{-1} at 308 nm can be reached.

In the absorption spectra for TP1 ($X = O$, $R_1 = (CH_2)_6$, $R_2 = CH_3$) as shown in Fig. 14.13, two distinct absorption maxima can be distinguished. The R_1 and R_2 substitutes change the properties such as T_g, film forming, and chromophore density. One maximum around 200 nm, which can be assigned to the aromatic system, and the other maximum around 330 nm that corresponds to the triazene unit [135]. These two well-separated absorption regions allow an excitation of different chromophores with different irradiation wavelengths such as 193, 248, and 308 nm and thereby to study their influence on the ablation behavior.

Higher ablation rates were measured for irradiation wavelengths that excite the triazene system (266, 308, and 351 nm) compared to the ablation rates for shorter wavelengths (248 and 193 nm) [6]. Also, a clear and well-defined ablation threshold fluence of 25 mJ/cm^2 ($\pm$5 mJ/cm^2) is observed for TP1 at an irradiation wavelength of 308 nm, while for irradiation with 248 nm a much broader range 16–28 mJ/cm^2 has been measured. For irradiation at 248 nm, carbonization of the polymer was detected upon irradiation, whereas the surface of the polymer remained unchanged after several laser pulses for irradiation with 308 nm [136]. This is also an indication for the different ablation mechanisms in the two absorption regions.

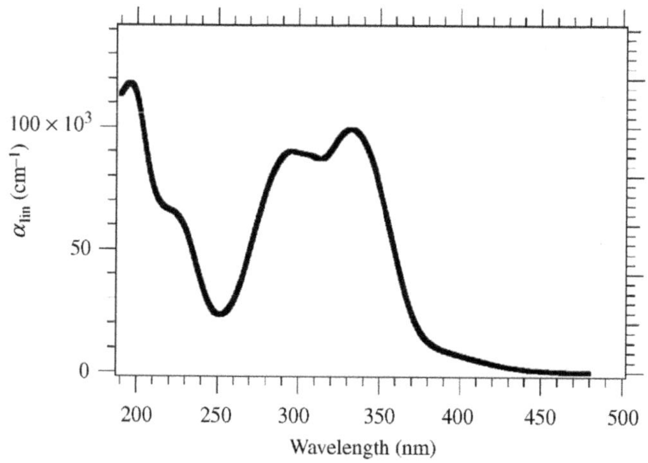

FIGURE 14.13 Linear absorption coefficient of TP1.

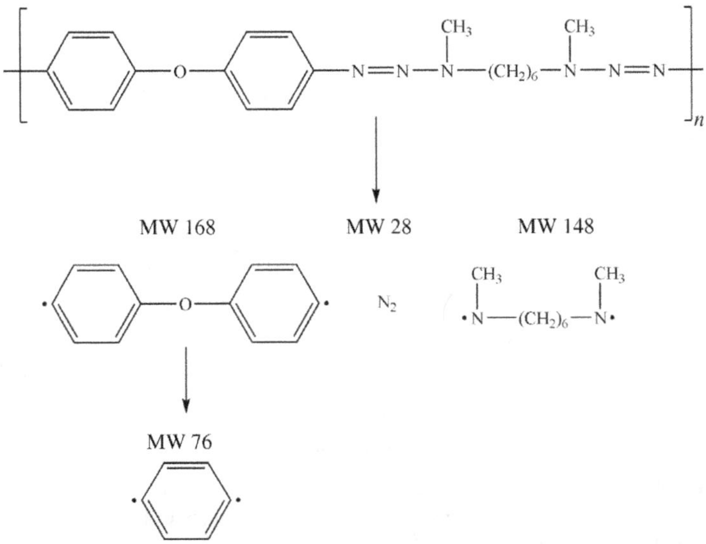

FIGURE 14.14 Decomposition pathway for TP1 measured by TOF-MS after irradiation with 308 nm.

The triazene polymers are also well suited as probes for the ablation mechanism. Mass spectrometry was used to study the ablation products and to determine the different ablation mechanisms at the different irradiation wavelengths [67,137,138]. All decomposition products were identified with time-resolved mass spectrometry for 248 and 308 nm irradiation. The proposed decomposition pathway for 308 nm irradiation is shown in Fig. 14.14, but similar products were observed also for a thermal decomposition [126]. A clearer indication for the presence of a photochemical mechanism for 308 nm irradiation was given by TOF-MS. Three different species of nitrogen were detected in the ablation plume: a very fast ground state neutral with up to 6 eV of kinetic energy, a slower ground state species with a broad energy distribution that is most probably a thermal product, and a metastable (excited) neutral N_2 species that can only be created by an electronic excitation [139].

The triazene polymers were also tested with mass spectrometry after 157 nm. These experiments showed a higher fragmentation of the polymer than for 308 or 248 nm, and even a complete fragmentation of the aromatic groups was observed [140].

The photochemical activity of the triazene group was also confirmed by irradiation at low fluences with excimer lamps, where one photon photochemistry is expected [137]. A decomposition of the triazene chromophore was observed below the ablation threshold fluence for irradiation at 308 and 222 nm. At 222 nm, an additional decomposition of the aromatic chromophores has been detected [141]. This suggests that the decomposition of the aromatic part is related to the carbonization. This selective decomposition of the triazene group by the less energetic wavelength (308 nm) clearly indicates that the triazene is the most sensitive unit in the triazene

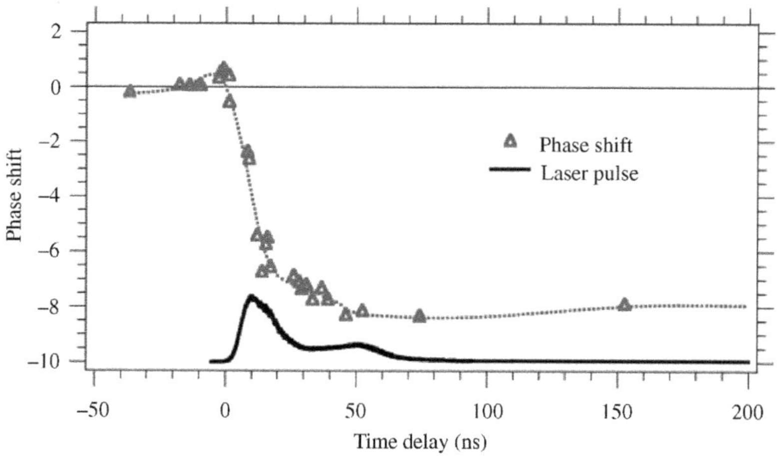

FIGURE 14.15 Interference measurement for TP1 during irradiation with 308 nm. The black curve represents the laser pulse, while the gray line corresponds to the phase shift, which is related to the ablation depth.

polymer and that a photochemical process is most probably also present during ablation.

Another method, which can be used to determine the ablation mechanism, is nanosecond interferometry. The ablation process could take place on a longer timescale (depending on the temperature) for a photothermal process than for a photochemical reaction. First, swelling is observed that is followed by etching [142,143], for example, discussed for a polyimide at 351 nm irradiation. This etching takes place on a microsecond timescale, which is much longer than the 30 ns excimer laser pulse. For the triazene polymer on the other hand, the etching starts and stops with the ablation laser pulse [141,144] (Fig. 14.15), which is again a clear indication for a photochemical process.

Irradiation experiments in the near-IR range at 800 nm with pulses in the pico- and femtosecond range were also performed. For femtosecond pulses, a lower ablation threshold fluence was found than for picosecond pulses, which indicates the presence of a thermal mechanism [145]. Also, no complete removal of a thin triazene polymer film from a glass substrate was possible with 100 fs pulses. These short pulses in the near-IR do not yield better results and are therefore no alternative to UV ablation [146].

The influence of the location of the predetermined "decomposition" site in the polymer has been tested by incorporating the triazene unit into the side chains. The obtained ablation structures were less defined, and stronger thermal effects were observed [147]. Investigation of the polymer "between" the individual triazene units suggests that a higher triazene density results in better ablation properties [7].

In Fig. 14.16, the ablation threshold fluences are plotted versus the polymer weight per triazene unit for TP1, two polyurethane polymers with the triazene unit in the polymer backbone (PUH-T1, PUH-T2 [148]), two polyacrylates with the

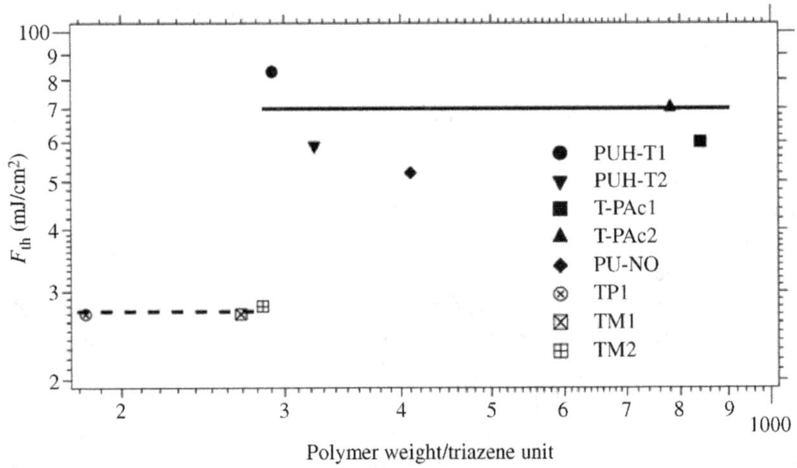

FIGURE 14.16 The ablation threshold fluence versus polymer weight per triazene unit is shown for various triazene polymers. The two lines in the plot are shown as guidelines.

triazene unit in the polymer side chain (T-PAc1, T-PAc2 [147]), two different triazene polymers with malonyl esther groups in the side chains (TM1 and TM2) [7], and a polyurethane polymer with the triazene unit in the side chain (PU-NO) [149]. A sudden increase in the ablation threshold fluence can be observed at about 285 amu/triazene group from ~25 to ~70 mJ/cm^2. Polymers above this jump have a higher ablation threshold fluence, as more bonds have to be broken to remove the larger remaining polymer fragments. Below or above this step, the ablation threshold fluence remains more or less constant, independent of the polymer weight per triazene unit. Why this sharp step is observed is not yet clear and must be studied in more detail.

14.6.2 Other Polymers Designed for Laser Ablation

Other designed polymers, such as *diazosulfidepolymer*, *pentazadienepolymer*, *diazophosphonatepolymer*, and *diazocopolyester*, have also been designed for ablation at 308 nm. All these polymers contain a photolabile chromophore based on diazo groups (–N=N–X) in the polymer main chain [150].

With all polymers, high-quality structures could be obtained, but not all polymers were stable enough to be analyzed by methods such as scanning electron microscopy. The diazosulfidepolymer could only be examined by optical microscopy, and also the pentazadienopolymer partially decomposed during sputtering prior to SEM analysis. The best results were obtained for diazocopolyester, where no incubation was observed for irradiation with 308 nm. By decreasing the triazene content below 35%, it became impossible to produce well-defined structures. The polymer started to form bubbles below the surface that were caused by a thermal decomposition of the ester into CO/CO_2 at a triazene contents above 90%.

14.7 COMPARISON OF DESIGNED AND COMMERCIALLY AVAILABLE POLYMERS

Compared to commercially available polymers such as polyimides or other designed polymers, for example, polyesters, the triazene polymers showed the highest ablation rates and the lowest ablation threshold fluence for selected wavelengths. The structure produced in TP (Fig. 14.17, top) with 308 nm irradiation are much sharper than those in Kapton (Fig. 14.17, bottom) and also no polymer debris is redeposited in and around the ablated structure in the case of the triazene polymer [141]. Kapton was chosen as commercially available reference because it has a similar α_{lin} at 308 nm. The absence of redeposited material for TP corresponds well with nanosecond-shadowgraphy measurements, where it was shown that no solid products are produced for 308 nm irradiation of TP [144].

All data obtained for TP strongly suggest that photochemical reactions play an important role during UV laser ablation, but also that photothermal processes are important. This is confirmed by the presence of the thermal N_2 products in the TOF curves. Photothermal processes will also always be present if the polymer decomposes exothermically during a photochemical decomposition and if the quantum yields of the photochemical reaction is not equal to one (which is most of time the case). The ablation of polymers will therefore always be a photophysical process (a mixture of photochemical and photothermal processes), where the ratio between the two mechanisms is a function of the irradiation wavelength and the polymer. In addition, photomechanical processes, such as pressure produced by trapped gaseous ablation products or shock and acoustic waves in the polymer, take place and can lead to a damage of the polymer and are most important for picosecond pulses.

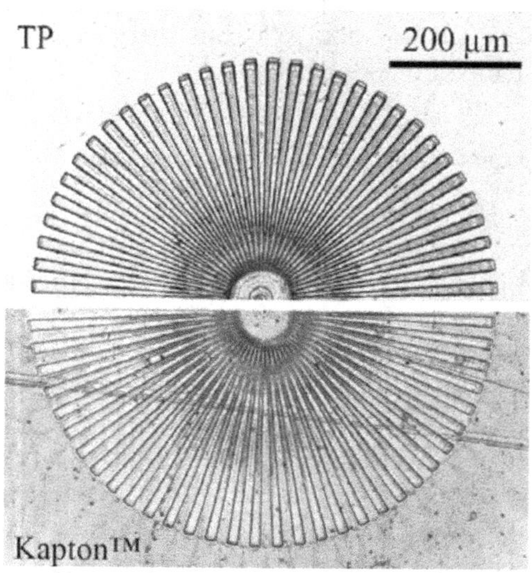

FIGURE 14.17 SEM of Siemens stars in TP (top) and Kapton (bottom), both produced with five laser pulses at 308 nm.

A more pronounced photochemical part is preferable for material structuring, as it leads to a more uniform decomposition of the polymer and results in less debris. In additional, large quantities of gaseous products are produced and less material is redeposited in and around the ablated area. The designed polymers such as the TP show a clear advantage over commercially available polymers.

REFERENCES

1. R. Srinivasan and V. Mayne-Banton *Appl. Phys. Lett.* **1982**, *41*, 576–578.
2. Y. Kawamura, K. Toyoda, and S. Namba, *Appl. Phys. Lett.* **1982**, *40*, 374–375.
3. R. Srinivasan and B. Braren *Chem. Rev.* **1989**, *89*, 1303–1316.
4. D. Bäuerle, *Laser Processing and Chemistry*, Springer Verlag, Berlin, 2000.
5. S. Lazare and V. Granier *Laser Chem.* **1989**, *10*, 25–40.
6. T. Lippert and J.T. Dickinson *Chem. Rev.* **2003**, *103*, 453–485.
7. T. Lippert, Laser application of polymers, *Polymers and Light*, Springer, Berlin, **2004**, pp. 51–246.
8. P.E. Dyer, Laser ablation of polymers in: I.W. Boyd and R.B. Jackman (Eds.), *Photochemical Processing of Electronic Materials*, Academic Press Limited, London, **1992**, pp. 360–385.
9. N. Bityurin, B.S. Luk'yanchuk, M.H. Hong, and T.C. Chong, *Chem. Rev.* **2003**, *103*, 519–552.
10. T. Lippert, *Plasma Processes Polym.* **2005**, *2*, 525–546.
11. N. Bityurin, *Annu. Rep. Prog. Chem. C,* **2005**, *101*, 216–247.
12. M. Karas, D. Bachmann, and F. Hillenkamp, *Anal. Chem.* **1985**, *57*, 2935–2939.
13. R. Zenobi and R. Knochenmuss *Mass Spectrom. Rev.* **1998**, *17*, 337–366.
14. L.J. Radziemski and D.A. Cremers (Eds.), *Laser-Induced Plasmas and Applications*, Marcel Dekker, New York, 1989.
15. G.B. Blanchet, *Chemtech* **1996**, *26*, 31–35.
16. W.B. Jiang, M.G. Norton, L. Tsung, and J.T. Dickinson, *J. Mater. Res.* **1995**, *10*, 1038–1043.
17. D.B. Chrisey and G.K. Hubler *Pulsed Laser Deposition of Thin Films*, John Wiley & Sons, New York, 1994.
18. R.W. Eason (Ed.), *Pulsed Laser Deposition of Thin Films: Applications-Led Growth of Functional Materials*, John Wiley & Son, Inc., New York, 2007.
19. H. Aoki, U.S. Patent 5736999, 1998.
20. R.S. Patel and T.A. Wassick *Proc. SPIE Int. Soc. Opt. Eng.* **2991**, *1997*, 217.
21. G. Kopitkovas, L. Urech, and T. Lippert, Fabrication of micro-optics in polymers and in UV transparent materials, in: E. Millon, J. Perriere, and E. Forgarassy (Eds.), *Recent Advances in Laser Processing of Materials*, Elsevier, Kidlington, 2006.
22. H. Klank, J.P. Kutter, and O. Geschke, *Lab Chip* **2002**, *2*, 242–246.
23. D. Snakenborg, H. Klank, and J.P. Kutter, *J. Micromech. Microeng.* **2004**, *14*, 182–189.
24. D. Gomez, F. Tekniker, I. Goenaga, I. Lizuain, and M. Ozaita, *Opt. Eng.*, **2005**, *44*.

25. D.F. Farson, H.W. Choi, C.M. Lu, and L.J. Lee, *J. Laser Appl.* **2006**, *18*, 210–215.
26. A. Pique, P. Wu, B.R. Ringeisen, D.M. Bubb, J.S. Melinger, R.A. McGill, and D.B. Chrisey, *Appl. Surf. Sci.* **2002**, *186*, 408–415.
27. T. Mito, T. Tsujita, H. Masuhara, N. Hayashi, and K. Suzuki, *Jpn. J. Appl. Phys. 2, Lett.* **2001**, *40*, L805–L806.
28. G.B. Blanchet, *Macromolecules* **1995**, *28*, 4603–4607.
29. E. Millon, J. Perriere, and E. Fogarassy (Eds.), *Recent Advances in Laser Processing of Materials*, Elsevier, Kidlington, 2006.
30. J.E. Andrew, P.E. Dyer, D. Forster, and P.H. Key, *Appl. Phys. Lett.* **1983**, *43*, 717–719.
31. R. Srinivasan and B. Braren *J. Polym. Sci. Pol. Chem.* **1984**, *22*, 2601–2609.
32. R. Srinivasan, B. Braren, and K.G. Casey, *J. Appl. Phys.* **1990**, *68*, 1842–1847.
33. S. Küper and M. Stuke *Appl. Phys. A: Mater. Sci. Process.* **1989**, *49*, 211–215.
34. S.V. Babu, G.C. Dcouto, and F.D. Egitto, *J. Appl. Phys.* **1992**, *72*, 692–698.
35. M. Himmelbauer, E. Arenholz, and D. Bauerle, *Appl. Phys. A: Mater. Sci. Process.* **1996**, *63*, 87–90.
36. G. Paraskevopoulos, D.L. Singleton, R.S. Irwin, and R.S. Taylor, *J. Appl. Phys.* **1991**, *70*, 1938–1946.
37. R.S. Taylor, D.L. Singleton, and G. Paraskevopoulos, *Appl. Phys. Lett.* **1987**, *50*, 1779–1781.
38. S. Küper, J. Brannon, and K. Brannon, *Appl. Phys. A: Mater. Sci. Process.* **1993**, *56*, 43–50.
39. T. Dumont, R. Bischofberger, T. Lippert, and A. Wokaun, *Appl. Surf. Sci.* **2005**, *247*, 115–122.
40. S.R. Cain, *J. Phys. Chem.* **1993**, *97*, 7572–7577.
41. S.R. Cain, F.C. Burns, and C.E. Otis, *J. Appl. Phys.* **1992**, *71*, 4107–4117.
42. G.C. D'Couto and S.V. Babu *J. Appl. Phys.* **1994**, *76*, 3052–3058.
43. B. Lukyanchuk, N. Bityurin, M. Himmelbauer, and N. Arnold, *Nucl. Instrum. Methods Phys. Res. B* **1997**, *122*, 347–355.
44. N. Arnold, B. Luk'yanchuk, and N. Bityurin, *Appl. Surf. Sci.* **1998**, *129*, 184–192.
45. T.F. Deutsch and M.W. Geis *J. Appl. Phys.* **1983**, *54*, 7201–7204.
46. E. Sutcliffe and R. Srinivasan *J. Appl. Phys.* **1986**, *60*, 3315–3322.
47. G.D. Mahan, H.S. Cole, Y.S. Liu, and H.R. Philipp, *Appl. Phys. Lett.* **1988**, *53*, 2377–2379.
48. V. Srinivasan, M.A. Smrtic, and S.V. Babu, *J. Appl. Phys.* **1986**, *59*, 3861–3867.
49. H. Schmidt, J. Ihlemann, B. Wolff-Rottke, K. Luther, and J. Troe, *J. Appl. Phys.* **1998**, *83*, 5458–5468.
50. B. Lukyanchuk, N. Bityurin, S. Anisimov, N. Arnold, and D. Bauerle, *Appl. Phys. A: Mater. Sci. Process.* **1996**, *62*, 397–401.
51. B. Lukyanchuk, N. Bityurin, S. Anisimov, and D. Bauerle, *Appl. Phys. A: Mater. Sci. Process.* **1993**, *57*, 367–374.
52. N. Bityurin, A. Malyshev, B. Luk'yanchuk, S. Anisimov, and D. Bäuerle, *Proc. SPIE.* **2802**, *1996*, 103.
53. N. Bityurin, *Appl. Surf. Sci.* **1999**, *139*, 354–358.

54. G.V. Treyz, R. Scarmozzino, and R.M. Osgood, *Appl. Phys. Lett.* **1989**, *55*, 346–348.
55. N. Arnold and N. Bityurin *Appl. Phys. A: Mater. Sci. Process.* **1999**, *68*, 615–625.
56. Y.G. Yingling and B.J. Garrison *J. Phys. Chem. B* **2005**, *109*, 16482–16489.
57. Y.G. Yingling and B.J. Garrison *J. Phys. Chem. B* **2004**, *108*, 1815–1821.
58. M. Prasad, P. Conforti, and B.J. Garrison, *J. Appl. Phys.* **2008**, *103*, 103114.
59. Y.G. Yingling, L.V. Zhigilei, and B.J. Garrison, *J. Photochem. Photobiol. A* **2001**, *145*, 173–181.
60. L.V. Zhigilei, E. Leveugle, B.J. Garrison, Y.G. Yingling, and M.I. Zeifman, *Chem. Rev.* **2003**, *103*, 321–347.
61. Y.G. Yingling and B.J. Garrison *Chem. Phys. Lett.* **2002**, *364*, 237–243.
62. Y.G. Yingling and B.J. Garrison *Nucl. Instrum. Methods Phys. Res. B*, **2003**, *202*, 188–194.
63. N. Bityurin and A. Malyshev *J. Appl. Phys.* **2002**, *92*, 605–613.
64. K. Suzuki, M. Matsuda, T. Ogino, N. Hayashi, T. Terabayashi, and K. Amemiya, *Proc. SPIE.* **2999**, *1997*, 98.
65. T. Lippert, M. Hauer, C.R. Phipps, and A. Wokaun, *Appl. Phys. A: Mater. Sci. Process.* **2003**, *77*, 259–264.
66. L.S. Bennett, T. Lippert, H. Furutani, H. Fukumura, and H. Masuhara, *Appl. Phys. A: Mater. Sci. Process.* **1996**, *63*, 327–332.
67. T. Lippert, S.C. Langford, A. Wokaun, G. Savas, and J.T. Dickinson, *J. Appl. Phys.* **1999**, *86*, 7116–7122.
68. O. Svelto, *Principles of Lasers*, Plenum Press, New York, **1998**.
69. F.W. Kneubühl and M.W. Sigrist *Laser*, Teubner Studienbücher Physik, Stuttgart, **1999**.
70. D. Basting (Ed.), *Excimer Laser Technology: Laser Sources, Optics, Systems and Applications*, Lambda Physics, Göttingen, **2001**.
71. R. Srinivasan, *J. Appl. Phys.* **1993**, *73*, 2743–2750.
72. R. Srinivasan, B. Braren, D.E. Seeger, and R.W. Dreyfus, *Macromolecules* **1986**, *19*, 916–921.
73. T. Efthimiopoulos, C. Kiagias, G. Heliotis, and E. Helidonis, *Can. J. Phys.* **2000**, *78*, 509–519.
74. A. Costela, J.M. Figuera, F. Florido, I. Garciamoreno, E.P. Collar, and R. Sastre, *Appl. Phys. A: Mater. Sci. Process.* **1995**, *60*, 261–270.
75. N. Bityurin, S. Muraviov, A. Alexandrov, and A. Malyshev, *Appl. Surf. Sci.* **1997**, *110*, 270–274.
76. H. Schmidt, J. Ihlemann, K. Luther, and J. Troe, *Appl. Surf. Sci.* **1999**, *139*, 102–106.
77. R.J. Lade, I.W. Morley, P.W. May, K.N. Rosser, and M.N.R. Ashfold, *Diamond Relat. Mater.* **1999**, *8*, 1654–1658.
78. T. Lippert, R.L. Webb, S.C. Langford, and J.T. Dickinson, *J. Appl. Phys.* **1999**, *85*, 1838–1847.
79. R. Srinivasan, *Appl. Phys. A: Mater. Sci. Process.* **1993**, *56*, 417–423.
80. T. Lippert, C. David, M. Hauer, T. Masubuchi, H. Masuhara, K. Nomura, O. Nuyken, C. Phipps, J. Robert, T. Tada, K. Tomita, and A. Wokaun, *Appl. Surf. Sci.* **2002**, *186*, 14–23.

81. R. Srinivasan, R.R. Hall, W.D. Loehle, W.D. Wilson, and D.C. Allbee, *J. Appl. Phys.* **1995**, *78*, 4881–4887.
82. D.L. Singleton, G. Paraskevopoulos, and R.S. Irwin, *J. Appl. Phys.* **1989**, *66*, 3324–3328.
83. S. Lazare, W.P. Guan, and D. Drilhole, *Appl. Surf. Sci.* **1996**, *96-8*, 605–610.
84. E.E. Ortelli, F. Geiger, T. Lippert, J. Wei, and A. Wokaun, *Macromolecules* **2000**, *33*, 5090–5097.
85. F. Bachman, *Chemtronics* **1989**, *4*, 149.
86. J.R. Lankard and G. Wolbold *Appl. Phys. A: Mater. Sci. Process.* **1992**, *54*, 355–359.
87. A. Beuhler, A. Tungare, and J. Savic, *Circuit World* **1998**, *24*, 36.
88. M. Inagaki, T. Takichi, Y. Hishiyama, and A. Obslei, *Chemistry and Physics of Carbon*, Marcel Dekker, New York, 1999.
89. Z. Ball and R. Sauerbrey *Appl. Phys. Lett.* **1994**, *65*, 391–393.
90. G.H. Wynn and A.W. Fountain *J. Electrochem. Soc.* **1997**, *144*, 3769–3772.
91. S. Klose, E. Arenholz, J. Heitz, and D. Bauerle, *Appl. Phys. A: Mater. Sci. Process.* **1999**, *69*, S487–S490.
92. N. Mansour and K.J. Ghaleh *Appl. Phys. A: Mater. Sci. Process.* **2002**, *74*, 63–67.
93. F. Wagner and P. Hoffmann *Appl. Surf. Sci.* **2000**, *154*, 627–632.
94. F. Wagner and P. Hoffmann *Appl. Phys. A: Mater. Sci. Process.* **1999**, *69*, S841–S844.
95. P.E. Dyer, G.A. Oldershaw, and D. Schudel, *J. Phys. D-Appl. Phys.* **1993**, *26*, 323–325.
96. T. Lippert, F. Zimmermann, and A. Wokaun, *Appl. Spectrosc.* **1993**, *47*, 1931–1942.
97. J.S. Rossier, P. Bercier, A. Schwarz, S. Loridant, and H.H. Girault, *Langmuir* **1999**, *15*, 5173–5178.
98. S. Küper and M. Stuke *Appl. Phys. Lett.* **1989**, *54*, 4–6.
99. N. Huber, J. Heitz, and D. Bauerle, *Eur. Phys. J.: Appl. Phys* **2004**, *25*, 33–38.
100. H. Niino, H. Okano, K. Inui, and A. Yabe, *Appl. Surf. Sci.* **1997**, *110*, 259–263.
101. H. Niino and A. Yabe *Appl. Phys. Lett.* **1993**, *63*, 3527–3529.
102. M. Okoshi, M. Murahara, and K. Toyoda, *J. Mater. Res.* **1992**, *7*, 1912–1916.
103. M. Murahara and K. Toyoda *J. Adhes. Sci. Technol.* **1995**, *9*, 1601–1609.
104. T. Gumpenberger, J. Heitz, D. Bauerle, and T.C. Rosenmayer, *Europhys. Lett.* **2005**, *70*, 831–835.
105. C. Hahn, T. Lippert, and A. Wokaun, *J. Phys. Chem. B*, **1999**, *103*, 1287–1294.
106. X.N. Wen, W.A. Tolbert, and D.D. Dlott, *Chem. Phys. Lett.* **1992**, *192*, 315–320.
107. H. Fujiwara, H. Fukumura, and H. Masuhara, *J. Phys. Chem.* **1995**, *99*, 11844–11853.
108. H. Kim, J.C. Postlewaite, T.H. Zyung, and D.D. Dlott, *Appl. Phys. Lett.* **1989**, *54*, 2274–2276.
109. T. Zyung, H. Kim, J.C. Postlewaite, and D.D. Dlott, *J. Appl. Phys.* **1989**, *65*, 4548–4563.
110. X.N. Wen, D.E. Hare, and D.D. Dlott, *Appl. Phys. Lett.* **1994**, *64*, 184–186.
111. D.E. Hare, J. Franken, D.D. Dlott, E.L. Chronister, and J.J. Flores, *Appl. Phys. Lett.* **1994**, *65*, 3051–3053.
112. D.E. Hare, J. Franken, and D.D. Dlott, *J. Appl. Phys.* **1995**, *77*, 5950–5960.
113. I.Y.S. Lee, X.N. Wen, W.A. Tolbert, D.D. Dlott, M. Doxtader, and D.R. Arnold, *J. Appl. Phys.* **1992**, *72*, 2440–2448.

114. R. Srinivasan, E. Sutcliffe, and B. Braren, *Appl. Phys. Lett.* **1987**, *51*, 1285–1287.

115. S. Küper and M. Stuke, *Appl. Phys. B: Photophys. Laser Chem.* **1987**, *44*, 199–204.

116. J.D. Bonlie, F. Patterson, D. Price, B. White, and P. Springer, *Appl. Phys. B: Lasers Opt.* **2000**, *70*, S155–S160.

117. S. Baudach, J. Bonse, and W. Kautek, *Appl. Phys. A*, **1999**, *69*, S395.

118. S. Baudach, J. Bonse, J. Kruger, and W. Kautek, *Appl. Surf. Sci.* **2000**, *154*, 555–560.

119. J. Bonse, S. Baudach, J. Krüger, and W. Kautek, *Proc. SPIE Int. Soc. Opt. Eng.* **2000**, *4065*, 161.

120. S. Baudach, J. Krüger, and W. Kautek, *Rev. Laser Engin.* **2001**, *29*, 705.

121. T. Lippert, A. Yabe, and A. Wokaun, *Adv. Mater.* **1997**, *9*, 105–119.

122. T. Lippert, A. Wokaun, J. Stebani, O. Nuyken, and J. Ihlemann, *Angew. Makromol. Chem.* **1993**, *213*, 127–155.

123. G. Bounos, A. Selimis, S. Georgiou, E. Rebollar, M. Castillejo, and N. Bityurin, *J. Appl. Phys.* **2006**. *100*.

124. G. Bounos, A. Athanassiou, D. Anglos, and S. Georgiou, *Chem. Phys. Lett.* **2006**, *418*, 317–322.

125. T. Lippert, J. Stebani, O. Nuyken, A. Stasko, and A. Wokaun, *J. Photochem. Photobiol. A*, **1994**, *78*, 139–148.

126. O. Nuyken, J. Stebani, T. Lippert, A. Wokaun, and A. Stasko, *Macromol. Chem. Phys.* **1995**, *196*, 751–761.

127. A. Stasko, V. Adamcik, T. Lippert, A. Wokaun, J. Dauth, and O. Nuyken, *Makromol. Chem. Macromol. Chem. Phys.* **1993**, *194*, 3385–3391.

128. C.R. Phipps, J.R. Luke, G.G. McDuff, and T. Lippert, *Proc. SPIE* **2002**, *4760*, 833–842.

129. T. Lippert, C. David, M. Hauer, A. Wokaun, J. Robert, O. Nuyken, and C. Phipps, *J. Photochem. Photobiol. A* **2001**, *145*, 87–92.

130. L. Urech, T. Lippert, C. Phipps, and A. Wokaun, *Appl. Surf. Sci.* **2007**, *253*, 7646.

131. J. Köhler and R. Meyer *Explosivstoffe*, Wiley-VCH, Weinheim, 1998.

132. M.B. Frankel, L.R. Grant, and J.E. Flanagan, *J. Propul. Power* **1992**, *8*, 560–563.

133. L. Urech, M. Hauer, T. Lippert, C.R. Phipps, E. Schmid, A. Wokaun, and I. Wysong, *Proc. SPIE* **2004**, *5448*, 52–64.

134. O. Nuyken, J. Stebani, T. Lippert, A. Wokaun, and A. Stasko, *Macromol. Chem. Phys.* **1995**, *196*, 739–749.

135. T. Lippert, L.S. Bennett, T. Nakamura, H. Niino, A. Ouchi, and A. Yabe, *Appl. Phys. A: Mater. Sci. Process.* **1996**, *63*, 257–265.

136. T. Lippert, T. Nakamura, H. Niino, and A. Yabe, *Appl. Surf. Sci.* **1997**, *110*, 227–231.

137. T. Lippert, C. David, J.T. Dickinson, M. Hauer, U. Kogelschatz, S.C. Langford, O. Nuyken, C. Phipps, J. Robert, and A. Wokaun, *J. Photochem. Photobiol. A,* **2001**, *145*, 145–157.

138. T. Lippert, A. Wokaun, S.C. Langford, and J.T. Dickinson, *Appl. Phys. A: Mater. Sci. Process.* **1999**, *69*, S655–S658.

139. M. Hauer, T. Dickinson, S. Langford, T. Lippert, and A. Wokaun, *Appl. Surf. Sci.* **2002**, *197*, 791–795.

140. M. Kuhnke, L. Cramer, J.T. Dickinson, T. Lippert, and A. Wokaun,TOF-MS study of photoreaktive polymers ablated by F_2 excimer laser (157 nm). *Poster at COLA*, 2003, Kreta.

141. T. Lippert, J.T. Dickinson, M. Hauer, G. Kopitkovas, S.C. Langford, H. Masuhara, O. Nuyken, J. Robert, H. Salmio, T. Tada, K. Tomita, and A. Wokaun, *Appl. Surf. Sci.* **2002**, *197*, 746–756.

142. H. Furutani, H. Fukumura, and H. Masuhara, *J. Phys. Chem.* **1996**, *100*, 6871–6875.

143. H. Furutani, H. Fukumura, H. Masuhara, S. Kambara, T. Kitaguchi, H. Tsukada, and T. Ozawa, *J. Phys. Chem. B,* **1998**, *102*, 3395–3401.

144. M. Hauer, D.J. Funk, T. Lippert, and A. Wokaun, *Appl. Surf. Sci.* **2003**, *208*, 107–112.

145. J. Bonse, S.M. Wiggins, J. Solis, and T. Lippert, *Appl. Surf. Sci.* **2005**, *247*, 440–446.

146. J. Bonse, S.M. Wiggins, J. Solis, T. Lippert, and H. Sturm, *Appl. Surf. Sci.* **2005**, *248*, 157–162.

147. E.C. Buruiana, T. Buruiana, H. Lenuta, T. Lippert, L. Urech, and A. Wokaun, *J. Polym. Sci. Pol. Chem.* **2006**, *44*, 5271–5282.

148. E.C. Buruiana, V. Melinte, T. Buruiana, B. Simonescu, T. Lippert, and L. Urech, *J. Photochem. Photobiol.* **2007**, *A186*, 270.

149. E.C. Buruiana, V. Melinte, T. Buruiana, T. Lippert, H. Yoshikawa, and H. Mashuhara, *J. Photochem. Photobiol. A,* **2005**, *171*, 261–267.

150. T. Lippert, T. Kunz, C. Hahn, and A. Wokaun, *Recent Res. Dev. Macromol. Res.* **1997**, *2*, 121.

15

PHOTODEGRADATION PROCESSES IN POLYMERIC MATERIALS

JEAN-LUC GARDETTE, AGNÈS RIVATON, AND SANDRINE THERIAS

15.1 INTRODUCTION

One major problem associated with the applications of polymers is their poor stability to weathering [1–8]. The formation of oxidation products appears as the main cause of this instability. Understanding the mechanisms by which these products are formed has been the subject of a considerable effort of the academic and industrial research for the past 30 years. In addition to the interest for understanding the nature and the

Photochemistry and Photophysics of Polymer Materials, Edited by Norman S. Allen

mechanistic behavior of these species in polymers, developing and adapting new stabilizers on rational basis and predicting the lifetime service of polymers are most valuable for practical applications of polymeric materials. One can consider that the mechanisms of photodegradation of most of the common polymers are fairly well understood.

The past decades have seen a rapid expansion in the use of polymers in many new fields of applications where they play an essential role. Unfortunately, most polymers are not inherently stable to light, so the studies of their photochemical behavior remain a subject of constant interest. Among the various fields that were the object of interest in the last few years, one has observed some of them emerging as the domains of organic solar cells, of coatings, and of nanocomposites.

This review focuses on these domains, which gives an overview of the recent progress that has been made in the field of polymer degradation and stabilization. The current understanding of stability/degradation is presented and the methods for studying and elucidating degradation are discussed.

15.2 POLYMERS FOR APPLICATIONS IN ORGANIC SOLAR CELLS

Organic solar cells, although less efficient than silicon cells, exhibit a unique combination of interesting properties: low cost, flexibility, and large surface processability. A crucial step toward the fabrication of efficient organic devices was the development of the so-called "bulk-heterojunction" concept based on an appropriate combination of a (p-type) electron donor and a (n-type) acceptor. The first interpenetrated network tested in solar cells was MDMO-PPV (poly[2-methoxy-5-(3′,7′-dimethyloctyloxy)-1,4-phenylenevinylene) blended with fullerene derivatives. Efficiency of 3.3% was obtained by Sariciftci and coworkers [9] with [60]PCBM (methano-fullerene[6,6]-phenyl C_{61}-butyric acid methyl ester). Hummelen and coworkers showed 3% with MDMO-PPV blended with [70]PCBM [10]. One of the main focuses of the research in the field of organic solar cells is to increase the power conversion efficiency of the devices. Aside from the efficiency, the requirement of long operational lifetimes of devices in real-life applications is a crucial factor for the development of viable devices. Several factors can contribute to the loss of the optoelectronic properties of the organic solar cells. Among them the photodegradation of the organic active layer could play an important role. Surprisingly, only few studies deal with the behavior of the active layer under continuous exposure to light. It is however well known that the exposure of polymers to UV–visible light or to moderate temperatures induces chemical modifications of their structure. These changes are likely to modify the macroscopic properties of the polymer and, as a consequence, are responsible for a deterioration of the performances of the structure in which the material is used. It is of first importance to identify these photochemical processes and to propose strategies of stabilization if required.

Usually, organic solar cells are encapsulated to prevent the degradation induced by oxygen and/or moisture. The protected devices present much longer lifetimes than "bare" devices [11,12]. To fulfill a very good encapsulation, glass substrate or expensive multilayer barrier coatings are required. As a consequence, organic devices lose some of their attractive properties such as flexibility (glass substrate) and/or low-cost fabrication (multilayer encapsulation). For this reason, the stability of these devices in presence and in absence of oxygen must be understood. Several factors can contribute to the loss of the electronic properties of the organic solar cells: changes in the morphology of the layer, problems at the electrodes interface [13,14], photodegradation of the organic active layer.

Characterizing the mechanisms involved in the photodegradation of the various polymers that compose organic cells offers the advisability of reviewing the majority of the mechanisms involved in the photodegradation of polymers. MDMO and MDMO-PPV can be used as an active layer, whereas polymers like polyethylene terephthalate (PET) are used as substrates, and various materials are sounded for encapsulation: PC, ceramic on PC, and nanocomposites.

15.2.1 Photodegradation Processes of Conjugated Polymers Used as Active Layers (MDMO-PPV)

In view of the chemical structure of MDMO-PPV, it is expected that this polymer is very sensitive to solar light. This is attested by several studies that clearly show that oxygen-dependent degradation of the polymer plays a significant role in the failure of devices based on MDMO-PPV [15–18]. Carbonyl defects were shown to have a major influence on devices like PLEDS and the increase of carbonyl defects was correlated to a dramatic reduction of the photoluminescence of the polymer [19,20]. Understandings about the origin and the formation of these defects seem relevant in order to improve the stability of the devices based on this kind of polymer.

15.2.1.1 Photooxidation Mechanisms Considering the degradation of poly (*p*-phenylene vinylene) (PPV)-type matrix, a mechanism of photooxidation involving singlet oxygen attack was proposed [21]. Absorption of light by the polymer is followed by an energy transfer from the triplet state of PPV to oxygen. The result is the formation of singlet oxygen. Singlet oxygen can then react with the double bonds of PPV. This reaction gives an unstable endoperoxide that decomposes to give two aromatic aldehydes accompanied by chain scission [22]. The mechanism of photodegradation of poly[2-methoxy-5-(2-ethylhexyloxy)-1,4-phenylenevinylene] (MEH-PPV) proposed by Cumpston et al. [22] is depicted in Scheme 15.1.

This phenomenon was detected by measuring the phosphorescence of singlet oxygen. It was also shown that in presence of diazabicyclooctane (DABCO), a well-known singlet oxygen quencher, the rate of degradation of PPV was considerably reduced. After this first step, oxidation of the ether substituents would occur to form ester. This simplified mechanism was proposed on the basis of infrared analysis of aged samples. More recently, some authors investigated the role of singlet oxygen in the photooxidation process of PPV oligomers [23]. On the one hand, DABCO was shown to deactivate the triplet state of PPV oligomer. On the other hand, it was shown that different singlet oxygen sensitizers such as Bengal rose or methylene blue had no effect on the oxidation rate of PPV oligomers. It was concluded that the triplet state of

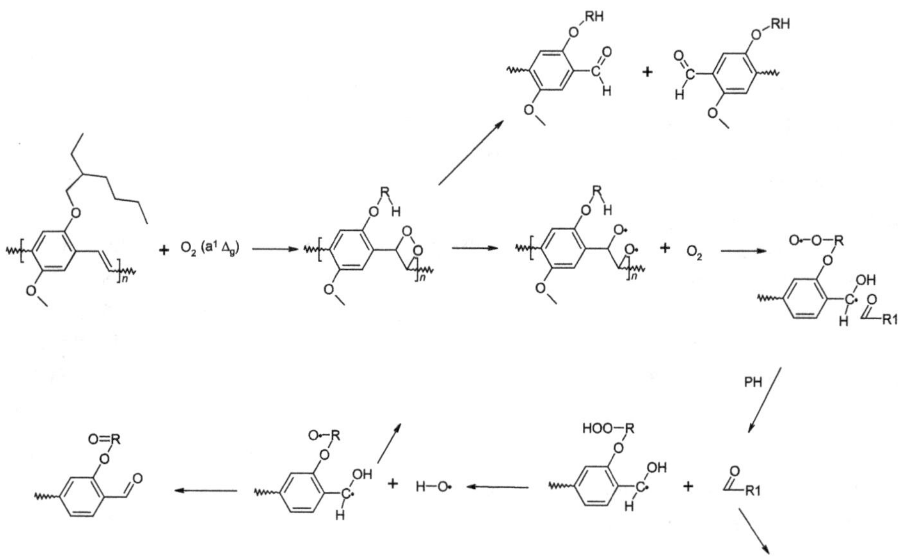

SCHEME 15.1 Mechanism of photodegradation of MEH-PPV proposed by Cumpson et al.

the PPV is involved in the degradation process, but not by generating singlet oxygen. At last, it was pointed out that the nature of the substituent and the length of the macromolecular chains are also likely to play a significant role in the mechanism of PPV-type polymers, as shown for oligomers of phenylene vinylene [24].

An alternative light-induced oxidative degradation of MDMO-PPV involving the ether side chains of MDMO-PPV was recently proposed by Chambon et al. [25] A combination of IR and UV techniques was used to elucidate the degradation mechanisms of the MDMO-PPV polymer. It was shown that oxidation of the ether side chains may predominate over direct attack of oxygen on the double bonds. The carbon atom next to the ether oxygen of the side chain is especially susceptible to radical-initiated oxidation. It is well known that polyethers are very reactive under exposure to UV–visible light in the presence of oxygen [26,27]. The sensitivity of polyethers toward photochemical oxidation results from the easy oxidation of the carbon atom in the α-position to the oxygen atom [28].

Irradiation under low pressure of oxygen permits two distinct processes to be discriminated:the oxidation of the ether groups and double bond saturation.

1. *Oxidation of Ether Functions*: Experimental results obtained with MDMO-PPV show that the ether part of the polymer is the most photooxidizable function and rapidly degrades. Abstraction of a hydrogen on the carbon atom in α-position of the oxygen leads to a macroalkyl radical [29]. By fixation of oxygen, a peroxy radical is formed giving an hydroperoxide by abstraction of a labile hydrogen atom and propagating the oxidation chain. Thermal and photochemical decomposition of these hydroperoxides give alkoxy and hydroxy radicals (Scheme 15.2). According to the literature, the main route of

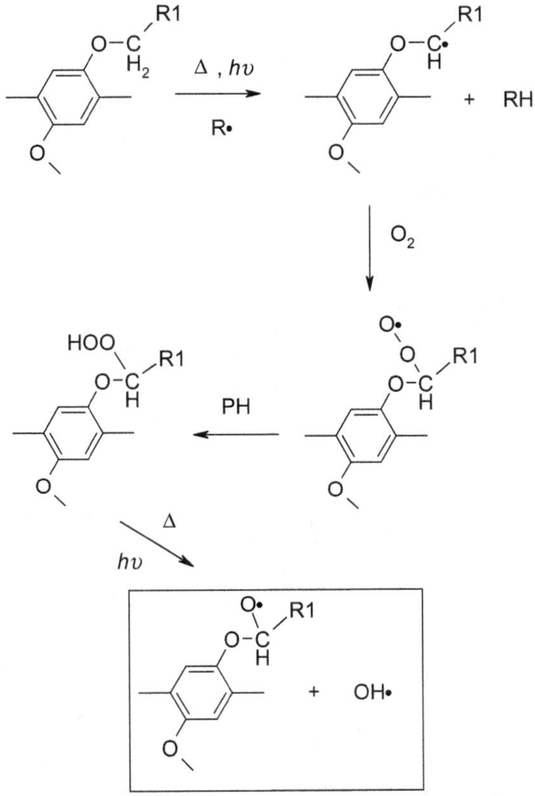

SCHEME 15.2 Thermal and photochemical decomposition of hydroperoxides.

evolution of the alkoxy radicals is a β-scission (Scheme 15.3) leading to the formation of formate [30] end groups and alkyl radicals (a). The alkoxy radical can also give an aromatic ester by cage reaction (b). The alkyl radicals produced by β-scission can also evolve following reaction (c) by fixation of oxygen followed by hydrogen abstraction. This leads to the formation of a primary hydroperoxide, which is photo- and thermo-unstable and gives a saturated carboxylic acid [31]. The third possible evolution of the macroalkoxy radical is an hydrogen abstraction on the polymeric backbone, leading to the formation of an hemiacetal (route (d)). This product is thermally unstable and decomposes to form a substituted phenol and an aliphatic carboxylic acid.

2. *Saturation and Oxidation of Double Bonds*: It is known that radicals can directly add on double bonds [32], which leads in turn to saturation of the double bonds and formation of a new radical. After fixation of oxygen, a peroxy radical is formed followed an hydroperoxide by the classical hydrogen abstraction. Thermal and photochemical decomposition of the hydroperoxides gives alkoxy radicals (Scheme 15.4).

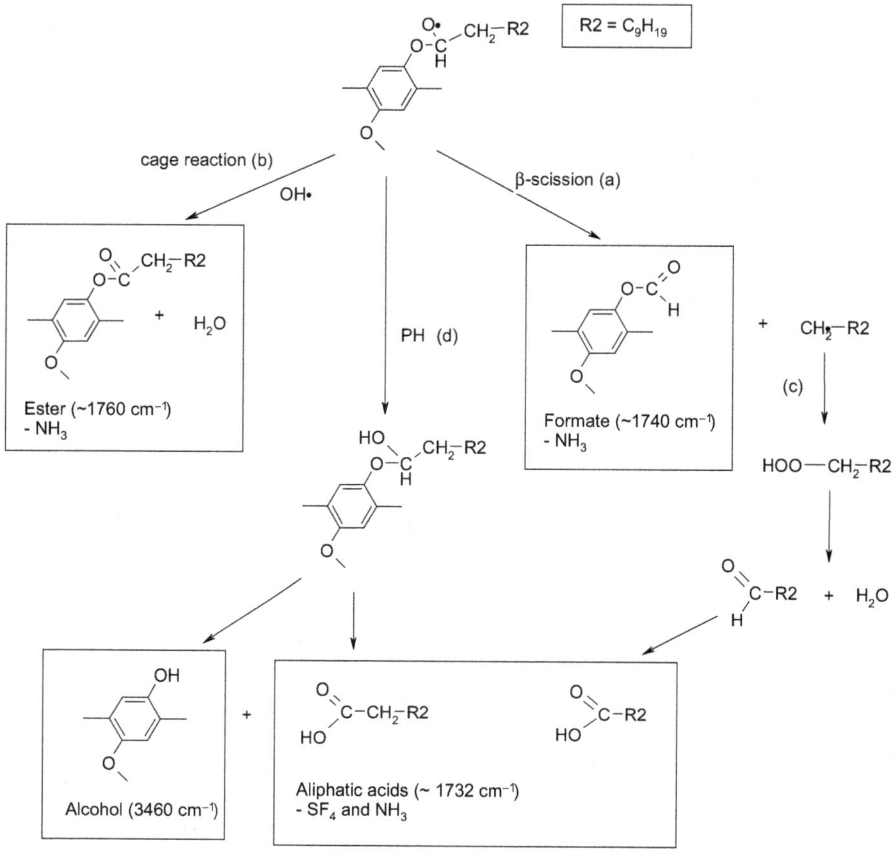

SCHEME 15.3 Oxidation of the ether functions.

After dissociation of the hydroperoxide giving a macroalkoxy radical and an hydroxy one, three different pathways are likely to occur:

- Cage reaction leading to the formation of an aromatic ketone (route (a))
- β -Scission leading to the formation of an aromatic aldehyde (route (b))
- Hydrogen abstraction leading to the formation of an alcohol (route (c))

Scheme 15.5 summarizes these various routes. According to the literature, when oxidation occurs on a secondary carbon, the cage reaction is favored compared to the β-scission [33]. The involvement of chain scissions in the photooxidation process has been confirmed by methanol treatment that showed that most of the oxidative species are low molecular weight products.

15.2.1.2 ***Photochemical Processes Under Irradiation in Absence of Oxygen*** The scientific community agrees to say that the active layer has to be protected from the

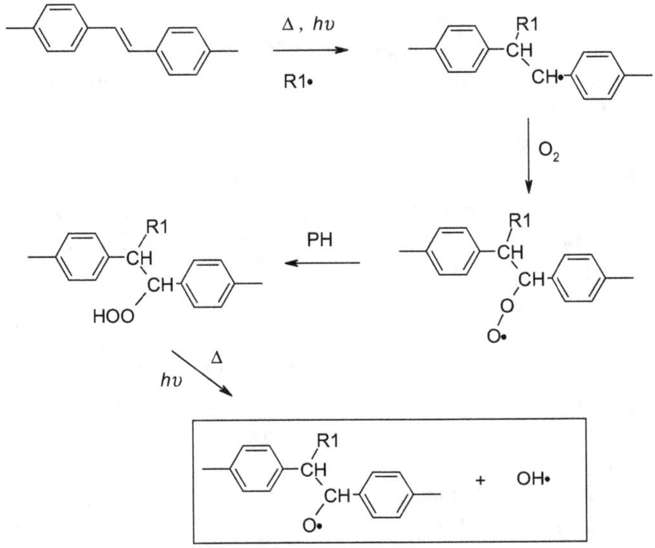

SCHEME 15.4 Thermal and photochemical decomposition of hydroperoxides following double bond saturation.

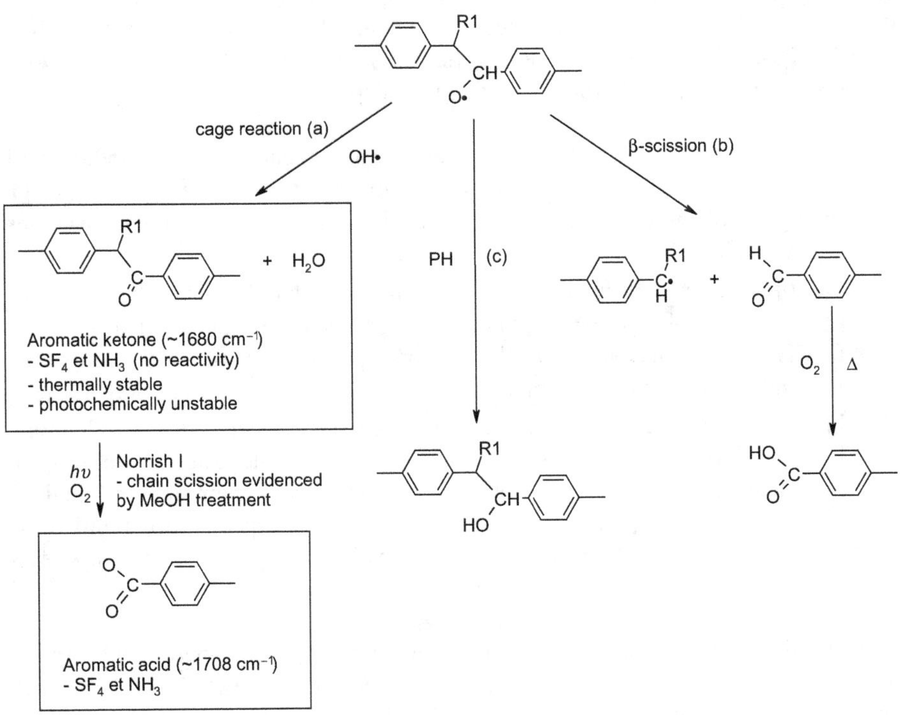

SCHEME 15.5 Saturation and oxidation of double bonds.

atmosphere [34]. However, providing an efficient protection to the solar cell against oxygen and humidity does not imply that the materials will be stable under exposure to light. The direct absorption of UV–visible light by the chromophores of polymeric matrix can be responsible for photophysical processes and direct photochemical reactions. The behavior of the active layer materials under illumination and in absence of oxygen has then to be investigated. It is well known that under the impact of light and without reacting with oxygen [35–39], aromatic polymers for instance can undergo various reactions, involving rearrangement (photo-Fries, Norrish, and so on), chain scissions, or cross-linking.

In the case of MDMO-PPV irradiated under vacuum, one observes that a reorganization of the chemical structure of the polymer occurs, leading to a loss of the conjugation [25]. In addition, the concomitant disappearance of the ether functions, the saturation of the double bonds, the formation of OH groups, and the modification in the aromatic ring substitution are observed. Concerning aryl–O–X components, the literature reports that a photo-Fries-type rearrangement can occur [40] under exposure without involvement of atmospheric oxygen. This reaction has been studied in details in the case of polycarbonate (PC) [41]. In the case of the photolysis of MDMO-PPV, such a rearrangement could account for the results mentioned above, disappearance of ether functions, and formation of hydroxyl products. The low intensity of the band that develops is due to the small amount of photoproduct formed and to the low molar extinction coefficient of hydroxyl groups ($\sim$90 L/(mol cm)) [42,43]. Scheme 15.6 summarizes the main reactions that occur under exposure in absence of oxygen. Absorption of UV–visible light by MDMO-PPV provokes the homolytic scission of the O–CH_2 bond. -Φ–O• and •CH_2–R macroradicals formed in this first step may react in two ways:

- Cage reaction according to photo-Fries rearrangement (a).•CH_2–R radical will substitute the cycle in *ortho*-position and hydrogen atom will be trapped by Φ–O• giving a phenolic group. The cycle, initially tetrasubstituted, becomes pentasubstituted.
- Out-of-cage reaction leading to the saturation of double bonds (b). Competitively to photo-Fries rearrangement, some •CH_2–R radicals formed by direct O–CH_2 scission may add to the double bonds provoking the loss of conjugation. It is indeed well know that radicals can directly add on unsaturations [32].
- From the infrared absorbance, the approximate concentration of alcoholic function formed on the one hand and of double bonds disappeared on the other hand is calculated using the molar extinction coefficient of both functions ($\sim$90 and $\sim$125 L/(mol cm) for -OH and =CH–, respectively). From these values, it was determined that the occurrence of the two photolytical processes is 60/40 for photo-Fries rearrangement over saturation reaction.

According to the mechanism of photodegradation proposed above, the absorption of light by the polymer induces the homolytical scission of the O–CH_2 bond, which generates the formation of two radicals that immediately recombine or add on

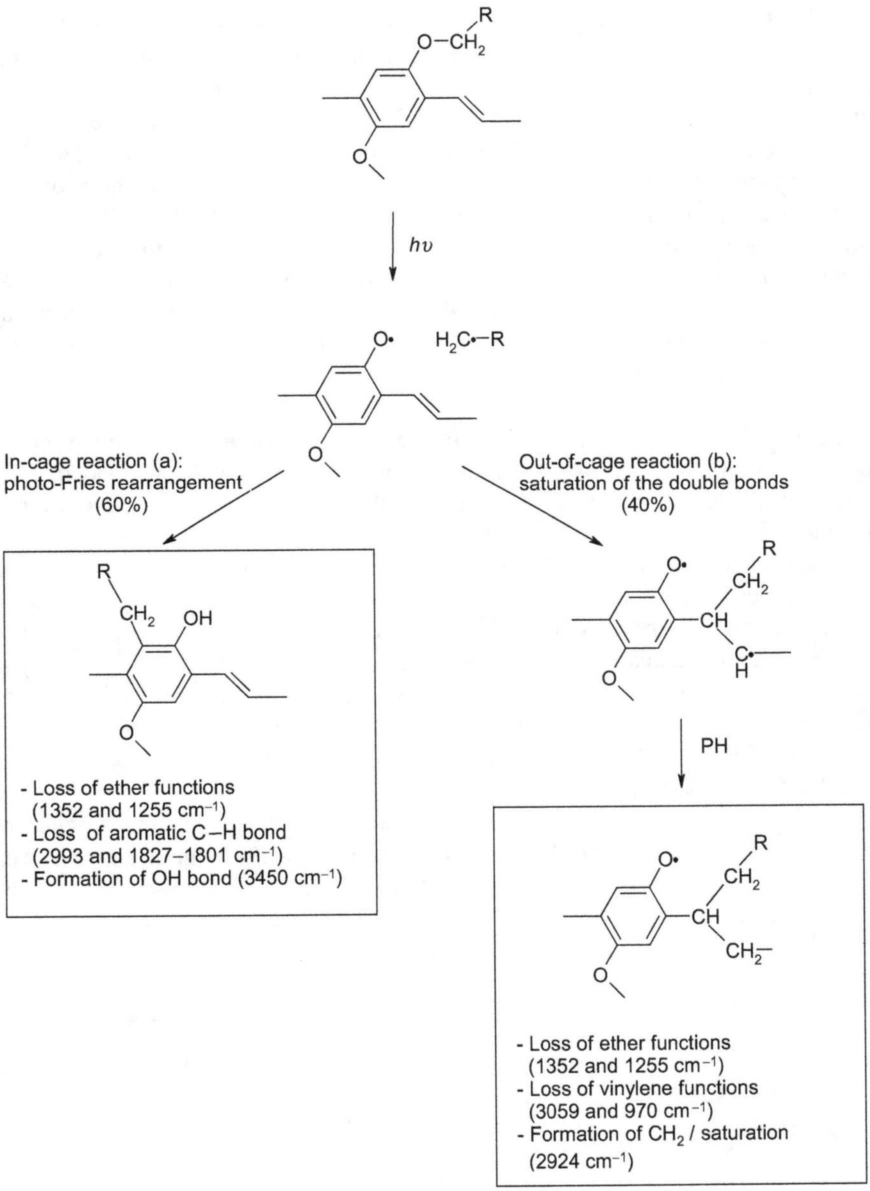

SCHEME 15.6 Vacuum photolysis of MDMO-PPV.

double bonds. Due to their low concentration and instability, it was not possible to evidence the formation of radicals by ESR. Nevertheless, a second pathway that could take part in the initiation step was suggested [34]. Charged radicals are likely to evolve and be considered as intermediates in the photodegradation of the polymer [32]. In the case of MDMO-PPV, transient absorption spectroscopy experiments that were carried out under nitrogen revealed the presence of transient species. Based on the study of Nogueira et al. [37], these intermediates were attributed to positive polarons MDMO-PPV$^{\bullet(+)}$ generated after excitation. Another segment of the polymer chain or an impurity would act as an electron acceptor. The radical cation can evolve by scission. If the positive charge is situated on the oxygen atom, the ether function could produce a cation Φ–O^+ and a radical •CH_2–R. The radicals thereby generated can add on double bonds or undergo photo-Fries reaction as shown in Scheme 15.1. Radical anion can also evolve by scission and generate an anion and a radical. The radical formed can enter the degradation mechanism by addition on double bonds. Moreover, radical anions are likely to saturate the vinylene function via nucleophilic addition.

15.2.1.3 Photodegradation of MDMO-PPV:PCBM Blends Chambon et al. recently published a paper [44] devoted to the study of the photo- and thermal ageing of MDMO-PPV blended with [60]PCBM, a commonly used active layer of organic solar cells. Thin films of MDMO-PPV, PCBM, and MDMO-PPV:PCBM were submitted to photo- and thermal ageing in presence or in absence of oxygen. The modifications of the chemical structure of the materials were analyzed by UV–visible and IR spectroscopies. The presence of PCBM in MDMO-PPV films was shown to dramatically reduce the degradation kinetic of the polymer not only in photooxidation but also in thermooxidation and photolysis (irradiation in the absence of oxygen) (Fig. 15.1). This effect was attributed to the radical scavenging properties of PCBM [45]. In addition, results showed that an efficient protection of MDMO-PPV:PCBM from the atmospheric oxygen was likely to decrease the degradation rate by a factor 1000. The authors concluded that the absence of oxygen combined to the radical scavenging property of PCBM significantly reduced the degradation rate of the polymer, and in this case, the active layer can be considered as relatively photostable.

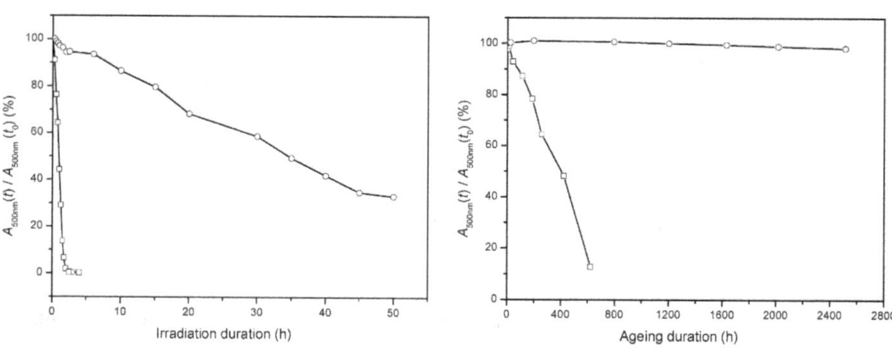

FIGURE 15.1 Normalized absorbance of UV–visible band at 500 nm during photooxidation (left) and thermooxidation (right) of MDMO-PPV (□) and MDMO-PPV blended with PCBM (○).

15.2.2 Photodegradation of the Substrates: Poly(butylene terephthalate)

Polyethylene terephthalate is a low-cost polymer that exhibits excellent optical and mechanical properties. PET is commonly used in the field of packaging, specifically plastic bottles. PET is also used as flexible substrate in organic solar cells. Exposing this polymer to environmental atmosphere changes its external appearance, modifies its surface, and degrades its properties.

The photoageing of PET [46–54] reported to be similar to that of PBT (poly (butylene terephthalate)) [55–59] has been the center of some interest in the literature. When sunlight falls onto PET, the monomer units absorb energy of short wavelength in the near-UV range: this is particularly so for the ester groups in α-position to the aromatic rings. The absorbed energy causes the rupture of covalent bonds initiating photolytic (without intervention of oxygen) and photooxidative (fixation of oxygen) reactions. In the environment, both mechanisms can take place and are closely related. Consequently, PET cannot be used in conditions of outdoor exposure without a stabilization agent to minimize the damage of photoaging.

15.2.2.1 Photochemical Behavior on Irradiation in Vacuum The absorption spectrum of the terephthalate chromophoric units in the UV domain extends up to 350 nm. This makes PBT directly accessible to UV light present in the terrestrial solar radiation. It was observed that photolysis of PBT is predominantly a surface effect: only the 35 μm top layer shows degradation while the subsequent layers of the sample are unaffected by UV irradiation. The changes in the UV–visible spectra show an increase in absorption without any defined maximum in the UV domain. FTIR spectroscopy coupled with various chemical derivatization treatments permits the identification of the various photoproducts that are formed. The increase in the hydroxyl absorption was attributed to the formation of alcohol and acid groups. In the carbonyl region, the broadening of the initial ester band was attributed to the formation of γ-lactone end groups, benzoic acid end groups, benzaldehyde end groups, and double bonds. In the region of the C–O stretching vibrations, the band that was observed at 1180 cm^{-1} has to be attributed to formates; the band that is observed to develop at 773 cm^{-1}, associated with the absorption at 1610 cm^{-1}, indicates the formation of *meta*-substituted rings. In parallel to the formation of the photoproducts, the intensity of several bands decreases, because photolysis involves hydrogen abstraction on CH_2 groups, substitution of aromatic cycles, loss of aromatic units, and scission of the ester bonds.

These experimental results may be explained as follows: by direct light absorption by the chromophoric groups of PBT units in the range 300–500 nm, the two alternative Norrish mechanisms shown in Scheme 15.7 may occur. The radicals generated by the various processes may react separately through hydrogen abstraction from neighboring molecules; this leads to the formation of benzoic acid, benzaldehyde, aliphatic alcohol, and butyl formate end group structures.

Double bonds that are formed through intramolecular Norrish type II process lead to the formation of γ-butyrolactone through further Norrish type I scission followed by intramolecular addition.

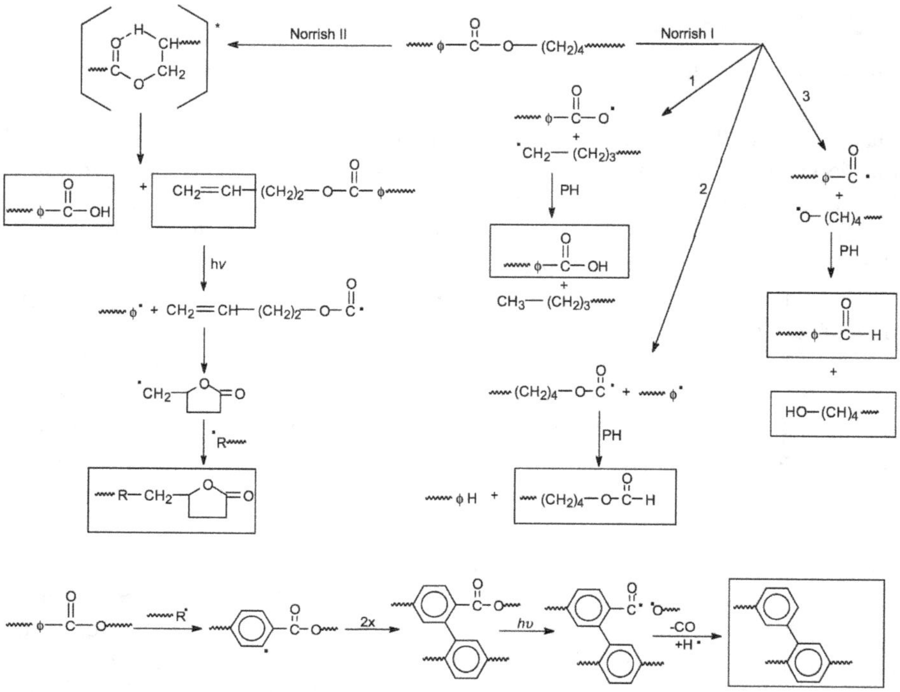

SCHEME 15.7 Vacuum photolysis of PBT at $\lambda > 300$ nm.

In parallel to hydrogen abstraction, the primary radicals formed by the Norrish type I processes may recombine together before or after decomposition through decarbonylation or decarboxylation. The hydrogen abstraction on the phenyl rings leads to *meta*-biphenyl groups and accounts for one of the recombination pathways.

15.2.2.2 Photochemical Behavior in the Presence of Oxygen Irradiations carried out in the presence of oxygen lead to a large increase of the absorbance in the range 300–500 nm. The direct scission of the aromatic ester chromophoric units results in the recombination of the radicals, which generates conjugated structures responsible for the yellowing of the samples. The rate of discoloration is more important in vacuum than in the presence of air: photooxidation leads to the bleaching of the photolysis products of PBT. The photooxidation of PBT causes drastic modifications of the IR spectra of the irradiated samples. Similar modifications of the carbonyl absorbance are observed in vacuum and in air, but the photoproducts are formed at a rate two times higher in the presence of oxygen. The photoproducts profiles have the same shape on exposure in vacuum and in air: photodegradation is confined to the first 35 μm layer. The distribution profile is rather related to the strong light absorption by the samples than to the low oxygen permeability of the polymer. Chemical derivatization reactions confirm that the photoproducts detected by IR analysis of the photolyzed films are also observed in photooxidation conditions. Photooxidation favors the consumption of CH_2 groups and accounts for the formation of hydroperoxides, esters, anhydrides, and peresters, photoproducts that are detected in fairly low concentration.

As a conclusion, the photooxidation of PBT involves a complex mechanism in which preponderant photolytic processes and minor photooxidative reactions interfere. Primary radicals formed in direct photoscission processes are able to induce the oxidation of the methylene groups in α-position to the ester bond by abstraction of labile hydrogen atoms. Hydroperoxides accumulate up to low concentration ($\leq$10 mol/kg); the homolysis of their O–O bond leads to the formation of macro-alkoxy and hydroxy radicals. The alkoxy radicals are converted into keto hemiacetal through hydrogen abstraction, anhydride through cage reaction, and aliphatic aldehyde and carboxy radicals through β-scission. These photoproducts are photo- and thermo-unstable, and their oxidation can occur. The detection of the IR bands of photoproducts and/or the characterization of derivatives of esters, peresters, and aliphatic acids, as well as the increase in the presence of oxygen of the quantum yield of formation of benzaldehyde, benzoic acid, and aliphatic alcohol permits the proposal of the mechanism shown in Scheme 15.8.

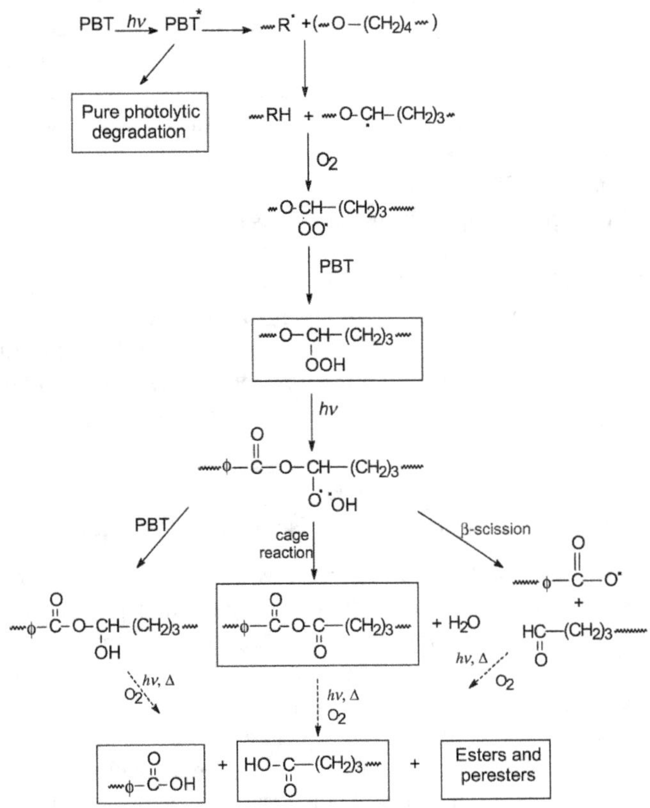

SCHEME 15.8 Photooxidation of PBT at wavelengths >300 nm.

15.3 PHOTODEGRADATION AND PHOTOSTABILIZATION OF CERAMIC COATED POLYMERS

The sensitivity of polymer solar cells toward oxygen and moisture makes it imperative to protect them with some sort of encapsulation membranes. The rate of transport across membranes is expressed in the oxygen transmission rate (OTR) and the water vapor transmission rate (WVTR). Upper limits of about 10^{-3} $cm^3/(m^2$ day atm) for OTR and 10^{-4} $g/(m^2$ day) for WVTR have been proposed for organic solar cells [60–62]. In general, the approach of a complex multilayer system is used to fulfill the requirements for these ultrabarrier applications. In addition to their barrier properties, these multilayer systems also have to exhibit high transparency for many applications. These layers may be inorganic oxides deposited by plasma-enhanced chemical vapor deposition (PECVD) reaching permeation levels 1000 times lower than that for the native polymer films. Multilayered barrier films consisting of alternating inorganic SiO_x and polymer PEN substrates were used to greatly increase the lifetime of MDMO-PPV/PCBM-type solar cells. Another layer system that can be used is based on sputtered inorganic layers that are separated by different types of intermediate smoothing layers [63].

Protection from photoaging of polymers with aromatic structure, such as PC, PET, or PEN, is a difficult challenge since these polymers strongly absorb UV light. Probably the most efficient method to prevent photoaging consists of covering the surface of an aromatic polymer with a thin ceramic coating. This outer layer has to be photochemically inert and transparent in the visible range. The coating absorbs the UV radiation that damages the polymer and also provides a barrier against oxygen diffusion. Ceramic thin films can be produced using several methods. Among them, radio-frequency magnetron sputtering techniques present several advantages such as low-temperature processing and good film–substrate adhesion. PET [64,65], PEEK [66], PC [67,68], and PEN [69] can be successfully coated with ZnO or TiO_2, depending on the polymer substrate. Ceramic coatings are deposited by sputtering from different argon–oxygen mixtures. This technique presents interesting advantages such as high ceramic density. In this review, we describe the concept on the stability improvement of sputtered inorganic films deposited on bisphenol-A polycarbonate.

15.3.1 Photodegradation Processes of Bisphenol-A Polycarbonate

Polycarbonate is increasingly used to replace various structural materials, such as metals and glasses, because of its high toughness, good transparency, and relatively low density. However, absorption of ultraviolet radiation from the exposure to sunlight induces chemical reactions that cause discoloration, significant changes in the surface mechanical properties, and embrittlement.

The reactions that are produced by exposure of PC to solar light have been described by a dual photochemistry [41,70–75]. Direct phototransformation and photoinduced oxidation occur, with a ratio largely dependent upon the spectral distribution of the excitation light source.

15.3.1.1 Direct Photochemistry of PC Excitation of PC at short wavelength (e.g., 254 nm) involves mainly two consecutive photo-Fries rearrangements of the aromatic carbonate units leading successively to the formation of phenylsalicylate (L_1) and dihydroxybenzophenone (L_2) units as shown in Scheme 15.9. Photo-Fries products have been well defined by definite maxima in the UV and in the carbonyl range of the IR domain: L_1 at 320 nm and 1689 cm^{-1} and L_2 at 355 nm and 1629 cm^{-1}.

As a minor pathway, L_3 units are formed competitively to photo-Fries rearrangements: some radicals formed in CO–O bond scissions may decarbonylate or decarboxylate before further radical recombination or hydrogen abstraction. This leads to the formation of hydroxy- and dihydroxybiphenyl units, aromatic ether structures, and phenol as end groups that further photolyzed into a mixture of species (in a convoluted absorption) that produces the yellowing of the PC film without any defined structure.

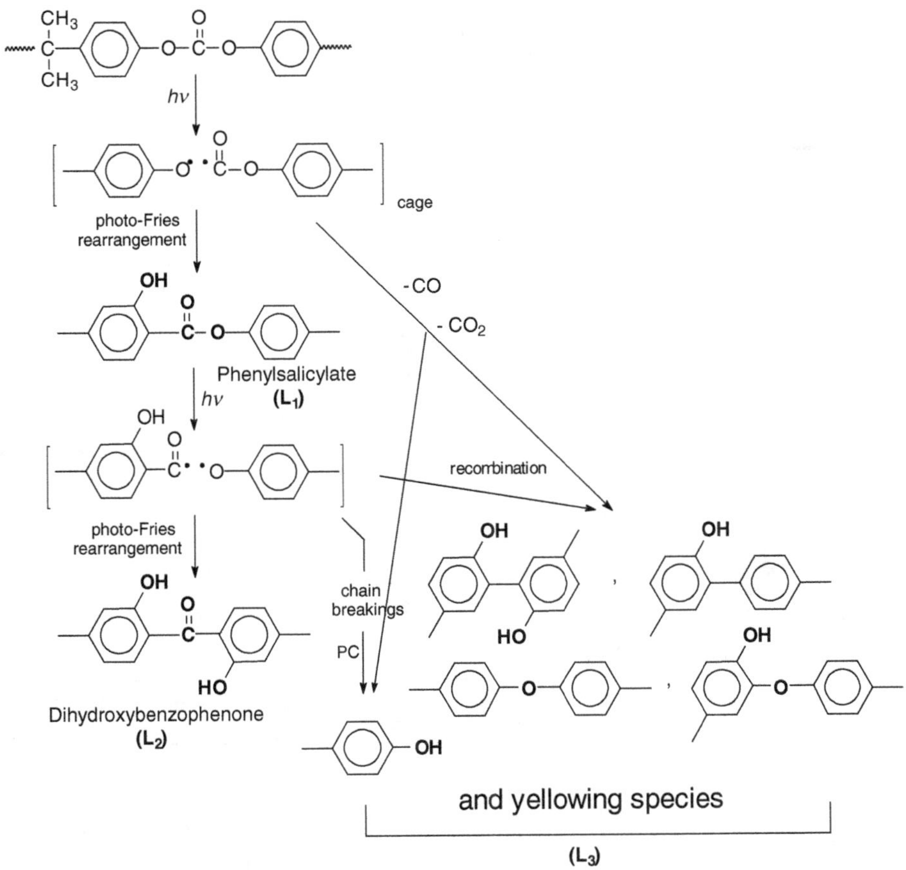

SCHEME 15.9 Direct phototransformation of PC at short wavelength in the absence of oxygen.

15.3.1.2 Photochemical Behavior at Long Wavelengths (λ > 300 nm) On irradiation at long wavelength (e.g., 365 nm) in the presence of oxygen, photoproducts have been shown to result mainly from the photoinduced oxidation on the gem dimethyl side chain and from the phenyl ring oxidation. The various steps of the gem dimethyl side-chain photooxidation, initiated by photo-Fries process, are reported in Scheme 15.10.

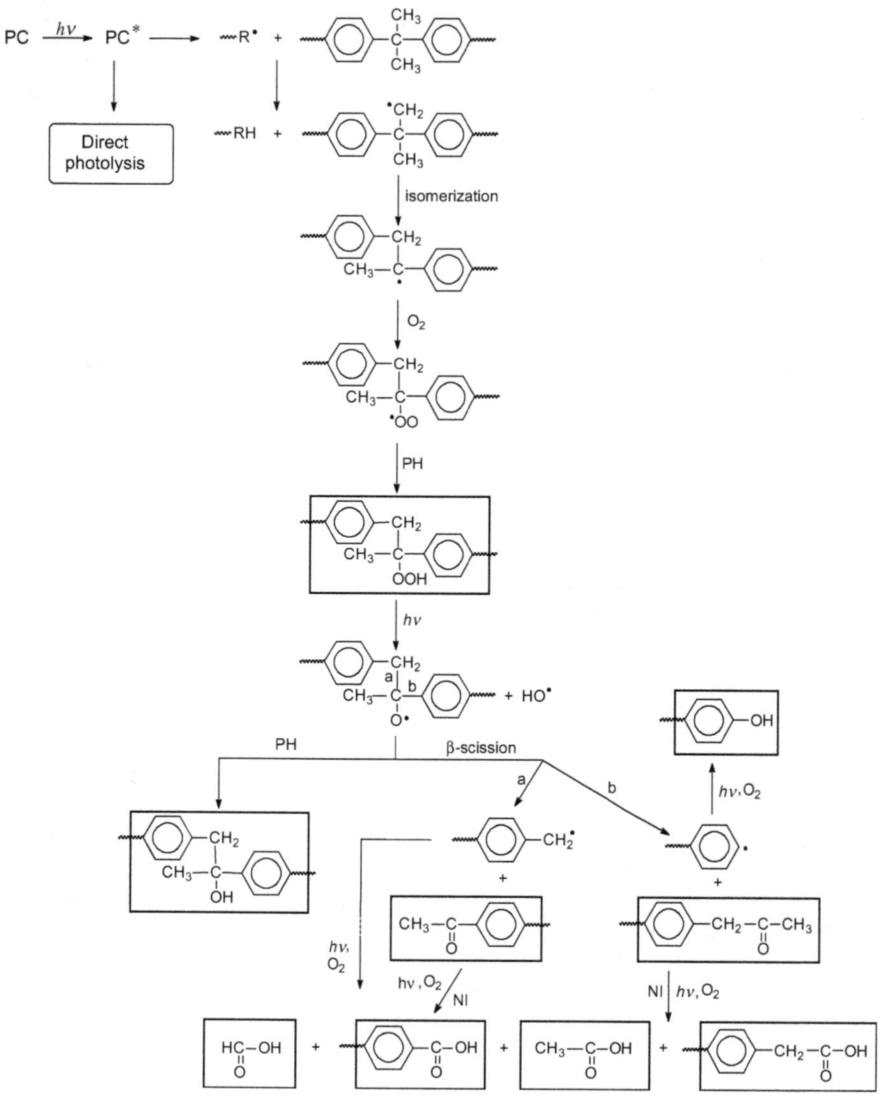

SCHEME 15.10 Gem dimethyl side-chain photooxidation of PC.

The first step of the oxidative reactions is a hydrogen abstraction on the polymeric backbone. Once formed, the primary methylene macroradicals (A) rearrange to yield the stable tertiary benzylic radicals (B):

$$^{\bullet}CH_2{-}C(CH_3)(Ph\sim)_2 \quad \mathbf{(A)} \xrightarrow{\text{isomerization}} CH_3{-}\dot{C}(Ph\sim)CH_2\sim \quad \mathbf{(B)}$$

The macroradical formed reacts with oxygen, leading to a peroxy radical that gives a hydroperoxide by abstraction of a labile hydrogen atom. Hydroperoxides decompose thermally or photochemically to give alkoxy and hydroxyl radicals (which can mainly resume the chain oxidation reaction) leading to the formation of the various photoproducts that have been identified, namely, aliphatic (1724 cm^{-1}) and aromatic (1690 cm^{-1}) chain ketones, aliphatic (1713 cm^{-1}) and aromatic (1696 cm^{-1}) chain acids, chain alcohols (3470 cm^{-1}), and formic and acetic acid that are able to migrate in the gas phase.

The opening of phenyl rings of PC has been assessed by several authors [73]. It has been suggested that cyclic anhydrides (1860/1840 cm^{-1}) could be formed in the thermal transformation of dicarboxylic acidic products (1713 cm^{-1}) that appear after phenyl ring scissions under long-wavelength irradiation [71]. A probable mechanism of formation is reported in Scheme 15.11. The initiation steps by OH• and $^1O_2$ have been proposed by Clark and Munro [76].

PC is photooxidized under polychromatic light, provided by sources emitting short- and long-wavelength radiations. The photochemical evolution of this polymer directly depends on the spectral distribution of the light. This means that the ratio of short to long wavelengths determines the occurrence of direct phototransformation and side-chain photooxidation. Photo-Fries rearrangement is the main process occurring by absorption of radiations below 330 nm, while gem dimethyl side-chain and phenyl ring oxidation are provoked by radiations above 330 nm.

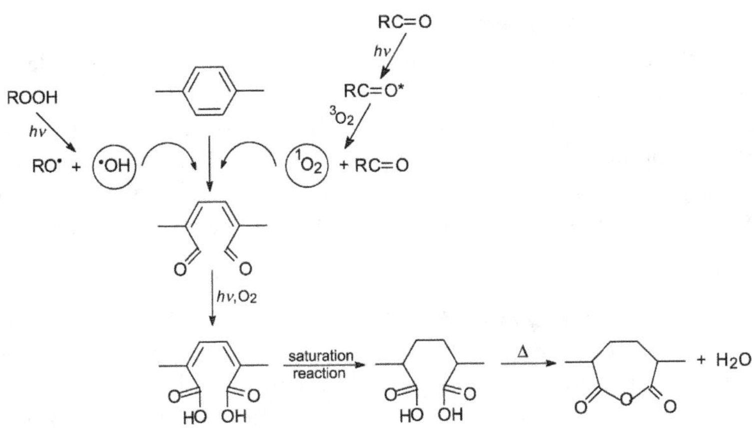

SCHEME 15.11 Probable mechanism of ring oxidation.

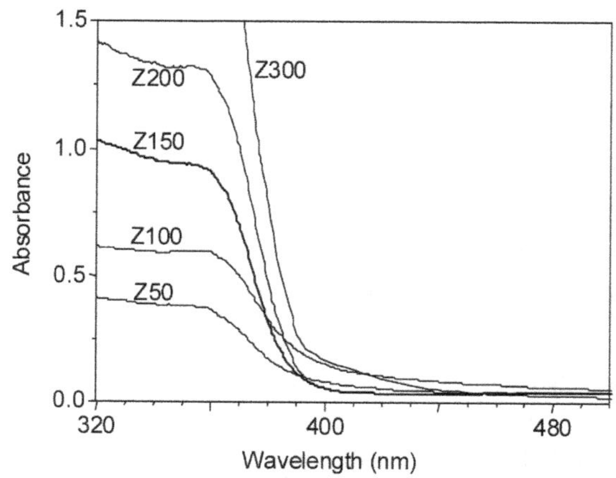

FIGURE 15.2 UV–vis absorbance spectra of ZnO coatings deposited on PC (the numbers indicate the corresponding layer thickness in nanometers) [67].

15.3.2 Photoprotection of Polycarbonate by Ceramic Coatings

15.3.2.1 ZnO Coatings on PC ZnO coatings can efficiently protect PC from photoageing. The efficacy of relatively thin ceramic coatings to reduce photodegradation and to enhance the surface nanomechanical properties of PC was evaluated in light of IR and UV–vis analysis [77,78]. It has been shown that a thin ZnO coating was effective in reducing the photodegradation rate of PC because of the screening effect of the coating in the UV range. Figure 15.2 shows the UV–visible spectra of PC coated with ZnO (thickness from 50 to 300 nm). ZnO exhibits a strong absorption in the region 300–400 nm that depends on the thickness of the layer. Thin ZnO coatings have a good transparency in the visible but absorb UV radiations.

The photochemical oxidation of PC can be characterized by the increase in absorbance at 3470 cm^{-1} versus irradiation time, as shown on Fig. 15.3. Oxidation occurred rapidly for uncoated PC. A decrease in the rate of oxidation was observed in the case of the coated samples. The more the ZnO layer is thick, the less is the oxidation rate. Similar conclusions were drawn when the absorbance was measured at 400 nm. The screening effect reduces therefore the PC photodegradation produced by the direct absorption of sunlight radiations. In addition, we have shown that the grain size and the density increase with the coating thickness. The coatings are denser and have lower microvoids. Consequently, they are more impermeable to oxygen and the oxidation is limited.

15.3.2.2 Multilayer Coatings on PC Despite the efficiency of ZnO and TiO_2 to protect polymers from irradiation damage, photocatalytic oxidation at the polymer/ceramic interface is a limiting factor [80]. The efficiency of the thin ZnO coating is then limited by the photocatalytic activity at the interface with the PC substrate. The photoactivity of these pigments could constitute a limitation to their use as coatings on polymeric substrates because they could induce a photocatalytic oxidation of the

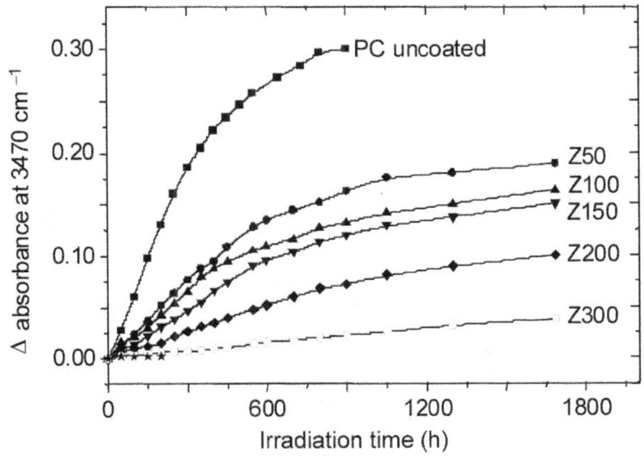

FIGURE 15.3 Increase in the IR absorbance of the band located at 3470 cm^{-1} versus the irradiation time for PC covered with ZnO coatings of different thicknesses [79].

polymer at the interface. The photoactivity results from the formation of active species that can initiate photooxidative degradation of the polymer by reaction with water and/or oxygen. These aspects are very well documented in the literature. Most authors agree that $HOO^{\bullet}$ and $HO^{\bullet}$ radicals are the predominant reactive species [81,82]. To overcome this deficiency, a thin Al_2O_3 coating was used as an intermediate layer [83]. It was found that the photocatalytic activity at the PC/ZnO interface was reduced as a result of the intermediate Al_2O_3 layer, which in addition limited the permeability of oxygen. The deposition of a thin Al_2O_3 layer between the ZnO layer and the PC substrate increased the resistance to photodegradation by suppressing the photocatalyzed degradation of the polymer at the interface. In addition, the Al_2O_3 and ZnO layers act as barriers to oxygen and limit the oxidative degradation caused by photoinduced aging. A higher photooxidation resistance was obtained with the three-layers coating (Al_2O_3/ZnO/Al_2O_3) than the two-layers coatings (Al_2O_3/ZnO and ZnOAl_2O_3) because of the increase in the impermeability to oxygen with the coating thickness. The ceramic coatings not only protect PC from photodegradation but also enhance the nanomechanical properties and lower the coefficient of friction [83]. The three-layers coatings exhibit the lowest coefficient of friction over the entire load range investigated. In contrast to the PC, the similar nanomechanical properties obtained before and after extensive irradiation indicate that photochemical reactions at the polymer interface are not detrimental to the adhesion of the Al_2O_3 and ZnO layers to the PC substrate. The results of this study demonstrated that thin ceramic coatings are effective inhibitors of photochemical reactions and deformation processes occurring in polymers, which lead to accelerated photodegradation, high friction, and excessive wear. Figure 15.4 shows the increase of absorbance at 3470 cm^{-1} and at 400 nm versus irradiation time for different polymer/ceramic assemblies.

15.3.2.3 Organic–Inorganic Hybrid Coatings Another elegant approach that permits the photoprotection of polycarbonate was recently published by Fabbri

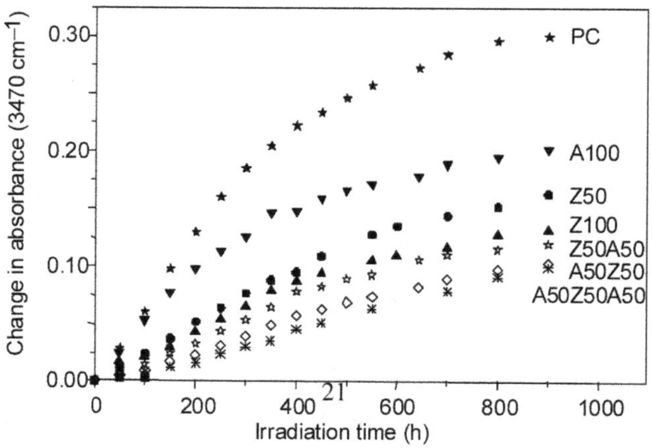

(a)

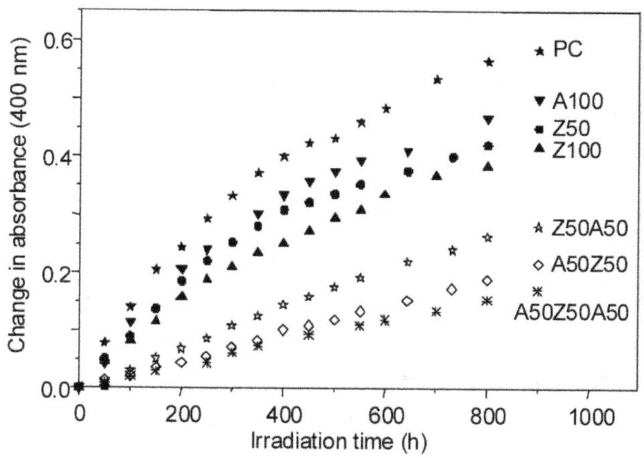

(b)

FIGURE 15.4 Change in absorbance at (a) 3470 cm^{-1} and (b) 400 nm of PC coated with single- and multilayer coatings of ZnO and Al_2O_3 of different thickness.

et al. [84] Organic–inorganic hybrid materials (ceramers) were prepared starting from triethoxysilane-terminated polymer chains (namely, poly(ethylene oxide) (PEO) or polycaprolactone (PCL)), both selected as an organic component for their good thermodynamic miscibility with PC [85–88] and tetraethoxysilane as inorganic network precursor; the hybrids were subsequently obtained as self-consistent materials or coated onto PC substrates. For the aforementioned significant advantages presented by the sol–gel approach, this method should be strongly suggested for the preparation of opportune hybrid coatings ensuring a barrier effect toward UV radiations. PCL was chosen rather than PEO because poly(ethylene oxide) is known as very sensitive to UV light-induced oxidation [30]. In this first approach of the protective effect of these ceramers, the attention was focused on the barrier effect of the coatings. The screen effect that these coatings can play when formulated with organic or mineral UV absorbers was studied. The oxidative degradation provoked by exposure to UV light of PC substrates coated with PCL ceramers was followed by IR and UV–visible spectroscopies. Photooxidation tests carried out on uncoated PC thin films and the same film coated with polycaprolactone/silica hybrids demonstrated a notable protective effect of the hybrid coatings against photooxidation.

15.4 PHOTOOXIDATIVE DEGRADATION OF POLYMER NANOCOMPOSITES

Finely dispersing particles in polymer matrices are nowadays a widely used method that allows for an improvement of properties such as stiffness, and mechanical properties of the composite materials. Polymer nanocomposites represent a new class of composite materials with fillers displaying at least one dimension in the nanometer range. Depending on the nature of the added filler, other properties can also been enhanced such as fire resistance and thermal and electrical conductive properties. These nanocomposites provide materials with significantly improved properties at lower filler content (about 3–5 wt%) than in conventional composites. Among the tremendous literature devoted to nanocomposite materials, most of the articles concern the synthesis, the processing and the characterization of nanocomposites, and the evaluation of their properties. Only few studies deal with the long-term behavior of these materials. One can question about the durability of the properties in the conditions of environmental ageing and only a little is known as far the lifetime of the materials is concerned.

This chapter reports the results of the literature that concerns the photooxidation of polymer nanocomposites. The published studies concern various polymers (PP, epoxy, ethylene–propylene–diene monomer (EPDM), PS, and so on) and different nanofillers such as organomontmorillonite or layered double hydroxides (LDH) were investigated. It is worthy to note that a specific attention was given to the interactions with various kinds of stabilizers and their efficiency to protect the polymer. One of the main objectives was to understand the influence of the nanofiller on the oxidation mechanism of the polymer and on the ageing of the nanocomposite material. Depending on the types of nanocomposite that were studied, the influence of several parameters such as morphology, processing conditions, and nature of the nanofiller was examined.

15.4.1 Polymer/Clay Nanocomposites

The most commonly studied polymer nanocomposites are clay-based nanocomposites, mainly with montmorillonite (MMt) as layered silicate filler (Scheme 15.12). Upon incorporation of organomodified clays (organoclays) into a polymer matrix, two nanomorphologies (Scheme 15.13) can be obtained, either intercalation of the polymer chain in between the clay platelets keeping the stacking of the sheets, or exfoliation of the clay platelets with a disordered dispersion of the inorganic sheets in the polymer.

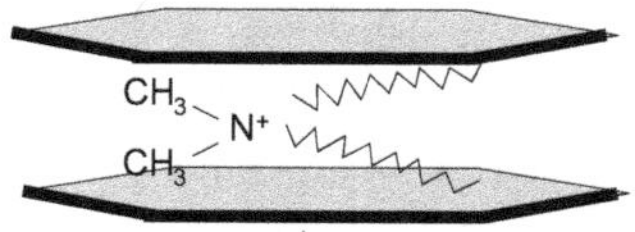

SCHEME 15.12 Structure of an organomontmorillonite.

15.4.1.1 Polypropylene/Montmorillonite A first approach of the photooxidation of polymer/clay nanocomposites [89] was published by Tidjani et al. In this paper, it was shown that the PP/MMt nanocomposites with PP-*g*-MA as compatibilizing agent had a higher propensity toward photooxidation. The authors suggested that this could arise from the clay and/or the structural form of the nanocomposite. Deeper investigations were carried within a general study dealing with the influence of the nanofiller on the behavior of nanocomposite materials in conditions of photolytic oxidation carried out on PP/MMt nanocomposites [90–92]. In a first article [91] the authors compared the oxidation mechanisms of both the pristine polypropylene (PP) and the polypropylene/montmorillonite nanocomposites and the influence of the compatibilizing agent was examined. The data indicated that the photooxidation mechanism of polypropylene was not modified, but that a dramatic shortening of the induction period of the oxidation was observed in the presence of the nanofiller, leading to a decrease in the durability of the nanocomposite. The comparison of the oxidation rates (the kinetic curves are presented in Fig. 15.5) of the four types of samples (PP, PP/MMt, and both with PP-*g*-MA) shows that the presence of the compatibilizer PP-*g*-MA had only a weak effect on the induction time whereas the length of this induction period was dramatically reduced in the presence of MMt. This unexpected result was attributed to the inhibition of the activity of the residual phenolic processing antioxidant (AO). In the case of PP/MMt, the oxidation was observed to start earlier than in the case of PP and PP/PP-*g*-MA, with a significant reduction in the induction period, which was even more accentuated in the case of the PP/PP-*g*-MA/MMt nanocomposites. At the opposite, no important modification of the oxidation rates at the permanent regime that follows the induction period was noted. The dramatic influence of the organoclay on the induction period suggested an interaction between the phenolic groups of the antioxidant and the montmorillonite.

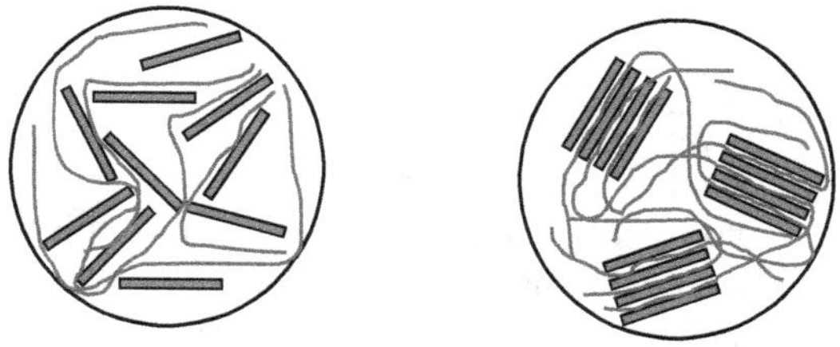

SCHEME 15.13 Nanocomposite structures.

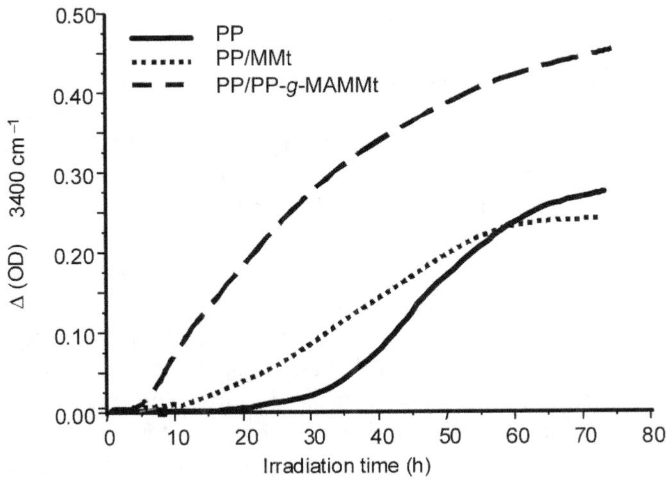

FIGURE 15.5 Variation of absorbance at 3400 cm^{-1} as a function of irradiation time for nanocomposite films.

The second part of this study [92] considered this peculiar point that has considerable consequences in terms of fundamental aspects of both the photooxidation of nanocomposites and the durability of these materials. A special attention was given to the interactions between the antioxidant, the clay (montmorillonite), and the compatibilizing agent (maleic anhydride), in terms of influence of each component on the rate of oxidation of the polymeric matrix. Two kinds of stabilizers were tested: a phenolic antioxidant (Scheme 15.14) generally used to process polypropylene and a redox antioxidant of the hindered amine light stabilizer (HALS) class (Scheme 15.15) that is used as long-term stabilizer for polypropylene.

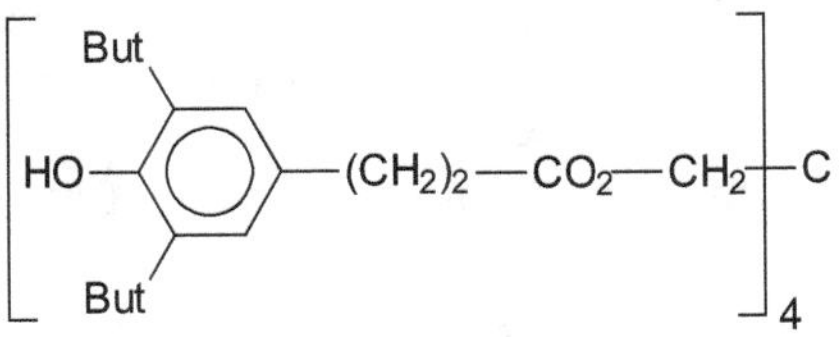

SCHEME 15.14 Chemical structure of the phenolic antioxidant.

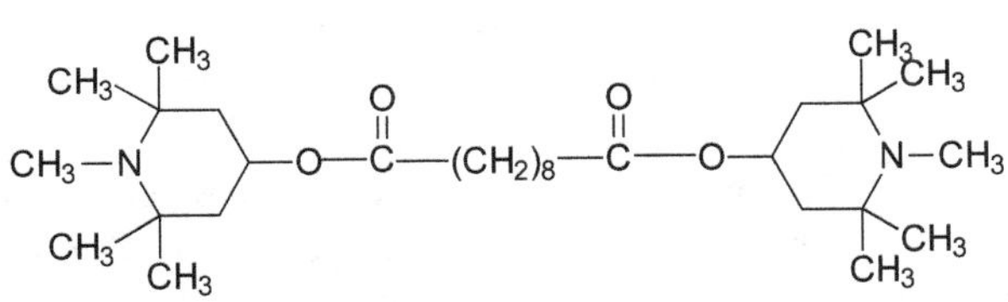

SCHEME 15.15 Chemical structure of the redox antioxidant (HALS).

With the two stabilizers tested, the main conclusion was that the presence of organomontmorillonite decreased the efficiency of the stabilizer. One possible explanation for this antagonism between the clay and the additives was that adsorption or intercalation phenomenon of the antioxidant with the silicate layered phase occurred. However, this intercalation would lead to a swelling of the clay. Complementary results based on X-ray diffraction pattern were obtained that compared the organomonmorillonite and a mixture of MMt and the phenolic antioxidant or the redox one. These results indicated quite clearly that the interlayer spacing was unchanged in both cases, which suggested that no swelling occurred with both stabilizers. If one considered the polarity of the phenol or amine functions of the stabilizers, and the facts that the polymeric matrix was hydrophobic whereas the clay had still an hydrophilic affinity (even though the presence of organic cations), the stabilizers might be inhomogeneously distributed in the nanocomposite material. This could favor the adsorption of the stabilizers on the clay. However, the experimental results showed that the additives can be easily removed by solvent extraction, which indicated that the interactions with the nanocomposite were not so strong.

Another explanation was the following. The organomontmorillonite used was a natural montmorillonite that contained iron. Chemical analysis of the clay confirmed the presence of a low amount of iron. It was recalled that iron and, in more general terms, metals are likely to induce the photochemical degradation of polymers. Iron at low concentration had a prooxidant effect that was due to the metal ion of iron that can initiate the oxidation of the polymer by the well-known redox reactions with hydroperoxides [93]. It was concluded that the transition metal ions, such as Fe, displayed a strong catalytic effect by redox catalysis of hydroperoxide decomposition, which was probably the most usual mechanism of filler accelerating effect on polymer oxidation. A characteristic of such catalytic effect was that it did not influence the steady-state oxidation rate, but it shortened the induction time.

The third point that was taken into account was the degradation of the alkylammonium cations. It has been shown that the irradiation of modified montmorillonite provoked an oxidation of the organic part of the exchanged clay [91]. Such a reaction is a supplementary way that can initiate the degradation of the polymer matrix. Another degradation pathway caused by the organoclay could result from the decomposition of alkylammonium during processing. It has been shown [94] that organoclays decompose during processing at high temperature by an Hofmann elimination mechanism to produce several by-products. The α-olefins formed in this reaction are sensitive toward photooxidation and could participate in the initiation of the polymer degradation. At this stage of the research on the long-term durability of nanocomposites in photooxidative conditions of ageing, it was however difficult to establish a hierarchy between the three plausible hypotheses that have been proposed to explain the loss of stability of stabilized nanocomposites.

There have been other studies dealing with the photooxidation of PP/MMt nanocomposites [95–100]. One of them concerned the consequences of photooxidation on the thermal stability and fire-retardant properties of PP-*g*-MA/MMt nanocomposites [99]. It was shown that the photodegradation dramatically affected the properties of the nanocomposites. A loss of thermal stability and fire retardancy performance was observed. This was ascribed to chain scission reactions occurring during oxidative degradation. Recently, a comparative study of photooxidation under natural and accelerated conditions of PP nanocomposites was carried out [100]. One of the objectives of running natural and accelerated tests was to determine the acceleration factor to accurately predict the service lifetime of polymers in use. The nanocomposites showed a higher propensity to photooxidation than the PP samples under both natural and accelerated conditions.

15.4.1.2 Polyethylene/Montmorillonite (PE/MMt) A photooxidative degradation study on PE/MMt was carried out [101] by Qin et al. on PE matrix that formed an intercalated nanocomposite with organomodified MMt and on conventional microcomposites with Na-MMt and Fe-MMt (no intercalation/exfoliation of the silicate layers). The rate of photooxidative degradation of PP/MMt nanocomposite was shown to be much faster than that of pure PE. The experiment results indicated that both ammonium ion and pure MMt could also accelerate the photooxidation of PE. The influence of Fe-MMt resulted in the catalysis effect of transition metal ions (Fe^{3+}). In a further paper [102], the authors reported a study of the influence of ammonium ions and transition metal ions on photooxidation of PE/MMt nanocomposites. It was said that the acceleration of degradation of nanocomposites was due to the decomposition of interlayer ammonium ions (leading to catalytic acidic sites created on the layers and the correspondingly olefin generated) and to the photoredox reactions of the transition metal ions.

Recently, La Mantia et al reported a study of LLDPE/clay nanocomposites with appropriate additives [103]. The photochemical behavior of nanocomposites without and with different UV stabilizers was studied under artificial accelerated conditions of ageing. Addition of a metal deactivator to the LLDPE–clay nanocomposites was also compared. Stabilizing effect on the physical properties such as elongation at break and tensile strength were described. The most effective photostabilization of these LLDPE/clay nanocomposites was achieved in presence of the metal deactivator.

15.4.1.3 Ethylene–Propylene–Diene Monomer/Montmorillonite The UV light-induced oxidation of EPDM/montmorillonite nanocomposites obtained with synthetic montmorillonite has been reported [104]. A specific attention was given to the dispersion of the clay. The nanocomposites were obtained by melt compounding EPDM-*g*-MA (EPDM grafted with maleic anhydride as compatibilizing agent) and organophilic synthetic clays. Four different samples were prepared and fully characterized: an EPDM grafted with maleic anhydride, an EPDM/Na-MMt microcomposite, and two EPDM/MMt nanocomposites, one with intercalated MMt and the other with exfoliated MMt. The results showed that the oxidation mechanism of the polymeric matrix was not modified by the presence of the clay. However, the influence of the organoclay on the durability of the nanocomposite material was observed through the induction time that was similar in the case of EPDM-*g*-MA and EPDM/MMt microcomposite. It was slightly reduced in the case of EPDM/MMt intercalated nanocomposite, and it almost disappeared in the case of the exfoliated EPDM/MMt nanocomposites. Conversely, no influence of the clay on the rate of oxidation of the polymeric matrix, once started, could be evidenced.

As the montmorillonite used was of synthetic origin and deprived of iron, the influence of the clay on the rate of degradation cannot be attributed to a catalytic effect of iron. Therefore, the adsorption of the antioxidant and additives was most likely to be the key factor behind the photodegradation of the exfoliated nanocomposites. The tremendous increase of accessible clay surface upon exfoliation enhanced the probability of additive adsorption on the solid surface and, hence, would help reduce dramatically its efficiency as antioxidant agent. Nevertheless, the organic moiety of the modified clay can be involved in the degradation, the effect being more pronounced in the case of the exfoliated sample that contained a larger amount of the organic cations. The exfoliated structure appeared to be more sensitive to the oxidation with higher dispersion of the organoclay and a higher amount of organic cations.

The photooxidation of vulcanized ethylene–propylene–diene monomer/montmorillonite nanocomposite as well as EPDM/nanocomposite with stabilizers was recently reported [105]. The photooxidation products were not changed in the presence of the nanofiller. However, the presence of MMt was observed to dramatically enhance the rate of photooxidation of EPDM with a shortening of the oxidation induction time, leading to a decrease in the durability of the nanocomposites. On the other side, it was observed that addition of stabilizers, either Tinuvin P or 2-mercaptobenzimidazole, was efficient in inhibiting the degradation effect of MMt.

A recent study dealt with photostabilization of EPDM/clay nanocomposites [106]. The EPDM/MMt hybrid nanocomposites were prepared by adding the phenolic stabilizers into the matrix prior to melt mixing and curing processes. A significant decrease in the rate of photodegradation was observed, and maximum stabilization was observed with antioxidant content up to 1.5%. This methodology might be able to prevent the sacrificial antioxidant loss and extend the lifetime of EPDM/clay nanocomposites when they are subjected to photooxidative degradation.

15.4.1.4 Polymer/Montmorillonite Recent studies reported the photooxidation of clay nanocomposites with different polymer matrices such as PA-6, PC, PMMA, epoxy, or conjugated polymers such as PPV. The nanocomposites demonstrated either a faster degradation than the pristine polymer in the case of PA-6 [107] or a faster rate of yellowing in the case of PC [108] or PMMA [109]. Concerning PPV nanocomposites, an improved environmental stability is reported, both with MMt [110] and with SiO2@Au nanoparticles [111]. The environmental degradation mechanisms of epoxy/clay nanocomposites due to accelerated UV and moisture exposure were recently reported [112]. The barrier characteristics of these nanocomposites were of interest with specific applications as protective coatings in mind. The photodegradation in dry conditions showed that the organic modifier and unexchangeable metal ions present in the organoclay aggravated the degradation of these nanocomposites. Moisture further accelerated the photodegradation process of the polymer; however, the deteriorating effect of moisture was significantly reduced by the excellent barrier characteristics of clay nanoplatelets that effectively restricted the movement of moisture and its interactions with radicals and oxygen, as suggested by the tortuous path models.

15.4.1.5 Polymer/Boehmite Nanocomposites Polymer containing clays are probably the most studied nanocomposites. However, for polyolefins the strategy of clay-containing nanocomposites preparation is relatively complex. A preliminary intercalation of onium ions between the clay layers by cation exchange is necessary to render them organophilic (organoclay). Furthermore, a compatibilizer is necessary (e.g., maleic anhydride grafted polypropylene) to melt compound the polyolefin matrix with modified organoclay and obtain diffusion of polyolefins in the interlamellar galleries. The use of nanofillers that can be dispersed in polyolefins without the use of compatibilizer is highly desirable in order to simplify the process. The possibility to produce boehmite with crystallite of nanometer size synthesized *via* sol–gel route and dispersed at nanometric level only using shear force of screw extruder makes this filler an ideal candidate to replace clays. Indeed, boehmite (AlOOH) is composed of Al–O double layers connected by hydrogen bonds between the hydroxyl groups that can be reacted with organic sulfates to improve compatibility with hydrophobic polymers. This allows the effect of a nanofiller on photooxidation to be studied without interfering with transition metal ions that are typical clay impurities known as photooxidation catalyzers.

The influence of boehmites on photooxidation of PP was studied in conditions of accelerated weathering, and samples containing inorganic boehmite and organomodified boehmite were compared [113,114]. It was shown that nanocomposites and pristine PP developed the same photoproducts, and boehmites did not modify the mechanism of photooxidation of the polymer [113]. The observed modifications were due to the consumption of the phenolic antioxidant present in commercial PP. This behavior was attributed to the inhomogeneous distribution of the antioxidant in presence of boehmites that were hydrophilic and could adsorb the stabilizers, thus preventing their antioxidant action. Organic modification might hinder migration of the antioxidant on the boehmite surface, showing a promising way to photostabilize nanocomposites. Attention should be paid to avoid effects due to the organic modifier in which oxidation of the alkyl chain of alkyl benzylsulfonate constituted a supplementary source of radicals that were likely to initiate the oxidation of PP, leading to an increase of the overall rate.

15.4.2 Polymer/LDH Nanocomposites

Mostly focused on cationic clays, and particularly on montmorillonite and hectorite, smectite-type layered silicates and clay-based nanofillers have recently been extended to the family of LDH. Hydrotalcite-like LDH materials are described according to the ideal formula, $[M^{II}_{1-x}M^{III}_{x}(OH)_2]^{x+}_{intra}\,[A^{m-}_{x/m}\cdot nH_2O]_{inter}$, where M^{II} and M^{III} are metallic cations, A the anions, and intra and inter denote the intralayer and interlayer domain, respectively. The structure consists of brucite-like layers constituted of edge-sharing octahedra. The presence of trivalent cations induces positive charges in the layers that are counterbalanced by interlamellar anions (Scheme 15.16).

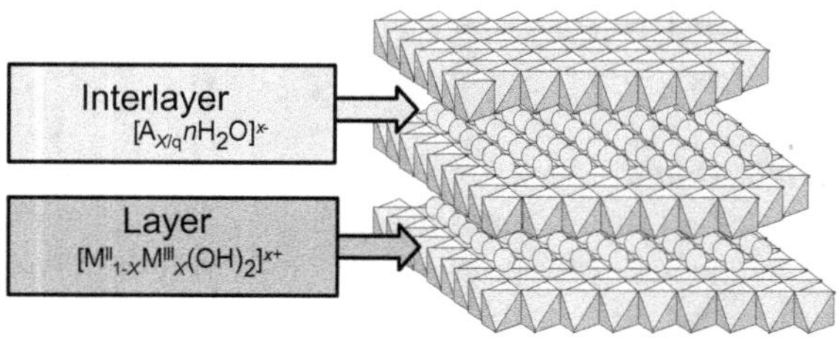

SCHEME 15.16 Structure of layered double hydroxides.

The interaction and subsequent oxidative behavior under UV light exposure of nanocomposite using poly(styrene) (PS) as polymer and LDH organomodified by a monomer surfactant as filler were recently investigated [115]. The photooxidation study revealed that the hybrid nanofiller did not modify the photooxidation mechanism of PS. The same products of oxidation were observed with the same proportions. A slightly higher oxidation rate was observed in the case of the sample with 5% of filler. The advantage of this system was its ability to be tailored in order to limit/control eventual interactions with photostabilizers and antioxidants.

The photooxidation of EPDM/LDH composites as well as EPDM/LDH with stabilizers was studied under accelerated UV irradiation [116]. EPDM filled LDH showed higher degradation rate than pristine EPDM. However, the results obtained in acidic medium showed the advantage of LDHs as an acid killer.

A comparison of the photooxidative behavior of PP/LDH and PP/MMt nanocomposites was reported [117], as well as structure and thermal properties of these nanocomposites. It appeared that the PP/LDH nanocomposites had an exfoliated structure and the LDH layers were well dispersed in the PP matrix. The photooxidation rate of PP/LDH was much lower than that of PP and PP/MMt samples, indicating a better UV stability for PP/LDH nanocomposites.

15.4.3 Polymer/Nanotube de Carbone (NTC) Nanocomposites

Carbon nanotubes (NTC) represent another family of potential nanofillers. Carbon nanotubes, first described by Iijima in 1991 [118], are a new allotropic form of carbon built-up of carbon atoms arranged in hexagons and pentagons, forming cylinders. Typically, carbon nanotubes consist of single-walled tubes (SWNTs) or multiwalled nanotubes (MWNTs) where several nanotubes of decreasing diameter are interlocked. These materials have potential applications in numerous fields such as electrical and thermal conductivity and hydrogen storage.

Results reported on photooxidation of EVA/NTC nanocomposites [119] showing that the global effect of NTs resulted mainly in a stabilizing effect on the nanocomposite degradation when the NTC amount was higher than 1% w/w. However, the two kinds of the NTC studied (thin MWNTs and DWNTs) have shown a higher degradation than the pristine EVA, which revealed a lower ability to act as antioxidant and/or a modification of their ability in dissipating the thermal energy. The size effect of carbon black particles on photodegradation previously reported [120] might be enhanced with the morphology and aspect ratio of carbon nanotubes.

Carbon nanotubes displayed a higher specific surface area, which could explain the influence of the carbon nanotube morphology, with mainly an effect of the diameter. Long carbon nanotubes were extremely difficult to disperse in EVA, actually forming well-known associated bundles.

15.4.4 Nanocomposites Obtained by UV Radiation Curing

Clay-based nanocomposite polymers have been synthesized by photoinitiated cross-linking polymerization of acrylate and epoxy-functionalized oligomers [121]. Both intercalated and exfoliated structures were present in the UV-cured nanocomposite photopolymer. The resistance to accelerated weathering of such UV-cured nanocomposites was evaluated in presence of light stabilizers such as UV absorber and HALS radical scavenger. These results, which showed no effect of the presence of the organoclay filler, were in marked contrast with almost all the results reported on the light stability of clay/polyolefin nanocomposites. Such UV-cured acrylic nanocomposites could therefore be used to protect organic materials against photodegradation upon outdoor exposure.

REFERENCES

1. B. Ranby and J.F. Rabek, *Photodegradation, Photooxidation and Photostabilization of Polymers*, Wiley, London, 1975, 493–500.
2. N. Grassie (Ed.), *Developments in Polymer Degradation*, Vols *1–7*, Elsevier Applied Science Publishers Ltd, London, 1978–1987.
3. J.F. McKellar and N.S. Allen, *Photochemistry of Man-Made Polymers*, Elsevier Applied Science Publishers Ltd, London, 1979.
4. N.S. Allen (Ed.), *Developments in Polymer Photochemistry*, Vols *1–3*, Elsevier Applied Science Publishers Ltd, London, 1980–1983.
5. J.F. Rabek, *Photodegradation of Polymers*, Springer-Verlag, Berlin Heidelberg, **1996**.
6. W. Schnabel, *Polymer Degradation*, Hansen, Munchen, 1981, pp 95–127.
7. K.S. Minsker, S.V. Kolesov, and G.E. Zaikov, *Degradation and Stabilization of Vinyl Chloride Based Polymers*, Pergamon Press, Oxford, 1988.
8. S.H. Hamid, M.B. Amin, and, A.G. Maadhah (Eds.), *Handbook of Polymer Degradation*, Marcel Dekker, New York, 1992.
9. C.J. Brabec, S.E. Shaheen, C. Winder, N.S. Sariciftci, and P. Denk, *Appl. Phys. Lett.* **2002**, *80*, 1288.
10. M.M. Wienk, J.M. Kroon, W.J.H. Verhees, J. Knol, J.C. Hummelen, P.A. van Hal, and R.A. Janssen, *J. Angew. Chem. Int. Ed.* **2003**, *42*, 3371.
11. M. Schaer, F. Nüesch, D. Berner, W. Leo, and L. Zuppiroli, *Adv. Funct. Mater.* **2001**, *11*, 116.
12. G. Dennler, C. Lungenschmied, H. Neugebauer, N.S. Sariciftci, M. Latreche, G. Czeremuszkin, and M.R. Wertheimer, *Thin Solid Films* **2006**, *511–512*, 349.
13. F.J.J. Janssen, L.J. van IJzendoorn, H.F.M. Schoo, J.M. Sturm, G.G. Andersson, A.W.D. van der Gon, H.H. Brongersma, and M.J.A. de Voigt, *Synth. Met.* **2002**, *131*, 167.

14. F.J.J. Janssen, J.M. Sturm, A.W. Denier van der Gon, L.J. van IJzendoorn, M. Kemerink, H.F.M. Schoo, M.J.A. de Voigt, and H.H. Brongersma, *Org. Electron.* **2003**, *4*, 209.
15. J.C. Scott, J.H. Kaufman, P.J. Brock, R. DiPietro, J. Salem, and J.A. Goitia, *J. Appl. Phys.* **1996**, *79*, 2745.
16. B.H. Cumpston, I.D. Parker, and K.F. Jensen, *J. Appl. Phys.* **1997**, *81*, 3716.
17. K. Norrman and F.C. Krebs, *Sol. Energy Mater. Sol. Cells* **2006**, *90*, 213.
18. F.C. Krebs, J.E. Carle, N. Cruys-Bagger, M. Andersen, M.R. Lilliedal, M.A. Hammond, and S. Hvidt, *Sol. Energy Mater. Sol. Cells* **2005**, *86*, 499.
19. F. Papadimitrakopoulos, K. Konstadinidis, M. Miller, R. Opila, E.A. Chandross, and M.E. Galvin, *Chem. Mater.* **1994**, *6*, 1563.
20. H. Antoniadis, L.J. Rothberg, F. Papadimitrakopoulos, M. Yan, M.E. Galvin, and M.A. Abkowitz, *Phys. Rev. B* **1994**, *50*, 14911.
21. R.D. Scurlock, B. Wang, P.R. Ogilby, J.R. Sheats, and R.L. Clough, *J. Am. Chem. Soc.* **1995**, *117*, 10194.
22. B.H. Cumpston and K.F. Jensen, *Synth. Met.* **1995**, *73*, 195.
23. L. Ma, X. Wang, B. Wang, J. Chen, J. Wang, K. Huang, B. Zhang, Y. Cao, Z. Han, S. Qian, and S. Yao, *Chem. Phys.* **2002**, *285*, 85.
24. N. Dam, R.D. Scurlock, B. Wang, L. Ma, M. Sundahl, and P.R. Ogilby, *Chem. Mater.* **1999**, *11*, 1302.
25. S. Chambon, A. Rivaton, J.-L. Gardette, M. Firon, and L. Lutzen, *J. Polym. Sci. A: Polym. Chem.* **2007**, *45*, 317.
26. F. Posada, J.-L. Philippart, P. Kappler, and J.-L. Gardette, *Polym. Degrad. Stab.* **1995**, *50*, 141.
27. P. Gauvin, J.L. Philippart, J. Lemaire, and D. Sallet, *Makromol. Chem.* **1985**, *186*, 1167.
28. P. Gauvin, J. Lemaire, and D. Sallet, *Makromol. Chem.* **1987**, *188*, 1815.
29. J.-L. Gardette, B. Mailhot, F. Posada, A. Rivaton, and C. Wilhelm, *Macromol. Symp.* **1999**, *143*, 95.
30. S. Morlat and J.-L. Gardette, *Polymer* **2001**, *42*, 6071.
31. J. Lacoste and D.J. Carlsson, *J. Polym. Sci. A: Polym. Chem.* **1992**, *30*, 493.
32. J. Fossey, D. Lefort, and J. Sorba, *Les radicaux libres en chimie organique*, Masson, Paris, 1993.
33. J.-L. Gardette and P. Delprat, *Science and Technology of Polymers and Advanced Materials– Emerging Technologies and Business Opportunities*, Plenum, New York, 1998, p. 587.
34. S.; Chambon, A.; Rivaton, J.-L.; Gardette, M. Firon, *Sol. Energy Mater. Sol. Cells*, **2008**, *92*, 785–792.
35. T. Grossetete, A. Rivaton, J.L. Gardette, C.E. Hoyle, M. Ziemer, D.R. Fagerburg, and H. Clauberg, *Polymer* **2000**, *41*, 3541.
36. J.-L. Philippart, C. Sinturel, and J.-L. Gardette, *Polym. Degrad. Stab.* **1997**, *58*, 261.
37. A.F. Nogueira, I. Montanari, J. Nelson, J.R. Durrant, C. Winder, N.S. Sariciftci, and C. Brabec, *J. Phys. Chem. B* **2003**, *107*, 1567.
38. A. Rivaton and J.L. Gardette, *Polym. Degrad. Stab.* **1999**, *66*, 385.
39. S. Chambon, A. Rivaton, J.-L. Gardette, M. Firon, and L. Lutsen, *J. Polym. Sci. A: Polym. Chem.* **2007**, *45*, 317.

40. D. Bellus, in: W.A. Noyes, G. Hammond, and J.N. Pitts,(Eds.), *Advances in Photochemistry*, Wiley–Interscience, New York, 1971.
41. A. Rivaton, *Polym. Degrad. Stab.* **1995**, *49*, 163.
42. J. Mallegol, D.J. Carlsson, and L. Deschenes, *Nucl. Instrum. Methods Phys. Res. Sect. B* **2001**, *185*, 283.
43. D.J. Carlsson, R. Brousseau, C. Zhang, and D.M. Wiles, *Polym. Degrad. Stab.* **1987**, *17*, 303.
44. S. Chambon, A. Rivaton, J.-L. Gardette, and M. Firon, *Sol. Energy Mater. Sol. Cells* **2007**, *91*, 394.
45. M. Seno, M. Maeda, and T. Sato, *J. Polym. Sci. A: Polym. Chem.* **2000**, *38*, 2572.
46. M. Day and D.M. Wiles, *J. Appl. Polym. Sci.* **1972**, *16*, 175.
47. M. Day and D.M. Wiles, *J. Appl. Polym. Sci.* **1972**, *16*, 191.
48. M. Day and D.M. Wiles, *J. Appl. Polym. Sci.* **1972**, *16*, 203.
49. M. Day and D.M. Wiles, *Polym. Lett.* **1971**, *9*, 665.
50. F.B. Marcotte, D. Campbell, J.A. Cleavaland, and D.T. Turner, *J. Polym. Sci. A1* **1967**, *5*, 481.
51. J.G. Pacifici and J.M. Straley, *Polym. Lett.* **1969**, *7*, 7.
52. G. Valk, M.L. Kehren, and I. Daamen, *Angew. Makromol. Chem.* **1970**, *13*, 97.
53. J.A. Dellinger and C.W. Roberts, *J. Polym. Sci., Polym. Chem. Ed.* **1980**, *18*, 97.
54. M.H. Tabankia and J.L. Gardette, *Polym. Degrad. Stab.* **1988**, *14*, 351.
55. A. Rivaton, *Polym. Degrad. Stab.* **1993**, *41*, 283.
56. A. Rivaton, *Polym. Degrad. Stab.* **1993**, *41*, 297.
57. H. Shah, I.B. Rufus, and C.E. Hoyle, *Macromolecules* **1994**, *27*, 553.
58. A. Rivaton, *Angew. Makromol. Chem.* **1994**, *216*, 155, 3882.
59. T. Grossetête, A. Rivaton, J.L. Gardette, C.E. Hoyle, M. Ziemer, D.R. Fagerburg, and H. Clauber, *Polymer* **2000**, *41*, 3541.
60. G. Dennler, C. Lungenschmied, H. Neugebauer, N.S. Sariciftci, and A. Labouret, *J. Mater. Res.* **2005**, *20*, 3224.
61. G. Dennler, C. Lungenschmied, H. Neugebauer, N.S. Sariciftci, M. Latréche, G. Czeremuzkin, and M.R. Wertheimer, *Thin Solid Films* **2006**, *511–512*, 349.
62. C. Lungenschmied, G. Dennler, H. Neugebauer, N.S. Sariciftci, M. Glatthaar, T. Meyer, and A. Meyer, *Sol. Energy Mater. Sol. Cells* **2007**, *91*, 379.
63. C. Charton, N. Schiller, M. Fahland, A. Holländer, A. Wedel, and K. Noller, *Thin Solid Films* **2006**, *502*, 99–103.
64. K.O., Awitor, A., Rivaton, J.-L., Gardette, A.J., Down, and M.B., Johnson, *Thin Solid Film*, **2008**, *516*, 2286–2291.
65. S. Ben Amor, G. Baud, M. Jacquet, and N. Pichon, *Surf. Coat. Technol.* **1998**, *102*, 63.
66. S. Giancaterina, S. Ben Amor, G. Baud, J.-L. Gardette, M. Jacquet, C. Perrin, and A. Rivaton, *Polymer* **2002**, *43*, 6397.
67. A. Moustaghfir, A. Rivaton, E. Tomasella, B. Mailhot, J.B. Cellier, M. Jacquet, and J.-L. Gardette, *J. Appl. Polym. Sci.* **2005**, *95*(2), 380.
68. A. Moustaghfir, E. Tomasella, M. Jacquet, A. Rivaton, B. Mailhot, J.L. Gardette, and E. Bêche, *Thin solid films* **2006**, *515*, 662–665.

69. L. Guedri, S. Ben Amor, J.-L. Gardette, M. Jacquet, and A. Rivaton, *Polym. Degrad. Stab.* **2005**, *88*(2), 199.
70. A. Factor and M.L. Chu, *Polym. Degrad. Stab.* **1980**, *2*, 203.
71. D. Freitag and U. Westeppe, *Makromol. Chem. Rapid Commun.* **1991**, *12*, 95.
72. A. Rivaton, *Angew. Makromol. Chem.* **1994**, *216*, 147–153 (3874).
73. J.E. Pickett, J.P. Barren, and R.J. Oliver, *Angew. Macromol. Chem.* **1997**, 252.
74. A. Rivaton, B. Mailhot, J. Soulestin, H. Varghese, and J.L. Gardette, *Polym. Degrad. Stab.* **2002**, *75*(1), 17–33.
75. A. Rivaton, B. Mailhot, J. Soulestin, H. Varghese, and J.L. Gardette, *Eur. Polym. J.* **2002**, *38*(7), 1349–1363.
76. D.T. Clark and H.S. Munro, *Polym. Deg. Stab.* **1982**, *4*, 441. D.T. Clark and H.S. Munro, *Polym. Deg. Stab.* **1984**, *8*, 195.
77. S. Ben Amor, G. Baud, M. Jacquet, and N. Pichon, *Surf. Coat. Technol.* **1998**, *102*, 63.
78. S. Giancaterina, S. Ben Amor, G. Baud, J.-L. Gardette, M. Jacquet, C. Perrin, and A. Rivaton, *Polymer* **2002**, *43*, 6397.
79. A. Moustaghfir, E. Tomasella, A. Rivaton, B. Mailhot, M. Jacquet, J.-L. Gardette, and J. Cellier, *Surf. Coat. Technol.* **2004**, *180–181*, 642.
80. G. Penot, R. Arnaud, and J. Lemaire, *Angew. Makromol. Chem.* **1983**, *117*, 71.
81. H.G. Völz, G. Kaempf, G.H. Fitzky, and A. Klaeren, *Am. Chem. Soc. Symp. Series* **1981**, *151*, 163.
82. R.B. Cundall, B. Hulme, V. Rudham, and M.S. Salim, *J. Oil Col. Chem.* **1978**, *61*, 351.
83. B. Mailhot, A. Rivaton, J.-L. Gardette, A. Moustaghfir, E. Tomasella, M. Jacquet, X.-G. Ma, and K. Komvopoulos, *J. Appl. Phys.* **2006**, *104310*, 99.
84. P. Fabbri, C. Leonelli, M. Messori, F. Pilati, M. Toselli, P. Veronesi, S. Thérias, A. Rivaton, and J.L. Gardette, *J. Appl. Polym. Sci.* **2008**, *108*(3), 1426–1436.
85. C.A. Cruz, D.R. Paul, and J.W. Barlow, *J. Appl. Polym. Sci.* **1979**, *23*(2), 589–600.
86. J.M. Jonza and R.S. Porter, *Macromolecules* **1986**, *19*(7), 1946–1951.
87. C. Li, Q. Kong, Q. Fan, and Y. Xia, *Mater. Lett.* **2005**, *59*, 773–778.
88. V. Balsamo, N. Calzadilla, G.A. Mora, and J. Müller, *J. Polym. Sci. B: Polym. Phys.* **2001**, *39*(7), 771–785.
89. A. Tidjani and C.A. Wilkie, *Polym. Degrad. Stab.* **2001**, *74*, 33–37.
90. B. Mailhot, S. Morlat, J.-L. Gardette, S. Boucard, J. Duchet, and J.-F. Gerard, *Polym. Degrad. Stab.* **2003**, *82*(2), 163–167.
91. S. Morlat, B. Mailhot, D. Gonzalez, and J.-L. Gardette, *Chem. Mater.* **2004**, *16*(3), 377–383.
92. S. Morlat -Therias, B. Mailhot, D. Gonzalez, and J.-L. Gardette, *Chem. Mater* **2005**, *17* (5),1072–1078.
93. G. Scott, H.H. Sheena, and A.M. Hariman, *Eur. Polym. J.* **1978**, *14*, 1071.
94. W. Xie, Z. Gao, W.-P. Pan, D. Hunter, A. Singh, and R. Vaia, *Chem. Mater.* **2001**, *13*, 2979.
95. H. Qin, S. Zhang, H. Liu, S. Xie, M. Yang, and D. Shen, *Polymer* **2005**, *46*, 3149–3156.
96. P. Ding and B. Qu, *Polym. Eng. Sci.* **2006**, *46*(9), 1153–1159.
97. S. Chmela, A. Kleinova, A. Fiedlerova, E. Borsig, D. Kaempfer, R. Thomann, and R. Muelhaupt, *J. Macromol. Sci., A: Pure Appl. Chem.* **2005**, *A42*(7), 821–829.

98. G. Kandilioti, A. Elenis, K.A. MacChiarola, and V.G. Gregoriou, *Appl. Spectrosc.* **2006**, *60*(11), 1285–1292.
99. M. Diagne, M. Gueye, A. Dasilva, L. Vidal, and A. Tidjani, *J. Mater. Sci.* **2006**, *41*(21), 7005–7010.
100. M. Diagne, M. Gueye, A. Dasilva, and A. Tidjani, *J. Appl. Polym. Sci.* **2007**, *105*(6), 3787–3793.
101. H. Qin, C. Zhao, S. Zhang, G. Chen, and M. Yang, *Polym. Degrad. Stab.* **2003**, *81*, 497–500.
102. H. Qin, Z. Zhang, M. Feng, F. Gong, S. Zhang, and M. Yang, *J. Polym. Sci. B: Polym. Phys.* **2004**, *42*, 3006–3012.
103. F.P. La Mantia, N. Tzankova Dintcheva, V. Malatesta, and F. Pagani, *Polym. Degrad. Stab.* **2006**, *91*(12), 3208–3213.
104. S. Morlat–Therias, B. Mailhot, J.-L. Gardette, C. Da Silva, B. Haidar, and A. Vidal, *Polym. Degrad. Stab.* **2005**, *90*, 78–85.
105. S. Morlat-Therias, E. Fanton, N.S. Tomer, S. Rana, R.P. Singh, and J.-L. Gardette, *Polym. Degrad. Stab.* **2006**, *91*, 3033–3039.
106. S.P. Lonkar, K.A. Pratheep, and R.P. Singh, *Polym. Adv. Technol.* **2007**, *18*(11), 891–900.
107. H.-L. Qin, S.-M. Zhang, M.S. Yang, and D.-Y. Shen, *Chem. J. Chin. Univ.* **2004**, *25*(1), 197–198.
108. J.M. Sloan, P. Patterson, and A. Hsieh, *Polym. Mater. Sci. Eng.* **2003**, *88*, 354–355.
109. P.H. Patterson, J.M. Sloan, and A.J. Hsieh, *ANTEC* **2003**, 3604–3607.
110. H.-C. Lee, T.W. Lee, Y.T. Lim, and O.O. Park, *Appl. Clay Sci.* **2002**, *21*, 287–293.
111. Y.T. Lim, T.-W. Lee, H.-C. Lee, and O.O. Park, *Opt. Mater.* **2002**, *21*, 585–589.
112. R.S.C. Woo, Y. Chen, H. Zhu, J. Li, J.K. Kim, and C.K.Y. Leung, *Compos. Sci. Technol.* **2007**, *67*, 3448–3456.
113. S. Bocchini, S. Morlat-Therias, J.-L. Gardette, and G. Camino, *Polym. Degrad. Stab.* **2007**, *92*, 1847–1856.
114. S. Chmela, A. Fiedlerova, E. Borsig, J. Erler, and R. Mülhaupt, *J. Macromol. Sci. A: Pure Appl. Chem.* **2007**, *44*, 1027–1034.
115. F. Leroux, L. Meddar, B. Mailhot, S. Morlat-Therias, and J.-L. Gardette, *Polymer* **2005**, *46*(11), 3571–3578.
116. B. Kumar, S. Rana, and R.P. Singh, *eXPRESS Polym. Lett.* **2007**, *1*(11), 748–754.
117. P. Ding and B. Qu, *Polym. Eng. Sci.* **2006**, *46*(9), 1153–1159.
118. S. Iijima, *Nature* **1991**, *354*, 56–58.
119. S. Morlat-Therias, E. Fanton, J.-L. Gardette, S. Peeterbroeck, M. Alexandre, and Ph. Dubois, *Polym. Degrad. Stab.* **2007**, *92*, 1873–1882.
120. A.R. Horrocks, J. Mwila, M. Miraftab, M. Liu, and S.S. Chohan, *Polym. Degrad. Stab.* **1999**, *65*, 25–36.
121. C. Decker, L. Keller, K. Zahouily, and S. Benfarhi, *Polymer* **2005**, *46*, 6640–6648.

16

PHOTODEGRADABLE POLYMERS

SAHAR AL-MALAIKA, C. HEWITT, AND H.H. SHEENA

16.1 SCOPE

Polymers are used in a wide range of fields either as replacement materials or for unique applications. Designing polymers for end-use target applications must take into account the entire life cycle of the product starting from raw materials to production, product lifetime(s), and waste. The environmental load engendered from such an activity cannot be overemphasized and efforts are currently being made to reduce the negative image and impact of polymers without undermining their usefulness and benefits to progress and society. The true impact of polymers on society, resources, and the environment is continually being reassessed and great strides have been made to address some of the issues and challenges relating to aspects of polymer durability and degradability.

This chapter deals with the photochemistry of photodegradable polymers, primarily hydrocarbon based. The discussion will be limited to that of the initial abiotic

Photochemistry and Photophysics of Polymer Materials, Edited by Norman S. Allen

degradation phase under the influence of UV light (the biodegradation phase and photodegradable hydrolytic polymers are not covered). This is critical to understanding the underlying science governing the conflicting demands of environmental issues of photodegradability of plastics, and their longevity and strength—the mainstay of conservation of resources in terms of materials and energy. Standardization and certification of degradable plastics including standard protocols for exposure testing of photodegradable polymers, which include Xenon arc (D5071-91) and fluorescent UV (D5208-91) have been reviewed elsewhere [1,2] are outside the scope of this chapter.

16.2 BACKGROUND

The photoxidative behavior of commercial polymers varies widely, the most photoxidizable of which are the unsaturated rubbers containing labile polymer–hydrogen bond. In the case of polyolefins, the photostability of different members of the family vary significantly depending on their chemical and morphological structures, defects and thermal history [3–9]. In general, polypropylene, PP, and high-density polyethylene, HDPE, are more susceptible to photoxidation than low-density polyethylene, LDPE [3,10]; polyvinyl chloride (PVC) and polystyrene (PS) should be relatively stable to photoxidation, but their thermal history exerts significant effects on their subsequent photoxidative behavior [11–13]. In view of this, antioxidants and stabilizers are used in order to achieve the target durability and longevity in-service gen. Consequently, many of the plastic products are generally over stabilized thereby prolonging their lifetime far beyond their intended end-use application, hence the associated issues relating to waste, resources, and the environment. The significant increase in the use of polymers in applications that only require a short or limited lifetime particularly in packaging and for single-use items, and the associated problems of litter and the ever increasing volume of plastics waste and its impact on landfills has triggered off a major change in attitude of the public to polymers because of their perceived high environmental burden. This has given rise to increasing pressures from governments and society to address the environmental issues associated with polymers including waste management and resource preservation as exemplified by recent efforts of the U.K. government to reducing the use of plastics shopping bags by imposing direct charges for their use [14], or by banning their use altogether, for example, in San Francisco, CA, USA, legislation have been passed to outlaw plastics bags from supermarkets and pharmacies [15]. Thus, the demand for positively promoting environmental degradation of plastics is on the increase especially for packaging, single-use disposables and hygiene products, to reduce their environmental load.

Environmental degradation of plastics may be promoted by different mechanisms initiated by the action of environmental factors, singly or in combination: oxygen (oxidatively degradable), sunlight normally in presence of oxygen (photodegradable), water (hydrolytically degradable), and microorganisms (biodegradable). The degradation process and its rate are affected directly by the chemical structure and physical

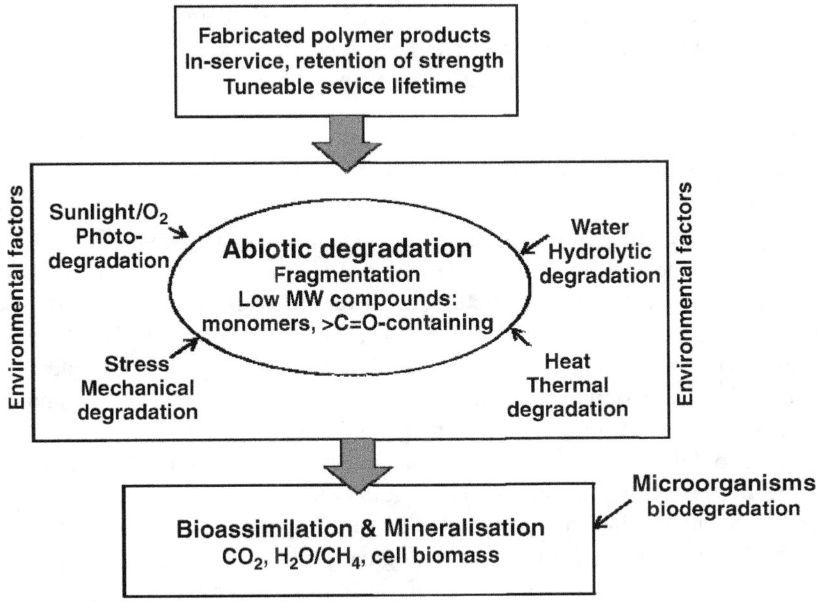

FIGURE 16.1 Schematic diagram of polymer degradation.

properties of the polymer [10]. Designing-in photodegradability necessitates addressing simultaneously the issues of use and lifetime performance, degradation rate and disposal [16–19] as described in the following list (see also Fig. 16.1). The main focus in this chapter, however, will be on aspects of abiotic degradation (a precursor stage to the process of biodegradation and fragmentation) brought about by the action of light.

(i) *Retention of Strength In-Use.* While the polymer is in-service it would remain strong and fit for purpose, and, ideally, the polymer would have a controlled tuneable service lifetime at this stage with well-defined performance characteristics before the onset of degradation in the second phase.

(ii) *Abiotic Degradation and Fragmentation.* At the end of its useful life, under the influence of one or more environmental factors, for example, oxygen, natural sunlight, and water (microorganisms in the latter case, i.e., biotic initiation), degradation of the polymer is triggered off causing changes in both physical and chemical characteristics leading to rapid and catastrophic failure in mechanical strength concomitant with fragmentation and formation of low molecular weight monomers, oligomers, carbonyl-containing compounds, and alcohols that will act as chemical nutrients for microorganisms. The extent and rate of degradation at this stage is very important and should lead to complete and rapid disintegration of the polymer, which is often accelerated by the addition of chemicals or species that would act as prodegradants.

(iii) *Biodegradation and Mineralization.* The polymer fragments, produced during the abiotic degradation stage and comprising low molecular weight hydrophilic metabolites, are metabolized and assimilated by microorganisms and converted ultimately to CO_2, H_2O, or methane and cell biomass, and hence returned back completely to the biological cycle.

16.3 PHOTODEGRADABLE POLYMERS

All polymers are, in general, unstable to heat and light. Polymers containing heteroatoms in the main chain or side and end chains are more photolabile than those based on C–C and C–H structural units due to their absorption of light and formation of photoexcited states and free radical formation involving bond breaking and transitions to unstable electronic states leading to energy transfer and photodecomposition. The subject of photoxidation is covered in Chapter 12 and only highlights of important processes will be given here.

In order to deliberately promote and enhance photodegradation, different strategies have been employed involving the use of additives or chemical moieties added to, or incorporated in, the polymers that are capable of photoinitiating and propagating free radical reactions. Technologically, photodegradable polymers, with many of the commercial ones based on polyolefins, are generally produced by either the chemical insertion of photoinitiator or photosensitizer groups, for example, carbonyl, into the main or side chains of a polymer backbone through copolymerization (intrinsically photodegradable polymers), or by the physical blending of prooxidant additives and fillers with polymers (photoinduced degradable polymers) and these would either induce photolytic initiation or promote the formation of free radical initiators by peroxidation under the influence of light, heat, and oxygen. These processes will give rise to cleavage of the polymer backbone thus lowering their molecular weight followed by extensive fragmentation of the polymer [3,17–21]. After the fragmentation process, the low molecular weight polar oxidation products produced from these photodegradable polyolefins become accessible to microorganisms and subsequent biodegradation [20,22–25]. Efforts have also been made [26] to synthesize photodegradable polymers where the degradation is triggered off by visible light through metal–metal bond insertion reactions along the polymer backbone.

The photodegradable polymers discussed below are grouped into two categories:

(i) Intrinsically photodegradable polymers due to their structural compositions (photolytic polymers).
(ii) Photoinduced oxo-degradable polymers affected by specialty additives or fillers.

16.3.1 Intrinsically Photodegradable Polymers by Copolymerization: Photolytic Polymers

In this approach, small amount of a photosensitizing group, typically based on a carbonyl-containing moiety, is directly incorporated into the main polymer chains or as short branches [21]. Carbonyl groups are unstable toward UV light, with absorption maxima around 280–290 nm, but are generally unaffected by visible light whereas most synthetic hydrocarbon polymers do not absorb light above 300 nm. The "erythmal region," 290–330 nm, of the solar spectrum coincides with the absorption region of the carbonyl group hence the unique switchable role of the carbonyl function in photodegradable polymers: it remains dormant indoors (since most artificial indoor lighting do not emit light in this region and glass filters it from sunlight) but is photoactivated outdoors. In this group of polymers, therefore, the chain scission and fragmentation is initiated by photochemical processes and starts as soon as the polymer, through the incorporated carbonyl groups, absorbs UV light without the need for molecular oxygen to be present initially. Two types of photodegradable polymers containing carbonyl groups (in-chain and side-chain) are available commercially.

16.3.1.1 Polymers with In-Chain Ketones In this family of polyethylene-based polymers the carbonyl function is built into the polymer chain. The synthesis of these polymers were first patented by Du Pont® in 1950 [27] with Dow® being currently the major producer of commercial ethylene-carbon monoxide systems. Typically, random ethylene-carbon monoxide (E-CO) polymers are manufactured by copolymerization of a small concentration of carbon monoxide in the range of 0.5–1.6 wt% with ethylene monomer under conditions similar to those used for making low density polyethylene [8]. When subjected to UV radiation in the 290 nm region at ambient temperatures, the carbonyl groups of the E-CO copolymers undergo rapid photolysis mainly by Norrish type II reaction [8,28] which gives rise to chain scission and occurs via the formation of a six-membered cyclic intermediate involving intramolecular γ-hydrogen-atom abstraction resulting in an end of chain ketone and a vinyl unsaturation, whereas at elevated temperatures the contribution of Norrish type I reaction that gives two free radicals starts to increase (Scheme 16.1) [8].

The photodegradation behavior of commercially produced films of photolytic polymers has been examined under laboratory accelerated photoaging conditions at 60 °C [29]. Figure 16.2 shows that exposure of the E-CO polymer (origin: Dupont) to light results in immediate and rapid photoxidation without an induction period and this is clearly seen to be associated with a very fast rate of reduction in its molar mass (MW drops from about 300,000 to 50,000–20,000) particularly at the early stages of photoxidation, followed by a much slower decrease to 6000–2000 on prolonged UV exposure well beyond polymer embrittlement [29]. It has been suggested that such changes in molecular weight [29–31] are due to the formation of cross-links, through the vinyl group, which competes with the chain scission at relatively early stage of the photolysis [32]. It was also shown [29] that there is also a good correlation between the reduction in molecular weight and the buildup of carbonyl-containing compounds as the main photoxidation products during accelerated UV exposure

In the same work [29], the nature of the low molecular weight degradation products of highly photoxidized and fragmented E-CO films (photofragmentation to a point where 96% reduction of original molecular weight has taken place) was

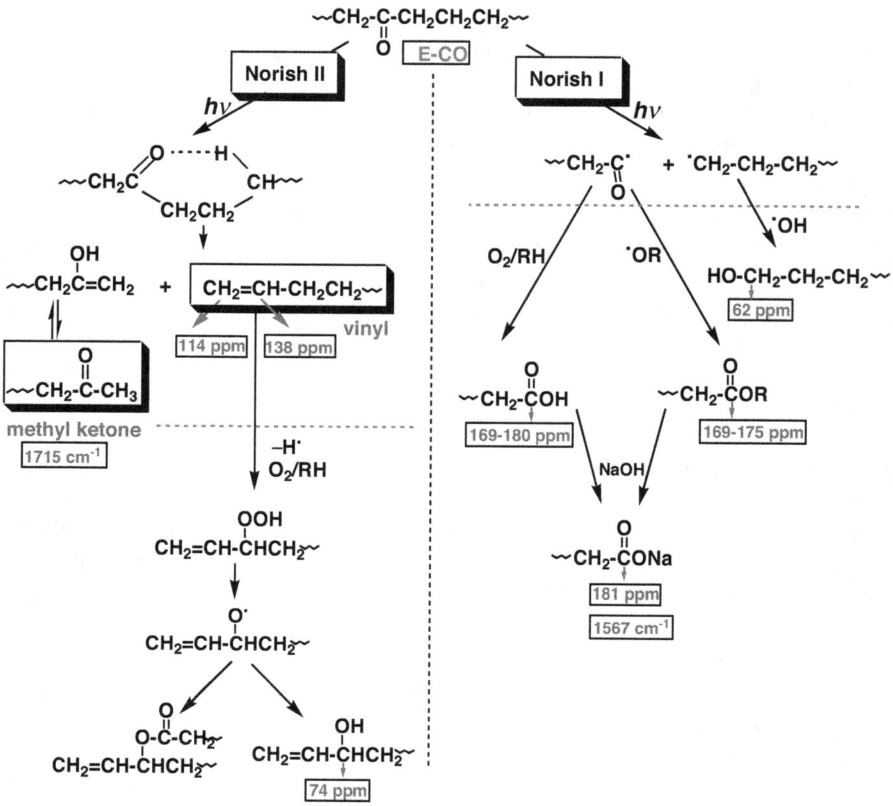

SCHEME 16.1 Photolysis, Norrish II and I processes, and further oxidation of E-CO copolymers. Numbers are the observed NMR ^{13}C chemical shifts (ppm) and FTIR wavenumber (cm^{-1}) of detected products.

examined by FTIR and NMR after extraction in aqueous alkaline solution. Both FTIR and ^{13}C NMR have shown that the photoxidation process has led to the formation of acids, esters, γ-lactones and ketones along with unsaturation of vinyl group formed via the Norrish type II photolysis reaction of the in-chain ketone groups in the E-CO polymer as well as secondary alcohols formed most likely from hydrogen abstraction and further oxidation of the active methylene next to the vinyl group (see Scheme 16.1).

In contrast to the random E-CO polymers that contain a small amount of CO, ethylene carbon monoxide copolymers containing much higher amounts of CO have been synthesized more recently to afford photodegradable alternating polyketone polymers, see structure below [33–35]. Photodegradation of these polymers does not occur through Norrish II chain scission reaction because of the absence of methylene group in the γ-position to the carbonyl group. Photodegradation of a commercial polyketone film (e.g., Shell "Carilon") was shown to occur initially via a Norrish I photolysis reaction whereby carbon monoxide is eliminated and two terminal

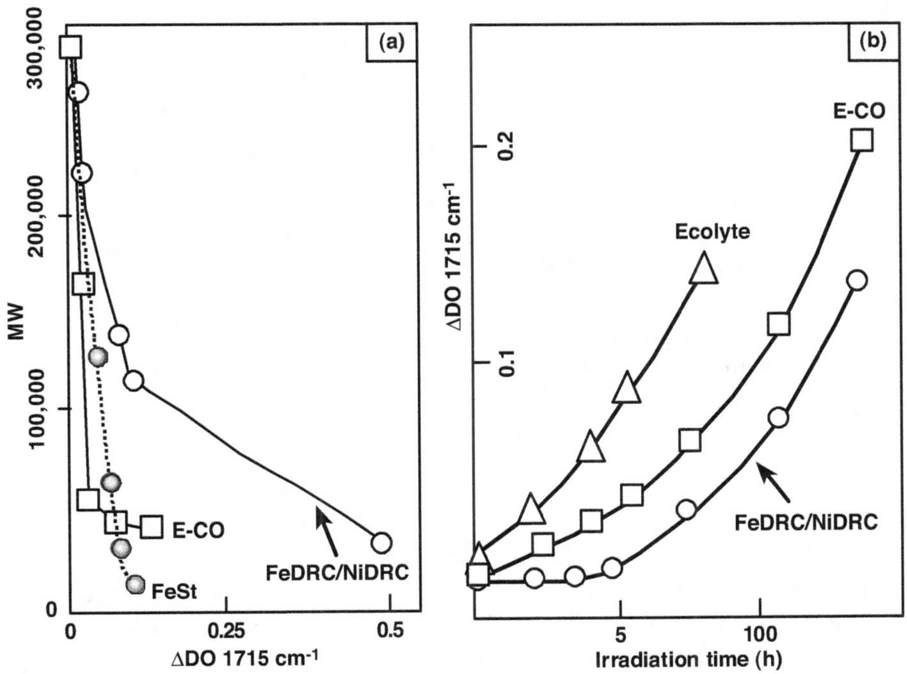

FIGURE 16.2 Rate of reduction in molecular weight as function of change in carbonyl absorption during UV exposure (SEPAP12-24) of different LDPE photodegradable films. (Reproduced with permission from Ref. 29; Published by Marcel Dekker, Inc., 1995.)

macroalkyl radicals, followed by photoxidation giving rise to various oxygenated products and a reduction in polymer molar mass [35,36].

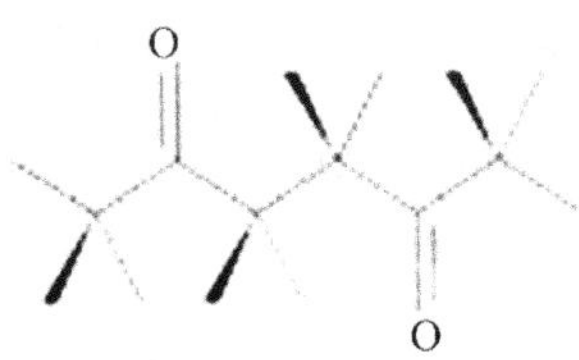

Alternating polyketones

16.3.1.2 Polymers with Side-Chain Ketones This family of photodegradable polymers are based on the copolymerization of a conventional monomer, for example, ethylene, styrene, propylene, with a small amount (e.g., 1%) of a vinyl ketone-containing comonomer of the general structure shown in Scheme 16.2 where the ketone is located in the side chain. Guillet, who had developed [21,37,38] these systems as photodegradable polymers and are marketed under the trade name Ecolyte™ (Eco Plastics), has shown that the rate of their photodegradation can be controlled by varying the nature of the substituent groups R′ and R. It was also shown

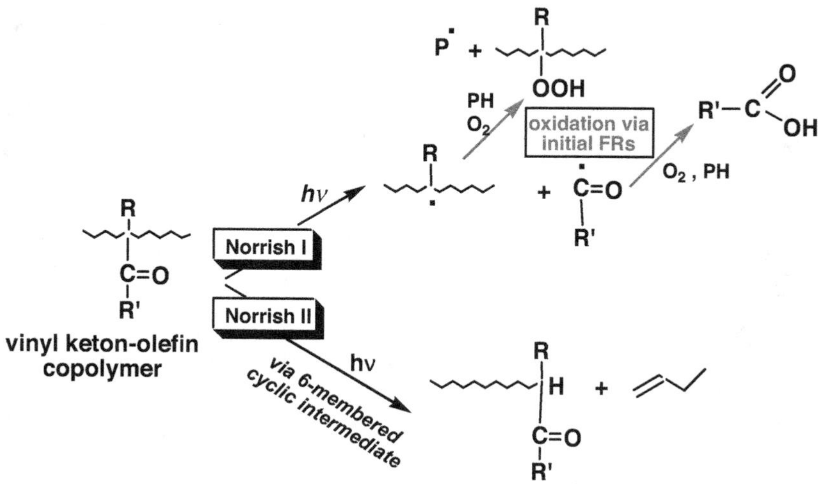

SCHEME 16.2 Photolysis and further oxidation of Ecolyte copolymers.

that their production is not limited to vinyl-based polymers but a wide range of commercial plastics can be made by this process including condensation polymers such as nylons, polyesters and even polymers that normally do not degrade by chain scission such as PVC and polyacrylonitrile, PAN [8,39–41].

These copolymers show accelerated photodegradation, without an induction period, due to the absorption of UV light by the side chain ketone groups through both radical (Norrish I) and nonradical (Norrish II) processes that leads to breaking of the main chains (see Scheme 16.2). The quantum yield for the chain scission Norrish II reaction was shown [21,42] to depend on both the structure and amount of the ketonic function and polymer chain length. For example, when the ketone monomer content (e.g., in methyl vinyl ketone-methyl methacrylate (MMA); phenyl vinyl ketone-MMA copolymers) is small the quantum yield of Norrish II reaction was low and photodegradation occurred mainly via Norrish I reaction [43,44]. On the other hand, the quantum yield for the Norrish II chain scission reaction of the Ecolyte-type copolymers was shown to be high and almost independent of the chain length, and occurs with a much higher efficiency than is the case of the in-chain ketones (E-CO polymers) where the quantum yield was very low and decreased with chain length (less efficient in polymers compared to small molecules) (see Table 16.1) [21]. The higher quantum efficiency of photolysis of Ecolyte compared to E-CO polymers is reflected by a faster rate of photoxidation of the former polymer (both polymers exposed under the same conditions and both give no photoinduction period) (see Fig. 16.2) [29]. These polymers are, therefore, useful for short-term applications such as disposable items.

The rate of photodegradation of Ecolyte polymers was found generally not to be affected by the presence of antioxidants in the polymer, as the process is photolytic and not photoxidative (unless if they absorb in wavelength range that initiate degradation), but was shown to be directly proportional to the concentration of the

TABLE 16.1 Changes in Quantum Yield of Norrish II (Φ_{II}) Reaction with Chain Length [21]

	Φ_{II}		
	In-chain Ketone	Side-Chain Ketone	
Chain Length *R* (Carbon Atoms)	$R\text{-}C(=O)\text{-}R$	$R\text{-}C(=O)\text{-}CH_3$	$R\text{-}C(=O)\text{-}C_6H_5$
4	0.11	0.25	0.31
5		0.20	0.31
6	0.092	0.20	0.25
7	0.080	0.20	0.30
8			0.29
9		0.20	
11	0.072		0.31
17		0.15	
21	0.059		
100*	0.025	~0.2	0.2–0.3

ketonic carbonyl groups for a given sample thickness although the overall rate of degradation is not affected by thickness since thicker samples would absorb more light, as is illustrated in Fig. 16.3 for polystyrene Ecolyte (Ecolyte S) films [21].

It is important to point out that the quantum yield for the Norrish II chain scission reaction (Φ_{CS}) is highly affected by the mobility of polymer chains. For example, the photolysis Φ_{CS} for a film of the copolymer poly(styrene-co-phenyl vinyl ketone) irradiated at 313 nm in the solid state was shown to be low (0.04–0.09) at temperatures below the copolymer T_g (glass transition temperature) but increased dramatically at,

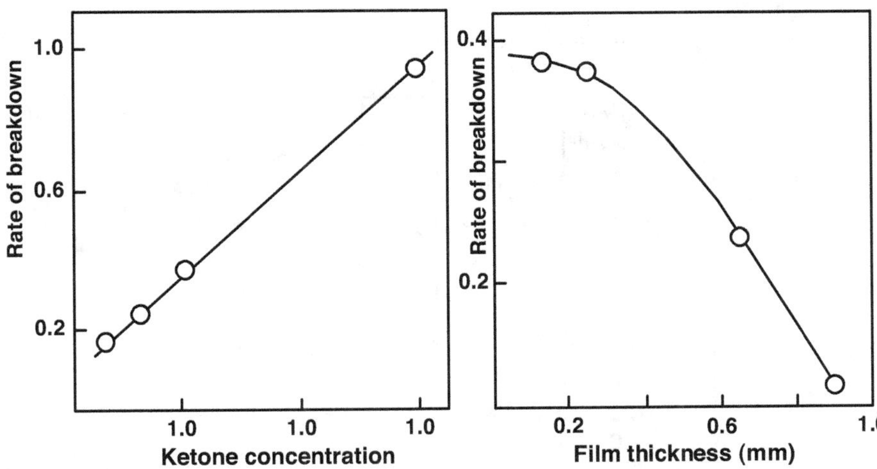

FIGURE 16.3 Effect of ketone concentration and film thickness of Ecolyte S on the rate of photodegradation (Reproduced with permission from Ref. 21; Published by Plenum Press, 1995.)

and above, its T_g (~100 °C) reaching a quantum yield value similar to that obtained from photolysis in solution (~0.3) after which the value remained constant [45,46]. Below T_g where the polymer chains mobility is frozen, the efficiency of the reaction depends on the ease of formation of the cyclic intermediate (Scheme 16.3) required for the occurrence of Norrish II reaction which is rather difficult under these conditions, whereas above T_g, the molecular mobility of the copolymer is sufficiently high such that the rate of formation of the intermediate is no longer rate determining and hence the reaction Φ_{CS} becomes similar to that observed in solution.

16.3.2 Photoinduced Oxo-Degradable Polymers

These polymers are prepared by blending commercially available polymers, for example, PE and PP, with special additives and fillers that can promote and accelerate phtodegradation under the influence of light and atmospheric oxygen. Unlike the photolytic polymers, molecular weight reduction and fragmentation in these polymers is initiated by an oxidation process through peroxidation (see Scheme 16.3) in the presence oxygen and light (photoxidation) or heat (thermal oxidation). This abiotic oxidative degradation process results in reduction in molecular weight and the formation of polar degradation products. When the molar mass of the polymer is sufficiently low, these polar oxidation products will be available for assimilation by microorganisms and ultimate biotic degradation (see Fig. 16.1). The special additives used here are typically photoactivators and prooxidants that are generally based on transition metal ions whereas the most widely used filler is generally based on starch. The different types of photoinduced degradable polymers are discussed below.

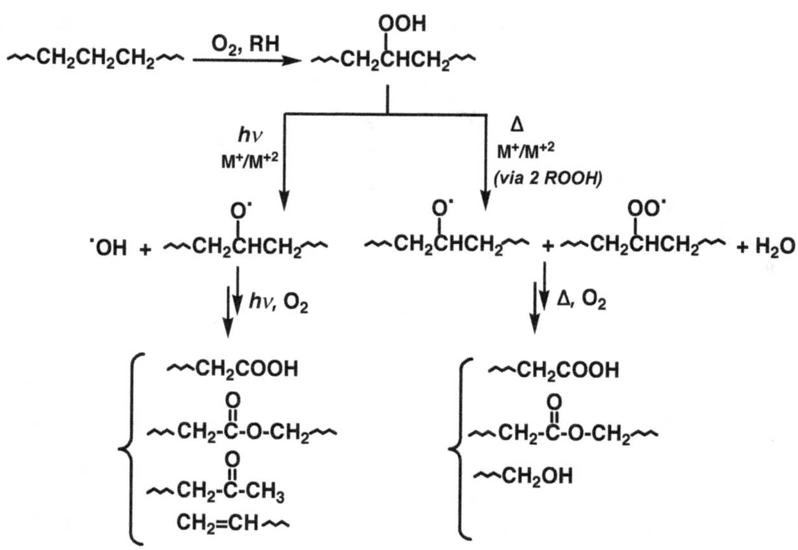

SCHEME 16.3 Peroxidation of polyethylene.

$$M^{+2} + ROOH \longrightarrow M^{+3} + RO^{\bullet} + HO^{-}$$

$$M^{+3} + ROOH \longrightarrow M^{+2} + ROO^{\bullet} + H^{+}$$

SCHEME 16.4 Redox reactions of iron.

16.3.2.1 Polymers Containing Metal Complexes as Photoactivators This is a family of polymers containing special additives based on transition metal ions introduced either as organosoluble (typically used as stearates) or as complexes containing sulfur ligands (e.g., dithiocarbamates), and in both cases are based typically on the metal ions Fe^{3+}, Co^{2+}, and Mn^{2+} [17,19,22,24]. These transition metal ions, particularly iron, afford efficient mechanisms for the photodegradation of polymers, for example, polyolefins, because of their dual roles as initiators of photoxidation and as catalysts for hydroperoxide decomposition through the well known redox reactions shown in Scheme 16.4 [5]. It is for this reason that a large number of commercial photodegradable polymers contain iron complexes that can accelerate their photodegradation.

It has been shown that transition metal carboxylate complexes of Mn^{2+}, Co^{2+}, and Fe^{+3} behave differently in that, whereas the Fe^{+3} complexes act as a source of metal ions producing free radicals that initiate photoxidation (Scheme 16.5) the Mn^{2+} and Co^{2+} complexes catalyze the decomposition of peroxides (associated with chain cleavage) without the influence of light [47]. The resulting chain scission leads to dramatic reduction in molecular weight and mechanical strength and formation of low molecular weight polar oxygenated products (see Scheme 16.3) that increase the hydrophilicity and wetability of the polymer surface, hence improved bioassimilation rates.

Polyethylene and polypropylene blended with iron carboxylate complexes, for example, acetylacetonate (FeAcAc) and stearates (FeSt), and irradiated by UV light under accelerated aging conditions were shown to act as effective phtoactivators giving rise to rapid photoxidation as shown from the rapid rate of carbonyl formation without any induction period (see Fig. 16.4a for FeAcAc in HDPE) and with a reduction in molar mass (see Fig. 16.2a for FeSt in LDPE). However, these complexes have been shown to cause considerable oxidation to both PE and PP during processing reflected in a sharp increase in the polymer's melt flow index (reflecting chain scission and drop in molar mass) (Fig 16.4b) and act, therefore, as thermal prooxidants and cannot be used without the use of additional antioxidants in the system [2,3,17–19,48,49].

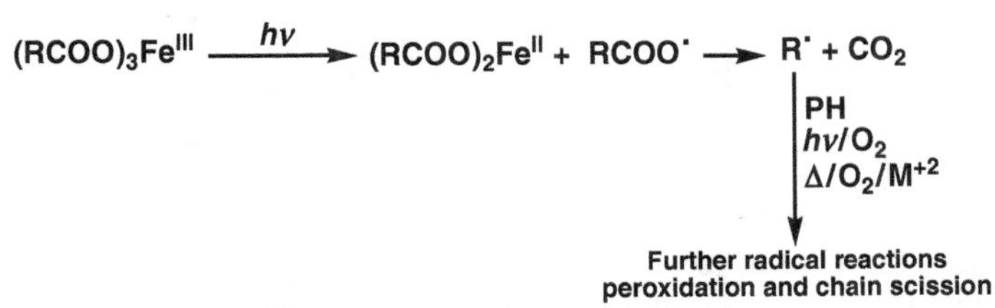

SCHEME 16.5 Reaction of Iron carboxylates in presence of UV light.

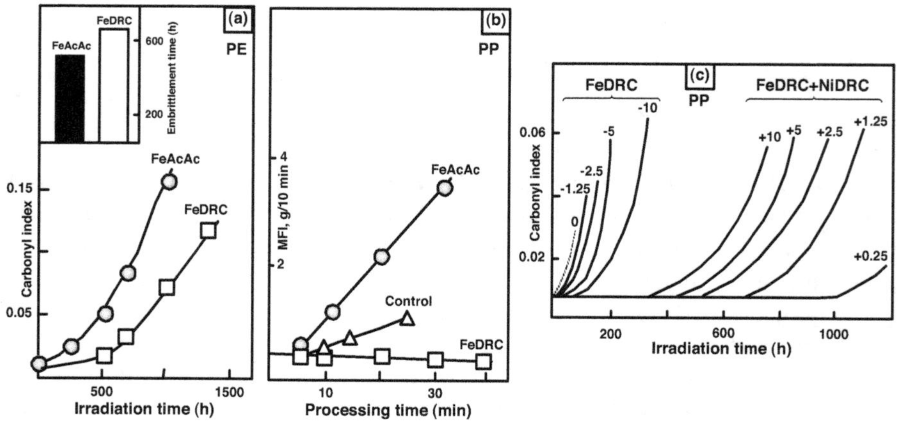

FIGURE 16.4 Photoxidation of PE containing iron complexes (a); melt flow index of PP containing metal complexes (b); and effect of FeDRC concentration on photoxidation of PP in presence (+) and absence (−) of 25×10^{-5} mol/100 g (numbers on curves are FeDRC concentration in mol/100 g) (c).

Iron complexes containing sulfur ligands, for example, iron dithiocarbamates, FeDRC, on the other hand, behave differently in that, unlike the FeAcAc and FeSt, they are highly effective processing stabilizers (Fig 16.4b) [48], but like the iron carboxylates they are photoxidized rapidly (rates were determined by monitoring carbonyl group formation) (Fig. 16.5) giving rise ultimately to similar photoproox-idant species in the form of iron carboxylates (see Scheme 16.5). The iron dithio-

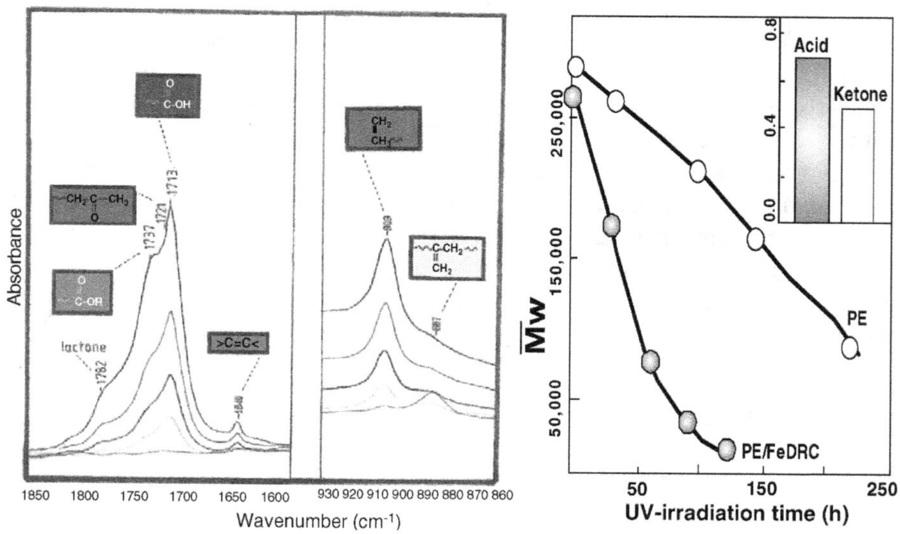

FIGURE 16.5 Evolution of carbonyl and unsaturation groups (from IR), reduction in MW with exposure time and amount of acids and ketones (calculated from SF_4 derivitization) formed during photoxidation of photodegradable PE film containing 0.05% FeDRC.

carbamates differs also from the carboxylates in that, unlike the latter which shows no induction period, FeDRC complexes act also as weak photostabilizers and, depending on their concentration, would give a variable but short photoinduction period, PIP (proportional to concentration) (see Fig. 16.4a) at the end of which they undergo a sharp inversion in activity and change from acting as weak photostabilizers to behaving as efficient photoxidation catalysts (photoactivators) (Fig. 16.4c) [17,19,48].

This special behavior of FeDRC with dual activity as antioxidant photoactivator, whereby it stabilizes the polymers during processing at all concentrations but causes rapid photoxidation in a concentration-dependent autoaccelerated process during exposure to UV light, has led to the first patent on photobiodegradable polymers containing sulfur complexed metal ions by Scott [49] in the 1970s. Like other peroxide decomposers, the antioxidant effect of FeDRC is due to the oxidation products of the thiocarbamyl ligand giving low molecular weight sulfur acids that are responsible for the nonradical decomposition of hydroperoxides [50,51] (see route A in Scheme 16.6); when the iron complex is destroyed by light, however, the free photoproxidant iron ion released (in the form of macromolecular carboxylate) is a powerful photodegradant and results in radical formation and reduction in molecular weight with consequent rapid physical disintegration of the polymer (route B in Scheme 16.6). The dependence of the useful lifetime (e.g., time to embrittlement) for such photodegradable PEs on the FeDRC concentration is shown in Fig 16.4c and is supported by the fact that higher concentrations of oxidation products (carboxylic acids, ketones, and esters) are formed in photofragmented samples containing higher initial [FeDRC] [48]. Further, the major final photoxidation product in FeDRC-containing polymers (both commercial and laboratory-produced) was found to be

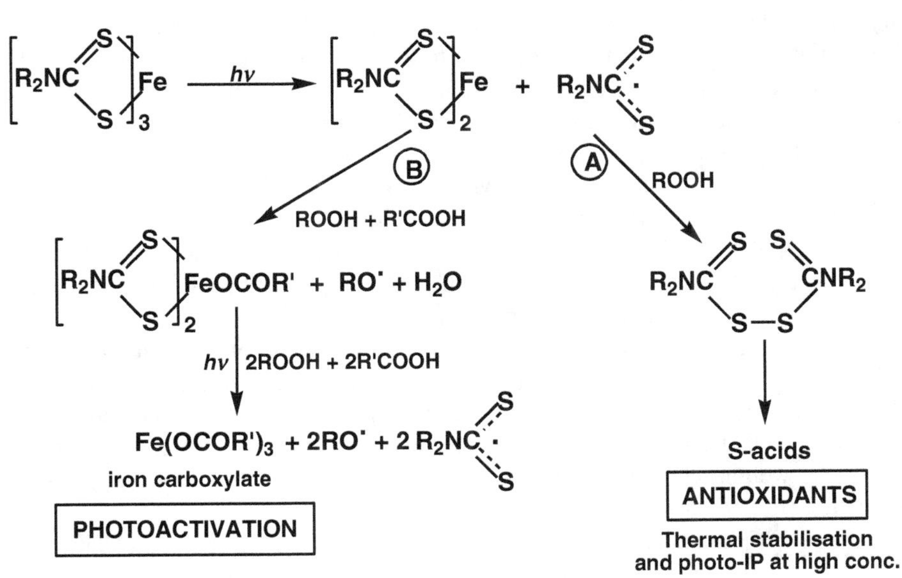

SCHEME 16.6 Mechanism of action of FeDRC as antioxidant photoactivator.

carboxylic acids present at higher concentrations compared to ketones and esters (calculated following SF4 treatment) (see Fig. 16.5b) and are formed mainly via Norrish I photolysis of intermediate ketones, which are themselves formed from thermolysis and photolysis of polymer hydroperoxides [29]. A photolytically induced reaction between ketones and hydroperoxides was proposed to explain the formation of carboxylic acids from ketones [52]. Norrish II photolysis (forming methyl ketones and vinyl) appears to play more important role when higher initial FeDRC concentrations are added to the polymer (see Scheme 16.1). However, both processes lead to chain scission (though only type I results in free radicals) and consequently a significant reduction in molecular weights and fragmentation (e.g., in a PE film containing 0.05 w/w% FeDRC a reduction of average MW from 250,000 to 20,000 took place in 150 h irradiation in SEPAP12-24) (Fig 16.5b) [29]. Compared to the photolytic E-CO in which the photodegradation was shown to be dominated by Norrish II photolysis of ketones (ninefold increase in vinyl products compared to that in FeDRC-PE for same thickness and under the same conditions), in the case of FeDRC-containing polymers the Norrish I is the more important photoxidation process resulting in higher amount of carboxylic acid (one third more than E-CO) and this was shown to give rise to a much faster biodegradation in the FeDRC polymers compared to the E-CO polymers [20,29].

The FeDRC gives a short PIP at high concentrations, however, the length of this PIP in polyolefins can be controlled much more effectively by using low concentration of FeDRC in combination with a very small concentration of the more photolytically more stable NiDRC which is also an effective thermal antioxidant [19,29,39], and this has been the basis for the Scott-Gilead process used for the preparation of time-controlled photodegradable polymers used for mulching films [17,53,54]. The effect of varying the concentration of FeDRC on its weak photoantioxidant effect (appearance of a short PIP) was highlighted earlier (increased PIP with increasing FeDRC concentration, see Fig. 16.4c). In contrast to the weak photostabilizing behavior of FeDRC when used alone, its combination with a fixed small amount of NiDRC behaves differently, in such combinations, FeDRC acts as a photosensitizer for the nickel as both the photoxidation (see Fig. 16.4c) and the rate of loss of the NiDRC complex from the polymer were shown to increase with increasing FeDRC concentration [19]. In the presence of the photostable NiDRC, therefore, adjusting the ratio of the FeDRC/NiDRC the length of the PIP can be predetermined and controlled precisely, and at its end the polymer system is rapidly photoxidized as seen from the rapid rate of increase in carbonyl formation paralleled by a significant reduction in molecular weight and viscosity (see Fig. 16.2) due to the photoreduction of the released trivalent iron carboxylate (forms initiating radicals) and oxidative breakdown of the polymer outlined above (Scheme 16.5). However, compared to the photolytic E-CO polymers that show very fast initial rate of reduction in their molecular weight, the FeDRC/NiDRC system gives a slower and more progressive reduction (Fig 16.2a).

Analysis of the final abiotic photodegradation products, by NMR and FTIR, and quantification of the different carbonyl compounds formed by derivitization has shown that the photoxidation of FeDRC-containing polymers (in presence or absence of NiDRC) occurs mainly by the Norrish I process that accounts for the formation of the major final photoproducts, the carboxylic acids. This contrasts with the photolytic E-CO polymers where their product analysis revealed higher concentrations of vinyl and methyl ketones in support of the more important role that Norrish II process plays in these polymers [29]. The identity of more than 200 abiotic oxidation products from various classes of degradable polymers have been examined extensively using a wide range of analytical techniques with the conclusion that the most dominant class of the final degradation products is aliphatic carboxylic acids [55–58].

Combinations of other metal ions such as Zn with FeDRC did not show such behavior but instead resulted in a synergistic stabilizing behavior [19]. The effect of the metal carboxylate complex FeAcAc in combinations with NiDRC was shown also to give an increased rate of photodestruction of the NiDRC in the polymer, but unlike the FeDRC/NiDRC system, it does not give variable PIP [19] and is therefore less useful as a "delayed action" photoactivator system. The above discussion indicates clearly that, compared to other metal ion combinations, the FeDRC/NiDRC system is the one that meets all the requirements for an ideal photodegradable polymers, where the additives protect the polymer initially from photoxidation giving a predetermined and precisely controlled lifetime (PIP) followed by a rapid reduction in molecular mass and complete physical fragmentation with formation of low molecular weight polar compounds that can be utilized by microorganisms leading ultimately to complete biodegradation (see Fig. 16.2). The importance of the initial abiotic process, that involves transition metal ion catalyzed thermal and photooxidation of the polymer, for the subsequent biodegradation cannot be overemphasized as hydrophobic polymers cannot be assimilated by microorganisms without this initial abiotic oxidation process [20].

***16.3.2.2 Polymers Containing TiO_2* as Photoactivators** TiO_2 pigment has been used traditionally as an additive in the polymer industry primarily to give white coloration and to improve the mechanical properties of the polymers [59–63]. More recently, it has been used to catalyze the photodegradation of polymers (polymer-TiO_2 composite systems) such as PE, PS, PVC [59,61,64–68]. It has been shown that the rate of photodegradation of PE, PS increased in the presence of TiO_2 doped with transition metal dopants based on V^{+5}, Mo^{+5}, and W^{+5}, whereas Mn^{+2} and Cr^{+3} reduced the photoactivity of the pigment. Moreover, the anatase form of titanium was shown to be more photoactive than the rutile form and the effect of increasing the calcination temperature of the pigment is to reduce the photoactivity by boosting the rutile fraction. Further, it was demonstrated that the concentration dependency of the photodegradation rates is directly related to the percentage of anatase achieved after calcination [59,60].

Further, TiO_2/dye (dye based on metal-phthalocyanine (MPc), e.g., CuPc) systems have also been used as composite photocatalysts to accelerate the degradation of PS and PE under solar radiation at ambient conditions [69,70]. For these TiO_2/dye polymer systems, it was demonstrated that, under outdoor solar exposure conditions, the amount of CO_2 released (the main product of photocatalytic degradation) and other volatile organic compounds, and the corresponding weight loss, of

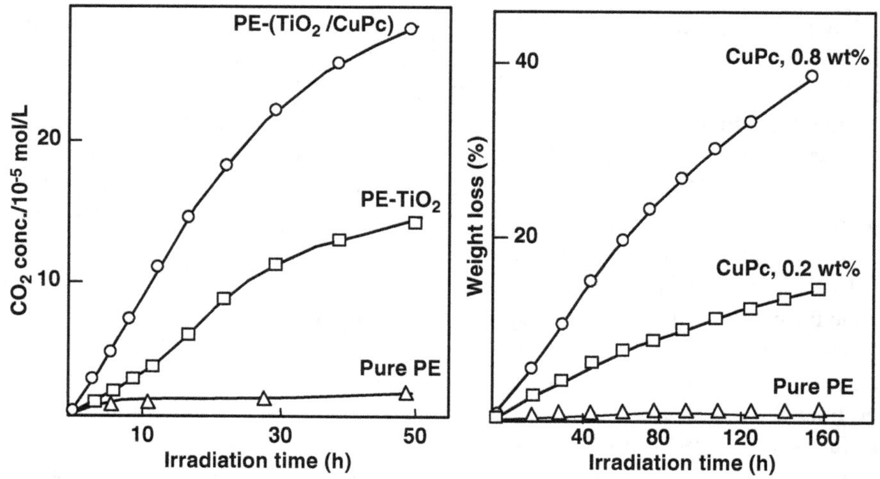

FIGURE 16.6 Concentration of CO_2 released and weight loss of PE, PE-TiO_2 and PE-(TiO_2/CuPc) films with solar irradiation time. (Reproduced with permission from Ref. 70; Published by Elsevier, 2008.)

PE-(TiO_2/CuPc) films were much higher than that of a PE-TiO_2 films, and both were very much higher than is the case for unmodified PE film exposed under the same conditions, indicating chain scission due to the photocatalytic action of TiO_2, which is further enhanced in the presence of the dye [70]. It was further suggested, from scanning electron microscopy, that the photocatalytic reaction starts first at the interface between the polymer and the exposed TiO_2 particles with the reaction rate of the PE-(TiO_2/CuPc) system being much higher than that of PE-TiO_2, (see Fig. 16.6) [70]. The mechanism suggested for the photocatalytic degradation of the PE-(TiO_2/CuPc) system is given in Scheme 16.7 [70]. The TiO_2 particles absorb UV light at wavelengths <390 nm giving mobile electrons (e^-) and holes (h^+) and a process of charge injection of the photogenerated holes from the valence band of TiO_2 to the ground state of CuPc that is thermodynamically permitted takes place (Equations 16.1 and 16.2 in Scheme 16.7). Further reactions of the electrons or holes generated from the UV excited TiO_2 with surface adsorbed oxygen and water results in the formation of various active oxygenated species, for example, $O_2^{\bullet -}$, $HO^{\bullet}$, and $HOO^{\bullet}$ that are important for the photolytic degradation caused by TiO_2 [71,72]. In the case of TiO_2/CuPc, the photocatalytic degradation of the polymer is not only

$$TiO_2 + h\nu \longrightarrow TiO_2(e^- + h^+) \quad (16.1)$$

$$CuPc + TiO_2(e^- + h^+) \longrightarrow CuPc^+ + TiO_2(e^-) \quad (16.2)$$

$$h^+ + \sim CH_2CH_2 \sim \longrightarrow \sim \dot{C}HCH_2 \sim + {}^{\bullet}OOH \quad (16.3)$$

$$^{\bullet}OOH + {}^{\bullet}OOH \longrightarrow H_2O_2 + O_2 \quad (16.4)$$

$$H_2O_2 \xrightarrow{h\nu} 2\ {}^{\bullet}OH \quad (16.5)$$

SCHEME 16.7 Photocatalytic degradation mechanism of PE-(TiO_2/CuPc) films [69].

due to the above-generated oxygenated radical species but also due to participation of the efficient holes generated in the ground state of CuPc in the oxidation process of PE, which is an energetically favorable process (Equations 16.3–16.5 in Scheme 16.7) [70].

For effective photocatalytic degradation to take place, uniformly dispersed TiO_2 nanoparticles in the polymer matrix give highly desirable. However, TiO_2 nanoparticles tend to aggregate significantly in low polarity hydrocarbon polymers such as polystyrene [73]. When these nanoparticles are directly incorporated into polymers, they aggregate into large agglomerates resulting in a significant reduction of their photodegradation efficiency. This was attributed to firstly to a reduction of the interface area between the photocatalytic TiO_2 particles and the polymer, and secondly, due to a rapid photoinduced whitening of the polymer that reduces the ability of light to penetrate into the composite polymer film thereby hindering the progress of photodegradation [66]. To improve this situation, it has been determined that a good dispersion of TiO_2 within the polymer can be achieved by grafting the nanoparticles onto the polymer, for example, PS-g-TiO_2 [64]. UV irradiation of PS-g-TiO_2 films which are characterized by a well dispersed TiO_2 have been shown to give a significant increase in the efficiency of the photodegradation of the polymer when compared to PS-TiO_2 films; after 300 h of UV irradiation the PS-g-TiO_2 film lost 29% of its weight concomitant with a decrease of the average molecular weight (MW) by one-fourth of its original value. The photocatalytic degradation reaction of PS is initiated by the active oxygen species formed through the absorption of light by TiO_2 (Scheme 16.8) [71–73,75], which react with the polystyryl radicals generating hydroxyl and carbonyl-containing intermediate products leading to the formation of carbon dioxide and chain scission of the polymer backbone [64,68]. An additional initiating activity of nano- and micro-TiO_2 particles was suggested to involve a direct interaction between the TiO_2 and the hydroperoxides resulting in the formation of further initiating radicals that give rise to reduced induction periods, as well as weakening of the inhibiting efficiency of stabilizers, in hydrocarbon model compounds, with the nanosized particles giving higher extent of interactions with hydroperoxides compared to the microparticles [76].

16.3.2.3 Photodegradable Polymers Containing Starch The incorporation of a biodegradable natural filler such as starch in a nondegradable hydrocarbon polymer

$$TiO_2 \xrightarrow{h\nu} TiO_2\,(e^- + h^+)$$

$$\left.\begin{array}{l} H_2O + h^+ \longrightarrow HO^\bullet + H^+ \\ O_2 + e^- \longrightarrow O_2^{\bullet -} \\ O_2^{\bullet -} + H_2O \longrightarrow {}^\bullet OOH + {}^-OH \\ 2\,HOO^\bullet \longrightarrow O_2 + H_2O_2 \\ h^+ + {}^-OH \longrightarrow {}^\bullet OH \end{array}\right\} \text{Active oxygen species}$$

SCHEME 16.8 Initiation of the photolytic degradation of PS-g-TiO_2 films [64].

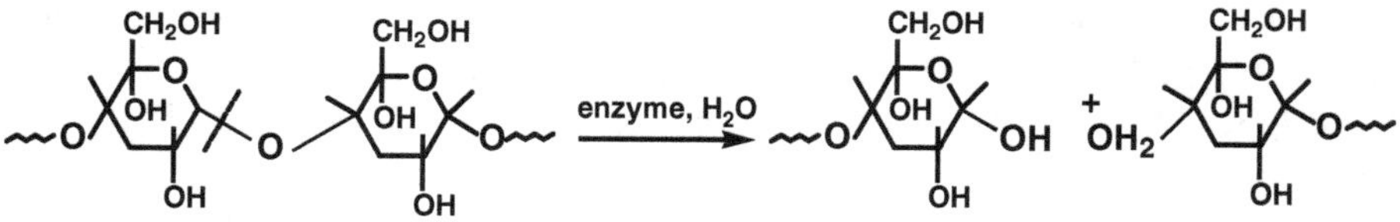

SCHEME 16.9 Enzymatic hydrolysis of starch.

such as PP, PS, PVC, and particularly, PE, that can ultimately give biodegradable starch-based plastics has generated much interest in recent years. Starch (mainly from potato, corn, and rice) that consists of about 20% linear amylase and 80% branched amylopectin polymers is digested enzymatically at the α-1,4-link (C–O–C) of both polymers and at the α-1,6-link in amylopectin (by glucosidases) (Scheme 16.9) [77,78].

Griffin [79,80] has developed starch-filled plastics (mainly PE-based) compositions in which the biodegradable starch filler is used at low concentrations of 6–15%. Starch-PE films have porous structure that enhances the accessibility of the polymer to microorganisms [81]. However, in such polymer-starch blends, the starch becomes encapsulated in the hydrocarbon polymer and would become unavailable to the usual starch digesting enzymes and the microorganisms can only remove the starch when the PE has been degraded [82]. As in the case of the other photodegradable polymers discussed above, it soon became clear that, in starch-filled polymers, the abiotic oxidation process resulting in chain scission reactions must take place first for biodegradation to take place. This has led to further developments by Griffin where a small amount of a readily oxidizable material, such as unsaturated fatty oils and diene rubbers, were used in the polymer formulation to promote oxidative degradation of the polymer and release the starch [80,82]. The oxidizability of starch-filled polyolefins is further enhanced through the use of metal ion prooxidants (particularly Fe^{+3}) that catalyze the thermal and photodegradation of the polymers (as discussed above) thus resulting in significant reduction in molecular weight to a level that can then be metabolized by microorganisms [49,53,83-85], hence many of the starch-based degradable polymers available commercially today contain iron ions, for example in the form of carboxylates. The photoproducts of the abiotic degradation of starch-filled PE have been identified and shown to be based on ketones, carboxylic acids, lactones, esters, alcohols, and alkanes and alkenes [86,87].

Comparison of the rate of photoxidation (outdoor exposure) of three commercial starch-filled PE samples (Ampacet) containing ferric stearate (FeSt), FeSt with C-black, and organic dye (no iron), has shown that the sample containing FeSt gave the highest rate of carbonyl formation (Fig. 16.7) similar to the rate of films containing FeDRC attributed to similar photoactivation reactions caused by the released prooxidant iron ion. In contrast, the starch sample containing an organic dye showed a much reduced rate of photoxidation not too dissimilar from the unmodified PE control, which indicates that the starch does not mediate in the photoxidation of the polymer matrix [29]. The sample containing FeSt and C-black showed fast rate of photoxidation (faster than the sample with organic dye, see Fig 16.7), an unexpected behavior since the highly UV-absorbing C-black is expected to prevent the

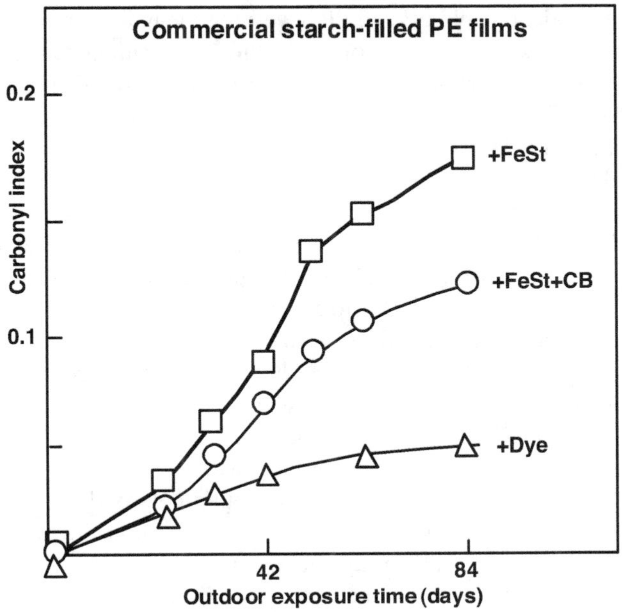

FIGURE 16.7 Rate of photoxidation of commercial starch-filled PE films during outdoor exposure. (Reproduced with permission from Ref. 29; Published by Marcel Dekker, Inc., 1995.)

photoexcitation of the iron stearate. The possibility of a synergistic effect involving both the oxidation of C-black and polymer surfaces was suggested as an explanation for the rapid photoxidation. A good correlation was observed between the rate of photoxidation, as measured from the increase in carbonyl absorption, and the reduction in molecular weight of these starch-filled polymers with the abiotic iron-catalyzed thermal and photoxidation process shown to be the rate limiting step in the subsequent bioassimilation process [20,29].

Other starch-based photobiodegradable polymers have been examined based on grafting of vinyl ketones on the starch backbone to produce graft copolymers, for example, styrene-grafted poly vinyl ketone, S-g-poly(PVK) or styrene grafted polymethyl vinyl ketone, S-g-poly(MVK). These polymers were shown to give rise to enhanced photodegradation. The extent of photodegradability was found to depend on the starch/ketone molar ratio; fastest photodegradation and subsequent biodegradation rates were obtained at 1:2 and 1:4 molar ratio for S-g-poly(MVK) and S-g-poly(PNK), respectively [88].

16.4 APPLICATIONS

The major market for photodegradable polymers has been in agriculture and packaging where the major group of polymers used are the polyolefins. In agriculture, the

Scott-Gilead photodegradable PE and PP have been the main commercial polymers used for over two decades in a variety of agricultural applications such as baler twines (e.g., Cleanfields™), control-release fertilizers (e.g., Nutricote™), and mulching films (e.g., Plastor™ in Europe), typically in the form of polyethylene sheets used to cover swathes of arable land, to provide cover to crops from seeding to germination and cropping that acts to protect the seedlings and crops from weeds and dehydration, and affording a controlled microclimate through a "greenhouse" effect and soil solarization [17,18,89,90]. The advantages and benefits in agricultural use are many but most important is that this family of polymers provide products, for example, films that are durable during use and are able to give an induction period (during which time the strength and properties of the film is intact) that is precisely controllable (both reproducible and capable of being varied) which is a very stringent requirement for mulching films and must be achieved if a dramatic reduction in crop yields or clogging of farm machinery is to be avoided [17,91].

The use of photodegradable polymers in disposable packaging applications has become even more important in recent years particularly due to the recognized threat that discarded plastics packaging can cause to marine life where deaths of birds and sea mammals can result through ingestion and strangulation [91]. The major class of commercial polymers targeted for packaging applications belong to the photolytic ketone-polyolefin copolymers based on the Guillet's process, which are marketed under the trade name Ecolyte along with the related E-CO polymers [21,37,38,92]. These polymers that give no induction period and photodegrade rapidly are mainly suitable for short-term and single-use packaging applications and are excellent for reducing plastics litter problem and are safer under marine environment, for example, their use in single-use six-collar pack beverage carriers (Hi-Cone™) which often ends up on sea shores would make them less hazardous to marine lives. For example, it was shown that an E-CO polymer containing about 1% carbonyl photodegrades after about 3 weeks of exposure to outdoor sunlight and breaks up into small fragments that have been suggested to then biodegrade completely after the first stage of abiotic photodegradation; the estimated total mass production for the six-pack carrier application exceeds 100 million pounds per annum [92].

Furthermore, in the marine environment, the entanglement hazard posed by plastics waste discarded on beaches depends on the time scale required for the material to degrade by any one of the degradation mechanisms, and definite embrittlement of the artifact shortly after it has been discarded is highly desirable as this may allow entangled sea animals to escape without permanent damage. Comparison of the photodegradability of films of three classes of commercial photodegradable polymers (E-CO with 1% CO, LDPE with prooxidant metal ion (Fe^{+3}) complex, and starch-LDPE containing metal ion prooxidant) under both outdoor (air) and marine floating exposure conditions conducted at the same location (in North Carolina, USA) has shown that all three samples became embrittled within 5–10 weeks under marine exposure conditions, significantly faster than unmodified PE control film [93]. However, it was also revealed that these polymers degraded less rapidly in seawater compared to land-based (air) exposure at the same location, although it is considered that the entanglement factor (relative to a control PE film) for marine exposure is still much larger than that for outdoor (dry) exposure due to the strong retardation of breakdown of the control film under marine conditions compared to its photofragmentation in outdoor air conditions [93]. The slower rate of degradation at sea was attributed both to the lower sample temperature at sea compared to dry outdoor conditions and to shielding from light due to surface fouling in samples exposed to seawater. E-CO polymers have also been used more recently as control-release packaging for water-soluble agricultural chemicals [94].

The use of photodegradable polymers in food contact applications has been limited because of risk of migration of additives from the polymer into the human environment. However, photodegrable polymers based on the Ecolyte process provide an attractive material for this purpose because of their immobilized ketonic groups thatare chemically attached to the polymer backbone [92]. Photodegradable polyketones, for example, Shell's Carillon, may also be used in food packaging applications due to their degradability, low gas permeability and chemical inertness [34,95]. Further, it has been shown that blending a small amount of polyketone with PP results in a significant reduction in the gas barrier property of PP too. However, polyketones are much more expensive than, for example, PE or PP [95], and unless this can be overcome they are unlikely to be economic for use on their own in these applications. Chemically bound iron, based on carboxylates, acrylates, and thiopropionates have been synthesized and used for the preparation of photodegradable polymers and shown to exhibit similar phtodegradation behavior to that of the traditional iron-carboxylate-containing PE (FeAcAc-PE or FeDRC-PE) [96]. Such photodegradable polymers with nonextractable constituents may offer also some potential for use in food packaging applications.

Starch-based degradable plastics are used mainly in packaging and also in hygienic products, for example, shopping bags, bread bags, over wrap, "flushable" sanitary product backing material. Foam loose fill packaging and injected moulded expanded products such as take-away containers have also been developed as a replacement to foam polystyrene. Further, granular starch has also been blended or mixed with petroleum-based polyesters or biologically derived polyesters, for example, polylactic acid and polyhydroxyalkanoate [97].

REFERENCES

1. R. Narayan, in: M. Vert, J. Feijen, A. Albertsson, G. Scott, and E. Chiellini (Eds.), *Biodegradable Polymers and Plastics*, Royal Society of Chemistry, Cambridge, 1992, p.176.
2. A. Krzan, S. Hemjinda, S. Miertus, A. Corti, and E. Chiellini, *Polym. Deg. Stability* **2006**, *91*, 2819.
3. S. Al-Malaika and G. Scott, in: N.S. Allen (Ed.), *Degradation and Stabilisation of Polyolefins*, Applied Science Publishers, London, 1983, p. 283.
4. G. Scott, *Atmospheric Oxidation and Antioxidants*, Elsevier, Amstrdam, 1965.
5. S. Al-Malaika, in: G. Scott (Ed.), *Atmospheric Oxidation and Antioxidants*, Vol. 1, Elsevier, 1993, p. 161.

6. D.J. Carlsson and D.M. Wiles, *Macromolecules* **1969**, *2*, 597.
7. S. Al-Malaika and G. Scott, in: N.S. Allen (Ed.), *Degradation and Stabilisation of Polyolefins*, Applied Science Publishers, 1983, p. 247.
8. G.H. Hartley and J. Guillet, *Macromolecules* **1968**, *1*, 165.
9. R. Arnaud, J.Y. Moisan, and J. Lemaire, *Macromolecules* **1984**, *17*, 332.
10. W.L. Hawkins, in: W.L. Hawkins (Ed.), *Polymer Stabilisation*, Wiley, New York, 1972, p. 9.
11. B.B. Cooray and G. Scott, in: G. Scott (Ed.), *Developments in Polymer Stabilisation-2*, Applied Science Publishers, London, 1980, p. 53.
12. G. Ayrey, B.C. Head, and R.C. Poller, *J. Polym. Sci., Macromol Rev.* **1974**, *8*, 1.
13. H.O. Wirthand and H. Andreas, *Pure and App. Chem.* **1977**, *49*, 627.
14. Reuters UK, 29 Feb. **2008**.
15. *San Francisco Cronicle*, 28 March 2007; *The New York Times*, 27 March 2007.
16. M. Vert, J. Feijen, A. Albertsson, G. Scott, and E. Chiellini (Eds.), *Biodegradable Polymers and Plastics,* Royal Society of Chemistry Cambridge **1992**.
17. D. Gilead and G. Scott, in: G. Scott (Ed.) *Development in Polymer Stabilisation-5*, Applied Science Publishers, London, 1982, p. 71.
18. G. Scott, *Polymers and the Environment*, Royal Society of Chemistry, Cambridge, **1999**.
19. S. Al-Malaika, A. Marogi, and G. Scott, *Polym. Degrad. Stab.* **1987**, *18*, 89.
20. R. Arnaud, P. Dabin, J. Lemair, S. Al-Malaika, S. Chohan, M. Coker, G. Scott, A. Fauve, and A. Maaroufi, *Polym. Degrad. Stab.* **1994**, *46*, 211.
21. J.E. Guillet, in: J. Guillet (Ed.), *Polymers and Ecological Problems*, Plenum Press, New York, 1973, p. 1.
22. M. Weiland, A. Daro, and C. David, *Polym. Degrd. Stab.* **1995**, *48*, 275.
23. S. Karlsson and A.C. Albertsson, *Polym. Eng. Sci.* **1998**, *38*, 1251.
24. I. Jakubowiz, *Polym. Degrad. Stab.* **2003**, *80*, 39.
25. E. Chiellini, A. Corti, and G. Swift, *Polym. Degrad. Stab.* **2003**, *81*, 341.
26. D.R. Tyler, *Coord. Chem. Rev.* **2003**, *246*, 291.
27. M.M. Brubaker, US Patent 2,495,286, **1950**.
28. J.E. Guillet, Plastics and the environment, in: G. Scott (Ed.), *Degradable Polymers: Principles and Applications*, Kluwer, Dordrecht, 2002, p. 413.
29. S. Al-Malaika, S. Chohan, M. Coker, G. Scott, R. Arnaud, P. Dabin, A. Fauve, and J. Lemaire, *J. M. S-Pure Appl. Chem*, **1995**, *A32*, 709.
30. G.M. Harlan and A. Nicholas, *Symposium on Degradable Plastics*, SPI, Washington, DC, **1987**, p. 14.
31. R.J. Statz and M.C. Dorris, *Symposium on Degradable Plastics*, SPI, Washington, DC, **1987**, p. 51.
32. G. Scott, *Polym. Degrd. Stab.* **1990**, *29*, 135.
33. F. Hollmanna, A.A. Mansourb, and B. Rieger, *Polymer* **2001**, *42*, 93–101.
34. F. Severini, R. Gallo, L. Brambilla, C. Castiglioni, and S. Ipsale, *Polym. Degrad. Stab.* **2000**, *69*, 133.
35. F. Severini, R. Gallo, L. Di Landro, M. Pegoraro, L. Brambilla, M. Tommasini, C. Castiglioni, and G. Zerbi, *Polymer* **2001**, *42*, 3609.
36. S.I. Kuzina, A.P. Pivovarov, A.I. Mikhailov, and G.P. Belov, *Eur. Polym. J.* (**2000**), *36*, 975.

37. J.E. Guillet, US Patent 3,753,952, **1973**.
38. S.K.L. Li and J.E. Guillet, *J. Polym. Sci. Polym. Chem.* **1980**, *18*, 2221.
39. M.M. Brubaker, US Patent 2,495,286, **1950**.
40. M. Heskins, W.J. Reid, D.J. Pinchin, and J.E. Guillet, *ACS Symp. Ser.* **1976**, *25*, 272.
41. L. Alexandru and J.E. Guillet, *Polym Sci., Polym. Chem.* **1975**, *13*, 483.
42. H.O.W. Eggins, J. Mills, A. Holt, and G. Scott, in: G. Sykes, and F.A. Skinner (Eds.), *Microbial Aspects of Pollution*, Academic Press, London, 1971.
43. S.G. Bond and J.R. Ebdon, *Polymer* **1994**, *35*, 451.
44. S.G. Bond and J.R. Ebdon, *Polymer* **1994**, *35*, 4079.
45. E. Dan and J.E. Guillet, *Macromolecules* **1973**, *6*, 228.
46. J.E. Guillet, S.K.L. Li, and H.C. Ng, *ACS Symp. Ser.* **1984**, *266*, 165.
47. M. Koutny, J. Lemaire, and A.-M. Delort, *Chemosphere* **2006**, *64*, 1243.
48. S. Al-Malaika, A. Marogi, and G. Scott, *J. Appl. Polym. Sci.* **1986**, *31*, 685.
49. G. Scott, British Patent 1,356,107, **1970**.
50. S. Al-Malaika, A. Marogi, and G. Scott, *J. Appl. Polym. Sci.* **1987**, *33*, 1455.
51. S. Al-Malaika, K.B. Chakraborty, and G. Scott, in: G. Scott (Ed.), *Development in Polymer Stabilisation-6*, Applied Science Publishers, London, 1983, p. 73.
52. G. Geuskens and M.S. Kabamba, *Polym. Degrd. Stab.* **1983**, *5*, 399.
53. G. Scott and D. Gilead, British Patent 1,586,344, **1978**.
54. D. Gilead, *Int. Polym. Mat.* **1978**, *6*, 185.
55. M. Hakkarainen, A.-C. Albertsson, and S. Karlsson, *J. Chromatogr. A* **1996**, *741*, 251.
56. M. Hakkarainen, A.-C. Albertsson, and S. Karlsson, *J. App. Polym. Sci* **1997**, *66*, 959.
57. S. Karlsson, M. Hakkarainen, and A.-C. Albertsson, *Macromolecules* **1997**, *30*, 7721.
58. M. Hakkarainen and A.-C. Albertssonin, in: A.-C. Albertsson (Ed.), *Advances in Polymer Science-169: Long Term Properties of Polyolefins*, Springer-Verlag, Berlin, 2004, p. 176.
59. T.J. Kemp and R.A. McIntyre, *Polym. Degrad. Stab.* **2006**, *91*, 3010.
60. T.J. Kemp and R.A. McIntyre, *Polym. Degrad. Stab.* **2006**, *91*, 3020.
61. T.J. Kemp and R.A. McIntyre, *Polym. Degrad. Stab.* **2006**, *91*, 165.
62. N.S. Allen, Y.S. Chow, F.F. Thompson, T.S. Jewitt, and M.R. Hornby, *Polym. Degrad. Stab.* **1991**, *32*, 119.
63. Y.S. Chow, N.S. Allen, F. Thompson, T.S. Jewitt, and M.R. Hornby, *Polym. Degrad. Stab.* **1991**, *34*, 243.
64. L. Zan, L. Tian, Z. Liu, and Z. Peng, *Appl. Catal. A Gen.* **2004**, *264*, 237.
65. L. Zan, S. Wang, W. Fa, Y. Hu, L. Tian, and K. Deng, *Polymer* **2006**, *47*, 8155.
66. S. Cho and W. Choi, *J. Photochem. Photobiol. A* **2001**, *143*, 221.
67. L. Zan, W.J. Fa, and S.G. Wang, *Environ. Sci. Technol.* **2006**, *40*, 1681.
68. H. Heidaka, Y. Suzuki, K. Nohara, S. Horikoshi, Y. Hissamatsu, E. Pelizzetti, and N. Serpone, *J. Polym. Sci. A Polym. Chem.* **1996**, *34*, 1311.
69. J. Shang, M. Chai, and Y.F. Zhu, *Environ. Sci. Technol.* **2003**, *37*, 4494.
70. X. Zhao, Z. Li, Y. Chen, L. Shi, and Y. Zhu, *Appl. Surf. Sci.* **2008**, *254*, 1825.
71. A.L. Linsebigler, G.Q. Lu, and J.T. Yates, *Chem. Rev.* **1995**, *95*, 735.
72. D.R. Park, J.L. Zhang, K. Ikeue, H. Yamashita, and M. Anpo, *J. Catal.* **1999**, *185*, 114.

73. R. Laible and K. Hamann, *Adv. Colloid. Interface Sci.* **1980**, *13*, 65.
74. M.R. Hoffmann, S.T. Martin, W. Choi, and D.W. Bahnemann, *Chem. Rev.* **1995**, *95*, 69.
75. C. Bauer, P. Jacques, and A. Kalt, *J. Photochem. Photobiol. A* **2001**, *14*, 87.
76. E.B. Zaynalov and N.S. Allen, *Polym. Degrad. Stab.* **2004**, *86*, 115.
77. R. Chandra and R. Rustgi, *Prog. Polym. Sci.* **1998**, *23*, 1273.
78. F.H. Otey and W.M. Doane, in: R.L. Whistler (Ed.), *Starch Chemistry and Technology*, Academic Press, 1984, p. 154, 397.
79. G.J.L. Griffin, British Patent, 1,485,833, **1972**.
80. G.J.L. Griffin, US Patent 4,983,651, **1991**.
81. G.J.L. Griffin, British Patent, 1,489,050, **1973**.
82. G.J.L. Griffin, in: G.J.L. Griffin (Ed.), *Chemistry and Technology of Biodegradable Polymers*, Blackie Academic & Professional, London, 1994, p. 18.
83. G.J.L. Griffin, Degradable plastic films, in: SPI Symposium on Degradable Plastics, Washington, DC, 1987, p. 47.
84. W.J. Maddever, in: S.H. Hamid, M.B. Amin, and A.G. Maadhah (Eds.), *Handbook of Polymer Degradation*, Marcel Dekker, New York, 1992.
85. C.R. Armistead in: G.J.L. Griffin (Ed.), *Chemistry and Technology of Biodegradable Polymers*, Blackie Academic & Professional, London 1994, ch.5.
86. A.C. Albertsson, and S. Carlsson, *Prog. Polym. Sci.* **1990**, *15*, 177.
87. F. Khabbaz, A.-C. Albertsson, and S. Karlsson, *Polym. Degrad. Stab.* **1998**, *61*, 329.
88. W.M. Choi, I.D. Jung, S.K. Kwon, C.S. Ha, and W.J. Cho, *Polym. Degrad. Stab.* **1998**, *61*, 15.
89. G. Scott, *Polym. Degrad. Stab.* **2000**, *68*, 1.
90. F. Kawai, in: A.-C. Alberttsson (Ed.), *5th International Symposium on Biodegradable Plastics and Polymers*, Stockholm, 9–13 June, 1998.
91. G. Scott, in: G. Scott and D. Gilead,(Eds.), *Degradable Polymers: Principles and Applications*, Chapman and Hall, London, 1995, p. 169.
92. J.E. Guillet, in: G. Scott, and D. Gilead (Eds.) *Degradable Polymers: Principles and Applications*, Chapman and Hall, London, 2002, p. 413.
93. A.J. Andrady, J.E. Pegram, and Y. Song, *J. Environ. Polym. Degrad.* **1993**, *1*, 117.
94. G. Harlan and C. Kmiec, in G. Scott, and D. Gilead (Eds.) *Degradable Polymers: Principles and Applications*, Chapman and Hall, London, 1995, p. 153.
95. E. Marklund, U.W. Gedde, M.S. Hednqvist, and G. Wiberg, *Polymer* **2001**, *42*, 3153.
96. G. Scott and S. Islam, *Polym. Degrad. Stab.* **1999**, *63*, 61.
97. K. Sriroth, and K. Sangseethong, in: *2nd International Symposium on Sweetpotato and Cassava: Innovative Technologies for Commercialisation*, ISHS Acta Horticultura, 704, February 2006.

17

PHOTOSTABILISATION OF POLYMER MATERIALS

Pieter Gijsman

Photochemistry and Photophysics of Polymer Materials, Edited by Norman S. Allen

17.1 INTRODUCTION

Synthetic polymers offer a wide range of attractive properties, which cause that they are widely used. In many of their applications, they are exposed to the outdoor environment, which might result in an undesired change of the mechanical or esthetical properties. The negative influence of the outdoor environment on polymer properties is in many cases called weathering or when there where the UV part of the solar radiation plays a major role, UV degradation. Polymer applications in which this type of degradation plays a major role can be found in different areas as the building and automotive industry as well as in agricultural areas. In building industry polyvinylchloride (PVC) window frames, polycarbonate (PC) roofing, and polypropylene (PP) stadium seats are important examples. Important polymer applications in cars are, for example, polypropylene bumpers and dashboards, polyamide or polyester mirror housings and rubber window sealings. The use of polymers in agricultural applications is nowadays very common; they are used instead of glass in greenhouses and in many cases as ground cover to increase the growing season. Other polymer applications that are subjected to the influence of outdoor environments are, for example, bottle crates, containers, and robes.

To make polymers useful for the above-mentioned applications, stabilizers of different types can be added to the polymers. As a result the lifetimes in the outdoor environments becomes long enough to make that polymers can be applied. In fact, without stabilizers the use of polymers in an outdoor environment would be very limited. Especially the discovery of hindered amine light stabilizer (HALS) led to a boost of the applicability of polymers in outdoor environments.

Although there are many similarities between the UV and the thermo-oxidative degradation of polymers there are many differences too. These differences cause that stabilization against UV or thermo-oxidative degradation is done with other types of stabilizer. A major reason that against thermo- and photo-oxidative degradation different stabilizers are applied is that one of the major requirements for UV stabilizers is that they are photostable. The majority of the phenolic antioxidants, which are the most applied stabilizers against the thermo-oxidative degradation, decompose under UV light [1,2], causing a low UV stabilizing activity. This is ascribed to the formation of phenoxyls by photochemically induced processes competing with radical scavenging, which was evidenced in pulse radiolysis experiments [3].

That UV light absorbers only are applied against UV and not against thermo-oxidative degradation is not surprising.

The most important classes of UV stabilizers are UV absorbers (UVAs), quenchers and hindered amine light stabilizers. In this chapter an overview is given about the types of UV stabilizers in use, their mechanisms of action as well as possible synergistic and antagonistic effects. In the last part for a selected number of polymers the effectiveness of different stabilizers is shown. The chemicals structures of all stabilizers mentioned in this chapter are shown in Appendix 17.9.

17.2 UV ABSORBERS

Sunlight degradation of polymers can be reduced by adding substances that absorb the harmful UV light more effectively than the polymer and transform the excess of energy to heat or less harmful radiation. In principle, this effect can be reached with dyes, pigments (colored), and UVAs (noncolored). The best UVA is carbon black, this pigment is able to absorb UV light very effectively causing that it increases the lifetime of polymers substantially. However, due to its color its applicability is limited. Organic UVAs are generally not (or very limited) colored, which makes their applicability much broader. In the following, the mechanism of action of several organic UVAs will be discussed.

17.2.1 Mechanism of UV-Absorption

To become an effective UV stabilizer UVAs must strongly absorb, for polymers, harmful UV light. They have to be able to transfer the absorbed light in harmless energy and they have to be active during the service life of the polymer. An overview of the currently in use classes of UVAs is given in Scheme 17.1. The UV absorption characteristics of these classes of UVAs are shown in several publications (for example, see Refs [4–6]). The UV absorption of these molecules depends, besides on the type, also on their substitution [4].

17.2.2 Mechanism of Deactivation

UVAs transform damaging radiation to less harmful energy such as heat or less energetic wavelengths. For hydroxybenzophenones and hydroxyphenylbenzotriazoles no significant amounts of phosphorescence of fluorescence was found [7]. Thus, the majority of the absorbed energy is transformed into heat. The heat energy generated in this way is small in comparison to the heat energy coming from absorbed visible, UV and IR wavelengths, which causes that it has no influence on the degradation rate.

On absorbing photons UVAs are excited to their first excited singlet state, UVAs with an intramolecular hydrogen bridge can undergo an excited-state intramolecular proton transfer (ESIPT). The excited proton transferred product, loses its energy radiationless as thermal energy, by fluorescence or phosphorescence, to form the ground-state proton transferred product, followed by a proton shift, which forms the UVA in the ground-state back [8,9]. This mechanism is shown for a

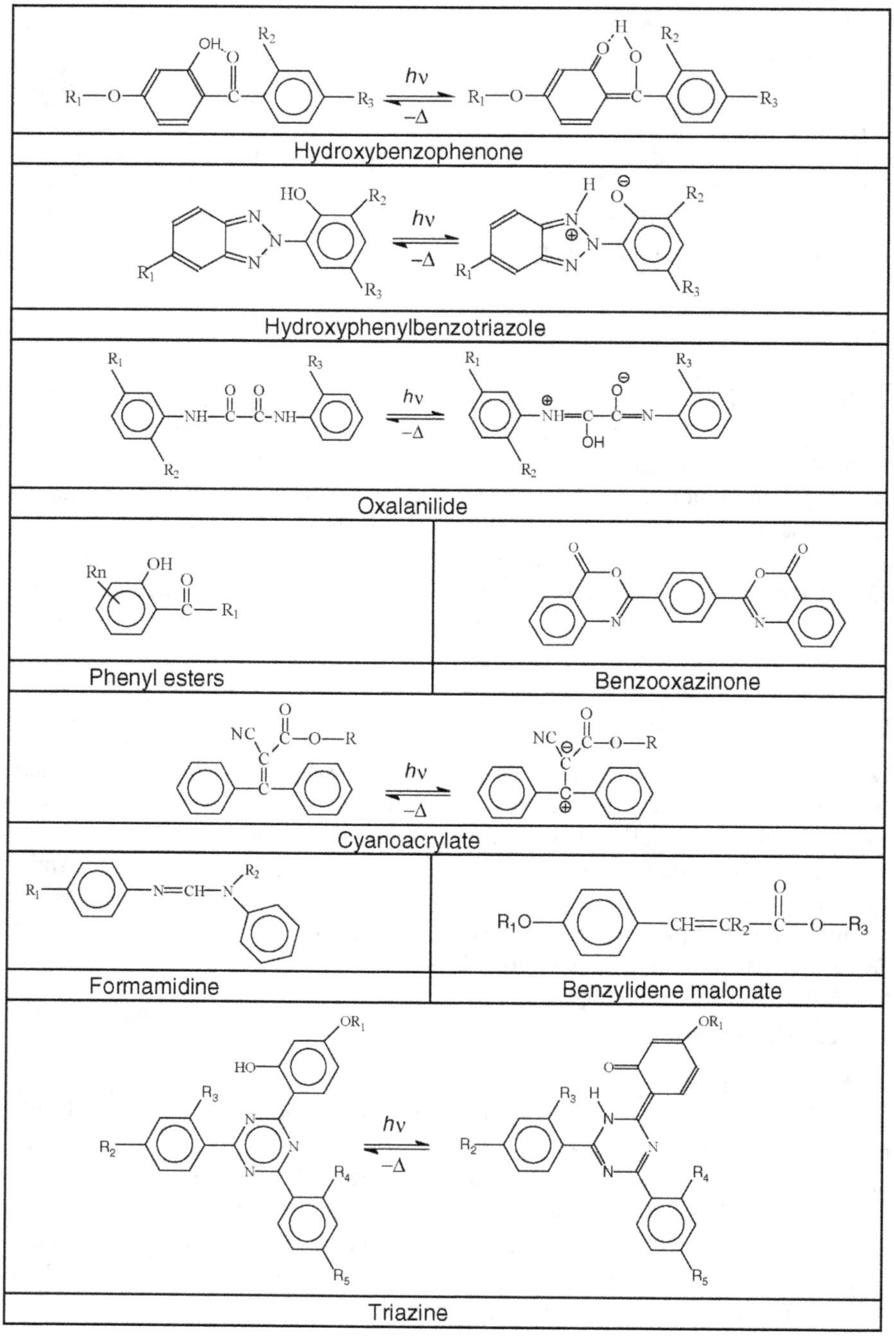

SCHEME 17.1 Currently in use classes of organic UVAs.

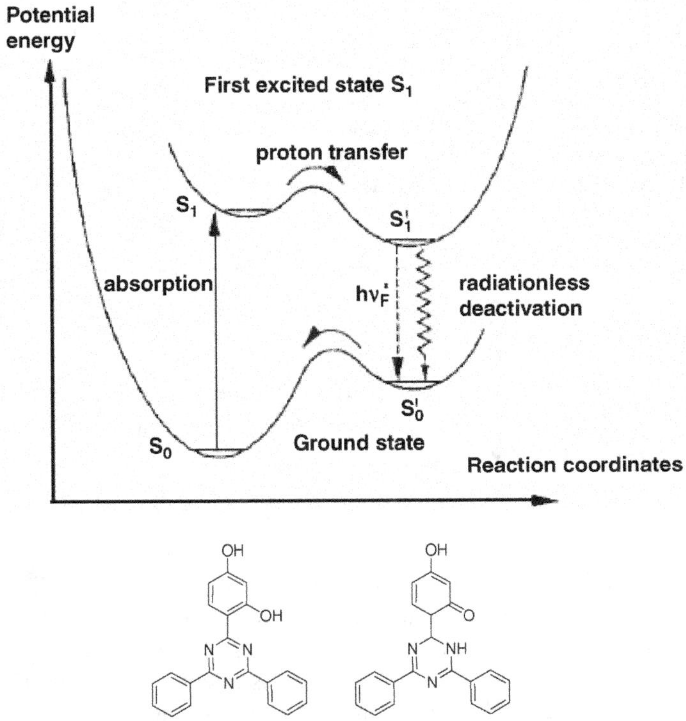

SCHEME 17.2 Simplified deactivation scheme and ESIPT for a 2-(2′-hydroxy-phenyl)-1,3,5-triazine [11].

2-(2′-hydroxy-phenyl)-1,3,5-triazine in Scheme 17.2. For this type of UVAs, the appearance of longer-wavelength fluorescence is observed that is assigned to the keto tautomer, formed following ESIPT [10].

17.2.2.1 Photostability of UVAs To be a good UV stabilizer the deactivation process has to be very effective, because destruction even with low quantum yields will cause undesired stabilizer concentration reduction. So it is estimated that even with a quantum yield for destruction of only 10^{-6} (one decomposition reaction upon 1 million photons absorbed), about 30% of the UV stabilizer is deactivated within 1 year of Florida exposure [12]. The energy dissipation mechanism of several UVAs is shown in Scheme 17.1. Because internal hydrogen bridges play a major role in the stability of UVAs, it is not surprising that the polarity of the matrix plays an important role in the stability of UVAs [13]. However, photolysis is not the only way UVAs degrade. Especially in easily degrading polymers radicals that are produced by the polymer matrix can attack the UVA, leading to inactive decomposition products.

The photolytic stability of UVAs can be determined in relatively stable polymers of which the loss by physical processes is known to be limited. Such a polymer is polymethylmethacrylate (PMMA). The photostability of several UVAs in PMMA was determined (Table 17.1). Of the evaluated UVAs, the triazine is the most stable.

TABLE 17.1 Rate of Loss of UVAs in PMMA When Exposed to Xenon Arc Weathering [8]

Absorber	Class	Rate (A/1000 kJ/m^2)
UVA-BP-1	Benzophenone	0.18
UVA-BZT-4	Benzotriazole	0.11
UVA-AN-2	Oxanalide	0.14
UVA-CA-1	Cyanoacrylate	0.14
UVA-HTZ-1	Triazine	0.085

17.2.3 Effectiveness of UVAs

Mechanism of action of UVAs is based on their ability to absorb harmful UV light. According to Lambert–Beer's law (see Equation 17.1), the light intensity at any depth (I_d) in a polymer is given by

$$I_d = I_0 \cdot 10^{-\Sigma(\varepsilon.c.d)} \tag{17.1}$$

where I_d is the intensity of the light at depth d, I_0 is the intensity of the light, ε is the extinction coefficient of the UVA, and c is the concentration of the UVA.

From this equation, the influence of UVA concentration and relative light intensity as a function of depth for a nonabsorbing polymer is calculated (see Fig. 17.1). From this figure, it is clear that UVAs mainly protect the bulk of the material. They are not very effective in preventing surface degradation. For absorbing polymers or when pigments are added, the influence of a UVA will be even smaller [14].

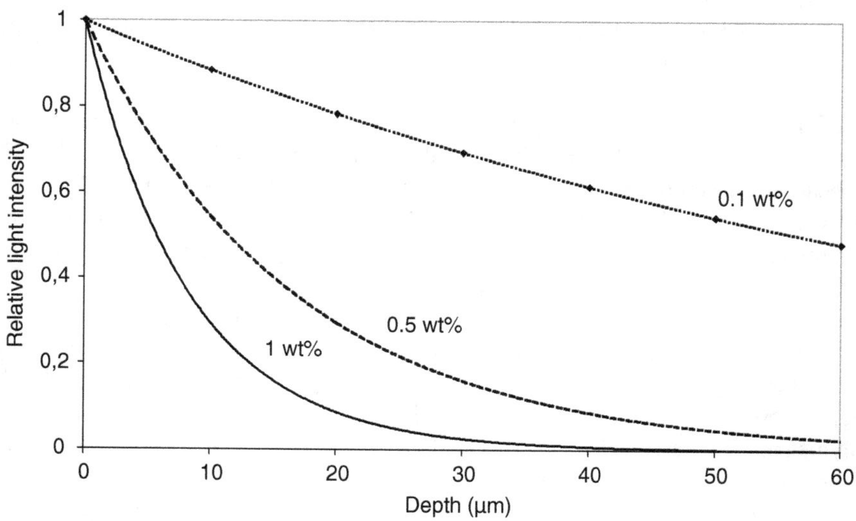

FIGURE 17.1 Calculated influence of the concentration of UVA (0.1, 0.5, and 1 wt%) and depth on the relative light intensity (nonabsorbing polymer, UVA: UVA-HTZ-2).

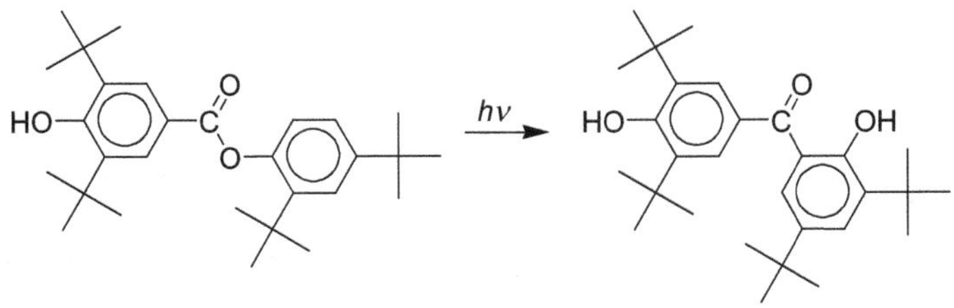

SCHEME 17.3 Photo-Fries reaction of phenyl substituted p-hydroxybenzoates.

The effectivity of UVAs will also depend on the influence of intensity on the degradation rate. If the degradation rate is linear in light intensity a 50% reduction will lead to a doubling of the lifetime. However, when the stability depend on the square root of the intensity a 50% light reduction only lead to a 30% increase in lifetime.

This means that UV absorption cannot be an effective mechanism for thin articles or if the degradation only appears at the surface. Nevertheless, UVAs are effective in thin articles and against surface degradation, which lead to the conclusion that they have to operate according to other mechanisms (e.g., quenching or radical scavenging) too.

Experiments in which an unstabilized PP film was shielded by a benzophenone or benzotriazole type UVAs showed that both could protect the unstabilized PP film by UV absorption. However, a film shielded by a benzophenone-type UVA-containing film showed a significantly faster degradation than a film containing this stabilizer. This clearly demonstrates that their protection is not only by UV absorption but also by other mechanisms. In contrast, a film shielded with a benzotriazole type UVA is considerably more stable than the shielding film itself, which clearly indicates the dominating role of UV absorption as main stabilization mechanism for this type of UVA [15,16].

Phenyl substituted p-hydroxybenzoates (e.g., UVA-BZ-2) do not absorb enough sunlight wavelengths to prevent UV degradation, but they are precursors for 2-hydroxybenzophenones, which are formed though a photo-Fries rearrangement (Scheme 17.3) [17]. Recently a polycarbonate polyester copolymer was developed that contains moieties that through this photo-Fries rearrangement forms 2-hydroxybenzophenone moieties, which act as UV shield and prevent the polymer from degradation [18].

17.3 QUENCHERS

Photodegradation is a result of the absorption of UV light that leads to molecules in the excited states, which undergo reaction. When molecules are present that take over the energy from the excited polymer and release this energy in a harmless form, an increased stability can be expected. This can be represented by a simple mechanism in which an excited chromophoric group (P^*) is deactivated by a stabilizer molecule (Q)

$$P^* + Q \longrightarrow P + Q^*$$

$$Q^* \xrightarrow{-h\nu/\Delta} Q$$

SCHEME 17.4 Deactivation of excited chromophoric groups in polymer (P*) by quenching molecules (Q).

that releases its energy as heat or luminescence, before the excited chromophoric leads to chain cleavage or initiating radicals (Scheme 17.4).

Quenching as a mechanism to stabilize polymers can only be successful if the quencher is within the lifetime of the excited chromophoric group within quenching distance. High diffusion coefficients and a long lifetime of the excited chromophoric group increase the quenching probability.

The energy transfer distance depends on the quenching mechanism. Different mechanisms to transfer the energy are known:

Energy Transfer by Collision. For this type of energy transfer very short distances are necessary, which can only be reached with quenchers with a very high diffusion coefficient.

Exchange Energy Transfer. In this case, the electron cloud of the donor and acceptor molecules should overlap; energy transfer can take place over distances of 1–1.5 nm.

Long-Range Energy Transfer. In this case the energy is transferred through a dipole-dipole interaction, this can happen over distances of 5–10 nm [19,20]. Requirements for this type of energy transfer are: donor and acceptor molecules should be immobile; their UV spectra should show a significant overlap, which is broad and not structured. For molecules with a molecular weight of 500, a concentration of 0.05–0.5% corresponds to distances of 5–10 nm, when the quencher is homogeneous distributed. For semi crystalline polymers, this concentration drops proportionally with crystallinity. These concentrations are for stabilizers quite realistic.

17.3.1 Quenching by Metal Complexes

In the class of quenchers, the nickel chelates are the most well known. For additives that are known to stabilize polyolefins, various attempts have been made to demonstrate energy-transfer to this type of molecules in liquid as well as in solid phase. For example, Briggs and McKellar [21] have shown that nickel (Ni(II)) chelates that are effective UV stabilizers for PP, are efficient acceptors for the excitation energy of triplet anthracene too. From this result it was concluded that the Ni chelates act as quencher in PP. However, this conclusion can be argued because anthracene cannot be expected to be an adequate model for the polypropylene carbonyl chromophores, because the excitation energy of triplet anthracene is −42 kcal/mole, whereas the value for an aliphatic ketone is −74 kcal/mole [22]. Efficient energy transfer usually occurs from a donor energy level to an acceptor level of lower energy than that of the donor. Hence, it is possible for an additive to quench ketone triplets but to be incapable in quenching anthracene triplets [22]. Chien and Conner [23] have also concluded that Ni(II) complexes can effectively prevent the diethyl ketone photosensitized oxidation of cumene in the liquid phase, possibly by a dipole–dipole type of energy transfer process. According to Guillory and Cook [24] excited singlet and triplet state quenching of benzophenone and 4-methyl-2-pentanone by nickel chelates was due to a diffusional exchange energy transfer mechanism and corresponds to their stabilization effectiveness.

Pivovarov and Lukovnikov [25] have attempted to make a more meaningful test of the energy-transfer ability of additives, which are present in a polymer. Their test is based on the efficiency of quenching of the fluorescence emission, which is detected from polymers, such as polyolefins, polyamides, and polystyrene, due to the presence of certain impurities. These fluorescent impurities are assumed to be responsible for the photoinstability of these polymers. Some effective polypropylene UV stabilizers efficiently quench the fluorescence emission of polypropylene observed at 330 nm. However, these results were disputed and it was augmented that in this case not quenching but UV absorption was the UV stabilization mechanism [22,24]. The effect of metal chelates on the yield of acetone formation from the photodecomposition of 2-pentanone was studied by Flood and Russell [26], although they found quenching, it was concluded that this was not enough to be the main stabilization mechanism of these metal chelates.

Although it was shown that metal chelates are possible quenchers, in more recent years their mechanism of action is more and more ascribed to other stabilization mechanisms. Ramani and Scott [27] evaluated a variety of metal complex stabilizers. All the metal complexes studied were found to be UVAs, although this was not their only stabilizing mechanism. In the case of the transition metal dibutyl dithiocarbamates the additional stabilization mechanism was consistent with their known peroxide decomposing behavior. In a series of papers Allen et al. [28–33] described the mechanism of action of metal chelates in polymers. According to them excited-state quenching and UV absorption only play little part in the stabilizing mechanism of metal chelates. Their mechanism of action is mainly ascribed to their ability to act as UV stable antioxidants, being able to scavenge radicals. Only one of the chelates, namely, Ni-1, behaved as a hydroperoxide decomposer.

According to Gugumus [34] the activity of nickel chelates is partly ascribed to their ability to quench excited polymer oxygen charge transfer complexes.

17.3.2 Quenching by Amines

Part of the stabilizing mechanism of Hindered amine light stabilizers in polyolefins was ascribed to the ability of the HALS oxygen charge transfer complex to quench initiating polymer–oxygen charge transfer complexes [35,36]. This reaction leads to excited amine oxygen CTCs (see Scheme 17.5). According to Davidson, excited amine–oxygen CTCs show the same configuration as the CTC complex of the amine with singlet oxygen [19]. Deactivation of this excited state is chemical or physical,

Initiation by charge transfer complexes

$$RH + O_2 \rightleftharpoons [RH, O_2] \xrightarrow{h\nu} [RH, O_2]^{\star} \longrightarrow \text{initiation}$$

CTC quenching mechanism

$$[RH, O_2]^{\star} + [NH, O_2] \longrightarrow [RH, O_2] + [NH, O_2]^{\star}$$

SCHEME 17.5 Schematic representation of the stabilization of amines through the quenching of excited polymer oxygen CTCs.

which depends on the chemical structure of the amine. On the analogy of singlet oxygen quenching [37], a mechanism is postulated for the quenching of polymer—oxygen CTCs by primary, secondary and tertiary amine–oxygen CTCs (Schemes 17.6 and 17.7).

According to these mechanisms, chemical quenching will lead to the destruction of the amine and the formation of hydrogen peroxide. Hydrogen peroxide can initiate the oxidation of polyolefins, so this mechanism does not lead to stabilization. However, chemical quenching will be found for primary and secondary amines containing an α-hydrogen. HALS does not contain an α-hydrogen and can therefore act as a physical quencher. Chemical quenching is also prevented when no unsaturation in the quenching molecule can be formed. According to Bredt's rule, amines with nitrogen in a bridgehead position cannot form a C=N bond [37,38], which means that these types of molecules will act by physical quenching. Thus, deactivation of excited oxygen-HALS and oxygen-bridged amines CTCs will mainly be physical and can

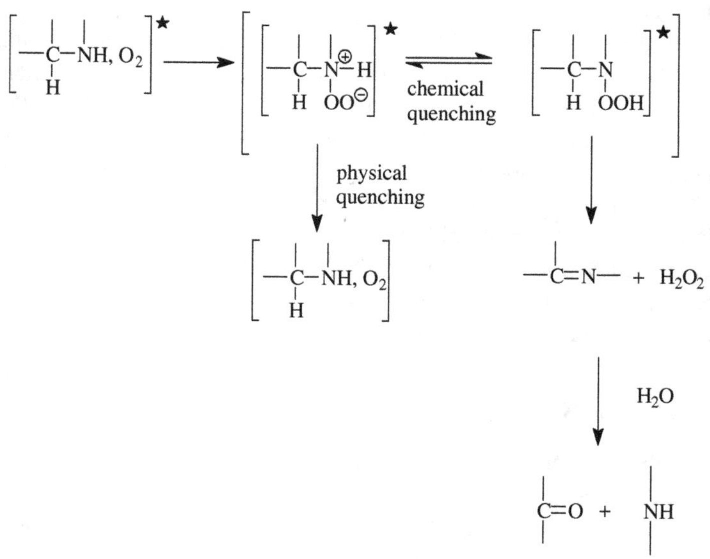

SCHEME 17.6 Proposed mechanism for the deactivation of excited primary and secondary amine oxygen CTCs.

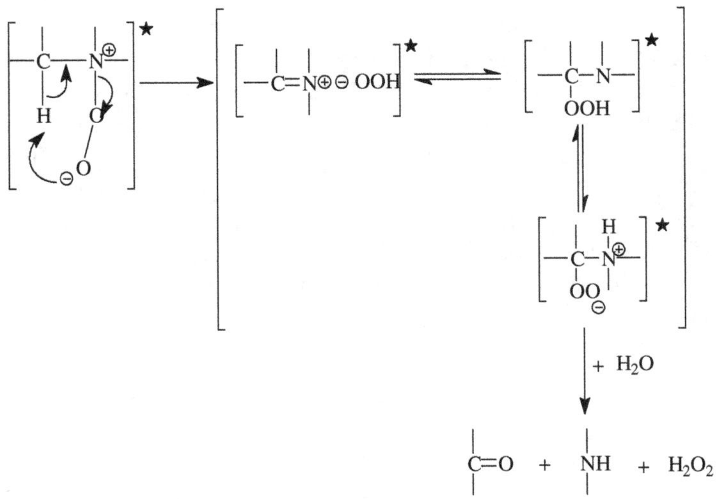

SCHEME 17.7 Proposed mechanism for the deactivation of excited tertiary amine oxygen CTCs.

lead to stabilization of the polymer. Indeed, it was found that the stability of PP could be increased with a factor of 2–3 by adding bridged amines [39–41].

17.4 RADICAL SCAVENGERS

Alkyl substituted *p*-hydroxybenzoates (e.g., UVA-BZ-1) were shown to operate as effective light stabilizers, particularly in the presence of pigments [42]. For these hydroxybenzoates, the Photo-Fries reaction as shown in Scheme 17.3 is not possible and therefore other reactions must cause the activity as UV stabilizer. Like many phenolic antioxidants, these hydroxybenzoates act as radical scavenger too. However, while antioxidants exhibit a weak absorption tail extending well above 300 nm, UVA-BZ-1 exhibits little, if any, absorption in this region. On this basis, it is expected to be more photostable under sunlight exposure conditions, which cause that they are ultraviolet stable radical scavengers. These UV stabilizers were shown to be more effective in PP than in PE, which was attributed to a longer propagation (kinetic chain) length for PP than for PE [16]. Due to the direct substitution of the carboxylic acid group onto the phenyl ring these stabilizers are resistant to dimerization and subsequent photoyellowing [17,43–45].

17.5 HALS

17.5.1 Introduction

After the introduction of the first HALS in the 1970s, this class of light stabilizers rapidly became the most important UV stabilizer for the majority of plastics. The invention of

HALS was based on the discovery that the 2,2,6,6-tetramethyl-1-piperidinyloxy, free radical (TEMPO) (**1**)), which already was known as an effective radical scavenger [46,47], was a very effective UV stabilizer too [48,49]. However, due to its physical and chemical properties TEMPO itself did not led to practical use. TEMPO is colored and will impart color to the to be stabilized polymer, it is thermally unstable and volatile [49]. Furthermore, it reacts with phenolic antioxidants present in many polymers leading to a reduction of processing and/or long-term heat stability. The discovery that compounds in which the *N*-oxyl functionality was replaced by a N–H functionality also showed good UV stabilization activity was the key finding that led to the development of HALS stabilizers [49].

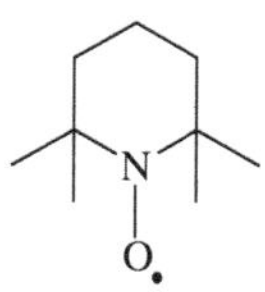

(1)

The introduction of HALS led to a large increase of the UV stability of polymeric materials. Without the discovery of HALS, the outdoor applicability of many polymers would be limited. So would the use of polypropylene in automotive application without the use of HALS be impossible.

Although HALS stabilizers were developed as UV stabilizer it is more and more recognized that these molecules could also impair long-term heat stability. Especially were phenolic antioxidants cannot be used due to their discoloration; HALS is used to protect the polymer against long-term heat degradation [50].

Nowadays, there are many HALS stabilizers commercial of which the majority is based on 2,2,6,6-tetra-methyl-4-piperidinyl moieties. The first commercial type (LMW-HALS-1) is relatively low in molecular weight, which caused that due to its high volatility it is not suitable for thin applications. To overcome this problem oligomeric HALS types were developed. One of the drawbacks of the piperidinyl moiety of HALS stabilizers is that it is basic causing that it can react with acids and forms a not stabilizing salt. Consequently, the effectiveness of HALS in systems where acids are present or can be formed is limited. To beat this problem less basic HALS types as, for example, N–O–R types were developed.

Since the discovery of HALS, a lot of research was done on their mechanism of action and many different mechanisms were proposed. At the moment, there is still no consensus, which might be related to the fact that the mechanism of action of HALS, as the UV degradation of polymers, is circumstances dependant. So are results of the mechanism of action in different polymers, UV degraded with different light sources and even thermo-oxidative compared. In the following an overview of the mechanistically work that was done on HALS as stabilizer as well as application data is given.

17.5.2 Synthesis

Almost all HALS stabilizers are derivatives of 2,2,6,6-tetramethylpiperidin-4-one, commonly known as triacetoneamine (TAA) (**2**):

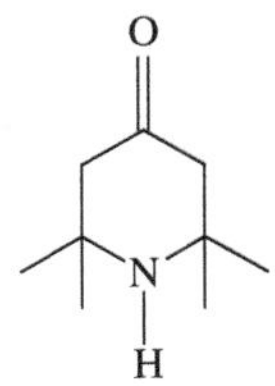

(2)

This product was already synthesized in the nineteenth century, although the yield at that time was low [51]. Since then the synthesis was continuously improved. TAA can be synthesized from acetone and ammonia. Although the mechanism has not been definitely proven, the mechanism postulated by Bradbury et al. [52] is generally accepted (Scheme 17.8).

As shown in Scheme 17.9, TAA derivatives can be made by hydrogenation or hydroamination with ammonia, butylamine or hexamethylene diamine leading to intermediates for the most commercial HALS stabilizers as, for example, HMW-HALS-2, HMW-HALS-4, or HMW-HALS-6 [53].

Up to my knowledge, only commercial HALSs that are not based on TAA derivatives are the piperazinone types (e.g., LMW-HALS-2). Their synthesis is more complicated (see Scheme 17.10).

17.5.3 Mechanism of Action of HAS

Proposed mechanisms of action of UV stabilizers are UV absorption, excited-state quenching, hydroperoxide decomposition and radical scavenging. For HAS the only mechanism that can be ruled out is UV absorption, because many very effective HAS stabilizers as, for example, LMW-HALS-1 do not absorb in the UV wavelength region

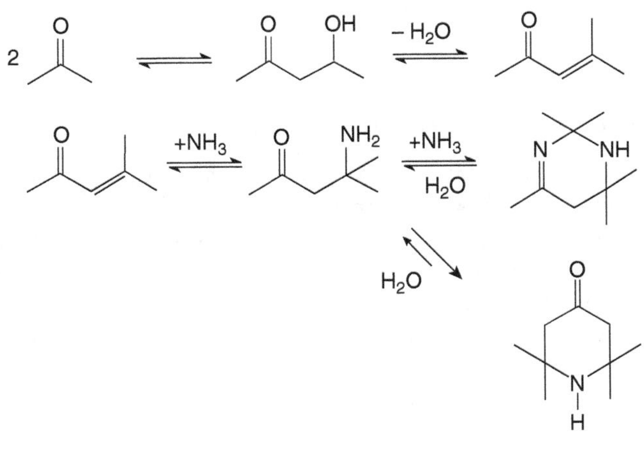

SCHEME 17.8 Reaction mechanism of triacetone amine from acetone and ammonia.

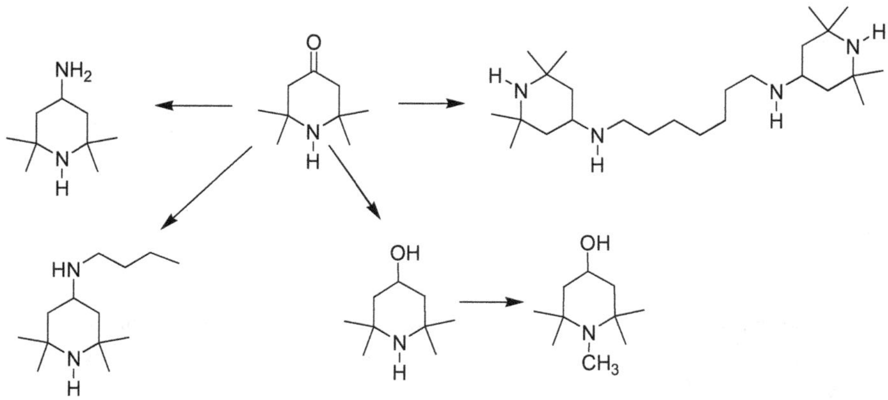

SCHEME 17.9 Synthesis of TAA derivatives.

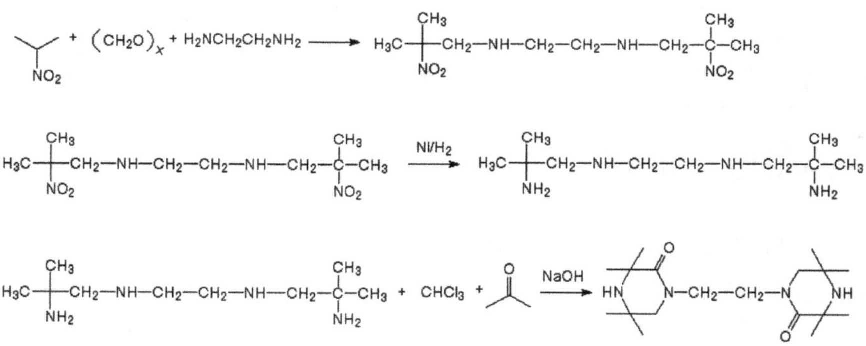

SCHEME 17.10 Synthesis of a piperazinone type HALS [54].

of terrestrial solar radiation. All other proposed mechanisms will be discussed in the following.

17.5.4 Stabilization Mechanism by the Piperidinoxyl Radical

ESR studies on the stabilizing activity of HALS showed that the amine is easily converted in a nitroxide [55,60]. Together with the knowledge that these radicals are stable and excellent scavengers of alkyl radicals, it was easily concluded that they should play an important role in their stabilizing activity. Shilov et al. [56] observed the consumption of nitroxide during the cumyl hydroperoxide initiated degradation of PP in an argon atmosphere, while in this experiment the addition of oxygen led to an increase of the nitroxide concentration leading to the conclusion that the nitroxide can be regenerated. Shilov and Denisov [57] proposed that the nitroxide is formed back in a reaction between the aminoether and a peroxy radical. From AIBN initiated model reactions in solution Kovtun et al. [58] suggested that this reaction leads to a

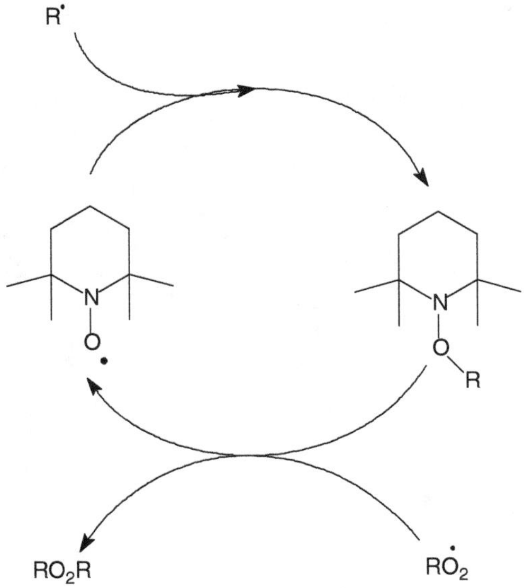

SCHEME 17.11 Nitroxide regeneration mechanism I.

dialkylperoxide. The high effectiveness of HALS can now be ascribed to the possibility to form the nitroxide back as shown in Scheme 17.11.

For PP the to the polymer chain grafted nitroxide was observed [59–61] and was ascribed to be the predominant stabilizer species [61].

According to Bolsman [62], the cyclic that is responsible for the catalytic scavenging of tertiary alkyl and peroxy radicals at 130°C involves the formation of a hydroxylamine as outlined in Scheme 17.12.

Based on the results of the degradation of unstabilized PP and the effectivity of HALS stabilizers in different model systems it is postulated that HALSs are mainly effective in systems in which peracids can be formed [50]. According to Step et al. [63,64] this is because in the nitroxide regeneration mechanism the peracyl radical plays a key role (Scheme 17.13).

In all the nitroxide regeneration mechanisms the nitroxide reacts with an alkyl radical, this reaction is in competition with the extremely fast reaction between an alkyl radical and oxygen (see Scheme 17.14)

Several groups investigated the rate constant for the reaction between the nitroxide and an alkyl radical [65–68]. The ratio between these reactions depends on the type of alkyl radical, but k_{NO}/k_{O_2} is about 0.1–0.25. For the reaction rate, besides the reaction constant, the concentration of reagents is important. For polyolefins, it was shown that the concentration of the nitroxide is an order of magnitude smaller than that of oxygen, which causes that only about 1–2.5% of the alkyl radicals can be scavenged by the nitroxide. This simple calculation raises doubts about the above postulated nitroxide regeneration mechanisms. However, it is believed that degradation is a heterogeneous process [69–74], if the HALS stabilizers concentrate themselves in these oxidizing

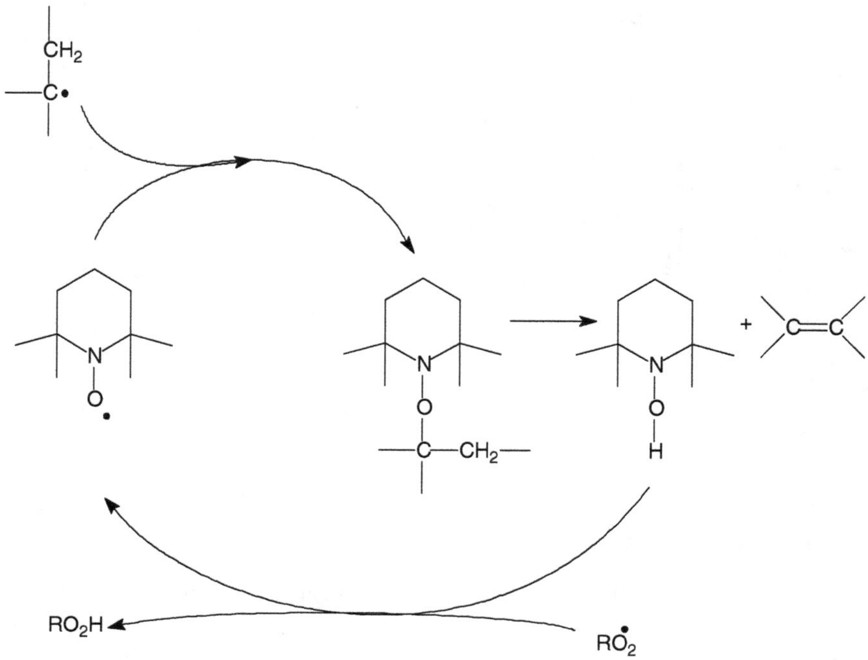

SCHEME 17.12 Nitroxide regeneration mechanism II.

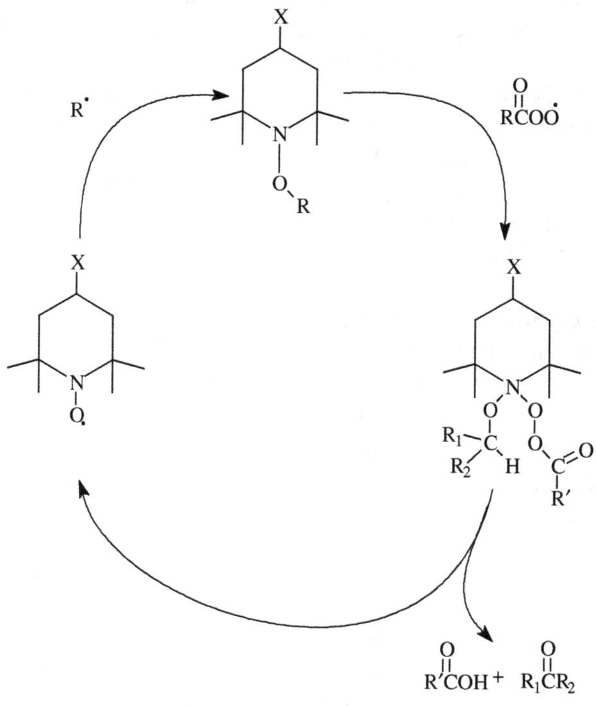

SCHEME 17.13 Nitroxide regeneration mechanism III [63,64].

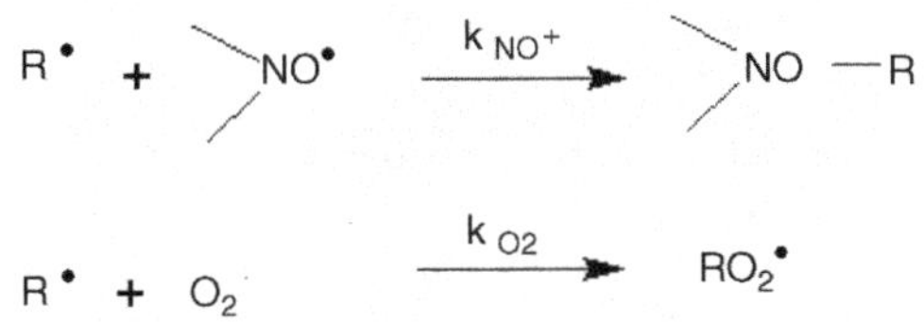

SCHEME 17.14 Competition reactions between an alkyl radical and oxygen or a nitroxide radical.

areas their concentration and that of their nitroxide increases causing an increased alkyl radical scavenging ability. One of the possibilities for an increased concentration in the oxidizing zones is the formation of associates between the hydroperoxides and the HALS. Oswald et al. [75] reported the formation of associates between non-hindered amines with hydroperoxides, Sedlar et al. [76] suggested that these associates are formed with HALS too. For model experiments Grattan et al. [77] found for zones being equimolar in hydroperoxide and alkane segments a 26-fold increase in the concentration of nitroxide radicals. From this result, it was suggested that association through hydrogen bonding, between the nitroxide and the hydroperoxides, causing a raise of the local nitroxide concentration to a level at which the reaction of the nitroxide with an alkyl radical could successfully compete with the reaction of an alkyl radical with oxygen.

Another possibility for the increased concentration of nitroxide in oxidized zones is the formation of charge transfer complexes between peroxy radicals and HALS [78].

17.5.4.1 Transformation of Piperidinyl to Piperidinoxyl Radical In the above-mentioned radical scavenging mechanisms, the nitroxide is the key intermediate. Several mechanisms have been suggested for the formation of the nitroxide from the parent amine. Sedlar et al. [76] proposed a reaction of a hydroperoxide with the amine to form an alkyloxyamine, which reacts with a peroxy radical to form a nitroxide (Scheme 17.15).

Carlsson et al. [79] published a mechanism involving an aminyl radical that is oxidized to a nitroxide (Scheme 17.16).

Geuskens and Nedelkos [80] suggested that the reaction of an amine with a peroxy radical leads to the nitroxide and an alcohol (Scheme 17.17).

The oxidation of alkanes leads to aldehydes. Toda et al. [81] and Felder [82] suggested that the formation of the nitroxide is related to the oxidation of the aldehydes formed.

SCHEME 17.15 Reaction of the hindered amine to an alkyloxyamine.

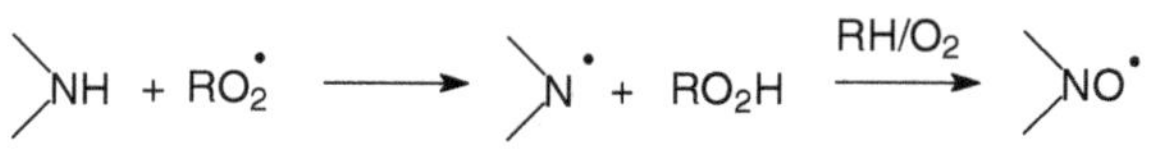

SCHEME 17.16 Formation of the nitroxide according to Carlsson et al. [79].

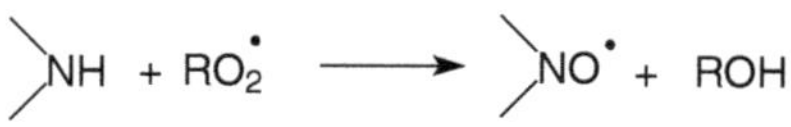

SCHEME 17.17 Formation of the nitroxide according to Geuskens and Nedelkos [80].

Toda et al. [81] assumed that the nitroxide is formed by a reaction of a peracid with the amine (Scheme 17.18) and Felder [82] suggested a reaction between an acylperoxy radical and the amine to form the nitroxide and an acid (Scheme 17.19).

Zahradnickova et al. [83] showed that the nitroxide can be formed by the reaction of the amine with oxidized polypropylene containing peracids.

17.5.4.2 Other Possible Stabilization Mechanisms

Transition Metal Complexation It is well known [84] that transition metals can initiate the photo-oxidation of polymers. Fairgrieve and McCallum [85] showed that HALS is capable in complexing transition metals. They postulated that the mechanism of action of HALS is related to this complexation, which could prevent the decomposition of hydroperoxides into initiating radicals. Moreover, in this case an increased concentration of HALS in the oxidizing zones can be expected.

Hydroperoxide Decomposition Piperidinyl derivatives have no influence on the photolysis rate of *t*-butylhydroperoxide in an inert environment [86,87] nor on the decomposition of cumyl hydroperoxide at 120°C in chlorobenzene [87]. From these results, it can be concluded that hydroperoxide decomposition cannot be the main stabilization mechanism of action of HALS.

Quenching of Excited Oxygen Polymer Charge Transfer Complexes At least a part of the stabilizing mechanism of HALSs in polyolefins was ascribed to the ability of the HALS oxygen charge transfer complex to quench initiating polymer-oxygen charge

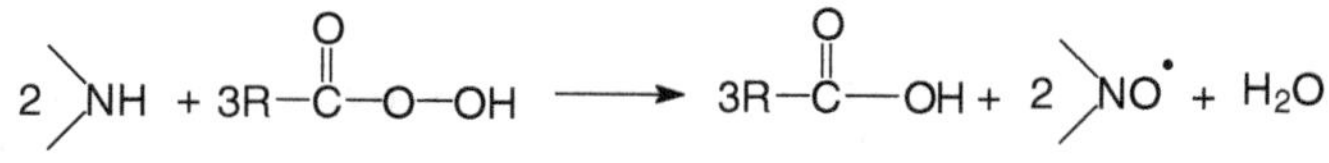

SCHEME 17.18 Formation of the nitroxide according to Toda et al. [81].

SCHEME 17.19 Formation of the nitroxide according to Felder [82].

transfer complexes [35,36,88]. Gugumus [88] based this conclusion on the high efficiency of HALS in preventing the reaction yielding trans-vinylene groups, while Gijsman based this conclusion on the difference in the ratio between during UV degradation formed carbonyl absorbance and oxygen uptake for unstabilized and HALS stabilized PE [35] and PP [36]. A possible quenching mechanism is described in Section 17.3.2.

17.5.5 Interactions of HALS with Other Stabilizers

In many stabilizer formulations, different types of stabilizers are used. As processing stabilizer phenolic antioxidants and phosphites are applied, for long-term heat stability phenolic antioxidants and thioethers are used and for UV stability, combinations of HALS with other types of UV stabilizers can be applied. HALS stabilizers show interactions with these types of stabilizers that can lead to synergisms as well as antagonisms.

17.5.5.1 Interactions Between HALS and Processing Stabilizers As the HALS stabilizers are for the majority of the polymers the most effective UV stabilizers, in many cases these stabilizers have to be combined with other stabilizers. Many polymers contain a phenolic antioxidant and/or a phosphite as processing stabilizer. Combinations of molecules can, for various reasons, give an adverse or a beneficial effect, which depends on the types used as well as on their application. For the combination of phenolic antioxidants with HALS there are antagonisms as well as synergisms reported [89–95]. Allen et al. [91] investigated the influence of different combinations of phenolic antioxidants and HALS on the UV stability of polypropylene and found, dependant on the specific combination, antagonisms as well as synergisms. Different reasons for the antagonism between HALS and phenolic antioxidants are postulated.

- Phenolic antioxidants inhibit the formation of hydroperoxides, which could cause that the transformation of the amine to the nitroxide is inhibited, which cause a loss of activity of the HALS.
- The formed nitroxide reacts with the phenolic antioxidant, causing an increased consumption of the phenolic antioxidant.
- Nitroxide reacts with decomposition products of the phenolic antioxidant (quinones, leading to deactivation of the nitroxide).

According to Yamashita and Ohkatsu [93] the antagonism of phenolic antioxidants and HALS can also be ascribed to the formation of a salt between both that accelerates the decomposition of hydroperoxide and in this way increases the degradation rate.

In many cases, phenolic antioxidants are used in combination with phosphites. Bauer et al. [96] showed an antagonism of P-2 with secondary as well as with tertiary HALS or the corresponding nitroxide radical. This antagonism is decreasing with increasing molar amount of the phosphite. George and Ghaemy [97] also found an increased photodegradation rate in the presence of P-2, they ascribed this to the inherent photoinstability of this phosphite leading to photoactivating species. In contrast to the aromatic, the aliphatic phosphite (P-1) showed a synergism with LMW-HALS-1. Allen et al. [92] showed a synergism between the aliphatic phosphite P-3 with a low molecular HALS (LMW-HALS-1) and an antagonism with the oligomeric HALSs (HMW-HALS-1, HMW-HALS-2, and HMW-HALS-4).

17.5.5.2 Interactions Between HALS and Sulfur Containing Stabilizers For application with a requirement on long-term heat stability at high temperatures (150°C) and on UV stability in many cases combinations of sulfur containing stabilizers and HALS are used. However, in many cases this combination behaves antagonistic. Using oxygen uptake experiments Bigger and Delatycki [98] showed the negative influence of BFS-1, S-1 and BFS-2 on the effectiveness of LMW-HALS-1 as UV stabilizer in HDPE (see Fig. 17.2). From these data, it is clear that at a concentration of 0.15 wt% of sulfur containing stabilizers, up to a concentration of 0.1–0.15 wt% the HALS is not effective at all. Only for higher concentrations, the HALS showed effectiveness. The difference between the different *S*-containing stabilizers is ascribed to their difference in mobility.

In the stabilizing mechanism of HALS the formation of the nitroxide plays a key role. The antagonistic effect between HALS and -*S*-containing compounds was ascribed to a reduced nitroxide formation rate due to decomposition of present hydroperoxides by the -*S*-containing compound [99] or to the formation of a salt between an acidic decomposition product of the -*S*-containing compound and the basic HALS that is not able to form the stabilizing nitroxide [100]. However, Kikkawa and Nakahara [101] showed that the nitroxide itself and sulfur-containing stabilizer showed an antagonism too. According to Luci et al. [102] the stabilizing reaction

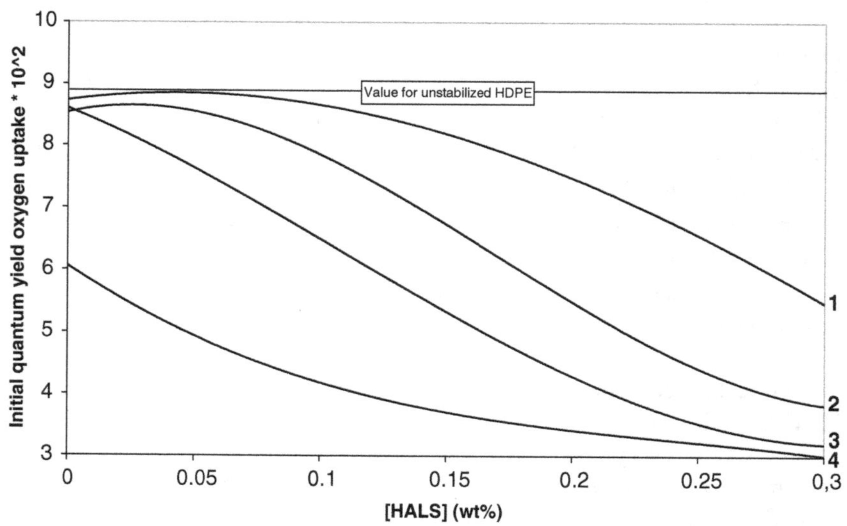

FIGURE 17.2 The initial quantum yield for oxygen uptake of HDPE as a function of the amount of LMW-HALS-1 for formulation with 0.15%wt/wt BFS-1 (1), S-1 (2), BFS-2 (3), and without a sulfur containing stabilizer (4) [98].

product is the nitroxide that is deactivated by a reaction with radicals formed on the -*S*-containing stabilizer. However, if a less basic HALS derivative is used the deactivation by a sulfur containing stabilizer (S-2) can be overcome, these types of HALS add stability to formulations containing S-2 [103].

17.5.5.3 Interactions Between HALS and Acids Stabilized polymers can be exposed to acids from pollution or from degradation products of halogen containing compounds. Due to their chemical structure HALS can be basic (see Table 17.2) and form ammonium salts with acids.

Carlsson et al. [104] showed for salts of HCl, HBr, and HNO_3 a large negative effect while sulfurous acid was somewhat less detrimental. The strongest carboxylic acid that can be formed during the oxidation of polymers (formic acid) had no effect at all. HCl treatment of with a NH or a $NCOCH_3$ HALS derivative stabilized PPs reduces the nitroxide formation rate [105]. This result was used to explain the detrimental effect. Besides this effect, it was shown that the nitroxide itself could form a salt too, causing that less basic HALSs will be deactivated after their conversion to the nitroxide too. The slight difference between secondary and tertiary amines is ascribed to the fact that the tertiary HALS types are first oxidized to the secondary (Scheme 17.20) to become effective [106].

17.5.5.4 Interactions Between HALS and Halogenated Flame-Retardants For polyolefins, brominated flame-retardants are still the most used. They may undergo decomposition leading to bromine radicals that in the presence of hydrocarbon polymer can lead to hydrobromic acid. The way the bromine radicals are formed depends on the structure of the flame-retardant. Aliphatic brominated flame-retardants are primarily decomposed thermally, mainly during processing. The more thermally stable aromatic brominated flame-retardant generates bromine radicals during weathering as a result of their UV absorption [108].

Since the levels of flame-retardants used in practice (up to 30%) exceeds many times those of HALS (up to 1%), even a small conversion of the flame-retardant to HBr can cause complete transformation of HALS to its ammonium salt [109]. This leads to total deactivation of HALS and ruins the HALS stabilizing performance [110].

Although, for the photolytically degrading halogenated flame-retardants combinations of absorbers and HALS might give sufficient stability [108], in combination with a flame-retardant several other UV stabilizer types can outperform HALS. The development of less basic NOR HALS types can change this situation. These types are less deactivated, causing that in combination with flame-retardants they can outperform all other stabilizers [103,108].

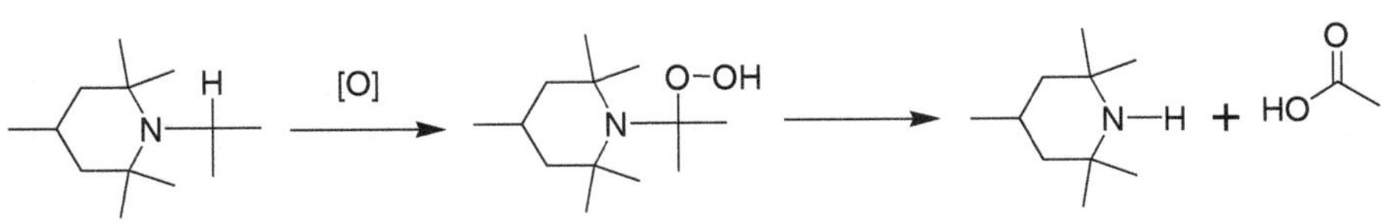

SCHEME 17.20 Formation of secondary from tertiary HALS.

TABLE 17.2 pK_a Values of Hindered Amine Stabilizers in Relation to Their Structure

Chemical Structure	pK_a	Reference(s)
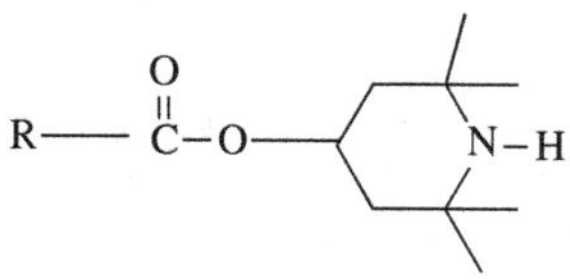	9.0	[107]
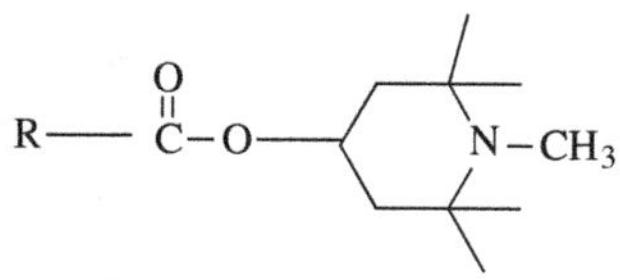	8.9	[107]
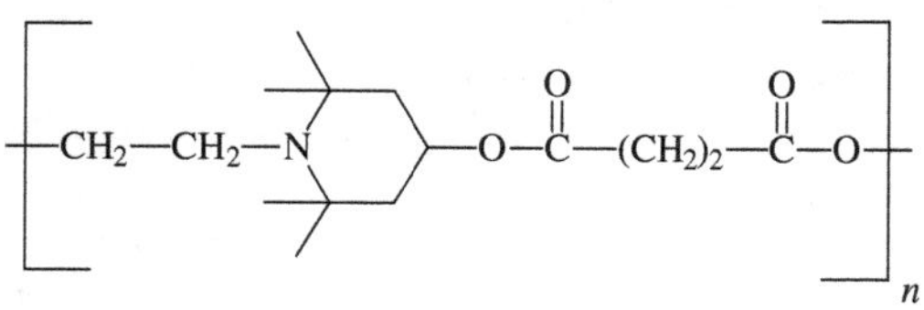	6.5	[103],[107]
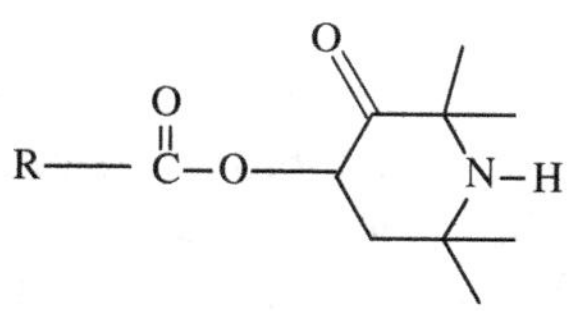	6.8	[107]
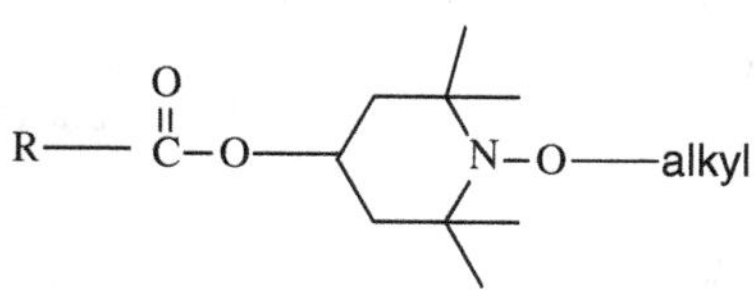	4.4	[107]
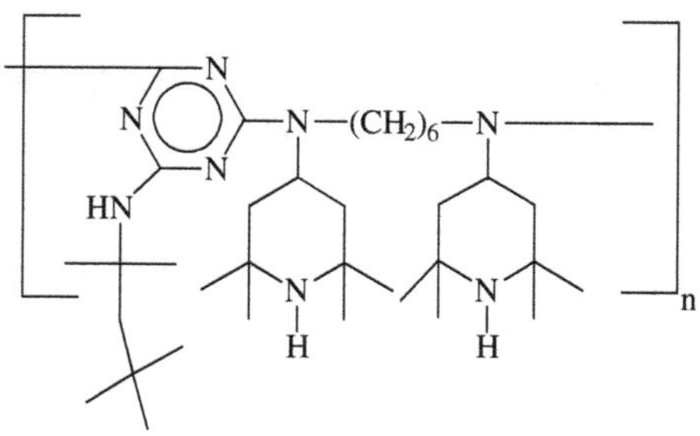	9.7	[103]

TABLE 17.2 (*Continued*)

Chemical Structure	pK_a	Reference(s)
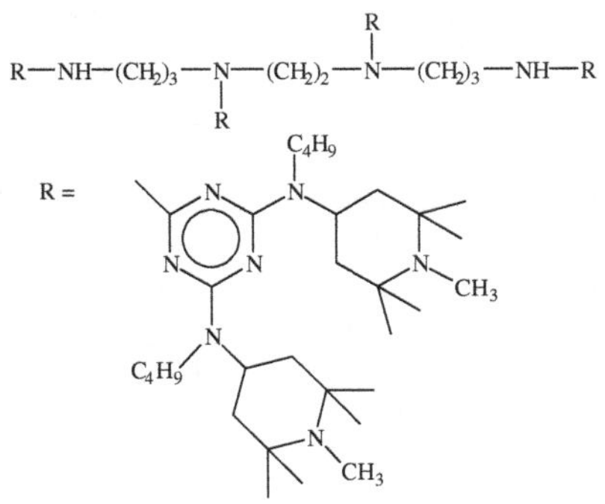	9.2	[103]
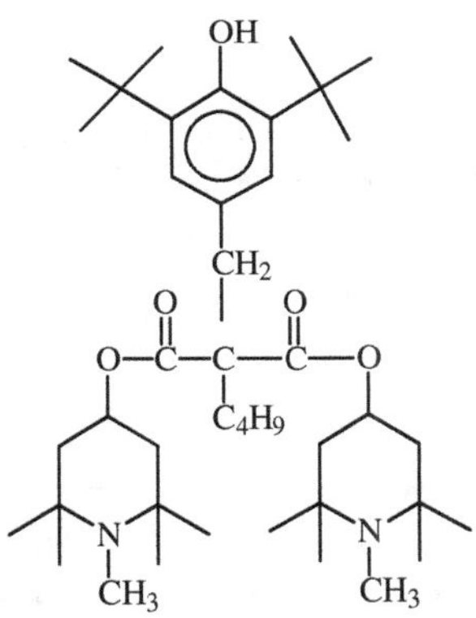	8.5	[103]

17.5.5.5 Interactions Between HALS and External Chemicals Besides reactions with on purpose in a polymer put additives, HALS can also interact with substances from the environment. UV stabilized polyethylene finds a large application in greenhouses, the interaction between agrochemicals and HALS is well investigated. As shown above sulfur and halogen containing additives have a negative effect on the UV stability of HALS stabilized polymers. Khan et al. [111] showed the same effect for sulfur and halogen-based insecticides and pesticides, by comparing the UV stability of HALS stabilized PE in a model greenhouse with weathering on a testing rack. Epacher and Pukanszky [112] evaluated the effect of different types of pesticides on the UV stability of HALS containing PE. They showed that the pesticides containing sulfur as active component are most harmful. Pauquet [113] stated that acidic species from pesticides and insecticides can interact with HALS and that this interaction depends on the pKa value of the HALS. He also suggested that chlorinated agrochemicals can deactivate HALS in two ways, first by abstraction of halogen radicals that are formed by decomposition of the in the

TABLE 17.3 Influence of *N*-Substitution on the UV Stability of Compression Moulded PP Films (0.1 mm) in a Xenontest 1200 [115])

	Time (h) to Carbonyl Absorbance = 0.1 HALS Concentration (mmol/kg^{-1})			
HALS	1.04	2.08	3.12	6.25
LMW-HALS-1 (NH)	1180	1840	2150	4000
Corresponding *N*-methyl derivative	1130	1830	2520	3080
Corresponding *N*-butyl derivative	980	1480	1340	2840
Corresponding *N*-allyl derivative	1160	1690	2100	3100
Corresponding *N*-benzyl derivative	1160	1650	2440	3700
Corresponding *N*-oxyl derivative	1100	1650	2280	3600

agrochemical present carbon halogen bond en secondly by salt formation with the formed HCl.

The susceptibility of LDPE with HMW-HALS-2 to chemical hydrolysis, photo-oxidation and composting was studied by Haider and Karlsson [114]. In this case, it was shown that nonconsumed stabilizer remained in the polymeric matrix after aging in compost, while a large loss of stabilizer was found in water at pH 5.

17.5.6 Alternatives for N–H

As shown above, one of the drawbacks of HALS is that they can be basic. Substitution on the *N* of the HALS has an influence on the basicity (see Table 17.2). However, this substitution might also have an influence on its activity as stabilizer. Several studies on the influence of *N*-substitution of HALS on the UV stability of PP were done. Gugumus [115] showed that N substitution of LMW-HALS-1 only had a marginal effect on the UV stability of PP (see Table 17.3).

In other studies the influence of N substitution on the effectivity of 4-benzoyloxy-2,2,6,6-tetramethylpiperidine was studied [116–118]. In this case the effectivity decreased in the order O > O-butyl > butyl > hydrogen > acetyl. It was shown that the tertiary hindered amines were oxidized and converted to the parent secondary amine.

One of the major new developments is the substitution of the NH HALS derivatives by their much less basic *N*-oxyl (NOR) analog (for example see Table 17.3). The relative effectivity between the NOR and the NH analogs depend on the application. Only in circumstances in which the NH derivative is deactivated the use of the NOR derivative can be justified.

17.6 INFLUENCE MOLECULAR WEIGHT OF THE STABILIZERS

In service life the stabilizer concentration is not only reduced by chemical consumption but also reduced by physical ways [119]. Especially for thin articles evaporation

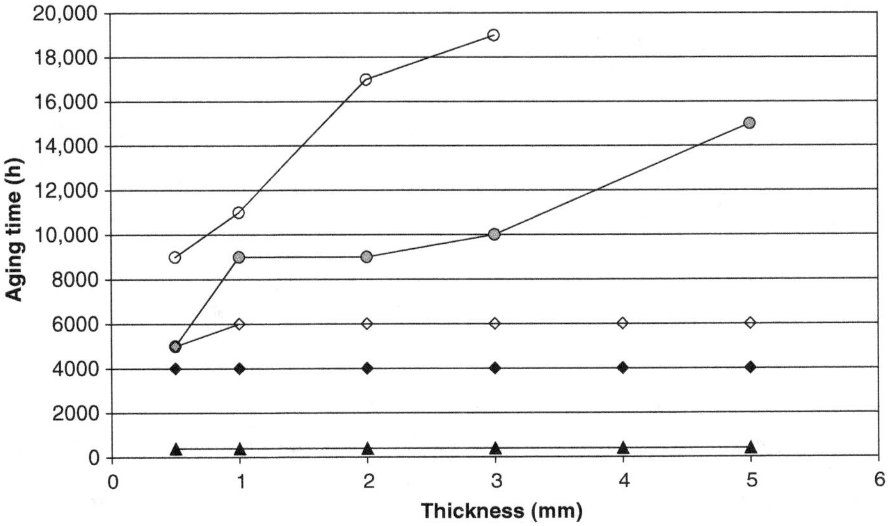

FIGURE 17.3 Influence of thickness on the stability (time to chalking) of medium impact dark gray-pigmented PP copolymer stabilized with a low molecular weight HALS (LMW-HALS-1) and a high molecular weight HALS (HMW-HALS-2) (Unstabilized (▲); 0.1% LMW-HALS(●); 0.2% LMW-HALS (○); 0.1% HMW-HALS (◆); 0.2% HMW-HALS (◇).

of the stabilizer can be the mechanism causing service life failure. For this reason, for thin articles generally relatively high molecular weight stabilizers perform better than their lower molecular weight analogs. For thicker articles this can be different, in this case the low molecular weight stabilizers performs in many cases better than their high molecular weight versions [120]. In Fig. 17.3 it is shown that for a relatively high molecular weight HALS (HMW-HALS), the influence of thickness on stability is limited, while for a low molecular weight stabilizer the stability increases with thickness. In this case the difference in effectivity between the low and high molecular weight HALS can be attributed to their difference in diffusion rate. For relatively thick articles, the UV degradation mainly takes place at the surface, while in the bulk nothing happens. Due to their relatively high diffusion coefficient LMW-HALS can (during weathering) diffuse to the surface and replace consumed or evaporated HALS, causing that in the case of thick articles, the bulk of the material acts as a reservoir of stabilizers that helps to protect the surface. This is not the case for the HMW-HALS their diffusion rate to the surface is to low to help to protect the surface. As a result, there is for LMW-HALS an influence of thickness on the stability, which is not the case for high molecular weight types

17.7 SYNERGISM BETWEEN UV STABILIZERS

To obtain an optimal stabilizing effect in many cases combinations of stabilizers are applied. Combination of stabilizers leads to an additive, an enhanced, or a reduced

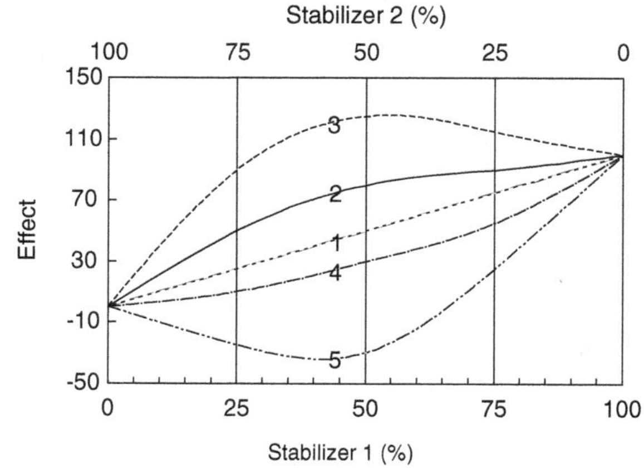

FIGURE 17.4 Definition of synergism and antagonism between two stabilizers.

effect. If the effect of the combination of two additives is larger than the sum of their separate effects when they are used alone, this is defined as synergism. If the effect is comparable to that of their sum, an additive effect is found and when the effect of the combination is less than the sum of the separate additives an antagonism is found. In many cases, these definitions are translated to a graphical presentation as shown in Fig. 17.4 [121,122]. In this figure, a synergism is defined when combination of two different stabilizers is more than additive (Fig. 17.4: lines 2 and 3) and when the effect is less than additive this is called antagonism (Fig. 17.4: lines 4 and 5). However, in this definition it is implicitly assumed that the relation between stability and concentration is linear. However, in many instances this is not true. According to Gugumus [6], the stability of PE and PP depend on the square root of the concentration of HALS. This cause that the additive effect of two types of HALS is always higher than the linear line in Fig. 17.4.

17.7.1 Synergisms with UVAs

To be active as a UV stabilizer, UVAs must strongly absorb, for polymers, harmful UV light and to be able to transfer the absorbed light in harmless energy. However, it was observed that photo-oxidation can occur by the fluorescence, which is emitted by a photoexcited UV-B absorber. In this case, it can be beneficial to add a second absorber that absorbs these wavelengths, which can lead to a synergism between two types of UVAs. Such a synergism was observed between hydrogen transfer-type UVAs and charge separation-type UVAs [123].

The interaction between UVAs and HALS is one of the most investigated interactions. In some cases, a synergism between these stabilizers is reported [16,122,124–127], although antagonisms are reported too [16,122,127].

These synergisms are ascribed to the difference in mechanism of action of UVAs and HALS; however, this by itself is not enough to explain the synergism. Kurumada et al. [124] investigated the synergism of a HALS (LMW-HALS-1) and two UVAs (UVA-BZT-1 and UVA-BZT-6) in PP, HDPE, ABS, and PS. A large synergism was

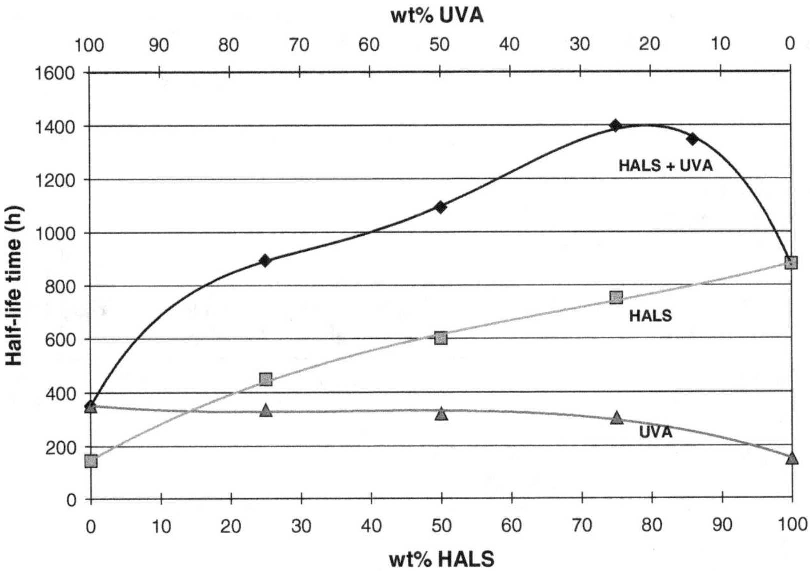

FIGURE 17.5 Half-life time (time to 50% retained elongation) of different HDPE plaques (2 mm thick) showing synergism of HALS (LMW-HALS-1) and UVA (UVA-BZT-1) (total amount UVA + HALS = 0.2 wt%) [124].

observed in polypropylene, high-density polyethylene (Fig. 17.5) and ABS resin. A moderate synergistic effect was obtained in polystyrene. They explained the synergism by the diffusion of very effective hindered amine light stabilizer from the polymer bulk, which is protected by the UVA, toward the surface layer where photooxidation proceeds.

The difference in interaction between HALS with UVA containing phenolic moieties and phenolic antioxidants is investigated by Mizakowa and Ohkatsu [125,126]. According to them Quinones, derived from UVA or phenolic antioxidants, were reduced to the corresponding hydroquinones by the action of HALS derivatives, strikingly faster and more easily for the UVA than for the phenolic antioxidants, which leads to a strong synergism of UVA and a weak synergism for phenolic antioxidants

In HDPE combination of HALS with UVA-HTZ-1 (at an overall level of 0.1% w/w) exposed in a Microscal Unit at about 50°C led to weak antagonism for LMW-HALS-3, LMW-HALS-4, and LMW-HALS-9. However, for LM-HALS-26 and HMW-HALS-4 a slight synergism was observed [127].

17.7.2 Synergisms between Different Types of HALS

17.7.2.1 Low and High Molecular Weight HALS It was shown that combinations of low and high molecular weight HALS led to an unexpected synergism in PP tapes (Table 17.4), although the effect of the combination on the UV stability is not

TABLE 17.4 Performance of Combinations of Low/High molecular Weight HALS in PP Tapes During Outdoor Weathering in Florida [122]

Light Stabilization	Energy to 50% Retained Tensile Strength (kJ/cm^2)
0.2% LMW-HALS-1	1925
0.2% HMW-HALS-2	795
0.1% LMW-HALS-1 + 0.1% HMW-HALS-2	1550

TABLE 17.5 Performance of Combinations of Low/High Molecular Weight HALS in 4 mm Thick Injection Moulded Gray-Pigmented PP/EPDM in Florida [122]

Light Stabilization	Time to Beginning of Chalking (Months)
0.45% LMW-HALS-1	18
0.2% LMW-HALS-1 + 0.2% HMW-HALS-1	24–30
0.2% LMW-HALS-1 + 0.2% HMW-HALS-2	36

higher than that of the LMW-HALS alone [6]. For injection moulded PP homo and copolymers, comparable results were found when the decrease of the mechanical properties is taken as aging criterion. However, for a PP/EPDM blend it was shown that the combination of a LMW-HALS and a HMW-HALS could perform better than the LMW-HALS alone (Table 17.5). For aging in Florida as well as in an accelerated test the same was found for PP tapes (Fig. 17.6). The synergistic effect was attributed

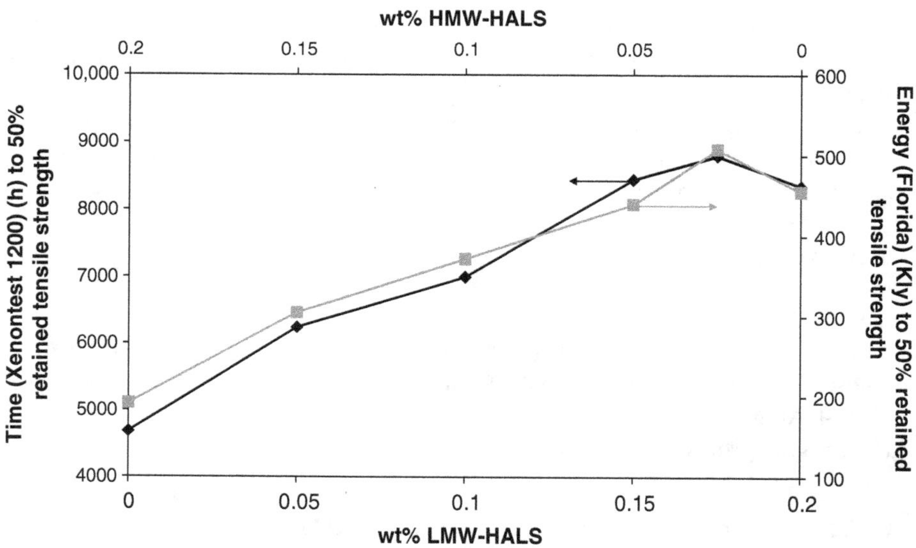

FIGURE 17.6 Performance of a combination of a low molecular weight HALS (LMW-HALS-1) and a high molecular weight HALS (HMW-HALS-2) in an accelerated weathering test (Xenon test 1200, BPT 53°C, no water spraying) and in Florida [122].

to the complimentary effects resulting from rapid diffusion of the LMW-HALS to the surface layers where the need for stabilizers is highest with the resistance to migration and extraction of HMW-HALS [122].

17.7.2.2 ***Two High Molecular Weight HALSs*** The synergistic effect of a LMW-HALS and a HMW-HALS can be ascribed to the difference in diffusion and evaporation rate. However, this is not the case for the combination of two HMW-HALSs, nevertheless synergisms were shown, which must be a result of very specific interactions or specific protection mechanisms of the components [122].

17.8 UV DEGRADATION AND EFFECTIVITY OF UV STABILIZERS FOR SELECTED POLYMERS

17.8.1 Polyethylene

17.8.1.1 ***Influence Aging Criterion*** Weathering leads to changes in the chemical composition of polymers. This change can lead to a decrease of the mechanical properties. In Figure 17.7 the development of the elongation at break, the dynstat impact energy, and the tensile impact energy of a white-pigmented UV stabilized HDPE during weathering in Florida is shown. It is shown that the polymer lifetime depends on the criterion used. It is shown that during exposure the drop of the

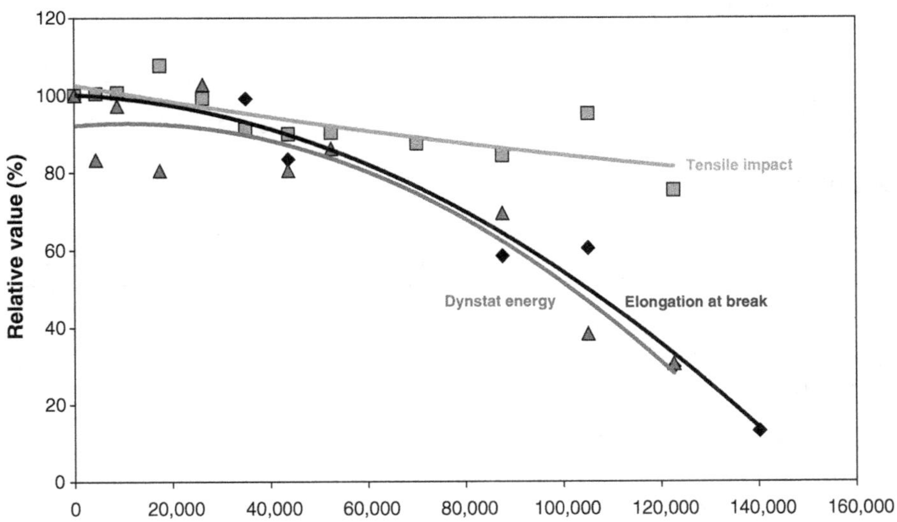

FIGURE 17.7 Development of the dynstat, elongation at break and tensile impact as a function of exposure time in Florida for a white-pigmented HDPE containing 750 ppm LMW-HALS-1.

elongation at break and the dynstat energy is comparable, while the drop in the tensile impact is slower.

17.8.1.2 Comparison Different UV Stabilizers The addition of UV stabilizers leads to a tremendous increase of the stability of LLDPE during outdoor weathering (Florida). The addition of a combination of a Nickel Quencher (Ni-1) and a UVA (UVA-BP-1) lead to a six times larger lifetime. The additions of a HMW-HALS (HMW-HALS-2) leads even to a larger (>10 times) increase in the lifetime (Fig. 17.8). In another comparison between different UV stabilizers (see Table 17.6) it is also shown that HALS is by far the best performing UV stabilizer. In this table it is also shown that UVA-BP-1 is the most effective UVA.

17.8.2 Polypropylene (Co) polymers

17.8.2.1 Appearance of Degradation As for PE the lifetime of PP depends on the criterion used. In the case of PP during weathering whitening (chalking) as well as a drop in the mechanical properties can be observed; which one is limiting depends on the toughness of the polymer. For homopolymers, the decrease in mechanical properties is limiting, while for copolymers it is chalking. For a homopolymer, a medium and a high impact copolymer the time to the appearance of chalking and the time until 50% of the original dynstat impact (mechanical property) was determined

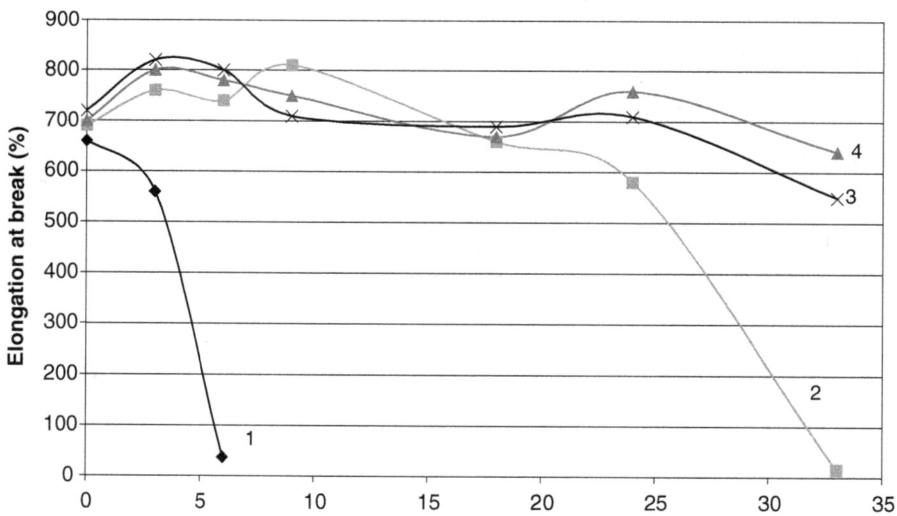

FIGURE 17.8 Drop in elongation at break during exposure in Florida of unstabilized LLDPE (1) and LLDPE containing 1% Ni-1 + 0.5% UVA-BP-1 (2), 0.5% HMW-HALS-2 (3) or 1.0% HMW-HALS-2 (4) [128].

TABLE 17.6 Effectivity of Different Classes of Stabilizers on the UV Stability of PP

Light Stabilizers	Energy (kLy) Received to 50% Retained Tensile Strength for LDPE Blown Films (200 μm) in Florida[a]	Time (h) Until Carbonyl Absorbance = 0.2. For Compression Moulded LDPE Plaques in a Weather–Ometer (Without Rain Cycle)[a]	Time (h) Until Carbonyl Absorbance = 0.1. For Compression Moulded LLDPE Plaques (0.2 mm) in a Weather–Ometer (BPT 63°C, Without Rain Cycle)[b]
0.2% UVA-BP-1			2,260
0.3% UVA-BP-1	75	950	
0.2% UVA-BZT-1			1,350
0.15% UVA-BZT-3	40		
0.15% UVA-AN-1		765	
0.15% UVA-BZ-1		720	
0.15%Ni-1		1,380	
0.15% HMW-HALS-1		10,500	8,580
0.15% HMW-HALS-2			10,510
0.15% UVA-BP-1 + 0.15% HMW-HALS-2	385		
0.15% UVA-BZT-3 + 0.15% HMW-HALS-2	365		

[a]From Ref. [34].
[b]From Ref. [129].

in an accelerated weathering test (Table 17.7). The different polymers contain different amounts of UV stabilizer, causing that their lifetimes cannot be compared. For the homopolymer, a decrease in the mechanical properties is observed far before chalking can be observed. For the medium impact copolymer the same is observed,

TABLE 17.7 Influence Criterion on the Lifetime for a Homopolymer, a Medium and a High Impact PP Copolymer

Polymer	UV Stabilization	HLT Dynstat (h)	Time to Visual Appearance of Chalking (h)
Homopolymer	0.3% LMW-HALS-1	2,300	>3000
Medium Impact Copolymer	0.08% UVA-BZT-2	550	1600
High Impact Copolymer	0.08% UVA-BZT-2	5,000	1500
High Impact Copolymer	0.1% UVA-BZT-2 + 0.1% LMW-HALS-1	8,000–10,000	2800–3100

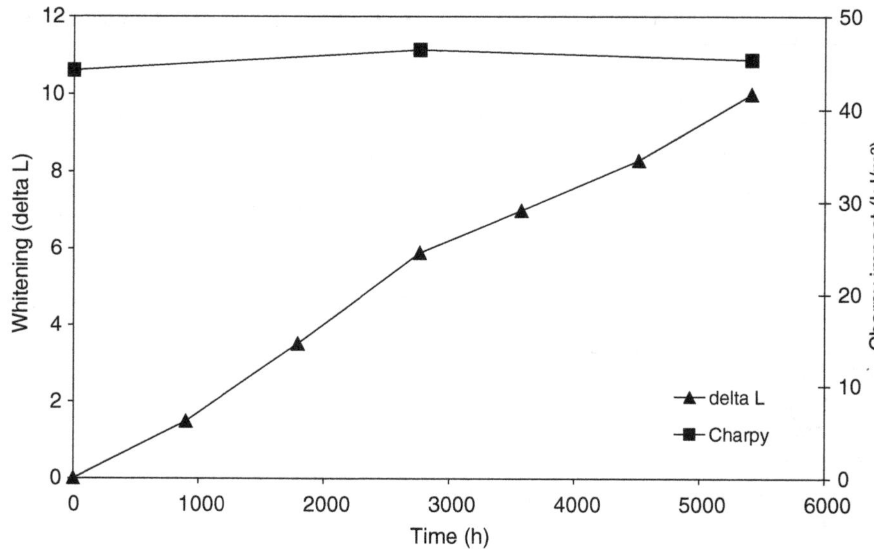

FIGURE 17.9 Whitening (delta L) and change of mechanical properties of a copolymer during artificial weathering.

however, independent on the stabilization system the high impact copolymers show chalking long before the mechanical properties decrease.

A comparison between the drop in mechanical properties (Charpy Impact) and the appearance of chalking or whitening during accelerated weathering of a high impact copolymer is shown in Fig. 17.9. From this figure, it is clear that the high impact copolymer first shows whitening before it looses its mechanical properties.

17.8.2.2 Comparison Different UV Stabilizers To increase the stability of PP different UVAs can be applied. From Table 17.8 it is clear that the HALS types are most effective and that the benzoates are more effective than the UVAs. Above it already was shown that especially in thicker applications LMW-HALS is more effective than HMW-HALS (Fig. 17.3).

A comparison between different LMW-HALS in a dark gray colored high impact polypropylene copolymer is shown in Table 17.9. In this case the HALSs with the lowest molecular weight performed best.

The influence of the concentration of different types of HMW-HALS on the UV stability of medium impact PP copolymer films was determined in an accelerated test (Fig. 17.10). In all cases an increase of the concentration of HALS led to an increase of the UV stability. This increase is up to 0.3% HALS about linear. The performance of HMW-HALS-5, HMW-HALS-2, and HMW-HALS-4 is about equal. HMW-HALS-3 performs a little bit less, especially at higher concentrations. In this case it is quite clear that HMW-HALS-1 is not as good as the other evaluated HALSs.

TABLE 17.8 Effectivity of different classes of stabilizers on the UV-stability of PP

Light Stabilizers	Energy (KLy) Received to 50% Retained Tensile Strength for PP Tapes in Florida[a]	Time (h) to 50% Retained Tensile Strength for PP Tapes in a Xenontest 1200 (BPT 50°C, Without Rain Cycle)[b]	Time (h) until Carbonyl Absorbance = 0.2. for PP Plaques in a Weather–Ometer (BPT 63°C, Without Rain Cycle)[c]
0.1% UVA-BP-1			1035
0.5% UVA-BP-1	70		
0.1% UVA-BZT-1			540
0.5% UVA-BZT-2	65	1070	
0.5% UVA-BZ-2	140	2500	
0.5% Ni-1	65	1690	
0.1% LMW-HALS-1	320	3500	2880
0.1% HMW-HALS-2	205		

[a]From Ref [16].
[b]From Ref [34].
[c]From Ref [129].

TABLE 17.9 Effectivity of Different LMW-HALSs (0.45%) at the UV Stability of a High Impact PP Copolymer [130]

HALS	Time (h) in a Weather–Ometer (BPT 63°C, Intensity 0.35 W (m^2.nm) at 340 nm 0.35 Rain Cycle 102 min Dry/8 min Rain) to a Gray Scale = 3
LMW-HALS-1	3500
LMW-HALS-5	5550
LMW-HALS-6	5550
LMW-HALS-7	4350
LMW-HALS-11	4350

As mentioned above for PP synergisms between low and high molecular weight HALS as well as between two high molecular weight HALSs were reported.

17.8.3 Polyamides

During photo-oxidation discoloration as well as a decrease of the mechanical properties can be observed, which of these two is most important highly depends on the thickness of the article and its application. To increase the stability of PA66 UVAs as well as HALS can be applied (Fig. 17.11). In the case of PA 6 and 12 the light

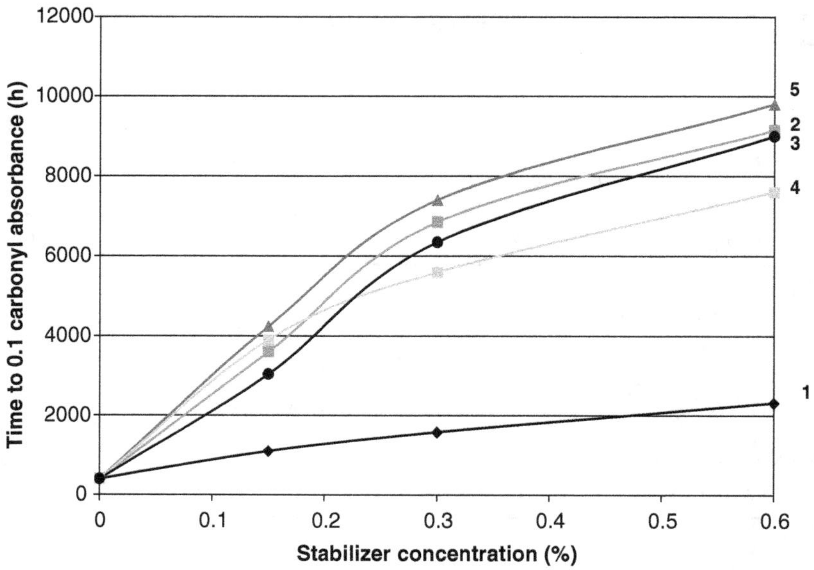

FIGURE 17.10 Influence of the concentration of different types of oligomeric HALS (HMW-HALS-1 (1), HMW-HALS-2 (2), HMW-HALS-4 (3), HMW-HALS-3 (4), HMW-HALS-5 (4)) on the time to reach a carbonyl absorbance of 0.1 of a medium impact PP copolymer.

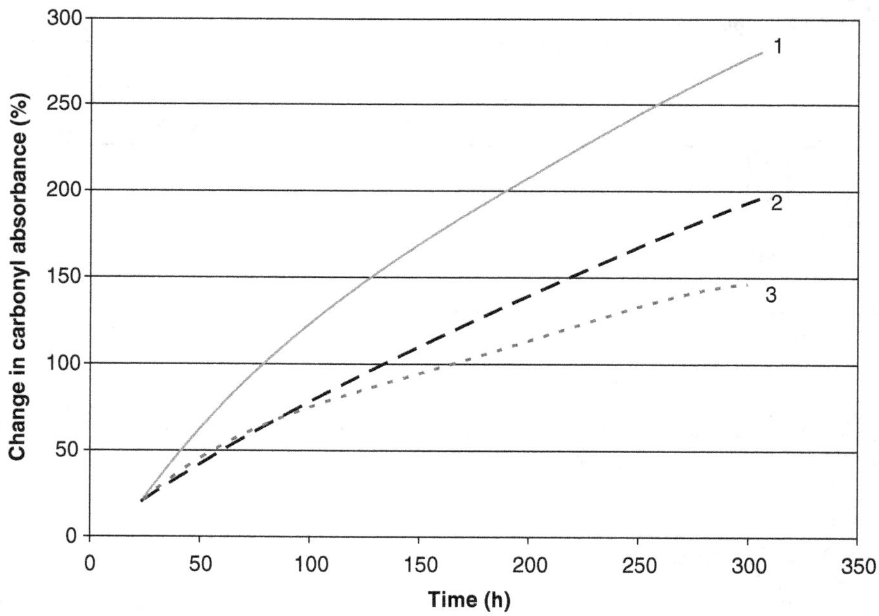

FIGURE 17.11 Photo-oxidation (in a SEPAP 12/24 containing four borosilicate filtered medium-pressure mercury vapor lamps (400 W) at 60°C of unstabilized Nylon 66 (1), with 0.5% UVA-BZT-1 (2), and 0.5% LMW-HALS-31 (3) [131].

TABLE 17.10 Influence of Phenolic Antioxidants and a Phosphites on the Light Stability (Energy or Time Reached to Reduce the Tensile Strength 50% in KJ/cm²) of PA6 and 12 [133,134]

	PA6		PA12
Stabilization	Central Europe (KJ/cm²)	Florida (Behind Glass) (KJ/cm²)	Florida (Direct) (KJ/cm²)
Unstabilized	175	190	315
0.25% AO-1			500
0.5% AO-1	375	545	515
0.25% AO-1/0.25% P-2	500	670	710
0.25% AO-2	525	500	710
0.5% AO-2			920
0.25% AO-2/0.25% P-2	690	630	965

stability can be increased with phenolic antioxidants and phosphites (see Table 17.10). However, the best protection can be reached by applying combinations of a thermo-oxidative stabilizers (such as phenols and phosphites) and UV stabilizers (such as UVAs and HALS). Leaving one of these components leads to a reduction in stability (see Table 17.11).

Although in many cases organic stabilizers are used to protect polyamides against photo-oxidation, in a recent study it was evidenced that a mixture of $CuCl_2$ and KI has a far better long-term stabilizing efficiency than a HALS Stabilizer (HMW-HALS-2) [132].

As for the unfilled PA6, for glass filled black PA6 the addition of just a phenolic antioxidants and a phosphite causes already a reduction of the UV degradation rate during accelerated aging. Addition of HALS leads to a further reduction of the rate of gloss loss during weathering (see Fig. 17.12). In this case, the combination of a low

TABLE 17.11 Influence the Combination of a Phenolic Antioxidant, a Phosphite, a UVA and a HALS on the UV Stability (Time Until a Color Change $\Delta E = 10$ in a Weather–Ometer (Black Standard Temperature: 80°C, Intensity: 0.55 W/m² (at 340 nm), and 18 min Rain/2 h) of Gray-Pigmented PA6)

AO-1	P-2	UVA-BZT-5	HMW-HALS-2	Time Until $\Delta E = 10$ (h)
—	—	—	—	150
			0.175%	150
0.175%			0.175%	150
0.175%	0.175%		0.175%	250
0.175%	0.175%	0.175%	0.175%	500

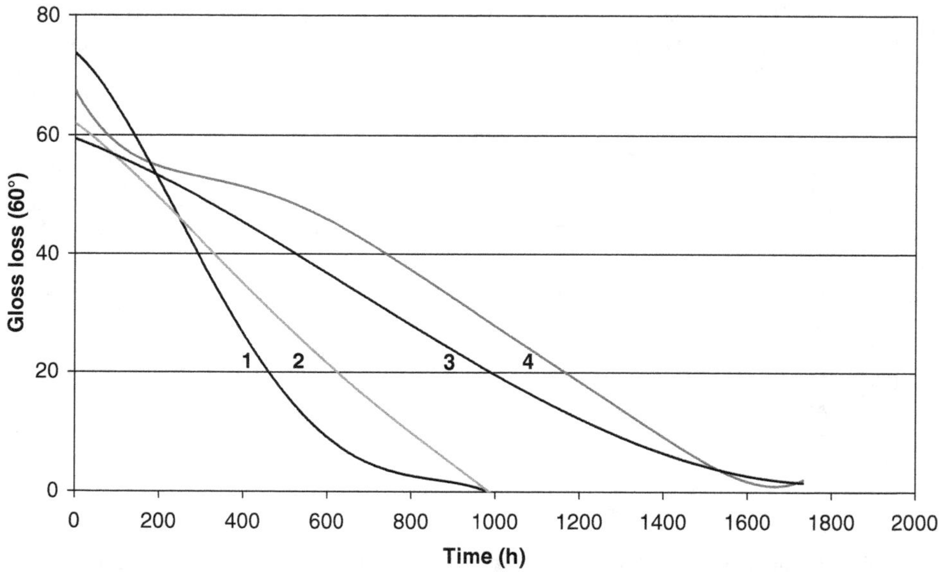

FIGURE 17.12 Gloss retention during exposure under exterior automotive conditions (SAE J1960) of black filled (33% glass) PA6. 1: Unstabilized; 2: 0.125% AO-1, 0.125% P-2; 3: 0.125% AO-1, 0.125% P-2, 0.125% HMW-HALS-2, 0.125% HMW-HALS-1; 4: 0.125% AO-1, 0.125%P-2, 0.125% HMW-HALS-2, 0.125% LMW-HALS-1 [135].

and a high molecular weight HALS performs better than the combination of two high molecular weight HALSs.

17.8.4 Polyesters

It was shown that the degradation of aromatic polyesters (such as polybutylene terephtalate) is mainly due to photolysis [136,137]. As a result for these polymers UVAs are much more effective than HALS (Fig. 17.13).

17.8.5 Polycarbonate

During weathering of polycarbonate (PC), yellowing is observed far before a drop of mechanical properties. It was shown that UVAs are more effective UV stabilizers than HALS [138]. The influence of a hydroxyphenylbenzotriazoles and oxalamide UVA compound on the discoloration rate was determined during outdoor exposure in Haifa [139]. At a concentration of 0.5% the hydroxyphenylbenzotriazoles outperformed the oxalamide; increasing of the concentration did not cause a remarkable effect.

The UV stability of polycarbonate can be increased by applying a UV absorbing coating on top [140].

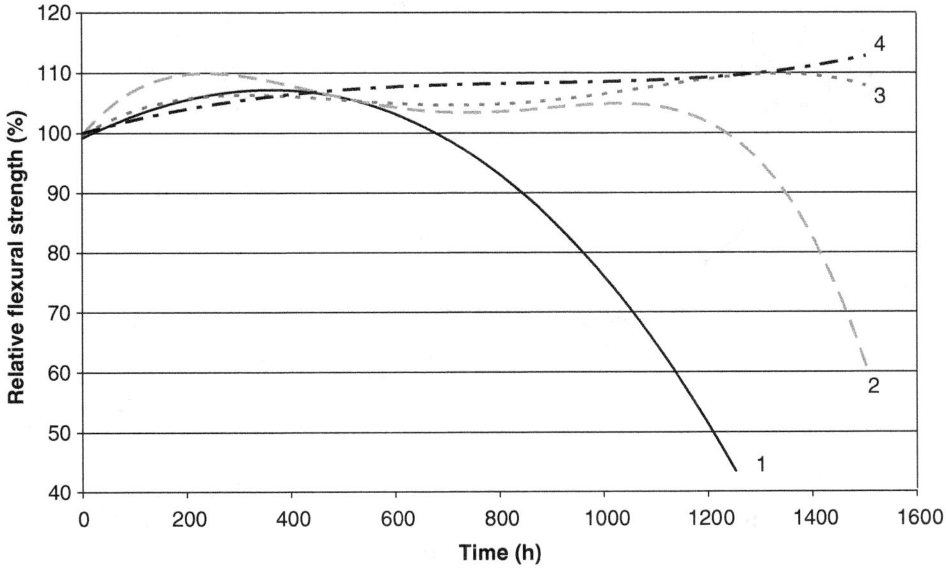

FIGURE 17.13 Relative drop in flexural strength as function of time in a Weather–Ometer (black standard temperature: 80°C, intensity 0.55 W/m^2 (at 340 nm), and 18 min rain/2 h) of Unstabilized PBT and PBT containing 0.5% LMW-HALS-1 (2), 0.5% UVA-BZT-5 (3) or 0.5% UVA-HTZ-2 (4).

17.8.6 Polystyrene

In contrast to polyamides in polystyrene phenolic antioxidants are not able to reduce the decrease in tensile strength during UV irradiation [141]. UVAs as well as HALS stabilizers are effective in polystyrene (PS). The best protection can be reached by using combinations of a UVA and a HALS. According to Gugumus [133,134] the synergism between UVA and HALS is distinct when the time until a $\Delta YI = 5$ is taken as failure criterion (see Table 17.12). However, when as criterion the yellowness index after 1600 h in a Weather–Ometer is taken, the synergism between a UVA and a HALS is much smaller (see Fig. 17.14).

TABLE 17.12 Accelerated Weathering Time (in a Weather–Ometer) Until Delta Yellowness Index = 5 [133,134]

Stabilization	Time until $\Delta YI = 5$ (h)
Unstabilized	700
0.2% UVA-BZT-6	2,700
0.2% LMW-HALS-1	4,500
0.1% UVA-BZT-6 + 0.1% LMW-HALS-1	11,000

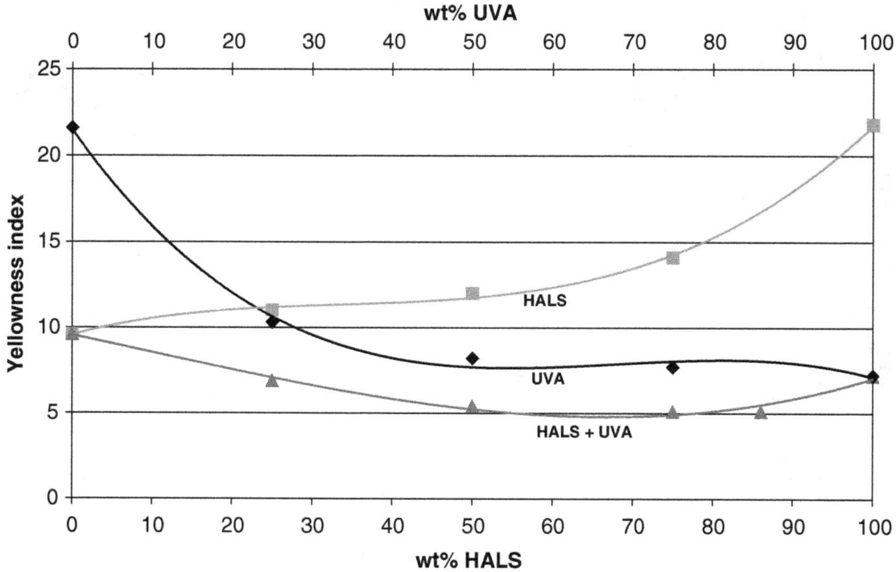

FIGURE 17.14 Yellowness index after 1600 h exposure in a Xenon Weather–Ometer (black panel temperature: 63 ± 3°C, without water spray) of HALS (LMW-HALS-1), UVA (UVA-BZT-6), and combinations (for combination total 0.3 wt%) [124].

17.9 APPENDIX

Abbreviation	Chemical structure
AO-1	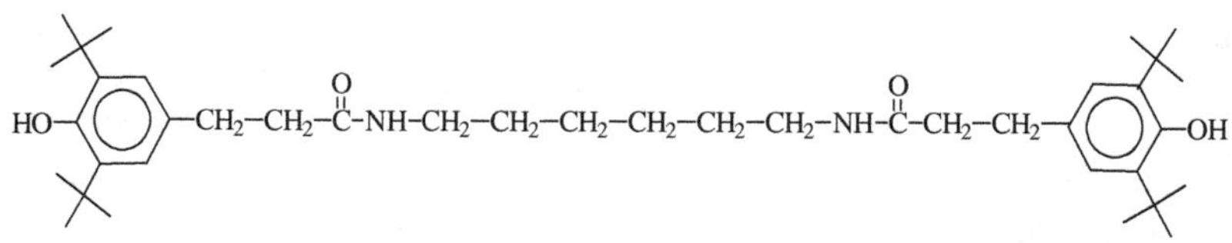
AO-2	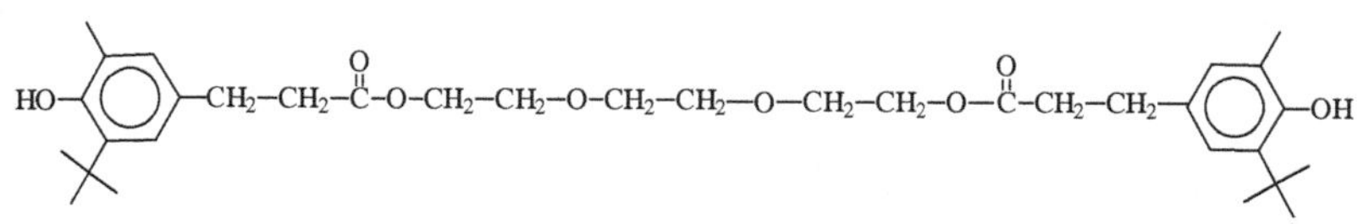
P-1	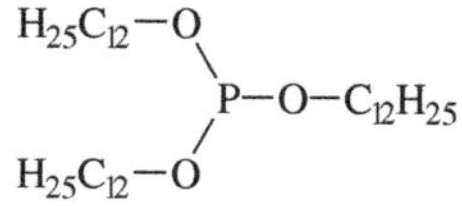

Abbreviation	Chemical structure
P-2	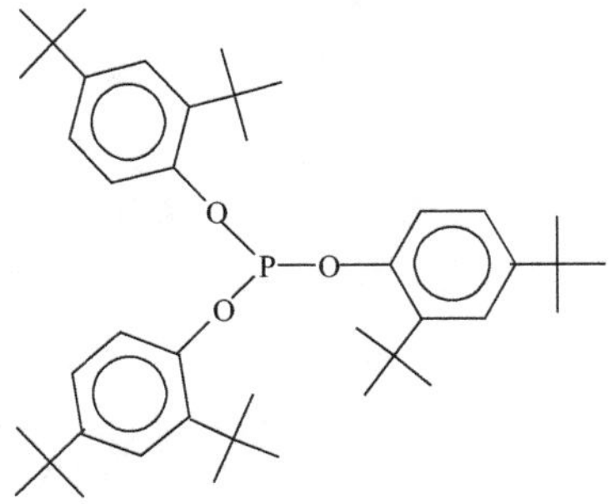
P-3	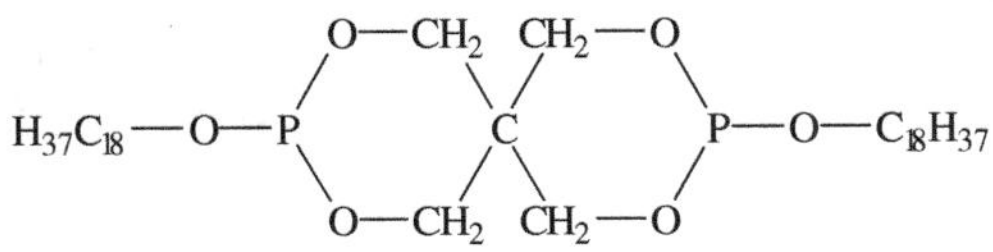
S-1	$H_{25}C_{12}-O-\overset{O}{\overset{\parallel}{C}}-CH_2-CH_2-S-CH_2-CH_2-\overset{O}{\overset{\parallel}{C}}-O-C_{12}H_{25}$
S-2	$H_{37}C_{18}-O-\overset{O}{\overset{\parallel}{C}}-CH_2-CH_2-S-CH_2-CH_2-\overset{O}{\overset{\parallel}{C}}-O-C_{18}H_{37}$
BFS-1	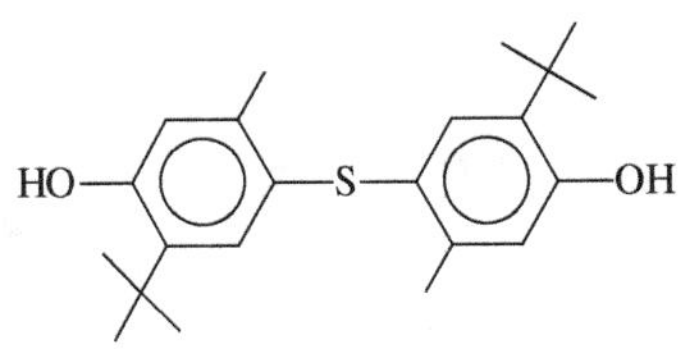
BFS-2	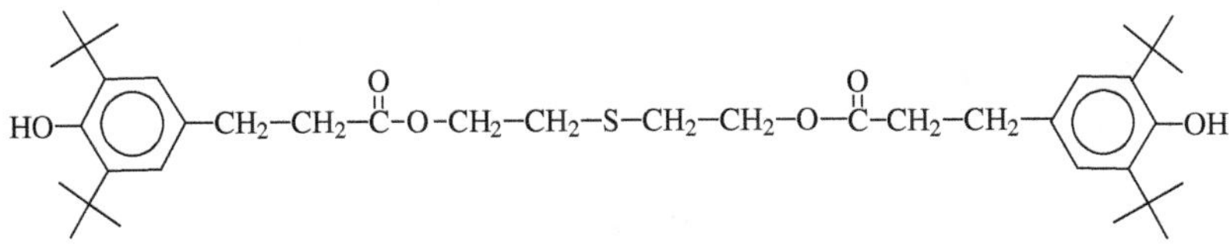
UVA-BP-1	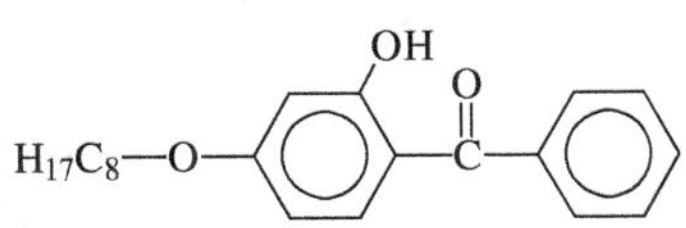
UVA-BZT-1	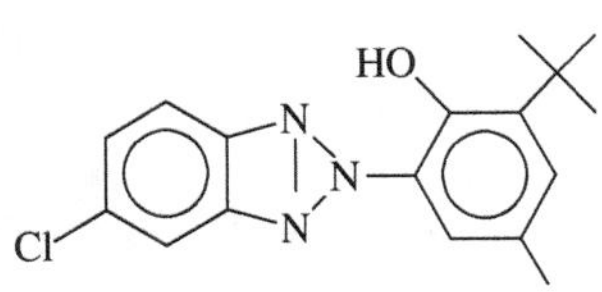

(*continued*)

Abbreviation	Chemical structure
UVA-BZT-2	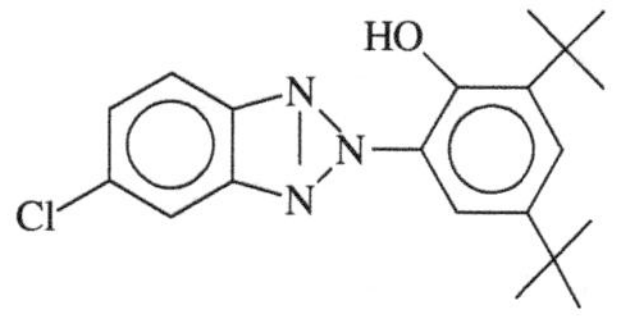
UVA-BZT-3	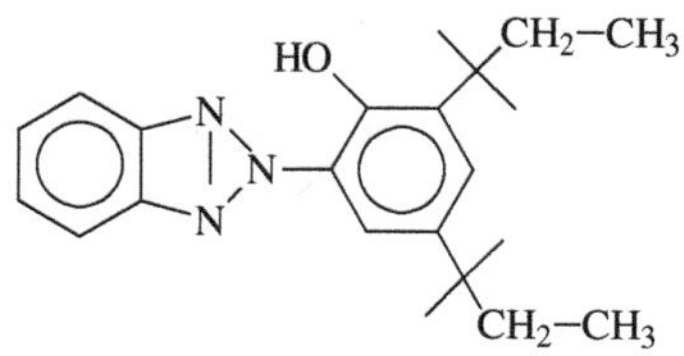
UVA-BZT-4	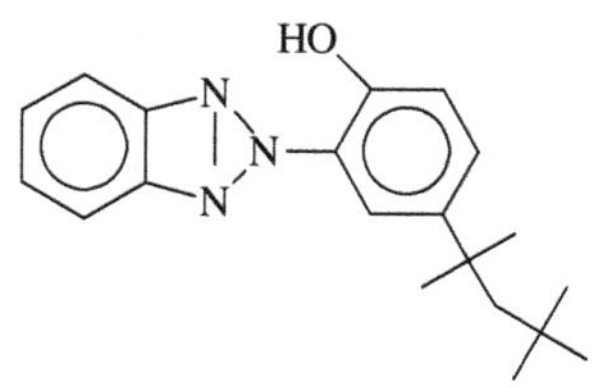
UVA-BZT-5	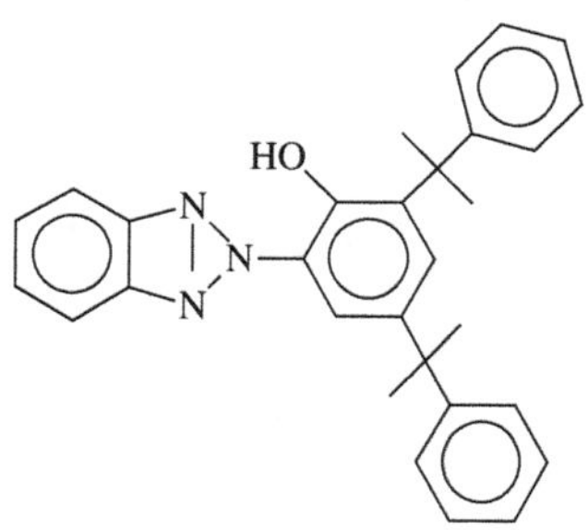
UVA-BZT-6	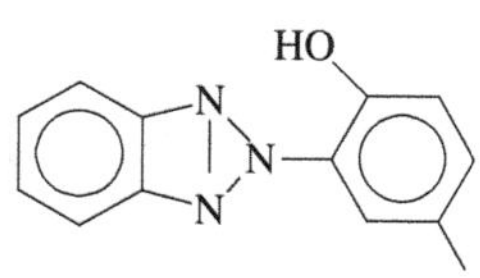
UVA-BZ-1	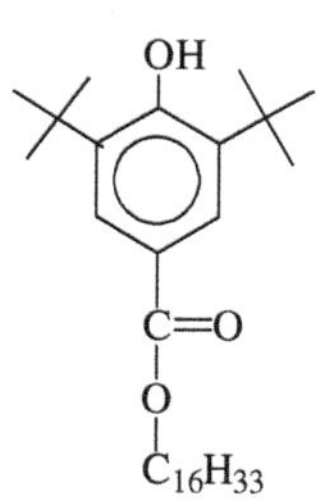

Abbreviation	Chemical structure
UVA-BZ-2	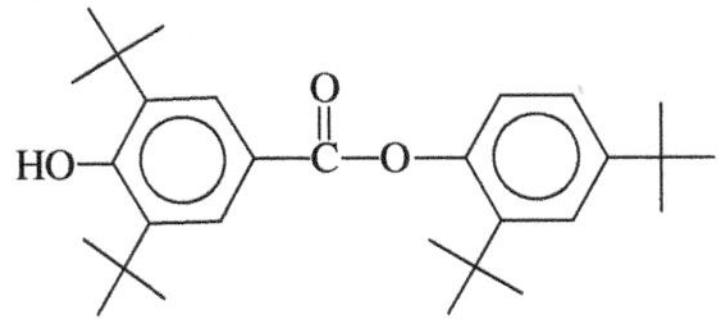
UVA-AN-1	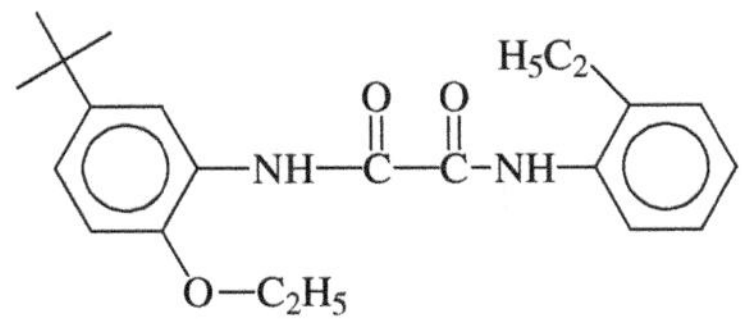
UVA-AN-2	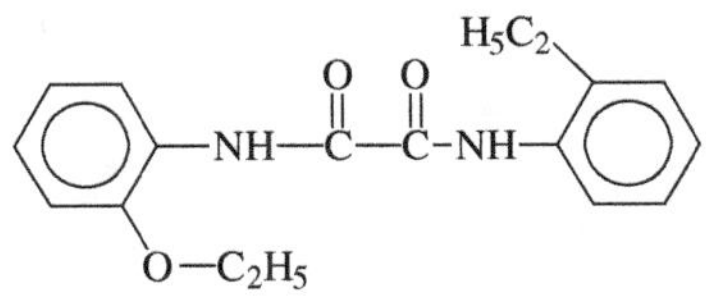
UVA-CA-1	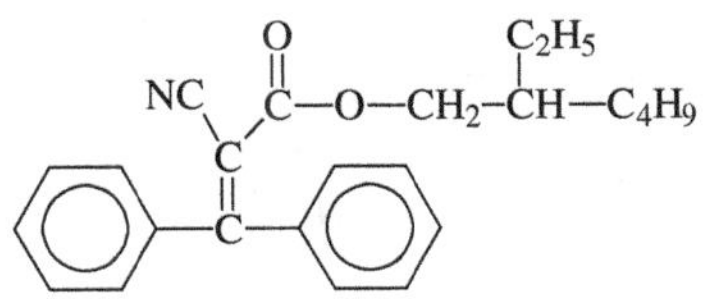
UVA-HTZ-1	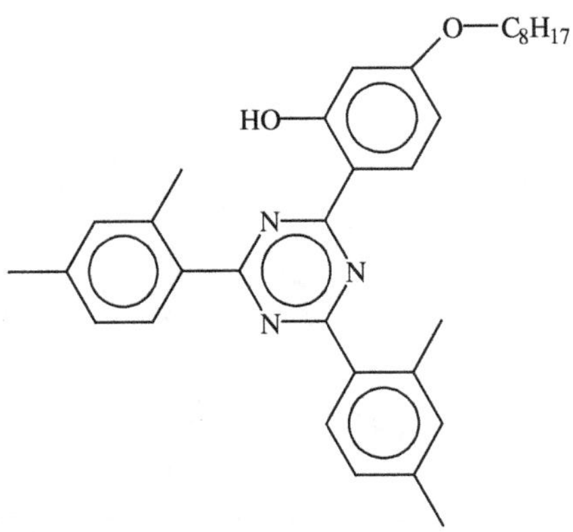
UVA-HTZ-2	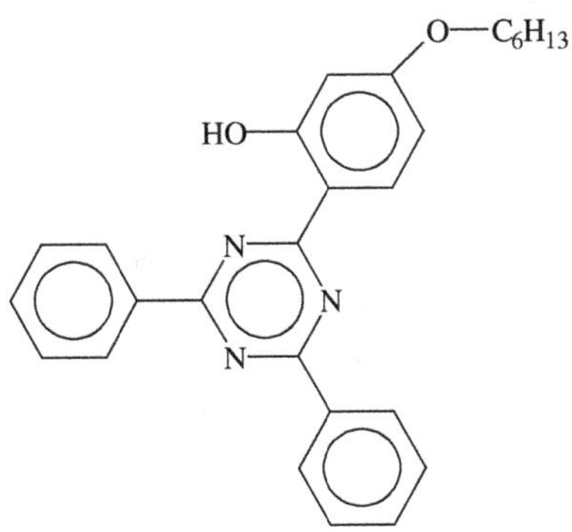

(continued)

Abbreviation	Chemical structure
Ni-1	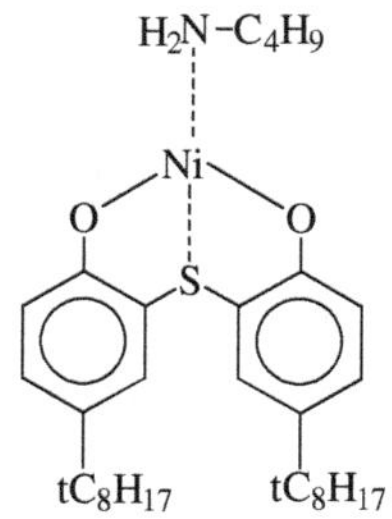
LMW-HALS-1	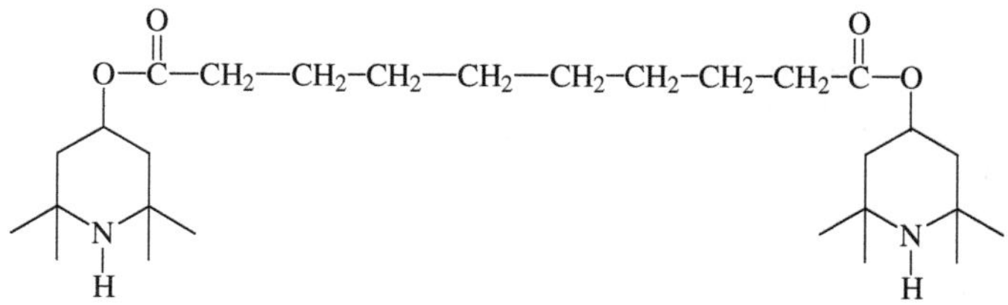
LMW-HALS-2	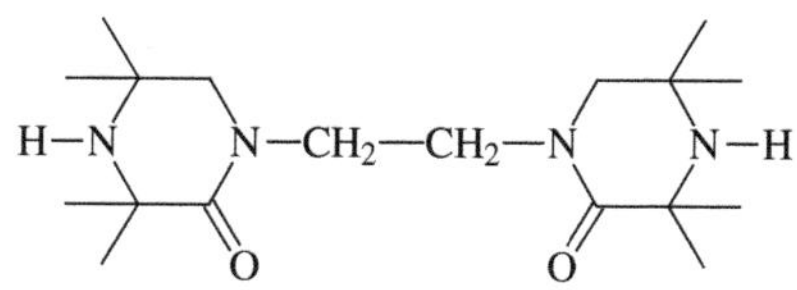
LMW-HALS-3	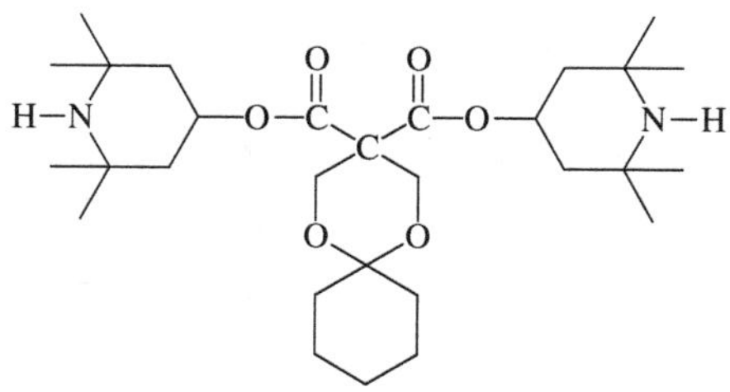
LMW-HALS-4	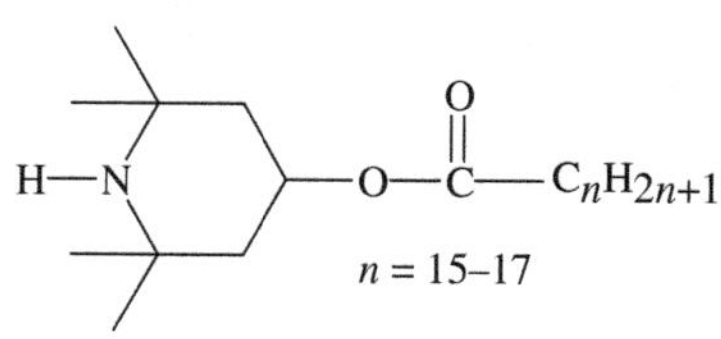
LMW-HALS-5	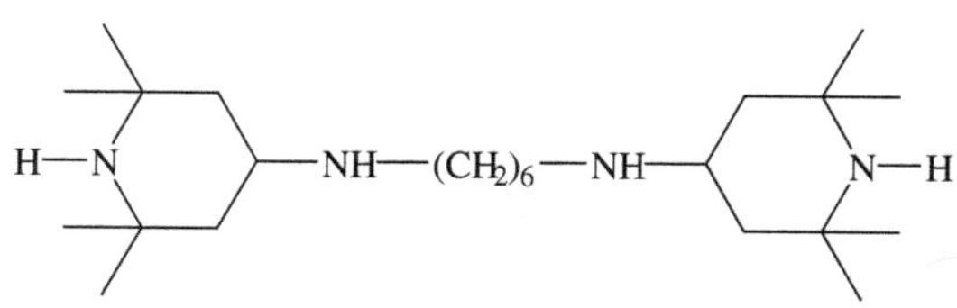

Abbreviation	Chemical structure
LMW-HALS-6	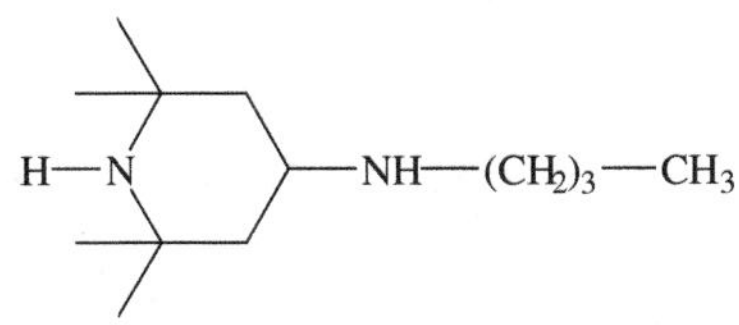
LMW-HALS-7	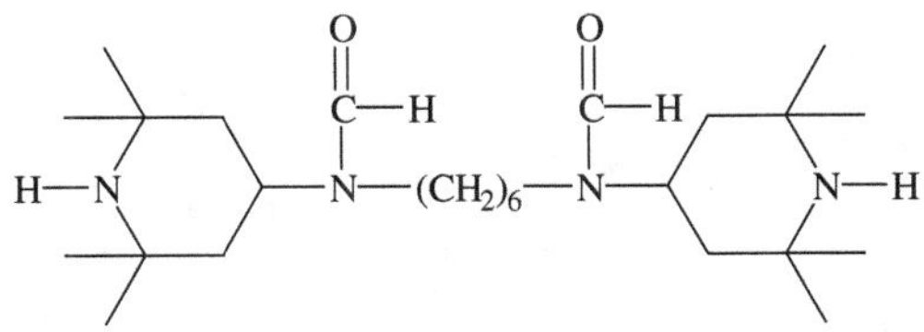
LMW-HALS-8	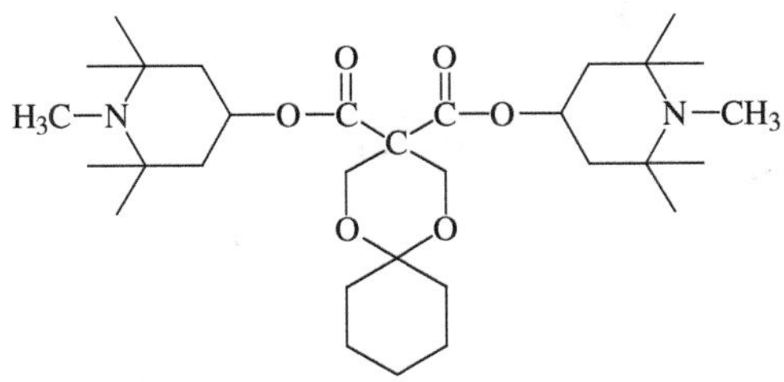
LMW-HALS-9	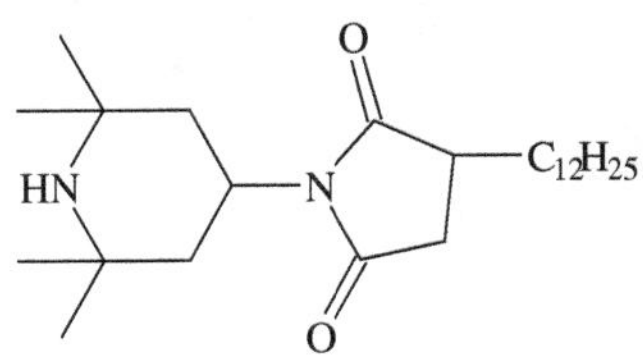
LMW-HALS-10	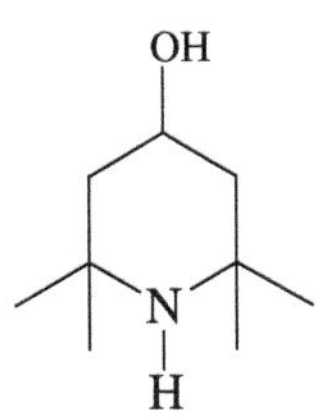

(continued)

Abbreviation	Chemical structure
LMW-HALS-11	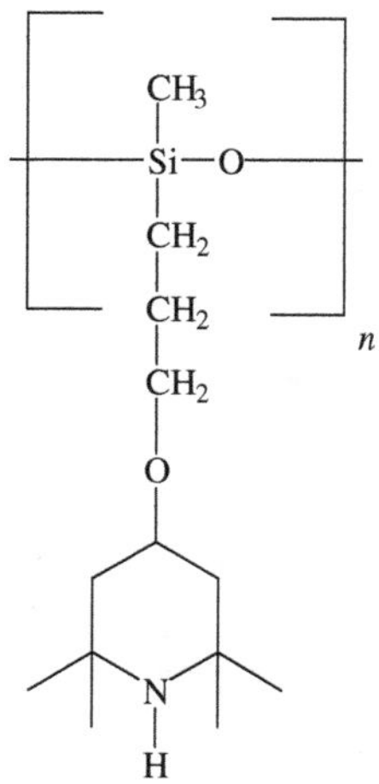
HMW-HALS-1	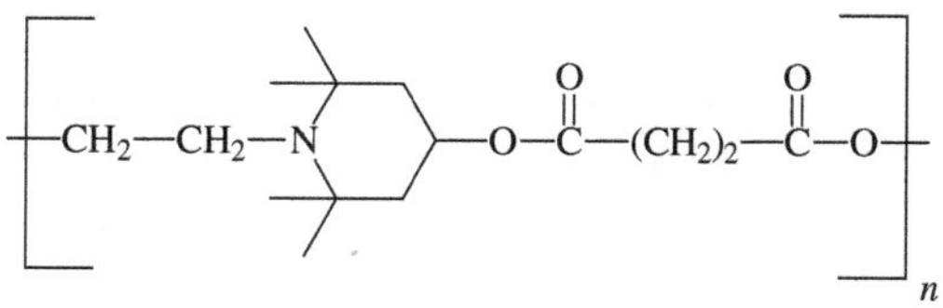
HMW-HALS-2	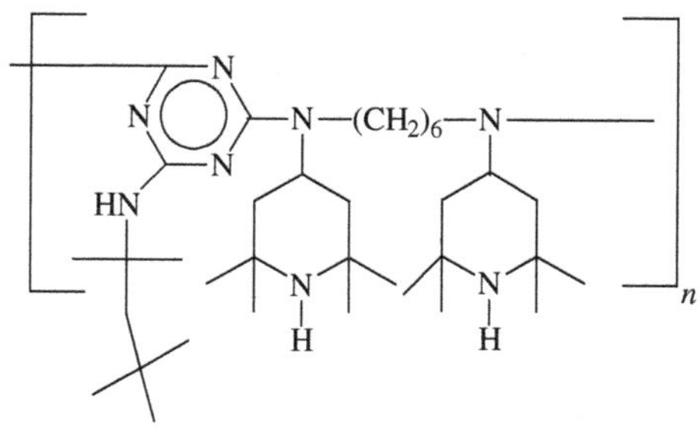
HMW-HALS-3	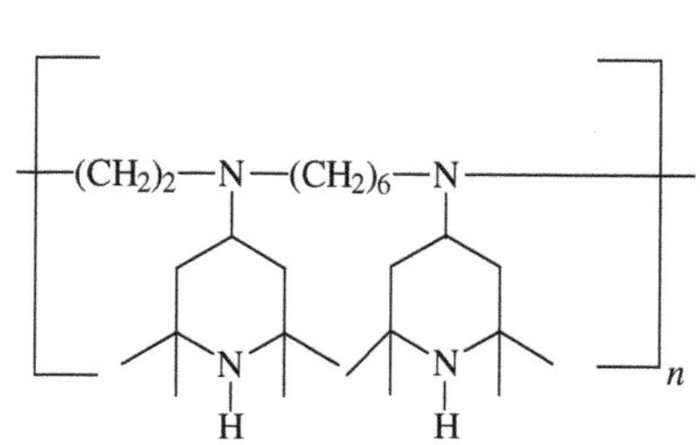

Abbreviation	Chemical structure
HMW-HALS-4	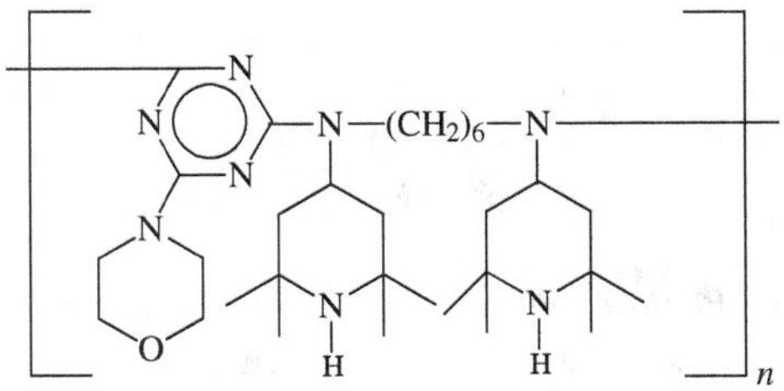
HMW-HALS-5	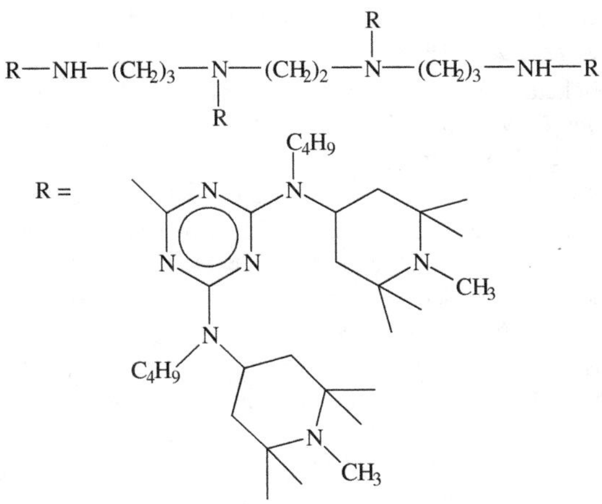
HMW-HALS-6	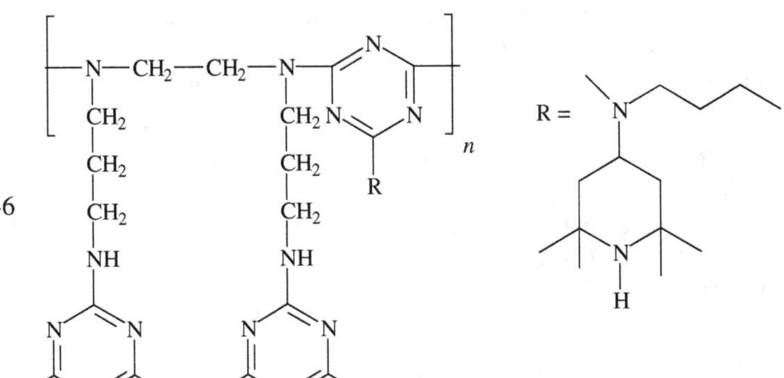

REFERENCES

1. J. Pospisil, Chemical and photochemical behavior of phenolic antioxidants in polymer stabilization-a state of the art report, Part I, *Polym. Degrad. Stab.* **1993**, *40*, 217–232.

2. J. Pospisil, Chemical and photochemical behaviour of phenolic antioxidants in polymer stabilization: a state of the art report, Part II, *Polym. Degrad. Stab.* **1993**, *39*, 103–115.
3. O. Brede, Pulse radiolysis investigations of reactions of phenolic stabilizers. Proceedings of the 12th International Conference on Advances in Controlled Degradation and Stabilization of Polymers, Luzern, **1990**, pp. 29–47.
4. H.J. Heller, Protection of polymers against light irradiation, *Eur. Polym. J. Suppl.* **1969**, 105–132.
5. J.F. Rabek, *Photostabilization of Polymers*, Elsevier Applied Science, London, 1990.
6. F. Gugumus and H. Zweifel (Ed.), *Light Stabilisers in Plastic Additives Handbook*, 5th edition, Hanser Publisher, Munich, **2001**, pp. 141–425.
7. H.J. Heller and H.R. Blattmann, Some aspects of stabilization of polymers against light, *Pure Appl. Chem.* **1973**, *36*, 141–161.
8. J.E. Pickett, Permanence of UV absorbers in plastics and coatings. in: *Handbook of Polymer Degradation*, Environmental Science and Pollution Control Series 21, 2nd Edition, 2000, pp. 163–190.
9. J. Keck, M. Roessler, C. Schroeder, G.J. Stueber, F. Waiblinger, M. Stein, D. LeGourrierec, H.E.A. Kramer, H. Hoier, S. Henkel, P. Fischer, H. Port, T. Hirsch, G. Rytz, and P. Hayoz, Ultraviolet Absorbers of the 2-(2-hydroxyaryl)-1,3,5-triazine class and their methoxy derivatives: fluorescence spectroscopy and X-ray structure analysis, *J. Phys. Chem. B* **1998**, *102*, 6975–6985.
10. S.W. Bigger, K.P. Ghiggino, I.H. Leaver, and A.D. Scully, Photophysics of 6-(2'-hydroxy-4'-methoxyphenyl)-*s*-triazine photostabilizers, *J. Photochem. Photobiol. A Chem.* **1987**, *40*, 391–399.
11. G. Rytz and R. Hilifiker, Introduction to a new class of high performance light stabilizers and the influence of light stabilizer structure on the polymer's life time. Eighteenth Annual International Conference on Advances in the Stabilization and Degradation of Polymers Luzern, Switzerland, **1996**.
12. J.E. Pickett and J.E. Moore, Photostability of UV screeners in polymers and coatings, *Adv. Chem. Ser.* **1996**, *249*, 287–301.
13. J.E. Pickett and J.E. Moore, Photodegradation of UV screeners, *Polym. Degrad. Stab.* **1993**, *42*, 231–244.
14. J.E. Pickett, Calculation of the efficiency of ultraviolet screeners in plastics, *J. Appl. Polym. Sci.* **1987**, *33*, 525–531.
15. J.P. Guillory and C.F. Cook, Mechanism of stabilization of polypropylene by ultraviolet absorbers, *J. Polym. Sci. Polym. Chem.* **1971**, *9*, 1529–1536.
16. F. Gugumus, The performance of light stabilizers in accelerated and natural weathering, *Polym. Degrad. Stab.* **1995**, *50*, 101–116.
17. N.S. Allen, A. Parkinson, F.F. Loffleman, and P.V. Susi, Mechanistic action of a new *p*-hydroxybenzoate light stabilizer in polypropylene film: a spectroscopic study, *Angew. Makromol. Chem.* **1983**, *116*, 203–219.
18. J.A. Suriano, T.M. Siclovan, J.E. Pickett, D.J. Brunelle, G.A. O'Neil, and H. Zhou, Weatherable polyarylate-CO-polycarbonate engineering thermoplastic, *Polym. Preprints* **2003**, *44*(1), 748–749.
19. R.S. Davidson and R. Foster (Ed.), *Molecular Associations*, Academic Press, 1975, pp. 216–334.

20. J.F. Rabek, *Photodegradation of polymers*, Springer, Berlin, Heidelberg, 1996.
21. B.J. Briggs and J.F. McKellar, Effects of some nickelchelates ultraviolet stabilizers on triplet anthracene, *Chem. Ind.* **1967**, *15*, 622–623.
22. D.J. Carlsson, T. Suprunchuk, and D.M. Wiles, Photooxidation of polypropylene films. VI. Possible UV-stabilization mechanisms, *J. Appl. Polym. Sci.* **1972**, *16*, 615–626.
23. J.C.W. Chien and W.P. Conner, Quenching of excited states of diethyl ketone by paramagnetic metal chelates, *J. Am. Chem. Soc.* **1969**, *90*, 1001–1006.
24. J.P. Guillory and C.F. Cook, Energy transfer processes involving ultraviolet stabilizers. Quenching of excited states of ketones, *J. Amer. Chem. Soc.* **1973**, *95*, 4885–4891.
25. A.P. Pivovarov and A.F. Lukovnikov, Mechanism of the protective effect of polymer light stabilizers, *Khimiya Vysokikh Energii* **1968**, *2*, 220–227.
26. J. Flood, K.E. Russell, D.J. Carlsson, and D.M. Wiles, Quenching of type II photodecomposition of 2-pentanone in *n*-hexane solution by metal chelates, *Can. J. Chem.* **1974**, *52*, 688–691.
27. P.R. Ramani and G. Scott, Mechanisms of antioxidant action. Metal complexes as UV stabilizers in polyethylene, *Eur. Polym. J.* **1976**, *12*, 591–597.
28. N.S. Allen, A. Chirinos-Padron, and J.H. Appleyard, Photo-stabilizing action of metal chelates in polypropylene. Part I. Excited state quenching versus UV antioxidant action under polychromatic irradiation, *Polym. Degrad. Stab.* **1982**, *4*, 223–237.
29. N.S. Allen, A. Chirinos-Padron, and J.H. Appleyard, Photo-stabilizing action of metal chelates in polypropylene. Part II. Photolysis versus photosensitized oxidation under monochromatic irradiation, *Polym. Degrad. Stab.* **1983**, *5*, 29–41.
30. N.S. Allen, A. Chirinos-Padron, and J.H. Appleyard, Photo-stabilizing action of metal chelates in polypropylene. Part III. Thermal antioxidant action and its relationship to photo-stabilization, *Polym. Degrad. Stab.* **1983**, *5*, 55–63.
31. N.S. Allen, A. Chirinos-Padron, and J.H. Appleyard, Photo-stabilizing action of metal chelate stabilizers in polypropylene. Part V. Light stability as a function of concentration and further studies in metal stearate-stabilizer systems, *Polym. Degrad. Stab.* **1983**, *5*, 323–338.
32. N.S. Allen, A. Chirinos-Padron, and J.H. Appleyard, Photostabilizing action of metal chelate stabilizers in polypropylene. Part VI. Importance of singlet ($^1\Delta_g$) oxygen quenching and UV screening, and flash photolysis studies, *Polym. Degrad. Stab.* **1984**, *6*, 31–45.
33. N.S. Allen, A. Chirinos-Padron, and J.H. Appleyard, The decomposition of cumene hydroperoxide by nickel(II), 2,2′-thiobis(4-*tert*-octyl-phenolato)-*n*-butylamine (NI-1) in *n*-hexane solution and its possible importance in the photostabilization of polypropylene, *Eur. Polym. J.* **1985**, *21*, 101–105.
34. F. Gugumus, Re-evaluation of the stabilization mechanisms of various light stabilizer classes, *Polym. Degrad. Stab.* **1993**, *39*, 117–135.
35. P. Gijsman, J. Hennekens, and D. Tummers, The mechanism of action of hindered amine light stabilizers, *Polym. Degrad. Stab.* **1993**, *39*, 225–233.
36. P. Gijsman and A. Dozeman, Comparison of the UV-degradation chemistry of unstabilized and HALS-stabilized polyethylene and polypropylene, *Polym. Degrad. Stab.* **1996**, *53*, 45–50.

37. D. Bellus, B. Ranby, and J.F. Rabek (Eds.), *Singlet Oxygen, Reactions with Organic Compounds and Polymers*, Wiley, New York, 1978, pp. 61–110.
38. R. Keese, Methods for preparing bridgehead olefins, *Angew. Chem.* **1975**, *87*, 568–578.
39. P. Gijsman, New synergists for hindered amine light stabilizers, *Polymer* **2002**, *43*, 1573–1579.
40. P. Gijsman,Plastics composition having improved weather resistance, WO9836023.
41. P. Gijsman and A. E. J. Schaafsma,Plastics composition with improved weather resistance, WO9948965.
42. J. A. Stretanski, SPE Regional Technical Conference, Houston, Texas, 23 February **1981**, p. 141.
43. N.S. Allen, A. Parkinson, F.F. Loffelman, and P.V. Susi, Photo-Stabilising action of a *p*-hydroxybenzoate compound in polyolefins. Part I. Thermal and photochemical behaviour in polypropylene film, *Polym. Degrad. Stab.* **1983**, *5*, 241–266.
44. N.S. Allen, A. Parkinson, F.F. Loffelman, and P.V. Susi, Photo-stabilizing action of a *p*-hydroxybenzoate compound in polyolefins. Part II. Thermal and photochemical behavior in high density polyethylene film, *Polym. Degrad. Stab.* **1984**, *6*, 65–79.
45. N.S. Allen, A. Parkinson, F.F. Loffelman, M.M. Rauhut, and P.V. Susi, Spectroscopic properties and mechanistic action of a new *p*-hydroxybenzoate light stabilizer: a comparative study in polypropylene and high density polyethylene and anti-oxidant interactions, *Polym. Degrad. Stab.* **1984**, *7*, 153–174.
46. E.G. Rozantsev, *Free Nitroxyl Radicals*, Plenium Press, New York London, 1970, p. 125.
47. M.B. Neiman, E.G. Rozantsev, and Yu. G. Mamedova, Free-radical reactions involving no unpaired electrons, *Nature* **1962**, *196*, 472–474.
48. K. Murayama and T. Yoshioka, Stable free radicals. IV. Decomposition of stable nitroxide radicals, *Bull. Chem. Soc. Jpn.* **1969**, *42*, 2307–2309.
49. T. Toda, T. Kurumada, and K. Murayama, Progress in the light stabilization of polymers, *Am. Chem. Soc. Symp. Ser.* **1985**, *280*, 37–54.
50. P. Gijsman and M. Gitton, Hindered amine stabilisers as long-term heat stabilisers for polypropylene, *Polym. Degrad. Stab.* **1999**, *66*, 365–371.
51. W. Heintz, *Lieb. Ann. Chem.* **1874**, *175*, 133.
52. R.B. Bradbury, N.C. Hancnox, and H.H. Hatt, Reaction between acetone and ammonia: the formation of pyrimidine compounds analogous to the aldoxanes of späth, *J. Chem. Soc.* **1947**, 1394–1399.
53. J. Kirchhoff and F. Kraushaar, Triacetoneamine derivatives Industrial applications and recent developments, *Polym. Polym. Compos.* **2000**, *8*, 245–254.
54. J.T. Lai, P.N. Son, and E. Jennings, Hindered diazacycloalkanones as ultraviolet stabilizers and antioxidants, *ACS Symp. Ser.* **1985**, *280*, 91–98.
55. G. Balint, A. Rockenbauer, T. Kelen, F. Tudos, and L. Jokay, Esr study of polypropylene photostabilisation by sterically hindered piperidine derivatives, *Polym. Photochem.* **1981**, *1*, 139–152.
56. Yu.B. Shilov, R.M. Battalova, and E.T. Denisov, Regeneration of iminoxyl radical during polypropylene oxidation, *Doklad. Acad. Nauk USSR* **1972**, *207*, 388–389.
57. Yu.B. Shilov and E.T. Denisov, Mechanism of the inhibiting activity of iminoxyl radicals during oxidation of polypropylene and polyethylene, *Vysokomol. Soyed. A* **1974**, *14*(10), 2313–2316.

58. G.A. Kovtun, A.L. Aleksandrov, and V.A. Golubev, Interaction of peroxide radicals with hydroxylamine ethers, *Izv. Akad. Nauk. SSSR Ser. Kim.* **1974**, 2197.

59. J. Durmis, D.J. Carlsson, K.H. Chan, and D.M. Wiles, Photostabilization of polypropylene by a hindered amine, *J. Polym. Sci. Polym. Lett. Ed.* **1981**, *19*, 549–554.

60. D.K. Hodgeman, Formation of polymer-bonded nitroxyl radicals in the UV stabilization of polypropylene by a bifunctional hindered amine light stabilizer, *J. Polym. Sci. Polym. Chem. Ed.* **1981**, *19*, 807.

61. D.J. Carlsson and D.M. Wiles, Polypropylene photo-stabilization by hindered secondary amines. A spectroscopic Investigation, *Polym. Degrad. Stab.* **1984**, *6*, 1–16.

62. T.A.B.M. Bolsman, A.P. Blok, and T.H.G. Frijns, Mechanism of the catalytic inhibition of hydrocarbon autoxidation by secondary amines and nitroxides, *Rec. J. R. Neth. Chem. Soc.* **1978**, *97*, 313–319.

63. E.N. Step, J. Turro, M.E. Gande, and P.P. Klemchuck, Flash photolysis and time-resolved electron spin resonance studies triplet benzophenone quenching by hindered amine light stbilizers (HALS). A comparison of HALS amines and aminoethers and hydrogen atom donors, *J. Photochem. Photobiool. A Chem.* **1993**, *74*, 203–210.

64. E.N. Step, J. Turro, M.E. Gande, and P.P. Klemchuck, Mechanism of polymer stabilization by hindered-amine light stabilizers (HALS). Model investigations of the interaction of peroxy radicals with HALS amines and amino ethers, *Macromolecules* **1994**, *27*, 2529–2539.

65. D.W. Grattan, D.J. Carlsson, and D.M. Wiles, Polyolefin photo-stabilisation mechanisms. Reaction of tetramethylpiperidine derivatives in model systems, *Polym. Degrad. Stab.* **1979**, *1*, 69–84.

66. V.Ya. Shlyapintokh, V.B. Ivanoc, Antioxidant action of sterically hindered amines and related compounds, in G. Scott, (Ed.), *Developments in Polymer Stabilisation*, Vol. *5*, Applied Science Publisher, London, **1982**, pp. 41–70.

67. A.L.J. Beckwith, V.W. Bowry, and K.U. Ingold, Kinetics of nitroxide radical trapping. 1. Solvent effects, *J. Am. Chem. Soc.* **1992**, *114*, 4983–4992.

68. V.W. Bowry and K.U. Ingold, Kinetics of nitroxide radical trapping. 2. Structural effects, *J. Am. Chem. Soc.* **1992**, *114*, 4992–4996.

69. P. Richters, Initiation process in the oxidation of polypropylene, *Macromolecules* **1970**, *3*, 262.

70. J.B. Knight, P.D. Calvert, and N.C. Billingham, Localization of oxidation in polypropylene, *Polymer* **1985**, *26*, 1713.

71. G.A. George, M. Celina, C. Lerf, G. Cash, and D. Wendell, A spreading model for the oxidation of polypropylene, *Macromol. Symp.* **1997**, *115*, 68–92.

72. M. Celina and G.A. George, Heterogeneous and homogeneous kinetic analysis of the thermal oxidation of polypropylene, *Polym. Degrad. Stab.* **1995**, *50*, 89–99.

73. I. Blakey, and G.A. George, Raman spectral mapping of photo/oxidised polypropylene, *Polym. Degrad. Stab.* **2000**, *70*, 269–275.

74. G. Ahlblad, P. Gijsman, B. Terselius, A. Jansson, and K. Möller, Thermo-oxidative stablility of PP waste films as studied by imaging chemiluminescence, *Polym. Degrad. Stab.* **2001**, *73*, 15–22.

75. A.A. Oswald, F. Noel, and A.J. Stephenson, Organic sulfur compounds. V. Alkylammonium thiolate and peroxide salts; possible intermediates in amine-catalyzed oxidation of mercaptans by hydroperoxides, *J. Org. Chem.* **1961**, *26*, 3969–3974.

76. J. Sedlar, J. Petruy, J. Pak, and M. Navratil, Polymer photostabilization by HALS derivatives: the role of piperidine-hydroperoxide associates, *Polymer* **1980**, *21*, 5–7.

77. D.W. Grattan, A.H. Reddoch, D.J. Carlsson, and D.M. Wiles, Polymer photostabilization by piperidine derivatives: the role of nitroxide-hydroperoxide complexing, *J. Polym. Sci. Polym. Lett. Ed.* **1978**, *16*, 143–148.

78. F. Gugumus, Developments in the U.V.-stabilisation of polymers, in: G. Scott, (Ed.), *Developments in Polymer Stabilisation*, Vol. *1*, Applied Science Publisher, London, 1979, pp. 261–308.

79. D.J. Carlsson, K.H. Chan, and D.M. Wiles, Polypropylene photostabilization by tetramethylpiperidine species. in S.R., Pappas, and F.H., Winslow, (Eds.), *Photodegradation and Photostabilisation of Coatings*, American Chemical Society Symposium Series No. 151, 1981, pp. 51–63.

80. G. Geuskens and G. Nedelkos, The oxidation of hindered amine light stabilizers to nitroxyl radicals in solution and in polymer, *Polym. Degrad. Stab.* **1987**, *19*, 365–378.

81. T. Toda, E. Mori, and K. Murayama, Stable free radicals. IX. Peroxy acid oxidation of hindered secondary amines to nitroxide radicals, *Bull. Chem. Soc. Jpn.* **1972**, *45*, 1904–1908.

82. B. Felder, R. Schumacher, and F. Sitek, Hindered amine light stabilizers: a mechanistic study, *ACS Symp. Ser.* **1985**, *280*, 69–90.

83. A. Zahradnickova, J. Sedlar, and D. Dastych, Peroxy acids in photo-oxidized polypropylene, *Polym. Deg. Stab.* **1991**, *32*, 155–176.

84. M.G. Chan, Metal deactivators, in: J. Pospisil and P.P. Klemchuck (Eds.), *Oxidation Inhibition in Organic Materials*, Vol. *1*, CRC Press Inc., Boca Raton, FL, 1990, pp. 225–246.

85. S.P. Fairgrieve and J.R. Mac Callum, Hindered amine light stabilizers: a proposed photostabilization mechanism, *Polym. Degrad. Stab.* **1984**, *8*, 107–121.

86. J. Sedlar, J. Petruj, J. Pac, and A. Zahradnicova, Photostabilizing action of sterically hindered piperidines-1, peroxide decomposing and hydrogen donating ability, *Eur. Polym. J.* **1980**, *16*, 659–662.

87. P.P. Klemchuk and M.E. Gande, Stabilization mechanisms of hindered amines, *Polym. Degrad. Stab.* **1988**, *22*, 241–274.

88. F. Gugumus, Findings pertaining to polyethylene photooxidation, *Makromol. Chem. Macromol. Symp.* **1989**, *25*, 1–22.

89. N.S. Allen and J.F. McKellar, Photostabilization of commercial polypropylene by hindered piperidine compounds, *Plast. Rubber Mat. Appl.* **1979**, 170–174.

90. N.S. Allen, Photo-stabilising performance of a hindered piperidine compound in polypropylene film: anti-oxidant/light stabiliser effects, *Polym. Degrad. Stab.* **1980**, *2*, 129–135.

91. J. Lucki, J.F. Rabek, and B. Ranby, Photostabilizing effect of hindered piperidine compounds: Interaction between hindered phenols and hindered piperidines, *Polym. Photochem.* **1984**, *5*, 351–354.

92. N.S. Allen, A. Hamid, F.L. Loffelman, P. Macdonald, M. Rauhut, and P.V. Sushi, Interactions of antioxidants with HALS in the thermal and photochemical oxidation of PP film, *Plast. Rubber Process. Appl.* **1985**, *5*, 259–265.

93. H. Yamashita and Y. Ohkatsu, A new antagonism between hindered amine light stabilizers and acidic compounds including phenolic antioxidants, *Polym. Degrad. Stab.* **2003**, *80*, 421–426.
94. J.E. Volponi, L.H.I. Mei, and D. dos Santos Rosa, The use of differential photocalorimetry to measure the oxidation induction time of isotatic polypropylene, *Polym. Testing* **2004**, *23*, 461–465.
95. P. Gijsman, J. Sampers, W. Bunge, and J. Vaassen, Stabilized polypropylene resin composition, EP1340789.
96. I. Bauer, W.D. Habicher, S. Korner, and S. Al-Malaika, Antioxidant interaction between organic phosphites and hindered amine light stabilizers: effects during photoxidation of polypropylene-II, *Polym. Degrad. Stab.* **1997**, *55*, 217–224.
97. G.A. George and M. Ghaemy, The effect of stabilizers on integrated chemiluminescence in the early stages of polypropylene photo-oxidation, *Polym. Degrad. Stab.* **1991**, *34*, 37–53.
98. S.W. Bigger and O. Delatycki, The effect of hindered amine light stabilizers on the photoxidative stability of high density polyethylene, *J. Polym. Sci. A Polym. Chem.* **1989**, *27*, 63–73.
99. K.B. Chakraborty and G. Scott, Mechanism of antioxidant action: the behavior of hindered amine U.V. stabilizers during the processing of LDPE, *Chem. Ind.* **1978**, 237–238.
100. L.Yu. Smoliak, N.R. Prokopchuck, Y.P. Losev, V.P. Prokopovich, and I.A. Klimotsova, The role of sulphur containing groups in inhibition of oxidative degradation of polyolefins by HALS, *Polym. Blends Polym. Compos.* **2002**, *1*, 151–159.
101. K. Kikkawa and Y. Nakahara, Antagonism between hindered amine light stabilizers and sulphur-containing compounds, *Polym. Degrad. Stab.* **1987**, *18*, 237–245.
102. J. Lucki, S.Z. Jian, J.F. Rabek, and B. Ranby, Antagonistic effects of hindered piperidines and organic sulphides in photostabilization of *cis*-1,4-polybutadiene, *Polym. Photochem.* **1986**, *7*, 27–47.
103. R.L. Gray, Additives: a novel non reactive HALS Boosts polyolefin stability, *Plast. Eng.* **1991**, *47*(6), 21–23.
104. D.J. Carlsson, K. Zhang, and D.M. Wiles, Polypropylene photostabilization by hindered amines in the presence of acidic species, *J. Appl. Polym. Sci.* **1987**, *33*, 875–884.
105. S. Chmela, D.J. Carlsson, and D.M. Wiles, Photo-stabilizing efficiency of *N*-substituted hindered amines in polypropylene: effects of processing conditions and exposure to a protonic acid, *Polym. Degrad. Stab.* **1989**, *26*, 185–195.
106. J.P.T. Jensen, D.J. Carlsson, and D.M. Wiles, Polypropylene photostabilization by a tertiary hindered amine, *Polym. Bull.* **1983**, *9*, 321–327.
107. H. Zweifel, *Stabilization of Polymeric Materials*, Springer-Verlag, Berlin, Heidelberg, New York, 1998.
108. R.L. Gray and R.E. Lee, The influence of flame retardant structure on UV stabilization approaches in polypropylene, *Angew. Makromol. Chem.* **1997**, *247*, 61–72.
109. J.-L. Gardette, C. Sinturel, and J. Lemaire, Photooxidation of fire retarded polypropylene, *Polym. Degrad. Stab.* **1999**, *64*, 411–417.
110. K. Antos and J. Sedlar, Influence of brominated flame retardant thermal decomposition products on HALS, *Polym. Degrad. Stab.* **2005**, *90*, 188–194.

111. J.H. Khan and S.H. Hamid, Durability of HALS-stabilized polyethylene film in a greenhouse environment, *Polym. Degrad. Stab.* **1995**, *48*, 137–142.

112. E. Epacher and B. Pukanszky, Interactions of pesticides and stabilizers in PE films for agricultural use, *Antec* **1999**, 3785–3790.

113. J.-R. Pauquet, Technological advances in the stabilization of polyethylene films, *Plast. Rubber Composit. Process. Appl.* **1998**, *27*, 19–24.

114. N. Haider and S. Karlsson, Loss of Chimassorb 944 from LDPE and identification of additive degradation products after exposure to water, air and compost, *Polym. Degrad. Stab.* **2001**, *74*, 103–112.

115. F. Gugumus, Mechanism and kinetics of photostabilisation of polyolefins with HALS, *Angew. Makromol. Chem.* **1990**, *176/177*, 241–289.

116. T. Kurumada, H. Ohsawa, T. Fujita, T. Toda, and T. Yoshioka, Photostabilizing activity of *N*-acylated hindered amine, *J. Polym. Sci. Polym. Chem. Ed.* **1985**, *23*, 2747–2756.

117. T. Kurumada, H. Ohsawa, O. Oda, T. Fujita, T. Toda, and T. Yoshioka, Photostabilizing activity of tertiary hindered amines, *J. Polym. Sci. Polym. Chem. Ed.* **1985**, *23*, 1477–1491.

118. T. Kurumada, H. Ohsawa, T. Fujita, and T. Toda, The effect of *N*-substituents of hindered amine on photo-oxidation of polypropylene, *J. Polym. Sci. Polym. Chem. Ed.* **1984**, *22*, 277–281.

119. N.C. Billingham, Physical phenomena in the oxidation and stabilization of polymers, in: J. Pospisil and P.P. Klemchuck (Eds.), *Oxidation Inhibition in Organic Materials*, Vol. 2, CRC Press, Inc., Boca Raton, FL, **1990**, pp. 249–297.

120. P. Delprat, X. Duteurtre, and J.-L. Gardette, Photo-oxidation of unstabilized and HALS-stabilized polyphasic ethylene-propylene polymers, *Polym. Degrad. Stab.* **1995**, *50*, 1–12.

121. B. Ranby and J.F. Rabek, *Photodegradation, photo-oxidation and photostabilization of polymers*, Wiley & Sons, London, 1975.

122. F. Gugumus, Possibilities and limits of synergism with light stabilizers in polyolefins 1. HALS in polyolefins, *Polym. Degrad. Stab.* **2002**, *75*, 295–308.

123. Y. Dobashi, T. Yuyama, and Y. Ohkatsu, Interaction of ultraviolet absorbers, *Polym. Degrad. Stab.* **2007**, *92*, 1227–1233.

124. T. Kurumada, H. Ohsawa, and T. Yamazaki, Synergism of hindered amine light stabilizers and UV-absorbers, *Polym. Degrad. Stab.* **1987**, *19*, 263–272.

125. S. Mizokawa and Y. Ohkatsu, Influence of ultraviolet absorbers on decomposition of hydroperoxide by hindered amine light stabilizers, *J. Jpn. Petrol. Inst.* **2007**, *50*, 1–7.

126. H. Takenaka, S. Mizokawa, and Y. Ohkatsu, Interaction of hindered amine light stabilizers and ultraviolet absorbers, *J. Jpn. Petrol. Inst.* **2007**, *50*, 8–15.

127. C.M. Liauw, A. Quadir, N.S. Allen, M. Edge, and A. Wagner, Effect of hindered piperidine light stabilizer molecular structure and UV-absorber addition on the oxidation of HDPE. Part 1. Long-term thermal and photo-oxidation studies, *J. Vinyl Add. Technol.* **2004**, *10*, 79–87.

128. J.H.W. Pouncy, How photostabilizers compare in LLDPE film weatherability, *Mod. Plast.* **1985**, 75–77.

129. F. Gugumus and N. Lelli, Light stabilization of metallocene polyolefins, *Polym. Degrad. Stab.* **2001**, *72*, 407–421.

130. P. Gijsman,Shaped polyolefin article stabilized by HALS compounds, WO 98/20065.

131. P.N. Thanki and R.P. Singh, Photostabilization of nylon 66 in presence of acid blue dyes, *Polym. Degrad. Stab.* **2002**, *75*, 423–430.

132. P. Cerruti, M. Lavorgna, C. Carfagna, and L. Nicolais, Comparison of photo-oxidative degradation of polyamide 6,6 films stabilized with HALS and $CuCl_2$ + KI mixtures, *Polymer* **2005**, *46*(13), 4571–4583.

133. F. Gugumus, Light Stabiliser, in: R. Gachter and H. Muller (Eds.), *Plastic Additives Handbook*, 3th edition, Hanser Publisher, Munich, pp. 129–270.

134. F. Gugumus, Photo-oxidation of polymers and its inhibition, in: J. Pospisil and P.P. Klemchuck (Eds.), *Oxidation Inhibition in Organic Materials*, Vol. 2, CRC Press, Inc., Boca Raton, FL, 1990, pp. 30–162.

135. H.H. Chin and F.H. Botkin, Light stabilization for black and white glass-filled nylon-6, *Ann. Tech. Conf. Soc. Plast. Eng.* **2000**, *58*, 2673–2677.

136. P. Gijsman and M. Diepens, Photolysis and photo-oxidation in engineering plastics, *Polym. Preprints* **2007**, *48*, 615–616.

137. P. Gijsman and M. Diepens, Photolysis and photo-oxidation in engineering plastics, *ACS Symp. Ser.* **2009**, *1004*, 287–306.

138. T. Thompson and P.P. Klemchuck, Light stabilization of bisphenol A polycarbonate, in: R. L. Clough, N.C. Billingham, and K.T. Gillen (Eds.), *Polymer Durability, Degradation and Life Time Prediction*, Advances in Chemistry Series 249, American Chemical Society, Washington, DC, 1996, pp. 303–317.

139. A. Ram, O. Zilber, and S. Kenig, Life expectation of polycarbonate, *Polymer Engineering and Science* **1985**, *25*, 535–540.

140. W. Nising, Twin-wall sheets from polycarbonate, *Kunststoffe* **1986**, *76*, 1095–1097.

141. G. Geuskens, P. Bastin, Q. Lu Vinh, and M. Rens, Photooxidation of polymers. Part IV. Influence of the processing conditions on the photooxidative stability of polystyrene, *Polym. Degrad. Stab.* **1981**, *3*, 295–306.

INDEX

Photochemistry and Photophysics of Polymer Materials, Edited by Norman S. Allen

www.ingramcontent.com/pod-product-compliance
Lightning Source LLC
LaVergne TN
LVHW011853200926
848484LV00002B/450